Lecture Notes in Computer Science 2387

Edited by G. Goos, J. Hartmanis, and J. van Leeuwen

Lecture Notes in Computer Science 2387

Edited by G. Goos, J. Hartmanis, and J. van Leeuwen

Springer

Berlin
Heidelberg
New York
Barcelona
Hong Kong
London
Milan
Paris
Tokyo

Oscar H. Ibarra   Louxin Zhang (Eds.)

Computing and Combinatorics

8th Annual International Conference, COCOON 2002
Singapore, August 15-17, 2002
Proceedings

Volume Editors

Oscar H. Ibarra
University of California, Department of Computer Science
Santa Barbara, California 93106, USA
E-mail: ibarra@cs.ucsb.edu

Louxin Zhang
Department of Mathematics, National University of Singapore
Singapore, Singapore 117543
E-mail: matzlx@nus.edu.sg

Cataloging-in-Publication Data applied for

Die Deutsche Bibliothek - CIP-Einheitsaufnahme

Computing and combinatorics : 8th annual international conference ; proceedings / COCOON 2002, Singapore, August 15 - 17, 2002. Oscar H. Ibarra ; Louxin Zhang (ed.). - Berlin ; Heidelberg ; New York ; Barcelona ; Hong Kong ; London ; Milan ; Paris ; Singapore ; Tokyo : Springer, 2002
(Lecture notes in computer science ; Vol. 2387)
ISBN 3-540-43996-X

CR Subject Classification (1998): F.2, G.2.1-2, I.3.5, C.2.3-4, E.1, E.4, E.5

ISSN 0302-9743
ISBN 3-540-43996-X Springer-Verlag Berlin Heidelberg New York

Springer-Verlag Berlin Heidelberg New York,
a member of BertelsmannSpringer Science+Business Media GmbH

http://www.springer.de

Printed in Germany

Typesetting: Camera-ready by author, data conversion by Olgun Computergrafik
Printed on acid-free paper SPIN: 10870538 06/3142 5 4 3 2 1 0

Preface

The abstract and papers in this volume were presented at the Eighth Annual International Computing and Combinatorics Conference (COCOON 2002), held on August 15-17 in Singapore. The topics cover various aspects of theoretical computer science and combinatorics related to computing.

Submissions to the conference this year were conducted electronically. The 60 papers were selected for presentation from a total of 106 submitted papers from Australia (6), Canada (3), China (6), Germany (9), India (5), Japan (11), Korea (10), Singapore (5), Taiwan (8), United States (29), and 11 other countries and regions (14). The papers were evaluated by an international program committee consisting of Mikhail Atallah, Jik Chang, Tim Ting Chen, Siu-Wing Cheng, Omer Egecioglu, Fan Chung Graham, Susanne Hambrusch, Sorin Istrail, Sampath Kannan, Ming-Yang Kao, Shlomo Moran, Koji Nakano, Takao Nishizeki, Steve Olariu, Gheorghe Paun, Pandu Rangan, Sartaj Sahni, Arto Salomaa, Igor Shparlinski, Janos Simon, Paul Spirakis, Chung Piaw Teo, Jan van Leeuwen, Paul Vitanyi, Peter Widmayer, and Hsu-Chun Yen. It is expected that most of the accepted papers will appear in a more complete form in scientific journals. In addition to the contributed papers, three invited lectures were presented by Eugene W. Myers, Sartaj Sahni, and Arto Salomaa.

We wish to thank all who have made this meeting possible: the authors for submitting papers, the program committee members and external referees (listed in the proceedings) for their excellent work, and the three invited speakers. Finally, we wish to express our sincere appreciation to the sponsors, local organizers, and our colleagues for their assistance and support.

August 2002 Oscar H. Ibarra, Louxin Zhang

Program Committee

Oscar H. Ibarra (Co-chair), UC Santa Barbara, USA
Louxin Zhang (Co-chair) Nat. U. of Singapore, Singapore

Mikhail Atallah, Purdue U., USA
Jik Chang, Sogang U., Korea
Tim Ting Chen, U. of Southern Calif., USA
Siu-Wing Cheng, HKUST, Hong Kong
Omer Egecioglu, UC Santa Barbara, USA,
Fan Chung Graham, UC San Diego, USA
Susanne Hambrusch, Purdue U., USA)
Sorin Istrail Celera Genomics Corp., USA
Sampath Kannan, U. of Penn, USA
Ming-Yang Kao, Northwestern U., USA
Shlomo Moran, Technion, Israel
Koji Nakano, JAIST, Japan
Takao Nishizeki, Tohuko, Japan
Steve Olariu, Old Dominion U., USA
Gheorghe Paun, Inst. of Math., Romania
Pandu Rangan, IIT Madras, India
Sartaj Sahni, U. of Florida, USA
Arto Salomaa, Turku U., Finland
Igor Shparlinski, Macquarie U., Australia
Janos Simon, U. of Chicago, USA
P. Spirakis, CTI, Greece
Chung Piaw Teo, NUS, Singapore
Jan van Leeuwen, U. of Utrecht, The Netherlands
Paul Vitanyi, CWI, The Netherlands
Peter Widmayer, ETHZ, Switzerland
Hsu-Chun Yen, Nat. Taiwan U., Taiwan

Organizing Committee

Khee Meng Koh (Co-chair), NUS, Singapore
Hon Wai Leong (Co-chair), NUS, Singapore

Fengming Dong, NTU, Singapore
Ee-Chien Chang, NUS, Singapore
Chung Piaw Teo, NUS, Singapore

Conference Secretary

Lynette M. L. Wong

Referees

Stephen Alstrup
Luzi Anderegg
Maria Andreou
Dan Archdeacon
Abdullah Arslan
Dorit Batler
Giuseppe Di Battista
Jacir Luiz Bordim
Ran Canetti
Alberto Caprara
Xin Chen
Sung-Woo Cho
Francisco Coelho
Barry Cohen
Zhe Dang
Mart de Graaf
Joerg Derungs
Stephan Eidenbenz
Panagiota Fatourou
Mike Fellows
Vladimir Filkov
Eldar Fischer
Dimitris Fotakis
Pierre Fraignaud
Jozef Gruska
Nicolas Hanusse
Tero Harju
Sariel Har-Peled
Joel Hass
Thomas Hofmeister
Ed Hong
Tao Jiang
Sungwon Jung
Michael Kaminski
George Karakostas
Dimitris Kavvadias
Daesan Kim
Spyros Kontogiannis
Jeff Lagarias
Donghoon Lee
Hanno Lefmann
Chin-Laung Lei
Stefano Lonardi
Hsueh-I Lu
Meena Mahajan
Ross McConnell
Janos Makowski
Pablo Moisset
Tal Mor
Matthias Mueller
Sotiris Nikoletseas
Roderic D. M. Page
Aris Pagourtzis
Vicky Papadopoulou
Jungheum Park
Kunsoo Park
Eynat Rafalin
Md. Saidur Rahman
S. Rajasekaran
B. Ravikumar
Hein Roehrig
Brigitte Servatius
Diane Souvaine
Mike Steel
Pavel Sumazin
Wing Kin Sung
Subhash Suri
Gabor Szabo
Laszlo Szekely
Arie Tamir
Joseph A. Thas
Takeshi Tokuyama
Nicholas Tran
John Tromp
Ming-Jer Tsai
Sam Wagstaff
Yuan-Fang Wang
Birgitta Weber
David Wei
Hongjun Wu
Jihoon Yang
Sheng Yu
Christos Zaroliagis
Shiyu Zhou
Xiao Zhou

Sponsoring Institutions

Department of Mathematics, NUS
Lee Foundation, Singapore

Organizing Institutions

Department of Mathematics, NUS
School of Computing, NUS
The Logistics Insitute - Asia Pacific, NUS

Table of Contents

Invited Lectures

Complexity Theory I

Discrete Algorithms I

Computational Biology and Learning Theory I

Coding Theory and Cryptography

Parallel and Distributed Architectures

Graph Theory

Radio Networks

Automata and Formal Languages

Internet Networks

Computational Geometry I

Computational Biology and Learning Theory II

Discrete Algorithms II

Computational Geometry II

Combinatorial Optimization

Complexity II

Quantum Computing

The Assembly of the Human and Mouse Genomes

Gene Myers, Ph.D

VP, Informatics Research
Celera Genomics
Gene.Myers@celera.com

Paired-read shotgun sequencing of a genome consists of randomly sampling segments of a fixed length, say 10,000 base pairs (10Kbp), and directly determining a "read" of 500 to 700bp at the two ends of each segment with an automated DNA sequencing instrument. Reconstructing or "assembling" a very large genome from such a data set was considered impossible at the time Jim Weber and I proposed it for the Human Genome (2.9Gbp) in 1996. Critics claimed that the computation would involve an impossible amount of computer time, that the size and repetitiveness of the genome would confound all attempts at assembly should sufficient computer efficiency be achieved, and that even if an assembly were produced it would be of an extremely poor quality and partial nature.

In mid-1998 Celera was formed and by the close of 1999 the informatics research team at Celera had assembled the 130Mbp Drosophila genome after producing a whole genome shotgun data set with enough reads to cover genome 13 times over, a 13X data set. An assembly of the Human genome followed in 2000 from a 5.1X data set and synthetic reads generated from public data. In April 2001, we produced an assembly of the Mouse genome from a 5.3X data set of three different mouse strains in equal proportions.

Our results from these projects prove unequivocally that whole genome shotgun sequencing is effective at delivering highly reliable reconstructions. The fact that assembly is achieved with only 5X data implies that a relatively complete picture of a large vertebrate genome can be obtained in six to nine months at very competitive cost with current technology. We demonstrate that the 5.3X mouse assembly is a solid substrate for annotation, and together with the human genome, many structural features become apparent through evolutionary conservation.

O.H. Ibarra and L. Zhang (Eds.): COCOON 2002, LNCS 2387, p. 1, 2002.

Data Structures for One-Dimensional Packet Classification Using Most-Specific-Rule Matching

Sartaj Sahni

CISE Department, University of Florida
Gainesville, FL 32611
sahni@cise.ufl.edu

We review the data structures that have been proposed for one-dimensional packet classification. Our review is limited to data structures for the case when ties among the rules that match an incoming packet are broken by selecting the matching rule that is most specific. For the case when the rule filters are destination-address prefixes or are nonintersecting ranges, this tie breaker corresponds to longest-prefix or shortest-range matching, respectively. When the rule filters are arbitrary ranges, this tie breaker resolves the tie only when the rule set is conflict free. Data structures for both static and dynamic rule tables are discussed.

O.H. Ibarra and L. Zhang (Eds.): COCOON 2002, LNCS 2387, p. 2, 2002.

DNA Complementarity and Paradigms of Computing

Arto Salomaa

Turku Centre for Computer Science
Lemminkäisenkatu 14, 20520 Turku, Finland
asalomaa@cs.utu.fi

Abstract. Watson-Crick complementarity is one of the central components of DNA computing, the other central component being the massive parallelism of DNA strands. While the parallelism drastically reduces (provided the laboratory techniques will become adequate) the computational complexity, the complementarity is the actual computational tool "freely" available. It is also the cause behind the Turing universality of models of DNA computing. This paper makes this cause explicit, reducing the matter to some previously known issues in computability theory. We also discuss some specific models.

1 Adleman's Experiment

Adleman, [2], demonstrated how standard methods of molecular biology could be used to solve a (small instance of a) computationally hard problem. The interest on "DNA-based computers" has been growing rapidly, both as regards the development of laboratory techniques and the study of theoretical models. This paper deals exclusively with the latter aspect.

There are still considerable obstructions to creating a practical molecular computer, and also very pessimistic views have been expressed. On the other hand, the possibilities in many fields such as cryptography seem quite amazing, see, for instance, [1].

Although the real practical feasibility of molecular computers remains still in doubt, the field has opened new vistas and important research problems both for computer scientists and biologists. The computer scientist and mathematician is looking for new models of computation. I would like to call such models "Watson-Crick machines" because of the Watson-Crick complementarity in DNA: the four bases A (adenine), T (thymine), C (cytosine) and G (guanine) form two complementary pairs (A,T) and (C,G). In a Watson-Crick machine, Turing's slow diligent clerk is replaced by DNA strands acting in a test tube. In this way the massive parallelism of DNA strands may render tractable some computational problems that were beyond the reach of the diligent clerk; not because the DNA strands are smarter but simply because they can make many tries at once. For the biologist, the unexpected results in DNA computing indicate that models of DNA computers could be significant also for the study of important biological problems

O.H. Ibarra and L. Zhang (Eds.): COCOON 2002, LNCS 2387, pp. 3–17, 2002.

such as evolution. Moreover, the techniques of DNA manipulation developed originally for computational purposes could find relevant applications in genetic engineering. However, because of the rather diverse research traditions, it will not be easy to establish common idioms, let alone common vocabulary, for the researchers in computer science and biology.

In the issue of Science containing Adleman's seminal work, David Gifford wrote, [7]: "If we are able to construct a universal machine out of biological macromolecular components, then we could perform any computation by means of biological techniques. There are certainly powerful practical motivations for this approach, including the information-encoding density offered by macromolecules and the high energy efficiency of enzyme systems. At present, there is no known way of creating a synthetic universal system based on macromolecules. Universal systems require the ability to store and retrieve information, and DNA is certainly up to the task if one could design appropriate molecular mechanisms to interpret and update the information in DNA. This ultimate goal remains elusive, but once solved, it will revolutionize the way we think about both computer science and molecular biology."

Although such an ultimate goal still remains elusive, there are by now already many DNA-based universal computational models. (For instance, see [3,4,6,10,11,9,22].) That there are many diverse ways of constructing DNA-based universal computers is due to the following overall observation. Watson-Crick complementarity guarantees universal computations in any model of DNA computers having sufficient capabilities of handling inputs and outputs. This is a consequence of the close similarity between the Watson-Crick complementarity and the *twin-shuffle language*. This observation was first made in [12], and developed further in [3,10,13,16,17]. The twin-shuffle language is a language over four letters, known, [5], to be powerful enough to serve as a basis for arbitrary computations. This state of affairs can be viewed also as a mathematical explanation to the number of nucleotides in DNA being four, rather than three or five.

It seems obvious that theoretical studies about DNA computing must make use of the following two advantages stemming from DNA molecules. (i) Watson-Crick complementarity which can be viewed as the actual computational tool, and (ii) the multitude of DNA molecules which brings massive parallelism to the computing scene. It seems that theoretical studies have so far concentrated on (i). It is still quite an open area to model (ii) mathematically, as well as to combine (i) and (ii) into one model of DNA computing.

The purpose of this paper is an overall discussion of the significance of the Watson-Crick complementarity in an arbitrary theoretical setup of DNA computing. Sections 2 and 3 are devoted to a general discussion. Specific setups, "case studies", are then presented in the following sections.

The paper assumes practically no background on the part of the reader. The necessary prerequisites about formal languages can be found, for instance, in [14] or [15].

2 Complementarity

DNA consists of polymer chains, usually referred to as *DNA strands.* A chain is composed of *nucleotides*, also referred to as *bases*. Sometimes the chains are also referred to as *oligonucleotides*, briefly *oligos*. The four DNA nucleotides or bases are customarily denoted by A, C, G and T.

According to a chemical convention, each strand has a $5'$ end and a $3'$ end, for instance,

$$5'CATTAG3' \quad or \quad 3'GTAATC5'.$$

Thus, the strands are oriented. The familiar double helix of DNA arises by the bondage of two separate strands. Actually, single strands are fragile and, consequently, information is stored in double strands.

The phenomenon known as *Watson-Crick complementarity* comes into the picture in the formation of double strands. Bonding happens by the pairwise attraction of bases: A bonds with T, and C bonds with G. Therefore, the two unordered pairs (A,T) and (C,G) are known as *complementary pairs* of bases. Bonding will only occur if the bases in the two strands are pairwise complementary and, moreover, the strands have opposite orientations: one of them extends from $5'$ to $3'$, and the other from $3'$ to $5'$. This is the case with the two strands mentioned above and, consequently, they form the double strand

$$5'CATTAG3'$$
$$3'GTAATC5'$$

Double strands can again be dissolved into single strands by heating the solution. This process is usually referred to as *melting.* The reverse process, referred to as *annealing*, is performed by cooling the solution. We do not try to give here any survey about the operations possible or feasible with DNA strands, or about the error-freeness of such operations. We will also ignore the orientation in the sequel. Studies of DNA computing have brought forward many significant new problem areas along these lines. However, for our purposes, the following observation is important. Watson-Crick complementarity gives us something important "for free": whenever bondage happens, we know that the bases at corresponding places must be complementary. If A appears in one of the strands, we know that T appears in the corresponding place in the other strand. If we know one member of a bond, we know also the other member "without looking". Otherwise, bondage would not have taken place. This observation is very basic for DNA computing. As we will see, it means that we have the *twin-shuffle language* freely available. This leads to universality of models of DNA computing satisfying certain requirements of handling inputs and outputs.

The observation mentioned is relevant also for simple tasks of DNA computing. It is also behind Adleman's fundamental experiment, [2], of solving an instance of the Hamiltonian path problem, HPP.

Given a directed graph, we want to solve its HPP. Each vertex is encoded by an oligonucleotide. (Adleman used strands of length 20.) Edges are encoded by oligonucleotides of the same length such that the first (resp. second) half of the oligonucleotide encoding an edge equals the (Watson-Crick) complement of the second (resp. first) half of the oligonucleotide encoding the outgoing (resp. incoming) vertex of the edge.

In symbols, let x and y, $|x| = |y| = 20$, be the oligonucleotides encoding two vertices such that there is an edge from x to y. Write

$$x = x_1x_2,\ y = y_1y_2,\ |x_1| = |x_2| = |y_1| = |y_2| = 10.$$

(As already mentioned, we ignore the orientation markers $5'$ and $3'$. Consider the alphabet $\Sigma_{DNA} = \{A, G, T, C\}$, referred to as the *DNA-alphabet* in the sequel. Define the letter-to-letter morphism $h_W : \Sigma^*_{DNA} \rightarrow \Sigma^*_{DNA}$ by

$$h_W(A) = T,\ h_W(T) = A,\ h_W(C) = G,\ h_W(G) = C.$$

(The morphism h_W will be called the *Watson-Crick morphism.*) Then the edge from x to y is encoded by the oligo $h_W(x_2y_1)$.

The experiment begins by forming a "DNA soup" containing, in large quantities, oligos of the vertices, as well as of the edges of the graph. Now the ligation reaction resulting from the Watson-Crick complementarity will link together compatible edges. This "domino game" continues, identifying longer and longer paths. In this way DNA molecules encoding random paths through the graph are formed. By a filtering procedure consisting of several operations possible for DNA strands one can check whether or not paths satisfying HPP are present.

It will be seen below that any model of DNA computing, where the filtering procedures available can simulate gsm mappings, is capable of Turing machine computations. Opinions may differ as to whether computational universality is of practical relevance. It is quite reasonable to argue that one should not aim to fit DNA models too tightly to Turing models, but should rather try to completely rethink the notion of computation. For some specific classes of hard computational problems, Turing universality might not at all be important. On the other hand, in a specific model of DNA computing, where computational universality is unknown, one could suddenly face a situation that numerous practical efforts of solving a particular problem are proven to have been futile.

3 Déjà vu: Complementarity and Universality

So far quite many and diverse theoretical models have been proposed for DNA-based computing. The early ones are discussed in [10], see also [3]. While the models have been based on different ideas and principles, Watson-Crick complementarity is somehow present in a computation or derivation step. This is natural, in view of the central role of complementarity in DNA operations. A typical model of DNA computing consists of augmenting a computational aspect of complementarity with some input-output format.

A property shared by most of the models is that they produce all recursively enumerable sets, that is, are universal in the sense of Turing machines. This property seems to be completely independent, for instance, of a model being grammatical or a machine model. Complementarity, augmented with adequate input-output facilities, seems to guarantee universality.

Why is this not surprising? Because this is something we have already seen before in theoretical computer science. We will now establish a link with certain fairly old results from computability theory, with the purpose of showing that complementarity is, in fact, a source of universality. Complementarity is such a powerful tool because it brings, in a certain sense, the universal *twin-shuffle language* to the computing scene. We are now ready for the formal details.

Consider the DNA-alphabet $\Sigma_{DNA} = \{A, G, T, C\}$, as well as the Watson-Crick morphism h_W. Following chemical terminology, A and G are called *purines*, and T and C *pyrimidines*. Clearly, the square of h_W is the identity. Words over Σ_{DNA} can be viewed as single strands. Two single strands x and y are complementary (and, thus, subject to bondage) if $x = h_W(y)$ or, equivalently, $y = h_W(x)$. The morphism h_W is denoted also by an upper bar: $h_W(x) = \overline{x}$. Thus, in this notation, the double bar will be the identity: $\overline{\overline{x}} = x$. Moreover, we will view the DNA-alphabet as an extended binary alphabet $\{0, 1, \overline{0}, \overline{1}\}$, with the conventions:

$$A = 0,\ G = 1,\ T = \overline{0},\ C = \overline{1}.$$

(Observe that this agrees with the bar notation for the Watson-Crick morphism.)

A generalization of the DNA-alphabet and our extended binary alphabet is the *DNA-like alphabet*

$$V_n = \{a_1, \ldots, a_n, \overline{a_1}, \ldots, \overline{a_n}\},\ n \geq 2.$$

The letters in the unordered pairs $(a_i, \overline{a_i})$, $1 \leq i \leq n$, are called *complementary.* Again, the morphism h_W mapping each letter to its complementary one is called the *Watson-Crick morphism* and also denoted by a bar. Extending the terminology concerning the DNA-alphabet, the non-barred letters are called *purines* and the barred ones *pyrimidines.*

We can now define the *twin-shuffle language* TS. Consider the binary alphabet $\{0, 1\}$, as well as its "complement" $\{\overline{0}, \overline{1}\}$. For a word x over $\{0, 1\}$, we denote by $\bar{x}$ the word over $\{\overline{0}, \overline{1}\}$, where every letter of x is replaced by its "barred version". (Observe that the bar defines in this fashion a letter-to-letter morphism of $\{0, 1\}^*$ onto $\{\overline{0}, \overline{1}\}^*$.) For instance, if $x = 001100$, then $\bar{x} = \overline{0}\overline{0}\overline{1}\overline{1}\overline{0}\overline{0}$.

For two words x and y, denote by $x \sqcup\!\sqcup\, y$ the set of words obtained by *shuffling* x and y, without changing the order of letters in x or y. For instance, each of the words

$$0\overline{0}0\overline{0}1\overline{1}1\overline{1}0\overline{0}0\overline{0},\ \overline{0}\overline{0}\overline{1}\overline{1}\overline{0}\overline{0}001100,\ \overline{0}001\overline{0}10\overline{1}\overline{1}\overline{0}0\overline{0}$$

is in $x \sqcup\!\sqcup\, \bar{x}$, where $x = 001100$, but $00\overline{1}\overline{1}\overline{0}\overline{0}0011\overline{0}\overline{0}$ is not in $x \sqcup\!\sqcup\, \bar{x}$).

Definition 1 *The* twin-shuffle language TS *consists of all words* $x \in w \sqcup\!\sqcup\, \bar{w}$ *over the alphabet* $\{0, 1, \overline{0}, \overline{1}\}$, *where* w *over* $\{0, 1\}$ *is arbitrary. The* generalized twin-shuffle language TS_n *over the DNA-like alphabet is defined exactly as* TS *except that now* w *ranges over the words over the alphabet* $\{a_1, \ldots, a_n\}$.

We now come to the *universality* of the twin-shuffle language TS. The universality is due to the following *basic representation result* for recursively enumerable languages.

Theorem 1 *For every recursively enumerable language L, a gsm-mapping g such that $L = g(TS)$ can be effectively constructed.*

Here "gsm" refers to "generalized sequential machine", a device obtained by providing a finite automaton with outputs. The result was established in [5]. For various proofs and the history of this result, the reader is referred to [15].

The basic representation result shows why TS is universal: It remains the same for all languages. Only the mapping g (that can be viewed to constitute the input-output format) has to be specified differently according to each particular L, in other words, according to the needs of each particular "task". The result is also highly invariant, which shows its fundamental character. The reader is referred to [13,15,16,17].

A further analysis of the mapping g leads to various strengthenings of the basic representation result. Strengthenings are needed for specific needs of particular models of DNA computing. We mention the following modifications.

Theorem 2 *An arbitrary recursively enumerable language L over the alphabet V can be represented as*

$$L = p_V(TS_n \cap R)$$

where TS_n is the generalized twin-shuffle language, R is a regular language and p_V is the projection on the alphabet V (leaving letters of V unchanged and erasing other letters.) The language L can also be represented in the forms

$$L = p_V(E(g,h) \cap R), \; L = p_V(g(E(g,h)) \cap R),$$

where g, h are morphisms and $E(g,h)$ is their equality set

$$E(g,h) = \{x | g(x) = h(x)\}.$$

The items R, p, n, g, h are effectively constructable, provided L is effectively given. We refer to [15] for a proof of this modified representation result. The modified version is stronger than the basic one, because it tells us that we may restrict the attention to a particular kind of gsm-mappings. Altogether the modified version fits very well to machine models of DNA computing, [6]. Many examples are given in [10].

The representation results presented above exhibit the universality of the twin-shuffle language TS. On the other hand, the interconnection between TS and Watson-Crick complementarity is rather obvious and will be discussed below.

The interconnection between the language TS and Watson-Crick complementarity can be presented in various ways, depending on the method of reading double strands as single strands. We now discuss some such methods. Instead of the DNA-alphabet $\{A, G, T, C\}$, we use the extended binary alphabet $\{0, 1, \overline{0}, \overline{1}\}$

in the way described above. Thus (disregarding orientation) the DNA double strands Z are of the form

$$\begin{matrix} x_1 \, x_2 \ldots x_n \\ \overline{x_1} \, \overline{x_2} \ldots \overline{x_n} \end{matrix}$$

where each x_i is a letter of the extended binary alphabet, and double bars can be ignored in the way described above. We will first construct a single strand (or a word) from the double strand Z by the *up-down* method, taking letters alternately from upper and lower strands, beginning from the upper strand. The result is

$$UD(Z) = x_1\overline{x_1}x_2\overline{x_2}\ldots x_n\overline{x_n}.$$

The word $UD(Z)$ is always in TS. Indeed, words of the form $UD(Z)$ constitute the regular subset $\{0\overline{0}, \overline{0}0, 1\overline{1}, \overline{1}1\}^*$ of TS.

Consider now the reverse problem of constructing a double strand from a word in TS. Let y be a nonempty word in TS. Necessarily, y is of even length, $|y| = 2m$. Moreover, the scattered subword y' (resp. y'') of y consisting of non-barred (resp. barred) letters is of length m. For $1 \leq i \leq m$, we denote by y'_i (resp. y''_i) the ith letter of y' (resp. y''). Because y is in TS, the unordered pair (y'_i, y''_i) equals either $(0, \overline{0})$ or $(1, \overline{1})$. When we speak of y'_i or y''_i, we have these particular occurrences in mind. The occurrences may lie far apart in y. However, one of them is always to the left of the other. The left occurrence is referred to as the *up-occurrence at position i*, the right occurrence is similarly referred to as the *down-occurrence at position i*.

Consider now the double strand of length m, where for $1 \leq i \leq m$, the ith letter in the upper (resp. lower) strand is the up- (resp. down-) occurrence at position i in y. This double strand is called the *left parse* of y and denoted $LP(y)$. Clearly, $LP(y)$ satisfies the complementarity requirement for DNA double strands. Observe that LP is not injective, for instance,

$$LP(\overline{1}010 = LP(\overline{1}1\overline{0}0).$$

On the other hand, for all double strands Z, we have $LP(UD(Z)) = Z$. The equation $UD(LP(y)) = y$ is valid if y belongs to the aforementioned subset $\{0\overline{0}, \overline{0}0, 1\overline{1}, \overline{1}1\}^*$ of TS.

We have shown how to go from words in TS to DNA double strands, and vice versa. Our observations can be summarized as follows.

Theorem 3 *For any nonempty word y in the twin-shuffle language TS, $LP(y)$ is a unique DNA double strand. For any DNA double strand Z, $UD(Z)$ is a unique word in the subset $\{0\overline{0}, \overline{0}0, 1\overline{1}, \overline{1}1\}^*$ of TS. When restricted to this subset, LP is the inverse of UD.*

The strength of the representation results (such as the basic and modified result presented above) is shown also by their invariance. For instance, the universality results are not affected if one assumes that one of the strands in the double strands contains only purines (non-barred letters).

Watson-Crick complementarity is a phenomenon provided for us "for free" by nature. When bondage takes place (under ideal conditions) between two single strands, we know that the bases opposite each other are complementary. This information is "free"; there is no need to check it in any way. At a first glance, it might seem that not much information is obtained: one just reads the same information twice when investigating a double strand. However, conclusions can be made from the *history* of a double strand, from the knowledge of how it came into being. The conclusions in Adleman's experiment are made in this way. If we know how information was encoded on the DNA strands subjected to bondage, we may learn much from the fact that bondage has actually taken place.

4 Watson-Crick Finite Automata

We will now briefly discuss some models of DNA-based computing, emphasizing computational universality. The reader is referred also to [6,10,16]. The proofs, omitted here, are based on the representation theorems in the preceding section. Some kind of interplay between complementarity and TS is always essential.

Consider the DNA-like alphabet $V_n = \{a_1, \ldots, a_n, \bar{a}_1, \ldots, \bar{a}_n\}$, $n \geq 1$. Observe that in complementary pairs one member is a purine and the other a pyrimidine – exactly as in case of DNA.

Generalizing the idea of a DNA double strand, we now introduce a data structure called a *double strand over* V_n. The set of all double strands over V_n will be denoted by $DS(V_n)$. By definition, $DS(V_n)$ is the free monoid generated by all ordered pairs $(a_i, \bar{a}_i)$ and $(\bar{a}_i, a_i)$, where i runs through the numbers $1, \ldots, n$.

Thus, an element Z of $DS(V_n)$ is a catenation of pairs of the form $(a_i, \bar{a}_i)$ and $(\bar{a}_i, a_i)$. The first (resp. second) members in the pairs define a word, referred to as the *upper* (resp. *lower*) *strand* of Z. We denote double strands similarly as before. For instance,

$$(a_2, \bar{a}_2)(\bar{a}_1, a_1)(a_1, \bar{a}_1)(a_1, \bar{a}_1) = \begin{matrix} a_2\bar{a}_1a_1a_1 \\ \bar{a}_2a_1\bar{a}_1\bar{a}_1 \end{matrix}$$

where $a_2\bar{a}_1a_1a_1$ (resp. $\bar{a}_2a_1\bar{a}_1\bar{a}_1$) is the upper (resp. lower) strand.

Definition 2 *A* Watson-Crick finite automaton *is a construct*

$$\mathcal{A} = (V_n,\ Q,\ q_0,\ F,\ \delta),$$

where V_n is a DNA-like alphabet, Q is a finite set (states)*, $q_0 \in Q$ (*initial state*), $F \subseteq Q$ (*final states*), and δ is a finite set of ordered quadruples $(q,\ x_u,\ x_d,\ q')$, where $q,\ q' \in Q$ and $x_u,\ x_d \in V_n^*$ (*transitions*).*

Intuitively, the automaton $\mathcal{A}$ consists of a control box, being at any moment in one of finitely many possible states, and two read-only heads, moving independently of each other from left to right. The inputs of $\mathcal{A}$ are double strands Z belonging to $DS(V_n)$. The overall behavior of $\mathcal{A}$ is nondeterministic. The reading heads H_u and H_d read the upper and lower strands of Z, respectively. Consider a quadruple $(q,\ x_u,\ x_d,\ q')$. Assume that $\mathcal{A}$ is in the state q and that x_u (resp. x_d)

appears in the upper (resp. lower) strand of Z, immediately to the right of the position scanned by H_u (resp. H_d). Then it is possible for $\mathcal{A}$ to pass H_u (resp. H_d) over x_u (resp. x_d) and go to the state q'. (Observe that δ might contain also the quadruple $(q,\ x'_u,\ x'_d,\ q'')$, where x'_u and x'_d are prefixes of x_u and x_d, respectively. Then $\mathcal{A}$ can also go to the state q'', by reading x'_u and x'_d.) Note that one or both of the middle elements in the quadruples may be the empty word. Thus, it is possible for the automaton $\mathcal{A}$ to read a portion of *one strand* only during a step of the computation.

A double strand Z belonging to $DS(V_n)$ is *accepted* by the automaton $\mathcal{A}$ if the following computation is possible, using the quadruples in δ. Initially, $\mathcal{A}$ is in the state q_0 and the head H_u (resp. H_d) is positioned to the left of the whole upper (resp. lower) strand of Z. At the end of the computation, $\mathcal{A}$ is in a state belonging to F and the head H_u (resp. H_d) is positioned to the right of the whole upper (resp. lower) strand of Z.

(This informal definition can be formalized in the usual way by introducing instantaneous descriptions and a yield relation for computational steps, based on δ. In such a formalization the elements of $DS(V_n)$ have to be decomposed further, because the heads move at different speeds. A similar further decomposition will be needed also below, in connection with matching systems.)

The language $L(\mathcal{A}) \subseteq V_n^*$ *accepted* by the Watson-Crick finite automaton $\mathcal{A}$ consists of all upper strands of the double strands Z accepted by $\mathcal{A}$.

We are now ready to state the universality result. Several modifications of it are possible. By a *weak coding* we mean a morphism mapping every letter either to a letter or to the empty word.

Theorem 4 *Every recursively enumerable language is a weak coding of a language accepted by a Watson-Crick finite automaton.*

A Watson-Crick finite automaton, as a construct, is essentially the same as a finite automaton with two one-way reading heads. (For such automata, though, we are not aware of any result analogous to the above theorem.) There is no difference if one reads twice the same single strand or once a double strand. However, in considerations involving DNA-based computing, there is a difference: something can be learned from the *history* of the double strands, how they came into being.

5 Matching Systems

As another case study, we consider a grammatical model. We need the set of *extended double strands* over V_n, in symbols, $EDS(V_n)$. Intuitively, elements of $EDS(V_n)$ are double strands having a "sticky" single strand at the end.

Define the alphabets

$$V_n^u = \{(x, \#) | x \in V_n\},\ V_n^d = \{(\#, x) | x \in V_n\}.$$

Intuitively, # is used to denote blank. The letters of V_n^u (resp. V_n^d) are used to construct "sticky" upper (resp. lower) strands which may become double strands by rules of complementarity.

By definition,

$$EDS(V_n) = DS(V_n)((V_N^u)^* \cup (V_n^d)^*).$$

Definition 3 *A* matching system *is a construct*

$$\mathcal{M} = (V_n,\ A,\ S_u,\ S_d),$$

where V_n is a DNA-like alphabet, $A \subseteq EDS(V_n)$ is a finite set (axioms)*, and $S_u \subseteq (V_n^u)^*$, $S_d \subseteq (V_n^d)^*$ are finite sets (*upper *and* lower strands, *respectively).* A matching system $\mathcal{M}$ *generates* a set

$$S(\mathcal{M}) \subseteq EDS(V_n)$$

as follows. Each element $Z \in S(\mathcal{M})$ possesses a *derivation*, a finite sequence of elements of $EDS(V_n)$. The derivation begins with an element of A and ends with Z. Every element of the derivation is obtained from the preceding element by extending it with some element of S_u or S_d. Elements of S_d (resp. S_u) have to be used to extend elements of $DS(V_n)(V_n^u)^*$ (resp. $DS(V_n)(V_n^d)^*$). (Thus, elements of $DS(V_n)$ can be extended by elements of both S_d and S_u.) The result of the extension has to be again an element of $EDS(V_n)$. Each element of S_d and S_u can be used arbitrarily many times in the same derivation.

The formal definition, based again on a yield relation, should be clear by the above informal remarks. Let us consider an example of a matching system $\mathcal{M}$. The alphabet will be the DNA alphabet $\{A, G, T, C\}$. The only axiom is $\begin{array}{c} CAT \\ G\#\# \end{array}$ and the set S_u has the strands $\begin{array}{c} GT \\ \#\# \end{array}$ and $\begin{array}{c} AT \\ \#\# \end{array}$ as its elements, whereas $\begin{array}{c} \#\#\#\# \\ TACA \end{array}$ is the only element of S_d.

Every derivation necessarily begins with the sequence

$$\begin{array}{c} CAT \\ G\#\# \end{array} \Rightarrow \begin{array}{c} CAT\#\# \\ GTACA \end{array} \Rightarrow \begin{array}{c} CATGT \\ GTACA \end{array}$$

After this, any of the three elements of $S_u \cup S_d$ can be applied. However, the application of $\begin{array}{c} GT \\ \#\# \end{array}$ leads to the element

$$\begin{array}{c} CATGTGT \\ GTACA\#\# \end{array}$$

after which the derivation necessarily blocks. The other two elements of $S_u \cup S_d$ lead to an extended double strand from which it is possible to continue. Altogether the possibilities are very limited, and the reader should have no difficulties in characterizing the set $S(\mathcal{M})$.

The elements of $S(\mathcal{M})$ are extended double strands. A language can be "squeezed out" of $S(\mathcal{M})$ in various ways. Only regular languages result in this way in the above setup. This is due to the unrestricted use of elements in $S_u \cup S_d$. To reach universality, some restrictions must be imposed on the derivations.

Consider a matching system $\mathcal{M}$. Assume that the sets S_u and S_d are of the same cardinality k. Label the elements of S_u and S_d by the integers $1, \ldots, k$. The *upper* (resp. *lower*) *control word* $C_u(D)$ (resp. $C_d(D)$) of a derivation D is the sequence of labels of elements of S_u (resp. S_d) used in the derivation D, in their order of application. A derivation D is termed *coherent* if $C_u(D) = C_d(D)$ and, moreover, the last element of D is a double strand, that is, an element in $DS(V_n)$. The *coherent set* $S_c(\mathcal{M}) \subseteq DS(V_n)$ generated by a matching system $\mathcal{M}$ (satisfying the assumption about the cardinalities of S_u and S_d) consists of the last elements of coherent derivations according to $\mathcal{M}$.

Theorem 5 *Every recursively enumerable language $L \subseteq \Sigma^*$ can be represented in the form*

$$L = f(g^{-1}(S_c(\mathcal{M}))),$$

for some matching system $\mathcal{M}$, morphism $g : \Sigma^* \to DS(V_n)$ and weak coding $f : \Sigma^* \to \Sigma^*$.

6 Lindenmayer Systems and Complementarity

Complementarity can be viewed also as a language-theoretic operation. As such h_W is only a morphism of a special kind. However, the operational complementarity can be considered also as a tool in a developmental model: undesirable conditions in a string *trigger* a transition to the complementary string. Thus, the class of "bad" strings is somehow specified. Whenever a bad string x is about to be produced by a generative process, the string $h_W(x)$ is taken instead of x. If the generative process produces a unique sequence of strings (words), the sequence continues from $h_W(x)$. The class of bad strings has to satisfy the following *soundness condition*: whenever x is bad, the complementary string $h_W(x)$ is not bad. This condition guarantees that no bad strings are produced.

While the operational complementarity can be investigated in connection with any generative process for words, it seems particularly suitable for *Lindenmayer systems*, the systems themselves being developmental models. The simplest L system, namely the D0L system, has been thoroughly investigated, [11]. A D0L system generates a sequence of words. When it is augmented with a trigger for complementarity transitions, as described above, the resulting sequences contain no bad words. However, even very simple triggers yield amazing computational possibilities. We will now mention some fundamental results concerning such *Watson-Crick D0L systems.* Because of their strength, especially in comparison with ordinary D0L systems, they have been quite widely studied, for instance, see [9,8,18,19,20,22,21,4]. For completeness, we begin with the definition of a D0L system.

Definition 4 *A* D0L system *is a triple* $G = (\Sigma, g, w_0)$, *where* Σ *is an alphabet,* $w_0 \in \Sigma^*$ *(the* axiom*) and* g *is an endomorphism of* Σ^*. *A D0L system defines the* sequence $S(G)$ *of words* w_i, $i \geq 0$, *where* $w_{i+1} = g(w_i)$, *for all* $i \geq 0$. *It defines also the* language $L(G)$, *consisting of all words in* $S(G)$, *the* length sequence $|w_i|$, $i \geq 0$, *as well as the* growth function $f(i) = |w_i|$.

A Watson-Crick D0L system is simply a D0L system augmented with a trigger. Below we consider just one trigger, namely, the language over the DNA-like alphabet, consisting of words, where the pyrimidines form a majority. Such Watson-Crick D0L systems are customarily referred to as *standard* in the literature. We will now give the formal definitions.

For a word x over a DNA-like alphabet V_n, we denote by $pur(x)$ (resp. $pyr(x)$) the scattered subword of x consisting of all purines (resp. pyrimidines) in x. (Recall that pyrimidines are the barred letters.) Thus, for any x,

$$|x| = |pur(x)| + |pyr(x)| \text{ and } |pur(x)| = |pyr(h_W(x))|.$$

These equations and the definition of the Watson-Crick morphism h_W give rise to the following basic lemma.

Lemma 1 *Let* x *be an arbitrary word over the DNA-like alphabet* V_n. *Then either* $|pur(x)| = \frac{|x|}{2}$ *or else exactly one of the inequalities*

$$|pur(x)| > \frac{|x|}{2} \text{ and } |pur(h_W(x))| > \frac{|x|}{2}$$

holds.

Definition 5 *A* Watson-Crick D0L system *is a construct* $G = (V_n,\ g,\ w_0)$, *where* V_n *is a DNA-like alphabet,* $g : V_n^* \to V_n^*$ *is a morphism and* $w_0 \in V_n^*$ *(the* axiom*) satisfies* $2|pur(w_0)| \geq |w_0|$.

The sequence $S_W(G)$ *generated by a Watson-Crick D0L system* G *is the sequence of words* $w_0,\ w_1,\ w_2, \ldots$, *defined by*

$$w_{i+1} = \begin{cases} g(w_i) \text{ if } |pur(g(w_i))| \geq \frac{|g(w_i)|}{2}, \\ h_W(g(w_i)), \text{ otherwise,} \end{cases}$$

where $i \geq 0$. *The* language, length sequence *and* growth function *of a Watson-Crick D0L system are defined as for D0L systems.*

The following result is now immediate.

Lemma 2 *Every word* x *in the sequence* $S_W(G)$ *generated by a Watson-Crick D0L system* G *satisfies the condition* $|pur(x)| \geq \frac{|x|}{2}$.

A celebrated problem concerning D0L systems, whose solution turned out to be rather involved, [11], is the *(sequence) equivalence problem*: given D0L systems G_1 and G_2, decide whether or not $S(G_1) = S(G_2)$. The same problem can be posed also when G_1 and G_2 are Watson-Crick D0L systems: decide whether or not $S_W(G_1) = S_W(G_2)$. It is not our purpose to enter the intricacies of this problem. However, it is easy to see that there are D0L systems G_1 and G_2 such that $S(G_1) = S(G_2)$ but $S_W(G_1) \neq S_W(G_2)$, and also D0L systems G_3 and G_4 such that $S(G_3) \neq S(G_4)$ but $S_W(G_3) = S_W(G_4)$.

Growth functions of Watson-Crick D0L systems can exhibit very weird behavior, even in the case of the 4-letter DNA-alphabet, [9,20].

We now come to the universality results. A Watson-Crick D0L *scheme* is a Watson-Crick D0L system without the axiom.

Definition 6 *Consider a Watson-Crick D0L scheme G_W. A partial recursive function f mapping a subset of the set of nonnegative integers into nonnegative integers is* computed *by G_W if its alphabet contains the letters B, b, E, e with the productions $E \to E$ and $e \to e$ and satisfying the following condition. For all $i \geq 0$, the equation $f(i) = j$ holds exactly in case there is a derivation according to G_W*

$$Bb^i \Longrightarrow^* Ee^j$$

and, moreover, the letters E and e appear in this derivation at the last step only. A function f is Watson-Crick computable *if it is computed by some Watson-Crick D0L scheme G_W.*

(We have used here the customary terminology involving productions and derivations.) The universality result, due originally to [22], can now be stated as follows.

Theorem 6 *Every partial recursive function is Watson-Crick computable.*

Language-theoretic counterparts will be expressed in the following theorem, [4]. The definitions of the Lindenmayer systems involved can be found in [11], whereas their Watson-Crick variants are defined analogously to Watson-Crick D0L systems.

Theorem 7 *Let L be a recursively enumerable language. Then there exists a Watson-Crick EDT0L system Σ such that $L = L(\Sigma)$. Moreover, the number of nonterminals of Σ is bounded by a constant not depending on L. There exists (effectively) also a Watson-Crick EDT0L system Σ with two tables such that $L = L(\Sigma)$, as well as a Watson-Crick E0L system Σ such that $L = L(\Sigma)$.*

7 Conclusion

We are still lacking a *killer app* of DNA-based computing, that is, an application which

- fits the DNA model,
- cannot be solved by the current or even future electronic machines,
- is of value (people are willing to pay for it).

Therefore, we do not want make any predictions about the ultimate fate of DNA-based computing, or its success or failure versus *quantum computing.*

We conclude by emphasizing some issues discussed above, as well as raising some questions.

- Watson-Crick complementarity amounts to the presence of the twin-shuffle language TS.
- The number *four* of the bases A, G, T, C is ideal: we get exactly the alphabet of TS.
- Universality results remain valid under several restrictions of the double strands. For instance, one strand may contain only purines A, G. What is the significance (computing, biology) of such restrictions?
- The pair (G, C) bonds stronger than the pair (A, T). What are the implications (computing, biology)?

References

1. L.M. Adleman, P.W.K. Rothemund, S. Roweiss and E. Winfree, On applying molecular computations to the Data Encryption Standard. In: E. Baum, D. Boneh, P. Kaplan, R. Lipton, J. Reif and N. Seeman (eds.), DNA Based Computers. Proc. of the Second Annual Meeting, Princeton (1996) 28-48.
2. L. M. Adleman, Molecular computation of solutions to combinatorial problems. Science 266 (1994) 1021–1024.
3. M. Amos, G. Paun, G. Rozenberg and A. Salomaa, DNA-based computing: a survey. To appear in Theoretical Computer Science.
4. J. Csima, E. Csuhaj Varjú and A. Salomaa, Power and size of extended Watson-Crick L systems, TUCS report 424, Turku Centre for Computer Science, Turku, 2001, to appear in Theoretical Computer Science.
5. J. Engelfriet and G. Rozenberg, Fixed-point languages, equality languages, and representations of recursively enumerable languages. J.Assoc.Comput.Mach. 27 (1980) 499-518.
6. R. Freund, G. Paun, G. Rozenberg and A. Salomaa, Watson-Crick finite automata. Proceedings of the 3rd DIMACS Conf. on DNA Based Computers, 1997, 305–317.
7. D. Gifford, On the path to computation with DNA. Science 266 (1994) 993–994.
8. J. Honkala and A. Salomaa, Watson-Crick D0L systems with regular triggers. Theoretical Computer Science 259 (2001) 689–698.
9. V. Mihalache and A. Salomaa, Language-theoretic aspects of DNA complementarity. Theoretical Computer Science 250 (2001) 163-178.
10. G. Păun, G. Rozenberg and A. Salomaa, DNA Computing. New Computing Paradigms. Springer-Verlag, Berlin, Heidelberg, New York (1998).
11. G. Rozenberg and A. Salomaa (eds.), Handbook of Formal Languages, Vol. 1–3. Springer-Verlag, Berlin, Heidelberg, New York, 1997.
12. G. Rozenberg and A. Salomaa, Watson-Crick complementarity, universal computations and genetic engineering. Leiden University, Computer Science Technical Report 28 (1996).
13. G. Rozenberg and A. Salomaa, DNA computing: new ideas and paradigms. Springer LNCS 1644 (1999) 106-118.

14. A. Salomaa, Formal Languages. Academic Press, New York, 1773.
15. A. Salomaa, Jewels of Formal Language Theory. Computer Science Press, Rockville, Md. (1981).
16. A. Salomaa, Turing, Watson-Crick and Lindenmayer. Aspects of DNA complementarity. In: C.S. Calude, J. Casti and M.J. Dinneen (eds.), Unconventional Models of Computation. Springer-Verlag, Singapore (1998) 94-107.
17. A. Salomaa, Computability paradigms based on DNA complementarity. In V. Keränen (ed.), Innovation in Mathematics, Proc. 2nd Intern. Mathematica Symposium, Computational Mechanics Publications, Southampton, Boston (1997) 15–28.
18. A. Salomaa, Watson-Crick walks and roads in D0L graphs. Acta Cybernetica 14 (1999) 179–192.
19. A. Salomaa, Uni-transitional Watson-Crick D0L systems. TUCS report 389, Turku Centre for Computer Science, Turku, 2001, to appear in Theoretical Computer Science.
20. A. Salomaa, Iterated morphisms with complementarity on the DNA alphabet. In M. Ito, G. Paun and S. Yu (eds.) Words, Semigroups, Transductions, World Scientific Publ. Co. (2001) 405–420.
21. A. Salomaa and P. Sosík, Watson-Crick D0L systems: the power of one transition. TUCS report 439, Turku Centre for Computer Science, Turku, 2002. Submitted for publication.
22. P. Sosík, D0L Systems + Watson-Crick Complement = Universal Computation. Springer LNCS 2055 (2001) 308–320.

On Higher Arthur-Merlin Classes

Jin-Yi Cai[1,*], Denis Charles[2,**], A. Pavan[3,**], and Samik Sengupta

[1] Computer Sciences Department, University of Wisconsin, Madison, WI 53706
jyc@cs.wisc.edu
[2] Computer Sciences Department, University of Wisconsin, Madison, WI 53706
cdx@cs.wisc.edu
[3] NEC Research Institute, 4 Independence Way, Princeton, NJ 08540
apavan@research.nj.nec.com
[4] Department of Computer Science and Engineering, University at Buffalo, Buffalo, NY 14260
samik@cse.buffalo.edu

Abstract. We study higher Arthur-Merlin classes defined via several natural probabilistic operators BP, R and coR. We investigate the complexity classes they define, and a number of interactions between these operators and the standard polynomial time hierarchy. We prove a hierarchy theorem for these higher Arthur-Merlin classes involving interleaving operators, and a theorem giving non-trivial upper bounds to the intersection of the complementary classes in the hierarchy.

1 Introduction

Arthur-Merlin Games were introduced by Babai [6,7] to study the power of randomization in interaction. Goldwasser and Sipser [13] proved soon afterwards that these classes are equivalent in power to Interactive Proof Systems introduced by [12]. In the last 15 years, this study has proved to be exceedingly successful in complexity theory [21,9,22,16,17]. Eventually the study of these proof systems (and multi prover systems) led to perhaps the most spectacular achievement in the Theory of Computing in the last decade [16,17,4,3].

It is well known that some traditional complexity classes can be characterized by operators. For example $\mathrm{NP} = \exists \cdot \mathrm{P}$, $\Sigma_2 = \exists \cdot \forall \cdot \mathrm{P}$, using the P-time bounded existential and universal operators $\exists$ and $\forall$ respectively (we omit the superscript P in this paper). They have been used fruitfully to prove (or to give simpler proofs of known) relations between complexity classes [22,18,20,19]. For example, Toda's Theorem was proved in this framework [18]. In this approach, Zachos and Fürer [22] showed that AM can also be characterized by the operator BP, i.e. $\mathrm{AM} = \mathrm{BP} \cdot \mathrm{NP}$. By successfully employing this operator machinery, they were able to give a simple and natural proof that the one-sided error version of AM coincides with the two-sided error version, and thus $\mathrm{AM} \subseteq \Pi_2$.

* Research supported in part by NSF grant CCR-0196197
** Work done while the author was at University at Buffalo

O.H. Ibarra and L. Zhang (Eds.): COCOON 2002, LNCS 2387, pp. 18–27, 2002.

In this paper, we define higher Arthur-Merlin classes using several natural probabilistic operators BP, R and co-R. We consider the hierarchy of classes $\mathrm{BP} \cdot \Sigma_k$, for $k \geq 1$, and prove some intricate relations among these classes and Σ_ℓ and ZPP. The class $\mathrm{BP} \cdot \Sigma_k$ can also be thought of as an Arthur-Merlin game where Arthur has access to Σ_{k-1} oracle. By investigating the interactions between the BP operators and the polynomial-time hierarchy, we are able to obtain some non-trivial containments; for example, we show that $\Sigma_k^{\mathrm{AM}} = \Sigma_{k+1}$ for $k \geq 2$, whereas the naïve argument gives $\Sigma_k^{\mathrm{AM}} \subseteq \Sigma_{k+2}$ (a special case of this is noted in [8] p. 253). Similarly, by interleaving n levels of Σ_k classes and n levels of $\mathrm{BP} \cdot \Sigma_\ell$ classes, we obtain results that improve trivial containments by n levels.

Recently classes $\mathrm{ZPP}^{\mathrm{NP}}, \mathrm{S}_2^{\mathrm{P}}$ and $\mathrm{AM} \cap \text{co-AM}$ have received much attention. Arvind and Köbler [1] and Goldreich and Zuckerman [14] proved that MA is contained in $\mathrm{ZPP}^{\mathrm{NP}}$. Köbler and Watanabe [15] and Bshouty *et al.* [10] proved that the polynomial-time hierarchy is in $\mathrm{ZPP}^{\mathrm{NP}}$, if NP has polynomial-size circuits. More recently Arvind and Köbler [2] proved that $\mathrm{AM} \cap \text{co-AM}$ is *low* for $\mathrm{ZPP}^{\mathrm{NP}}$, i.e., $\mathrm{ZPP}^{\mathrm{NP}^{\mathrm{AM} \cap \text{co-AM}}} \subseteq \mathrm{ZPP}^{\mathrm{NP}}$. Cai [11] has shown that $\mathrm{S}_2^{\mathrm{P}} \subseteq \mathrm{ZPP}^{\mathrm{NP}}$, this combined with the observation of Sengupta that if NP has polynomial-size circuits then $\mathrm{PH} \subseteq \mathrm{S}_2^{\mathrm{P}}$ gives the strongest version of the Karp-Lipton theorem.

We consider a generalization of the class $\mathrm{AM} \cap \text{co-AM}$. Let E_k be $\mathrm{BP} \cdot \Sigma_k \cap \mathrm{BP} \cdot \Pi_k$. We prove $\mathrm{NP}^{E_1} \subseteq \mathrm{ZPP}^{\mathrm{NP}}$. We then prove a far reaching extension of this upper bound. Arvind and Köbler's result follows as a corollary, and thereby, we obtain an alternative proof of their lowness theorem. By interleaving n levels of E_k and n levels of Σ_l classes, we obtain an upper bound that improves over the trivial bound by n levels.

Our proofs illustrate the power of the operator machinery, when properly deployed. The apparent ease (even a tautness) with which these theorems on the hierarchy are revealed speaks to its effectiveness. While the formalism of this operator machinery may appear austere and less vivid intuitively, it should also be pointed out that only through such a formalism can we even begin to state some of the delicate relations in this hierarchy. While lacking intuitive appeal, there is also a certain beauty in its succinctness. We also note that a certain amount of care must be exercised in giving the correct definition of these classes by operators. For instance in [8] on page 243, the authors claim that $\mathrm{MA} = \mathrm{NP} \cdot \mathrm{BPP}$, but this is incorrect. The predicate which is used to define the BPP language need not exhibit a BPP type computation, if we have a "Yes" instance but the non-deterministic guess is not the "correct" Merlin proof.

2 Preliminaries

We give formal definitions in this section. An Arthur-Merlin game is a combinatorial game, played by Arthur–a probabilistic polynomial-time machine (with public coins), and Merlin–a computationally unbounded Turing machine.

Given a string x, Merlin tries to convince Arthur that x belongs to some language L. The game consists of a finite number of moves. In the end Arthur either accepts or rejects x. Babai [6,7] defines a language L to be in AM as follows. For every string x of length n, the game consists of a random move by Arthur and a reply by Merlin. If $x \in L$, then the probability that there exists a move by Merlin that leads to acceptance by Arthur is at least $\frac{3}{4}$; on the other hand, if $x \notin L$, then the probability that there exists a move by Merlin that leads to acceptance by Arthur is at most $\frac{1}{4}$.

This definition has error on both sides. One can define the one-sided error version, temporarily denoted as AM_1, which is defined the same way as AM, except that when x is in the language then *for every* random choice by Arthur there exists a move by Merlin that leads to the acceptance of x by Arthur.

Babai [6,7] also defined the class MA. Here Merlin, instead of Arthur, moves first.

Proposition 1. *We know the following relationships about* AM. *i)* $\mathrm{AM} = \mathrm{AM}_1$ *[22], ii)* $\mathrm{AM} \subseteq \Pi_2$ *[7], iii)* $\mathrm{AM} \cap \text{co-AM} \subseteq \mathrm{ZPP}^{\mathrm{NP}}$*, and iv)* $\mathrm{AM} \subseteq \text{co-RP}^{\mathrm{NP}} \subseteq \mathrm{BPP}^{\mathrm{NP}}$.

The following inclusions are known about MA: i) $\mathrm{MA} \subseteq \mathrm{S}_2^{\mathrm{P}} \subseteq \mathrm{ZPP}^{\mathrm{NP}}$ [1,14,11], and ii) $\mathrm{NP}^{\mathrm{BPP}} \subseteq \mathrm{MA}$ [22].

The probabilistic operator BP is the following:

Definition 1. *Given a class $\mathcal{C}$, a language L is in* $\mathrm{BP}\cdot\mathcal{C}$ *if there exists a language L' in $\mathcal{C}$ and a polynomial $p(\cdot)$ such that the following holds:*

$$x \in L \Longrightarrow \Pr_{r\in\{0,1\}^{p(n)}}[\langle x,r\rangle \in L'] \geq \frac{3}{4}$$

$$x \notin L \Longrightarrow \Pr_{r\in\{0,1\}^{p(n)}}[\langle x,r\rangle \in L'] \leq \frac{1}{4}.$$

Zachos and Fürer [22] showed that $\mathrm{AM} = \mathrm{BP}\cdot\mathrm{NP}$. We can define the one-sided error version of $\mathrm{BP}\cdot\mathcal{C}$ using the operator co-R. Again, the difference is if $x \in L$ then for every $r \in \{0,1\}^{p(n)}$ the tuple $\langle x,r\rangle$ is in L'. Then, by using the fact that $\mathrm{AM} = \mathrm{AM}_1$ we obtain $\mathrm{BP}\cdot\mathrm{NP} = \text{co-R}\cdot\mathrm{NP}$. Similarly we can define the operator R. Note that for any complexity class $\mathcal{C}$, $\text{co-}(\mathrm{R}\cdot\mathcal{C}) = \text{co-R}\cdot\text{co-}\mathcal{C}$.

We define higher Arthur-Merlin classes using these operators. Observe that we do not obtain any new classes by applying the operator R (co-R) to Σ_k (Π_k respectively). Therefore, only $\text{co-R}\cdot\Sigma_k$ and $\mathrm{R}\cdot\Pi_k$ are possibly new classes, and since $\mathrm{AM} = \text{co-R}\cdot\mathrm{NP}$, $\text{co-R}\cdot\Sigma_k$ will be a generalization of AM. However, as the following proposition shows, we can equivalently consider $\mathrm{BP}\cdot\Sigma_k$ as an extension of AM.

Proposition 2. *Proposition 1 can be easily extended as follows: i)* $\mathrm{BP}\cdot\Sigma_k = \text{co-R}\cdot\Sigma_k$*, ii)* $\text{co-R}\cdot\Sigma_k \subseteq \Pi_{k+1}$*, iii)* $\text{co-R}\cdot\Sigma_k \cap \mathrm{R}\cdot\Pi_k \subseteq \mathrm{ZPP}^{\Sigma_k}$*, and iv)* $\mathrm{BP}\cdot\Sigma_k \subseteq \text{co-RP}^{\Sigma_k} \subseteq \mathrm{BPP}^{\Sigma_k}$.

Now, we consider the relativizations of the class AM. Given a class $\mathcal{C}$, $\mathrm{AM}^{\mathcal{C}}$ is defined similar to the class AM where Arthur has access to some $L \in \mathcal{C}$ as the oracle. This generalization coincides with the classes we obtained by operators; namely, we can prove the following, given a class of languages $\mathcal{C}$,

Proposition 3. $\mathrm{AM}^{\mathcal{C}} = \text{co-R} \cdot \mathrm{NP}^{\mathcal{C}} = \mathrm{BP} \cdot \mathrm{NP}^{\mathcal{C}}$.

Hence, in particular $\mathrm{AM}^{\Sigma_k} = \mathrm{BP} \cdot \Sigma_{k+1}$. Similarly, given class $\mathcal{C}$, we can define the class $\mathrm{MA}^{\mathcal{C}}$. Arvind and Köbler [2] proved a non-trivial result about the class $\mathrm{AM} \cap \text{co-AM}$. They showed that $\mathrm{AM} \cap \text{co-AM}$ is *low* for $\mathrm{ZPP}^{\mathrm{NP}}$, which means $\mathrm{ZPP}^{\mathrm{NP}^{\mathrm{AM} \cap \text{co-AM}}} \subseteq \mathrm{ZPP}^{\mathrm{NP}}$. Note that a naïve attempt, that just uses the fact $\mathrm{AM} \cap \text{co-AM} \subseteq \mathrm{ZPP}^{\mathrm{NP}}$, yields $\mathrm{ZPP}^{\mathrm{NP}^{\mathrm{AM} \cap \text{co-AM}}} \subseteq \mathrm{ZPP}^{\Sigma_2}$.

The rest of the paper is organized as follows. In section 3, we shall prove results about $\mathrm{BP} \cdot \Sigma_k$, culminating in Theorem 2 about the hierarchy with interleaving levels of $\mathrm{BP} \cdot \Sigma_l$ and Σ_k. In section 4, we concentrate on $\mathrm{BP} \cdot \Sigma_k \cap \mathrm{BP} \cdot \Pi_k$. We extend the lowness results about $\mathrm{AM} \cap \text{co-AM}$ and prove Theorem 5 about interleaving levels of $\mathrm{BP} \cdot \Sigma_l \cap \mathrm{BP} \cdot \Pi_l$ and Σ_k. Finally, in section 5, we give examples of languages in $\mathrm{BP} \cdot \Sigma_k$, for every $k \geq 2$.

3 On $\mathrm{BP} \cdot \Sigma_k$

We begin with the following result and generalize it later.

Lemma 1. $\mathrm{NP}^{\mathrm{AM}} \subseteq \mathrm{MA}^{\mathrm{NP}}$.

Proof. (**Sketch.**) Let $L \in \mathrm{NP}^{\mathrm{AM}}$. Then there is a language $A \in \mathrm{AM}$ and a polynomial-time bounded non-deterministic oracle Turing machine N^A, that accepts the language L. We will assume (without loss of generality) that the predicate that we use to decide A has 1-sided error, and our proof will yield a predicate that has 2-sided error. However the 2-sided error version of $\mathrm{MA}^{\mathrm{NP}}$ is the same as the 1-sided error version. Consider the following $\mathrm{MA}^{\mathrm{NP}}$ protocol: Merlin provides an accepting computation of N^A with query answers filled in. Arthur, here, is a $\mathrm{BPP}^{\mathrm{NP}}$ machine. Since both AM and co-AM are in $\mathrm{BPP}^{\mathrm{NP}}$, Arthur can verify that the path given by Merlin is indeed a correct accepting computation. □

The following theorem is a natural extension of the above Lemma.

Theorem 1.

$$\mathrm{NP}^{\mathrm{BP} \cdot \Sigma_l} \subseteq \mathrm{MA}^{\Sigma_l} \subseteq \mathrm{S}_2^{\mathrm{P}^{\Sigma_l}} \subseteq \mathrm{ZPP}^{\Sigma_{l+1}}, \textit{ for } l \geq 0.$$

Proof. Note that when $l = 0$, the theorem is saying $\mathrm{NP}^{\mathrm{BPP}} \subseteq \mathrm{MA} \subseteq \mathrm{S}_2^{\mathrm{P}} \subseteq \mathrm{ZPP}^{\mathrm{NP}}$. All these inclusions are known [1,14,22,11]. The proof of the case $l \geq 1$ is similar to the proof of previous lemma. For the first containment, Merlin gives an accepting computation of the NP machine with query answers filled in. Arthur, now a BPP^{Σ_l} machine can verify the all the query answers are correct, as $\mathrm{BP} \cdot \Sigma_l$ and $\text{co-BP} \cdot \Sigma_l$ are subsets of BPP^{Σ_l}. The second inclusion follows from relativizing the proof of $\mathrm{MA} \subseteq \mathrm{S}_2^{\mathrm{P}} \subseteq \mathrm{ZPP}^{\mathrm{NP}}$. □

The following two equalities can be easily proved using the fact that $\mathrm{NP}^{\mathrm{ZPP}} = \mathrm{NP}$. Here $k \geq 1$ and $l \geq 0$.

Lemma 2. *i)* $\Sigma_k^{\mathrm{ZPP}^{\Sigma_l}} = \Sigma_{k+l}$, *ii)* $\mathrm{BP} \cdot \Sigma_k^{\mathrm{ZPP}^{\Sigma_l}} = \mathrm{BP} \cdot \Sigma_{k+l}$.

Next, we consider a hierarchy with interleaving levels of $\mathrm{BP}\cdot\Sigma_l$ and Σ_k. The upper bound in this theorem is an improvement of n levels over the trivial bounds.

Theorem 2. *1. For $k_1,\ldots,k_n \geq 1$ and $l_1,\cdots l_n \geq 0$, where $n \geq 1$*

$$\Sigma_{k_1}^{\mathrm{BP}\cdot\Sigma_{l_1}^{\cdot^{\cdot^{\Sigma_{k_n}^{\mathrm{BP}\cdot\Sigma_{l_n}}}}}} \subseteq \mathrm{ZPP}^{\Sigma_{k_1+k_2+\cdots+k_n+l_1+l_2+\cdots+l_n}}.$$

2. For $k_1 \geq 2, k_2,\ldots,k_n \geq 1$, and $l_1,\cdots l_n \geq 0$, where $n \geq 1$

$$\Sigma_{k_1}^{\mathrm{BP}\cdot\Sigma_{l_1}^{\cdot^{\cdot^{\Sigma_{k_n}^{\mathrm{BP}\cdot\Sigma_{l_n}}}}}} = \Sigma_{k_1+k_2+\cdots+k_n+l_1+l_2+\cdots+l_n}.$$

Proof. We prove the first equality, by induction on n, the number of levels. Our base case is

$$\Sigma_k^{\mathrm{BP}\cdot\Sigma_l} \subseteq \mathrm{ZPP}^{\Sigma_{k+l}}, \text{ for } k \geq 1, l \geq 0,$$

which we shall prove by induction on k. We know, by Theorem 1, that $\mathrm{NP}^{\mathrm{BP}\cdot\Sigma_l} \subseteq \mathrm{ZPP}^{\Sigma_{l+1}}$. So when $k=1$ the base case is true. Now consider the case $k \geq 2$. Assume $\Sigma_{k-1}^{\mathrm{BP}\cdot\Sigma_l} \subseteq \mathrm{ZPP}^{\Sigma_{k+l-1}}$, for $k \geq 2$. Now,

$$\Sigma_k^{\mathrm{BP}\cdot\Sigma_l} = \mathrm{NP}^{\Sigma_{k-1}^{\mathrm{BP}\cdot\Sigma_l}} \subseteq \mathrm{NP}^{\mathrm{ZPP}^{\Sigma_{k+l-1}}} = \Sigma_{k+l} \subseteq \mathrm{ZPP}^{\Sigma_{k+l}}$$

The second inclusion is by assumption and the last equality is by Lemma 2. Let us assume that, for $k_2,\ldots,k_n \geq 1$ and $l_2,\ldots l_n \geq 0$

$$\Sigma_{k_2}^{\mathrm{BP}\cdot\Sigma_{l_2}^{\cdot^{\cdot^{\Sigma_{k_n}^{\mathrm{BP}\cdot\Sigma_{l_n}}}}}} \subseteq \mathrm{ZPP}^{\Sigma_{k_2+\cdots+k_n+l_2+\cdots+l_n}}.$$

The following inclusions prove item 1.

$$\begin{aligned}
\Sigma_{k_1}^{\mathrm{BP}\cdot\Sigma_{l_1}^{\cdot^{\cdot^{\Sigma_{k_n}^{\mathrm{BP}\cdot\Sigma_{l_n}}}}}} &\subseteq \Sigma_{k_1}^{\mathrm{BP}\cdot\Sigma_{l_1}^{\mathrm{ZPP}^{\Sigma_{k_2+\cdots+k_n+l_2+\cdots+l_n}}}} && \text{(by assumption)} \\
&= \Sigma_{k_1}^{\mathrm{BP}\cdot\Sigma_{k_2+\cdots+k_n+l_1+l_2+\cdots+l_n}} && \text{(by Lemma 2)} \\
&\subseteq \mathrm{ZPP}^{\Sigma_{k_1+k_2+\cdots+k_n+l_1+l_2+\cdots+l_n}} && \text{(By base case)}
\end{aligned}$$

We omit the proof of second item due to space constraints. □

4 On $\mathrm{BP}\cdot\Sigma_k \cap \mathrm{BP}\cdot\Pi_k$

Definition 2. *Let $E_k \equiv \mathrm{BP}\cdot\Sigma_k \cap \mathrm{BP}\cdot\Pi_k$, and if $\mathcal{C}$ is a class of languages we define $E_k^{\mathcal{C}} \equiv \mathrm{BP}\cdot\Sigma_k^{\mathcal{C}} \cap \mathrm{BP}\cdot\Pi_k^{\mathcal{C}}$.*

Note that $E_k^{\Sigma_l} = E_{k+l}$. In this section we study the interaction of the class E_k with Σ_l classes. Recall that AM∩co-AM ⊆ $\mathrm{ZPP}^{\mathrm{NP}}$; in fact Arvind and Köbler [2] proved that AM ∩ co-AM is low for $\mathrm{ZPP}^{\mathrm{NP}}$. This lowness result is a corollary of the following theorem. Thus we offer a different proof of their result, which we believe is intuitively clearer.

Theorem 3.

$$\mathrm{NP}^{\mathrm{E}_l} \subseteq \mathrm{ZPP}^{\Sigma_l}, \textit{ for } l \geq 1.$$

The proof is by induction. The base case is the following lemma.

Lemma 3.

$$\mathrm{NP}^{\mathrm{AM}\cap\mathrm{co\text{-}AM}} \subseteq \mathrm{ZPP}^{\mathrm{NP}}$$

Proof. Let $L \in E_1 = \mathrm{AM} \cap \mathrm{coAM}$. By definition there exist two deterministic polynomial-time computable predicates $A(\cdot,\cdot,\cdot)$ and $B(\cdot,\cdot,\cdot)$ and a polynomial $p(\cdot)$ such that the following holds: $\forall x \in \{0,1\}^* :$ let $m = p(|x|)$

$$\begin{aligned} x \in L \quad &\Rightarrow \quad \forall r \in \{0,1\}^m : \exists w \in \{0,1\}^m : A(x,w,r) \\ &\text{AND} \quad \Pr_{r\in\{0,1\}^m}[\forall w \in \{0,1\}^m : \neg B(x,w,r)] \geq \frac{3}{4} \qquad (1) \\ x \notin L \quad &\Rightarrow \quad \Pr_{r\in\{0,1\}^m}[\forall w \in \{0,1\}^m : \neg A(x,w,r)] \geq \frac{3}{4} \\ &\text{AND} \quad \forall r \in \{0,1\}^m : \exists w \in \{0,1\}^m : B(x,w,r) \end{aligned}$$

We will show that $\mathrm{NP}^L \subseteq \mathrm{ZPP}^{\mathrm{NP}}$. Let $S \in \mathrm{NP}^L$ and let N be an oracle Turing machine which witnesses this membership.

Let $T(\cdot)$ be the bound on running time of N. Hence, N can query strings of length at most $T(n)$ for input size n. We first amplify the success probability for A and B such that the following holds for each string q, with $|q| \leq T(n)$:

$$q \in L \Rightarrow \Pr_{r\in\{0,1\}^{l(n)}}[\forall w \in \{0,1\}^{p(|q|)} : \neg B(q,w,r)] \geq 1 - \tfrac{1}{2^{T(n)+n+1}}$$
$$q \notin L \Rightarrow \Pr_{r\in\{0,1\}^{l(n)}}[\forall w \in \{0,1\}^{p(|q|)} : \neg A(q,w,r)] \geq 1 - \tfrac{1}{2^{T(n)+n+1}}.$$

To achieve this amplification it suffices to take $l(n) = n^d$ for some large enough constant $d > 0$. Thus we have

$$\Pr_{r\in\{0,1\}^{l(n)}}\left[\forall q,\ |q| \leq T(n), \left[q \in L \Rightarrow \left(\forall w \in \{0,1\}^{p(|q|)} : \neg B(q,w,r)\right)\right]\right] \geq 1 - \frac{1}{2^n}$$
$$\Pr_{r\in\{0,1\}^{l(n)}}\left[\forall q,\ |q| \leq T(n), \left[q \notin L \Rightarrow \left(\forall w \in \{0,1\}^{p(|q|)} : \neg A(q,w,r)\right)\right]\right] \geq 1 - \frac{1}{2^n}.$$

Now we define the non-deterministic Turing machine N'. Let $|x| = n$ and r be a string of length $l(n)$.

N' on input $\langle x,r\rangle$ operates as follows. It simulates N on input x. Whenever N poses a query q to the oracle L, N' non-deterministically guesses an answer $a_q \in \{\text{Yes}, \text{No}\}$ and then uses r, as random bits and also guesses a "proof" w and verifies that if $a_q = \text{Yes}$, then $A(q,w,r)$ holds and if $a_q = \text{No}$, then $B(q,w,r)$ holds.

Now we construct a probabilistic oracle Turing machine M with oracle access to **SAT** that recognizes S. Given an x with $|x| = n$, it generates a random string r of length $l(n)$. Then, M poses the query to the **SAT** oracle asking (using Cook reduction) whether there is an accepting path of N' on input $\langle x,r\rangle$ and if so, M finds such a path C using the self-reducibility of **SAT**. If the answer was "No" then M rejects x. If the answer was "Yes" then M checks for contradictions to the query answers on the path C. In other words, for every query answered "Yes" we ask the **SAT** oracle if there is a w' such that $B(q,w',r)$ holds, and similarly for every query answered "No" we ask the **SAT** oracle if there is a w'' such that $A(q,w'',r)$ holds. If for every query along C there were no contradictions then we accept x, otherwise we output "?".

Clearly M rejects x only if the **SAT** oracle did not find any accepting path of N'. If $x \in L$ there always is an accepting path of N' regardless of r so we are correct in rejecting the string. Now if M accepts x then we obtained a path C of N' where there were no contradictions to the queries. If any of the queries were answered wrongly there always is a contradiction. This is because if for any query q such that $q \in L$, but the **SAT**-oracle found a w' such that $B(q,w',r)$ then by (1) there is also a w'' such that $A(q,w'',r)$ regardless of r. Similarly for $q \notin L$ if the **SAT**-oracle found a w'' such that $A(q,w'',r)$ then again by (1), irrespective of r there is also a w' such that $B(q,w',r)$. Thus $x \in L$ if M accepts x. Hence M is always right in accepting or rejecting a string.

The proof is complete if we can argue that we output a "?" with low probability. Now a "?" is reached only if there was a contradiction to some query along the path C. By our amplification the probability that there exists a contradiction to any query of length bounded by $T(n)$ is at most $\frac{1}{2^n}$. We note that a contradiction happens at a query only if either $q \in L$ and $\exists w' : B(q,w',r)$ or $q \notin L$ and $\exists w'' : A(q,w'',r)$. □

Proof. (of theorem 3) It is clear that the above lemma relativizes. So we have

$$\mathrm{NP}^{\mathrm{E}_l} = \mathrm{NP}^{\mathrm{E}_1^{\Sigma_{l-1}}} \subseteq \mathrm{ZPP}^{\mathrm{NP}^{\Sigma_{l-1}}} = \mathrm{ZPP}^{\Sigma_l}.$$

□

Now $\mathrm{ZPP}^{\mathrm{NP}^{\mathrm{AM}\cap\text{co-AM}}} \subseteq \mathrm{ZPP}^{\mathrm{ZPP}^{\mathrm{NP}}} = \mathrm{ZPP}^{\mathrm{NP}}$. Thus we obtain an alternate proof of Arvind and Köbler's result.

Corollary 1 *[2]* AM $\cap$ co-AM *is low for* $\mathrm{ZPP}^{\mathrm{NP}}$.

We will generalize Theorem 3 further culminating in Theorem 6. First we have

Theorem 4.

$$\Sigma_k^{\mathrm{E}_l} = \Sigma_{k+l-1} \textit{ for } k \geq 1, l \geq 1.$$

Proof. We again prove this by induction. Base case is $\mathrm{NP}^{\mathrm{NP}^{\mathrm{E}_l}} = \Sigma_{l+1}$.

$$\begin{aligned}\mathrm{NP}^{\mathrm{NP}^{\mathrm{E}_l}} &\subseteq \mathrm{NP}^{\mathrm{ZPP}^{\Sigma_l}} \text{ (by Theorem 3)}\\ &= \Sigma_{l+1} \text{ (by Lemma 2)}.\end{aligned}$$

Let us assume $\Sigma_{k-1}^{\mathrm{E}_l} \subseteq \mathrm{ZPP}^{\Sigma_{k+l-2}}$, for $k \geq 3$. Now,

$$\begin{aligned}\Sigma_k^{\mathrm{E}_l} &= \mathrm{NP}^{\Sigma_{k-1}^{\mathrm{E}_l}}\\ &\subseteq \mathrm{NP}^{\mathrm{ZPP}^{\Sigma_{k+l-2}}} \text{ (by assumption)}\\ &= \Sigma_{k+l-1} \text{ (by Lemma 2)}.\end{aligned}$$

Since E_l clearly contains Σ_{l-1} the other inclusion also holds. □
We obtain the following corollary.

Corollary 2 $\mathrm{AM} \cap \text{co-AM}$ *is low for* Σ_k, $k \geq 2$.

Theorem 5.

$$E_k^{\mathrm{ZPP}^{\Sigma_l}} = E_{k+l} \text{ for } k \geq 1, l \geq 1$$

Proof.

$$E_k^{\mathrm{ZPP}^{\Sigma_l}} = \mathrm{BP}\cdot\Sigma_k^{\mathrm{ZPP}^{\Sigma_l}} \cap \mathrm{BP}\cdot\Pi_k^{\mathrm{ZPP}^{\Sigma_l}} = \mathrm{BP}\cdot\Sigma_{k+l} \cap \mathrm{BP}\cdot\Pi_{k+l}$$

The last equality is by Lemma 2. □

Theorem 6. *1. For* $l_1, l_2, \ldots, l_n, k_1, k_2, \ldots, k_n \geq 1$, $n \geq 1$,

$$\Sigma_{l_1}^{E_{k_1}^{\Sigma_{l_2}^{\cdot^{\cdot^{\Sigma_{l_n}^{E_{k_n}}}}}}} \subseteq \mathrm{ZPP}^{\Sigma_{l_1+l_2+\cdots+l_n+k_1+k_2+\cdots+k_n-n}}.$$

2. For $l_1 \geq 2$ *and* $l_2, \ldots, l_n, k_1, k_2, \ldots, k_n \geq 1$, $n \geq 1$,

$$\Sigma_{l_1}^{E_{k_1}^{\Sigma_{l_2}^{\cdot^{\cdot^{\Sigma_{l_n}^{E_{k_n}}}}}}} = \Sigma_{l_1+l_2+\cdots+l_n+k_1+k_2+\cdots+k_n-n}.$$

Note the presence of "$-n$" appearing on the right hand side. This is a drastic improvement over the naïve bound.

Proof. We prove the first inclusion by induction. Our base case $\Sigma_l^{E_k} \subseteq \mathrm{ZPP}^{\Sigma_{k+l-1}}$ is due to Theorem 4 and Theorem 3. Now assume

$$\Sigma_{l_2}^{E_{k_2}^{\Sigma_{l_3}^{\cdot^{\cdot^{\Sigma_{l_n}^{E_{k_n}}}}}}} \subseteq \mathrm{ZPP}^{\Sigma_{l_2+\cdots+l_n+k_2+\cdots+k_n-(n-1)}}.$$

Now,

$$
\begin{aligned}
\Sigma_{l_1}^{E_{k_1}^{\Sigma_{l_2}^{\cdot^{\cdot^{\Sigma_{l_n}^{E_{k_n}}}}}}} &\subseteq \Sigma_{l_1}^{E_{k_1}^{\mathrm{ZPP}^{\Sigma_{l_2+\cdots+l_n+k_2+\cdots+k_n-(n-1)}}}} \text{(by assumption)} \\
&= \Sigma_{l_1}^{E_{l_2+\cdots+l_n+k_1+k_2+\cdots+k_n-(n-1)}} \text{(by Theorem 5)} \\
&\subseteq \mathrm{ZPP}^{\Sigma_{k_1+k_2+\cdots+k_n+l_1+\cdots l_n-n}} \text{(by Theorem 4 and Theorem 3).}
\end{aligned}
$$

We omit the proof of second item due to space constraints. □

5 Examples of Languages in BP $\cdot \Sigma_k$

We know that Graph non-isomorphism is in AM $\cap$ co-NP and is not known to be in NP. Agrawal and Thierauf [5] proved that the complement of boolean isomorphism is in BP $\cdot \Sigma_2 \cap \Pi_2$. We generalize their result (the proof is omitted here) to give examples of languages in BP $\cdot \Sigma_k \cap \Pi_k$. We consider *circuit isomorphism* problem with oracle gates which is similar to boolean isomorphism problem. Recall that two circuits C_1 and C_2 (with n inputs) are *equivalent* if for every $x \in \{0,1\}^n$ $C_1(x) = C_2(x)$. The following definition holds for any $k \geq 0$.

Definition 3. Circuit Isomorphism with Σ_k-oracle gates (CI_k) : $\{\langle C_1, C_2\rangle : C_1$ *and* C_2 *are circuits with* n *inputs and have* Σ_k *oracle gates, and there exists permutation* π *of* $\{x_1, x_2, \ldots, x_n\}$ *such that for all* $\hat{x} = (x_1, x_2, \cdots x_n)$ $C_1(\hat{x}) = C_2(\pi(\hat{x}))$ *holds*$\}$.

Clearly, $\overline{CI}_0$ is same as $\overline{CI}$ and $\overline{CI}_k \in \Pi_{k+2}$. We claim that $\overline{CI}_k \in \mathrm{BP} \cdot \Sigma_{k+2}$. Given C, a circuit with n inputs and Σ_k-oracle gates, and an input $\hat{x} \in \{0,1\}^n$, $C(\hat{x})$ can be determined in P^{Σ_k}. Secondly, given circuits C_1, C_2 (also with Σ_k-oracle gates), a P^{Σ_k+1} machine can determine an $\hat{x}$ such that $C_1(\hat{x}) \neq C_2(\hat{x})$ (if such an $\hat{x}$ exists). To see this, define $L = \{\langle C_1, C_2\rangle : \exists \hat{x} \in \{0,1\}^n\ C_1(\hat{x}) \neq C_2(\hat{x})\}$. Clearly $L \in \Sigma_{k+1}$ and by self-reducibility, $\hat{x}$ can be determined in P^{Σ_k+1}. Again, note that C_1 and C_2 are equivalent if and only if $\langle C_1, C_2\rangle \notin L$ and so one call to L determines the equivalence. Given these two facts, we can argue as in [5] to get the following theorem.

Theorem 7. $\overline{CI}_k \in \mathrm{BP} \cdot \Sigma_{k+2}$*, for* $k \geq 0$.

6 Conclusions

We have considered natural extensions of the class AM using the BP operator. We have established several non-trivial relations involving these classes and the polynomial hierarchy. However, it would be interesting to give more natural examples of languages in BP $\cdot \Sigma_k \cap \Pi_k$.

Another possible extension is to study the relationship of the class $\mathrm{S}_2^{\Sigma_k}$ and E_{k+1}. Even the case $k = 0$ is interesting: we know that AM $\cap$ co-AM $\subseteq \mathrm{ZPP}^{\mathrm{NP}}$ and $\mathrm{S}_2^{\mathrm{P}} \subseteq \mathrm{ZPP}^{\mathrm{NP}}$, but we do not yet know how the classes AM $\cap$ co-AM and $\mathrm{S}_2^{\mathrm{P}}$ are related.

References

1. V. Arvind and J. Köbler, *On Pseudorandomness and Resource-Bounded Measure*, Proc. 17th FST and TCS, Springer-Verlag, LNCS 1346, 235-249, 1997.
2. V. Arvind and J. Köbler, *Graph isomorphism is low for ZPP^{NP} and other lowness results*, STACS 2000.
3. S. Arora, C. Lund, R. Motwani, M. Sudan and M. Szegedy, *Proof verification and hardness of approximation problems*. Proceedings of the 33rd IEEE Symposium on Foundations of Computer Science, 14–23, 1992.
4. S. Arora and S. Safra, *Approximating clique is NP-complete.*, Proceedings of the 33rd IEEE Symposium on Foundations on Computer Science, 2–13, 1992.
5. M. Agrawal and T. Thierauf, *The Boolean isomorphism problem*, Proc. 37th Annual Symposium on Foundations of Computer Science, 422–430.
6. L. Babai, *Trading group theory for randomness*, STOC 17:421–429(85).
7. L. Babai and S. Moran, *Arthur-Merlin Games : a randomized proof system, and a hierarchy of complexity classes*, Journal of Computer and System Sciences, 36:254-276, 1988.
8. J. L. Balcázar, J. Díaz, J. Gabarró, *Structural Complexity II*, EATCS Monographs on Theoretical Computer Science, Springer-Verlag, 1988.
9. R. Boppana, J. Hastad and S. Zachos, *Does* co-NP *have short interactive proofs?*, Information Processing Letters, 25:127-132, 1987.
10. N. Bshouty, R. Cleve, S. Kannan and C. Tamon, *Oracles and Queries that are sufficient for Exact Learning*, Proceedings of the 17th Annual ACM conference on Computational Learning Theory, 130–19 (1994).
11. Jin-Yi Cai, $S_2^P \subseteq \mathrm{ZPP}^{\mathrm{NP}}$, ECCC Tech-report TR-02-30, also to appear in FOCS 2001.
12. S. Goldwasser, S. Micali and C. Rackoff, *The Knowledge Complexity of Interactive Proofs*, Proc. 17th ACM Symp. om Computing, Providence, RI, 1985, pp. 291-304.
13. S. Goldwasser and M. Sipser, *Private coins versus public coins in interactive proof systems*, STOC 18:59–68(1986).
14. O.Goldreich and D.Zuckerman, *Another Proof that $BPP \subseteq PH$ (and more)*, ECCC, TR97-045, October 1997.
15. J. Köbler and O. Watanabe, *New collapse consequences of NP having small circuits* ICALP, LNCS 944:196–207(1995).
16. C. Lund, L. Fortnow, H. Karloff and N. Nisan, *Algebraic Methods for Interactive Proof Systems*, Journal of the ACM, 39(4):859-868, October 1992.
17. A. Shamir, *IP = PSPACE*, Journal of the ACM, 39(4):869-877, October 1992.
18. S. Toda, *PP is as hard as polynomial-time hierarchy*. SIAM Journal on Computing, 20(5):865–877, 1991.
19. H. Vollmer and K. Wagner, *The complexity of finding middle elements*. International Journal of Foundations of Computer Science, 4:293–307, 1993.
20. O. Watanabe and S. Toda, *Polynomial time 1-Turing reductions from #PH to #P*. Theoritical Computer Science, 100(1):205–221, 1992.
21. S. Zachos and H. Heller, *A Decisive characterization of BPP*, Information and Control, 69:125–135(1986).
22. S. Zachos and M. Fürer, *Probabilistic quantifiers vs Distrustful adversaries*, FSTTCS 1987, LNCS-287:449–455.

$(2 + f(n))$-SAT and Its Properties*

Xiaotie Deng[1], C.H. Lee[1], Yunlei Zhao[1,2], and Hong Zhu[2]

[1] Department of Computer Science
City University of Hong Kong, Hong Kong
{csdeng,cschlee,csylzhao}@cityu.edu.hk
[2] Department of Computer Science
Fudan University, Shanghai, China

Abstract. Consider a formula which contains n variables and m clauses with the form $\Phi = \Phi_2 \wedge \Phi_3$, where Φ_2 is an instance of 2-SAT which contains m_2 2-clauses and Φ_3 is an instance of 3-SAT which contains m_3 3-clauses. $\Phi$is an instance of $(2 + f(n))$-SAT if $\frac{m_3}{m_2+m_3} \leq f(n)$. We prove that $(2 + f(n))$-SAT is in $\mathcal{P}$ if $f(n) = O\left(\frac{\log n}{n^2}\right)$, and in $\mathcal{NPC}$ if $f(n) = \frac{1}{n^{2-\varepsilon}}(\forall \varepsilon : 0 < \varepsilon < 2)$. Most interestingly, we give a candidate $\left(2 + \frac{(\log n)^k}{n^2}\right)$-SAT $(k \geq 2)$, for natural problems in $\mathcal{NP} - \mathcal{NPC} - \mathcal{P}$ (denoted as $\mathcal{NPI}$) with respect to this $(2+f(n))$-SAT model. We prove that the restricted version of it is not in $\mathcal{NPC}$ under the assumption $\mathcal{P} \neq \mathcal{NP}$. Actually it is indeed in $\mathcal{NPI}$ under some stronger but plausible assumption, specifically, the Exponential-Time Hypothesis (ETH) which was introduced by Impagliazzo and Paturi.

1 Introduction

In 1975, Lander had shown that there exist some languages in $\mathcal{NP} - \mathcal{NPC} - \mathcal{P}$ (denoted as $\mathcal{NPI}$) under the assumption $\mathcal{P} \neq \mathcal{NP}$ [1]. But the language constructed there is not a natural one because the construction needs to run all Turing machines. So far, no natural problems have been proven to be in $\mathcal{NPI}$ under $\mathcal{P} \neq \mathcal{NP}$ and finding such a natural problem is considered an important open problem in complexity theory [2, 3]. The problems of GI (Graph Isomorphism) and Factoring, which were suggested by Karp, are regarded as two most likely candidates [2, 3].

The satisfiability problem of Boolean formula (SAT) has played a central role in the field of computational complexity theory. It is the first $\mathcal{NP}$-Complete problem. And up to now, all known algorithms to find a solution for 3-SAT require exponential time in problem size in the worst case. In practice, the time complexity of the fastest algorithm for 3-SAT is $(\frac{4}{3})^n$, where n is the variable number in the formula [4]. It is also an important open question whether sub-exponential time algorithms exist. The plausibility of such a sub-exponential

* This research is supported by a research grant of City University of Hong Kong 7001023.

O.H. Ibarra and L. Zhang (Eds.): COCOON 2002, LNCS 2387, pp. 28–36, 2002.

time algorithm for 3-SAT was investigated in [5], using sub-exponential time reduction. It was shown there that linear size 3-SAT is complete for the class $\mathcal{SNP}$ (Strict$-\mathcal{NP}$) with respect to such reduction. It implies that if there exists a sub-exponential time algorithm for 3-SAT then all the languages in $\mathcal{SNP}$ can be decided in sub-exponential time. Note that some well-studied problems, such as k-SAT, k-Colorability, for any $k \geq 3$, and so on, have been proven to be $\mathcal{SNP}$-Complete. In light of both the practical and theoretical supports, Impagliazzo and Paturi introduced the ETH (Exponential-Time Hypothesis) for 3-SAT: 3-SAT does not have a sub-exponential-time algorithm [6]. Although ETH is stronger than $\mathcal{NP} \neq \mathcal{P}$, it is still quite reasonable. In recent advances of cryptography, many important cryptographic primitives and protocols were constructed under the ETH for the one-way functions: DLP or RSA, e.g., verifiable pseudorandom functions [7], verifiable pseudorandom generator [8] and resettable zero-knowledge arguments systems for $\mathcal{NP}$ [9, 10] and so on.

On the other hand, recently there has been a growth of interests to study the link between the hardness of computational complexity of decision problems and the phase boundaries in physical systems [11, 12]. It was observed that, similar to physical systems, across certain phase boundaries dramatic changes occur in the computational difficulty and solution character. $\mathcal{NP}$-Complete problems become easier to solve away from the boundary and the hardest problems occur at the phase boundary [11, 13].

To understand the onset of exponential complexity that occurs when going from a problem in $\mathcal{P}$(2-SAT) to a problem that is $\mathcal{NP}$-Complete (3-SAT), the $(2{+}p)$-SAT model was introduced in [14, 11], where p is a constant and $0 \leq p \leq 1$. An instance of $2{+}p$-SAT is a formula with m clauses, of which $(1-p)\cdot m$ contain two variables (2-clauses) and pm contain three variables (3-clause). $2 + p$-SAT smoothly interpolates between 2-SAT ($p = 0$) and 3-SAT ($p = 1$) when the instances are generated randomly. The median computation cost scales linearly with n (the number of variables) when $p < p_0$ and exponentially for $p > p_0$, where p_0 lies between 0.4 and 0.416 [11]. However, for the worst case complexity, $(2 + p)$-SAT is $\mathcal{NP}$-Complete for any constant p, $p > 0$ [11, 12].

In this work, we further explore the worst case complexity boundary of $\mathcal{P}$ and $\mathcal{NPC}$ when p is further reduced (not a constant but a function of n). Somewhat surprisingly, such an extension allows us to suggest another candidate for natural problems in $\mathcal{NPI}$ under $\mathcal{NP} \neq \mathcal{P}$. In fact, we present a natural problem in $\mathcal{NPI}$ under ETH . In Section 2, we present the necessary definitions and the related important properties for our study. In Section 3, we present a candidate for natural problems in $\mathcal{NPI}$ and prove it not in $\mathcal{NPC}$ under $\mathcal{NP} \neq \mathcal{P}$. In Section 4, we prove it is not in $\mathcal{P}$ under ETH. We conclude with discussions in Section 5.

2 Properties of $(2 + f(n))$-SAT

In this section, we introduce the $(2{+}f(n))$-SAT model. We are mainly concerned with the boundary of $f(n)$ that separates the problems between $\mathcal{P}$ and $\mathcal{NPC}$.

Let Φ is an formula and denoted $|\Phi|$ as the number of clauses in Φ. We introduce the definition of $(2+f(n))$-SAT:

Definition 1. *Consider a formula which contains n variables and m clauses with the form $\Phi = \Phi_2 \wedge \Phi_3$, where Φ_2 is an instance of 2-SAT which contains m_2 2-clauses, and Φ_3 is an instance of 3-SAT which contains m_3 3-clauses.An instance of $(2+f(n))$-SAT is one satisfying the condition:*

$$\frac{|\Phi_3|}{|\Phi|} = \frac{m_3}{m} = \frac{m_3}{m_2+m_3} \leq f(n).$$

Throughout the paper, we restrict our discussion to instances with $f(n) = \frac{|\Phi_3|}{|\Phi|}$. Indeed, all our claims hold if they hold under this restriction. Note that $m_2 \leq 4n_2^2$, $m_3 \leq 8n_3^3$, $n_2 \leq 2m_2$, $n_3 \leq 3m_3$, $n \leq 3m$, and that the variables which appear in Φ_2 may appear in Φ_3, and vice versa, i. e., $n \leq n_2 + n_3 \leq 2n$.

Theorem 1. *For any constant $k > 0$, $\left(2+\frac{k\log n}{n^2}\right)$-SAT is in $\mathcal{P}$.*

Proof. Consider any instance of $\left(2+\frac{k\log n}{n^2}\right)$-SAT ($k > 0$), a formula $\Phi = \Phi_2 \wedge \Phi_3$, where $\frac{m_3}{m_2+m_3} = \frac{k\log n}{n^2}$. We get

$$\begin{aligned} m_3 &= \frac{k\log n \cdot m_2}{n^2 - k\log n} \leq \frac{km_2\log n + k\log n}{n^2} \leq \frac{(k\cdot 4n^2+k)\cdot\log n}{n^2} \\ &= (4k+\frac{k}{n^2})\cdot\log n \leq 5k\cdot\log n. \end{aligned}$$

Note that the variables which appear in Φ_2 may appear in Φ_3, and vice versa. For the $5k\cdot\log n$ variables which appear in Φ_3, we can enumerate all the at most n^{5k} truth assignments and then for each truth assignment we can determine Φ_2 in polynomial time of n, and thus the $\left(2+\frac{k\log n}{n^2}\right)$-SAT ($k \geq 0$) is in $\mathcal{P}$. □

Claim 1. Given n variables, we can construct a satisfiable formula Φ, where Φ is an instance of 2-SAT and $|\Phi| \leq \frac{3}{2}n^2 - \frac{3}{2}n$.

Proof. We construct 2-clauses as follows: $(\frac{1}{2}n^2 - \frac{1}{2}n)$ clauses with the form $(x_i \vee x_j)(i \neq j, 1 \leq i,j \leq n)$, $(n^2 - n)$ clauses with the form $(x_i \vee \neg x_j)$, $(i \neq j, 1 \leq i,j \leq n)$. From all these 2-clauses, we select $k, 1 \leq k \leq \frac{3}{2}n^2 - \frac{3}{2}n$, clauses to construct the formula Φ we need, then Φ is satisfiable when all these n variable are assigned the value "true". □

Theorem 2. $(2+\frac{1}{n^{2-\varepsilon}})$*-SAT ($\forall\varepsilon$, $0 < \varepsilon < 2$) is in $\mathcal{NPC}$.*

Proof. We show that there is a many-one reduction from 3-SAT to $(2+\frac{1}{n^{2-\varepsilon}})$-SAT ($0 < \varepsilon < 2$). Let Φ_3 be an instance of 3-SAT that contains n_3 variables and m_3 3-clauses. Without loss of generality, we assume that $m_3 \geq 2$. Then we add $n_2 = m_3^{\frac{8}{\varepsilon}}$ new variables and using these new variables to construct a satisfiable formula Φ_2 which contains m_2 2-clauses.

Let $\frac{m_3}{m_2+m_3} = \frac{1}{n^{2-\varepsilon}} (0 < \varepsilon < 2)$ then

$$\frac{m_3}{m_2 + m_3} = \frac{1}{n^{2-\varepsilon}} \geq \frac{1}{(n_2 + n_3)^{2-\varepsilon}}$$

$$m_2 \leq \left((n_2 + n_3)^{2-\varepsilon} - 1\right) \cdot m_3 \leq (n_2 + n_3)^{2-\varepsilon} \cdot m_3 \leq \left(m_3^{\frac{8}{\varepsilon}} + 3m_3\right)^{2-\varepsilon} \cdot m_3.$$

But note that $m_3 \geq 2$, we get

$$\left(m_3^{\frac{8}{\varepsilon}} + 3m_3\right)^2 \cdot m_3 \leq \left[\frac{3}{2}\left(m_3^{\frac{8}{\varepsilon}}\right)^2 - \frac{3}{2} m_3^{\frac{8}{\varepsilon}}\right] \cdot (m_3)^8$$
$$= \left(\frac{3}{2} n_2^2 - \frac{3}{2} n_2\right) \cdot m_3^8$$
$$\leq \left(\frac{3}{2} n_2^2 - \frac{3}{2} n_2\right) \cdot \left(m_3^{\frac{8}{\varepsilon}} + 3m_3\right)^{\varepsilon}.$$

That is,

$$m_2 \leq (m_3^{\frac{8}{\varepsilon}} + 3m_3)^{2-\varepsilon} \cdot m_3 \leq \frac{3}{2} n_2^2 - \frac{3}{2} n_2 \Rightarrow m_2 \leq \frac{3}{2} n_2^2 - \frac{3}{2} n_2.$$

The satisfiable formula Φ_2 can be constructed according to Claim 1 .

Let $\Phi = \Phi_2 \wedge \Phi_3$, then Φ is an instance of $(2 + \frac{1}{n^{2-\varepsilon}})$-SAT $(0 < \varepsilon < 2)$ and Φ is satisfiable if and only if Φ_3 is satisfiable.

Note that the above many-one reduction indeed can be constructed in polynomial time of m_3 (also in polynomial time of n_3, since $n_3 \leq 3m_3$, $m_3 \leq 8n_3^3$).

Obviously, $(2 + \frac{1}{n^{2-\varepsilon}})$-SAT $(0 < \varepsilon < 2)$ is in $\mathcal{NP}$, so the theorem does hold. □

One open problem related to our $(2 + f(n))$-SAT model is:

Open Problem Does there exist some $f(n)$, s.t. $\frac{k \log n}{n^2} < f(n) < \frac{1}{n^{2-\varepsilon}}$, where $k \geq 0$ and $0 < \varepsilon < 2$, so that $(2 + f(n))$-SAT is in $(\mathcal{NP} - \mathcal{NPC}) - \mathcal{P}$ (denoted as $\mathcal{NPI}$) under the assumption $\mathcal{P} \neq \mathcal{NP}$?

Note that $\left(2 + \frac{k \log n}{n^2}\right)$-SAT is in $\mathcal{P}$, $k \geq 0$ and $(2 + \frac{1}{n^{2-\varepsilon}})$-SAT $(0 < \varepsilon < 2)$ is in $\mathcal{NP}$-Complete according to the above theorems.

Now, we give another candidate and also another open problem with regard to our $(2 + f(n))$-SAT for natural problems in $\mathcal{NPI}$ under $\mathcal{P} \neq \mathcal{NP}$:

Open Problem In the $(2 + f(n))$-SAT model, is $\left(2 + \frac{(\log n)^k}{n^2}\right)$-SAT $(k \geq 2)$ in $(\mathcal{NP}$-$\mathcal{NPC})$-$\mathcal{P}$ under the assumption $\mathcal{NP} \neq \mathcal{P}$?

Note that $\frac{k_1 \log n}{n^2} < \frac{(\log n)^k}{n^2} (k \geq 2) < \frac{1}{n^{2-\varepsilon}}$, where $k_1 \geq 0$ and $0 < \varepsilon < 2$.

3 A Candidate for Natual Problems in $\mathcal{NPI}$ Under $\mathcal{NP} \neq \mathcal{P}$

Now, we give another candidate for natural problems in $\mathcal{NPI}$ under $\mathcal{P} \neq \mathcal{NP}$ which is a restricted version of $\left(2 + \frac{(\log n)^k}{n^2}\right)$-SAT $(k \geq 2)$. We will prove that it is not $\mathcal{NP}$-Complete under the assumption $\mathcal{P} \neq \mathcal{NP}$. Actually it is indeed in $\mathcal{NPI}$ under some stronger but reasonable assumptions.

Theorem 3. *In the $(2+f(n))$-SAT model, if the variables which appear in Φ_2 do not appear in Φ_3, and vice versa, then $\left(2+\frac{(\log n)^k}{n^2}\right)$-SAT is not in $\mathcal{NPC}$ under the assumption $\mathcal{NP} \neq \mathcal{P}$, $k \geq 2$.*

Proof. Clearly this problem is in $\mathcal{NP}$. We prove this theorem by showing that 3-SAT can not be reduced to $\left(2+\frac{(\log n)^k}{n^2}\right)$-SAT by many-one reduction, where $k \geq 2$.

Assume that there exists a many-one reduction(denoted as F) from 3-SAT to $\left(2+\frac{(\log n)^k}{n^2}\right)$-SAT ($k \geq 2$). It means that for any instance of 3-SAT, a formula Φ_0 which contains n_0 variables and m_0 3-clauses, we can construct the $F(\Phi_0)$ which is an instance of $\left(2+\frac{(\log n)^k}{n^2}\right)$-SAT ($k \geq 2$) in polynomial time of n_0, where $F(\Phi_0)$ contains n variables amd m clauses, and $F(\Phi_0)$ is satisfiable if and only if Φ_0 is satisfiable. Let $F(\Phi_0) = \Phi_2 \wedge \Phi_3$, where Φ_2 is an instance of 2-SAT which contains m_2 2-clauses and n_2 variables and Φ_3 is an instance of 3-SAT which contains m_3 3-clauses and n_3 variables, then $\frac{(\log n)^k}{n^2} = \frac{|\Phi_3|}{|\Phi|} = \frac{m_3}{m} = \frac{m_3}{m_2+m_3}$, $k \geq 2$.

We consider the relation between m_3 and m_0. there are two cases:

Case 1. $m_3 \geq m_0$.

Claim 2. $m = m_2 + m_3$ *can not be expressed as a polynomial of* m_3.

Proof. (of Claim 2) Firstly, for sufficiently large n, $\frac{(\log n)^k}{n^2} = \frac{m_3}{m} \leq \frac{1}{2}$ (i.e. $m \geq 2m_3$), where $k \geq 2$. Secondly,

$$m = m_2 + m_3 \leq 4n^2 + m_3 \Longrightarrow n^2 \geq \frac{m - m_3}{4}.$$

Then, for sufficiently large n, the following holds

$$\begin{aligned} \frac{m_3}{m} = \frac{(\log n)^k}{n^2} &\leq \frac{4(\log 3m)^k}{m - m_3} \\ \Longrightarrow 4(\log 3m)^k &\geq m_3 \cdot \frac{m - m_3}{m} \geq \frac{1}{2} \cdot m_3 \\ \Longrightarrow m &\geq \frac{1}{3} \cdot 2^{\left(\frac{m_3}{8}\right)^{\frac{1}{k}}} \end{aligned}$$

□

According to Claim 2, in Case 1, we get the fact that m can not be expressed as a polynomial of m_3, and since $m_3 \geq m_0$, so m also can not be expressed as a polynomial of m_0 (of course m also can not be expressed as a polynomial of n_0 since $m_0 \leq 8n_0^3$). It's absurd since the many-one reduction $F(\Phi_0)$ must be done in polynomial time of n_0.

Case 2 $m_3 < m_0$.

Since we assume $F(\Phi_0)$ can be constructed in polynomial time of n_0, then m_2 must can be expressed as $P(n_0)$, where $P(\cdot)$ is a polynomial. So, if $m_3 < m_0$ it means that we can decrease the 3-clause number in Φ_0 by adding $P(n_0)$ 2-clauses (by imposing F on Φ_0). However, note that we assume the variables which appear in Φ_2 do not appear in Φ_3, and vice versa, then we can impose F on Φ_3, and so on. Repeat the above process at most m_0 times we can eliminate all 3-clauses in $F(\Phi_0)$ to get a formula Φ' and guarantee that Φ' is satisfiable if and only if $F(\Phi_0)$ is satisfiable if and only if Φ_0 is satisfiable, where Φ' contains only 2-clauses and $\mid \Phi' \mid$ is at most $m_0 P(n_0)$, or at most $8n_0^3 \cdot P(n_0)$, another polynomial of n_0. This means that there exists a many-one reduction from 3-SAT to 2-SAT , which contradicts our assumption $\mathcal{P} \neq \mathcal{NP}$.

So, from the arguments above, we can conclude that $\left(2 + \frac{(\log n)^k}{n^2}\right)$-SAT ($k \geq 2$) is not in $\mathcal{NP}$-Complete under the assumption $\mathcal{P} \neq \mathcal{NP}$. □

4 Can the Candidate Be in $\mathcal{P}$?

In this section, we further show that the candidate presented in the previous section is indeed in $\mathcal{NPI}$ under *ETH*.

Definition 2. (*SE*) *A language $L \in SE$ if for any $x \in L$ there exists an algorithm to find a y so that $|y| \leq m(x)$ and $R(x, y)$ in time $poly(|x|)2^{\varepsilon m(x)}$ for every fixed ε, $1 > \varepsilon > 0$, where R is a polynomial time relation called the constraint, and m is a polynomial-time computable and polynomial bounded complexity parameter.*

Definition 3. (*SERF*) *The sub-exponential reduction family SERF from A_1 with parameter m_1 to A_2 with parameter m_2 is defined as a collection of Turing reduction $M_\varepsilon^{A_2}$, such that for each ε, $1 > \varepsilon > 0$:*

1. *$M_\varepsilon^{A_2}(x)$ runs in time at most $poly(|x|)2^{\varepsilon m_1(x)}$*
2. *If $M_\varepsilon^{A_2}(x)$ queries A_2 with the input x', then $m_2(x') = O(m_1(x))$ and $|x'| = |x|^{O(1)}$.*

If such a reduction family exists, A_1 is *SERF*-reducible to A_2. If each problem in $\mathcal{SNP}$ is *SERF*-reducible to a problem A, then A is $\mathcal{SNP}$-Hard under *SERF*-reduction. And if A is also in $\mathcal{SNP}$ then we say A is $\mathcal{SNP}$-Complete under *SERF*-reductions. Note that the *SERF*-reducibility is transitive, and, if (A_1, m_1) *SERF*-reduces to (A_2, m_2), and $(A_2, m_2) \in SE$, then $(A_1, m_1) \in SE$ [5].

Definition 4. *(Strong many-one reduction) Let A_1 be a problem with complexity parameter m_1 and constraint R_1 and A_2 be a problem with complexity parameter m_2 and constraint R_2. A many-one reduction f from A_1 to A_2 is called a strong many-one reduction if $m_2(f(x)) = O(m_1(x))$. Strong many-one reduction is a special case of SERF-reduction [5].*

Lemma 1. *3-SAT with complexity parameter n, the number of variables, is SERF-reducible to 3-SAT with complexity parameter m, the number of clauses [5].*

Lemma 2. *3-SAT is $\mathcal{SNP}$-Complete under SERF-reductions, with either clauses or variables as the parameter. [5]*

Definition 5. *(3-ESAT) 3-ESAT is a variant of 3-SAT, for any instance of 3-ESAT, say a formula Φ, in which the clause number is equal to the number of variables which appear in Φ.*

Claim 3. *Given n ($n \geq 5$) variables, we can construct a satisfiable formula Φ in polynomial time of n, where Φ is an instance of 3-SAT and $|\Phi| \leq 2n$.*

Proof. We construct $2n$ 3-clauses with the form $x_i \vee x_j \vee x_k$, where $1 \leq i, j, k \leq n$, $i \neq j$, $i \neq k$, $j \neq k$. This can be done since there are $C_n^3 \geq 2n$ 3-clauses with such form. Then we select k, $1 \leq k \leq 2n$, 3-clauses to construct the formula Φ. Φ is satisfiable when all these n variables are assigned the value "true". □

Theorem 4. *3-ESAT is $\mathcal{SNP}$-Hard under SERF-reductions, with either clauses or variables as the parameter. Consequently, 3-ESAT $\in$ SE implies $\mathcal{SNP} \subseteq SE$.*

Proof. According to Lemma 1, Lemma 2 and the definition of strong many-one reduction, we only need to show there exists a strong many-one reduction from 3-SAT with m (the clause number) as complexity parameter to 3-ESAT with m as complexity parameter.

For any given instance of 3-SAT, a formula Φ_0 which contains n_0 variables and m_0 clauses, we construct the many-one reduction respectively according to whether $m_0 > n_0$ or not.

Firstly, if $m_0 > n_0$, we add $\frac{3}{2}(m_0 - n_0)$ new variables and use them to construct a formula Φ_1 which contains $\frac{1}{2}(m_0 - n_0)$ clauses, in which each of all those $\frac{3}{2}(m_0 - n_0)$ new variables appears once and only once. This means that Φ_1 is always satisfiable. Let $\Phi = \Phi_1 \wedge \Phi_0$ then we get the instance of 3-ESAT since $m_0 + \frac{1}{2}(m_0 - n_0) = n_0 + \frac{3}{2}(m_0 - n_0)$, and Φ is satisfiable if and only if Φ_0 is satisfiable, and the reduction can be done in polynomial time of n_0.

Note that $m_0 + \frac{1}{2}(m_0 - n_0) < 2m_0$.

In the second case, we add n_1 new variables, where $n_1 = max\{n_0 - m_0, 5\}$ and construct a satisfiable formula Φ_1, with the size $(n_1 + n_0 - m_0)$. This can be done according to Claim 3 since $n_1 + n_0 - m_0 \leq 2n_1$. Then similar to the first case, let $\Phi{=}\Phi_1 \wedge \Phi_0$, we get the instance of 3-ESAT with parameter $n_1 + n_0$ and Φ is satisfiable if and only if Φ_0 is satisfiable. Thus the reduction is done in polynomial time of n_0.

Note that

$$(n_1 + n_0 - m_0) + m_0 = max\{2n_0 - m_0, 5 + n_0\} \leq max\{5m_0, 3m_0 + 5\}.$$

Then according to the properties of *SERF* reduction, the theorem does hold. □

From the above proof, it is also easy to see that 3-ESAT is also $\mathcal{NP}$-Complete.

Definition 6. *Define s to be the infimum of* $\{\delta$*: there exists an* $O(2^{\delta n})$ *algorithm for solving 3-ESAT*$\}$. *Define ETH (Exponential-Time Hypothesis) for 3-ESAT to be that:* $s > 0$. *In other words, 3-ESAT does not have sub-exponential time algorithm.*

Note that this hypothesis is stronger than $\mathcal{NP} \neq \mathcal{P}$ but yet plausible according to both theoretical and practical arguments presented in Section 1. Under this assumption, we have the following result.

Theorem 5. *In the* $(2+f(n))$*-SAT model, if the variables which appear in* Φ_2 *do not appear in* Φ_3*, and vice versa, then the* $\left(2+\frac{(\log n)^k}{n^2}\right)$*-SAT is indeed in* $\mathcal{NPI}$ *under ETH for 3-ESAT,* $k \geq 2$.

Proof. Consider the special case of $\left(2+\frac{(\log n)^k}{n^2}\right)$-SAT, where Φ_3 is an instance of 3-ESAT and $n_3 = m_3 = (\log n)^k$ and Φ_2 is always satisfiable. That is,

$$\frac{m_3}{m} = \frac{m_3}{m_2 + m_3} = \frac{(\log n)^k}{n^2} = \frac{m_3}{(n_2+n_3)^2}.$$

$$m_2 = (n_2+n_3)^2 - m_3 \leq (n_2+n_3)^2.$$

Note that $n_2 = n - n_3 = n - (\log n)^k$, $n_3 = (\log n)^k$, for sufficiently large n we get

$$(n_2+n_3)^2 \leq \frac{3}{2}n_2^2 - \frac{3}{2}n_2.$$

This means the special case of $\left(2+\frac{(\log n)^k}{n^2}\right)$-SAT indeed exists according to Claim 1.

Then for this special case of $\left(2+\frac{(\log n)^k}{n^2}\right)$-SAT ($k \geq 2$), Φ_3 can not be solved in polynomial time of n under *ETH* for 3-ESAT since there are $(\log n)^k$ variables in Φ_3, so does $\Phi = \Phi_2 \wedge \Phi_3$ since the variables which appear in Φ_2 do not appear in Φ_3, and vice versa.

Thus, $\left(2+\frac{(\log n)^k}{n^2}\right)$-SAT is indeed not in $\mathcal{P}$ under *ETH* for 3-ESAT, $k \geq 2$, and according to theorem 3 the theorem does hold. □

The more general case of $\left(2+\frac{(\log n)^k}{n^2}\right)$-SAT ($k \geq 2$), where the variables which appear in Φ_2 may appear in Φ_3, and vice versa, is currently under investigation.

5 Remarks and Conclusion

In this work, we study the boundary between $\mathcal{P}$ and $\mathcal{NPC}$ for the model of $(2+p)$-SAT when p is considered as a function of n, the number of variables in the Boolean formula. The model allows us to obtain a natural problem in $\mathcal{NPI}$ under the ETH assumption. It is an interesting open problem whether this can be further shown to be in $\mathcal{NPI}$ under the weaker assumption $\mathcal{NP} \neq \mathcal{P}$.

Acknowledgments

The authors are indebted to Shirley Cheung for her many valuable helps.

References

1. R. E. Lander. On the Structure of Polynomial Time Reducibility. Jour. of ACM, 22, 1975, pp. 155-171.
2. H. Papadimitriou. Computational Complexity, 329-332, Addison-Wesley (1994)
3. O. Goldreich. Introduction to Complexity, 23-25 (1999). Also available from http://theory.lcs.mit.edu/~oded/
4. Uwe Schoning. A Probabilistic Algorithm for k-SAT and Constraint Satisfaction Problems. In FOCS 1999: 410-420.
5. Russell Impagliazzo and Ramamohan Paturi. Which Problems Have Strongly Exponential Complexity? In FOCS 1998: 653-664.
6. Russell Impagliazzo and Ramamohan Paturi. Complexity of k-SAT. JCSS 62 (2): 367-375, 2001.
7. S. Micali, M. Rabin and S. Vadhan. Verifiable random functions. In FOCS 1999: 120-130.
8. C. Dwork and M. Naor. Zaps and their applications. In FOCS 2000: 283-293. Also available from http://www.wisdom.weizmann.ac.il/~naor/
9. R. Canetti, O. Goldreich, S. Goldwasser and S. Micali. Resettable zero-knowledge. In STOC 2000: 235-244.
10. S. Micali and L. Reyzin. Soundness in the public-key model. In Crypt 2001:542-565. Also available from http://www.cs.bu.edu/~reyzin/
11. R. Monasson, R. Zecchina, S. Kirkpatrick, B. Selman and L. Troyansky. Determining computational complexity from characteristic 'phase transitions', Nature 400, 133-137 (1999).
12. W. Anderson. Solving problems in finite time, Nature 400, 115-116 (1999).
13. S. Kirkpatrick and B. Selman. Critical behavior in the satisfiability of random Boolean expressions, Science, (264): 1297-1301 (1994).
14. R. Monasson and R. Zecchina. Tricritical points in random combinatorics: the 2+p SAT case, J. Phys. A (31): 9209-9217 (1998).

On the Minimal Polynomial of a Matrix*

Thanh Minh Hoang and Thomas Thierauf

Abt. Theoretische Informatik
Universität Ulm
89069 Ulm, Germany
{hoang,thierauf}@informatik.uni-ulm.de

Abstract. We investigate the complexity of the degree and the constant term of the minimal polynomial of a matrix. We show that the degree of the minimal polynomial behaves as the matrix rank.
We compare the constant term of the minimal polynomial with the constant term of the characteristic polynomial. The latter is known to be computable in the logspace counting class **GapL**. We show that this holds also for the minimal polynomial if and only if the *logspace exact counting class* $\mathbf{C_=L}$ is closed under complement. The latter condition is one of the main open problems in this area.
As an application of our techniques we show that the problem to decide whether a matrix is diagonalizable is complete for $\mathbf{AC^0(C_=L)}$, the $\mathbf{AC^0}$-*closure of* $\mathbf{C_=L}$.

1 Introduction

A rule of thumb says that *Linear Algebra is in* $\mathbf{NC^2}$. However, if we look more closely, we see that this is a very rough statement. In particular, we are not able to show that the various problems in Linear Algebra are equivalent under, say, logspace many-one reductions.

It seems to be more appropriate to express the complexity of problems in Linear Algebra in terms of *logspace counting classes*. The initial step in this direction was done by Damm [5], Toda [17], Vinay [19], and Valiant [18]. They showed that the determinant of an integer matrix characterizes the complexity class **GapL** (see [13] for more details on the history). Toda [17] showed more problems to be complete for **GapL**, including matrix powering, and the inverse of a matrix. There are also graph theoretic problems related to counting the number s-t-paths in a graph.

The verification of **GapL** functions is captured by the class $\mathbf{C_=L}$. An example of a complete problem is to decide whether an integer matrix A is singular, i.e., whether $\det(A) = 0$. More general, the decision problem, whether the rank of A is less than some given number k, is complete for $\mathbf{C_=L}$. The problem whether the rank of A *equals* k can be expressed as the conjunction of problems in $\mathbf{C_=L}$ and in $\mathbf{coC_=L}$, a class that we denote by $\mathbf{C_=L} \wedge \mathbf{coC_=L}$. The problem to determine the rank of a matrix is captured by the $\mathbf{AC^0}$-closure of $\mathbf{C_=L}$, which we denote

* This work was supported by the Deutsche Forschungsgemeinschaft

O.H. Ibarra and L. Zhang (Eds.): COCOON 2002, LNCS 2387, pp. 37–46, 2002.

by $\mathbf{AC}^0(\mathbf{C_=L})$. Finally, the problem to decide whether two matrices have the same rank is complete for $\mathbf{AC}^0(\mathbf{C_=L})$. The results on the rank were shown by Allender, Beals, and Ogihara [2].

The complexity of the minimal polynomial has been studied before [11](see also [9,10]). In this paper, we show that there is a strong relationship between the *degree of the minimal polynomial* of a matrix and the matrix rank problem. Namely, the problems to decide whether the degree of the minimal polynomial is less than k or equal k, for some given k, are complete for $\mathbf{C_=L}$ and $\mathbf{C_=L} \wedge \mathbf{coC_=L}$, respectively. To decide whether the degrees of the minimal polynomials of two matrices are equal is complete for $\mathbf{AC}^0(\mathbf{C_=L})$.

We also investigate the complexity of the constant term of the minimal polynomial. The constant term of the characteristic polynomial is **GapL**-complete. By analogy, we ask whether the constant term of the minimal polynomial can be computed in **GapL**, too. We show that this question is strongly connected with another open problem: *the constant term of the minimal polynomial can be computed in* **GapL** *if and only if* $\mathbf{C_=L}$ *is closed under complement*. This connection is a consequence of a hardness result: to decide whether the constant terms of the minimal polynomials of two matrices are equal is complete for $\mathbf{AC}^0(\mathbf{C_=L})$.

Whether $\mathbf{C_=L}$ is closed under complement is one of the big open questions in this area. Recall that many related classes have this property: **NL** [12,16], **SL** [14], **PL** (trivially), and nonuniform **UL** [15]. Thus our results on the constant term of the minimal polynomial might offer some new points to attack this problem.

A final observation is about the diagonalizability of matrices. In [10] it is shown that this decision is hard for $\mathbf{AC}^0(\mathbf{C_=L})$. We show that this class also is an upper bound for this problem. It follows that diagonalizability is complete for $\mathbf{AC}^0(\mathbf{C_=L})$. We extend the result to *simultaneous diagonalizability* where one has to decide whether *all* of k given matrices are diagonalizable by the same diagonalizing matrix.

2 Preliminaries

We assume familiarity with some basic notions of complexity theory and linear algebra. We refer the readers to the papers [2,3] for more details and properties of the considered complexity classes, and to the textbooks [6,8,7] for more background in linear algebra.

Complexity Classes. For a nondeterministic Turing machine M, we denote the number of accepting and rejecting computation paths on input x by $acc_M(x)$ and by $rej_M(x)$, respectively. The difference of these two quantities is gap_M, i.e., for all $x : gap_M(x) = acc_M(x) - rej_M(x)$. The function class **GapL** is defined as the class of all functions $gap_M(x)$ such that M is a nondeterministic logspace bounded Turing machine. **GapL** has many closure properties: for example it is closed under addition, subtraction, and multiplication (see [3]). In [1] (Corollary 3.3) it is shown that **GapL** is closed under composition in a very strong sense:

if each entry of an $n \times n$ matrix A is **GapL**-computable, then the determinant of A is still computable in **GapL**.

A set S is in $\mathbf{C_=L}$, if there exists a function $f \in \mathbf{GapL}$ such that for all x we have $x \in S \iff f(x) = 0$. Since it is open whether $\mathbf{C_=L}$ is closed under complement, it makes sense to consider the *Boolean closure of* $\mathbf{C_=L}$, i.e., the class of sets that can be expressed as a Boolean combination of sets in $\mathbf{C_=L}$. For our purposes, it suffices to consider the following two classes: a) $\mathbf{coC_=L}$ is the class of complement sets $\overline{L}$ where $L \in \mathbf{C_=L}$, b) $\mathbf{C_=L} \wedge \mathbf{coC_=L}$ [2] is defined as the class of intersections of sets in $\mathbf{C_=L}$ with sets in $\mathbf{coC_=L}$, i.e.,

$$L \in \mathbf{C_=L} \wedge \mathbf{coC_=L} \iff \exists L_1 \in \mathbf{C_=L},\ L_2 \in \mathbf{coC_=L}:\ L = L_1 \cap L_2.$$

For sets S_1 and S_2, we say that S_1 is $\mathbf{AC}^0$*-reducible to* S_2, if there is a logspace uniform circuit family of polynomial size and constant depth that computes S_1 with unbounded fan-in AND- and OR-gates, NOT-gates, and oracle gates for S_2. In particular, we consider the classes $\mathbf{AC}^0(\mathbf{C_=L})$ and $\mathbf{AC}^0(\mathbf{GapL})$: the sets that are $\mathbf{AC}^0$-reducible to a set in $\mathbf{C_=L}$, respectively a function in **GapL**. The known relationships among these classes are as follows:

$$\mathbf{C_=L} \subseteq \mathbf{C_=L} \wedge \mathbf{coC_=L} \subseteq \mathbf{AC}^0(\mathbf{C_=L}) \subseteq \mathbf{AC}^0(\mathbf{GapL}) \subseteq \mathbf{TC}^1 \subseteq \mathbf{NC}^2.$$

Furthermore, we say that S_1 is *(logspace many-one) reducible to* S_2, if there is a function $f \in L$ (deterministic logspace) such that for all x we have $x \in S_1 \iff f(x) \in S_2$. In an analogous way one can define $\mathbf{AC}^0$- or $\mathbf{NC}^1$-many-one reductions. Unless otherwise stated, all reductions in this paper are logspace many-one.

Linear Algebra. Let $A \in \boldsymbol{F}^{n\times n}$ be a matrix over the field $\boldsymbol{F}$. The *characteristic polynomial* of A is the polynomial $\chi_A(x) = \det(xI - A)$. A nonzero polynomial $p(x)$ over $\boldsymbol{F}$ is called an *annihilating polynomial* for A if $p(A) = \mathbf{0}$. The Cayley-Hamilton Theorem states that $\chi_A(x)$ is an annihilating polynomial for A. The characteristic polynomial is a *monic polynomial*: its highest coefficient is one. The *minimal polynomial* of A, denoted by $\mu_A(x)$, is the unique monic annihilating polynomial for A with minimal degree. Note that if A is an integer matrix, then all coefficients of $\chi_A(x)$ and of $\mu_A(x)$ are also integer. Let's denote the degree of a polynomial p by $\deg(p)$. Then we have $1 \leq \deg(\mu_A(x)) = m \leq n$.

Two matrices $A, B \in \boldsymbol{F}^{n\times n}$ are called *similar* if there is a nonsingular matrix $P \in \boldsymbol{F}^{n\times n}$ such that $A = PBP^{-1}$. Furthermore, A is called *diagonalizable* if A is similar to a diagonal matrix. The matrices $A_1, \ldots, A_k$ are called *simultaneously diagonalizable* if there is a nonsingular matrix P such that $PA_1P^{-1}, \ldots, PA_kP^{-1}$ are diagonal.

Problems. Unless otherwise stated the domain for the algebraic problems are the integers. By DETERMINANT we denote the problem to compute the determinant of a given $n \times n$ matrix A. In POWERELEMENT there is additionally given an integer m and have to compute $(A^m)_{1,n}$, the element of A^m at position $(1, n)$. Both POWERELEMENT and DETERMINANT are complete for **GapL** [4,5,17,18,19].

Various decision problems are based on **GapL**-functions. The *verification* of a **GapL**-function is captured by the class $\mathbf{C_=L}$. A **GapL**-complete function yields a $\mathbf{C_=L}$-complete verification problem. For example, to verify whether the determinant is zero, i.e., testing singularity, is complete for $\mathbf{C_=L}$. Similarly, to verify whether A^m at position $(1,n)$ is zero, is complete for $\mathbf{C_=L}$. The latter problem we denote by PowerElement$_=$.

With respect to the minimal polynomial, MinPolynomial is the problem to compute the i-th coefficient d_i of $\mu_A(x)$ for given A and i. MinPolynomial is computable in $\mathbf{AC^0(GapL)}$ and is hard for **GapL** [10,11]. With respect to the degree of the minimal polynomial, DegMinPol is the set of all triple (A,k,b), where b is the k-th bit of $\deg(\mu_A(x))$.

There is a bunch of decision problems related to MinPolynomial and DegMinPol: Given two matrices A and B, and $k \geq 1$,

- EqMinPolynomial is to decide whether $\mu_A(x) = \mu_B(x)$,
- EqCTMinPol is to decide whether the minimal polynomials of A and B have the same constant term,
- EqDegMinPol is to decide whether the minimal polynomials of A and B have the same degree,
- DegMinPol$_=$ is to decide whether $\deg(\mu_A(x)) = k$,
- DegMinPol$_\leq$ is to decide whether $\deg(\mu_A(x)) \leq k$.

Finally, the set of all diagonalizable matrices is denoted by Diagonalizable. The set of all simultaneously diagonalizable matrices is denoted by SimDiagonalizable.

3 The Minimal Polynomial

In this section we investigate the complexity of the degree and the constant term of the minimal polynomial of a matrix. The upper bounds on the complexity of these problems follow easily from the predecessor paper [10,11]. The main contributions here are the lower bounds for these problems. In particular, we want to point out that the degree of the minimal polynomial has essentially the same complexity as the matrix rank.

3.1 Upper Bounds

In [10] it is shown that the minimal polynomial of a matrix A can be computed in $\mathbf{AC^0(GapL)}$. The algorithm was based on the following observation. Define $\boldsymbol{a}_i = vec(A^i)$, where $vec(A^i)$ is the vector of length n^2 that is obtained by putting the columns of A^i below each other, for $i = 0,1,2,\ldots,n$. Then the minimal polynomial $\mu_A(x)$ with degree m is characterized by the following two properties:

(i) $\mu_A(A) = \mathbf{0}$. Equivalently we can say that $\boldsymbol{a}_0, \boldsymbol{a}_1, \ldots, \boldsymbol{a}_m$ are *linearly dependent*, and

(ii) for every monic polynomial $p(x)$ with degree $m-1$, we have $p(A) \neq \mathbf{0}$.
Equivalently we can say that $\boldsymbol{a}_0, \boldsymbol{a}_1, \ldots, \boldsymbol{a}_{m-1}$ are *linearly independent.*

Note that $\boldsymbol{a}_m, \ldots, \boldsymbol{a}_n$ linearly depend on $\boldsymbol{a}_0, \boldsymbol{a}_1, \ldots, \boldsymbol{a}_{m-1}$ in this case. Define the $n^2 \times j$ matrices C_j and the symmetric $j \times j$ matrices D_j as

$$C_j = (\boldsymbol{a}_0\, \boldsymbol{a}_1\, \cdots\, \boldsymbol{a}_{j-1}),\ D_j = C_j^T\, C_j,\quad \text{for } j = 1, \ldots, n.$$

Then $C_m, \ldots, C_n$ and $D_m, \ldots, D_n$ all have the same rank m, which is precisely the degree of $\mu_A(x)$. Hence we have $\deg(\mu_A(x)) = \operatorname{rank}(D_n)$.

Let $\chi_{D_n}(x) = x^n + c_{n-1}x^{n-1} + \cdots + c_1 x + c_0$. Since D_n is symmetric, we have $\operatorname{rank}(D_n) = n - l$, where l is the smallest index such that $c_l \neq 0$. Because **GapL** is closed under composition [1], each of the coefficients $c_{n-1}, \ldots, c_0$ is computable in **GapL**. Therefore, in $\mathbf{C_=L}$ we can test whether one or several of the c_i's are zero (note that $\mathbf{C_=L}$ is closed under conjunction). In particular, we get a method to verify the degree of the minimal polynomial.

Proposition 1. *1.* DEGMINPOL$_\leq$ *is in* $\mathbf{C_=L}$.
2. DEGMINPOL$_=$ *is in* $\mathbf{C_=L} \wedge \mathbf{coC_=L}$,
3. DEGMINPOL, EQDEGMINPOL *are in* $\mathbf{AC^0}(\mathbf{C_=L})$,

Part 1 and 2 of the proposition follow directly from the discussion above. The problems in part 3 can be solved with some extra $\mathbf{AC^0}$-circuitry.

Next, we consider the coefficients of $\mu_A(x) = x^m + d_{m-1}x^{m-1} + \cdots + d_0$. The vector $(d_0, d_1, \ldots, d_{m-1})^T$ is the unique solution of the system of linear equations $C_m \boldsymbol{x} = -\boldsymbol{a}_m$. Hence we get

$$(d_0, d_1, \ldots, d_{m-1})^T = -D_m^{-1} C_m^T \boldsymbol{a}_m. \tag{1}$$

Notice that D_m nonsingular for $m = \deg(\mu_A(x))$, and each element of D_m^{-1} can be computed in **GapL** [1].

Let B be another matrix and we want to know whether A and B have the same minimal polynomial, or, whether their minimal polynomials have the same constant term. We can express the coefficients of $\mu_B(x)$ analogously as for A in equation (1). It follows that we can compare the coefficients in $\mathbf{AC^0}(\mathbf{C_=L})$.

Proposition 2. EQMINPOLYNOMIAL *and* EQCTMINPOL *are in* $\mathbf{AC^0}(\mathbf{C_=L})$.

3.2 Lower Bounds

Allender, Beals, and Ogihara [2] showed that the decision problem *Feasible Systems of Linear Equations*, FSLE for short, is complete for $\mathbf{AC^0}(\mathbf{C_=L})$. More precisely, an input for FSLE are an $m \times n$ matrix A and a vector $\boldsymbol{b}$ of length m over the integers. One has to decide whether the system of linear equations $A\boldsymbol{x} = \boldsymbol{b}$ has a rational solution. We use FSLE as reference problem to show the hardness results.

Theorem 1. EQDEGMINPOL, EQMINPOLYNOMIAL, *and* EQCTMINPOL *are hard for* $\mathbf{AC^0}(\mathbf{C_=L})$.

Proof. Let A and $\boldsymbol{b}$ be an input for FSLE. Define the symmetric matrix $B = \begin{pmatrix} \mathbf{0} & A \\ A^T & \mathbf{0} \end{pmatrix}$ and vector $\boldsymbol{c} = (\boldsymbol{b}^T, \mathbf{0})^T$ of length $m+n$. We prove that

$$
\begin{aligned}
(A, \boldsymbol{b}) \in \text{FSLE} &\iff (B, \boldsymbol{c}) \in \text{FSLE} & (2)\\
&\iff C = \begin{pmatrix} B & \mathbf{0} \\ 0 \cdots 0 & 0 \end{pmatrix} \text{ is similar to } D = \begin{pmatrix} B & \boldsymbol{c} \\ 0 \cdots 0 & 0 \end{pmatrix} & (3)\\
&\iff D \in \text{DIAGONALIZABLE} & (4)\\
&\iff \mu_C(x) = \mu_D(x) & (5)\\
&\iff \deg(\mu_C(x)) = \deg(\mu_D(x)) & (6)\\
&\iff \text{ct}(\mu_{C_\alpha}(x)) = \text{ct}(\mu_{D_\alpha}(x)), & (7)
\end{aligned}
$$

where $\text{ct}(\mu_M(x))$ denotes the constant term of $\mu_M(x)$, and $C_\alpha = C + \alpha I$ and $D_\alpha = D + \alpha I$ for an appropriate positive integer α to be chosen later.

Equivalences (2), (3), and (4) were shown in [10]. For completeness, we include a proof.

Equivalence (2). Note that the system $A^T x = \mathbf{0}$ is always feasible.

Equivalence (3). Let x_0 be a solution of the system $B\boldsymbol{x} = \boldsymbol{c}$. Define the nonsingular matrix $T = \begin{pmatrix} I & x_0 \\ \mathbf{0} & -1 \end{pmatrix}$. It is easy to check that $CT = TD$, therefore C is similar to D. Conversely, if the above system is not feasible, then C and D have different ranks and can therefore not be similar.

Equivalence (4). Observe that matrix C is symmetric. Therefore, C is always diagonalizable, i.e., C is similar to a diagonal matrix, say C'. Now, if C is similar to D, then D is similar to C' as well, because the similarity relation is transitive. Hence D is diagonalizable as well. Conversely, if D is diagonalizable, then D has only elementary divisors of the form $(x - \gamma_i)$ where γ_i is any of its eigenvalues. Since C is diagonalizable, its elementary divisors are also linear. Note furthermore that C and D have the same characteristic polynomial. Therefore, they must have the same system of elementary divisors, i.e., they are similar.

Equivalence (5). If C is similar to D, then it is clearly that $\mu_C(x) = \mu_D(x)$. Conversely, if $\mu_C(x) = \mu_D(x)$, then $\mu_D(x)$ contains only linear irreducible factors, because $\mu_C(x)$ has this property (since C is symmetric matrix). Therefore D is diagonalizable.

Equivalence (6). Recall that $\deg(\mu_C(x))$ is exactly the number of all distinct eigenvalues of C. Since C and D have the same characteristic polynomial, they have the same eigenvalues, and therefore $\deg(\mu_C(x)) \leq \deg(\mu_D(x))$. These degrees are *equal* iff every root of $\mu_D(x)$ has multiplicity 1. The latter holds iff D is diagonalizable.

Equivalence (7). Observe that equivalences (2) to (6) still hold when we replace C_α and D_α for C and D, respectively, for any α. For an appropriate choice of α we show: if the constant terms of $\mu_{C_\alpha}(x)$ and $\mu_{D_\alpha}(x)$ are equal, then these polynomials are equal.

Fix any α. Let $\lambda_1, \ldots, \lambda_k$ be the distinct eigenvalues of C. Then the distinct eigenvalues of C_α are $\lambda_1 + \alpha, \ldots, \lambda_k + \alpha$. Since C_α is symmetric and since C_α and D_α have the same eigenvalues, we can write

$$\mu_{C_\alpha}(x) = \prod_{i=1}^{k}(x - (\lambda_i + \alpha)) \text{ and } \mu_{D_\alpha}(x) = \prod_{i=1}^{k}(x - (\lambda_i + \alpha))^{t_i},$$

where $t_i \geq 1$ for $i = 1, 2, \ldots, k$. In order to prove that $\mu_{C_\alpha}(x) = \mu_{D_\alpha}(x)$, we have to show that all $t_i = 1$, for an appropriate α.

Note that the constant terms of these polynomials are the product of the eigenvalues (in the case of D_α, with multiplicities t_i each). Hence it suffices to choose α such that all eigenvalues of C_α are greater than 1. This is done as follows. By $\rho(C)$ we denote the *spectral radius* of C, i.e. $\rho(C) = \max_{1 \leq i \leq k} |\lambda_i|$. The *maximum column sum matrix norm* of $C = (c_{i,j})$ is defined as

$$||C|| = \max_{1 \leq j \leq 2n+1} \sum_{i=1}^{2n+1} |c_{i,j}|.$$

It is well known that $\rho(C) \leq ||C||$. Therefore, if we choose (in logspace) $\alpha = ||C|| + 2$, then we have $\lambda_i + \alpha > 1$, for $i = 1, 2, \ldots, k$.

□

Corollary 1. EqDegMinPol, EqMinPolynomial, *and* EqCTMinPol *are complete for* $\mathbf{AC^0(C_=L)}$.

Recall that the constant term of the characteristic polynomial can be computed in **GapL**. Now assume for a moment, that the constant term of the minimal polynomial is in **GapL** as well. It follows that EqCTMinPol is in $\mathbf{C_=L}$, because this is asking whether the difference of two constant terms (a **GapL**-function) is zero. By Corollary 1, it follows that $\mathbf{AC^0(C_=L)} = \mathbf{C_=L}$. This argument is a proof of the following corollary:

Corollary 2. *If the constant term of the minimal polynomial of a matrix is computable in* **GapL**, *then* $\mathbf{C_=L}$ *is closed under complement.*

Theorem 2. *1.* $\text{DegMinPol}_{\leq}$ *is hard for* $\mathbf{C_=L}$, *and*
2. $\text{DegMinPol}_{=}$ *is hard for* $\mathbf{C_=L} \wedge \mathbf{coC_=L}$.

Proof. 1) To show the first claim, we reduce $\text{PowerElement}_{=}$ to $\text{DegMinPol}_{\leq}$. Let A be a $n \times n$ matrix and $m \geq 1$ be an input for $\text{PowerElement}_{=}$. One has to decide whether $(A^m)_{1,n} = 0$. In [11] (see also [10]) it is shown how to construct a matrix B in logspace such that

$$\mu_B(x) = x^{2m+2} - ax^{m+1}, \text{ where } a = (A^m)_{1,n}.$$

Let C be the companion matrix of the polynomial x^{2m+2}, that is, a $(2m+2) \times (2m+2)$ matrix, where all the elements on the first sub-diagonal are 1 and all the other elements are 0. Then we have $\chi_C(x) = \mu_C(x) = x^{2m+2}$.

Define $D = \begin{pmatrix} B & \mathbf{0} \\ \mathbf{0} & C \end{pmatrix}$. It is known that the minimal polynomial of D is the *least common multiple* (for short: lcm) of the polynomials $\mu_B(x)$ and $\mu_C(x)$. Therefore we have

$$\begin{aligned} \mu_D(x) &= \operatorname{lcm}\{x^{m+1}(x^{m+1} - a), x^{2m+2}\} \\ &= \begin{cases} x^{2m+2}, & \text{if } a = 0, \\ x^{2m+2}(x^{m+1} - a), & \text{if } a \neq 0. \end{cases} \end{aligned}$$

It follows that $(A^m)_{1,n} = 0 \iff \deg(\mu_D(x)) = 2m + 2$.

2) To show the second claim, we reduce an arbitrary language $L \in \mathbf{C_{=}L} \wedge \mathbf{coC_{=}L}$ to DegMinPol$_{=}$. Namely, we compute (in logspace) matrices A_1 and A_2 of order n_1 and n_2, respectively, and integers $1 \leq m, l \leq n$ such that for all w

$$w \in L \iff (A_1^m)_{1,n_1} = 0 \text{ and } (A_2^l)_{1,n_2} \neq 0.$$

We show in Lemma 1 below that we may assume w.l.o.g. that $m > l$. Let $a_1 = (A_1^m)_{1,n_1}$ and $a_2 = (A_2^l)_{1,n_2}$. As explained in the first part of the proof, we can compute matrices B_1 and B_2 such that

$$\begin{aligned} \mu_{B_1}(x) &= x^{2m+2} - a_1 x^{m+1}, \\ \mu_{B_2}(x) &= x^{2l+2} - a_2 x^{l+1}. \end{aligned}$$

By C we denote again the companion matrix of x^{2m+2}. For the diagonal block matrix $D = \begin{pmatrix} B_1 & & \\ & B_2 & \\ & & C \end{pmatrix}$, we get (for $m > l$)

$$\begin{aligned} \mu_D(x) &= \operatorname{lcm}\{\mu_{B_1}(x),\ \mu_{B_2}(x),\ \mu_C(x)\} \\ &= \operatorname{lcm}\{x^{m+1}(x^{m+1} - a_1),\ x^{l+1}(x^{l+1} - a_2),\ x^{2m+2}\} \\ &= \begin{cases} 2m + l + 3, & \text{for } a_1 = 0,\ a_2 \neq 0, \\ 3m + 3, & \text{for } a_1 \neq 0,\ a_2 = 0, \\ 2m + 2, & \text{for } a_1 = 0,\ a_2 = 0, \\ 3m + 3 + r, & \text{for } a_1 \neq 0,\ a_2 \neq 0, \text{ where } r > 0. \end{cases} \end{aligned}$$

In summary, we have

$$w \in L \iff a_1 = 0 \text{ and } a_2 \neq 0 \iff \deg(\mu_D(x)) = 2m + l + 3.$$

□

The following lemma completes the proof of Theorem 2

Lemma 1. *Let A be an $n \times n$ matrix and $m \geq 1$. For any $k \geq 1$ there is a matrix $\widetilde{A}$ of order $p = n(mk+1)$ such that $(A^m)_{1,n} = (\widetilde{A}^{km})_{1,p}$.*

Proof. Define the following $(mk+1) \times (mk+1)$ *block matrix* $\widetilde{A}$

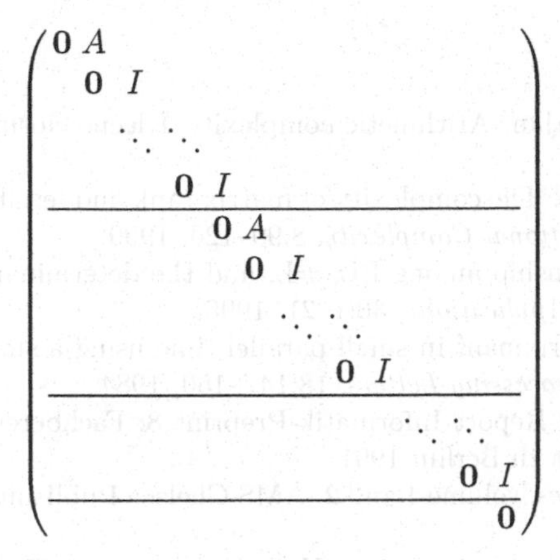

Each block of $\widetilde{A}$ is a matrix of order n. All blocks are zero except for the ones on the first block super-diagonal. Here we start with A followed by $(k-1)$-times I. This pattern occurs m-times in total.

An elementary calculation shows that $\widetilde{A}^{mk}$ has A^m as its upper right block at position $(1, mk+1)$. All other blocks are $\mathbf{0}$. This proves the lemma.

□

4 Diagonalizability

In [9] it is shown that the decision whether two matrices are similar is complete for $\mathbf{AC}^0(\mathbf{C_=L})$. It is well known that DIAGONALIZABLE is hard for $\mathbf{AC}^0(\mathbf{C_=L})$ (see Theorem 1) and is contained in $\mathbf{AC}^0(\mathbf{GapL})$ [10]. In this section we show that DIAGONALIZABLE and SIMDIAGONALIZABLE are complete for $\mathbf{AC}^0(\mathbf{C_=L})$.

Theorem 3. DIAGONALIZABLE *is complete for* $\mathbf{AC}^0(\mathbf{C_=L})$.

Proof. It remains to prove that DIAGONALIZABLE is in $\mathbf{AC}^0(\mathbf{C_=L})$. Given matrix A. In Section 3.1 we shown how to construct a matrix C_n such that $\deg(\mu_A(x)) = \text{rank}(C_n)$.

Matrix A is diagonalizable iff its minimal polynomial contains only linear irreducible factors. This is the case iff $\deg(\mu_A(x))$ equals the number of distinct eigenvalues of A. The latter number can be determined as the rank of the *Hankel matrix* H_A associated with A (see Chapter XV. in [6]). Therefore, we have

$$\begin{aligned} A \text{ is diagonalizable } &\iff \deg(\mu_A(x)) = \#\text{ of distinct eigenvalues of } A \\ &\iff \text{rank}(C_n) = \text{rank}(H_A). \end{aligned} \tag{8}$$

Since each element of C_n and H_A can be computed in **GapL**, equation (8) can be checked in $\mathbf{AC}^0(\mathbf{C_=L})$. □

We consider the problem SIMDIAGONALIZABLE. Given matrices $A_1, \ldots, A_k$ of order n and $k \geq 1$. We have to test whether there is a nonsingular matrix S such that SA_iS^{-1} are diagonal, for all $1 \leq i \leq k$. If all matrices A_i are diagonalizable then they are simultaneously diagonalizable iff they are pairwise commutable, i.e. $A_i\, A_j = A_j\, A_i$ for all i, j. The latter test can be done in $\mathbf{NC}^1$. Therefore the main part is to test whether $A_i \in$ DIAGONALIZABLE, for all i. By Theorem 3 we get the following:

Corollary 3. SIMDIAGONALIZABLE *is complete for* $\mathbf{AC}^0(\mathbf{C_=L})$.

References

1. E. Allender, V Arvind, and M. Mahajan. Arithmetic complexity, Kleene closure, and formal power series, 1999.
2. E. Allender, R. Beals, and M. Ogihara. The complexity of matrix rank and feasible systems of linear equations. *Computational Complexity*, 8:99–126, 1999.
3. E. Allender and M. Ogihara. Relationship among PL, #L, and the determinant. *RAIRO-Theoretical Informatics and Applications*, 30:1–21, 1996.
4. S. Berkowitz. On computing the determinant in small parallel time using a small number of processors. *Information Processing Letters*, 18:147–150, 1984.
5. C. Damm. DET = $L^{(\#L)}$. Technical Report Informatik-Preprint 8, Fachbereich Informatik der Humboldt Universitaet zu Berlin, 1991.
6. F. Gantmacher. *The Theory of Matrices*, volume 1 and 2. AMS Chelsea Publishing, 1977.
7. R. Horn and C. Johnson. *Matrix Analysis*. Cambridge University Press, 1985.
8. R. Horn and C. Johnson. *Topics in Matrix Analysis*. Cambridge University Press, 1991.
9. T. M. Hoang and T. Thierauf. The complexity of verifying the characteristic polynomial and testing similarity. In *15th IEEE Conference on Computational Complexity (CCC)*, pages 87–95. IEEE Computer Society Press, 2000.
10. T. M. Hoang and T. Thierauf. The complexity of the minimal polynomial. In *26th International Symposium, MFCS 2001*, pages 408–420. Springer, 2001.
11. T. M. Hoang and T. Thierauf. The complexity of the characteristic and the minimal polynomial. Invited paper to the special issue in *Theoretical Computer Science* of the 26th MFCS conference 2001, to appear, 2002.
12. N. Immerman. Nondeterministic space is closed under complement. *SIAM Journal on Computing*, 17:935–938, 1988.
13. M. Mahajan and V Vinay. Determinant: Combinatorics algorithms, and complexity. *Chicago Journal of Theoretical Computer Science*, 5, 1997.
14. N. Nisan and A. Ta-Shma. Symmetric logspace is closed under complement. *Chicago Journal of Theoretical Computer Science*, 1995.
15. K. Reinhardt and E. Allender. Making nondeterminism unambiguous. *SIAM Journal on Computing*, 29:1118–1131, 2000.
16. R. Szelepcsényi. The method of forced enumeration for nondeterministic automata. *Acta Informatica*, 26(3):279–284, 1988.
17. S. Toda. Counting problems computationally equivalent to the determinant. Technical Report CSIM 91-07, Dept. of Computer Science and Information Mathematics, University of Electro-Communications, Chofu-shi, Tokyo 182, Japan, 1991.
18. L. Valiant. Why is Boolean complexity theory difficult. In M.S. Paterson, editor, *Boolean Function Complexity*, London Mathematical Society Lecture Notes Series 169. Cambridge University Press, 1992.
19. V Vinay. Counting auxiliary pushdown automata and semi-unbounded arithmetic circuits. In *6th IEEE Conference on Structure in Complexity Theory*, pages 270–284, 1991.

Computable Real Functions of Bounded Variation and Semi-computable Real Numbers (Extended Abstract)

Robert Rettinger[1], Xizhong Zheng[2,*], and Burchard von Braunmühl[2]

[1] FernUniversität Hagen, 58084 Hagen, Germany
[2] BTU Cottbus, 03044 Cottbus, Germany
zheng@informatik.tu-cottbus.de

Abstract. In this paper we discuss some basic properties of computable real functions of bounded variation (CBV-functions for short). Especially, it is shown that the image set of semi-computable real numbers under CBV-functions is a proper subset of the class of weakly computable real numbers; Two applications of CBV-functions to semi-computable real numbers produce the whole closure of semi-computable real numbers under total computable real functions, and the image sets of semi-computable real numbers under monotone computable functions and CBV-functions are different.

1 Introduction

Continuity of a real function is one of the most important property in analysis. The effective counterpart of a continuous real function is the computable real function which can be computed by some algorithm with the respect to the effectively convergent Cauchy representation of real numbers. Computable real functions are widely discussed in literature, e.g., [4,5,11]. There are many problems, especially in applications to physical science, where more precise information about a function than continuity or computability are required. For example, it is very useful to be able to measure how rapidly a real function f oscillates on some interval $[a;b]$. However, the oscillatory character of a function is not easily determined from its continuity or even its computability. For this reason, the notion of the variation of a function was introduced in mathematics by Camille Jordan (1838–1922). Concretely, the variation $V_a^b(f)$ of f on the interval $[a;b]$ is defined as the supremum $\sup\left(\sum_{i<k}|f(x_i) - f(x_{i+1})|\right)$ which is taken over all possible subdivision $a = x_0 < x_1 < x_2 < \cdots < x_k = b$ of the interval (cf. [7]). This quantity turns out to be very useful for problems in physics, engineering, probability theory, Fourier series, and so forth.

A function f is called of *bounded variation* (BV for short) on an interval $[a;b]$, if the variation $V_a^b(f)$ of f on this interval is finite. Denote by $\mathbb{BV}[a;b]$ ($\mathbb{CBV}[a;b]$) the class of all (computable) real functions $f : [a;b] \to [a;b]$ which are of bounded

* Corresponding author

O.H. Ibarra and L. Zhang (Eds.): COCOON 2002, LNCS 2387, pp. 47–56, 2002.

variation on $[a;b]$. Especially, $\mathbb{BV}[0;1]$ and $\mathbb{CBV}[0;1]$ are denoted simply by $\mathbb{BV}$ and $\mathbb{CBV}$, respectively. The class $\mathbb{BV}$ is widely discussed in classical mathematics. In this paper, we are more interested in the class $\mathbb{CBV}$. For example, we will discuss which classes of real numbers are closed under $\mathbb{CBV}$ and clarify the relationships among the image sets of $\mathbb{CBV}$ for different classes of real numbers.

Let's remind the definition of several interesting classes of real numbers discussed in effective analysis. A real number x is called *computable* if there is a computable sequence (x_s) of rational numbers which converges effectively to x in the sense that $|x - x_s| \leq 2^{-s}$ for any $s \in \mathbb{N}$; x is *left (right) computable* if there is an increasing (decreasing) computable sequence of rational numbers which converges to x; Left and right computable real numbers are called *semi-computable*; x is *weakly computable* if there are two left computable real numbers y and z such that $x = y - z$, and x is *recursively approximable* if there exists a computable sequence of rational numbers which converges to x. We denote by $\mathbf{EC}, \mathbf{LC}, \mathbf{RC}, \mathbf{SC}, \mathbf{WC}$ and $\mathbf{RA}$ the classes of computable, left computable, right computable, semi-computable, weakly computable and recursively approximable real numbers, respectively. These classes have been widely discussed in literature ([1,3,8,12,14]). Their relationship can be summarized as follows: $\mathbf{EC} = \mathbf{LC} \cap \mathbf{RC} \subsetneq \mathbf{SC} = \mathbf{LC} \cup \mathbf{RC} \subsetneq \mathbf{WC} \subsetneq \mathbf{RA}$. Besides, the classes $\mathbf{EC}, \mathbf{WC}$ and $\mathbf{RA}$ are algebraic fields, i.e., they are closed under the arithmetical operations $+, -, \times$ and $\div$. For $\mathbf{WC}$, another characterization is shown in [12] that, $x \in \mathbf{WC}$ iff there is a computable sequence (x_s) of rational numbers which converges weakly effectively to x in the sense that $\sum_{s\in\mathbb{N}} |x_s - x_{s+1}|$ is finite.

Obviously, the classes $\mathbf{EC}$ and $\mathbf{RA}$ are closed under the CBV-functions. Furthermore, the function f defined by $f(x) := 1 - x$ is a CBV-function which maps left computable real numbers to right computable ones and vice versa. Moreover, $g \circ f \in \mathbb{CBV}$ iff $g \in \mathbb{CBV}$ for any g. This observation implies that $\mathbb{CBV}(\mathbf{LC}) = \mathbb{CBV}(\mathbf{RC}) = \mathbb{CBV}(\mathbf{SC})$, where $\mathbb{CBV}(\mathbf{C}) := \{g(x) : x \in \mathbf{C} \ \& \ g \in \mathbb{CBV}\}$ denotes the image set of $\mathbf{C}$ under functions of $\mathbb{CBV}$. In [6], it is shown that both the classes $\mathbf{SC}$ and $\mathbf{WC}$ are not closed under $\mathbb{CBV}$ and that the image of a semi-computable real number under a CBV-function is weakly computable. In other words, $\mathbf{SC} \subsetneq \mathbb{CBV}(\mathbf{SC}) \subseteq \mathbf{WC}$ and $\mathbf{WC} \subsetneq \mathbb{CBV}(\mathbf{WC})$ hold. On the other hand, it is also shown in [6] that $\mathbf{WC} \subsetneq \mathbb{CTF}(\mathbf{SC}) = \mathbb{CTF}(\mathbf{WC}) \subsetneq \mathbf{RA}$, where $\mathbb{CTF}$ is the set of all computable total real functions $f : [0;1] \to [0;1]$. Namely, the image sets of semi-computable and weakly computable real numbers under total computable real functions are the same and they locate strictly between the classes $\mathbf{WC}$ and $\mathbf{RA}$. Two interesting questions remain open. That is, whether $\mathbb{CBV}(\mathbf{SC}) = \mathbf{WC}$? and $\mathbb{CBV}(\mathbf{WC}) = \mathbb{CTF}(\mathbf{WC})$? For the first question, we will show a negative answer. A positive answer to the second question follows from the stronger result $\mathbb{CBV}^2(\mathbf{SC}) = \mathbb{CTF}(\mathbf{SC})$. This result shows that any application of a total computable real function to a semi-computable real number can be realized by two consecutive applications of the CBV-functions to some (possibly different) semi-computable real number. Finally, we will show that the image sets of weakly computable real numbers under computable monotone functions and under usual total computable real functions are different.

2 Preliminaries

In this section we will recall some known results, notions and notations which will be used later.

Let Σ be any alphabet. Σ^* and Σ^ω are the sets of all finite strings and infinite sequences of Σ, respectively. For $u, v \in \Sigma^*$, denote by uv the concatenation of v after u. u is an initial segment of w (denoted by $u \sqsubseteq w$) if $w = uv$ for some v and $u \sqsubset w$ means $u \sqsubseteq w$ & $w \neq w$. If $w \in \Sigma^* \cup \Sigma^\omega$, then $w[n]$ denotes its n-th element. Thus, $w = w[0]w[1]\cdots w[n-1]$, if $|w|$, the length of w, is n, and $w = w[0]w[1]w[2]\cdots$, if $|w| = \infty$. The unique string of length 0 is denoted by λ (so-called empty string). For any finite string $w \in \{0;1\}^*$, and number $n \leq |w|$, the restriction $w \upharpoonright n$ is defined by $(w \upharpoonright n)[i] := w[i]$ if $i < n$ and $(w \upharpoonright n)[i] := \uparrow$, otherwise. Then the length $|w \upharpoonright n| = n$.

We denote by $\mathbb{N}, \mathbb{Q}$ and $\mathbb{R}$ the sets of all natural, rational and real numbers, respectively. $[0;1]_{\mathbb{Q}}$ is the set of all rational numbers $x \in [0;1]$. For any sets A and B, $f :\subseteq A \to B$ is a partial function such that $\mathrm{dom}(f) \subseteq A$ and $\mathrm{range}(f) \subseteq B$, while $f : A \to B$ denotes a total function from A to B, i.e., $\mathrm{dom}(f) = A$. If $I \subset \mathbb{R}$ is an interval, then its length is denoted by $l(I)$.

The computability notions on subsets $A \subseteq \mathbb{N}$ and functions $f :\subseteq \mathbb{N}^k \to \mathbb{N}$ are well defined and developed in classical computability theory (cf. [9,10]). For other countable set, say, $\mathbb{Q}$, the corresponding notions of computability can be defined accordingly by means of some coding. For example, if $\langle \cdot, \cdot \rangle : \mathbb{N}^2 \to \mathbb{N}$ is a computable pairing function, then we can define a coding $\sigma : \mathbb{N} \to \mathbb{Q}$ by $\sigma(\langle \langle n, m \rangle, k \rangle) := (n - m)/(k+1))$ for any $n, m, k \in \mathbb{N}$ and call a set $A \subseteq \mathbb{Q}$ recursive or recursively enumerable if the set $\sigma^{-1}(A) := \{n \in \mathbb{N} : \sigma(n) \in A\}$ is recursive or recursively enumerable, respectively. A function $f :\subseteq \mathbb{N} \to \mathbb{Q}$ is computable if there is a computable function $g :\subseteq \mathbb{N} \to \mathbb{N}$ such that $f(n) = \sigma \circ g(n)$ for any $n \in \mathrm{dom}(\sigma \circ g)$, and so forth. Especially, we call a sequence (x_s) of rational numbers computable if there is a computable function $f : \mathbb{N} \to \mathbb{Q}$ such that $x_s = f(s)$ for any s.

The computability of real functions can be defined by type-2 Turing machines (Weihrauch [10,11]). A type-2 Turing machine M extends the classical Turing machine in such a way that it accepts infinite sequences as well as finite strings as inputs and outputs. For any input p, $M(p)$ outputs a finite string q if $M(p)$ halts after finite steps with q in its write-only output tape, while $M(p)$ outputs an infinite sequence q means that $M(p)$ will never halt and keep writing the sequence q on its output tape. A real function $f :\subseteq \mathbb{R} \to \mathbb{R}$ is *computable*, if there is a type-2 Turing machine M which computes f in the sense that, for any $x \in \mathrm{dom}(f)$ and any sequence p (of rational numbers) which converges effectively to x, $M(p)$ outputs a sequence q (of rational numbers) which converges effectively to $f(x)$. Therefore, any computable function is continuous on its domain.

The closure properties of the real number classes mentioned in the last section under the class $\mathbb{CPF}$ of computable partial real functions are first discussed in [13]. These discussions are extended to the case of $\mathbb{CTF}$ in [6], where the divergence bounded computability is also introduced. For any sequence (x_s) and $n \in \mathbb{N}$. The n-divergence of (x_s) is defined as the maximal m such that, for some chain $i_0 < j_0 \leq i_1 < j_1 \leq \cdots \leq i_m < j_m$ of natural numbers, $|x_{i_s} - x_{j_s}| \geq 2^{-n}$ holds for any $s \leq m$. A real number x is called *divergence bounded computable* if there is a recursive function h and a computable sequence (x_s) of rational numbers converging to x such that the n-divergence of (x_s) is bounded by $h(n)$ for any n. The class of all divergence bounded computable real numbers is denoted by **DBC**. We can summarize some of the main results of [13,6] as follows.

Theorem 2.1 (Zheng [13] and Rettinger et al. [6]).
1. $\mathbb{CPF}(\mathbf{LC}) = \mathbb{CPF}(\mathbf{RC}) = \mathbb{CPF}(\mathbf{RA}) = \mathbf{RA}$.
2. $\mathbb{CTF}(\mathbf{LC}) = \mathbb{CTF}(\mathbf{RC}) = \mathbb{CTF}(\mathbf{WC}) = \mathbf{DBC}$.
3. $\mathbf{WC} \subsetneq \mathbb{CBV}(\mathbf{WC}) \subseteq \mathbb{CTF}(\mathbf{WC}) \subsetneq \mathbf{RA}$.
4. $\mathbf{SC} \subsetneq \mathbb{CBV}(\mathbf{LC}) = \mathbb{CBV}(\mathbf{RC}) = \mathbb{CBV}(\mathbf{SC}) \subseteq \mathbf{WC}$.

Let $\delta : \mathbb{N} \to \mathbb{N}^+$ and $\mathbb{N}_\delta^* := \{w \in \mathbb{N}^* : \forall n < |w|\,(w[n] < \delta(n))\}$. We define a *$\delta$-interval tree* ($\delta$-i.t., for short) on $[0;1]$ as a function $I : \mathbb{N}_\delta^* \to \mathbb{I}$, where $\mathbb{I}$ is the set of all rational intervals on $[0;1]$, such that $I(\lambda) = [0;1]$; $\bigcup_{i<\delta(|w|)} I(wi) = I(w)$, for any $w \in \mathbb{N}_\delta^*$ and $\lim_{s\to\infty} l(I(w_s)) = 0$ for any sequence (w_s) of $\mathbb{N}_\delta^*$ with $w_s \sqsubset w_{s+1}$. For any δ-i.t. I and $w \in \mathbb{N}_\delta^*$, the interval $I(w)$ is denoted by $I(w) := [a_w^\delta; b_w^\delta]$. A δ-i.t. I is called *computable* if the functions $a, b : \mathbb{N}_\delta^* \to [0;1]_\mathbb{Q}$ defined by $a(w) := a_w^\delta$ and $b(w) := b_w^\delta$, respectively, are computable. A δ-i.t. I is called *canonical* if, for any $w \in \mathbb{N}_\delta^*$, the interval $I(w)$ is divided into subintervals $I(w0), I(w1), \cdots, I(w(\delta(|w|)-1))$ disjunctively and equally, in other words, $a_w^\delta := \sum_{i<|w|}\left(w[i] \cdot \prod_{j\leq i} \delta(j)^{-1}\right)$ and $b_w^\delta := a_w^\delta + \prod_{j<|w|} \delta(j)^{-1}$.

Furthermore, for any $\delta_1, \delta_2 : \mathbb{N} \to \mathbb{N}^+$, a function $\iota :\subseteq \mathbb{N}_{\delta_1}^* \to \mathbb{N}_{\delta_2}^*$ is called *(δ_1, δ_2)-compatible* if the domain $\mathrm{dom}(\iota)$ of ι is infinite and alternate in the sense that, $w(i-1), w(i+1) \notin \mathrm{dom}(\iota)\ \&\ i \neq 0, \delta_1$ for any $w \in \mathbb{N}_{\delta_1}^*$ and $i < \delta_1(|w|)$ such that $wi \in \mathrm{dom}(\iota)$; $\forall w, v \in \mathbb{N}_{\delta_1}^*$ $(w \in \mathrm{dom}(\iota)\ \&\ v \sqsubseteq w \Longrightarrow v \in \mathrm{dom}(\iota)\ \&\ \iota(v) \sqsubseteq \iota(w))$ and $\forall u, v \in \mathrm{dom}(\iota)\ (|u| = |v| \Longrightarrow |\iota(u)| = |\iota(v)|)$.

The most important application of this notion is the following technical lemma which is very useful to construct some computable real functions.

Lemma 2.2 (Rettinger et al. [6]). *Let $\delta_1, \delta_2, e : \mathbb{N} \to \mathbb{N}^+$ be computable functions, I_1 a canonical δ_1-i.t., and I_2 a computable δ_2-i.t. with $l(I(w)) \leq 2^{-e(|w|)}$ for all $w \in \mathbb{N}_{\delta_2}^*$. If $\iota :\subseteq \mathbb{N}_{\delta_1}^* \to \mathbb{N}_{\delta_2}^*$ is a (δ_1, δ_2)-compatible computable function, then there is a computable function $f : [0;1] \to [0;1]$ such that $f(I_1(w)) \subseteq I_2(\iota(w))$ and $f(a_w^{\delta_1}) = a_{\iota(w)}^{\delta_2}$ for all $w \in \mathrm{dom}(\iota)$.*

3 Computable Functions of Bounded Variation

In this section, we will discuss some basic properties of $\mathbb{CBV}$. Especially, we investigate which properties of continuous functions of bounded variation can be extended to that of $\mathbb{CBV}$ accordingly. For the BV-functions, we have at first the following simple properties which hold obviously for the CBV-functions too.

Proposition 3.1. *1. If $f, g \in \mathbb{BV}$, then $f+g, f-g, f \cdot g \in \mathbb{BV}$. If, in addition, $(\exists c > 0)(\forall x \in [0;1])(|g(x)| \geq c)$ holds, then $f/g \in \mathbb{BV}$;*

2. Let $L_a^b(f)$ denote the length of the graph of function f on the interval $[a;b]$, then $V_a^b(f) + (b-a) \geq L_a^b(f) \geq [(V_a^b(f))^2 + (b-a)^2]^{1/2}$. Therefore, $f \in \mathbb{BV}[a;b]$ iff the graph of f on $[a;b]$ has finite length.

3. There are $f, g \in \mathbb{BV}$ such that $f \circ g \notin \mathbb{BV}$, i.e., $\mathbb{BV}$ is not closed under composition.

Some other properties of $\mathbb{CBV}$ are summarized in the following lemma.

Lemma 3.2. *1. If $f \in \mathbb{CBV}$, then $V_0^1(f) = \sup_{\mathbb{Q}} (\sum_{i<m} |f(r_i) - f(r_{i+1})|)$, where the supremum $\sup_{\mathbb{Q}}$ is taken over all rational subdivision $0 = r_0 < r_1 < r_2 < \cdots < r_m = 1$ for $r_i \in \mathbb{Q}$.*

2. If both $f : [0;1] \to [0;1]$ and its first order derivative f' are computable, then $f \in \mathbb{CBV}$ and v_f is a computable function, where $v_f(x) := V_0^x(f)$.

*3. For $f \in \mathbb{CBV}$, the variation $V_0^1(f)$ is a left computable real number. And for any $y \in$ **LC**, there is a function $f \in \mathbb{CBV}$ such that $V_0^1(f) = y$.*

Classically, for any $f \in \mathbb{BV}$, there are nondecreasing functions g, h such that $f(x) = g(x) - h(x)$ for all $x \in [0;1]$. Moreover, if f is continuous, then g, h can also be chosen to be continuous. Unfortunately, the result cannot be extended immediately to $\mathbb{CBV}$as shown in [15]. However this claim can still be true if we require that $V_0^1(f)$ is computable as well. This observation belongs essentially to Douglas Bridges [2].

Theorem 3.3 (Bridges [2]). *Let $f \in \mathbb{CBV}$. If $V_0^1(f)$ is computable, then there are two computable nondecreasing function $g, h : [0;1] \to [0;1]$ such that $f(x) = g(x) - h(x)$ for any $x \in [0;1]$.*

4 $\mathbb{CBV}$(SC) and WC

By Theorem 2.1.4, the image of a semi-computable real number under a CBV-function is weakly computable. In this section we will show that not every weakly computable real number is such an image. To this end, let's look at an important property of CBV-functions.

Given an interval $J \subseteq [0;1]$ of length δ and a continuous function $f : [0;1] \to [0;1]$, a pair (x_1, x_2) of real numbers is called a crossing of f over J if $f(x_1)$ and $f(x_2)$ locate on different sides of the interval J. Denoted by $z(f, J)$ the number of crossings of f over J, namely

$$\begin{aligned} z(f,J) := \max\{n \in \mathbb{N} : (\exists (x_i)_{i \leq n})(0 \leq x_0 < x_1 < \cdots < x_n \leq 1 \ \& \\ (\forall i < n)((x_i, x_{i+1}) \text{ is a crossing of } f \text{ over } J))\}. \end{aligned}$$

If $f \in \mathbb{CBV}$ and $[0;1]$ is divided equally into n-subintervals J_i of length $1/n$ for $i < n$. Then $\sum_{i<n} z(f, J_i)/n \leq V_a^b(f)$ holds for any $n \in \mathbb{N}$. This implies that, $(\forall e \in \mathbb{N})(\exists n \in \mathbb{N})(\exists i < n)(z(f, J_i)/n \leq 2^{-e})$. Because this observation is very essential for the proof of the following Theorem 4.2, we state it as a separate lemma.

Lemma 4.1. *Let $f \in \mathbb{CBV}$ and $I \subseteq [0;1]$. For any $e \in \mathbb{N}$, there are $\delta > 0$ and $a_0, a_1, a_2, a_3 \in I$ such that the intervals $I_i := (a_i; a_{i+1})$ have the same length δ for $i < 3$ and $z(f, I_1) \cdot 3 \cdot \delta \leq 2^{-e}$.*

Theorem 4.2. $\mathbb{CBV}(\mathbf{LC}) \subsetneq \mathbf{WC}$.

Proof. (sketch) The inclusion part is quite straightforward. Here we prove only the inequality part. Namely we show that there is a weakly computable real number y such that $y \neq f(x)$ for any $x \in \mathbf{LC}$ and any $f \in \mathbb{CBV}$.

Let (φ_s) and (γ_s) be effective enumerations of computable functions $\varphi_s :\subseteq \mathbb{R} \to \mathbb{R}$, and $\gamma_s :\subseteq \mathbb{N} \to \mathbb{Q}$, respectively. It suffices to construct effectively a computable sequence (y_s) of rational numbers which converges weakly effectively to some y and y satisfies, for any $i, j \in \mathbb{N}$, the requirement

$$R_{\langle i,j\rangle}: \quad \varphi_i \in \mathbb{CBV} \;\&\; \forall s(\gamma_j(s) \leq \gamma_j(s+1)) \Longrightarrow y \neq \lim_{s\to\infty} \varphi_i(\gamma_j(s))).$$

The strategy to satisfy a single requirement R_e ($e = \langle i,j\rangle$) is simple. For example, we can fix a rational interval $I \subseteq [0;1]$ as a base interval and choose arbitrarily two open subintervals I_1 and I_2 of I such that they have at least a positive distance. Then, one of these subintervals is a witness interval of R_e in the sense that each element of this interval satisfies R_e. Actually, this witness interval can be determined in finite steps, if $\varphi_i \in \mathbb{CBV}$ & $\forall s(\gamma_j(s) \leq \gamma_j(s+1))$ holds. Namely, we choose at first I_1, then change to I_2 if some $\varphi_i(\gamma_j(s))$ enters I_1, and change to I_1 again whenever $\varphi_i(\gamma_j(s))$ enters I_2 for a larger s, and so on. If the limit $\lim_{s\to\infty} \varphi_i(\gamma_j(s))$ exists, we can change the interval only finitely often and the last interval we have chosen is the witness interval of R_e. Otherwise, if the limit does not exist, then both I_1 and I_2 are witness intervals of R_e.

To satisfy all requirements R_e simultaneously, we will try to find a sequence (I_s) of nested open intervals such that $I_{s+1} \subsetneq I_s$. For each s, I_s and I_{s+1} are base and witness interval of R_s, respectively. If we require in addition that $\lim_{s\to\infty} l(I_s) = 0$, and define $y_s := \mathrm{mid}(I_s)$ (the middle point of I_s), then the sequence (y_s) converges to a limit y which belongs to all intervals I_e and hence satisfies all requirements R_e. To ensure that the sequence (y_s) converges weakly effectively, we choose the witness intervals in such a way that $l(I_s) \leq 2^{-s}$ for any $s \in \mathbb{N}$. This implies that $|y_s - y_{s+1}| \leq 2^{-s}$ and hence $\sum_{s\in\mathbb{N}} |y_s - y_{s+1}| \leq 2$.

Unfortunately, the sequence (y_s) mentioned above is not computable, because the sequence (I_s) of witness intervals is not computable. However we can construct one of its effective approximation $(I_{e,s})_{e<d_s, s\in\mathbb{N}}$ such that $\lim_{s\to\infty} d_s = \infty$, $I_{e,s}$ and $I_{e+1,s}$ are current base and witness intervals at stage s, respectively, of the requirement $R_{\langle i,j\rangle}$ for $(\varphi_{i,s}, \gamma_{j,s})$ instead of (φ_i, γ_j). At the same time, define $y_s := \mathrm{mid}(I_{d_s,s})$. Of course, the "injury" phenomenon could appear in this construction. For example, given $(I_{e,s})_{e\leq d_s}$, we might define a new witness interval $I_{e_1,s+1}$ for some $e_1 < d_s$ at stage $s+1$. In this case, all $I_{e,s}$ for $e_1 < e \leq d_s$ are destroyed and have to be redefined later again. We say that the corresponding requirements R_e is injured (by R_{e_1}). Fortunately, any requirement R_e can be injured finitely often and its witness interval $I_e := \lim_{s\to\infty} I_{e,s}$ exists.

Nevertheless, the injury in the above construction introduces also extra jumps of the sequence (y_s). To guarantee that the sum $\sum_{s\in\mathbb{N}} |y_s - y_{s+1}|$ is still finite, more efforts are needed. Concretely, given a current base interval $I_{e,s}$ of R_e, we choose at stage $s+1$, according to Lemma 4.1, three subintervals I_0, I_1 and I_2 of length δ such that $z(f, I_1)\cdot 3\delta \leq 2^{-(e+b_{e,s})}$ where $b_{e,s}$ is the number of injuries that R_e received up to stage s and let I_0 or I_2 be the witness interval of R_e. This implies that, R_e contributes to the jumps of (y_s) hereafter at most $2^{-(e+b_{e,s})}$ whenever it is not injured again. Generally, let S_e be the set of all e-stages s at which y_s is defined according to R_e. Then, we have $\sum_{s+1\in S_e} |y_s - y_{s+1}| \leq 2^{-e}$. In other words, the e-stages contribute at most 2^{-e} to the sum $\sum_{s+1\in\mathbb{N}} |y_s - y_{s+1}|$. Therefore, $\sum_{s\in\mathbb{N}} |y_s - y_{s+1}| = \sum_{e\in\mathbb{N}} \sum_{s+1\in S_e} |y_s - y_{s+1}| \leq \sum_{e\in\mathbb{N}} 2^{-e} \leq 2$. That is $y := \lim_{s\to\infty} y_s$ is a weakly computable real number which satisfies all requirements R_e. Thus, $x \in \mathbf{WC}/\mathbb{CBV}(\mathbf{LC})$.

Corollary 4.3. $\mathbb{CBV}(\mathbf{LC}) \subsetneq \mathbb{CTF}(\mathbf{LC})$.

5 $\mathbb{CBV}^2(\mathbf{LC})$ and $\mathbb{CTF}(\mathbf{LC})$

In the last section we have shown that one application of CBV-functions to **SC** produces a proper subset of **WC**. However, we will show in this section that two applications of CBV-functions suffice to produce the set $\mathbb{CTF}(\mathbf{LC})$. In the following, let $\mathbb{CBV}^2 := \{f \circ g : f, g \in \mathbb{CBV}\}$.

Theorem 5.1. $\mathbb{CBV}^2(\mathbf{LC}) = \mathbb{CTF}(\mathbf{LC}) = \mathbf{DBC}$

Proof. (sketch) The inclusion $\mathbb{CBV}^2(\mathbf{LC}) \subseteq \mathbb{CTF}(\mathbf{LC})$ is trivial because $\mathbb{CBV}^2 \subseteq \mathbb{CTF}$. We prove now that $\mathbf{DBC} \subseteq \mathbb{CBV}^2(\mathbf{LC})$.

Given a $y \in \mathbf{DBC}$, there is a recursive function $b : \mathbb{N} \to \mathbb{N}$ and a computable sequence (y_s) of rational numbers converging to y such that, for any $n \in \mathbb{N}$, the n-divergence of (y_s) is bounded by $b(n)$. Assume w.l.o.g. that $b(n) \geq 1$. We will construct two computable functions $g, h \in \mathbb{CBV}$ and an increasing computable sequence (x_s) of rational numbers converging to x such that $g \circ h(x) = y$.

By definition, for any $f \in \mathbb{CBV}$, f can only have few big jumps or many small jumps. Since $\mathbb{CBV}(\mathbf{LC}) \neq \mathbb{CTF}(\mathbf{LC})$, the composition $g \circ h$ cannot be of bounded variation. That is, to satisfy $gh(x) = y$, the function $g \circ h$ should have a lot of big jumps but g and h should not. The essential idea is that, we let h have a lot of small jumps and then let g amplify them to the big ones.

Let $\delta_1(n) := 2b(3n) + 1$, $\delta_2(n) := 2^{n+1} \cdot \prod_{i\leq n}(b(3i) + 1) + 1$ and $\delta_3(n) := 2$ be computable functions. I_1 and I_2 are canonical δ_1- and δ_2-interval trees, respectively. We define I_3 as a δ_3-interval tree in such a way that, for any $w \in \{0,1\}^*$ and $i \in \{0,1\}$, the interval $I_3(w)$ is covered by intervals $I_3(w0)$ and $I_3(w1)$ which are overlapped in the middle of the interval $I_3(w)$ for a length of $2^{-3|w|+1}$. More precisely, the interval $I_3(w) := [a_w^{\delta_3}; b_w^{\delta_3}]$ has the length of $l_w^{\delta_3} := \prod_{i<|w|}(2^{-1} + 2^{-3i})$ for $a_w^{\delta_3}, b_w^{\delta_3}$ defined by

$$a_w^{\delta_3} := \sum_{i<|w|} w[i] \cdot (2^{-1} + 2^{-3(i+1)}) \cdot \prod_{j<i}(2^{-1} + 2^{-3j}) \quad \text{and}$$

$$b_w^{\delta_3} := \sum_{i<|w|} w[i] \cdot (2^{-1} + 2^{-3(i+1)}) \cdot \prod_{j<i} (2^{-1} + 2^{-3j}) + \prod_{i<|w|} (2^{-1} + 2^{-3i}).$$

Furthermore, we define two functions $\iota_1 : \mathbb{N}^*_{\delta_1} \to \mathbb{N}^*_{\delta_2}$ and $\iota_2 : \mathbb{N}^*_{\delta_2} \to \mathbb{N}^*_{\delta_3}$ inductively by $\iota_1(\lambda) := \lambda$, $\iota_2(\lambda) := \lambda$ and

$$\begin{aligned} \iota_1(wi) &:= (\iota_1(w)1 \text{ if } \exists j(i = 4j+1); \iota_1(w)3 \text{ if } \exists j(i = 4j+3); \uparrow \text{ otherwise}) \\ \iota_2(wi) &:= (\iota_2(w)0, \text{ if } i = 1; \iota_2(w)1 \text{ if } i = 3; \uparrow \text{ otherwise }) \end{aligned}$$

for any $w \in \mathbb{N}^*$. Obviously, both ι_1 and ι_2 are computable functions. They are also (δ_1, δ_2)- and (δ_2, δ_3)-compatible, respectively. By Lemma 2.2, there are computable real functions $g, h : [0;1] \to [0;1]$ such that

$$h(I_1(w)) \subseteq I_2(\iota_1(w)) \;\&\; g(I_2(u)) \subseteq I_3(\iota_2(u)) \tag{1}$$

for any $w \in \mathrm{dom}(\iota_1)$ and $u \in \mathrm{dom}(\iota_2)$. It is not difficult to see that both g and h are of finite variations. It remains to construct an increasing computable sequence (x_s) of rational numbers. Notice that, for any $w \in \mathrm{dom}(\iota_1)$, $\iota_1(w) \in \mathrm{dom}(\iota_2)$ and hence $gh(a_w^{\delta_1}) = g(a_{\iota_1(w)}^{\delta_2}) = a_{\iota_1\iota_2(w)}^{\delta_3}$. We define $x_s := a_{w_s}^{\delta_1}$ where (w_s) is a computable sequence in $\mathbb{N}^*_{\delta_1}$ which is defined as follows.

Stage $s = 0$. Define simply $w_0 = \lambda$ and hence $x_0 := 0$.

Stage $s+1$. Given $w_s \in \mathrm{dom}(\iota_1)$. If $y_s \in I_3(\iota_2\iota_1(w_s))$, then define $w_{s+1} := w_s1$ if $y_s \in I_3(\iota_2\iota_1(w_s1))$ and $w_{s+1} := w_s3$ otherwise.

Suppose now that $y_s \notin I_3(\iota_1\iota_2(w_s))$. Then choose an $n \leq |w_s|$ such that $y_s \in I_3(\iota_1\iota_2(w_s) \upharpoonright n)$, $y_s \notin I_3(\iota_1\iota_2(w_s) \upharpoonright (n+1))$ and let $w_{s+1} = (w_s \upharpoonright n)(w_s[n] + 2)$. Notice that, $\iota_1(w_{s+1})[n] = 1$ if $\iota_1(w_s)[n] = 3$ and $\iota_1(w_{s+1})[n] = 3$ otherwise. This implies that $\iota_2\iota_1(w_{s+1})[n] = 1 \dot{-} \iota_2\iota_1(w_s)[n]$, and hence $y_s \in I_3(\iota_2\iota_1(w_{s+1})[n])$.

We can show that (x_s) constructed above is an increasing computable sequence and $\lim_{s\to\infty} gh(x_s) = \lim_{s\to\infty} y_s = y$.

Corollary 5.2. $\mathbb{CBV}(\mathbf{WC}) = \mathbb{CTF}(\mathbf{WC})$

Proof. This follows immediately from Theorem 2.1 and Theorem 5.1.

6 $\mathbb{CMF}(\mathbf{WC})$ and $\mathbb{CTF}(\mathbf{WC})$

In this section we will show that the image sets of **WC** under $\mathbb{CTF}$ and $\mathbb{CMF}$ are different, where $\mathbb{CMF}$ is the class of computable monotone functions $f : [0;1] \to [0;1]$. Notice first that, $y \in \mathbb{CMF}(\mathbf{WC})$ iff there is an $x \in \mathbf{WC}$ and a strictly monotone and computable function f such that $y = f(x)$. For strictly monotone functions, we have the following useful lemma.

Lemma 6.1. *Let $f : [0;1] \to [0;1]$ be a strictly monotone function, $J \subseteq [0;1]$ a non-empty rational interval. There is a $t_0 \in \mathbb{N}$ such that, for any $r \geq t_0$, there are rational numbers $a_1 < a_2 < a_3 < a_4$ which belong to the interval J and satisfy the following condition:*

$$a_3 - a_2 \geq 2^{-(2r+2)} \;\&\; |f(a_1) - f(a_4)| \leq 2^{-(r+1)}. \tag{2}$$

Theorem 6.2. $\mathbb{CMF}(\mathbf{WC}) \subsetneq \mathbb{CBV}(\mathbf{WC}) = \mathbf{DBC}$

Proof. It suffices to prove the inequality part. We will construct a recursive function h and a computable sequence (y_s) of rational numbers converging to y so that the n-divergence of (y_s) is bounded by $h(n)$ for any n. Thus $y \in \mathbf{DBC} = \mathbb{CTF}(\mathbf{WC})$. Furthermore, y satisfies, for all $i, j \in \mathbb{N}$, the requirement

$$R_{\langle i,j\rangle}: \quad \begin{array}{l}\text{If } \varphi_i \text{ is a strictly monotone total function and } \gamma_j \text{ is total such} \\ \text{that } \sum_{s\in\mathbb{N}} |\gamma_j(s) - \gamma_j(s+1)| \leq 1, \text{ then } \lim_{s\to\infty} \varphi_i(\gamma_j(s)) \neq y,\end{array}$$

where (φ_e) and (γ_e) are effective enumerations of all computable functions $\varphi_e :\subseteq [0;1] \to [0;1]$ and $\gamma_e :\subseteq \mathbb{N} \to [0;1]_{\mathbb{Q}}$. Thus, $y \notin \mathbb{CMF}(\mathbf{WC})$.

Given a strictly monotone function φ_i and a weakly convergent sequence $(\gamma_j(s))$ with $\sum_{s\in\mathbb{N}} |\gamma_j(s) - \gamma_j(s+1)| \leq 1$, we consider a base interval $J \subseteq [0;1]$. Let r and a_1, a_2, a_3, a_4 satisfy Lemma 6.1 and define $I^1 := [a_1; a_2]$, $I^2 := [a_3; a_4]$, $J^1 := \varphi_i(I^1)$ and $J^2 := \varphi_i(I^2)$. Now, if $\varphi_i(\gamma_j(s))$ enters J^1 (hence $\gamma_j(s)$ enters I^1), then we define y_{s+1} as the middle point of J^2. Similarly, if $\varphi_i(\gamma_j(s))$ enters J^2 (hence $\gamma_j(s)$ enters I^2), then we define y_{s+1} to be the middle point of J^1. This guarantees that the limits $y := \lim_{s\to\infty} y_s$ and $\lim_{s\to\infty} \varphi_i(\gamma_j(s))$ have at least a distance of $|\varphi_i(a_2) - \varphi_i(a_3)|$, hence y satisfies the requirement $R_{\langle i,j\rangle}$. Notice that the y_s's can be redefined according to this strategy at most 2^{2r+2} times because $a_3 - a_2 \geq 2^{-(2r+2)}$ and $\sum_{s\in\mathbb{N}} |\gamma_j(s) - \gamma_j(s+1)| \leq 1$. On the other hand, every redefinition of y_s contributes only a jump which is bounded by $2^{-(r+1)}$ because of (2).

To satisfy all requirements R_e simultaneously, let's begin with the base interval $I_0 := [0; 1]$ and search for the minimal $e := \langle i, j\rangle$ such that we can apply Lemma 6.1 for the function φ_i. Choose r_1 and a_1, a_2, a_3, a_4 which satisfy Lemma 6.1 and let $I_e^1 := [a_1; a_2]$, $I_e^2 := [a_3; a_4]$ and $J_e^u := \varphi_i(I_e^u)$ for $u := 1, 2$. By default, let $I_1 := I_e^1$ be a new base interval, define y_{s_1} to be the middle point of J_e^1. If at a later stage $s_2 > s_1$, $\varphi_i(\gamma_j(s_2))$ enters the interval J_e^1, then set $I_1 := I_e^2$. If there is another $s_3 > s_2$ such that $\varphi_i(\gamma_j(s_2))$ enters the interval J_e^2, then redefine $I_1 := I_e^1$, and so on. In each case, we will define a new value of (y_s) as the middle point of I_1. Of course, this redefinition can appear at most $2^{(2r_1+1)}$ times if $\sum_{s\in\mathbb{N}} |\gamma_j(s) - \gamma_j(s+1)| \leq 1$.

Now on the base interval I_1 we will look for another minimal $e_1 := \langle i_1, j_1\rangle > e$ such that Lemma 6.1 can be applied to φ_{i_1}. Define r_2, $I_{e_1}^u$, $J_{e_1}^u$ $(u := 1, 2)$, I_2 and new y_s similarly. This procedure can be carried out further. By the above strategy, we can see that, first, the limit $y := \lim_{s\to\infty} y_s$ exists. In fact it is the unique common point of a nested interval sequence $(I_e)_{e\in\mathbb{N}}$; Second, every requirement $R_{\langle i,j\rangle}$ is satisfied, because $\lim_{s\to\infty} \varphi_i(\gamma_j(s))$ and y have at least the distance $|\varphi_i(a_2) - \varphi_i(a_3)|$ (for some $a_2 < a_3$), if φ_i and γ_j satisfy the premise of $R_{\langle i,j\rangle}$; Third, the n-divergence of (y_s) is bounded by a recursive function h defined by $h(n) := \sum_{m<n} 2^{2m+2}$. Here the third claim follows from the observation that we define new y_s only according to some requirement and some natural number r which satisfies Lemma 6.1 and any jump which is related to this r is not greater than $2^{-(r+1)}$. Different requirements relate to different such r and, for any fixed r, there are at most 2^{2r+2} jumps related to this r.

Unfortunately, the construction above is not effective, because, first, we cannot decide whether φ_i is a monotone total function and, second, we can't calculate the value $\varphi_i(\gamma_j(s))$ in finite steps, even if it is defined. To solve this problem, let $\beta_i :\subseteq [0;1]_{\mathbb{Q}} \times \mathbb{N} \to [0;1]_{\mathbb{Q}}$ be an approximation of φ_i such that $|\varphi_i(x) - \beta_i(x,n)| \leq 2^{-n}$ and use the function pair $(\beta_{i,s}, \gamma_{j,s})$ in the construction instead of (φ_i, γ_j). In this case, the finite injury priority method should be applied.

References

1. K. Ambos-Spies, K. Weihrauch, and X. Zheng. Weakly computable real numbers. *Journal of Complexity*, 16(4):676–690, 2000.
2. D. Bridges. A constructive look at functions of bounded variation. *Bull. London Math. Soc.*, 32(3):316–324, 2000.
3. C. S. Calude. A characterization of c.e. random reals. *Theoretical Computer Science*, 217:3–14, 2002.
4. K.-I. Ko. *Complexity Theory of Real Functions*. Progress in Theoretical Computer Science. Birkhäuser, Boston, 1991.
5. M. B. Pour-El and J. I. Richards. *Computability in Analysis and Physics*. Perspectives in Mathematical Logic. Springer, Berlin, 1989.
6. R. Rettinger, X. Zheng, R. Gengler, and B. von Braunmühl. Weakly computable real numbers and total computable real functions. In J. Wang, editor, *Computing and Combinatorics*, volume 2108 of *Lecture Notes in Computer Science*, pages 586–595, Berlin, 2001. Springer. 7th Annual International Conference, COCOON 2001, Guilin, China, August 20-23, 2001.
7. H. L. Royden. *Real Analysis*. The Macmillan Company, New York, 1963.
8. R. Soare. Recursion theory and Dedekind cuts. *Trans. Amer. Math. Soc.*, 140:271–294, 1969.
9. R. I. Soare. *Recursively enumerable sets and degrees. A study of computable functions and computably generated sets*. Perspectives in Mathematical Logic. Springer-Verlag, Berlin, 1987.
10. K. Weihrauch. *Computability*, volume 9 of *EATCS Monographs on Theoretical Computer Science*. Springer, Berlin, 1987.
11. K. Weihrauch. *Computable Analysis*. Springer, Berlin, 2000.
12. K. Weihrauch and X. Zheng. A finite hierarchy of the recursively enumerable real numbers. In L. Brim, J. Gruska, and J. Zlatuška, editors, *Mathematical Foundations of Computer Science 1998*, volume 1450 of *Lecture Notes in Computer Science*, pages 798–806, Berlin, 1998. Springer. 23rd International Symposium, MFCS'98, Brno, Czech Republic, August, 1998.
13. X. Zheng. Closure properties on real numbers under limits and computable operators. In D.-Z. Du, P. Eades, V. Estivill-Castro, X. Lin, and A. Sharma, editors, *Computing and Combinatorics*, volume 1859 of *Lecture Notes in Computer Science*, pages 170–179, Berlin, 2000. Springer. 6th Annual Conference, COCOON'2000, Sydney, Australia, July 2000, full version to appear in *Theor. Comput. Sci.*
14. X. Zheng. Recursive approximability of real numbers. *Mathematical Logic Quarterly*, 48, 2002. (to appear).
15. X. Zheng, R. Rettinger and B. von Braunmühl. On the Jordan decomposability for computable functions of bounded variation. Manuscript, 2002.

Improved Compact Routing Tables for Planar Networks via Orderly Spanning Trees

Hsueh-I Lu*

Academia Sinica, Taiwan
http://www.iis.sinica.edu.tw/~hil/
hil@iis.sinica.edu.tw

Abstract. We address the problem of designing compact routing tables for an unlabeled connected n-node planar network G. For each node r of G, the designer is given a routing spanning tree T_r of G rooted at r, which specifies the routes for sending packets from r to the rest of G. Each node r of G is equipped with ports $1, 2, \ldots, d_r$, where d_r is the degree of r in T_r. Each port of r is supposed to be assigned to a neighbor of r in T_r in a one-to-one manner. For each node v of G with $v \neq r$, let $\text{port}_r(v)$ be the port to which r should forward packets with destination v. Under the assumption that the designer has the freedom to determine the label and the port assignment of each node in G, the *routing table design problem* is to design a compact routing table R_r for r such that $\text{port}_r(v)$ can be determined *only* from R_r and the label of v.
Compact routing tables for various network topologies have been extensively studied in the literature. Planar networks are particularly important for routing with geometric metrics. Based upon four-page decompositions of G, Gavoille and Hanusse gave the best previously known result for this problem: Each $\text{port}_r(v)$ is computable in $O(\log^{2+\epsilon} n)$ bit operations for any positive constant ϵ; and the number of bits required to encode their R_r is at most $8n + o(n)$. We give a new design that improves the code length of R_r to at most $7.181n + o(n)$ bits without increasing the time required to compute $\text{port}_r(v)$.

1 Introduction

We address the problem of designing compact routing tables for an unlabeled connected planar network G over an n-node set V. For each node r of G, we are given a *routing spanning tree* T_r of G rooted at r, which specifies the routes for sending packets from r to the rest of G. For example, T_r could be a precomputed shortest-path tree rooted at r for some weighted version of G. It is reasonable to assume that those n routing spanning trees are *consistent*, i.e., if s and v are two nodes such that s is on the path of T_r between r and v, then the path of T_s between s and v is identical to the path of T_r between s and v. Each node r of G is equipped with ports $1, 2, \ldots, d_r$, where d_r is the degree of r in T_r. Each port

* Research supported in part by NSC grant NSC 90-2213-E-001-018. Institute of Information Science, Academia Sinica, 128 Academia Road, Sect. 2, Taipei 115, Taiwan.

O.H. Ibarra and L. Zhang (Eds.): COCOON 2002, LNCS 2387, pp. 57–66, 2002.

of r is supposed to be assigned to a neighbor of r in T_r in a one-to-one manner. We assume the freedom to determine the label and the port assignment of each node in G. For each neighbor s of r in T_r, let $\text{port}_r(s)$ denote the port of r that is assigned to s. For each node v of G with $v \neq r$, define $\text{port}_r(v) = \text{port}_r(s)$, where s is the neighbor of r in T_r whose removal disconnects v and r in T_r. That is, if r receives a packet whose destination is v, then r should pass the packet to its port with index $\text{port}_r(v)$. The *routing table design problem* is to come up with a compact routing table R_r for each node r of G such that $\text{port}_r(v)$ can be determined *only* from R_r and the label of v. Natural objectives of this problem include

- minimizing the number $\lambda_r(n)$ of bits to encode the label of node r;
- minimizing the number $\beta_r(n)$ of bits to encode R_r for node r; and
- minimizing the time $\tau_r(n)$ to obtain $\text{port}_r(v)$ from R_r and the label of v.

Based upon four-page decompositions of G [39], Gavoille and Hanusse [20] gave the best formerly known result for this problem, which outperformed several previous trade-offs among the above objectives. Specifically, their design achieves $\lambda_r(n) = \lceil \log_2 n \rceil$, $\beta_r(n) \leq 8n + o(n)$, $\sum_{r \in V} \beta_r(n) \leq 8n^2 + o(n^2)$, and $\tau_r(n) = O(\log^{2+\epsilon} n)$ for any positive constant ϵ, where $\tau_r(n)$ is measured in terms bit operations[1]. In [20], they also mentioned a lower bound $\beta_r(n) \geq n - O(\log n)$ for the routing table design problem on G. In the present paper, we improve the code length to $\beta_r(n) < 7.181n + o(n)$ while maintain the same bounds on $\lambda_r(n)$ and $\tau_r(n)$. Moreover, our design has the feature that T_r can be recovered from R_r.

Planar networks are important in routing with geometric metrics [3, 23, 27]. For example, some of the best network topology maps used by ISPs and Internet Backbone Networks can be modeled as planar or almost planar graphs [17, 26, 28]. The near-shortest routes in wireless *ad hoc* networks are obtained through various types of planar subnetworks with low stretch factors [4, 5, 28, 32]. Other results related to routing in planar networks can be found in [2, 15, 30].

Compactness of routing tables has been extensively studied in the literature [1, 8, 9, 12–14, 18, 19, 21, 22, 33, 34, 37]. Our improved compact routing tables are based upon succinct encodings for planar graphs with efficient query support [6, 7, 24, 31]. The best of such results for an n-node m-edge planar graph without multiple edges and self-loops, due to Chiang, Lin, and Lu [6], takes $2m+2n+o(m+n)$ bits, is obtainable in $O(m+n)$ time, and supports $O(1)$-time adjacency and degree queries. Chiang et al.'s encoding is based on an algorithmic tool called *orderly spanning tree*, which generalizes the concept of realizers [36] and canonical orderings [10, 25] for planar graphs. Besides graph encoding, orderly spanning tree also finds applications in graph drawing [6] and VLSI floor planning [29]. Our improved routing tables rely heavily on Chiang et al.'s techniques for encoding planar graphs via orderly spanning trees. Therefore, our results can be regarded as an application of orderly spanning tree in computer networks.

[1] The base of each logarithm is two for the rest of the paper.

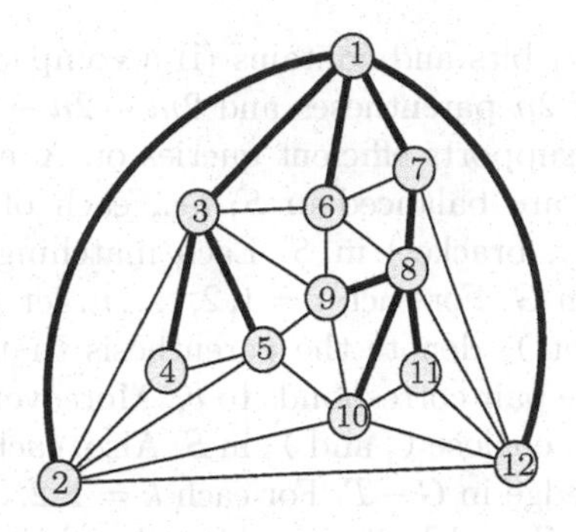

Fig. 1. A plane graph G with it orderly spanning tree T rooted at node 1 drawn in dark. The node labels show the counterclockwise preodering of the nodes in T.

The rest of the paper is organized as follows. Section 2 gives the preliminaries. Section 3 describes our design of routing tables. Section 4 explains how to efficiently obtain the correct port from the routing table and the label of the destination. Section 5 concludes the paper with a couple of open questions.

2 Preliminaries

All graphs in this paper do not contain multiple edges and self-loops. Let $|X|$ denote the number of bits in a binary string X. Let $|S|$ denote the number of elements in a set S. Unless explicitly stated otherwise, the rest of the paper sticks with the $\log n$-bit word model of computation [16, 38] that assumes operations read, write, and add on $O(\log n)$ consecutive bits take $O(1)$ time. Clearly, $O(t)$ time in $\log n$-bit word model implies $O(t \log n)$ time in terms of bit operations. For brevity of notation, let $\tilde{O}(t)$ denote $O(t) \cdot O((\log \log n)^{O(1)})$. Clearly, $\tau_r(n) = \tilde{O}(t)$ implies $\tau_r(n) = O(t \log^\epsilon n)$ for any positive constant ϵ.

Let T be a rooted spanning tree of a plane graph G. Two nodes are *unrelated* in T if they are distinct and neither of them is an ancestor of the other in T. Let $v_1, v_2, \ldots, v_n$ be the counterclockwise preordering of the nodes in T. A node v_i is *orderly* in G with respect to T if the neighbors of v_i in G form the following four blocks in counterclockwise order around v_i, where each block could be empty: the parent $p(v_i)$ of v_i in T; $V_1(v_i)$ consisting of the unrelated neighbors v_j of v_i with $j < i$; $V_2(v_i)$ consisting of the children of v_i in T; and $V_3(v_i)$ consisting of the unrelated neighbors v_j of v_i with $j > i$. T is an *orderly spanning tree* of G if v_1 is on the boundary of G's exterior face, and each $v_i, 1 \leq i \leq n$, is orderly in G with respect to T. An example of orderly spanning tree is shown in Figure 1. For any connected planar graph G, Chiang et al. [6] gave a linear-time algorithm that computes a plane embedding H of G, and an orderly spanning tree of H.

Let us review the technical details of Chiang et al.'s encoding [6] that are relevant to our improved routing tables. A key step of their encoding algorithm is to find an orderly spanning tree T for a plane embedding of the input n-node m-edge planar graph G. For each $i = 1, 2, \ldots, n$, let v_i be the i-th node in the counterclockwise preordering of the nodes in T. Chiang et al.'s encoding for G

has $2n + 2m + o(m + n)$ bits and contains (i) a compact representation X for a string S consisting of $2n$ parentheses and $2m - 2n + 2$ brackets, and (ii) an auxiliary string χ that supports efficient queries on X and S. The parentheses (respectively, brackets) are balanced in S, i.e., each of them has a matching parenthesis (respectively, bracket) in S. Each matching parenthesis pair in S corresponds to a node in G. For each $i = 1, 2, \ldots, n$, let $(_i$ denote the i-th open parenthesis in S; and let $)_i$ denote the parenthesis that matches $(_i$ in S. The matching $(_i$ and $)_i$ is the pair corresponds to v_i. Moreover, v_i is an ancestor of v_j in T if and only $(_i$ and $)_i$ enclose $(_j$ and $)_j$ in S. Also, each matching bracket pair in S corresponds to an edge in $G - T$. For each $k = 1, 2, \ldots, m - n + 1$, let $[_k$ be the k-th open bracket in S, and $]_k$ its matching close bracket in S. The matching pair $[_k$ and $]_k$ corresponds the edge $e_k = (v_i, v_j)$, where $)_i$ (respectively, $(_i$) is the rightmost parenthesis that precedes $[_k$ (respectively, $]_k$) in S. We call k the *rank* of edge (v_i, v_j). A naive representation of S requires $4m + O(1)$ bits, since S has length $2m + O(1)$ and contains four distinct symbols. However, due to some property of orderly spanning tree, S can be encoded into a string X with $2m + 2n + O(1)$ bits. Moreover, the first $2n$ bits of X is the above string of balanced parentheses. The auxiliary string χ has $o(m + n)$ bits, supporting several efficient queries of T and G as summarized in the following lemma.

Lemma 1 (See [6]). *There is an $O(n)$-time computable $2m + 2n + o(n)$-bit encoding for G and T from which each of the following queries can be answered in $O(1)$ time:*

- *$p(v_i)$, $|V_1(v_i)|$, $|V_2(v_i)|$, and $|V_3(v_i)|$;*
- *the j-th neighbor $\mathrm{nbr}(v_i, j)$ of v_i in $G - T$;*
- *the rank $\mathrm{rank}(v_i, v_j)$ of each edge $(v_i, v_j) \in G - T$; and*
- *the endpoints of the edge with rank k.*

The result stated in Lemma 1 takes advantage of auxiliary strings for strings of parentheses, as summarized in the next two lemmas, which are also needed by our routing tables.

Lemma 2 (see [11]). *Let $Z_1, \ldots, Z_k$ be binary strings. If $\sum_{i=1}^{k} |Z_i| \le n^{O(1)}$ and $k = O(1)$, then there exists an $o(n)$-bit string ζ, obtainable in $O(n)$ time, such that given the concatenation of $\zeta, Z_1, Z_2, \ldots, Z_k$, the position of the first bit of each Z_i in the concatenation can be computed in $O(1)$ time.*

Let $Z_1 \oplus Z_2 \oplus \cdots \oplus Z_k$ denote the concatenation of $\zeta, Z_1, Z_2, \ldots, Z_k$ as in Lemma 2.

Lemma 3 (See [31]). *Let P be a string of n balanced parentheses. There is an $o(n)$-bit string $\mu(P)$, computable in $O(n)$ time from P, such that each of the following queries can be answered in $O(1)$ time from P and $\mu(P)$ for each $i = 1, 2, \ldots, n$:*

- *the position of the parenthesis that matches the parenthesis $P[i]$ in P;*
- *the position of the i-th open parenthesis in P; and*
- *the number of open parentheses in $P[1], P[2], \ldots, P[i]$.*

For any binary string B, let $\text{rank}(B, i)$ denote the number of one bits in $B[1], B[2], \ldots, B[i]$; and $\text{select}(B, i)$ denote the position j such that $B[j]$ is the i-th one bit in B. Let e denote the base of the natural logarithm. Our routing tables also need the following information-theoretically optimal representation of a binary string with efficient query support.

Lemma 4 (See [35]). *Let B be an n-bit string with k one bits. Then B can be encoded in expected $O(k)$ time into a compressed string $Z(B)$ with $\log \binom{n}{k} + o(n) = k \log \frac{en}{k} + o(n)$ bits such that each $B[i]$ and $\text{select}(B, i)$ can be determined directly from $Z(B)$ in $O(1)$ time. Moreover, if $B[i] = 1$, then $\text{rank}(B, i)$ can also be obtained from $Z(B)$ in $O(1)$ time.*

3 The Improved Design of Routing Tables

In this section, we present a design of routing table satisfying $\lambda_r(n) = \lceil \log n \rceil$ and $\beta_r(n) < 7.181n + o(n)$. We first compute in $O(n)$ time a plane embedding G of the input n-node planar graph and an orderly spanning spanning T for G [6]. Let $v_1, v_2, \ldots, v_n$ be the counterclockwise preordering of the nodes in T. For each $i = 1, 2, \ldots, n$, let $i - 1$ be the label of v_i, which can be encoded into $\lceil \log n \rceil$ bits. As for the port assignment, for each $j = 1, 2, \ldots, d_r$, we simply assign port j of r to the neighbor of r in T_r with the j-th smallest label. It remains to describe the content of R_r.

It is not difficult to see that the edges of G not in $T \cup T_r$ are irrelevant to the information to be stored in R_r. The rest of the section focuses on the m_r-edge planar graph $G_r = T \cup T_r$. Clearly, both T and T_r are spanning trees of G_r. Moreover, T remains an orderly spanning tree of G_r. Let $p_r(v_i)$ denote the parent of v_i in T_r. An edge (v_i, v_j) of T_r with $i < j$ is *forward* (respectively, *backward*) in G_r if $v_i = p_r(v_j)$ (respectively, $v_j = p_r(v_i)$). Clearly, each edge in $G_r - T$ is either forward or backward. However, an edge of T may be neither forward nor backward, since some edge of T might not be in T_r. Define

$$L_1 = \lceil (\log n)(\log \log n) \rceil;$$
$$L_2 = \lceil (\log \log n)^2 \rceil.$$

Clearly, $L_1 = \tilde{O}(\log n)$ and $L_2 = \tilde{O}(1)$. A node is *popular* if it has at least L_1 neighbors in G_r. A node is *lonely* if it is has at most L_2 neighbors in G_r. Since G_r is planar, the number of popular nodes is $O(n/L_1)$. Similarly, G_r contains $O(n/L_2)$ nodes that are not lonely. To define R_r, we still have to choose $O(n/L_1)$ *special* nodes from T_r as follows. Choose an arbitrary leaf u with maximum depth in T_r. Let w be the highest ancestor of u in T_r whose distance from u in T_r is no more than L_1. (Observe that if $w \neq r$, then w is the ancestor of u whose distance from u in T_r is exactly L_1.) We then choose w as a special node, delete the subtree of T_r rooted at w from T_r, and then repeat the above steps until all nodes are deleted from T_r. By the above choice of special nodes, each node in T_r either has depth less than L_1 or has a special ancestor whose distance from u is at most L_1. Define

$$R_r = S_1 \oplus S_2 \oplus S_3 \oplus S_4 \oplus S_5 \oplus S_6 \oplus S_7 \oplus S_8,$$

where $S_1, \ldots, S_8$ are defined as follows.

1. The encoding S_1 for G_r and T with $2m_r+2n+o(n)$ bits as stated in Lemma 1.
2. An $(m_r - n + 1)$-bit string S_2 that encodes the direction of each edge in $G_r - T$. Specifically, $S_2[k] = 1$ if and only if the edge of $G_r - T$ with rank k is forward.
3. A string $S_3 = Z(S_3')$, where S_3' encodes whether each edge in $T \cap T_r$ is backward or not. Specifically, for each $i = 2, \ldots, n$, $S_3'[i] = 1$ if only if the edge $(v_i, p(v_i))$ is in T_r and is backward.
4. A string $S_4 = Z(S_4') \oplus S_4''$, where S_4' encodes whether each node is popular, and S_4'' stores the parent of each popular node in T_r. Specifically, for each $i = 1, 2, \ldots, n$, $S_4'[i] = 1$ if and only if v_i is popular. $S_4''[i]$ stores the $O(\log n)$-bit index of the parent of the i-th popular node in T_r.
5. A string $S_5 = Z(S_5') \oplus S_5''$, where S_5' encodes whether each node is lonely, and S_5'' is a table storing the rank of $p_r(v)$ among the neighbors of v in G_r for each node v that is not lonely. Specifically, for each $i = 1, 2, \ldots, n$, $S_5'[i] = 0$ if and only if v_i is lonely. If $S_5'[i]$ is the j-th one bit in S_5' (i.e., select$(S_5', j) = i$), then $S_5''[j]$ keeps the number k_j such that $p_r(v_i)$ is the neighbor of v_i in G_r with the k_j-th smallest label.
6. A string $S_6 = Z(S_6') \oplus S_6'' \oplus \mu(S_6'')$, where S_6' is a string consisting of three kinds of distinct characters (,), and 0 defined as follows, and S_6'' is the string obtained from S_6' by deleting its 0 characters. Recall that the first $2n$ bits of S_1 keeps a string of $2n$ balanced parentheses. For each v_i with $p_r(v_i) \neq p(v_i)$ that is not lonely, let $S_6'[j_1] =$ (and $S_6'[j_2] =$), where
   - j_1 is the index with $S_1[j_1 - 1] = (_i$, and
   - j_2 is the index with $S_1[j_2] =)_\ell$ and $v_\ell = p_r(v_i)$.

   Let the remaining characters of S_6' be 0. Clearly, the parentheses in S_6' are balanced. Moreover, the parentheses at $S_6'[j_1]$ and $S_6'[j_2]$ as defined above match each other in S_6'.
7. A string $S_7 = Z(S_7') \oplus S_7''$, where S_7' encodes whether a node is special, and S_7'' stores port$_r(v)$ for each special node v. Specifically, for each $i = 1, 2, \ldots, n$, let $S_7'[i] = 1$ if and only if v_i is special. The i-th $\lceil \log n \rceil$-bit word of S_7'' keeps port$_r(v_j)$, where v_j is the i-th special node, i.e., select$(S_7', i) = j$.
8. A string $S_8 = S_8' \oplus S_8''$. For each $j_1 = 1, 2, \ldots, \lceil n/L_1 \rceil$, the j_1-th $\lceil \log n \rceil$-bit word of S_8' stores the number of r's neighbors in T_r from v_1 to $v_{j_1 L_1}$. For each $j_2 = 1, 2, \ldots, \lceil n/L_2 \rceil$, the j_2-th $\lceil \log L_1 \rceil$-bit word of S_8'' stores the number of r's neighbors in T_r from v_j to $v_{j_2 L_2}$, where $j = 1 + j_2 L_2 - \lfloor j_2 L_2 / L_1 \rfloor L_1$.

Lemma 5. *R_r is computable in expected $O(n)$ time and $|R_r| < 7.181n + o(n)$.*

Proof. (Sketch) One can verify that $|S_4|+|S_5|+|S_6|+|S_7|+|S_8| = o(n)$ and R_r is computable in expected $O(n)$ time. Therefore, $\beta_r(n) = |S_1|+|S_2|+|S_3|+o(n) = 3m_r+n+(2n-m_r)\log\frac{en}{2n-m_r}+o(n) = \beta_r(n) = 7n+t\left(\log\frac{en}{t}-3\right)+o(n)$, where $t = 2n - m_r$. By verifying that the maximum of $\beta_r(n)$ occurs at $t = \frac{1}{8}n$, we have $\beta_r(n) \le \left(7 + \frac{1}{8}\log e\right)n + o(n) < 7.181n + o(n)$. □

4 Determining the Correct Port Efficiently

In this section, we demonstrate how to determine $\text{port}_r(v_i)$ from the routing table R_r and the number i in $\tilde{O}(\log n)$ time.

Lemma 6. *It takes $\tilde{O}(\log n)$ time to determine* $\text{port}_r(v)$ *from R_r and v's label.*

Proof. (Sketch) The task of determining $\text{port}_r(v)$ in $\tilde{O}(\log n)$ time can be reduced to the task of determining each $p_r(v)$ from R_r and the label of v in $\tilde{O}(1)$ time as follows. If $p_r(v)$ is obtainable from R_r and the label of v in $\tilde{O}(1)$ time, then we can traverse T_r from v_i toward r in $\tilde{O}(1)$ time per step. Clearly, the traversal always reaches a node v_j that is either special or a neighbor of r in $O(L_1) = \tilde{O}(\log n)$ steps, i.e., $\tilde{O}(\log n) \cdot \tilde{O}(1) = \tilde{O}(\log n)$ time. Observe that whether v_j is special can be determined from S_7 in $O(1)$. Also, by Lemma 1, whether v_j is a neighbor of r can be determined in $O(1)$ time from S_1. If v_j is special, then $\text{port}_r(v_i) = \text{port}_r(v_j) = S''_7[\text{rank}(S'_7, j)]$, which, by Lemma 4, is obtainable from S_7 in $\tilde{O}(1)$ time. If v_j is a neighbor of r, then let $j_1 = \lfloor j/L_1 \rfloor$ and $j_2 = \lfloor j/L_2 \rfloor$. By the definition of S_8, $\text{port}_r(v_j)$ equals $S'_8[j_1] + S''_8[j_2]$ plus the number of r's neighbors in T_r from $v_{j_2 L_2+1}$ to v_j. By Lemmas 1 and 4, $\text{port}_r(v_i) = \text{port}_r(v_j)$ is obtainable from S_1 and S_8 in $\tilde{O}(1)$ time.

It remains to show how to determine $p_r(v_i)$ in $\tilde{O}(1)$ time from R_r and the number i. First of all, we look up S_4 to determine if v_i is popular. If v_i is popular, then we can obtain $p_r(v_i)$ from S''_4 in $O(1)$ time. Therefore, the rest of the proof assumes that v_i is not popular.

Clearly, $p_r(v_i)$ is a neighbor of v_i in either one of the edge sets $G_r - T$ and T. Before looking for $p_r(v_i)$ among the neighbors of v_i in G_r, we look up S_5 in $O(1)$ time (by Lemma 4) to see whether v_i is lonely.

Case 1: v_i is not lonely. By Lemma 4, we can compute $k_i = S''_5[\text{rank}(S'_5, i)]$ from S_5 in $\tilde{O}(1)$ time. Since T is an orderly spanning tree of G_r, we know which edge set to look for $p_r(v_i)$ by k_i, $|V_1(v_i)|$, $|V_2(v_i)|$, and $|V_3(v_i)|$, which, by Lemma 1, are obtainable in $O(1)$ time from S_1.

Case 2: v_i is lonely. Observe that $p_r(v_i) \in V_1(v_i)$ if and only if exactly one of the edges between v_i and $V_1(v_i)$ is forward. Also, $p_r(v_i) \in V_3(v_i)$ if and only if exactly one of the edges between v_i and $V_3(v_i)$ is backward. Since v_i is lonely, we can afford checking the direction of each incident edges of v_i in $G_r - T$: By Lemma 1, the direction of each edge in $G_r - T$ can be determined from S_1 and S_2 in $O(1)$ time, and, thus, we know whether $p_r(v_i)$ is a neighbor of v_i in $G_r - T$ in $\tilde{O}(1)$ time. Moreover, if we find out that $p_r(v_i)$ is a neighbor of v_i in $G_r - T$, $p_r(v_i)$ can also be computed from S_1 in $O(1)$ time.

After figuring out whether $p_r(v_i)$ is a neighbor of v_i in $G_r - T$ or T, we can compute $p_r(v_i)$ in $\tilde{O}(1)$ time as follows. We first consider the case that $p_r(v_i)$ is a neighbor of v_i in $G_r - T$. If v_i is not lonely, then $p_r(v_i)$ is either $\text{nbr}(v_i, k_i)$ or $\text{nbr}(v_i, k_i - |V_2(v_i)|)$; otherwise, $p_r(v_i)$ can be computed from S_1 as stated above.

It remains to consider the case that $p_r(v_i)$ is a neighbor of v_i in T. Clearly, $p_r(v_i)$ is either $p(v_i)$ or a child of v_i in T. We first determine whether v_i is a leaf of T by the value of $|V_2(v_i)|$, which, by Lemma 1, is computable in $O(1)$

time. It suffices to consider the case $|V_2(v_i)| > 0$, since otherwise $p_r(v_i) = p(v_i)$. If v_i is lonely, then we can afford checking the direction of the edge between v_i and each child of v_i in T. Since the direction of each edge can be determined from S_1 and S_3 in $O(1)$ time, one can figure out whether $p_r(v_i)$ is a child of v_i in T in $O(L_2) = \tilde{O}(1)$ time. If $p_r(v_i)$ turns out to be a child of v_i in T, then $p_r(v_i)$ can also be determined from S_1 and S_3 in $O(1)$ time. Otherwise, we know $p_r(v_i) = p(v_i)$, which can be obtained from S_1 in $O(1)$ time. As for the case that v_i is not lonely, one can compute $p_r(v_i)$ in $\tilde{O}(1)$ time from S_6 as follows. Let j be the index such that $S_1[j-1] = (_i$ in S_1. Let j' be the rank of (in S'_6. Let $S''_6[j'']$ be the close parenthesis that matches the j'-th open parenthesis in S''_6. Suppose j^* is the rank of) in S'_6. Let ℓ be the index such that $)_\ell = S_1[\ell']$, where $S'_6[\ell']$ is the j^*-th close parenthesis in S'_6. One can verify $v_\ell = p_r(v_i)$. □

The main theorem of the paper follows immediately from Lemmas 5 and 6.

Theorem 1. *Given an n-node planar network G, there is a design of routing tables for G such that $\lambda_r(n) = \lceil \log n \rceil$, $\beta_r(n) < 7.181n + o(n)$, $\tau_r(n) = \tilde{O}(\log n)$, and R_r is computable in expected $O(n)$ time for each node r of G.*

5 Concluding Remarks

Some questions might worth further investigation. For example, closing the gap between the lower and upper bounds on $\beta_r(n)$ is an interesting one. Also, if $\lambda_r(n) > \lceil \log n \rceil$, e.g., $10\lceil \log n \rceil$ or $O((\log n)^{O(1)})$, can one significantly reduce the upper bound on $\beta_r(n)$? If $\lambda_r(n) > \lceil \log n \rceil$, is there still an $\Omega(n)$-bit lower bound on $\beta_r(n)$?

Acknowledgments

We thank several anonymous referees for their helpful comments which significantly improve the presentation of the paper. We thank Cyril Gavoille for bringing [20] to our attention at SODA 2001. We thank Kunihiko Sadakane, Rasmus Pagh, Rajeev Raman, and Venkatesh Raman for helping with Lemma 4. We also thank Kuan-Ling Chen for discussion and Nen-Fu Huang and Jia-Shung Wang for interesting comments.

References

1. B. Awerbuch, A. Bar-Noy, N. Linial, and D. Peleg. Improved routing strategies with succinct tables. *Journal of Algorithms*, 11(3):307–341, 1990.
2. M. Becker and K. Mehlhorn. Algorithms for routing in planar graphs. *Acta Informatica*, 23(2):163–176, 1986.
3. P. Bose and P. Morin. Competitive online routing in geometric graphs. In *Proceedings of the 8th International Colloquium on Structural Information and Communication Complexity*, pages 35–44, 2001.
4. P. Bose, P. Morin, I. Stojmenovic, and J. Urrutia. Routing with guaranteed delivery in ad hoc wireless networks. *Wireless Networks*, 7(6):609–616, 2001.

5. L. P. Chew. There are planar graphs almost as good as the complete graph. *Journal of Computer and System Sciences*, 39:205–219, 1989.
6. Y.-T. Chiang, C.-C. Lin, and H.-I. Lu. Orderly spanning trees with applications to graph drawing and graph encoding. In *Proceedings of the 12th Annual ACM-SIAM Symposium on Discrete Algorithms*, pages 506–515, Washington, DC, 7–9 Jan. 2001. A revised and extended version can be found at `http://xxx.lanl.gov/abs/cs.DS/0102006`.
7. R. C.-N. Chuang, A. Garg, X. He, M.-Y. Kao, and H.-I. Lu. Compact encodings of planar graphs via canonical ordering and multiple parentheses. In K. G. Larsen, S. Skyum, and G. Winskel, editors, *Proceedings of the 25th International Colloquium on Automata, Languages, and Programming*, Lecture Notes in Computer Science 1443, pages 118–129, Aalborg, Denmark, 1998. Springer-Verlag.
8. L. Cowen and C. G. Wagner. Compact roundtrip routing in directed networks. In *Prooceedings of the 19th Annual ACM Symposium on Principles of Distributed Computing*, pages 51–59. ACM PRESS, 2000.
9. L. J. Cowen. Compact routing with minimum stretch. *Journal of Algorithms*, 38(1):170–183, 2001.
10. H. de Fraysseix, J. Pach, and R. Pollack. How to draw a planar graph on a grid. *Combinatorica*, 10:41–51, 1990.
11. P. Elias. Universal codeword sets and representations of the integers. *IEEE Transactions on Information Theory*, IT-21:194–203, 1975.
12. P. Fraigniaud and C. Gavoille. Routing in trees. In F. Orejas, P. G. Spirakis, and J. v. Leeuwen, editors, *Proceedings of the 28th International Colloquium on Automata, Languages and Programming*, Lecture Notes in Computer Science 2076, pages 757–772. Springer, July 2001.
13. P. Fraigniaud and C. Gavoille. A space lower bound for routing in trees. In *Proceedings of the 19th Annual Symposium on Theoretical Aspects of Computer Science*, Lecture Notes in Computer Science 2285, pages 65–75. Springer, Mar. 2002.
14. G. N. Frederickson and R. Janardan. Designing networks with compact routing tables. *Algorithmica*, 3(1):171–190, 1988.
15. G. N. Frederickson and R. Janardan. Efficient message routing in planar networks. *SIAM Journal on Computing*, 18:843–857, 1989.
16. M. L. Fredman and D. E. Willard. Trans-dichotomous algorithms for minimum spanning trees and shortest paths. *Jouranl of Computer and System Sciences*, 48(3):533–551, June 1994.
17. J. Gao, L. J. Guibas, J. Hershburger, L. Zhang, and A. Zhu. Geometric spanner for routing in mobile networks. In *Proceedings of the ACM Symposium on Mobile Ad Hoc Networking & Computing (MobiHoc)*, 2001.
18. C. Gavoille. A survey on interval routing. *Theoretical Computer Science*, 245(2):217–253, 2000.
19. C. Gavoille and M. Gengler. Space-efficiency of routing schemes of stretch factor three. *Journal of Parallel and Distributed Computing*, 61:679–687, 2001.
20. C. Gavoille and N. Hanusse. Compact routing tables for graphs of bounded genus. In J. Wiedermann, P. van Emde Boas, and M. Nielsen, editors, *26th International Colloquium on Automata, Languages and Programming*, Lecture Notes in Computer Science 1644, pages 351–360. Springer, July 1999. A full version is available at http://dept-info.labri.fr/~gavoille/article/ GH99_up.ps.gz.
21. C. Gavoille and D. Peleg. The compactness of interval routing. *SIAM Journal on Discrete Mathematics*, 12(4):459–473, Oct. 1999.

22. C. Gavoille and D. Peleg. The compactness of interval routing for almost all graphs. *SIAM Journal on Computing*, 31(3):706–721, 2001.
23. Y. Hassin and D. Peleg. Sparse communication networks and efficient routing in the plane. *Distributed Computing*, 14(4):205–215, 2001.
24. G. Jacobson. Space-efficient static trees and graphs. In *Proceedings of the 30th Annual Symposium on Foundations of Computer Science*, pages 549–554, Research Triangle Park, North Carolina, 30 Oct.–1 Nov. 1989. IEEE.
25. G. Kant. Drawing planar graphs using the canonical ordering. *Algorithmica*, 16(1):4–32, 1996.
26. B. N. Karp. *Geographic Routing for Wireless Networks.* PhD thesis, Harvard University, Cambridge, MA, Oct 2000.
27. B. N. Karp and H. T. Kung. GPSR: Greedy perimeter stateless rouring for wireless networks. In *Proceedings of the Sixth Annual ACM/IEEE International Conference on Mobile Computing and Networking (MobiCom)*, pages 243–254, Boston, 2000.
28. X.-Y. Li, G. Calinescu, and P.-J. Wan. Distributed construction of a planar spanner and routing for ad hoc wireless networks. In *Proceedings of the 21st Annual Joint Conference of the IEEE Computer and Communications Societies (InfoCom)*, New York City, 2002. To appear.
29. C.-C. Liao, H.-I. Lu, and H.-C. Yen. Floor-planning via orderly spanning trees. In *Proceedings of the 9th International Symposium on Graph Drawing*, Lecture Notes in Computer Science 2265, pages 367–377, Vienna, Austria, 2001. Springer.
30. G. Lin. Fault tolerant planar communication networks. In *Proceedings of the 24th Annual ACM Symposium on the Theory of Computing*, pages 133–139, 1992.
31. J. I. Munro and V. Raman. Succinct representation of balanced parentheses, static trees and planar graphs. *SIAM Journal on Computing*, 31(3):762–776, 2001.
32. G. Narasimhan and M. Smid. Approximating the stretch factor of euclidean graphs. *SIAM Journal on Computing*, 30(3):978–989, 2000.
33. D. Peleg. *Distributed Computing: A Locality-Sensitive Approach.* Monographs on Discrete Mathematics and Applications. SIAM, 2000.
34. D. Peleg and E. Upfal. A trade-off between space and efficiency for routing tables. *Journal of the ACM*, 36(3):510–530, 1989.
35. R. Raman, V. Raman, and S. S. Rao. Succinct indexable dictionaries with applications to representations of k-ary trees and multisets. In *Proceedings of the 13th Annual ACM-SIAM Symposium on Discrete Algorithms*, pages 233–242, San Francisco, 6–8 Jan. 2002.
36. W. Schnyder. Embedding planar graphs on the grid. In *Proceedings of the First Annual ACM-SIAM Symposium on Discrete Algorithms*, pages 138–148, 1990.
37. M. Thorup and U. Zwick. Compact routing schemes. In *Proceedings of the 13th Annual ACM Symposium on Parallel Algorithms and Architectures*, pages 1–10. ACM PRESS, 2001.
38. P. van Emde Boas. Machine models and simulations. In J. van Leeuwen, editor, *Handbook of Theoretical Computer Science*, volume A, chapter 1, pages 1–60. Elsevier, Amsterdam, 1990.
39. M. Yannakakis. Embedding planar graphs in four pages. *Jouranl of Computer and System Sciences*, 38(1):36–67, Feb. 1989.

Coloring Algorithms on Subcubic Graphs

Harold N. Gabow and San Skulrattanakulchai

Department of Computer Science, University of Colorado at Boulder,
Boulder CO 80309 USA
{hal,skulratt}@cs.colorado.edu

Abstract. We present efficient algorithms for three coloring problems on subcubic graphs (ones with maximum degree 3). These algorithms are based on a simple decomposition principle for subcubic graphs. The first algorithm is for 4-edge coloring, or more generally, 4-list-edge coloring. Our algorithm runs in linear time, and appears to be simpler than previous ones. As evidence we give the first randomized EREW PRAM algorithm that uses $O(n/\log n)$ processors and runs in $O(\log n)$ time with high probability, where n is the number of vertices of the input graph. The second algorithm is the first linear-time algorithm to 5-total-color subcubic graphs. The third algorithm generalizes this to the first linear-time algorithm to 5-list-total-color subcubic graphs.

1 Introduction

We present efficient algorithms for three coloring problems on subcubic graphs. (A subcubic graph has maximum degree 3.) The algorithms are based on a simple decomposition principle for subcubic graphs. The problems we attack are by now well-studied generalizations of standard vertex and edge coloring. Our first algorithm is for 4-edge coloring, or more generally, 4-list-edge coloring. Our algorithm runs in linear time, and appears to be simpler than previous ones. As evidence we give the first randomized EREW PRAM algorithm that uses $O(n/\log n)$ processors and runs in $O(\log n)$ time with high probability. The second algorithm is the first linear-time algorithm to 5-total-color subcubic graphs. The third algorithm generalizes this to the first linear-time algorithm to 5-list-total-color subcubic graphs. We now give relevant definitions and discuss the problems and previous work in detail.

We follow the terminology of [1]. Let $G = (V, E)$ be a (multi)graph having n vertices and m edges. A *total coloring* is a map $\varphi : V \cup E \to \mathbb{N}$ satisfying (i) no adjacent vertices or edges have the same image, and (ii) the image of each vertex is distinct from the images of edges incident with it. To *k-total-color* G is to find a total coloring map whose image is included in $[k] = \{1, 2, \ldots, k\}$. The graph is *k-total-colorable* when such a map exists. The *total chromatic number* χ'' is the least k for which G is k-total-colorable.

Let $X \in \{V, E, V \cup E\}$ and let $\lambda : X \to 2^{\mathbb{N}}$ be an assignment of lists of colors to elements of X. A *λ-coloring on* X is a map $\varphi : X \to \mathbb{N}$ such that $\varphi(x) \in \lambda(x)$ for all $x \in X$, and $\varphi(x_1) = \varphi(x_2)$ implies x_1 is neither adjacent nor incident

O.H. Ibarra and L. Zhang (Eds.): COCOON 2002, LNCS 2387, pp. 67–76, 2002.

to x_2. A *λ-vertex-coloring* is a λ-coloring on V. A *λ-edge-coloring* is a λ-coloring on E. A *λ-total-coloring* is a λ-coloring on $V \cup E$.

A graph is *k-choosable* if there exists a λ-vertex-coloring for any λ satisfying $|\lambda(v)| = k$ for all $v \in V$. The notions of *k-edge-choosability* and *k-total-choosability* are defined similarly. The *choice number* $\mathrm{ch}(G)$ is the least k for which G is k-choosable. The *list chromatic index* χ'_ℓ is the least k for which G is k-edge-choosable. The *total choosability* χ''_ℓ is the least k for which G is k-total-choosable.

By Vizing's Theorem [2] the chromatic index χ' of any simple graph with maximum degree Δ is either Δ or $\Delta + 1$. A simple graph can be edge-colored using $\Delta + 1$ colors in $O(m\sqrt{n \log n})$ time [3]. However, Holyer [4] shows that deciding whether the chromatic index of a given simple cubic graph equals 3 or 4 is **NP**-complete [5]. Reference [6] gives an algorithm to 4-edge-color any (not necessarily simple) subcubic graph in $O(n)$ time. The **List Coloring Conjecture** (LCC) states that the list chromatic index χ'_ℓ of a graph equals its chromatic index χ'. (See [7,8].) Only a handful of special families of graphs are known to satisfy the LCC. (See [9].) Clearly $\chi'_\ell \geq \chi'$. A simple graph would satisfy $\chi'_\ell \leq \Delta + 1$ if the LCC were true. Even this upper bound has not been established. Independently, Vizing [10] and Erdős et al. [11] prove the following list version of Brooks' theorem [12]: *the choice number of any connected simple graph that is not complete or an odd cycle does not exceed its maximum degree.* Reference [13] gives an $O(m+n)$-time algorithm to Δ-list-vertex-color any graph satisfying the hypotheses of the above theorem. Suppose G is a subcubic graph. By reducing the problem of list-edge-coloring G to that of list-vertex-coloring its line graph and using the above theorem we see that G is 4-edge-choosable. Thus G can be 4-list-edge-colored in $O(n)$ time by executing the algorithm of [13] on the line graph of G. The authors of reference [14] introduce the term *subcubic graphs* and show that they are 4-edge-choosable. They examine subcubic graphs with *halfedges*, i.e., edges with only one endpoint. They obtain their result through complicated case-by-case analysis of 4-edge-coloring of paths and cycles with halfedges and of some special types of graphs; reduction of the input graph to a specific form; and coloring procedures that avoid known obstruction. Even though the algorithm derived from their proof has a linear time bound, it appears to be too complicated for practical use. We will present a direct and simple $O(n)$-time sequential algorithm, and a work-optimal, $O(\log n)$-time with high probability, randomized EREW PRAM algorithm to 4-list-edge-color subcubic graphs.

The **Total Coloring Conjecture** (TCC) states that the total chromatic number χ'' of a simple graph is at most $\Delta + 2$. Clearly $\chi'' \geq \Delta + 1$. However, deciding whether or not a given simple graph satisfies $\chi'' = \Delta + 1$ is **NP**-complete [15,16]. The TCC has been shown to hold for some families of graphs [17]. Rosenfeld [18] shows that the TCC holds for subcubic graphs. Vijayaditya [19] shows that the TCC holds for simple subcubic graphs. The proof of [18] yields a super-linear time algorithm because it has to find a shortest cycle in each recursive step. The algorithm in [19] requires a routine to find a perfect matching in a bridgeless, cubic graph. The current best known algorithm [20] to find such a matching runs in $O(n \log^4 n)$ time. We will present the first-known $O(n)$-time algorithm to 5-total-color subcubic graphs.

The authors of reference [21] (and also those of reference [22]) conjecture that the total choosability χ''_ℓ of any graph equals its total chromatic number χ''. They also show that subcubic graphs are 5-total-choosable. They claim that their proof gives a polynomial-time algorithm, without specifying the degree of the polynomial. Their algorithm has a super-linear running time because it has to find a shortest cycle in each recursive step. We will present the first-known $O(n)$-time algorithm to 5-list-total-color subcubic graphs.

Definitions. A *cycle* is a connected graph every vertex of which has degree 2. It is *even* if it has an even number of vertices, and *odd* otherwise. A vertex of a tree is *pendant* if its degree is one. A tree edge is *pendant* if it is incident with a pendant vertex. Let C be a cycle in G. A *chord* is an edge of G joining two vertices of C but is itself not an edge of C. A *triple bond* is a graph consisting of 2 vertices and 3 parallel edges. We use the term *available color* at several places in this paper. Let x be either a vertex or an edge and let $\lambda(x)$ be its list of colors. By a *neighbor* of x we mean any vertex/edge adjacent/incident to x. During the execution of an algorithm, a color $\alpha \in \lambda(x)$ is *available for* x if no neighbor of x has yet been assigned color α by the algorithm; it is *unavailable* otherwise.

2 Decomposition Theorem & List Coloring Lemmas

This decomposition theorem (see also [6]) is at the heart of our coloring method.

Theorem 1. *A subcubic graph G can be decomposed in linear time into edge-disjoint subgraphs $\mathcal{C}$ and $\mathcal{T}$, where $\mathcal{C}$ is a collection of vertex-disjoint cycles, and $\mathcal{T}$ is a forest of maximum degree at most 3. Furthermore, G admits a decomposition without chords unless it contains a triple bond.*

Proof. Let $\mathcal{C}$ be a maximal collection of edge-disjoint cycles in G. Any two cycles of $\mathcal{C}$ are vertex-disjoint since G is subcubic. The forest $\mathcal{T} = G - E(\mathcal{C})$ obviously has maximum degree at most 3. If G contains no triple bond then we can choose the cycles of $\mathcal{C}$ to be chordless. □

We can classify the edges as *cycle edges* or *tree edges* for a given decomposition.

Lemma 1. *Let C be a cycle with color lists $L(\cdot)$ assigned to its vertices. If every list has at least 2 colors, then C is L-vertex-colorable in linear time unless C is odd and all lists are the same list of size 2.*

Proof. The following procedure produces an L-vertex-coloring of C if it runs to completion. Its input consists of a labeled cycle $C = \langle v_1, v_2, \ldots, v_k, v_1 \rangle$, color lists $L(\cdot)$, and a color $\alpha \in L(v_1)$.

1. assign color α to v_1
2. **for** $i \leftarrow 2$ **to** $k-1$ **do**
3. assign to v_i any color in $L(v_i)$ distinct from that of v_{i-1}
4. assign to v_k any color in $L(v_k)$ distinct from that of v_{k-1} or v_1

Statements 1 & 3 can always be carried out since all v_i have $|L(v_i)| \geq 2$. We show Statement 4 too can always be carried out by judiciously labeling C and picking the color α. First suppose some adjacent vertices x, y satisfy $L(x) \neq L(y)$; say that $L(y) \setminus L(x) \neq \emptyset$. Label C so that $x = v_k$ and $y = v_1$ and pick $\alpha \in L(v_1) \setminus L(v_k)$. Next suppose $|L(x)| > 2$ for some x. Label C so that $x = v_k$ and pick any $\alpha \in L(v_1)$. The remaining possibility is for every vertices x, y to satisfy $L(x) = L(y)$ and $|L(x)| = 2$. Assume C is even or else there is nothing to prove. Choose any valid labeling of C and pick any $\alpha \in L(v_1)$. □

Lemma 2. *Let $C = \langle v_1, v_2, \ldots, v_k, v_1 \rangle$ be an odd cycle. For each i let $L(v_i)$ be a list of colors for vertex v_i such that $|L(v_1)| = 1$, $|L(v_2)| = |L(v_k)| = 2$, $L(v_i) \supseteq L(v_2)$ for $3 \leq i < k$, $L(v_1) \subset L(v_2)$, and $L(v_k) \neq L(v_2)$. Then C is L-vertex-colorable in linear time.*

Proof. By throwing away all excess colors, we may assume $L(v_i) = L(v_2)$ for all $3 \leq i < k$. Color $v_1, v_2, \ldots, v_k$ in that order, using any color available for each vertex. It is easy to see that this coloring procedure never fails. □

To edge-color a graph is the same as to vertex-color its line graph. Since a cycle is isomorphic to its line graph, these corollaries follow from the above Lemmas.

Corollary 1. *Let C be a cycle with color lists $L(\cdot)$ assigned to its edges. If every list has at least 2 colors, then C is L-edge-colorable in linear time unless C is odd and all lists are the same list of size 2.* □

Corollary 2. *Let $C = \langle v_1, e_1, \ldots, v_k, e_k, v_1 \rangle$ be an odd cycle. For each i let $L(e_i)$ be a list of colors for edge e_i such that $|L(e_1)| = 1$, $|L(e_2)| = |L(e_k)| = 2$, $L(e_i) \supseteq L(e_2)$ for $3 \leq i < k$, $L(e_1) \subset L(e_2)$, and $L(e_k) \neq L(e_2)$. Then C is L-edge-colorable in linear time.* □

3 List Edge Coloring Algorithms

3.1 Sequential Algorithm

Let G be a subcubic graph every edge e of which has a list $\lambda(e)$ of 4 colors. The algorithm has two stages. The first stage obtains, by Theorem 1, a decomposition of G into a forest $\mathcal{T}$ and a collection $\mathcal{C}$ of cycles. The second stage does the coloring. The trees are colored first; the cycles are colored later. For each tree T in $\mathcal{T}$, start a depth-first search [23] from some arbitrary vertex in T. Assign some available color to each tree edge when it is first discovered by the search. Since $\Delta(\mathcal{T}) \leq 3$ and $|\lambda(e)| = 4$ for every edge e, when a tree edge e is discovered by the search no more than two colors in $\lambda(e)$ are unavailable for e. So coloring of $\mathcal{T}$ can be carried out to completion.

Each cycle $C \in \mathcal{C}$ is colored by invoking Corollary 1. For each $e \in E(C)$, let $L(e)$ be the set of colors available for e. Any edge of C has at least 2 available colors. So Corollary 1 applies unless C is odd and every edge in it has the same 2 colors available. Take any vertex in C. It has degree 3, so we can change the color of its pendant edge. This changes the available colors for some but not all edges of C (since C has ≥ 3 edges). So again Corollary 1 applies.

3.2 Parallel Algorithm

We now describe how to implement the 2 stages of the sequential algorithm to obtain a randomized EREW PRAM algorithm that uses $O(n/\log n)$ processors and runs in time $O(\log n)$ with high probability.

To decompose G into cycles and trees, take a spanning tree T of G. Each nontree edge e has an associated fundamental cycle $C(e)$. Let $\mathcal{C}$ be the direct sum, i.e., the mod-2 sum, of all fundamental cycles $C(e_1), C(e_2), \ldots, C(e_{m-n+1})$. Then $\mathcal{C}$ is a collection of edge-disjoint cycles. In fact, $\mathcal{C}$ is a collection of vertex-disjoint cycles since G is subcubic. Also, $\mathcal{C}$ contains all the nontree edges e_i. So we need only compute $\mathcal{C}$ and output the decomposition consisting of cycles $\mathcal{C}$ and forest $\mathcal{T} = T - \{\text{edges of } \mathcal{C}\}$.

To find $\mathcal{C}$, we have to compute, for each tree edge f, the parity of the number of fundamental cycles $C(e_i)$ containing f. Edge f is in $\mathcal{C}$ if and only if this number is odd. This can be done as follows. First make T into a rooted out-tree. Take any tree edge $f = wv$, where w is the parent of v. Then compute the parity of the sum, over all descendants u of v, of the vertex degrees $d_{G-T}(u)$ in the graph $G - T$. To find a spanning tree T of G, use the randomized $O(\log n)$-time, $O(m+n)$-work, EREW PRAM algorithm of [24]. Computing the desired parity of the sum can be done on an EREW PRAM in $O(\log n)$ time using $O(m+n)$ work by parallel prefix computation on the Euler tour of T [25,26].

To color a tree, number every vertex by its depth. Each vertex r of even depth will color the subtree of ≤ 6 edges that descend from it and no other even-depth vertex. Let the subtree consist of edges ra_i for $i = 1, 2$, and $a_i a_{ij}$ for $i = 1, 2$ and $j = 1, 2$. Some of these edges may not exist. The coloring is in 2 parallel steps.

Step 1 There are two substeps.

1. Assign each edge ra_i 2 colors from its list, such that the 4 colors assigned to these 2 edges are distinct. This can be done since all lists have length 4.
2. For $i = 1, 2$ do
   (a) Let c_1, c_2 be the colors on ra_i.
   (b) Assign color d_j to $a_i a_{ij}$ for $j = 1, 2$, where d_j is in the edge's list and c_1, c_2, d_1, d_2 are 4 distinct colors. This can be done since all lists have length 4.

Step 2 Color ra_i with c_1 or c_2, whichever is distinct from the color of r's parent edge.

Clearly this tree coloring is an EREW algorithm that can be made work-optimal by counting the number of vertices on even levels, which can be done.

To color a cycle, first determine the first edge e_1 and its appropriate color as in the proof of Corollary 1. The remaining edges can then be colored by pointer jumping.

4 Total Coloring Lemmas

Lemma 3. *A tree of maximum degree ≤ 3 is 4-list-total-colorable in linear time.*

Proof. Let us be given a tree of maximum degree at most 3, every vertex/edge of which has its own list of 4 colors. Start a depth-first search from some arbitrary vertex. Assign an available color for each vertex/edge when it is discovered by the search. (While scanning an edge $e = \{v, w\}$ out of a vertex v, if both e and w are discovered for the first time, then we'll say edge e is discovered before vertex w.) It is easy to check that this greedy procedure gives a list-total coloring. □

Lemma 4. *A cycle $C = \langle v_1, e_1, \ldots, v_k, e_k, v_1 \rangle$ is L-total-colorable in linear time if the sizes of the lists $L(v_i)$ of vertices v_i and lists $L(e_i)$ of edges e_i satisfy*

$$|L(v_i)| \geq \begin{cases} 5, & \text{if } i = 1, \\ 2, & \text{if } i = 2, \\ 3, & \text{if } 3 \leq i \leq k, \end{cases} \quad \text{and} \quad |L(e_i)| \geq \begin{cases} 1, & \text{if } i = 1, \\ 3, & \text{if } 2 \leq i < k, \\ 4, & \text{if } i = k. \end{cases}$$

Proof. Color vertices/edges e_1, v_2, e_2, v_3, ..., e_{k-1}, v_k, e_k, v_1 in that order, using any color available for each. It's easy to check that when a vertex/edge x is about to be colored, some color is available for it because fewer than $|L(x)|$ of its neighbors are already colored. □

5 Total Coloring Algorithms

Let G be a connected subcubic graph every vertex/edge x of which has its own list $\lambda(x)$ of 5 colors. We have to find a λ-total-coloring. To specialize the following description to the 5-total-coloring problem, simply set $\lambda(x) = [5]$ for all x. The only difference between the general λ-total-coloring algorithm and the 5-total-coloring algorithm is in Case 2 of the ensuing description.

A triple bond is clearly λ-total-colorable; so assume G is not a triple bond. By Theorem 1, let G be decomposed into trees $\mathcal{T}$ and chordless cycles $\mathcal{C}$. For each $T \in \mathcal{T}$, let T' be the vertex-deleted tree $T - \{\, v : v$ is a pendant vertex of T on some cycle in $\mathcal{C}\}$. The coloring algorithm has two steps.

Step 1 for each $T \in \mathcal{T}$, total-color T' using Lemma 3.
Step 2 for each $C \in \mathcal{C}$, color those pendant edges incident with C whose non-incident endpoints are already colored, and then total-color C.

The algorithm maintains the invariant that no two neighbors receive the same color. How Step 2 performs its task depends on each cycle C. Let us adopt the following naming convention. Let $C = \langle v_1, e_1, \ldots, v_k, e_k, v_1 \rangle$ be a labeled cycle. If v_i is incident with a pendant tree edge, name that edge f_i, and name f_i's other endpoint w_i. Note that no w_i is the same as any v_i since the decomposition is chordless. However, it is possible that $w_i = w_j$ for some distinct i, j. For each vertex/edge x of G, write $L(x)$ for the set of colors still available for x. If x is colored, let $\varphi(x)$ denote its color. When $i = 1$ (resp. $i = k$), vertex v_{i-1} (resp. v_{i+1}) refers to v_k (resp. v_1); similarly for vertex w_{i-1} (resp. w_{i+1}) and edge f_{i-1} (resp. f_{i+1}). There are two possibilities for each cycle C.

Case 1. Some vertex x on C is incident with no pendant tree edge, or its pendant edge $\{x, y\}$ has y still uncolored. First label C so that $x = v_1$. Then, maintaining the invariant, color all existing f_i whose w_i are already colored. All the hypotheses of Lemma 4 are satisfied; so C can be total-colored.
Case 2. Every vertex x on C is incident with a pendant edge $\{x, y\}$ with y already colored. Choose any valid labeling of C. Observe that each f_i has at most 3 colored neighbors; and thus $|L(f_i)| \geq 2$ in the current coloring φ.

5.1 Case 2 of the 5-Total-Coloring Algorithm

Subcase 2.1 Some index i satisfies $\varphi(w_i) \neq \varphi(w_{i+1})$. By relabeling if necessary, we may assume $\varphi(w_k) \neq \varphi(w_1)$. We will color all the f_i and e_i so that, for all i, either $\varphi(e_i) = \varphi(w_i)$ or $\varphi(e_{i-1}) = \varphi(w_i)$. We do this as follows.

for $i \leftarrow 1$ **to** k **do** {
  $\varphi(f_i) \leftarrow$ any color in $L(f_i) \setminus \{\varphi(w_{i-1})\}$;
  if $\varphi(w_i) \in L(e_i)$ **then** $\varphi(e_i) \leftarrow \varphi(w_i)$ }
for $i \leftarrow 1$ **to** k **do if** e_i is still uncolored **then** $\varphi(e_i) \leftarrow$ any color in $L(e_i)$

It is easy to see the above procedure never gets stuck and it extends φ to cover all f_i and e_i while maintaining the invariant. It also results in $|L(v_i)| = 2$ for every i. The last statement follows from the fact that 4 neighbors of v_i, viz. w_i, f_i, e_{i-1}, and e_i, are colored, with exactly 2 of them colored the same. If all the hypotheses of Lemma 1 are satisfied, then φ can be extended to a total-coloring of C and we are done. So suppose C is odd and every $L(v_i)$ contains the same 2 colors. Consider edge e_k. Since $L(v_k) = L(v_1)$, we have $\{\varphi(e_{k-1}), \varphi(f_k)\} = \{\varphi(e_1), \varphi(f_1)\}$. Hence there exists some color $\alpha \in L(e_k)$ different from the current color $\varphi(e_k)$ of e_k. Change the color of e_k to α. This affects $L(v_k)$ and $L(v_1)$ but leaves the remaining $L(v_i)$ intact. Following this color change (and relabeling of C if necessary), either all the hypotheses of Lemma 1 or all the hypotheses of Lemma 2 are satisfied. Therefore, φ can be extended to a total-coloring of C.
Subcase 2.2 All w_i receive the same color. Color all the f_i and v_i as follows.

Procedure A
  $\varphi(f_1) \leftarrow$ any color in $L(f_1)$;
  for $i \leftarrow 2$ **to** k **do** {
    $\varphi(f_i) \leftarrow$ any color in $L(f_i) \setminus \{\varphi(f_{i-1})\}$; $\varphi(v_i) \leftarrow \varphi(f_{i-1})$ }
  if $\varphi(f_k) \neq \varphi(f_1)$ **then** $\varphi(v_1) \leftarrow \varphi(f_k)$ **else** $\varphi(v_1) \leftarrow$ any color in $L(v_1)$

It is easy to see that the above procedure never gets stuck and it extends φ to cover all f_i and v_i while maintaining the invariant. It results in either
  (i) $\varphi(f_i) = \varphi(v_{i+1})$ and $\varphi(f_i) \neq \varphi(f_{i+1})$ for all $1 \leq i \leq k$, or
  (ii) $\varphi(f_k) = \varphi(f_1)$, but $\varphi(f_i) = \varphi(v_{i+1})$ and $\varphi(f_i) \neq \varphi(f_{i+1})$ for all $1 \leq i < k$.
In both cases we have $|L(e_i)| \geq 2$ for all i. If all the hypotheses of Corollary 1 are satisfied, then φ can be extended to a total-coloring of C and we are done. So suppose C is odd and every $L(e_i)$ contains the same 2 colors, say α and β. Letting γ, δ, and ϵ be the colors of f_1, f_2, and v_1 respectively, we see that $3 \mid k$ and for all $1 \leq i \leq k$ we have

$$\varphi(f_i) = \varphi(v_{i+1}) = \begin{cases} \gamma, & \text{if } i \equiv 1 \pmod 3, \\ \delta, & \text{if } i \equiv 2 \pmod 3, \\ \epsilon, & \text{if } i \equiv 0 \pmod 3. \end{cases} \quad (1)$$

Change the color of v_2 to the only remaining color in $[5] \setminus \{\gamma, \delta, \epsilon, \varphi(w_2)\}$. This changes $L(e_1)$ and $L(e_2)$ but leaves the remaining $L(e_i)$ intact. Following this color change and appropriate relabeling of C, all the hypotheses of Corollary 2 are satisfied. Therefore, φ can be extended to a 5-total-coloring of C.

5.2 Case 2 of the λ-Total-Coloring Algorithm

First some definitions. An edge e_i is *safe* if $|L(e_i)| \geq 2$. A vertex v_i is *safe* if any of the following holds (and *unsafe* otherwise): **(i)** $\varphi(w_i) \notin \lambda(v_i)$ **(ii)** f_i is colored and $\varphi(f_i) \notin \lambda(v_i)$ **(iii)** for $j = i-1$ or $j = i$, e_j is colored and ($\varphi(e_j) \notin \lambda(v_i)$ or $\varphi(e_j) = \varphi(w_i)$). Let $\sigma(i) = (\lambda(e_i) \setminus \lambda(v_i)) \cup (\lambda(e_i) \cap \{\varphi(w_i)\})$ for all $1 \leq i \leq k$. (A color in $\sigma(i)$ makes v_i safe.) These 3 facts easily follow from the definitions.

Fact 1. Suppose that for all $1 \leq i \leq k$ no e_i is colored but all f_i, v_i are colored. Then edge e_i is safe if and only if $\{\varphi(f_i), \varphi(v_i), \varphi(f_{i+1}), \varphi(v_{i+1})\} \setminus \lambda(e_i) \neq \emptyset$ or $\varphi(f_i) = \varphi(f_{i+1})$ or $\varphi(f_i) = \varphi(v_{i+1})$ or $\varphi(v_i) = \varphi(f_{i+1})$.

Fact 2. Suppose none of $v_1, \ldots, v_k$ is colored. Then vertex v_i is safe if and only if $|L(v_i)| \geq 2$, with equality holding only when f_i, e_{i-1}, e_i are colored and exactly one of the conditions in **(i)–(iii)** holds.

Fact 3. If $\varphi(w_i) \in \lambda(v_i)$ then $\sigma(i)$ is nonempty. In particular, $\sigma(i)$ is nonempty if v_i is unsafe. Furthermore, $\sigma(i) \subseteq \lambda(e_i)$ and assigning any color in $\sigma(i)$ to e_i makes v_i safe.

There are 3 subcases to consider. In subcases 2.1 & 2.2, we'll first color all the f_i and e_i so as to make all v_i safe, and then extend φ to cover all the v_i as well, after making any necessary color changes on some edges. In subcase 2.3, we'll first color all the f_i and v_i so as to make all e_i safe, and then extend φ to cover all the e_i as well, after making any necessary color change on a vertex.

Subcase 2.1 Some index i satisfies $\varphi(w_i) \notin \lambda(v_i)$ or $L(f_i) \setminus \lambda(v_i) \neq \emptyset$. By relabeling if necessary, we may assume $i = k$ satisfies the condition. Color all the f_i and e_i as follows.

```
for i ← 1 to k do {
   if L(f_i) \ λ(v_i) ≠ ∅ then φ(f_i) ← any color in L(f_i) \ λ(v_i)
   else φ(f_i) ← any color in L(f_i);
   if v_i is unsafe /* always false if i = k */ then
      φ(e_i) ← any color in σ(i) }
for i ← 1 to k do if e_i is still uncolored then φ(e_i) ← any color in L(e_i)
```

Subcase 2.2 Subcase 2.1 does not apply, and some valid labeling of C has some index i satisfying $\sigma(i) \neq \sigma(i-1)$ or $|\sigma(i)| > 1$. By relabeling if necessary, we may assume $i = 1$ satisfies the condition. Let $\alpha \in \sigma(1)$ and $\beta \in \sigma(k)$ be distinct colors. Define a function c by setting $c(1) = \alpha$, $c(k) = \beta$, and setting $c(i)$ to be any color in $\sigma(i)$ for each $1 < i < k$. Color all the f_i and e_i as follows.

$\varphi(f_1) \leftarrow$ any color in $L(f_1) \setminus \{c(k)\}$; $\varphi(e_1) \leftarrow c(1)$;
for $i \leftarrow 2$ **to** k **do** {
  $\varphi(f_i) \leftarrow$ any color in $L(f_i) \setminus \{c(i-1)\}$;
  if $c(i) \in L(e_i)$ **then** $\varphi(e_i) \leftarrow c(i)$ }
for $i \leftarrow 2$ **to** k **do if** e_i is still uncolored **then** $\varphi(e_i) \leftarrow$ any color in $L(e_i)$

It is easy to see that the coloring procedures in subcases 2.1 & 2.2 extend φ to cover all f_i and e_i while maintaining the invariant. They also result in safe v_i for all i. If all the hypotheses of Lemma 1 are satisfied, then φ can be extended to a λ-total-coloring of C and we are done. So suppose C is odd and every $L(v_i)$ consists of the same 2 colors, say α and β. Consider edge e_i such that either $(\varphi(e_i) \in \lambda(v_i)$ and $\varphi(e_i) \neq \varphi(w_i))$ or $(\varphi(e_i) \in \lambda(v_{i+1})$ and $\varphi(e_i) \neq \varphi(w_{i+1}))$. Existence of such an edge is guaranteed because $k \geq 3$ and $|L(v_i)| = 2$ for all i. By relabeling if necessary, we may assume $\varphi(e_i) \in \lambda(v_i)$ and $\varphi(e_i) \neq \varphi(w_i)$.

First suppose $L(e_i) \supset \{\varphi(e_i)\}$. Say that the current color of e_i is $\gamma \notin \{\alpha, \beta\}$. Change the color of e_i to some other color in $L(e_i)$. This color change affects only lists $L(v_i)$ and $L(v_{i+1})$ in the following manner. Suppose the new color is either α or β, say wlog that it is α. Then $L(v_i)$ becomes $\{\gamma, \beta\}$; and $L(v_{i+1})$ becomes either $\{\gamma, \beta\}$ or $\{\beta\}$. Thus either Lemma 1 or Lemma 2 applies. Now suppose the new color is $\delta \notin \{\alpha, \beta\}$. Then $L(v_i)$ becomes $\{\gamma, \alpha, \beta\}$; and $L(v_{i+1})$ becomes either $\{\gamma, \alpha, \beta\}$ or stays $\{\alpha, \beta\}$ like before. Thus Lemma 1 applies.

Next suppose $L(e_i) = \{\varphi(e_i)\}$, but $L(e_{i+1}) \supset \{\varphi(e_{i+1})\}$. Say γ is the current color of e_{i+1} and δ is some other color in $L(e_{i+1})$. Recolor e_{i+1} by δ. If either Lemma 1 or Lemma 2 applies, then we are done. Otherwise we have $(\gamma \notin \lambda(v_{i+1}) \cup \lambda(v_{i+2}))$ or $(\gamma = \varphi(w_{i+1}) \notin \lambda(v_{i+2}))$ or $(\gamma = \varphi(w_{i+2}) \notin \lambda(v_{i+1}))$ or $(\gamma = \varphi(w_{i+1}) = \varphi(w_{i+2}))$. In that case, also recolor e_i by γ. Now Lemma 2 applies.

Lastly suppose $L(e_i) = \{\varphi(e_i)\}$ and $L(e_{i+1}) = \{\varphi(e_{i+1})\}$. Then we can switch the colors of e_i and e_{i+1} and still maintain the invariant! Lemma 1 applies after the color switch.

Subcase 2.3 Subcase 2.1 does not apply, and every valid labeling of C has all $\sigma(i)$ as the same one-color set, say $\{\alpha\}$. This can happen only when there exist distinct colors $\beta, \gamma, \delta, \epsilon$ all different from α such that for each i we have $\lambda(e_i) = \{\alpha, \beta, \gamma, \delta, \epsilon\}$ and either $\lambda(v_i) = \lambda(e_i)$ or $(\{\alpha\} = \lambda(e_i) \setminus \lambda(v_i)$ and $\{\varphi(w_i)\} = \lambda(v_i) \setminus \lambda(e_i))$. Color all the f_i and v_i using Procedure A. It is easy to see that Procedure A extends φ to cover all f_i and v_i while maintaining the invariant. It also results in safe e_i for all i. If all the hypotheses of Corollary 1 are satisfied, then φ can be extended to a λ-total-coloring of C and we are done. So suppose C is odd and every $L(e_i)$ consists of the same 2 colors, say α and β. By the way Procedure A works, this can happen only when $3 \mid k$ and there exist distinct colors γ, δ, ϵ all different from α, β such that Equation (1) holds for all i. Change the color of v_2 to the only color in $\lambda(v_2) \setminus \{\gamma, \delta, \epsilon, \varphi(w_2)\}$. This changes $L(e_1)$ and $L(e_2)$ but leaves the remaining $L(e_i)$ intact. After the color change and appropriate relabeling of C, all the hypotheses of Corollary 2 are satisfied. So φ can be extended to a λ-total-coloring of C.

References

1. Bondy, J.A., Murty, U.S.R.: Graph Theory with Applications. Macmillan (1976)
2. Vizing, V.G.: On an estimate of the chromatic class of a p-graph. Metody Diskret. Analiz. **3** (1964) 25–30 In Russian.
3. Gabow, H.N., Nishizeki, T., Kariv, O., Leven, D., Terada, O.: Algorithms for edge-coloring graphs. Technical Report TRECIS-8501, Tohoku University (1985)
4. Holyer, I.J.: The NP-completeness of edge-coloring. SIAM Journal on Computing **10** (1981) 718–720
5. Gary, M.R., Johnson, D.S.: Computers and Intractability: A Guide to the Theory of NP-Completeness. W.H. Freeman & Co., San Francisco, CA (1979)
6. Skulrattanakulchai, S.: 4-edge-coloring graphs of maximum degree 3 in linear time. Information Processing Letters **81** (2002) 191–195
7. Bollobás, B., Harris, A.J.: List-colourings of graphs. Graphs and Combinatorics **1** (1985) 115–127
8. Chetwynd, A.G., Häggkvist, R.: A note on list-colorings. Journal of Graph Theory **13** (1989) 87–95
9. Jensen, T.R., Toft, B.: Graph Coloring Problems. John Wiley & Sons (1995)
10. Vizing, V.G.: Coloring the vertices of a graph in prescribed colors. Metody Diskret. Anal. v Teorii Kodov i Schem **29** (1976) 3–10
11. Erdős, P., Rubin, A.L., Taylor, H.: Choosability in graphs. In: Proceedings of the West-Coast Conference on Combinatorics, Graph Theory and Computing. Volume XXVI of Congressus Numerantium., Arcata, California (1979) 125–157
12. Brooks, R.L.: On colouring the nodes of a network. Proceedings of the Cambridge Philosophical Society. Mathematical and Physical Sciences **37** (1941) 194–197
13. Skulrattanakulchai, S.: Δ-list vertex coloring in linear time. In: Proc. SWAT '02. LNCS (2002) To appear.
14. Juvan, M., Mohar, B., Škrekovski, R.: On list edge-colorings of subcubic graphs. Discrete Mathematics **187** (1998) 137–149
15. Sánchez-Arroyo, A.: Total colourings and complexity. Master's thesis, University of Oxford (1989)
16. Sánchez-Arroyo, A.: Determining the total colouring number is NP-hard. Discrete Mathematics **78** (1989) 315–319
17. Yap, H.P.: Total Colourings of Graphs. LNM Volume 1623. Springer (1996)
18. Rosenfeld, M.: On the total coloring of certain graphs. Israel Journal of Mathematics **9** (1971) 396–402
19. Vijayaditya, N.: On total chromatic number of a graph. Journal of the London Mathematical Society **3** (1971) 405–408
20. Biedl, T.C., Bose, P., Demaine, E.D., Lubiw, A.: Efficient algorithms for Petersen's Matching Theorem. Journal of Algorithms **38** (2001) 110–134
21. Juvan, M., Mohar, B., Škrekovski, R.: List total colorings of graphs. Combinatorics, Probability & Computing **7** (1998) 181–188
22. Borodin, O.V., Kostochka, A.V., Woodall, D.R.: List edge and list total colourings of multigraphs. Journal of Combinatorial Theory Series B **71** (1997) 184–204
23. Cormen, T.H., Leiserson, C.E., Rivest, R.L., Stein, C.: Introduction to Algorithms. Second edn. McGraw-Hill, New York (2001)
24. Halperin, S., Zwick, U.: Optimal randomized EREW PRAM algorithms for finding spanning forests. Journal of Algorithms **39** (2001) 1–46
25. Tarjan, R.E., Vishkin, U.: An efficient parallel biconnectivity algorithm. SIAM Journal on Computing **14** (1985) 862–874
26. Reif, J.H., ed.: Synthesis of Parallel Algorithms. Morgan Kaufmann, CA (1993)

Efficient Algorithms for the Hamiltonian Problem on Distance-Hereditary Graphs

Sun-yuan Hsieh[1], Chin-wen Ho[2], Tsan-sheng Hsu[3], and Ming-tat Ko[3]

[1] Department of Computer Science and Information Engineering, National Cheng Kung University, Tainan, Taiwan
hsiehsy@mail.ncku.edu.tw

[2] Department of Computer Science and Information Engineering, National Central University, Chung-Li, Taiwan
hocw@csie.ncu.edu.tw

[3] Institute of Information Science, Academia Sinica, Taipei, Taiwan
{tshsu,mtko}@iis.sinica.edu.tw

Abstract. In this paper, we first present an $O(|V|+|E|)$-time sequential algorithm to solve the Hamiltonian problem on a distance-hereditary graph $G=(V,E)$. This algorithm is faster than the previous best result which takes $O(|V|^2)$ time. Let $T_d(|V|,|E|)$ and $P_d(|V|,|E|)$ denote the parallel time and processor complexities, respectively, required to construct a decomposition tree of a distance-hereditary graph on a PRAM model M_d. We also show that this problem can be solved in $O(T_d(|V|,|E|) + \log |V|)$ time using $O(P_d(|V|,|E|) + (|V|+|E|)/\log |V|)$ processors on M_d. Moreover, if G is represented by its decomposition tree form, the problem can be solved optimally in $O(\log |V|)$ time using $O((|V|+|E|)/\log |V|)$ processors on an EREW PRAM.

1 Introduction

A graph is *distance-hereditary* [2,11] if the distance stays the same between any of two vertices in every connected induced subgraph containing both (where the *distance* between two vertices is the length of a shortest path connecting them). Distance-hereditary graphs form a subclass of perfect graphs [6,10,11] that are graphs G in which the maximum clique size equals the chromatic number for every induced subgraph of G [8]. Two well-known classes of graphs, trees and cographs, both belong to the class of distance-hereditary graphs. Properties of distance-hereditary graphs are studied by many researchers [2,3,4,5,6,7,9,10,11,12,13,14,15,16,17,18] which resulted in sequential or parallel algorithms to solve several interesting graph-theoretical problems on this special class of graphs.

A cycle in a graph G is called a *Hamiltonian cycle* if it contains every vertex of G exactly once. A graph is said to be *Hamiltonian* if it contains a Hamiltonian cycle. The *Hamiltonian problem* is to determine whether there exists a Hamiltonian cycle in a given graph and find one if such a cycle does exist. Previous

O.H. Ibarra and L. Zhang (Eds.): COCOON 2002, LNCS 2387, pp. 77–86, 2002.

related works on distance-hereditary graphs are summarized below. By investigating the neighborhood of the last pendant vertex in a one-vertex extension-sequence, Müller and Nicolai developed an $O(|V|(|V|+|E|))$-time sequential algorithm to solve the Hamiltonian problem on bipartite distance-hereditary graphs, where $|V|$ (respectively, $|E|$) is the number of vertices (respectively, edges) of the given graph [17]. In [18], Nicolai presented an $O(|V|^3)$-time sequential algorithm to solve the Hamiltonian problem on distance-hereditary graphs. Hung et al. [16] further reduced the above sequential complexity to $O(|V|^2)$. To the best of our knowledge, there is no parallel algorithm to solve the problem on distance-hereditary graphs in the literature.

In this paper, we first present an $O(|V|+|E|)$-linear-time algorithm to solve the Hamiltonian problem on distance-hereditary graphs. Let $T_d(|V|,|E|)$ and $P_d(|V|,|E|)$ denote the parallel time and processor complexities, respectively, required to construct a decomposition tree representation of a distance-hereditary graph on a PRAM model M_d. We show that the Hamiltonian problem can be solved in $O(T_d(|V|,|E|)+\log|V|)$ time using $O(P_d(|V|,|E|)+(|V|+|E|)/\log|V|)$ processors on M_d. The best known result to construct a decomposition tree needs $O(\log^2|V|)$ time using $O(|V|+|E|)$ processors on a CREW PRAM [15]. If G is given in its decomposition tree form, the problem can be solved in $O(\log|V|)$ time using $O((|V|+|E|)/\log|V|)$ processors on an EREW PRAM. This time-processor complexity matches the best sequential algorithm.

2 Preliminaries

This paper considers finite, simple and undirected graphs $G=(V,E)$, where V and E are the vertex and edge sets of G, respectively. Let $n=|V|$ and $m=|E|$. For two graphs $G_1=(V_1,E_1)$ and $G_2=(V_2,E_2)$, the *union of* G_1 *and* G_2, denoted by $G_1\cup G_2$, is the graph $(V_1\cup V_2, E_1\cup E_2)$. Let $G[X]$ denote the subgraph of G induced by $X\subseteq V$. For graph-theoretic terminologies and notations not mentioned here, see [8].

Definition 1 [4] A graph consisting of a single vertex v is a *primitive distance-hereditary graph with the twin set* $\{v\}$. Let G_1 and G_2 be distance-hereditary graphs with the twin sets S_1 and S_2, respectively. Then, (1) The graph obtained from G_1 and G_2 by connecting every vertex of S_1 to all vertices of S_2 is a distance-hereditary graph with the twin set $S_1\cup S_2$; (2) The graph obtained from G_1 and G_2 by connecting every vertex of S_1 to all vertices of S_2 is a distance-hereditary graph with the twin set S_1; (3) The union of G_1 and G_2 is a distance-hereditary graph with the twin set $S_1\cup S_2$.

A distance-hereditary graph G is said to be formed from G_1 and G_2 by the *true twin* (respectively, *attachment*) operation if G is obtained through (1) (respectively, (2)) of Definition 1, and by the *false twin operation* if G is obtained through (3) of Definition 1.

For a rooted tree T, we use *root*(T) to denote the root of T. A distance-hereditary graph can be represented by a binary tree form, called a *decomposition tree* D_G, which is defined as follows.

Definition 2 [4] (1) The tree consisting of a single vertex v is a decomposition tree of a primitive distance-hereditary graph $G = (\{v\}, \emptyset)$.
(2) Let D_{G_1} and D_{G_2} be decomposition trees of distance-hereditary graphs G_1 and G_2, respectively. (a) If G is formed from G_1 and G_2 by the true twin operation, then a tree D_G with $root(D_G)$ being represented by $\otimes$ and with $root(D_{G_1})$ and $root(D_{G_2})$ being the two children of $root(D_G)$ is a decomposition tree of G. (b) If G is formed from G_1 and G_2 by the attachment operation, then a tree D_G with $root(D_G)$ being represented by $\oplus$ and with the roots of D_{G_1} and D_{G_2} being the left and right children of $root(D_G)$, respectively, is a decomposition tree of G. (c) If G is formed from G_1 and G_2 by the false twin operation, then a tree D_G with $root(D_G)$ being represented by $\odot$ and with the roots of D_{G_1} and D_{G_2} being the two children of $root(D_G)$ is a decomposition tree of G.

For a node v in D_G, let $D_G(v)$ be the subtree of D_G rooted at v, and let G_v be a subgraph of G induced by the leaves of $D_G(v)$. Also let S_v be the twin set of G_v.

Note that a decomposition tree of a distance-hereditary graph can be constructed in $O(n+m)$ sequential time[4], and constructed in parallel in $O(\log^2 n)$ time using $O(n+m)$ processors on a CREW PRAM.

A *closed integer interval* is an ordered pair of integers $[t_1, t_2]$, with $t_1 \leq t_2$. The interval $[t_1, t_2]$ represents the set $\{t \in \mathbf{Z} |\ t_1 \leq t \leq t_2\}$. A *path partition* of a graph $G = (V, E)$ is a set of pairwise vertex disjoint paths such that the union of the vertices of these paths equals V. Given a distance-hereditary graph G with the twin set S, a path partition of G is said to be *crucial* if the end-vertices of each path are in S. Let $\mathcal{P}_S(G)$ denote a crucial path partition of G. Furthermore, a *crucial k-path partition of G*, $\mathcal{P}_S^k(G) = \{P_1, P_2, \ldots, P_k\}$ is a crucial path partition of G composed of exactly k paths $P_1, P_2, \ldots, P_k$. Since all path partitions under consideration throughout this paper are crucial, the subscripts S in the notations $\mathcal{P}_S(G)$ and $\mathcal{P}_S^k(G)$ are omitted if no ambiguity arises. Let $N(G) = (l_1, l_2, \ldots, l_t)$ denote the set of integers in the increasing order, i.e., $l_1 < l_2 < \cdots < l_t$, such that G has a crucial l_i-path partition. As we will show later, the elements of $N(G)$ form a segment of consecutive integers, that is, $l_{i+1} = l_i + 1$ for all $1 \leq i \leq t-1$. Thus $N(G)$ can be represented by a closed integer interval $[l_1, l_t]$. In particular, let l_1 (respectively, l_t) be the *left* (respectively, *right*) *end-point* of $N(G)$, denoted by $l(N(G))$ (respectively, $r(N(G))$). We also define $N(G) = [0, 0]$ if G does not contain any crucial path partition.

3 A Sequential Algorithm

In the following sections, let G be a distance-hereditary graph with the twin set S, which is formed from two distance-hereditary graphs G_1 and G_2 with the twin sets S_1 and S_2, respectively.

Lemma 1 *Let* $G = G_1 \otimes G_2$. *A crucial q-path partition* $\mathcal{P}^q(G)$ *can be constructed from a crucial i-path partition* $\mathcal{P}^i(G_1)$ *and a crucial j-path partition* $\mathcal{P}^j(G_2)$, *where* $\text{MAX}\{1, \text{MAX}\{i,j\} - \text{MIN}\{i,j\}\} \leq q \leq i + j$.

Lemma 2 *Let $G = G_1 \oplus G_2$. A crucial path partition $\mathcal{P}^{i-j}(G)$ can be constructed from $\mathcal{P}^i(G_1)$ and $\mathcal{P}^j(G_2)$ if $i > j$.*

According to the above lemmas, we can further obtain the following results.

Lemma 3 *Assume that $G = G_1 \otimes G_2$. Let a_i and b_i be positive integers for $i \in \{1,2\}$. If $N(G_1) = [a_1, b_1]$ and $N(G_2) = [a_2, b_2]$, then $N(G)$ equals (1) $[1, b_1 + b_2]$ when $[a_1, b_1] \cap [a_2, b_2] \neq \emptyset$; (2) $[a_2 - b_1, b_1 + b_2]$ when $[a_1, b_1] \cap [a_2, b_2] = \emptyset$ and $a_2 > b_1$; (3) $[a_1 - b_2, b_1 + b_2]$ when $[a_1, b_1] \cap [a_2, b_2] = \emptyset$ and $a_1 > b_2$.*

Corollary 1 *Assume that $G = G_1 \otimes G_2$. Let a_i and b_i be positive integers for $i \in \{1,2\}$. If $N(G_1) = [a_1, b_1]$ and $N(G_2) = [a_2, b_2]$, then $N(G) = [\text{MAX}\{1, a_2 - b_1, a_1 - b_2\}, b_1 + b_2]$.*

Lemma 4 *Assume that $G = G_1 \oplus G_2$. Let a_i and b_i be positive integers for $i \in \{1,2\}$. If $N(G_1) = [a_1, b_1]$ and $N(G_2) = [a_2, b_2]$, then $N(G)$ equals (1) $[0, 0]$ when $a_2 \geq b_1$; (2) $[1, b_1 - a_2]$ when $[a_1, b_1] \cap [a_2, b_2] \neq \emptyset$ and $a_2 < b_1$; (3) $[a_1 - b_2, b_1 - a_2]$ when $[a_1, b_1] \cap [a_2, b_2] = \emptyset$ and $a_2 < b_1$.*

Corollary 2 *Assume that $G = G_1 \oplus G_2$. Let a_i and b_i be positive integers for $i \in \{1,2\}$. If $N(G_1) = [a_1, b_1]$ and $N(G_2) = [a_2, b_2]$, then $N(G) = [\text{MIN}\{\text{MAX}\{0, b_1 - a_2\}, \text{MAX}\{1, a_1 - b_2\}\}, \text{MAX}\{0, b_1 - a_2\}]$.*

By the structure characterization that each vertex of G_1 is not adjacent to any vertex of G_2, we have the following result.

Lemma 5 *Assume that $G = G_1 \odot G_2$. Let a_i and b_i be positive integers for $i \in \{1,2\}$. If $N(G_1) = [a_1, b_1]$ and $N(G_2) = [a_2, b_2]$, then $N(G) = [a_1 + a_2, b_1 + b_2]$.*

Theorem 1 *Let D_G be a decomposition tree of a distance-hereditary graph G. If $N(G) \neq [0, 0]$, then $N(G_v) = [a_v, b_v]$ for all nodes v in D_G, where $a_v \leq b_v \in \mathbf{Z}^+$.*

Theorem 2 *Let $G = G_1 \otimes G_2$ or $G = G_1 \oplus G_2$ and let $|V(G)| \geq 3$. Then, G has a Hamiltonian cycle if and only if $N(G_1) \cap N(G_2) \neq \emptyset$ and $N(G_i) \neq [0, 0]$, $i \in \{1, 2\}$.*

In the remaining of this section, we develop an $O(n+m)$-linear-time sequential algorithm to solve the Hamiltonian problem on distance-hereditary graphs. Consider a Hamiltonian distance-hereditary graph G. Let x be an arbitrary non-root node of D_G. Given a Hamiltonian cycle C of G, it is clear that this cycle in G_x forms a crucial k-path partition of G_x for some $1 \leq k \leq |V(G_x)|$. In particular, we refer such a k to be the *target number* $tar_C(x)$ *of* G_x (the subscript C can be omitted from the notation if no ambiguity aries). Note that $tar(l) = 1$ if l is a leaf of D_G. For a non-root internal node $v \in V(D_G)$ with the two children u and

w, $\mathcal{P}^{tar(v)}(G_v)$ can be constructed from $\mathcal{P}^{tar(u)}(G_u)$ and $\mathcal{P}^{tar(w)}(G_w)$ by augmenting some specified edges in $S_u \times S_w$. These edges are called *marked edges of* $\mathcal{P}^{tar(v)}(G_v)$, denoted by $M(\mathcal{P}^{tar(v)}(G_v))$. Clearly, $\mathcal{P}^{tar(u)}(G_u)$ and $\mathcal{P}^{tar(w)}(G_w)$ interleaved with $M(\mathcal{P}^{tar(v)}(G_v))$ form $\mathcal{P}^{tar(v)}(G_v)$.

Assume that $N(G_1) = [a_1, b_1]$ and $N(G_2) = [a_2, b_2]$. Given $N(G)$ and a fixed non-zero number $q \in N(G)$, we can summarize the values of q_1 and q_2 in $O(1)$ time such that $\mathcal{P}^q(G)$ can be constructed from $\mathcal{P}^{q_1}(G_1)$ and $\mathcal{P}^{q_2}(G_2)$. We now describes our algorithm in details.

PHASE 1: Compute $N(G_x)$ for each non-root node x of D_G. Initially, let $N(G_l) = [1, 1]$ for each leaf l. By Lemmas 3–5, the desired values can be computed by a bottom-up evaluation process. During the progress of computation, if $N(G_v) = [0, 0]$ for some node v, we terminate the algorithm and exit. Otherwise, all nodes of D_G are associated with the desired values. Clearly, this phase takes $O(n)$ time.

PHASE 2: Compute the target number $tar(x)$ for each non-root node x. Assume that y and z are the left and right children of $root(D_G)$, respectively. Select an arbitrary number $k \in N(G_y) \cap N(G_z)$ and let $tar(y) = tar(z) = k$. For each non-root internal node v with the two children u and w, select $tar(u)$ and $tar(w)$ such that $\mathcal{P}^{tar(v)}(G_v)$ can be constructed from $\mathcal{P}^{tar(u)}(G_u)$ and $\mathcal{P}^{tar(w)}(G_w)$. This phase takes $O(n)$ time.

PHASE 3: Find a set of edges to form a Hamiltonian cycle of G in $O(n + m)$ time using the information obtained in the previous two phases.

4 A Parallel Algorithm

4.1 Computing $N(G_v)$

Let u and w be the left and right children, respectively, of an internal node v in D_G. Also let $N(G_u) = [a_1, b_1]$ and $N(G_w) = [a_2, b_2]$. If v is $\otimes$ (respectively, $\odot$), then the formulas in Corollary 1 (respectively, Lemma 5) are used to compute $N(G_v)$, after relaxing the original constraint of requiring two numbers a_i and b_i, $i \in \{1, 2\}$, being positive integers to integers. For ease of parallel implementation, we adopt the following formula to compute $N(G_v)$ instead of that in Corollary 2:

$$N(G_v) = [\mathrm{MAX}\{1, a_1 - b_2\}, b_1 - a_2]. \quad (1)$$

Recall that the input graph G considered in this paper is connected with $n > 2$. In fact, the type of $root(D_G)$ (that is, $\otimes$-node or $\oplus$-node) does not effect the correctness of our algorithms. From now on, we further assume that $root(D_G)$ is a $\otimes$-node for convenience. By a bottom-up evaluation process, the following result can be obtained.

Lemma 6 *Assume that $N(G_u)$ is computed for each non-leaf node u using Equation 1 and the formulas described in Corollary 1 and Lemma 5. Let y and z be the two children of $root(D_G)$. Then, G is Hamiltonian if and only if $r(N(G_v)) > 0$ for all nodes v, and $N(G_y) \cap N(G_z) \neq \emptyset$.*

In the rest of this section, we apply the binary tree contraction technique described in [1] to compute $N(G_v)$ using Equation 1 and the formulas described in Corollary 1 and Lemma 5. This technique recursively applies the two operations, *prune* and *bypass*, to a given binary tree. $Prune(u)$ is the operation which removes a leaf node u from the current tree, and $bypass(v)$ is the operation that removes a node v with exactly one child w and then let the parent of v become the new parent of w.

Lemma 7 [1] *If the prune operation and the bypass operation can be performed by one processor in a constant time, then the binary tree contraction algorithm can be implemented in $O(\log n)$ time using $O(n/\log n)$ processors on an EREW PRAM, where n is the number of nodes in the input binary tree.*

Definition 3 A pair of binary functions $F(x,y) = [f(x,y), g(x,y)]$, where $f, g : \mathbf{Z}^2 \mapsto \mathbf{Z}$, possess the *closed form* if the following two conditions hold. (1) $f(x,y) = \text{MAX}\{x+c_1, -y+c_2, d\}$, where $c_i, d \in \mathbf{Z} \cup \{-\infty\}$ and c_1, c_2, d cannot be $-\infty$ at the same time; (2) $g(x,y) = e - \text{MAX}\{x+h_1, -y+h_2, j\}$, where $h_i, j \in \mathbf{Z} \cup \{-\infty\}$, $e \in \mathbf{Z}$, and h_1, h_2, j cannot be $-\infty$ at the same time.

For a pair of binary functions $F(x,y) = [f(x,y), g(x,y)]$, let $l(F)$ and $r(F)$ be $f(x,y)$ and $g(x,y)$, respectively.

Lemma 8 *Let F_1 and F_2 be two arbitrary function pairs possessing the closed form. Then, $F_1 \circ (l(F_2), r(F_2))$ also possess the closed form.*

Given a decomposition tree D_G, we next develop a parallel algorithm to compute $N(G_v)$ for each node v in D_G. For a node u in the current tree H, let $par_H(u)$ (respectively, $child_H(u)$) denote the *parent* (respectively, *children*) of u, and let $sib_H(u)$ denote the *sibling* of u. The subscript H can be omitted if no ambiguity arises. Recall that $l(N(G_u))$ (respectively, $r(N(G_u))$) is the *left* (respectively, *right*) *end-point* of the $N(G_u)$.

During the process of executing the tree contraction, we aim at constructing a pair of binary functions $[f_v(x,y), g_v(x,y)]$ associated with each node v of the current tree such that both functions possess the closed form and satisfies the invariant described below. Let v be an internal node in the current tree whose left and right children are u and w, respectively. Also let $\delta(u,v)$ and $\delta(w,v)$ be the left and right children of v in the original tree, respectively. Note that $\delta(u,v)$ and $\delta(w,v)$ are ancestors of u and w in the original tree, respectively. From now on, we call $\delta(u,v)$ (respectively, $\delta(w,v)$) *replacing ancestors of* u (respectively, w) *with respect to* v, and abbreviate it to $\delta(u)$ (respectively, $\delta(w)$) if no ambiguity arises.

INVARIANT: Once $l(N(G_u))$, $r(N(G_u))$, $l(N(G_w))$, and $r(N(G_w))$ are computed and provided as the inputs of $[f_u(x,y), g_u(x,y)]$ and $[f_w(x,y), g_w(x,y)]$, the following three statements hold:
S1: $N(G_{\delta(u)}) = [f_u(l(N(G_u)), r(N(G_u))), g_u(l(N(G_u)), r(N(G_u)))]$;

S2: $N(G_{\delta(w)}) = [f_w(l(N(G_w)), r(N(G_w))), g_w(l(N(G_w)), r(N(G_w)))]$;
S3: $N(G_v)$ can be computed using $N(G_{\delta(u)})$ and $N(G_{\delta(w)})$.

For a node v in the current tree, we call the above functions $f_v(x, y)$ and $g_v(x, y)$ *the crucial functions of* v. We next describe the details of our algorithm. Initially, for each node v in the given tree we construct $f_v(x, y) = \text{MAX}\{x + 0, -y + (-\infty), -\infty\}$ and $g_v(x, y) = 0 - \text{MAX}\{x + (-\infty), -y + 0, -\infty\}$. Moreover, if v is a leaf, then let $l(N(G_v)) = r(N(G_v)) = 1$. In the execution of the tree contraction, assume that *prune*(u) and *bypass*$(par(u))$ are performed consecutively. Let $par(u) = v$ and $sib(u) = w$ in the current tree. Assume that $[f_u(x, y), g_u(x, y)]$ and $[f_w(x, y), g_w(x, y)]$ are crucial functions of u and w in the current tree, respectively. Thus we have $N(G_{\delta(u)}) = [f_u(l(N(G_u)), r(N(G_u)))$, $g_u(l(N(G_u)), r(N(G_u)))]$; $N(G_{\delta(w)}) = [f_w(l(N(G_w)), r(N(G_w))), g_w(l(N(G_w))$, $r(N(G_w)))]$. Since u is a leaf, $l(N(G_u)) = r(N(G_u)) = 1$. Therefore, $N(G_{\delta(u)})$ can be obtained through the function evaluation. On the other hand, since w may not be a leaf in the current tree, $l(N(G_w))$ and $r(N(G_w))$ are indeterminate values represented by the variables x and y, respectively. Hence, $N(G_{\delta(w)})$ can be represented by the interval $[f_w(x, y), g_w(x, y)]$. Let $s = f_u(l(N(G_u)), r(N(G_u)))$ $= f_u(1, 1)$; $t = g_u(l(N(G_u)), r(N(G_u))) = g_u(1, 1)$; $l(N(G_{\delta(w)})) = f_w(l(N(G_w))$, $r(N(G_w))) = f_w(x, y) = \text{MAX}\{x + c_1, -y + c_2, d\}$; $r(N(G_{\delta(w)})) = g_w(l(N(G_w))$, $r(N(G_w))) = g_w(x, y) = e - \text{MAX}\{x + h_1, -y + h_2, j\}$. We then construct the following two intermediate functions in order to form $N(G_v)$ from those of $\delta(u)$ and $\delta(w)$:

CASE A: v is a $\otimes$-node. According to Corollary 1, we construct the following functions.

CASE A1: w is the left child of v. Then, $N(G_v) = [\text{MAX}\{1, l(N(G_{\delta(u)})) - r(N(G_{\delta(w)}))$, $l(N(G_{\delta(w)})) - r(N(G_{\delta(u)}))\}$, $r(N(G_{\delta(w)})) + r(N(G_{\delta(u)}))]$ $= [\text{MAX}\{x + \text{MAX}\{s - e + h_1, c_1 - t\}, -y + \text{MAX}\{s - e + h_2, c_2 - t\}, \text{MAX}\{1, s - e + j, d - t\}\}, (e + t) - \text{MAX}\{x + h_1, -y + h_2, j\}]$

CASE A2: w is the right child of v_i. The construction is similar to that of Case A1.

CASE B: v is a $\oplus$-node. According to Equation 1, we construct the following functions.

CASE B1: w is the left child of v. Then, $N(G_v) = [\text{MAX}\{1, l(N(G_{\delta(w)})) - r(N(G_{\delta(u)}))\}$, $r(N(G_{\delta(w)})) - l(N(G_{\delta(u)}))] = [\text{MAX}\{x + (c_1 - t), -y + (c_2 - t), \text{MAX}\{1, d - t\}\}, (e - s) - \text{MAX}\{x + h_1, -y + h_2, j\}]$.

CASE B2: w is the right child of v_i. The construction is similar to Case B1.

CASE C: v is a $\odot$-node. According to Lemma 5, $N(G_v) = [l(N(G_{\delta(w)})) + l(N(G_{\delta(u)})), r(N(G_{\delta(w)})) + r(N(G_{\delta(u)}))] = [\text{MAX}\{x + (c_1 + s), -y + (c_2 + s), d + s\}, (e + t) - \text{MAX}\{x + h_1, -y + h_2, j\}]$.

Therefore, the functions representing $N(G_v)$ in Cases A–C all possess the closed form. Let H denote the current tree. We construct the above functions after executing *prune*(u). Given the two functions $l(N(G_v))$ and $r(N(G_v))$ constructed above, the contribution to the left and right end-points of $N(G_{par_H(v)})$

can be obtained using $f_v(l(N(G_v)), r(N(G_v)))$ and $g_v(l(N(G_v)), r(N(G_v)))$. The functions are constructed for w after executing $bypass(par_H(u)) = bypass(v)$. By Lemma 8, the above functions possess the closed form. Therefore, during the process of executing the binary tree contraction, the crucial functions constructed after executing $prune(u)$ and $bypass(par(u))$ can be implemented in $O(1)$ time using one processor. This implies that the interval $N(G_v)$ for each node $v \in V(D_G)$, can be computed in $O(\log n)$ time using $O(n/\log n)$ processors on an EREW PRAM.

After computing $N(G_v)$, we check whether G is Hamiltonian by Lemma 6. This can be implemented in $O(1)$ time using $O(n)$ processors on an EREW PRAM. If G is Hamiltonian, then a Hamiltonian cycle of G can be generated using the method described in Section 4.2.

4.2 Computing $tar(x)$

Given a decomposition tree D_G associated with $N(G_x)$, $x \in V(D_G)$, we present an algorithm to compute $tar(x)$ in $O(\log n)$ time using $O(n/\log n)$ processors on an EREW PRAM based on the binary tree contraction technique.

For an internal node v with the two children u and w in D_G, recall that G_v is constructed from G_u and G_w using one of the three operations defined in Definition 1. For a positive integer $q_v \in N(G_v)$, we call $q_u \in N(G_u)$ and $q_w \in N(G_w)$ *contributing numbers of* q_v if a crucial q_v-path partition $\mathcal{P}^{q_v}(G_v)$ can be constructed from crucial path partitions $\mathcal{P}^{q_u}(G_u)$ and $\mathcal{P}^{q_w}(G_w)$. Two functions $f_u : N(G_v) \mapsto N(G_u)$ and $f_w : N(G_v) \mapsto N(G_w)$ are said to be *contributing functions of* v if $f_u(q_v) = q_u$ and $f_w(q_v) = q_w$.

On the other hand, consider a distance-hereditary graph G obtained from the two distance-hereditary graphs G_1 and G_2. Recall that, for a fixed $q \in N(G)$, we can return the contributing numbers $q_1 \in N(G_1)$ and $q_2 \in N(G_2)$ of q in $O(1)$ time. For ease of parallel implementation, we further observe that given $q \in N(G), N(G_1) = [a_1, b_1]$ and $N(G_2) = [a_2, b_2]$, two values of q_1 (respectively, q_2) can be obtained using the closed formula.

Observation 1 *For each non-root internal node* $v \in D_G$ *with the two children* u *and* w*, we can construct contributing functions of* v *having the unique form* $\textsc{Min}\{\textsc{Max}\{\pm x + c, a\}, b\}$*, where* x *is a variable drawn from* $\mathbf{Z}^+$*,* $a \in \mathbf{Z} \cup \{-\infty\}$*,* $b \in \mathbf{Z} \cup \{\infty\}$*, and* $c \in \mathbf{Z}$.

We call the form *min-max form* and call the function with the min-max form *min-max function.*

Lemma 9 *The class of min-max functions is closed under composition.*

In the rest of this section, we assume that the input of our algorithm is a decomposition tree D_G satisfying the following condition. For each non-root internal node v with the two children u and w, the edges (u, v) and (w, v) are associated with the contributing functions f_u and f_w of v, respectively. In particular, two edges incident with $root(D_G)$ are associated with two identity functions. Note that the identity functions clearly possess the min-max form. Our parallel algorithm consists of two stages, called the *tree contraction stage* and the *unwrapping stage*. In the tree contraction stage, we use the binary tree contraction technique to contract the given tree into a tree-node tree T_3. During the contraction, we also construct min-max functions (described latter) associated with the remaining edges of the current tree. In the unwrapping stage, we restore T_3 into the original tree to compute the target values progressively.

The Tree Contraction Stage. In the execution of the tree contraction, assume that $prune(u)$ and $bypass(par(u))$ are performed consecutively. Without loss of generality, assume that u is the left child of $par(u) = v$ (the case of u being the right child can be handled similarly). Let $sib(u) = w$ and v' be the parent of v in the current tree. Also assume that f_v, f_u and f_w are three contributing min-max functions associated with $(v, v'), (u, v)$ and (w, v), respectively. After executing $prune(u)$ and then $bypass(v)$, the edge (w, v') is associated with the function $f_w \circ f_v$. Note that $f_w \circ f_v$ also possesses the min-max form by Lemma 9.

After executing the binary tree contraction, a three-node tree T_3 is obtained. Note that the two functions associated with two edges of T_3 are both constant functions 1. Assume that y and z be the left and right children of D_G. Let $N(G_y) = [a_y, b_y]$ and $N(G_z) = [a_z, b_z]$. We first set $tar(root(D_G)) = tar(root(T_3)) = q \in [a_y, b_y] \cap [a_z, b_z]$ and then go on the next stage.

The Unwrapping Stage. This stage is to restore T_3 into D_G together with some function evaluations. Define two operations *arcprune* and *arcbypass*, denoted by $prune^{-1}$ and $bypass^{-1}$, respectively. $Prune^{-1}$ (respectively, $bypass^{-1}$) is the operation that restores the node deleted by *prune* (respectively, *bypass*), that is, $prune^{-1}(prune(u)) = u$ (respectively, $bypass^{-1}(bypass(v)) = v$). Consider an internal node w of the current tree from which two nodes u and v will be restored as the two children of w in the next step. Let $T(w)$ be the subtree rooted at w in the current tree T and $par_T(w) = t$. Without loss of generality, assume that $bypass^{-1}(bypass(v)) = v$ and $prune^{-1}(prune(u)) = u$ are restored consecutively such that in the current tree T', u is the leaf-child of v and w is the other child of v in T'. Also assume that g_v and g_u are two min-max functions associated with (v, t) and (u, v), respectively. Note that $tar(t)$ is obtained in the previous step. We evaluate $tar(v) = g_v(tar(t))$ and let $tar(u) = 1$ because g_u must be the constant function 1.

Lemma 10 *Given a decomposition tree D_G, the target values for all nodes can be computed in $O(\log n)$ time using $O(n/\log n)$ processors on an EREW PRAM.*

Given a decomposition tree D_G with $tar(x)$ being associated with each node $x \in V(D_G)$, we can find those edges to form a Hamiltonian cycle in $O(\log n)$ time using $O(m/\log n)$ processors on an EREW PRAM. The details are omitted due to limited space. Let $T_d(n, m)$ and $P_d(n, m)$ denote the parallel time and processor complexities, respectively, required to construct a decomposition tree of a distance-hereditary graph $G = (V, E)$ on a PRAM model M_d.

Theorem 3 *The Hamiltonian problem on distance-hereditary graphs can be solved in $O(T_d(n, m) + \log n)$ time using $O(P_d(n, m) + (n + m)/\log n)$ processors on M_d. Moreover, if a decomposition tree is given, the problem can be solved in $O(\log n)$ time using $O((n + m)/\log n)$ processors on an EREW PRAM.*

References

1. K. Abrahamson, N. Dadoun, D. G. Kirkpatrick, and T. Przytycka, A simple parallel tree contraction algorithm, *Journal of Algorithms*, 10:287–302, 1989.
2. H. J. Bandelt and H. M. Mulder, Distance-hereditary graphs, *Journal of Combinatorial Theory Series B*, 41(1):182–208, 1989.
3. A. Brandstädt and F. F. Dragan, A linear time algorithm for connected γ-domination and Steiner tree on distance-hereditary graphs, *Networks*, 31:177–182, 1998.
4. M. S. Chang, S. Y. Hsieh, and G. H. Chen, Dynamic programming on distance-hereditary graphs, *Proceedings of 7th International Symposium on Algorithms and Computation (ISAAC'97)*, LNCS 1350, pp. 344–353, 1997.
5. B. Courcelle, J. A. Makowsky, and U. Rotics, Linear time solvable optimization problems on graphs of bounded clique-width, *Theory of Computing Systems*, 33:125–150, 2000.
6. A. D'atri and M. Moscarini, Distance-hereditary graphs, steiner trees, and connected domination, *SIAM Journal on Computing*, 17(3):521–538, 1988.
7. F. F. Dragan, Dominating cliques in distance-hereditary graphs, *Algorithm Theory-SWAT'94-4th Scandinavian Workshop on Algorithm Theory, LNCS 824, Springer, Berlin*, pp. 370–381, 1994.
8. M. C. Golumbic, *Algorithmic Graph Theory and Perfect Graphs*, Academic press, New York, 1980.
9. M. C. Golumbic and U. Rotics, On the clique-width of perfect graph classes, *WG'99, LNCS 1665*, pp. 135–147, 1999.
10. P. L. Hammer and F. Maffray, Complete separable graphs, *Discrete Applied Mathematics*, 27(1):85–99, 1990.
11. E. Howorka, A characterization of distance-hereditary graphs, *Quarterly Journal of Mathematics (Oxford)*, 28(2):417–420, 1977.
12. S.-y. Hsieh, C. W. Ho, T.-s. Hsu, M. T. Ko, and G. H. Chen, Efficient parallel algorithms on distance-hereditary graphs, *Parallel Processing Letters*, 9(1):43–52, 1999.
13. S.-y. Hsieh, C. W. Ho, T.-s. Hsu, M. T. Ko, and G. H. Chen, Characterization of Efficiently Solvable Problems on Distance-Hereditary Graphs, *Proceedings of 9th International Symposium on Algorithms and Computation (ISAAC'98)*, LNCS 1533, pp. 257–266, 1998.
14. S.-y. Hsieh, C. W. Ho, T.-s. Hsu, M. T. Ko, and G. H. Chen, A faster implementation of a parallel tree contraction scheme and its application on distance-hereditary graphs, *Journal of Algorithms*, 35:50–81, 2000.
15. S.-y. Hsieh, Parallel decomposition of distance-hereditary graphs, *Proceedings of the 4th International ACPC Conference Including Special Tracks on Parallel Numerics (ParNum'99) and Parallel Computing in Image Processing, Video Processing, and Multimedia (ACPC'99)*, LNCS 1557, pp. 417–426, 1999.
16. R. W. Hung, S. C. Wu, and M. S. Chang, Hamiltonian cycle problem on distance-hereditary graphs, manuscript.
17. H. Müller and F. Nicolai, Polynomial time algorithms for Hamiltonian problems on bipartite distance-hereditary graphs, *Information Processing Letters*, 46:225–230, 1993.
18. Falk Nicolai, Hamiltonian problems on distance-hereditary graphs, Technique report, Gerhard-Mercator University, Germany, 1994.

Extending the Accommodating Function*

Joan Boyar, Lene M. Favrholdt, Kim S. Larsen, and Morten N. Nielsen

Department of Mathematics and Computer Science
University of Southern Denmark, Odense
http://www.imada.sdu.dk/~{joan,lenem,kslarsen,nyhave}

Abstract. The applicability of the accommodating function, a relatively new measure for the quality of on-line algorithms, is extended. If a limited amount n of some resource is available, the accommodating function $\mathcal{A}(\alpha)$ is the competitive ratio when input sequences are restricted to those for which the amount αn of resources suffices for an optimal off-line algorithm. The accommodating function was originally used only for $\alpha \geq 1$. We focus on $\alpha < 1$, observe that the function now appears interesting for a greater variety of problems, and use it to make new distinctions between known algorithms and to find new ones.

1 Introduction

The Accommodating Function. The accommodating function $\mathcal{A}$ is a new performance measure for on-line optimization problems with a limited amount n of some resource. Informally, $\mathcal{A}(\alpha)$ is the competitive ratio when input sequences are restricted to those for which an optimal off-line algorithm does not benefit from having more than the amount αn of resources. The accommodating function was recently defined in [6], and it was applied to various problems in [6] and [1], but only for $\alpha \geq 1$. In this paper, values of $\alpha < 1$ are considered for the first time. The accommodating function is formally defined in Section 2.

Background and Motivation. The original motivation for considering this type of restriction of request sequences is from the Seat Reservation Problem [5], the problem of assigning seats to passengers in a train on-line, in a "fair" manner, to maximize earnings. For the unit price version, the competitive ratio for this problem is $\Theta(\frac{1}{k})$, where k is the number of stations where the train stops. This very discouraging performance cannot occur, however, for realistic request sequences. Since the management is often able to judge approximately how many cars are necessary to accommodate all requests, it is more realistic to consider only request sequences that can be fully accommodated by an optimal off-line algorithm. Such sequences, corresponding to $\alpha = 1$, are called accommodating sequences. For the unit price problem, $\mathcal{A}(1) \geq \frac{1}{2}$ [5].

* Supported in part by the Future and Emerging Technologies program of the EU under contract number IST-1999-14186 (ALCOM-FT) and in part by the Danish Natural Science Research Council (SNF).

O.H. Ibarra and L. Zhang (Eds.): COCOON 2002, LNCS 2387, pp. 87–96, 2002.

The idea of restricting the adversary to only giving accommodating sequences carries over to any optimization problem with some limited resource, such as the seats in a train. Thus, for instance, it can be used on the k-Server Problem, where the servers constitute a limited resource since there are only k of them, or scheduling problems, where there is only a fixed number of machines.

In addition to giving rise to new interesting algorithmic and analytical problems, the accommodating function, compared to just one ratio, contains more information about the on-line algorithms. This information can be exploited in several ways. The shape of the function, for instance, can be used to warn against critical scenarios, where the performance of the on-line algorithm compared to the off-line can suddenly drop rapidly when fewer resources are available.

In [6] a variant of bin packing is investigated, for which the number of bins is fixed and the goal is to maximize the number of items packed. The bins have height k and the items are integer sized. It is shown that, in general, Worst-Fit has a strictly better competitive ratio than First-Fit, while, in the special case of accommodating sequences, First-Fit has a strictly better competitive ratio than Worst-Fit. In this case, the competitive ratio on accommodating sequences seems the more appropriate measure, since it is constant while, in the general case, the competitive ratio is $\Theta(\frac{1}{k})$, basically due to some sequences which seem very contrived. This shows that in addition to giving more realistic performance measures for some problems, the competitive ratio on accommodating sequences can be used to distinguish between algorithms, showing, not surprisingly, that the decision as to which algorithm should be used depends on what sort of request sequence is expected. The obvious question at this point was: Where is the cross-over point? When does First-Fit become better than Worst-Fit? This is another motivation for considering the accommodating function.

Main Contribution. In [6], the accommodating function was investigated for $\alpha \geq 1$. In this paper we extend the definition of the accommodating function to include $\alpha < 1$. As example problems, we first consider two maximization problems. For Unrestricted Bin Packing, we show upper and lower bounds for First-Fit and investigate a new variant of Unfair-First-Fit [1], called Unfair-First-Fit$_\alpha$, which turns out to be better than First-Fit for α close to 1. For Seat Reservation, we consider three deterministic algorithms, which asymptotically have the same competitive ratio on accommodating sequences, and show that these algorithms can be distinguished using the proposed extension of the accommodating function. Finally, we consider well known on-line minimization problems to emphasize the broad scope of the accommodating function for $\alpha < 1$.

The Accommodating Function for Two Maximization Problems. For the Seat Reservation Problem, considering the accommodating function for $\alpha < 1$ corresponds to the situation where the management has provided more cars than needed by an optimal off-line algorithm to accommodate all passenger requests. This seems to be a desirable situation, since the only way the train company can hope to accommodate all requests on-line is by having more resources than would be necessary if the problem was solved optimally off-line.

In this paper, we obtain positive and negative results on the accommodating function for the problem in general and for First-Fit and Worst-Fit in particular. The results for First-Fit and Worst-Fit are depicted in Figure 1 to the left of the line $\alpha = 1$. To the right of this line, general results from [6] are shown.

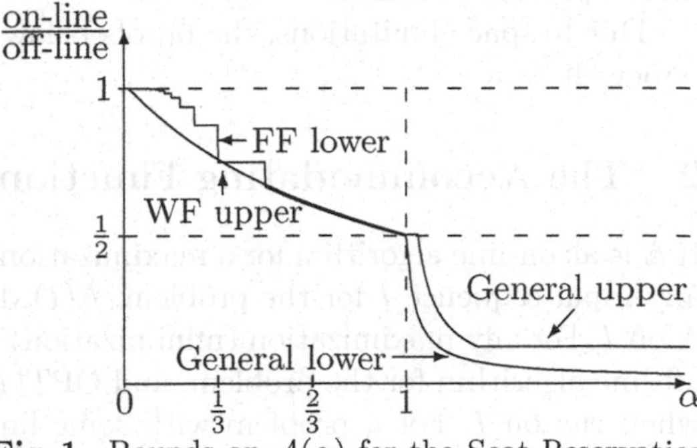

Fig. 1. Bounds on $\mathcal{A}(\alpha)$ for the Seat Reservation Problem.

The competitive ratio on accommodating sequences fails to distinguish between fair algorithms for the Seat Reservation Problem (they all have ratio $\frac{1}{2}$ in the limit). In contrast, we investigate the accommodating function at $\alpha = \frac{1}{3}$ for three different algorithms and discover that Worst-Fit is the worst there, First-Fit is better, and Kierstead-Trotter has competitive ratio 1 at $\alpha = \frac{1}{3}$.

We also consider another maximization problem, a variant of bin packing called Unrestricted Bin Packing. We obtain positive and negative results on the accommodating function for the problem in general and for First-Fit in particular. The results for First-Fit are depicted in Figure 2 to the left of the line $\alpha = 1$. The results to the right of the line are from [6]. Moreover, we show how positive results for accommodating sequences can be extended in a fairly general manner to better bounds on the competitive ratio for more restricted sets of input sequences. Thus, accommodating sequences play a more important role than originally anticipated. One can also see from the graphs in Figures 1 and 2 that accommodating sequences have unique properties; the shape of the curve seems to often change significantly at $\alpha = 1$. In addition, for Unrestricted Bin Packing, we investigate a new algorithm, Unfair-First-Fit$_\alpha$, showing that knowledge of α can be exploited to obtain a better performance ratio.

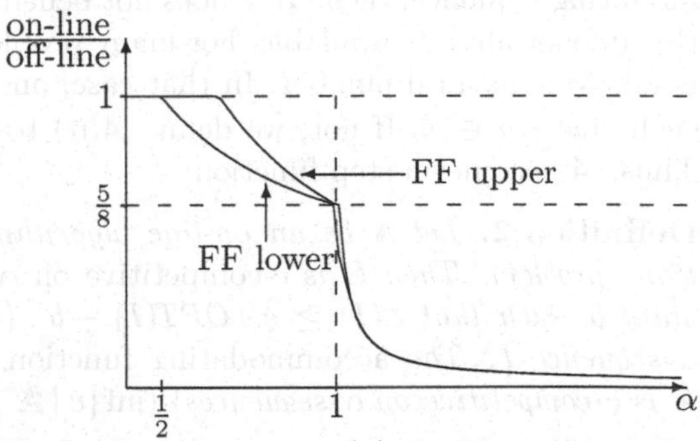

Fig. 2. Bounds on $\mathcal{A}_{\mathrm{FF}}(\alpha)$ for Unrestricted Bin Packing.

The Connection to Resource Augmentation. Resource augmentation is another technique which is used to give more realistic results when the standard competitive ratio seems too negative [12]. With resource augmentation analysis, the on-line algorithm is given more resources than the off-line algorithm, but the performance ratio is still the worst case over all request sequences. There is clearly a similarity between resource augmentation and the accommodating function with $\alpha < 1$. The similarities and differences are discussed in section 5, which includes examples showing that the accommodating function can give more optimistic results.

Due to space limitations, the proofs below have been eliminated. See the full paper [4].

2 The Accommodating Function

If $\mathbb{A}$ is an on-line algorithm for a maximization (minimization) problem, then, for any input sequence I for the problem, $\mathbb{A}(I)$ denotes the value (cost) of running $\mathbb{A}$ on I. For any maximization (minimization) problem, OPT denotes an optimal off-line algorithm for the problem, and OPT(I) denotes the value (cost) of OPT when run on I. For a problem with some limited resource, OPT_n denotes the value (cost) of OPT when the amount n of the limited resource is available.

Definition 1. *Let P be an on-line problem with a fixed amount n of resources. For any $\alpha > 0$, an input sequence I is said to be an α-sequence, if $OPT_{\alpha n}(I) = OPT_{n'}(I)$, for all $n' \geq \alpha n$. 1-sequences are also called* accommodating sequences.

If an input sequence is an α-sequence, then OPT does not benefit from having more than the amount αn of resources. Thus, if an input sequence is an accommodating sequence, then OPT does not benefit from having more resources than the amount already available. For many problems, the amount n of the resource is always a natural number. In that case, one should only consider values of α such that $\alpha n \in \mathbb{N}$. If not, we define $\mathcal{A}(\alpha)$ to have the same value as $\mathcal{A}(\frac{\lfloor \alpha n \rfloor}{n})$. Thus, $\mathcal{A}$ becomes a step function.

Definition 2. *Let $\mathbb{A}$ be an on-line algorithm for a maximization (minimization) problem. Then $\mathbb{A}$ is* c-competitive *on α-sequences, if there exists a constant b, such that $\mathbb{A}(I) \geq c \cdot OPT(I) - b$ $(\mathbb{A}(I) \leq c \cdot OPT(I) + b)$, for any α-sequence I. The* accommodating function *is defined as $\mathcal{A}_{\mathbb{A}}(\alpha) = \sup\{c \mid \mathbb{A}$ is c-competitive on α-sequences$\}$ $(\inf\{c \mid \mathbb{A}$ is c-competitive on α-sequences$\})$.*

3 Unrestricted Bin Packing

In this section we consider a maximization variant of Bin Packing, called Unrestricted Bin Packing, in which the objective is to maximize the number of items accepted within n bins all of the same size $k \in \mathbb{N}$. An input sequence consists of integer-sized items, and an α-sequence can be packed (by an optimal off-line algorithm) within αn bins.

We use the following notation for a given $\alpha \leq 1$. Assume that we have a numbering of the bins from 1 to n. The first αn bins, denoted B_A, are called *internal* bins, and the remaining $n - \alpha n$ bins, denoted B_X, are called *external* bins. Whenever we consider a fixed request sequence, we let A denote the set of items accepted within internal bins, and let X denote the set of items accepted within the external bins. The set R contains the remaining (rejected) items. The bounds on First-Fit found in this section are depicted in Figure 2.

Definition 3. *An algorithm for Unrestricted Bin Packing is called* fair *if it never rejects an item that it is able to pack.*

Theorem 1. *Any fair algorithm* $\mathbb{A}$ *for Unrestricted Bin Packing has* $\mathcal{A}_{\mathbb{A}}(\alpha) \geq \frac{1}{1+\alpha-\frac{1}{k}}$*, when* $\alpha \leq 1$.

This performance guarantee is asymptotically tight due to Worst-Fit's behavior.

The proof of Theorem 1 shows that fair algorithms will reject less than αn items. When considering First-Fit, a stronger fact can be shown, namely that αn is an upper bound on the number of items not packed in the internal bins, i.e., $|R| + |X| < \alpha n$. Similar results can be shown for a broader class of algorithms.

Definition 4. *Consider any* α*-sequence* I *with* $\alpha \leq 1$ *and an algorithm* $\mathbb{A}$*. Let* X' *be the set of items in* I *which* $\mathbb{A}$ *places as the first in some external bin, so* $X' \subset X$*. Define* $f(I)$ *to be the subsequence of* I *consisting of those requests in* $A \cup X' \cup R$*. An algorithm* $\mathbb{A}$ *is said to be* uniform *if, for any sequence* I*, when* $\mathbb{A}$ *processes* $f(I)$ *using* αn *bins (and any* α *the algorithm may use internally set to 1), it accepts exactly those items in* A.

Lemma 1. *For any fair, uniform algorithm* $\mathbb{A}$ *and any* α*-sequence, with* $\alpha \leq 1$, $|R| + |X'| < \alpha n$ *and whenever* $|R| > 0$, $|X'| = n - \alpha n$.

Note that, by Lemma 1, any fair, uniform algorithm is optimal when $\alpha \leq \frac{1}{2}$. This can be seen in the following way. For $\alpha \leq \frac{1}{2}$, $|R| + |X'| < \alpha n$ and $|X'| = n - \alpha n$ cannot both be fulfilled. Thus, for any fair, uniform algorithm, $|R| = 0$ when $\alpha \leq \frac{1}{2}$.

First-Fit is a fair, uniform algorithm, as are other algorithms, such as Best-Fit. For such algorithms, the theorem below extends good performance guarantees obtained on accommodating sequences to results for the accommodating function for $\alpha < 1$. The theorem can also be applied to some algorithms which are not fair.

Theorem 2. *Suppose the following four conditions hold for an on-line algorithm* $\mathbb{A}$*: (1)* $\mathcal{A}_{\mathbb{A}}(1) \geq \beta$*, (2)* $\mathbb{A}$ *is uniform, (3)* $|R| + |X'| \leq \alpha n$*, (4) whenever* $|R| > 0$, $|X'| = n - \alpha n$*. Then for* $\alpha \leq 1$*, we have* $\mathcal{A}_{\mathbb{A}}(\alpha) \geq \min\left(\frac{1-\alpha-\beta+2\alpha\beta}{\alpha}, 1\right)$.

The first corollary to this theorem concerns the fair algorithms First-Fit and Best-Fit.

Corollary 1. *For* $\alpha \leq 1$, $\mathcal{A}_{FF}(\alpha) \geq \min\left(\frac{3+2\alpha}{8\alpha}, 1\right)$ *and* $\mathcal{A}_{BF}(\alpha) \geq \min\left(\frac{3+2\alpha}{8\alpha}, 1\right)$.

The second corollary to the theorem above concerns a slight variation of the algorithm Unfair-First-Fit [1]. Unfair-First-Fit behaves just like First-Fit unless the given item has size larger than $\frac{k}{2}$. In this case, the item is rejected on purpose, if the number of currently accepted items is at least $\frac{2}{3}$ of the number of items in the entire sequence seen so far. The new algorithm given in Figure 3, called Unfair-First-Fit$_\alpha$ (UFF$_\alpha$), assumes that α is known in advance. Using this knowledge, the algorithm divides the bins into B_A and B_X. Note that the ratio considered in the if-statement of the algorithm is the number of items in the internal bins, compared to all items given. If this ratio is at least $\frac{2}{3}$, the item is rejected if it does not fit in an external bin.

```
Input: S = ⟨o_1, o_2, ..., o_n⟩
while S ≠ ⟨⟩
    o := hd(S); S := tail(S)
    if size(o) ≤ k/2 or |A|/(|A|+|X|+|R|+1) < 2/3
        Try to pack o using First-Fit
    else
        Try to pack o in B_X using First-Fit
    Update A, X, R accordingly
```

Fig. 3. The algorithm Unfair-First-Fit$_\alpha$.

Corollary 2. *$\mathcal{A}_{UFF_\alpha}(\alpha) \geq \min\left(\frac{1+\alpha}{3\alpha} - \epsilon, 1\right)$, for $\alpha \leq 1$ and $\alpha n \geq 9$, where $\epsilon = \frac{4\alpha-2}{(4n+1)\alpha} < \frac{1}{n}$.*

The theorem below extends a result in [6] and it shows that any (even randomized) on-line algorithm $\mathbb{A}$ for Unrestricted Bin Packing has $\mathcal{A}_{\mathbb{A}}(\alpha) < 1$, for $\alpha > \frac{4}{5}$.

Theorem 3. *Any on-line algorithm $\mathbb{A}$ for Unrestricted Bin Packing has $\mathcal{A}_{\mathbb{A}}(\alpha) \leq \frac{2\alpha+4}{7\alpha} + O(\frac{1}{n})$, when $\frac{4}{5} \leq \alpha \leq 1$ and $k \geq 5$.*

The following theorem extends a hardness result for First-Fit on accommodating sequences in [1] and shows that, when n is sufficiently large, $\mathcal{A}_{\mathrm{UFF}}(\alpha) \geq \mathcal{A}_{\mathrm{FF}}(\alpha)$, for $\frac{7+\sqrt{85}}{18} < \alpha \leq 1$.

Theorem 4. *For Unrestricted Bin Packing, the accommodating function for First-Fit is at most $\mathcal{A}_{FF}(\alpha) \leq \frac{5}{9\alpha-1} + O(\frac{1}{\sqrt{n}})$, when $\frac{2}{3} + \frac{1}{3n} \leq \alpha \leq 1$.*

4 Unit Price Seat Reservation

A train with n seats travels from a start station to an end station, stopping at $k \geq 2$ stations, including the first and last. Reservations can be made for any trip from a station s to a station t. The passenger is given a single seat number when the ticket is purchased, which can be any time before departure. For political reasons, the problem must be solved in a *fair* manner, i.e., the ticket agent may not refuse a passenger if it is possible to accommodate him when he attempts to make his reservation. For this problem an α-sequence can be packed by an optimal off-line algorithm using αn seats. The algorithms (ticket agents) attempt to maximize income, i.e., the sum of the prices of the tickets sold. In this paper, we consider only the pricing policy in which all tickets have the same price, the *unit price problem.*

The performance guarantee for fair algorithms on accommodating sequences found in [5] can be extended to a general performance guarantee for $\alpha \leq 1$.

Theorem 5. *For $\alpha \leq 1$, any fair algorithm for Unit Price Seat Reservation is $\frac{1}{1+\alpha}$-competitive on α-sequences.*

This result is asymptotically tight due to Worst-Fit's behavior:

Theorem 6. $\mathcal{A}_{WF}(\alpha) \leq \frac{1}{1+\alpha-\frac{1}{k-1}-\frac{1}{n}}$, *for $\alpha \leq 1$.*

It was shown in [5] that First-Fit's competitive ratio on accommodating sequences is not strictly better than $\frac{1}{2}$, the performance guarantee for any fair algorithm. In contrast, for $\alpha \leq \frac{1}{2}$, it is possible to prove that First-Fit's performance is better than Worst-Fit's.

Theorem 7. $\mathcal{A}_{FF}(\alpha) \geq \frac{2^{l-1}(1-l\alpha)+(2^l-1)\alpha}{2^{l-1}(1-l\alpha)+(2^l)\alpha} \geq 1 - \frac{1}{2^l}$, *where $l = \lfloor \frac{1}{\alpha} \rfloor$ and $\alpha \leq 1$.*

This performance guarantee is not tight. Kierstead and Qin [13] have shown that First-Fit colors every interval graph using at most 25.72 times as many colors as necessary, so the competitive ratio on α-sequences is 1 when $\alpha \leq \frac{1}{25.72} \approx 0.039$.

The following theorem shows that for any fair algorithm $\mathbb{A}$, for k sufficiently large, $\mathcal{A}_{\mathbb{A}}(\alpha) < 1$, for $\alpha > \frac{4}{5}$. The proof is based on Theorem 3.1 in [3], which is based on a proof from [5].

Theorem 8. *For $\alpha \leq 1$ and $k \geq 9$, no fair on-line algorithm for Unit Price Seat Reservation is more than $(\frac{8-\alpha}{9\alpha} + O(\frac{1}{k}))$-competitive, even on α-sequences.*

We consider a class of fair algorithms called Any-Fit which will use an empty seat, only if there is not enough space on partially used seats. Any-Fit includes both First-Fit and Best-Fit.

Theorem 9. *For an Any-Fit algorithm $\mathbb{A}$, $\mathcal{A}_{\mathbb{A}}(\alpha) \leq \frac{1}{3\alpha-1} + O(\frac{1}{k})$ for $\frac{1}{3} < \alpha \leq 1$.*

Kierstead and Trotter's algorithm [14] has a competitive ratio of 1 on $\frac{1}{3}$-sequences. Their algorithm solves the problem of minimizing the number of seats (colors) used; it is undefined for values of $\alpha > \frac{1}{3}$ when applied to the maximization problem where the number of seats is limited. However, their algorithm can obviously be extended in many ways so that it solves this maximization problem.

Chrobak and Ślusarek [8] have shown that there exists a sequence where First-Fit uses more than 4.4 times as many seats (colors) as OPT, and thus is not 1-competitive on $\frac{1}{3}$-sequences. Thus, Kierstead and Trotter's algorithm has a better competitive ratio on $\frac{1}{3}$-sequences than First-Fit.

Kierstead and Trotter's algorithm is not, however, better than First-Fit on 1-sequences, since it has been shown [5] that no fair algorithm has a competitive ratio on accommodating sequences which is strictly better than $\frac{1}{2}$ asymptotically.

In conclusion, we have that the competitive ratio on α-sequences, where $\alpha < 1$, can be useful in distinguishing between algorithms.

Theorem 10. *Kierstead and Trotter's algorithm, First-Fit, and Worst-Fit, all have asymptotic competitive ratio $\frac{1}{2}$ on accommodating sequences, but have different competitive ratios on $\frac{1}{3}$-sequences.*

5 Comparison with Resource Augmentation

The concept of the accommodating function should not be confused with resource augmentation introduced in [12]. Resource augmentation analysis gives the on-line algorithm more resources than the optimal off-line algorithm that it is compared to, but there is no restriction on the input sequences.

Note 1. In the resource augmentation setting, the on-line algorithm has the amount m of resources and the optimal off-line algorithm has the amount $n \leq m$ of resources. Positive results (upper bounds for minimization problems and lower bounds for maximization problems) obtained in this setting are also valid for the case where all input sequences are $\frac{n}{m}$-sequences. This is because, even though the optimal off-line algorithm in our setting has the same amount of resources as the on-line algorithm, the result it obtains cannot be improved with extra resources beyond n when considering $\frac{n}{m}$-sequences.

The contrapositive of the observation above gives that when considering negative results, it is the other way around: negative results for the accommodating function carry over to the resource augmentation setting.

Resource augmentation does not always give as realistic or optimistic results as the accommodating function for $\alpha < 1$. In order to obtain the same results, one needs to restrict to accommodating sequences while also doing resource augmentation. To see that resource augmentation alone can be insufficient, consider the algorithm First-Fit for the Seat Reservation Problem. Theorem 7 gives a lower bound on First-Fit's accommodating function for $\alpha \leq 1$: $\mathcal{A}_{\mathrm{FF}}(\alpha) \geq 1 - \frac{1}{2^l}$, where $l = \lfloor \frac{1}{\alpha} \rfloor$. In contrast, resource augmentation cannot give a constant lower bound on First-Fit's accommodating function.

Theorem 11. *For $\alpha > \frac{2}{n}$, the competitive ratio of First-Fit for the Seat Reservation Problem is at most $\frac{1+\alpha}{(\alpha - \frac{2}{n})(k-1)}$, when First-Fit has at most $\frac{1}{\alpha}$ times as many seats as OPT.*

For Unrestricted Bin Packing, one can also show that resource augmentation analysis alone gives results which are much more pessimistic than the corresponding accommodating function results. Recall that the performance guarantee from Corollary 1 for First-Fit for $\alpha \leq 1$ is $\mathcal{A}_{\mathrm{FF}}(\alpha) \geq \frac{3+2\alpha}{8\alpha}$.

Theorem 12. *For Unrestricted Bin Packing, the competitive ratio of First-Fit is at most $\frac{1}{\alpha k}$, when First-Fit has at most $\frac{1}{\alpha}$ times as many bins as OPT.*

5.1 Paging

We consider the Paging Problem in the page fault model, i.e., the algorithm maintains a fast memory (cache) consisting of k pages of memory and the input is a sequence of page requests. If a page in cache is requested, no cost is incurred; otherwise the requested page must be transferred from the slow memory at a cost of 1 and another page must be evicted from the cache. The goal is to choose an eviction strategy which minimizes cost. Before the first page request is served, the cache is empty. In this problem, the limited resource is the cache.

When an optimal off-line algorithm serves an α-sequence, it will have the same cost for every cache size $k' \geq \alpha k$. To see why such a sequence can consist of more than αk different pages, consider the following example: For $k = 6$, the following sequence is a $\frac{1}{2}$-sequence, $S = \langle 1, 2, 3, 4, 5, 6, 7, 1, 2, 8, 9, 10, 1, 2, 11, 12 \rangle$. Keeping the two pages 1 and 2 in the cache will give a cost of 12 with a cache of size k' for all $k' \geq 3$.

Using Note 1 and a result from [16], we find that FIFO (First-In/First-Out) and LRU (Least-Recently-Used) are $\frac{k}{(1-\alpha)k+1}$-competitive. Sleator and Tarjan [16] have shown that their performance guarantee is tight, and a very similar proof shows that it is also tight for α-sequences.

Theorem 13. *Any deterministic Paging algorithm* $\mathbb{A}$ *has* $\mathcal{A}_{\mathbb{A}}(\alpha) \geq \frac{k}{(1-\alpha)k+1}$, *when* $\alpha \leq 1$.

Using Note 1 and a result from [17], we find that when $\alpha < 1$, the randomized algorithm MARK is 2-competitive when $\frac{1}{1-\alpha} < e$ and $2(\ln \frac{1}{1-\alpha} - \ln \ln \frac{1}{1-\alpha} + \frac{1}{e-1})$-competitive when $\frac{1}{1-\alpha} \geq e$. Young's hardness result [17] for any randomized algorithm can be used directly for the competitive ratio on α-sequences for $\alpha < 1$, since his sequences are α-sequences. Thus, any randomized algorithm has a competitive ratio of at least $\ln \frac{1}{1-\alpha} - \ln \ln \frac{1}{1-\alpha} - \frac{3}{k(1-\alpha)}$ when $\frac{1}{1-\alpha} \geq e$, so MARK is within a factor of approximately 2 of optimal.

5.2 Other Problems

The k-server problem and machine scheduling minimizing makespan are considered in the full paper [4]. Resource augmentation results for these problems are used to give accommodating function results.

6 Conclusion

The accommodating function for $\alpha \leq 1$ seems to be interesting for a variety of on-line problems, possibly for a greater variety than when $\alpha > 1$. For example, the accommodating function for deterministic algorithms for the Paging Problem has a very uninteresting shape for $\alpha > 1$; the value is constant at k, while for $\alpha \leq 1$, $\mathcal{A}(\alpha) = \frac{k}{(1-\alpha)k+1}$.

The study of the accommodating function in general and in particular for $\alpha \leq 1$ has given rise to new algorithms, Unfair-First-Fit and Unfair-First-Fit$_\alpha$.

In addition, the Seat Reservation Problem demonstrates the utility of the accommodating function with $\alpha < 1$ in distinguishing between different algorithms. Three algorithms, which all have competitive ratio close to $\frac{1}{2}$ on accommodating sequences, have different competitive ratios on $\frac{1}{3}$-sequences.

The proofs of positive results seem, in general, to be more interesting than the proofs of negative results. This appears less true for those problems where positive results concerning resource augmentation can be used directly, giving positive results for the accommodating function with $\alpha \leq 1$. However, the original proofs in the resource augmentation setting tend to be interesting, and those results become even more interesting given their application to this setting where the set of request sequences is limited. On the other hand, the accommodating function with $\alpha \leq 1$ sometimes gives much more useful information than resource augmentation. Examples of this were given for the Seat Reservation Problem and Unrestricted Bin Packing, using First-Fit.

References

1. Y. Azar, J. Boyar, L. Epstein, L. M. Favrholdt, K. S. Larsen, and M. N. Nielsen. Fair versus Unrestricted Bin Packing. *Algorithmica*. To appear. Preliminary version in SWAT 2000.
2. Y. Azar, L. Epstein, and R. van Stee. Resource Augmentation in Load Balancing. In *SWAT 2000*, volume 1851 of *LNCS*, pages 189–199, 2000.
3. E. Bach, J. Boyar, L. Epstein, L. M. Favrholdt, T. Jiang, K. S. Larsen, G.-H. Lin, and R. van Stee. Tight Bounds on the Competitive Ratio on Accommodating Sequences for the Seat Reservation Problem. *Journal of Scheduling*. To appear. Preliminary version in COCOON 2000.
4. J. Boyar, L. M. Favrholdt, K. S. Larsen, and M. N. Nielsen. Extending the Accommodating Function. Technical report PP-2002-02, Department of Mathematics and Computer Science, University of Southern Denmark, Odense, 2002.
5. J. Boyar and K. S. Larsen. The Seat Reservation Problem. *Algorithmica*, 25:403–417, 1999.
6. Joan Boyar, Kim S. Larsen, and Morten N. Nielsen. The Accommodating Function: a generalization of the competitive ratio. *SIAM Journal on Computing*, 31(1):233–258, 2001.
7. M. Brehop, E. Torng, and P. Uthaisombut. Applying Extra Resource Analysis to Load Balancing. *Journal of Scheduling*, 3:273–288, 2000.
8. M. Chrobak and M. Ślusarek. On Some Packing Problems Related to Dynamic Storage Allocation. *RAIRO Informatique Théoretique et Applications*, 22:487–499, 1988.
9. E. Koutsoupias. Weak Adversaries for the k-Server Problem. In *FOCS*, pages 444–449, 1999.
10. R. L. Graham. Bounds for Certain Multiprocessing Anomalies. *Bell Systems Technical Journal*, 45:1563–1581, 1966.
11. T. R. Jensen and B. Toft. *Graph Coloring Problems*. John Wiley & Sons, 1995.
12. B. Kalyanasundaram and K. Pruhs. Speed is as powerful as clairvoyance. In *FOCS*, pages 214–221, 1995.
13. H. A. Kierstead and J. Qin. Coloring Interval Graphs with First-Fit. *Discrete Mathematics*, 144:47–57, 1995.
14. H. A. Kierstead and W. T. Trotter. An Extremal Problem in Recursive Combinatorics. *Congressus Numerantium*, 33:143–153, 1981.
15. M. S. Manasse, L. A. McGeoch, and D. D. Sleator. Competitive Algorithms for Server Problems. *Journal of Algorithms*, 11(2):208–230, June 1990.
16. D. D. Sleator and R. E. Tarjan. Amortized Efficiency of List Update and Paging Rules. *Communications of the ACM*, 28(2):202–208, 1985.
17. N. Young. On-Line Caching as Cache Size Varies. In *SODA*, pages 241–250, 1991.

Inverse Parametric Sequence Alignment*

Fangting Sun[1], David Fernández-Baca[1], and Wei Yu[2]

[1] Department of Computer Science, Iowa State University, Ames, IA 50010
{ftsun,fernande}@cs.iastate.edu

[2] Department of Computer Science, Washington University, St. Louis, MO 63130
weiyu@ccrc.wustl.edu

Abstract. We consider the inverse parametric sequence alignment problem, where a sequence alignment is given and the task is to determine parameter values such that the given alignment is optimal at that parameter setting. We describe a $O(mn \log n)$-time algorithm for inverse global alignment without gap penalty and a $O(mn \log m)$ time algorithm for global alignment with gap penalty, where m, n $(n \leq m)$ are the lengths of input strings. We then discuss algorithms for local alignment.

1 Introduction

Finding the best alignment of two DNA, RNA or amino acid sequences has become a standard technique for determining similarity between biological sequences. There are hundreds of papers written on this topic and its applications to biology. The review [1] gives relevant references.

Given two sequences S and T of lengths n and m, $n \leq m$, an alignment is obtained by inserting special *space* characters into the two sequences in such a way as to build sequences S' and T' of equal length, denoted by $\mathcal{A} = (S', T')$. A *match* is a position where S' and T' have the same characters. A *mismatch* is a position in which S' and T' have different characters, neither of which is a space. An *indel* is a position in which one of S' and T' has a space. A *gap* is a sequence of one or more consecutive spaces in S' and T'.

An alignment $\mathcal{A}$ can be characterized by its number of matches, mismatches, indels and gaps, denoted w, x, y, z, respectively. In scoring an alignment matches are rewarded, while mismatches, indels and gaps are penalized. Let α, β and γ denote the mismatch, indel and gap penalties. Then the *score* of $\mathcal{A}$ is

$$score_{\mathcal{A}} = w - \alpha x - \beta y - \gamma z$$

The case where the weight of the matches is a parameter is ignored since we can divide all the parameters by this value and reduce it to the above case. The *optimal alignment problem* is to find a maximum-score alignment $\mathcal{A}$ between two strings. For fixed weights, this problem can be solved in $O(mn)$ time [11]. The problem we have just defined is often referred to as *global* alignment, so as to distinguish it from its *local* version, which is defined later (see Section 4).

* Research partially supported by grant CCR-9988348 from the National Science Foundation.

O.H. Ibarra and L. Zhang (Eds.): COCOON 2002, LNCS 2387, pp. 97–106, 2002.

The *parametric sequence alignment* problem is to compute optimal alignments for two fixed sequences as a function of varying penalties. The value of an alignment is a linear function of the parameters; thus the parameter space can be partitioned into *optimal regions* such that in every region one alignment is optimal throughout and the regions are maximal for this property. This approach was proposed by Fitch and Smith [4] at first. Later, both mathematical formulations and algorithms for *parametric sequence alignment* were obtained by Gusfield et al. [7,8]. Additional work is found in [3,12,13,14].

In *inverse parametric optimization* [2] one is given a parametric optimization problem and a desired optimal solution and the task is to determine parameter settings such that the given solution is optimal for those values. The *inverse parametric sequence alignment* problem is to find parameter values such that reference alignment is optimal for those values or, if no such settings exist, find a parameter setting minimizing the numerical difference between the score of the optimal alignment and the score of the reference alignment. These parameter values define an *inverse optimal point* on the parameter space. Inverse parametric computation is useful for deducing parameter settings where the optimal alignment is likely to reconstruct correct alignments that have been determined by other methods.

One way to locate the "correct" parameter settings is to first construct the entire decomposition of the parameter space and then choose the correct values. Alternatively, one can try to find the parameter settings directly. This can be done by gradient descent [8], although it is not clear how to obtain bounds on the worst-case performance of this method. Megiddo's method of *parametric search* [9,10] can be used instead, leading to a $O(m^2n^2)$ method for the case where only one parameter is varied. While powerful, Megiddo's method has the drawback that it leads to complex algorithms. Improvements in the running time are possible by relying on the existence of a *parallel* algorithm for the problem, but this only complicates the results further. Here we give an approach that is much simpler than Megiddo's method and exploits the integer nature of the scoring of sequence alignments.

Since the optimal regions are bounded by the intersection of hyperplanes, all regions are convex polygons [4,7,8,3]. Hence, the inverse optimal parameter setting(s) must occur at a single vertex (intersection point of three or more optimal regions), at a single edge (intersection line between two optimal regions), or at a single complete polygon of the polygonal decomposition of the parameter space. The main idea of our algorithms is to find the inverse-optimal point in the parameter space using binary search. Our main contribution is a proof that this simple algorithm converges quickly.

The rest of this paper is organized as follows. Section 2 gives a $O(mn \log n)$ algorithm for global alignment without gap penalty. A $O(mn \log m)$ algorithm for global alignment with gap penalty is described in Section 3. Further results are discussed in Section 4.

2 Global Alignment without Gap Penalty

In this section we consider global alignment where the gap penalties are ignored ($\gamma = 0$). Then, the score function is $score = w - \alpha x - \beta y$. Given a reference alignment $\mathcal{A}_0$ with w_0 matches, x_0 mismatches and y_0 indels, we must find an inverse-optimal point for $\mathcal{A}_0$ in the α, β plane. We need some preliminary results.

Theorem 1 (Gusfield et al. [7]). *Any line forming a boundary between two regions is of the form $\beta = c + (c + 0.5)\alpha$, for some $c > -1/2$.*

Corollary 1. *Suppose (α_0, β_0) is an inverse-optimal point for reference alignment $\mathcal{A}_0$ in the α, β space. Then all points on the line that goes through $(-1, -1/2)$ and (α_0, β_0) are inverse-optimal for $\mathcal{A}_0$.*

Lemma 1 ([7]). *The positive β-axis intersects all the region boundaries. Let $\mathcal{A}_1, \mathcal{A}_2, \ldots, \mathcal{A}_k$ be the optimal alignments encountered by β axis in order of increasing β-value. Then $y_{i+1} < y_i$ for all $\mathcal{A}_i (i < k)$.*

Let $\mathcal{A}_i, \mathcal{A}_j$ be the optimal alignments in two neighboring optimal regions encountered by β axis, with score $w_i - \alpha x_i - \beta y_i$ and $w_j - \alpha x_j - \beta y_j$. Then the equation of the boundary line between the regions is:

$$\beta = \frac{w_i - w_j}{y_i - y_j} + \frac{x_j - x_i}{y_i - y_j}\alpha \tag{1}$$

A *breakpoint* along any given line is the point where the line moves between two adjacent optimal regions.

Lemma 2. *The length of the interval between any two successive breakpoints along the β-axis is greater than $1/n^2$.*

Proof. According to Equation (1), the boundary line of two neighboring optimal regions where $\mathcal{A}_i, \mathcal{A}_j$ are optimal respectively intersects the β axis at $(0, \frac{w_i - w_j}{y_i - y_j})$.

Let $\mathcal{A}_i, \mathcal{A}_j, \mathcal{A}_k$ $(i < j < k)$ be the optimal alignments in three consecutive optimal regions when going along the β axis and let $\Delta w_1 = w_j - w_k, \Delta w_2 = w_i - w_j, \Delta y_1 = y_j - y_k, \Delta y_2 = y_i - y_j$. Then the interval between two breakpoints on the β axis is:

$$\Delta\beta = \frac{w_j - w_k}{y_j - y_k} - \frac{w_i - w_j}{y_i - y_j} = \frac{\Delta w_1 \Delta y_2 - \Delta w_2 \Delta y_1}{\Delta y_1 \Delta y_2}$$

Since $\Delta w_1, \Delta w_2, \Delta y_1, \Delta y_2$ are all integers and $\Delta\beta > 0$, then $\Delta w_1 \Delta y_2 - \Delta w_2 \Delta y_1 \geq 1$. Notice that $m - n \leq y \leq m + n$, then $\Delta y_1 + \Delta y_2 = y_i - y_k \leq (m + n) - (m - n) \leq 2n$. Therefore $\Delta y_1 \Delta y_2 \leq ((\Delta y_1 + \Delta y_2)/2)^2 \leq n^2$. It follows that $\Delta\beta > 1/n^2$. □

The main idea of our algorithm is to use binary search on the β axis. The details are given below:

Algorithm 1

1. $high = n; low = 0;$
2. **while** $((high - low) > 1/n^2)$
3. $\quad mid = low + (high - low)/2;$
4. $\quad$ compute the optimal global alignment $\mathcal{A}(w, x, y)$ at point $(0, mid)$;
5. $\quad$ **if** $(y == y_0)$ **then**
6. $\qquad$ **return** the line passing through $(-1, -\frac{1}{2})$ and $(0, mid)$;
7. $\quad$ **else**
8. $\qquad$ **if** $(y > y_0)$ **then** $low = mid$ **else** $high = mid$ **end if**;
9. $\quad$ **end if**
10. **end while**
11. compute optimal global alignments $\mathcal{A}_{high}$, $\mathcal{A}_{low}$ for points $(0, high)$, $(0, low)$;
12. $mid = low + (high - low)/2;$
13. **if** ($\mathcal{A}_{high}$ is the same as $\mathcal{A}_{low}$) **then**
14. $\quad$ **return** the line passing through $(-1, -\frac{1}{2})$ and $(0, mid)$;
15. **else**
16. $\quad$ compute β_0 such that $w_{high} - \beta_0 y_{high} = w_{low} - \beta_0 y_{low}$;
17. $\quad$ compute optimal alignment $\mathcal{A}$ for point $(0, \beta_0)$;
18. $\quad$ **if** $(w_0 - \beta_0 y_0 == w - \beta_0 y)$ **then**
19. $\qquad$ **return** the line passing through $(-1, -\frac{1}{2})$ and $(0, \beta_0)$;
20. $\quad$ **else**
21. $\qquad$ **return** the line passing through $(-1, -\frac{1}{2})$ and $(0, mid)$;
22. $\quad$ **end if**
23. **end if**

Theorem 2. *Algorithm 1 correctly solves the inverse parametric alignment problem for global alignment without gaps in $O(mn \log n)$ time.*

Proof. From equation (1), all breakpoints on the β axis lie below $(0, n)$, so it is correct to restrict the search space to the portion on the β-axis between $(0, 0)$ to $(0, n)$. Lemma 1 guarantees that binary search works for this problem, since the algorithm can decide to go up or down along the β axis according to the number of indels.

If the algorithm finds a point $(0, \beta_0)$ such that the optimal alignment $\mathcal{A}$ for that point has the same number of indels as the reference alignment, then $(0, \beta_0)$ is either inverse-optimal $(w = w_0)$ or it is approximately inverse-optimal $(w \neq w_0)$. Following Corollary 1, the algorithm returns a line.

Lemma 2 shows that when the length of the search interval is smaller than $1/n^2$, it cannot contain a complete optimal region. It includes either part of one optimal region ($\mathcal{A}_{high}$ equals $\mathcal{A}_{low}$) or one breakpoint. In the first case, all points in the remaining search space are approximately inverse-optimal. In the second case, the breakpoint is inverse-optimal or all points in the remaining search space are approximately inverse-optimal. Thus Algorithm 1 gives the correct answer.

Steps 1, 3, 5-9, 12 need $O(1)$ time; step 11 and 13-22 need $O(mn)$ time; step 4 needs $O(mn)$ time, and the while statement can loop at most $3 \log n$ times. Therefore, the total time is $3 \log n \cdot O(mn) = O(mn \log n)$. □

3 Global Alignment with Gap Penalty

In this section we solve the inverse global alignment problem with gap penalty. There are now three parameters to consider, α, β and γ. Given a reference alignment $\mathcal{A}_0$ with w_0 matches, x_0 mismatches, y_0 indels and z_0 gaps, we need to find a point on the α, β, γ space where $\mathcal{A}_0$ is optimal or approximately optimal.

First, let us describe the boundary lines of the optimal regions in the α, β, γ and an important property about the centroid of the convex polyhedra.

Theorem 3 (Gusfield et al. [7]). *Any line forming a boundary between three or more regions is of the form $\beta = c + (c + 1/2)\alpha, \gamma = d + d\alpha$.*

Corollary 2. *All region boundaries intersect with either the positive β, γ coordinate plane or with the positive α, γ coordinate plane.*

Theorem 4 (Grunbaum [5]). *Let S be a convex body of volume 1 in R^d. Let v_1 be the larger of the two volumes in a division of S by a hyperplane through its centroid. Then $v_1 \leq 1 - (\frac{d}{d+1})^d$.*

Finally, as in Equation (1), the boundary line between optimal regions on the β, γ plane associated with alignments $\mathcal{A}_i$ and $\mathcal{A}_j$ has the form

$$\beta = \frac{w_i - w_j}{y_i - y_j} - \frac{z_i - z_j}{y_i - y_j}\gamma \tag{2}$$

A *vertex* is the intersection point of three or more optimal regions. Suppose vertex v is intersection point of optimal regions whose optimal alignments are $\mathcal{A}_1, \mathcal{A}_2, \mathcal{A}_3$. Let $\Delta w_1 = w_1 - w_2$, $\Delta w_2 = w_2 - w_3$, etc. Then $v = (\beta_v, \gamma_v)$, where

$$\beta_v = \frac{\Delta w_1 \Delta z_2 - \Delta w_2 \Delta z_1}{\Delta y_1 \Delta z_2 - \Delta y_2 \Delta z_1} \quad \text{and} \quad \gamma_v = \frac{\Delta w_1 \Delta y_2 - \Delta w_2 \Delta y_1}{\Delta z_1 \Delta y_2 - \Delta z_2 \Delta y_1} \tag{3}$$

According to Corollary 2, if there exists inverse-optimal point(s) in the search space, then there exists inverse-optimal point(s) on either the positive β, γ or positive α, γ coordinate plane. Thus we can search on the β, γ ($\alpha = 0$) coordinate plane first. If an inverse-optimal point is found on the β, γ coordinate plane, algorithm terminates; otherwise, continue to search on the α, γ coordinate plane. If an inverse-optimal point is found on the α, γ coordinate plane, then return it. If there is no inverse-optimal point on the α, γ coordinate plane either, then return an approximate inverse-optimal point.

The algorithm uses the following idea to reduce the search space. Let $v = (0, \beta_v, \gamma_v)$ be a point in the current search space on the β, γ plane and let $\mathcal{A}_v$ be the optimal alignment at v, with w_v matches, x_v mismatches, y_v indels and z_v gaps. If reference alignment $\mathcal{A}_0$ is optimal at v, then v is an inverse-optimal point. Otherwise, suppose $\mathcal{A}_0$ is optimal at point $(0, \beta, \gamma)$. By the optimality of $\mathcal{A}_0$ and $\mathcal{A}_v$, it follows that:

$$w_0 - \beta y_0 - \gamma z_0 \geq w_v - \beta y_v - \gamma z_v \quad \text{and} \quad w_v - \beta_v y_v - \gamma_v z_v > w_0 - \beta_v y_0 - \gamma_v z_0.$$

Therefore,

$$(y_v - y_0)\beta + (z_v - z_0)\gamma > (y_v - y_0)\beta_v + (z_v - z_0)\gamma_v \quad (4)$$

The boundary line l of the halfplane defined by inequality (4) passes through v and divides the remaining search space into two regions. The region whose points do not satisfy inequality (4) can be discarded, thereby reducing the search space. Line l becomes a new boundary line for the remaining search space; we say that this boundary line is *defined by* point v and alignment $\mathcal{A}_v$. According to Theorem 4, if the *centroid* of the present search space is selected as point v, the search space will reduce by a factor of at least 4/9.

If $\mathcal{A}_0$ is not an optimal alignment at any point on the β, γ plane, then suppose $(0, \beta, \gamma)$ is an approximately inverse-optimal point that minimizes the numerical difference between the score of the optimal alignment and the score of $\mathcal{A}_0$, and that $\hat{\mathcal{A}}$ is optimal at that point. Thus, we have: $(w_v - \beta_v y_v - \gamma_v z_v) - (w_0 - \beta_v y_0 - \gamma_v z_0) \geq (\hat{w} - \beta\hat{y} - \gamma\hat{z}) - (w_0 - \beta y_0 - \gamma z_0)$ and $(\hat{w} - \beta\hat{y} - \gamma\hat{z}) \geq (w_v - \beta y_v - \gamma z_v)$. Therefore $(y_v - y_0)\beta + (z_v - z_0)\gamma \geq (y_v - y_0)\beta_v + (z_v - z_0)\gamma_v$, which is the same boundary line as inequality (4).

Using the centroid as described above, we repeatedly reduce the search region until its area is smaller than $\frac{1}{2m^7}$. By Lemma 3, which is proved later, when the region is smaller than this size, it cannot include a complete optimal region.

After the binary search terminates, if there exists an inverse-optimal point, there must exist an inverse-optimal *vertex* in the remaining search space. From Lemma 5, proved later, if there exists inverse-optimal vertex (β^*, γ^*) in the remaining search space, there exist two distinct boundary lines l_1 and l_2, defined by $\mathcal{A}_1$ and by $\mathcal{A}_2$, such that $\mathcal{A}_1, \mathcal{A}_2$ are optimal at (β^*, γ^*); that is, (β^*, γ^*) is the intersection of the scores of $\mathcal{A}_0$, $\mathcal{A}_1$, and $\mathcal{A}_2$, and we say (β^*, γ^*) is *located* (*determined*) by l_1 and l_2 or by $\mathcal{A}_1$ and $\mathcal{A}_2$. From Lemma 6, if there exists inverse-optimal vertex, when m is big, we can use the two longest boundary lines to locate that vertex; when m is small enough, we need check different pairs of boundary lines to locate the inverse-optimal *vertex*.

We now describe the algorithm:

Algorithm 2

1. set the search space $R = \{(0, \beta, \gamma) | 0 \leq \beta \leq m^2, 0 \leq \gamma \leq m^2\}$;
2. **while** $(Area(R) > 1/2m^7)$
3. compute the centroid $v(\beta_v, \gamma_v)$ of search space R;
4. compute the optimal global alignment $\mathcal{A}_v$ at point v on β, γ space;
5. **if** ($\mathcal{A}_v$ is the same as $\mathcal{A}_0$) **then**
6. **return** the line passing through $(-1, -\frac{1}{2}, 0)$ and $(0, \beta_v, \gamma_v)$;
7. **else**
8. $R \leftarrow R \cap$ the halfplane defined by Equation (4);
9. **end if**
10. **end while**
11. $d = max\{|uv| |$u,v are points on the boundary of R$\}$;
12. **if** $(d < 1/m^3$ or $m > 80)$ **then**

13. select the two longest boundary lines defined by alignments $\mathcal{A}_1, \mathcal{A}_2$;
14. compute $(\hat{\beta}, \hat{\gamma})$ so that $\mathcal{A}_0, \mathcal{A}_1, \mathcal{A}_2$ have same score at $(\hat{\beta}, \hat{\gamma})$;
15. **if** $((\hat{\beta}, \hat{\gamma})$ is in $R)$ **then**
16. compute the optimal global alignment $\hat{\mathcal{A}}$ at point $(\hat{\beta}, \hat{\gamma})$;
17. **if** $(\hat{\mathcal{A}}$ and $\mathcal{A}_0$ have same score at $(\hat{\beta}, \hat{\gamma}))$ **then**
18. **return** the line passing through $(-1, -\frac{1}{2}, 0)$ and $(0, \hat{\beta}, \hat{\gamma})$
19. **end if**
20. **end if**
21. **else**
22. store all pairs of different boundary lines of R into stack S;
23. **while** (S is not empty)
24. pop a pair of boundary lines defined by alignments $\mathcal{A}_1, \mathcal{A}_2$ from S;
25. compute $(\hat{\beta}, \hat{\gamma})$ so $\mathcal{A}_0, \mathcal{A}_1, \mathcal{A}_2$ have same score at $(\hat{\beta}, \hat{\gamma})$;
26. **if** $((\hat{\beta}, \hat{\gamma})$ is in $R)$ **then**
27. compute the optimal global alignment $\hat{\mathcal{A}}$ at point $(\hat{\beta}, \hat{\gamma})$;
28. **if** $(\hat{\mathcal{A}}$ and $\mathcal{A}_0$ have same score at $(\hat{\beta}, \hat{\gamma}))$ **then**
29. **return** the line passing through $(-1, -\frac{1}{2}, 0)$ and $(0, \hat{\beta}, \hat{\gamma})$
30. **end if**
31. **end if**
32. **end while**
33. **end if**
34. continue to search on α, γ plane;

To show that Algorithm 2 is correct, we first need to prove some results.

Lemma 3. *The area of any complete optimal region on the β, γ plane is greater than $\frac{1}{2m^6}$.*

Proof. According to Equation (2) and (3), select a boundary line l of optimal regions and a vertex $v = (\beta_v, \gamma_v)$ as:

$$l : \beta = \frac{\Delta w_i}{\Delta y_i} - \frac{\Delta z_i}{\Delta y_i}\gamma \text{ and}$$

$$\beta_v = \frac{\Delta w_1 \Delta z_2 - \Delta w_2 \Delta z_1}{\Delta y_1 \Delta z_2 - \Delta y_2 \Delta z_1}, \gamma_v = \frac{\Delta w_1 \Delta y_2 - \Delta w_2 \Delta y_1}{\Delta z_1 \Delta y_2 - \Delta z_2 \Delta y_1}$$

Then the distance between v and l is:

$$d = \left| \beta_v + \frac{\Delta z_i}{\Delta y_i}\gamma_v - \frac{\Delta w_i}{\Delta y_i} \right| \cdot \frac{|\Delta y_i|}{\sqrt{\Delta y_i^2 + \Delta z_i^2}}$$

Since $-m < \Delta w_i, \Delta y_i, \Delta z_i < m$, when $d > 0$, $d > 1/m^3$

Since the distance between any two vertices should be greater than the distance between a vertex and a boundary line of an optimal region, the distance between any two vertices is greater than $1/m^3$.

A complete optimal region is composed of at least three vertices. Suppose the minimal complete optimal region is made with 3 vertices, then the base and the height of this triangle are both greater than $1/m^3$. Therefore, the area is greater than $\frac{1}{2m^6}$. □

The precondition for the following results is that the area of the remaining search space is smaller than $\frac{1}{2m^7}$.

Lemma 4. *Suppose reference alignment $\mathcal{A}_0(w_0, y_0, z_0)$ is optimal at vertex (β^*, γ^*) in the remaining search space. Let l be a boundary line of the remaining search space defined by (β_v, γ_v) and alignment $\mathcal{A}_v$. If $\mathcal{A}_v$ is not optimal at (β^*, γ^*), then the distance between (β^*, γ^*) and l is greater than $\frac{1}{m^3}$.*

Proof. According to inequality (4), the boundary line l is

$$(y_v - y_0)\beta + (z_v - z_0)\gamma = (y_v - y_0)\beta_v + (z_v - z_0)\gamma_v \qquad (*)$$

Suppose $\mathcal{A}_v$ is not optimal at (β^*, γ^*). From the optimality of $\mathcal{A}_0$,$\mathcal{A}_v$, we have:

$$w_0 - \beta^* y_0 - \gamma^* z_0 = w_v - \beta^* y_v - \gamma^* z_v + \Delta c_1, \quad \Delta c_1 > 0 \qquad (**)$$

$$w_v - \beta_v y_v - \gamma_v z_v = w_0 - \beta_v y_0 - \gamma_v z_0 + \Delta c_2, \quad \Delta c_2 > 0 \qquad (***)$$

Adding (**) and (***) we obtain

$$(y_v - y_0)\beta^* + (z_v - z_0)\gamma^* = (y_v - y_0)\beta_v + (z_v - z_0)\gamma_v + \Delta c_1 + \Delta c_2 \qquad (****)$$

Equations (*) and (****) define two parallel lines, and (****) passes through (β^*, γ^*). Thus the distance between (β^*, γ^*) and line l equals the distance between line (****) and line l. Hence the distance is:

$$d = \left| \frac{\Delta c_1 + \Delta c_2}{y_v - y_0} \cdot \frac{y_v - y_0}{\sqrt{(y_0 - y_v)^2 + (z_0 - z_v)^2}} \right| = \left| \frac{\Delta c_1 + \Delta c_2}{\sqrt{(y_0 - y_v)^2 + (z_0 - z_v)^2}} \right|$$

According to (**), $\Delta c_1 = w_0 - w_v - (y_0 - y_v)\beta^* - (z_0 - z_v)\gamma^* > 0$. Since (β^*, γ^*) is a vertex, according to Equation (3), it is clear that $\Delta c_1 > 1/m^2$, then $d > |\frac{1}{m^2} \frac{1}{\sqrt{(y_0 - y_v)^2 + (z_0 - z_v)^2}}| > \frac{1}{m^3}$. □

Lemma 5. *If there exists an inverse-optimal vertex (β^*, γ^*) in the remaining search space, then there exist two boundary lines that are defined by (β_1, γ_1), $\mathcal{A}_1$, and (β_2, γ_2), $\mathcal{A}_2$ such that alignments $\mathcal{A}_1, \mathcal{A}_2$ are optimal at (β^*, γ^*).*

Proof. Suppose reference alignment $\mathcal{A}_0$ is optimal at vertex (β^*, γ^*) in the remaining search space. Assume that no boundary line that is defined by (β_i, γ_i), alignment $\mathcal{A}_i$ such that $\mathcal{A}_i$ is optimal at (β^*, γ^*). Then from Lemma 4, the distance between (β^*, γ^*) and any point on the boundary of the remaining search space is greater than $\frac{1}{m^3}$. Hence the area of the remaining search space is greater than $\frac{1}{m^6}$. But the area of the remaining search space is smaller than $\frac{1}{2m^7}$, a contradiction. So there exists at least one boundary line that is defined by (β_1, γ_1), alignment $\mathcal{A}_1$ and $\mathcal{A}_1$ is optimal at (β^*, γ^*).

If there is only one boundary line that satisfies the above requirements, we can find a contradiction from similar reasoning. Thus there exist at least two boundary lines that satisfy above requirements. □

Lemma 6. *If there exists an inverse-optimal vertex in the remaining search space, then when $m > 80$, it can be located using the two longest boundary lines.*

Proof. Suppose that there exists an inverse-optimal *vertex* in the remaining search space. Notice that if the distance of the two farthest points on the boundary is smaller than $1/m^3$, then any two boundary lines can be used to locate the inverse-optimal *vertex* (according to Lemma 4).

Since the search space can be initially restricted to $0 \leq \beta \leq m^2, 0 \leq \gamma \leq m^2$, the area of remaining search space is smaller than $1/2m^7$, and every iteration reduces the search by at least 4/9 and increases the number of boundary lines by at most one, there are at most $\log_{\frac{9}{5}} 2 + 11 \log_{\frac{9}{5}} m$ boundary lines of the remaining search space.

If the length of a boundary line l is more than $1/m^4$, then l can be used to locate that inverse-optimal *vertex*. Thus if the lengths of the two longest boundary lines are greater than $1/m^4$, then the inverse-optimal *vertex* can be located by them. If there are no two boundary lines whose lengths are greater than $1/m^4$, then the distance of two farthest points on the boundary is smaller than $d = (\log_{\frac{9}{5}} 2 + 11 \log_{\frac{9}{5}} m) \cdot \frac{1}{m^4}$. When $m > 80$, $d < \frac{1}{m^3}$. Thus if $m > 80$, the two longest boundary lines can be used to locate the inverse-optimal *vertex*. □

Now we prove the correctness of Algorithm 2 and analyze its running time.

Theorem 5. *If there exists an inverse-optimal point on the β, γ coordinate plane, then Algorithm 2 can find it in $O(mn \log m)$ time.*

Proof. According to Equation 3, the maximum coordinate for a vertex is (m^2, m^2). Thus, we can restrict the search space to $0 \leq \beta \leq m^2, 0 \leq \gamma \leq m^2$. If the algorithm finds an inverse-optimal point (β_v, γ_v) during the binary search, it returns a line. Lemma 3 shows when the area of the remaining search space is smaller than $1/2m^7$, it cannot include a complete optimal region. So binary search terminates and the algorithm begins to check the vertices in the remaining region. If there exists an inverse-optimal vertex in the remaining search space, from Lemma 5 and Lemma 6, step 11-33 can find it. So if there exists inverse-optimal point on β, γ coordinate plane, Algorithm 2 can find that point.

The initial area of search space is m^4; the binary search terminates when the area is less than $1/2m^7$. Every iteration reduces the area of search space by at least 4/9 and increases the number of boundary lines by at most 1, so there are $O(\log m)$ iterations and $O(\log m)$ boundary lines. In every iteration, the algorithm computes one optimal alignment which takes $O(mn)$ time. Thus steps 1-10 need $O(mn \log m)$ time. Step 11 needs $O(\log m)$ time, since there are $O(\log m)$ intersection points on the boundary and the two farthest points must both be intersection points. Step 12-20 need $O(mn)$ time. In step 22-32, since $m < 80$, we can consider the number of different pairs of boundary lines as constant, then the time that step 22-32 need is $O(mn)$. Thus the total time need by Algorithm 2 is $O(mn \log m)$. □

The above algorithm also works for searching on the α, γ plane, only that, in step 34, when we have found that the reference alignment cannot be optimal on the α, γ coordinate plane, we need to return the centroid of the remaining search space as an approximately inverse-optimal point.

4 Further Results and Open Problems

Given two sequences S and T, a *local alignment* is obtained by finding substrings S' and T' of S and T, respectively, whose optimal global alignment score is maximum over all pairs of substrings from S and T [6]. This problem can be solved in $O(mn)$ time [6], where m, n are the lengths of S and T.

The inverse local alignment problem without gap penalty ($\gamma = 0$) asks to find an inverse-optimal or approximately inverse-optimal point on the α, β coordinate plane. This problem can be solved by slightly modifying Algorithm 2 of Section 3 in $O(mn \log m)$ time.

An open problem is to extend our binary search strategy into fixed-dimensional space. For example, this could lead to an efficient algorithm for inverse local alignment with gap penalty, which is a search problem in the α, β, γ space.

References

1. A. Apostolico and R. Giancarlo. Sequence alignment in molecular biology. *Journal of Computational Biology*, 5(2):173–196, 1998.
2. D. Eppstein. Setting parameters by example. *Proc. 40th Symp. Foundations of Computer Science, IEEE*, pages 309–318, 1999.
3. D. Fernández-Baca, T. Seppäläinen, and G. Slutzki. Bounds for parametric sequence comparison. *Discrete Applied Mathematics*, 2002, to appear.
4. W. Fitch and T. F. Smith. Optimal sequence alignments. *Proceedings of the National Academy of Sciences of the USA*, 80:1382–1386, 1983.
5. B. Grunbaum. Partitions of mass distributions and of convex bodies by hyperplanes. *Pacific Journal of Mathematics*, 10:1257–1261, 1960.
6. D. Gusfield. Algorithms on strings, trees, and sequences: Computer science and computational biology. Cambridge University press, Cambridge, New York, Melbourne, 1997.
7. D. Gusfield, K. Balasubramanian, and D. Naor. Parametric optimization of sequence alignment. *Algorithmica*, 12:312–326, 1994.
8. D. Gusfield and P. Stelling. Parametric and inverse-parametric sequence alignment with XPARAL. *Methods in Enzymology*, 226:481–494, 1996.
9. N. Megiddo. Combinatorial optimization with rational objective functions. *Math. Oper. Res.*, 4:414–424, 1979.
10. N. Megiddo. Applying parallel computation algorithms in the design of serial algorithms. *Journal of the ACM*, 30(4):852–865, 1983.
11. D. Sankoff and E. J. Kruskal. Time warps, string edits, and macromolecules: the theory and practice of sequence comparison. *Addison-Wesley*, 1983.
12. M. Vingron and M. Waterman. Sequence alignment and penalty choice: review of concepts, case studies, and implications. *J. of Molecular Biology*, 235:1–12, 1994.
13. M. S. Waterman. Parametric and ensemble sequence alignment. *Bulletin of Mathematical Biology*, 56(4):743–767, 1994.
14. M. S. Waterman, M. Eggert, and E. Lander. Parametric sequence comparisons. *Proceedings of the National Academy of Sciences of the USA*, 89:6090–6093, 1992.

The Full Steiner Tree Problem in Phylogeny

Chin Lung Lu[1], Chuan Yi Tang[2], and Richard Chia-Tung Lee[3]

[1] National Center for High-Performance Computing, P.O. Box 19-136, Hsinchu, Taiwan 300, R.O.C., cllu@nchc.gov.tw
[2] Department of Computer Science, National Tsing Hua University, Hsinchu, Taiwan 300, R.O.C., cytang@cs.nthu.edu.tw
[3] Department of Computer Science and Information Engineering, National Chi-Nan University, Puli, Nantou Hsien, Taiwan 545, R.O.C., rctlee@ncnu.edu.tw

Abstract. Motivated by the reconstruction of phylogenetic tree in biology, we study the full Steiner tree problem in this paper. Given a complete graph $G = (V, E)$ with a length function on E and a proper subset $R \subset V$, the problem is to find a full Steiner tree of minimum length in G, which is a kind of Steiner tree with all the vertices of R as its leaves. In this paper, we show that this problem is NP-complete and MAX SNP-hard, even when the lengths of the edges are restricted to either 1 or 2. For the instances with lengths either 1 or 2, we give a $\frac{5}{3}$-approximation algorithm to find an approximate solution for the problem.

1 Introduction

Given a graph $G = (V, E)$, a subset $R \subseteq V$ of vertices, and a length (or distance) function $d : E \to \mathbb{R}^+$ on the edges, a *Steiner tree* is a connected and acyclic subgraph of G which spans all vertices in R. The vertices in R are usually referred to as *terminals* and the vertices in $V \setminus R$ as *Steiner* (or *optional*) vertices. Note that a Steiner tree might contain the Steiner vertices. The *length* of a Steiner tree is defined to be the sum of the lengths of all its edges. The so-called *Steiner tree problem* is to find a *Steiner minimum tree* (i.e., a Steiner tree of minimum length) in G. The Steiner tree problem has been extensively studied in the past years because it has many important applications in VLSI design, network routing, wireless communications, computational biology and so on [1,2,3,4]. This problem is well known to be NP-complete [5], even in the Euclidean metric [6] or rectilinear metric [7]. However, it has many approximation algorithms with constant performance ratios [4,8].

Motivated by the reconstruction of phylogenetic (or evolutionary) tree in biology, we study a variant of the Steiner tree problem, called the full Steiner tree problem, in this paper. A Steiner tree is *full* if all terminals are the leaves of the tree [4]. The *full Steiner tree problem* is to find a full Steiner tree with minimum length. If we restrict the lengths of edges to be either 1 or 2, then the problem is called the *(1,2)-full Steiner tree problem.* From the viewpoints of biologists, the terminals of a full Steiner tree $\mathcal{T}$ can be regarded as the extant taxa (or species, morphological features, biomolecular sequences), the internal

O.H. Ibarra and L. Zhang (Eds.): COCOON 2002, LNCS 2387, pp. 107–116, 2002.

vertices of $\mathcal{T}$ as the extinct ancestral taxa, and the length of each edge in $\mathcal{T}$ as the evolutionary time along it. Then $\mathcal{T}$ might correspond to an evolutionary tree of the extant species, which trends to minimize the tree length according to the principle of parsimony (i.e., nature always finds the paths that require a minimum evolution) [9]. Hence, the problem of reconstruction of such kind of phylogenetic tree can be considered as the full Steiner tree problem. We refer the readers to [10,11] for other models of evolutionary trees and time-complexities of their constructions.

To our knowledge, little work has been done on the full Steiner tree problem. In [12], Hwang gave a linear-time algorithm for constructing a relatively minimal full Steiner tree $\mathcal{T}$ with respect to $\mathcal{G}$ in the Euclidean metric, where $\mathcal{G}$ is the given topology of $\mathcal{T}$. In this paper, we show that the full Steiner tree problem is NP-complete and MAX SNP-hard, even when the lengths of edges are restricted to be either 1 or 2. However, we give a $\frac{5}{3}$-approximation algorithm for the (1,2)-full Steiner tree problem.

2 Preliminaries

To make sure that a full Steiner tree exists, we restrict the given graph $G = (V, E)$ to be complete and R to be a proper subset of V (i.e., $R \subset V$) in the full Steiner tree problem (FSTP for short).

FSTP (FULL STEINER TREE PROBLEM)
INSTANCE: A complete graph $G = (V, E)$, a length function $d : E \to \mathbb{R}^+$ on the edges, a proper subset $R \subset V$, and a positive integer bound B.
QUESTION: Is there a full Steiner tree $\mathcal{T}$ in G such that the length of $\mathcal{T}$ is less than or equal to B?

The length function d is called a *metric* if it satisfies the following three conditions: (1) $d(x, y) \geq 0$ for any x, y in V, where equality holds if and only if $x = y$, (2) $d(x, y) = d(y, x)$ for any $x, y \in V$, and (3) $d(x, y) \leq d(x, z) + d(z, y)$ for any x, y, z in V (triangle inequality). If we restrict that all edge lengths are either 1 or 2 (i.e., $d : E \to \{1, 2\}$), then we call this restricted FSTP as *(1,2)-full Steiner tree problem* (FSTP(1,2) for short), where such length function is a metric. For convenience, we use MIN-FSTP and MIN-FSTP(1,2) to be referred as the optimization problems of FSTP and FSTP(1,2), respectively.

Given two optimization problems Π_1 and Π_2, we say that Π_1 *L-reduces* to Π_2 if there are polynomial-time algorithms f and g and positive constants α and β such that for any instance I of Π_1, the following conditions are satisfied. (1) Algorithm f produces an instance $f(I)$ of Π_2 such that $\mathsf{OPT}(f(I)) \leq \alpha \cdot \mathsf{OPT}(I)$, where $\mathsf{OPT}(I)$ and $\mathsf{OPT}(f(I))$ stand for the optimal solutions of I and $f(I)$, respectively. (2) Given any solution of $f(I)$ with cost c_2, algorithm g produces a solution of I with cost c_1 in polynomial time such that $|c_1 - \mathsf{OPT}(I)| \leq \beta \cdot |c_2 - \mathsf{OPT}(f(I))|$. A problem is said to be *MAX SNP-hard* if a MAX SNP-hard problem can be L-reduced to it. In [13], Arora et al. showed that if any MAX SNP-hard problem has a PTAS (POLYNOMIAL TIME APPROXIMATION SCHEME), then P=NP, where a problem has a PTAS if for any fixed $\epsilon > 0$, the problem can be approximated within a factor of $1+\epsilon$ in polynomial time [14]. In other words, it is very unlikely for a MAX SNP-hard problem to have a PTAS. On the other hand, if Π_1 L-reduces to Π_2 and Π_2 has a PTAS, then Π_1 has a PTAS [15].

3 NP-Completeness Result

In this section, we will show that FSTP(1,2) is an NP-complete problem by a reduction from the exact cover by 3-sets problem (X3C for short), which is well known to be NP-complete [16].

X3C (Exact Cover by 3-Sets Problem)
Instance: A finite set X with $|X| = 3n$ and a collection $\mathcal{S}$ of 3-element subsets of X with $|\mathcal{S}| = m$.
Question: Does $\mathcal{S}$ contain an exact cover for X, i.e., a subcollection $\mathcal{S}' \subseteq \mathcal{S}$ such that every element of X occurs in exactly one member of $\mathcal{S}$?

Theorem 1. *FSTP(1,2) is an NP-complete problem.*

Proof. In the following, we only show that FSTP(1,2) is NP-hard by reducing X3C to it. Let $X = \{x_1, x_2, \cdots, x_{3n}\}$ and $\mathcal{S} = \{S_1, S_2, \cdots, S_m\}$ be an instance of X3C. Without loss of generality, we assume that $m > n$. Then we transform X and $\mathcal{S}$ into an instance of FSTP(1,2) as follows.

- A complete graph $G = (V, E)$ with $V = X \cup \mathcal{S} \cup \{y\}$, $R = X$ and $B = 4n$.
- For each edge e, $d(e) = \begin{cases} 1, \text{ if } e \in \{(x_i, S_j) | x_i \in S_j\} \cup \{(y, S_j) | 1 \leq j \leq m\}, \\ 2, \text{ otherwise.} \end{cases}$

See Figure 1 for an example of the reduction with $X = \{x_1, x_2, x_3, x_4, x_5, x_6\}$ and $\mathcal{S} = \{S_1, S_2, S_3, S_4\} = \{\{x_1, x_3, x_5\}, \{x_2, x_3, x_5\}, \{x_2, x_4, x_5\}, \{x_2, x_4, x_6\}\}$, where only the edges of length 1 in G are shown and the white circles and the black boxes denote the Steiner vertices and the terminals, respectively.

Next, we claim that X3C has a positive answer (i.e., $\mathcal{S}$ has an exact cover) if and only if there is a full Steiner tree in G with length less than or equal to B,

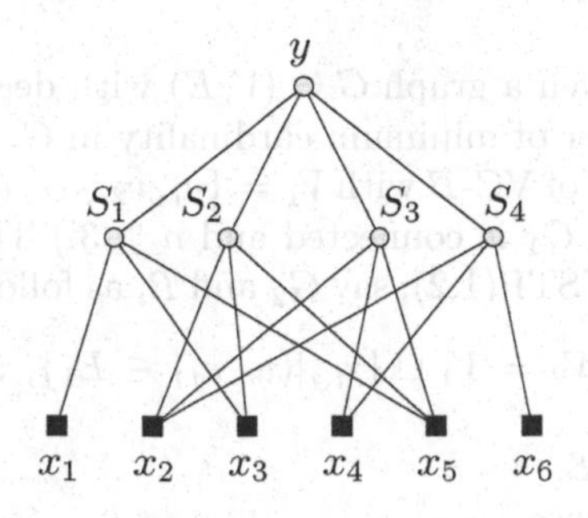

Fig. 1. A reduction of X3C to FSTP(1,2).

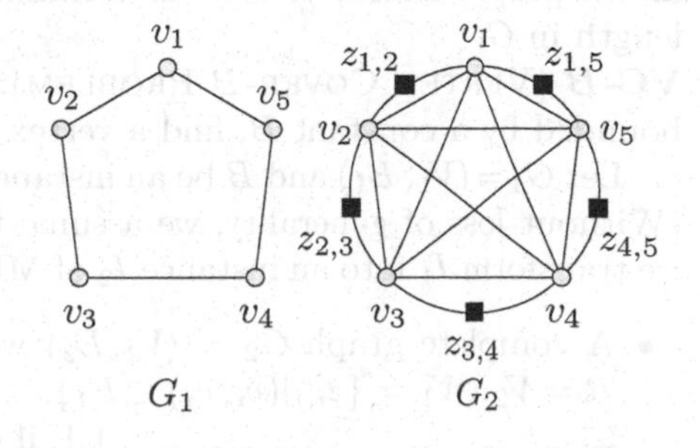

Fig. 2. An L-reduction of VC-B to MIN-FSTP(1,2).

where $B = 4n$. First, suppose that $\mathcal{S}$ has an exact cover $\mathcal{S}'$. Then we can build a tree $\mathcal{T}$ with the set $E(\mathcal{T})$ of edges, where $E(\mathcal{T}) = \{(x_i, S_j), (y, S_j) | S_j \in \mathcal{S}', x_i \in S_j\}$. It is not hard to see that $\mathcal{T}$ is a full Steiner tree in G and its length is $4n$.

Conversely, suppose that there is a full Steiner tree $\mathcal{T}$ in G with length less than or equal to $4n$. For convenience, we call the edges in $\{(S_i, S_j) | 1 \leq i < j \leq m\}$ as the *type-1* edges and the edges in $\{(y, x_i) | 1 \leq i \leq 3n\}$ as the *type-2* edges. Note that the length of each type-1 or type-2 edge in G is 2 by the reduction. Then we can transform $\mathcal{T}$ into $\mathcal{T}'$ using the following two methods such that $\mathcal{T}'$ contains no type-1 or type-2 edge. (1) Replace each type-1 edge (S_i, S_j) of $\mathcal{T}$ with the edges (y, S_i) and (y, S_j). (2) Replace each type-2 edge (y, x_i) of $\mathcal{T}$ with the edges (y, S_j) and (x_i, S_j), where $x_i \in S_j$. Clearly, $\mathcal{T}'$ is still a full Steiner tree of G, which can be obtained from $\mathcal{T}$ in polynomial time, and its length is less than or equal to that of $\mathcal{T}$ (i.e., $\leq 4n$) since we always use two edges of length 1, which might be in $\mathcal{T}$ already, to replace an edge of length 2. Since $\mathcal{T}'$ contains no type-1 or type-2 edge, its leaves are adjacent to some vertices in $\mathcal{S}$, all of which are then adjacent to y. For convenience, we use $\mathcal{Y}$ to denote the set of internal vertices of $\mathcal{T}'$ which are adjacent to y. Then the length of $\mathcal{T}'$ is greater than or equal to $|\mathcal{X}| + |\mathcal{Y}| = 3n + |\mathcal{Y}|$, which is less than or equal to $4n$. This implies that $|\mathcal{Y}| \leq n$. On the other hand, since each vertex in $\mathcal{Y}$ is adjacent to at most 3 leaves, $3 \cdot |\mathcal{Y}| \geq 3n$ and hence $|\mathcal{Y}| \geq n$. In other words, we have $|\mathcal{Y}| = n$ and clearly, $\mathcal{Y}$ corresponds to a vertex cover of $\mathcal{S}$. □

The proof of Theorem 1 can be used to prove the NP-completeness of FSTP, even when the length function is metric.

4 MAX SNP-Hardness Result

In this section, we will show that the optimization problem of FSTP(1,2), referred to as MIN-FSTP(1,2), is MAX SNP-hard by an L-reduction from the vertex cover-B problem (VC-B for short), which was shown to be MAX SNP-hard by Papadimitriou and Yannakakis [15].

MIN-FSTP(1,2) (Minimum (1,2)-Full Steiner Tree Problem): Given a complete graph $G = (V, E)$ with a length function $d : E \to \{1, 2\}$ on the edges and a proper subset $R \subset V$ of terminals, find a full Steiner tree of minimum length in G.

VC-*B* (Vertex Cover-B Problem): Given a graph $G = (V, E)$ with degree bounded by a constant B, find a vertex cover of minimum cardinality in G.

Let $G_1 = (V_1, E_1)$ and B be an instance I_1 of VC-B with $V_1 = \{v_1, v_2, \cdots, v_n\}$. (Without loss of generality, we assume that G_1 is connected and $n \geq 3$.) Then we transform I_1 into an instance I_2 of MIN-FSTP(1,2), say G_2 and R, as follows.

- A complete graph $G_2 = (V_2, E_2)$ with $V_2 = V_1 \cup \{z_{i,j} | (v_i, v_j) \in E_1\}$, and $R = V_2 \setminus V_1 = \{z_{i,j} | (v_i, v_j) \in E_1\}$.
- For each edge $e \in E_2$, $d(e) = \begin{cases} 1, \text{ if } e \in \mathcal{E}, \\ 2, \text{ otherwise}, \end{cases}$
  where $\mathcal{E} = \{(v_i, v_j) | 1 \leq i < j \leq n\} \cup \{(v_i, z_{i,j}), (z_{i,j}, v_j) | (v_i, v_j) \in E_1\}$.

See Figure 2 for an example of the reduction, where G_1 is a C_5 (i.e, a cycle of length 5) and only the edges of length 1 in G_2 are shown.

Lemma 1. *Let $\mathcal{T}$ be a solution of length c to MIN-FSTP(1,2) on the instance I_2 which is obtained from a reduction of an instance I_1 of VC-B. Then in polynomial time, we can find another solution $\mathcal{T}'$ of length no more than c to MIN-FSTP(1,2) on instance I_2 such that $\mathcal{T}'$ contains no edge of length 2.*

Proof. In the following, we only show how to replace an edge of length 2 from $\mathcal{T}$ with some edges of length 1 in polynomial time without increasing the length of the resulting $\mathcal{T}$. Then by repeatedly applying this procedure to $\mathcal{T}$, we will finally obtain $\mathcal{T}'$ in polynomial time. Let (x, y) be an edge of length 2 in $\mathcal{T}$. Since both x and y cannot belong to R or V_1 at the same time, one of them must be a terminal and the other must be a Steiner vertex according to the rules to construct G_2. Without loss of generality, we assume that x is a terminal and y is a Steiner vertex. Since we assume that G_1 is connected and $n \geq 3$, x must be connected to some one terminal z with a path of two edges of length 1, say (x, v) and (v, z). Let u be the Steiner vertex of $\mathcal{T}$ which is adjacent to z. Then we consider the following two possibilities. Case 1: $(x, u) \in E_1$. This means that the length of (x, u) is 1. Then we replace (x, y) with (x, u). Case 2: $(x, u) \notin E_1$. This means that the length of (x, u) is 2. Then we replace (x, y) with (x, v) and (v, u) of length 1. It is easy to see that the resulting $\mathcal{T}$ is still a full Steiner tree of G_2, which can be obtained in polynomial time, without increasing the length. □

Theorem 2. *MIN-FSTP(1,2) is a MAX SNP-hard problem.*

Proof. Let f denote the polynomial-time algorithm (as described in the beginning of this section) to transform an instance I_1 of VC-B to the instance I_2 of MIN-FSTP(1,2) (i.e., $f(I_1) = I_2$). We design another polynomial-time algorithm g as follows. Given a full Steiner tree $\mathcal{T}$ in G_2 of length c, we transform it into another full Steiner tree $\mathcal{T}'$ using the method described in the proof of Lemma 1. Clearly, $\mathcal{T}'$ contains no edge of length 2 and its length is no more than c, which implies that the number of vertices in $\mathcal{T}'$ is less than or equal to $c+1$ (since $\mathcal{T}'$ is a tree). Then the collection of those internal vertices of $\mathcal{T}'$ which are adjacent to the leaves of $\mathcal{T}'$ corresponds to a vertex cover of G_1 whose size is less than or equal to $c - |E_1| + 1$. Next, we prove that algorithms f and g are an L-reduction from VC-B to MIN-FSTP(1,2) by showing the following two inequalities.

(1) $\mathsf{OPT}(f(I_1)) \leq \alpha \cdot \mathsf{OPT}(I_1)$, where $\alpha = 2B$. Note that $B \cdot \mathsf{OPT}(I_1) \geq |E_1|$ since each vertex in G_1 covers at most B edges. Let u be a vertex in G_1 whose degree is B. Then we can build a star $\mathcal{T}$ with u as its center and R as its leaves. Clearly, $\mathcal{T}$ is a feasible solution of MIN-FSTP(1,2) on $f(I_1)$ whose length is $B + 2 \cdot (|E_1| - B) = 2 \cdot |E_1| - B$. Hence, $\mathsf{OPT}(f(I_1)) \leq 2 \cdot |E_1| - B \leq 2 \cdot |E_1| = 2B \cdot \frac{|E_1|}{B} \leq 2B \cdot \mathsf{OPT}(I_1)$.

(2) $|c_1 - \mathsf{OPT}(I_1)| \leq \beta \cdot |c_2 - \mathsf{OPT}(f(I_1))|$, where $\beta = 1$. Given a vertex cover $\mathcal{C}$ in G_1 of size c, we can create a full Steiner tree $\mathcal{T}$ in G_2 of length $c + |E_1| - 1$ in the following way. Connect each edge of E_1 (corresponding a

terminal in G_2) to an arbitrary vertex in $\mathcal{C}$ (corresponding a Steiner vertex in G_2) and connect all vertices of $\mathcal{C}$ by $c-1$ edges of length 1 in G_2. Hence, $\mathsf{OPT}(f(I_1)) \leq \mathsf{OPT}(I_1)+|E_1|-1$. Conversely, by algorithm g, a full Steiner tree $\mathcal{T}$ of G_2 with length c_2 can be transformed into a vertex cover of G_1 of size c_1 less than or equal to $c_2-|E_1|+1$ (i.e., $c_1 \leq c_2-|E_1|+1$). Then $c_1-\mathsf{OPT}(I_1) \leq (c_2-|E_1|+1)-\mathsf{OPT}(I_1) = c_2-(\mathsf{OPT}(I_1)+|E_1|-1) \leq c_2-\mathsf{OPT}(f(I_1))$. Hence, $|c_1-\mathsf{OPT}(I_1)| \leq 1 \cdot |c_2-\mathsf{OPT}(f(I_1))|$. □

Clearly, the proof of Theorem 2 can be applied to show that MIN-FSTP is still MAX SNP-hard, even though the length function is metric.

5 A $\frac{5}{3}$-Approximation Algorithm for MIN-FSTP(1,2)

For MIN-FSTP(1,2), it is not hard to see that any star with an arbitrary Steiner vertex as its center and all terminals as its leaves is an approximate solution with performance ratio within 2 of the optimal one. In this section, we give a $\frac{5}{3}$-approximation algorithm for MIN-FSTP(1,2) using the so-called average length (or distance) heuristics [17,18].

Let *Steiner star* be a star $\mathcal{T}$ with a Steiner vertex as its $\mathsf{center}(\mathcal{T})$ and the terminals as its $\mathsf{leaves}(\mathcal{T})$, where $\mathsf{center}(\mathcal{T})$ and $\mathsf{leaves}(\mathcal{T})$ denote the center and the leaves of $\mathcal{T}$, respectively. For a Steiner star $\mathcal{T}$ with $|\mathsf{leaves}(\mathcal{T})| \geq 2$, we define its *average length* to be $f(\mathcal{T}) = \frac{\sum_{v \in \mathsf{leaves}(\mathcal{T})} d(\mathsf{center}(\mathcal{T}),v)}{|\mathsf{leaves}(\mathcal{T})|-1}$. For convenience, we use $\mathcal{X}_k$*-star* to denote a Steiner star $\mathcal{T}$ with k leaves and $d(\mathsf{center}(\mathcal{T}),v)=1$ for each v of $\mathsf{leaves}(\mathcal{T})$. By definition, we have the following lemmas immediately.

Lemma 2. *Let $\mathcal{T}$ be a Steiner star with k terminals, where $k \geq 2$. If $\mathcal{T}$ contains no leaf at distance 1 from* $\mathsf{center}(\mathcal{T})$, *then $f(\mathcal{T}) = 2+\frac{2}{k-1}$.*

Lemma 3. *Let $\mathcal{T}$ be a Steiner star with k terminals, where $k \geq 2$. If $\mathcal{T}$ contains only one leaf at distance 1 from* $\mathsf{center}(\mathcal{T})$, *then $f(\mathcal{T}) = 2+\frac{1}{k-1}$.*

Lemma 4. *Let $\mathcal{T}$ be a Steiner star with k terminals, where $k \geq 2$. If $\mathcal{T}$ contains exactly two leaves at distance 1 from* $\mathsf{center}(\mathcal{T})$, *then $f(\mathcal{T}) = 2$.*

Lemma 5. *Let $\mathcal{T}$ be a $\mathcal{X}_k$-star with $k \geq 3$. Then $f(\mathcal{T}) = 1+\frac{1}{k-1}$.*

Lemma 6. *Let $\mathcal{T}_1$ be an $\mathcal{X}_k$-star with $k \geq 3$ and let $\mathcal{T}_2$ be the Steiner star obtained from $\mathcal{T}_1$ by adding a new terminal z with $d(\mathsf{center}(\mathcal{T}_1),z)=2$. Then $f(\mathcal{T}_1) < f(\mathcal{T}_2)$.*

Next, we describe our approximation algorithm for MIN-FSTP(1,2) in the following algorithm APX-FSTP(1,2). Without loss of generality, we assume that $|R| \geq 8$, since for $|R| < 8$, the optimal solution can be found by an exhaustive search in polynomial time. According to Lemmas 2 to 6, our algorithm APX-FSTP(1,2) always selects an $\mathcal{X}_k$-star with maximum k, $k \geq 3$, to do the reduction if it exists, since its average length must be minimum. If only $\mathcal{X}_2$-stars are found in the (resulting) instance, then the average length of the minimum Steiner star

APX-FSTP(1,2)
Input: A complete graph $G = (V, E)$ with $d : E \to \{1, 2\}$ and a subset $R \subset V$.
Output: A full Steiner tree $\mathcal{T}_{\mathsf{APX}}$ in G.
1: Let $\mathcal{E}$ be an empty set;
2: /* Choose a Steiner star with the minimum average length */
if there are two or more remaining Steiner vertices **then**
Find a Steiner star $\mathcal{T}$ with minimum average length;
if $f(\mathcal{T}) = 2$ **then /* Transform $\mathcal{T}$ into an $\mathcal{X}_2$-star */**
Remove from $\mathcal{T}$ those leaves at distance 2 from $\mathsf{center}(\mathcal{T})$ if they exist;
else Let $\mathcal{T}$ be the Steiner star with the only Steiner vertex as its center and all remaining terminals as its leaves;
3: Let $\mathcal{E} = \mathcal{E} \cup \{(\mathsf{center}(\mathcal{T}), v) | v \in \mathsf{leaves}(\mathcal{T})\}$; **/* Perform a reduction */**
Replace the Steiner star $\mathcal{T}$ by a single new terminal, say z;
Let $d(z, u) = d(\mathsf{center}(\mathcal{T}), u)$ for each remaining vertex u;
4: if there is still more than one terminal **then** Go to Step 2;
else Let $\mathcal{T}_{\mathsf{APX}}$ be the full Steiner tree induced by $\mathcal{E}$;

must be 2 by Lemmas 2 to 4. In this case, the minimum Steiner star selected by APX-FSTP(1,2) might contain some leaves at distance 2 from the center. To avoid this situation, APX-FSTP(1,2) will transform it into an $\mathcal{X}_2$-star without changing its average length by Lemma 4. If the (resulting) instance does not contain a $\mathcal{X}_k$-star with $k \geq 2$, then APX-FSTP(1,2) will perform only one reduction by Lemmas 2 and 3. As discussed above, we can find that APX-FSTP(1,2) will always select an $\mathcal{X}_k$-star, $k \geq 2$, to do the reduction except the last one.

We analyze the time-complexity of APX-FSTP(1,2) as follows. Let n and m be the numbers of the terminals and the Steiner vertices in G, respectively (i.e., $n = |R|$ and $m = |V \setminus R|$). Clearly, the time-complexity of APX-FSTP(1,2) is dominated by the cost of Step 2, which needs to find a Steiner star with minimum average length. It can be implemented by first finding an optimal Steiner star with each Steiner vertex as the center and then selecting the best one among these optimal Steiner stars. For each Steiner vertex v, we can find an optimal Steiner star with v as its center in $\mathcal{O}(n')$ time, where n' denotes the number of the resulting terminals. The reason is that we just calculate the number of terminals at distance 1 from v and then we are able to know what its optimal Steiner star is by Lemmas 2 to 6. Suppose that there are m' Steiner vertices in each reduction. Then Step 2 can be done in $\mathcal{O}(n'm' + m')$ time. Since each reduction eliminates one Steiner vertex and at least one terminal, the number of the iterations is at most $\min\{n, m\}$ and hence the total time-complexity of APX-FSTP(1,2) is polynomial.

Let the *performance ratio* of our approximation algorithm APX-FSTP(1,2) for instance I be $\mathsf{ratio}(I) = \frac{\mathsf{APX}(I)}{\mathsf{OPT}(I)}$, where $\mathsf{OPT}(I)$ denotes the length of an optimal full Steiner tree for I and $\mathsf{APX}(I)$ denotes the length of $\mathcal{T}_{\mathsf{APX}}$ obtained by APX-FSTP(1,2). In the following, we assume that I is a worst-case instance among all instances. That is, $\mathsf{ratio}(I') \leq \mathsf{ratio}(I)$ for each $I' \neq I$.

Lemma 7. *If instance I contains an $\mathcal{X}_k$-star for $k \geq 5$, then* $\mathsf{ratio}(I) \leq \frac{5}{3}$.

Proof. Let $\mathcal{T}$ be an arbitrary $\mathcal{X}_k$-star in I whose k is maximum and let $E(\mathcal{T})$ be the set of its edges. Then by Lemmas 2 to 6, the first iteration of our algorithm APX-FSTP(1,2) will reduce $\mathcal{T}$ since its average length $f(\mathcal{T})$ is minimum. Let I' be the resulting instance of APX-FSTP(1,2) after $\mathcal{T}$ is reduced. Clearly, we have $\mathsf{APX}(I') = \mathsf{APX}(I) - k$. Let $\mathcal{T}_{\mathsf{OPT}}$ be an optimal full Steiner tree of I, and let $\mathcal{H}$ be the resulting graph obtained by adding the k edges of $E(\mathcal{T})$ to $\mathcal{T}_{\mathsf{OPT}}$. Then by removing from $\mathcal{H}$ some edges not in $E(\mathcal{T})$ and adding some one edge if possible, we can build a full Steiner tree $\mathcal{T}'$ of I such that it contains all edges of $E(\mathcal{T})$. In the worst case, the k edges of $E(\mathcal{T})$ and the $\mathsf{center}(\mathcal{T})$ are not in $\mathcal{T}_{\mathsf{OPT}}$. Then we need to add edge $(\mathsf{center}(\mathcal{T}), v)$ to build a full Steiner tree $\mathcal{T}'$, where v is a Steiner vertex in $\mathcal{T}_{\mathsf{OPT}}$. Clearly, the length of $\mathcal{T}'$ is less than or equal to $\mathsf{OPT}(I)+2$ since $d(\mathsf{center}(\mathcal{T}), v) \leq 2$. If we reduce $\mathcal{T}$ in $\mathcal{T}'$, then we obtain a full Steiner tree $\mathcal{T}''$ of instance I' whose length is less than or equal to $\mathsf{OPT}(I) - k + 2$. In other words, $\mathsf{OPT}(I') \leq \mathsf{OPT}(I) - k + 2$. Hence, $\mathsf{ratio}(I') = \frac{\mathsf{APX}(I')}{\mathsf{OPT}(I')} \geq \frac{\mathsf{APX}(I)-k}{\mathsf{OPT}(I)-k+2}$. Recall that $\mathsf{ratio}(I') \leq \mathsf{ratio}(I)$. Then we have $\frac{\mathsf{APX}(I)-k}{\mathsf{OPT}(I)-k+2} \leq \frac{\mathsf{APX}(I)}{\mathsf{OPT}(I)} \iff k \cdot \mathsf{OPT}(I) \geq (k-2) \cdot \mathsf{APX}(I) \iff \frac{k}{k-2} \geq \frac{\mathsf{APX}(I)}{\mathsf{OPT}(I)} = \mathsf{ratio}(I)$. Hence, we have $\mathsf{ratio}(I) \leq \frac{5}{3}$. □

In the following, we assume that I contains no such an $\mathcal{X}_k$-star with $k \geq 5$ and we will then show that $\mathsf{ratio}(I) \leq \frac{5}{3}$. Given an instance I consisting of $G = (V, E)$ and $R \subset V$, we say that a vertex $v \in V$ *1-dominates* (or *dominates* for simplicity) itself and all other vertices at distance 1 from v. For any $D \subseteq V$, we call it as a *1-dominating set* (or *dominating set*) of R if every terminal in R is dominated by at least one vertex of D. A dominating set of R with minimum cardinality is called as a *minimum* dominating set of R.

Lemma 8. *Given an instance I of MIN-FSTP(1,2), let D be a minimum dominating set of R. Then* $\mathsf{OPT}(I) \geq n + |D| - 1$, *where* $n = |R|$.

Proof. Let $\mathcal{T}_{\mathsf{OPT}}$ be an optimal full Steiner tree of I (i.e., $\mathsf{OPT}(I) = |\mathcal{T}_{\mathsf{OPT}}|$) and let $R' \subseteq R$ be the set of terminals that are dominated by the vertices of D', where $D' \subseteq V \setminus R$ is the set of Steiner vertices in $\mathcal{T}_{\mathsf{OPT}}$. Note that for those vertices in D', $\mathcal{T}_{\mathsf{OPT}}$ needs to contain at least $|D'| - 1$ edges to connect them. Clearly, the length of $\mathcal{T}_{\mathsf{OPT}}$ is $|\mathcal{T}_{\mathsf{OPT}}| \geq |R'| + (|D'| - 1) + 2 \cdot |R \setminus R'| = |R| + |D'| + |R \setminus R'| - 1$. Since the union of D' and $R \setminus R'$ is a dominating set of R and they are disjoint, $|D| \leq |D' \cup (R \setminus R')| = |D'| + |R \setminus R'|$. In other words, we have $\mathsf{OPT}(I) \geq |R| + |D| - 1 = n + |D| - 1$. □

Let D be a minimum dominating set of R. Then we can partition R into many subsets in a way as follows. Assign each terminal z of R to a member of D which dominates it. If two or more vertices of D dominate z, then we arbitrarily assign z to one of them. Let $\mathcal{C}_1, \mathcal{C}_2, \cdots, \mathcal{C}_q$ be the partitions consisting of exactly 4 terminals.

Lemma 9. $\mathsf{OPT}(I) \geq \frac{4n}{3} - \frac{q}{3} - 1$.

Proof. According to the partition of R, we have $4 \cdot q + 3(|D| - q) \geq n$, which means that $|D| \geq \frac{n-q}{3}$. Recall that $\mathsf{OPT}(I) \geq n + |D| - 1$ by Lemma 8. Hence, $\mathsf{OPT}(I) \geq n + \frac{n-q}{3} - 1 = \frac{4n}{3} - \frac{q}{3} - 1$. □

Lemma 10. *If instance I contains no $\mathcal{X}_k$-star with $k \geq 5$, then* $\mathsf{ratio}(I) \leq \frac{5}{3}$.

Proof. Assume that APX-FSTP(1,2) totally reduces j $\mathcal{X}_{k_i}$-stars, where $1 \leq i \leq j$ and $k_i \geq 2$. Note that $\mathcal{X}_{k_i}$ is a subtree of the full Steiner tree $\mathcal{T}_{\mathsf{APX}}$ produced by APX-FSTP(1,2) and its length is k_i. Since the reduction of $\mathcal{X}_{k_i}$ merges k_i old terminals into a new one, the number of the terminals is decreased by $k_i - 1$. After reducing $\mathcal{X}_{k_j}$, the number of the remaining terminals is $n - \sum_{i=1}^{j}(k_i - 1)$. To reduce these terminals, APX-FSTP(1,2) creates a Steiner star with length less than or equal to $2 \cdot (n - \sum_{i=1}^{j}(k_i - 1))$. Hence, the total length of $\mathcal{T}_{\mathsf{APX}}$ is less than or equal to $(\sum_{i=1}^{j} k_i) + (2 \cdot (n - \sum_{i=1}^{j}(k_i - 1))) = 2n - \sum_{i=1}^{j}(k_i - 2)$. In other words, we have $\mathsf{APX}(I) \leq 2n - p$, where $p = \sum_{i=1}^{j}(k_i - 2)$.

Recall that we partition R into many disjoint subsets in which $\mathcal{C}_1, \mathcal{C}_2, \cdots, \mathcal{C}_q$ are the partitions with each consisting of exactly 4 terminals. In other words, there are at least q disjoint $\mathcal{X}_4$-stars in I. Next, we claim that $p > \frac{5q}{9}$. The best situation is that each partition $\mathcal{C}_i$, $1 \leq i \leq q$, corresponds to an $\mathcal{X}_4$-star which will be reduced by APX-FSTP(1,2). In this case, each such an $\mathcal{X}_4$-star contributes 2 to p and hence we have $p \geq 2q > \frac{5q}{9}$. Otherwise, we consider the case with the following four properties, where for simplicity of illustration, we assume that $q_2 \equiv 0 \pmod 2$, $q_3 \equiv 0 \pmod 3$ and $q_4 \equiv 0 \pmod 4$, and $q_1 + q_2 + q_3 + q_4 = q$.

(1) There are q_1 partitions $\mathcal{C}_{i_1}, \cdots, \mathcal{C}_{i_{q_1}}$ in which each partition $\mathcal{C}_{i_h}$ $1 \leq h \leq q_1$, corresponds to an $\mathcal{X}_4$-star reduced by APX-FSTP(1,2);
(2) There are q_2 partitions $\mathcal{C}_{i_{q_1+1}}, \cdots, \mathcal{C}_{i_{q_1+q_2}}$ in which every other two consecutive partitions $\mathcal{C}_{i_{q_1+h+1}}$ and $\mathcal{C}_{i_{q_1+h+2}}$, $0 \leq h \leq q_2 - 2$ and $h \equiv 0 \pmod 2$, correspond to an $\mathcal{X}_4$-star reduced by APX-FSTP(1,2);
(3) There are q_3 partitions $\mathcal{C}_{i_{q_1+q_2+1}}, \cdots, \mathcal{C}_{i_{q_1+q_2+q_3}}$ in which every other three consecutive partitions $\mathcal{C}_{i_{q_1+q_2+h+1}}, \mathcal{C}_{i_{q_1+q_2+h+2}}$ and $\mathcal{C}_{i_{q_1+q_2+h+3}}$, $0 \leq h \leq q_3 - 3$ and $h \equiv 0 \pmod 3$, correspond to an $\mathcal{X}_4$-star reduced by APX-FSTP(1,2);
(4) There are q_4 partitions $\mathcal{C}_{i_{q_1+q_2+q_3+1}}, \cdots, \mathcal{C}_{i_{q_1+q_2+q_3+q_4}}$ in which every other four consecutive partitions $\mathcal{C}_{i_{q_1+q_2+q_3+h+1}}, \mathcal{C}_{i_{q_1+q_2+q_3+h+2}}, \mathcal{C}_{i_{q_1+q_2+q_3+h+3}}$ and $\mathcal{C}_{i_{q_1+q_2+q_3+h+4}}$, $0 \leq h \leq q_4 - 4$ and $h \equiv 0 \pmod 4$, correspond to an $\mathcal{X}_4$-star reduced by APX-FSTP(1,2).

It is not hard to see that the reduction of $\mathcal{X}_4$-stars of property (1) (respectively, (2), (3) and (4)) will contribute $2q_1$ (respectively, $\frac{2q_2}{2}$, $\frac{2q_3}{3}$ and $\frac{2q_4}{4}$) to p and in the worst case, produce 0 (respectively, 0, $\frac{2q_3}{3}$ and q_4) $\mathcal{X}_3$-star in the remaining instance. In the worst case, the $(0 + 0 + \frac{2q_3}{3} + q_4 = \frac{2q_3+3q_4}{3})$ produced $\mathcal{X}_3$-stars will further contribute $\frac{\frac{2q_3+3q_4}{3}}{3}$ to p. Hence, we have $p \geq 2q_1 + \frac{2q_2}{2} + \frac{2q_3}{3} + \frac{2q_4}{4} + \frac{2q_3+3q_4}{9} = q_1 + \frac{18(q_1+q_2+q_3+q_4)-2q_3-3q_4}{18} \geq q_1 + \frac{13q}{18}$ (since $q_1 + q_2 + q_3 + q_4 = q$ and $2q_3 + 3q_4 \leq 5q$) $> \frac{5q}{9}$. As discussed above, we have $p > \frac{5q}{9}$ (i.e., $q < \frac{9p}{5}$). Recall that $\mathsf{OPT}(I) \geq \frac{4n}{3} - \frac{q}{3} - 1$ by Lemma 9. Then we have $\mathsf{ratio}(I) = \frac{\mathsf{APX}(I)}{\mathsf{OPT}(I)} \leq \frac{2n-p}{\frac{4n}{3}-\frac{q}{3}-1} = \frac{6n-3p}{4n-q-3} \leq \frac{6n-3p}{4n-\frac{9p}{5}-3}$. Clearly, $\frac{6n-3p}{4n-\frac{9p}{5}-3} \leq \frac{5}{3}$ if $n \geq 8$. Hence, $\mathsf{ratio}(I) \leq \frac{5}{3}$ for $n \geq 8$. Note that for $n < 8$, the optimal solution can be found by an exhaustive search in polynomial time. □

According to Lemmas 7 and 10, we have the following theorem immediately.

Theorem 3. *APX-FSTP(1,2) is a $\frac{5}{3}$-approximation algorithm for the minimum (1,2)-full Steiner tree problem (MIN-FSTP(1,2)).*

References

1. Cheng, X., Du, D.: Steiner Trees in Industry. Kluwer Academic Publishers, Dordrecht, Netherlands (2001)
2. Du, D., Smith, J., Rubinstein, J.: Advances in Steiner Trees. Kluwer Academic Publishers, Dordrecht, Netherlands (2000)
3. Foulds, L., Graham, R.: The Steiner problem in phylogeny is NP-complete. Advances in Applied Mathematics **3** (1982) 43–49
4. Hwang, F., Richards, D., Winter, P.: The Steiner Tree Problem. Annals of Discrete Mathematics 53. Elsevier Science Publishers B. V., Amsterdam (1992)
5. Karp, R.: Reducibility among combinatorial problems. In Miller, R.E., Thatcher, J.W., eds.: Complexity of Computer Computations. Plenum Press, New York (1972) 85–103
6. Garey, M., Graham, R., Johnson, D.: The complexity of computing Steiner minimal trees. SIAM Journal on Applied Mathematics **32** (1977) 835–859
7. Garey, M., Johnson, D.: The rectilinear Steiner problem is NP-complete. SIAM Journal on Applied Mathematics **32** (1977) 826–834
8. Gröpl, C., Hougardy, S., Nierhoff, T., Prömel, H.: Approximation algorithms for the Steiner tree problem in graphs. In Cheng, X., Du, D., eds.: Steiner Trees in Industry. Kluwer Academic Publishers, Dordrecht, Netherlands (2001) 235–279
9. Graur, D., Li, W.: Fundamentals of Molecular Evolution. 2nd edition, Sinauer Publishers, Sunderland, Massachusetts (2000)
10. Kim, J., Warnow, T.: Tutorial on phylogenetic tree estimation. Manuscript, Department of Ecology and Evolutionary Biology, Yale University (1999)
11. Wareham, H.T.: On the computational complexity of inferring evolutionary trees. Technical Report 93-01, Department of Computer Science, Memorial University of Newfoundland (1993)
12. Hwang, F.: A linear time algorithm for full Steiner trees. Operations Research Letters **4** (1986) 235–237
13. Arora, S., Lund, C., Motwani, R., Sudan, M., Szegedy, M.: Proof verification and the hardness of approximation problems. Journal of the Association for Computing Machinery **45** (1998) 501–555
14. Ausiello, G., Crescenzi, P., Gambosi, G., Kann, V., Marchetti-Spaccamelai, A., Protasi, M.: Complexity and Approximation—Combinatorial Optimization Problems and Their Approximability Properties. Springer Verlag, Berlin (1999)
15. Papadimitriou, C., Yannakakis, M.: Optimization, approximization and complexity classes. Journal of Computer and System Sciences **43** (1991) 425–440
16. Garey, M., Johnson, D.: Computers and Intractability—A Guide to the Theory of NP-Completeness. San Francisco, Freeman (1979)
17. Bern, M., Plassmann, P.: The Steiner problem with edge lengths 1 and 2. Information Processing Letters **32** (1989) 171–176
18. Rayward-Smith, V.: The computation of nearly minimum Steiner trees in graphs. International Journal of Mathematical Education in Science and Technology **14** (1983) 15–23

Inferring a Union of Halfspaces from Examples

Tatsuya Akutsu[1,2] and Sascha Ott[3]

[1] Bioinformatics Center, Institute for Chemical Research, Kyoto University, Uji-city, Kyoto 611-0011, Japan
takutsu@kuicr.kyoto-u.ac.jp

[2] Graduate School of Informatics, Kyoto University, Sakyo-ku, Kyoto 606-8501, Japan

[3] Human Genome Center, Institute of Medical Science, University of Tokyo, Minato-ku, Tokyo 108-8639, Japan
ott@ims.u-tokyo.ac.jp

Abstract. We consider the following problem which is motivated by applications in Bioinformatics: given positive and negative points in d-dimensions, find a minimum cardinality set of halfspaces whose union covers all positive points and no negative points. We prove that approximation of this problem is at least as hard as approximation of graph coloring. On the other hand, we show that the two-dimensional case of the problem can be solved in polynomial time. Other related results are shown, too.

1 Introduction

Separation of *positive examples* from *negative examples* using a *hyperplane* is an important problem in both *machine learning* and *statistics* [2,3,6,13]. Though it is a classic problem, extensive studies are still being done motivated by the invention of the *support vector machine* [6]. Recent studies focus on finding large margin classifiers [6] and minimizing the number of misclassified examples [3].

Learning a *mixture of models* is another important problem in machine learning and statistics, because a single model (or a single set of parameters) is not always enough to characterize the given data (the given examples). Using multiple sets of parameters is also important in Bioinformatics, because parameter sets (for example, amino acid scores) sometimes depend on environments [7,16]. For example, *Dirichlet mixture* was effectively applied to sequence analysis [7]. Many techniques have been proposed for deriving a mixture of probability distributions such as the EM algorithm and local search heuristics [4,7]. Recently, Arora and Kannan proposed approximation algorithms for deriving a mixture of Gaussian distributions [4].

However, to our knowledge, there had been no algorithmic study on deriving a union or a mixture of halfspaces from examples. Therefore, we study the computational complexity of this problem and derive inapproximability as well as approximability results. Furthermore, we show that this problem has a close relationship to the problem of deriving a PSSM (*Position Specific Score Matrix*)

O.H. Ibarra and L. Zhang (Eds.): COCOON 2002, LNCS 2387, pp. 117–126, 2002.

from examples [1], where PSSMs are widely used in Bioinformatics [7]. Usually, PSSMs are derived from positive examples using simple statistical methods based on residue frequencies or local search algorithms (such as the EM algorithm) [7]. However, almost no theoretical results had been known. Recently, we proved that derivation of a PSSM (resp. derivation of a mixture of PSSMs) from both positive and negative examples is NP-hard [1].

We define the problem of deriving a union of halfspaces in the following way (see also Fig. 1), where a halfspace means a *closed halfspace*.

Problem 1. Given point sets POS and NEG in d-dimensional Euclidean space, find a minimum cardinality set of halfspaces $\{h_1, \ldots, h_k\}$ such that $POS \subseteq \cup_{i=1,\ldots,k} h_i$ and $NEG \cap (\cup_{i=1,\ldots,k} h_i) = \emptyset$.

Along with the above, we consider the following decision problem.

Problem 2. Given point sets POS and NEG in d-dimensional Euclidean space and an integer K, find a set of halfspaces $\{h_1, \ldots, h_K\}$ such that $POS \subseteq \cup_{i=1,\ldots,K} h_i$ and $NEG \cap (\cup_{i=1,\ldots,K} h_i) = \emptyset$.

It should be noted that there is a solution for K in Problem 2 if there is a solution for $K-1$. Hereafter, we let $n = |POS|$ and $m = |NEG|$.

In this paper, we prove that approximation of Problem 1 is at least as hard as approximation of graph coloring. We also prove that Problem 2 is NP-hard even for $K = 2$. This result is interesting because Problem 2 can be trivially solved in polynomial time using *linear programming* if $K = 1$. On the other hand, we show that the two-dimensional case of Problem 1 can be solved in polynomial time. We present approximation algorithms for a special case of Problem 1 and variants (maximization of the number of correctly classified examples) of Problem 2.

Though the *Boosting* technique [10] might be applied to these problems, it would not solve the problems optimally or near optimally because we derive strong hardness results.

We can consider the *dual problems* in which a union of halfspaces is replaced by an intersection of halfspaces. We can obtain analogous results for these problems by exchanging POS and NEG.

The organization of the paper is as follows. First, we show hardness results and a close relationship between a halfspace and a PSSM. Next, we present a polynomial time algorithm for the two-dimensional case of Problem 1. Then, we present approximation algorithms. Finally, we conclude with future work.

2 Hardness Results

We briefly introduce the problem of deriving a mixture of PSSMs [1]. Let POS_{pssm} and NEG_{pssm} be sets of strings of length l over an alphabet Σ. For a string S, $S[i]$ denotes the i-th letter of S. A PSSM is a function $f_i(a)$ from $[1, \ldots, l] \times \Sigma$ to the set of real numbers, where $i \in [1, \ldots, l]$ and $a \in \Sigma$. For a string S and a PSSM $f_i(a)$, we define $f(S)$ (the score of S) by $f(S) = \sum_{i=1,\ldots,l} f_i(S[i])$.

Problem 3. (Derivation of a Mixture of PSSMs)
Given Σ, POS_{pssm}, NEG_{pssm} and a positive integer K, find a set of K PSSMs

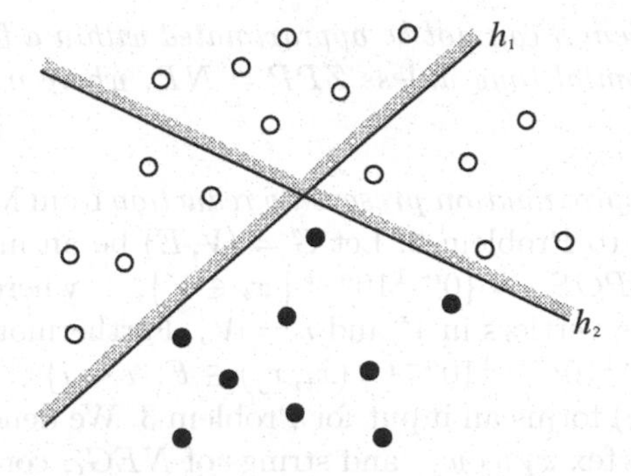

Fig. 1. Separation of positive points from negative points using two halfspaces. White circles and black circles denote positive points and negative points, respectively.

and a threshold Θ which satisfy the following conditions: (i) for all $S \in POS_{pssm}$, $f^k(S) \geq \Theta$ for some $k \in [1, \ldots, K]$, (ii) for all $S \in NEG_{pssm}$ and for all $k \in [1, \ldots, K]$, $f^k(S) < \Theta$, where f^k denotes the k-th PSSM.

It was proven in [1] that Problem 3 is NP-hard even for $K = 2$ and $\Sigma = \{0, 1\}$.

Theorem 1. *Problem 2 is NP-hard even for* $K = 2$.

Proof. We use a polynomial time reduction from Problem 3.

We identify Σ with $\{1, 2, \ldots, |\Sigma|\}$. For a point $\boldsymbol{p}$ in $(l|\Sigma|)$-dimensional Euclidean space, $\boldsymbol{p}[i]$ denotes the i-th coordinate value of $\boldsymbol{p}$. From each string S of length l in $POS_{pssm} \cup NEG_{pssm}$, we create a point $\boldsymbol{s}$ by $\boldsymbol{s}[|\Sigma|(i-1) + S[i]] = 1$ for $i = 1, \ldots, l$, $\boldsymbol{s}[j] = 0$ for the other j.

We identify a PSSM $f_i(a)$ with an $(l|\Sigma|)$-dimensional vector $\boldsymbol{a}$ by $\boldsymbol{a}[|\Sigma|(i - 1) + a] = f_i(a)$ for $i = 1, \ldots, l$ and $a = 1, \ldots, |\Sigma|$.

Then, we have the following relation: $f(S) \geq \Theta$ iff. $\boldsymbol{a} \cdot \boldsymbol{s} \geq \Theta$, where $\boldsymbol{a} \cdot \boldsymbol{s}$ is the *inner product* between $\boldsymbol{a}$ and $\boldsymbol{s}$. That is, each PSSM corresponds to a halfspace. It should be noted that each halfspace $\boldsymbol{a} \cdot \boldsymbol{s} \geq \theta$ can be normalized so that the righthand side value takes Θ.

From this relation, it is straight-forward to see the following:

there exists a solution $(\{f^1, \ldots, f^K\}, \Theta)$ for Problem 3 iff.
there exists a solution $\{\ \boldsymbol{a}^1 \cdot \boldsymbol{x} \geq \Theta,\ \ldots,\ \boldsymbol{a}^K \cdot \boldsymbol{x} \geq \Theta\ \}$ for Problem 2.

Therefore, the theorem follows from the NP-hardness of Problem 3 [1]. □

On the minimization of the number of PSSMs, we have a strong inapproximability result as shown below, where ZPP (Zero-error Probabilistic Polynomial time) is the class of languages that have Las Vegas algorithms running in expected polynomial time [15]. We say that a minimization problem (resp. a maximization problem) can be approximated within a factor of $f(n)$ if there is an algorithm for which $\max(\frac{APR}{OPT}, \frac{OPT}{APR}) \leq f(n)$ holds where APR and OPT are the scores of an approximate solution and an optimal solution, respectively.

Theorem 2. *Problem 3 can not be approximated within a factor of $O(n^{1-\epsilon})$ for any $\epsilon > 0$ in polynomial time unless $ZPP = NP$, where n denotes the number of positive strings.*

Proof. We use an *approximation preserving reduction* from Minimum Graph Coloring (GT4 in [11]) to Problem 3. Let $G = (V, E)$ be an undirected graph. We define POS_G as $POS_G = \{0^{i-1}10^{n-i} \mid x_i \in V\}$, where we choose an arbitrary ordering of the vertices in V and $n = |V|$. Furthermore, we define NEG_G as $NEG_G = \{0^{i-1}10^{j-i-1}10^{n-j} \mid (x_i, x_j) \in E,\ i < j\} \cup \{0^n\}$.
$I = (POS_G, NEG_G)$ forms an input for Problem 3. We denote strings of POS_G corresponding to vertex x_i as w_{x_i} and strings of NEG_G corresponding to $e \in E$ as w_e.

It suffices to show, that there is a solution for I with K matrices, iff G can be colored with K colors. Let $g : V \to \{1, \ldots, K\}$ be a coloring for G. For $k \in \{1, \ldots, K\}$, we define a PSSM f^k. Let $i \in \{1, \ldots, n\}$ and $a \in \{0, 1\}$.

$$f_i^k(a) = \begin{cases} 0 & \text{if } a = 0 \\ 1 & \text{if } a = 1,\ g(x_i) = k \\ \text{-1} & \text{if } a = 1,\ g(x_i) \neq k \end{cases}$$

With $\Theta = 1$, f^k accepts all strings of POS_G, which correspond to a vertex colored with color k. Since every string in NEG_G corresponds to an edge of G and all edges of G connect vertices with different colors, all negative strings are rejected by all PSSMs. Therefore, $f^1, \ldots, f^K$ is a solution for I with K PSSMs.

For the proof of the opposite direction, suppose there is a solution for I with K PSSMs. For every vertex x_i, we choose a PSSM f^k accepting w_{x_i} and color x_i with color k. If there were an edge $(x_i, x_j) \in E$ with both x_i and x_j colored with the same color k, we could conclude

$$f^k(w_{(x_i,x_j)}) + f^k(0^n) = f^k(w_{x_i}) + f^k(w_{x_j}) \geq 2\Theta,$$

which is a contradiction to f^k rejecting both $w_{(x_i,x_j)}$ and 0^n.

Thus, the above construction of an input for Problem 3 yields an approximation preserving reduction from Minimum Graph Coloring to Problem 3. Since Minimum Graph Coloring can not be approximated within $O(|V|^{1-\epsilon})$ unless $ZPP = NP$ ([9]), we have the theorem. □

Using the reduction in the proof of Theorem 1, we have:

Corollary 1. *Problem 1 can not be approximated within a factor of $O(n^{1-\epsilon})$ for any $\epsilon > 0$ in polynomial time unless $ZPP = NP$.*

In [1], it was shown that a set of $|POS|$ PSSMs can be computed by linear programming. Though it is a trivial solution to compute one PSSM for each positive string, this simple algorithm is surprisingly nearly optimal, since Theorem 2 implies that the minimal number of PSSMs can not be approximated within a factor of $O(|POS|^{1-\epsilon})$.

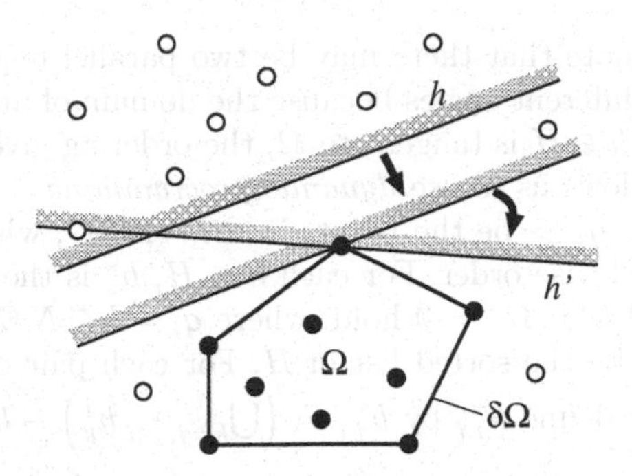

Fig. 2. Explanation of Proposition 1.

3 A Polynomial Time Algorithm in Two-Dimensions

In this section, we show that the two-dimensional case of Problem 1 can be solved in polynomial time.

We assume without loss of generality that all points in $POS \cup NEG$ are in general positions (i.e., no three points are collinear), where we can modify the algorithm for a general case without increasing the order of the time complexity.

For sets X and Y, $X - Y$ denotes the subset of X obtained by removing $X \cap Y$ from X. Let Ω denote the *convex hull* [8] of NEG, where the boundary is included in Ω. Let $\delta\Omega$ denote the boundary of Ω and $\Omega^- = \Omega - \delta\Omega$. Similarly, δh denotes the boundary of a halfplane h.

Proposition 1. *For each halfplane h such that $h \cap NEG = \emptyset$ and $h \cap POS \neq \emptyset$, there exists a halfplane h' satisfying the following conditions: (i) $h' \cap \Omega^- = \emptyset$, (ii) $(h \cap POS) \subseteq (h' \cap POS)$, (iii) $|h' \cap NEG| = 1$, (iv) $|\delta h' \cap POS| = 1$.*

Proof. First, simply translate h so that the intersection with NEG consists of exactly one point. Next, rotate the halfplane around the point so that the intersection of the boundary with POS consists of exactly one point. □

It follows that there exists a solution (a set of halfplanes) for Problem 1 and Problem 2 in which each halfplane is obtained by slightly perturbing a halfplane satisfying the conditions of Proposition 1.

Let H be the set of halfplanes satisfying the conditions of Proposition 1, where the number of such halfplanes is $O(n)$. We fix an arbitrary element h_0 in H. For each halfplane h, we define *angle*(h) (the angle of h to h_0) as follows. Let $\boldsymbol{q}$ be the intersection point of δh and δh_0. Assume that h_0 coincides with h if h_0 is rotated clockwise around $\boldsymbol{q}$ by A radian $(0 \le A < 2\pi)$. Then, we define *angle*$(h) = A$.

We define the order on halfplanes by

$$h \prec h' \text{ iff. } angle(h) < angle(h') .$$

For example, $h_0 \prec h_j^* \prec h_j \prec h_i^* \prec h_i$ holds for halfplanes in Fig. 3. It should be noted that *angle*(h) is a bijection from H to $[0, 2\pi)$ if the domain of *angle*(h)

is restricted to H (note that there may be two parallel hyperplanes tangent to $\delta\Omega$, but they have different angles because the domain of angles is defined to be $[0, 2\pi)$). Since each $h \in H$ is tangent to Ω, the ordering gives the *total order* on H. This ordering allows us to use *dynamic programming.*

Let $< \boldsymbol{q}_1, \boldsymbol{q}_2, \ldots, \boldsymbol{q}_h >$ be the points in $\delta\Omega \cap NEG$, where these points are arranged in the clockwise order. For each $h \in H$, h^* is the halfplane such that $h^* \supset \{\boldsymbol{q}_i, \boldsymbol{q}_{i-1}\}$ and $h^* \cap \Omega^- = \emptyset$ hold, where $\boldsymbol{q}_i = h \cap NEG$ and $\boldsymbol{q}_0 = \boldsymbol{q}_h$. Let $< h_0, h_1, \ldots, h_N >$ be the sorted list of H. For each pair of halfplanes (h_j, h_i) such that $j < i$, we define $h^*_{j,i}$ by $h^*_{j,i} = \left(\bigcup_{k=j,\ldots,i} h^*_k\right) - h^*_j$, where $h^*_{j,i} = \emptyset$ if $h^*_j = h^*_i$.

Algorithm 1 shown below (see also Fig. 3) outputs the minimum number of halfplanes, where it can be easily modified for computing the halfplanes. Though we assume that $h^*_0 \neq h^*_{i_k}$ holds for the optimal solution $< h_{i_1} = h_0, h_{i_2}, \ldots, h_{i_k} >$, Algorithm 1 can be modified for the case of $h^*_0 = h^*_{i_k}$ without increasing the order of the time complexity.

```
ALGORITHM 1
  Construct the convex hull Ω of NEG;
  if POS ∩ Ω ≠ ∅ then output "no solution" and halt;
  k ← +∞;
  for all h_0 ∈ H do k ← min(k,SUBPROC(h_0));
  Output k;

SUBPROC(h_0)
  Let < h_0, h_1, ..., h_N > be the sorted list of H;
  if h_0 ∩ POS = POS return 1;
  T[0] ← 1;
  for i = 1 to N do T[i] ← +∞;
  for i = 1 to N do
    for j = 0 to i − 1 do
      if (h*_{j,i} ∩ POS) ⊆ h_j ∪ h_i then T[i] ← min(T[i], T[j] + 1);
  for i = 1 to N do
    if (POS − h*_{0,i}) − (h_0 ∪ h_i) ≠ ∅ then T[i] ← +∞;
  return min{T[i] | i = 1, ..., N, h*_i ≠ h*_0};
```

Theorem 3. *The two-dimensional case of Problem 1 can be solved in polynomial time.*

Proof. First, we show the correctness of Algorithm 1. Clearly, Problem 1 has a solution if and only if $POS \cap \Omega = \emptyset$.

From Proposition 1, it suffices to find a minimum cardinality set of halfplanes satisfying $POS \subseteq \cup h_i$ and the conditions of Proposition 1. Let H_{opt} be such a set.

Then, we can put the total order on H_{opt} by choosing an arbitrary element h_0 in H_{opt}. Let $< h_0 = h_{i_1}, h_{i_2}, \ldots, h_{i_k} >$ be the ordered list of H_{opt}. Since all the

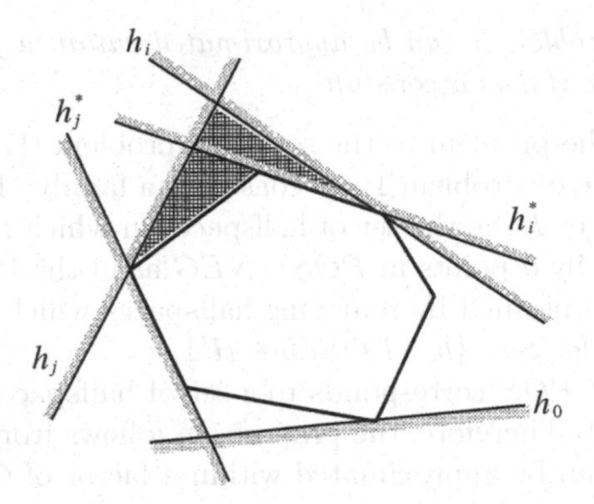

Fig. 3. Explanation of Algorithm 1. $T[i]$ is updated if the halftone region $(h_{j,i}^* - h_j - h_i)$ does not contain a point of POS.

elements in POS must be covered by $\cup h_{i_k}$, $\left(h_{i_j,i_{j+1}}^* \cap POS\right) \subseteq h_{i_j} \cup h_{i_{j+1}}$ must hold for $j = 1, \ldots, k-1$ and $(POS - h_{0,i_k}^*) \subseteq (h_0 \cup h_{i_k})$ must hold. Moreover, any h_k such that $k > i > j$ can not cover $h_{j,i}^* - h_j - h_i$. Therefore, we can use dynamic programming technique for computing the minimum number of halfspaces required to cover $\cup_{j=0}^{i} h_j$ for each i, where the number will be stored in $T[i]$. Since all h_0 are examined, Algorithm 1 will find H_{opt}.

Next we analyze the time complexity. The convex hull of NEG can be constructed in $O(m \log m)$ time [8]. Before applying the dynamic programming, we compute $h_{i,j}^* \cap POS$ for all i, j. It takes $O((n+m)^2 n)$ time using an incremental procedure, where we omit details. Since both $|H|$ and $|POS|$ are $O(n)$, the dynamic programming procedure takes $O(n^3)$ time per h_0.

Therefore, Algorithm 1 takes $O(n^4 + m^2 n)$ time in total. □

4 Approximation Algorithms in *d*-Dimensions

In this section, we assume that all points in $POS \cup NEG$ are in general positions. It is easy to see that Problem 2 can be solved in polynomial time if both d and K are fixed.

Proposition 2. *Problem 2 can be solved in $O((n+m)^{dK+1})$ time for constants d and K.*

Proof. It is easy to see that if Problem 2 has a solution, then there exists a solution in which the boundary of each halfspace is defined by d points in $POS \cup NEG$, where we slightly perturb it so that each halfspace does not contain the point(s) in NEG on the boundary. Therefore, it is enough to examine all combinations of K such halfspaces. Since the number of halfspaces mentioned in the above is $O((n+m)^d)$, the number of the combinations is $O((n+m)^{dK})$. For each combination, $O(n+m)$ time is required for checking the conditions of Problem 2. □

Proposition 3. *Problem 1 can be approximated within a factor of $O(\log n)$ in $O((n+m)^{d+1})$ time if d is a constant.*

Proof. We reduce the problem to the set cover problem [11,12].

From an instance of Problem 1, we construct a family of sets $\mathcal{S}$ over POS in the following way. Let H be the set of halfspaces in which the boundary of each halfspace is defined by d points in $POS \cup NEG$ as in the Proposition 2. Let H' be the subset of H obtained by removing halfspaces which contain at least one point in NEG. We let $\mathcal{S} = \{h \cap POS | \ h \in H'\}$.

Then, a cover of POS corresponds to a set of halfspaces satisfying the conditions of Problem 1. Therefore, the proposition follows from a fact [12] that the set cover problem can be approximated within a factor of $O(\log n)$. □

From a practical viewpoint, it is more important to develop algorithms for minimizing the number of misclassified examples or maximizing the number of correctly classified examples, where the number of halfspaces is bounded by a constant K. We consider the latter case in this paper, because it seems difficult to develop a good approximation algorithm for the former problem [3]. Let H be a set of halfspaces. $\boldsymbol{p} \in POS$ (resp. $\boldsymbol{p} \in NEG$) is called a *true positive* (resp. a *false positive*) if $\boldsymbol{p} \in h$ for some $h \in H$. $\boldsymbol{p} \in NEG$ (resp. $\boldsymbol{p} \in POS$) is called a *true negative* (resp. *false negative*) if $\boldsymbol{p} \notin h$ for any $h \in H$. Let $\#TP$ and $\#FP$ denote the numbers of true positives and false positives, respectively. Let $\#TN$ and $\#FN$ denote the numbers of true negatives and false negatives, respectively. The following proposition is almost trivial (or directly follows from the result of Amaldi and Kann [2]).

Proposition 4. *The maximization problem of $\#TP + \#TN$ can be approximated within a factor of 2 in polynomial time.*

Proof. We can use the following simple algorithm: output a halfspace which covers all elements in $POS \cup NEG$ if $|POS| \geq |NEG|$, otherwise output a halfspace which covers no element in $POS \cup NEG$. □

We are also interested in maximizing $\#TP$ under the condition that $\#FP = 0$ since it is sometimes important not to output false positives. We call it Problem 4. Clearly, Problem 4 is NP-hard even for $K = 2$. We propose a simple randomized approximation algorithm (Algorithm 2) for the case of $K = 2$.

ALGORITHM 2
Let R be a set of r points randomly selected from POS;
$H_{max} \leftarrow \emptyset$;
for all partition $R_1 \cup R_2$ of R **do**
  $H \leftarrow \emptyset$;
  if there is a halfspace h_1 such that $h_1 \cap R_1 = R_1$ and $h_1 \cap NEG = \emptyset$
  then $H \leftarrow H \cup h_1$;
  if there is a halfspace h_2 such that $h_2 \cap R_2 = R_2$ and $h_2 \cap NEG = \emptyset$
  then $H \leftarrow H \cup h_2$;
  if $|H \cap POS| > |H_{max} \cap POS|$ **then** $H_{max} \leftarrow H$;
Output H_{max};

The performance of Algorithm 2 is poor if $\#TP$ in the optimal solution is small (e.g., $\#TP$ is $O(\log n)$). However, such a case is meaningless since we need a classifier with small errors. Therefore, we are interested in the case where $\#TP$ is large enough in the optimal solution. In order to analyze Algorithm 2 for such a case, we use β-center points. A point $c \in \mathcal{R}^d$ is called a *β-center point* of a point set P if every closed halfspace containing c contains at least βn points of P [5]. Clarkson *et al.* showed that $\left(\frac{1}{d+1} - \epsilon\right)$-center point can be found in $O((d/\epsilon)^2 \log(d/\epsilon))^{d+O(1)} \log(1/\delta))$ time with probability $1-\delta$ by using $O((d/\epsilon) \cdot \log(d/\epsilon) \cdot \log(1/\delta))$ elements randomly sampled from P.

Theorem 4. *Let $K = 2$. Suppose that $\#TP > n/2$ holds in the optimal solution for Problem 4. For any constants d, $\delta > 0$ and $\epsilon > 0$, Algorithm 2 outputs a pair of hyperplanes such that $\#FP = 0$ and $\#TP \geq \frac{n}{4} \cdot \left(\frac{1}{(d+1)} - \epsilon\right)$ in $O(n+m)$ time with probability $1-\delta$.*

Proof. Let (h_a, h_b) be an optimal solution, where we assume without loss of generality that $|h_a \cap POS| \geq |h_b \cap POS|$.

Using a sufficiently large constant r, there exists R_1 with high probability which can be considered as a random sample of size $O((d/\epsilon) \cdot \log(d/\epsilon) \cdot \log(1/\delta))$ from $h_a \cap POS$. Since h_1 contains R_1 and $|h_a \cap POS| \geq \frac{n}{4}$ holds, h_1 contains $\frac{n}{4} \cdot \left(\frac{1}{d+1} - \epsilon\right)$ points of POS with high probability.

Since r is a constant and linear programming can be done in linear time for fixed d [14], Algorithm 2 works in linear time. □

It should be noted that Problem 4 can be solved exactly in $O((n+m)^{dK+1})$ time for fixed d and K as in Proposition 2.

5 Concluding Remarks

In this paper, we studied the complexity of inferring a union of halfspaces from positive and negative examples. Though this problem was motivated by applications in Bioinformatics, we believe that it has applications in other areas, too. We presented a polynomial time algorithm for Problem 1 in two-dimensions. However, it is left as an open problem whether or not Problem 1 is NP-hard in three-dimensions (or higher dimensions).

From a practical viewpoint, maximization of the number of correctly classified examples (or, minimization of the number of misclassified examples) is more important. We presented approximation algorithms for this purpose. However, the performances of the algorithms are far from enough to be practical. Development of better algorithms is important future work as well as deriving lower bounds on the performance ratios.

Acknowledgement

Tatsuya Akutsu was supported in part by a Grant-in-Aid for Scientific Research on Priority Areas (C) for "Genome Information Science" and Grant-in-Aid #13680394 from the Ministry of Education, Culture, Sports, Science and Technology (MEXT) of Japan. We thank Jean-Phillipe Vert for helpful discussions.

References

1. Akutsu, T., Bannai, H., Miyano, S., Ott, S.: On the complexity of deriving position specific score matrices from examples. Proc. CPM 2002. Lecture Notes in Computer Science (to appear)
2. Amaldi, E., Kann, V.: The complexity and approximability of finding maximum feasible subsystems of linear relations. Theoretical Computer Science **147** (1995) 181–210
3. Amaldi, E., Kann, V.: On the approximability of minimizing nonzero variables or unsatisfied relations in linear systems. Theoretical Computer Science **209** (1998) 237–260
4. Arora, S., Kannan, R.: Learning mixtures of arbitrary gaussians. Proc. 33rd ACM Symp. Theory of Computing (2001) 247–257
5. Clarkson, K.L., Eppstein, D., Miller, G.L., Sturtivant, C., Teng S-H.: Approximating center points with iterated radon points. Proc. 9th ACM Symp. Computational Geometry (1993) 91–98
6. Cortes, C., Vapnik, V.: Support-vector networks. Machine Learning **20** (1995) 273–297
7. Durbin, R., Eddy, S., Krogh, A., Mitchison, G.: Biological Sequence Analysis. Probabilistic Models of Proteins and Nucleic Acids. Cambridge University Press, New York (1998)
8. Edelsbrunner, H.: Algorithms in Combinatorial Geometry. Springer-Verlag, Berlin Heidelberg New York (1987)
9. Feige, U., Kilian, J.: Zero knowledge and the chromatic number. J. Computer and System Sciences **57** (1998) 187–199
10. Freund, Y., Schapire, R.E.: A decision-theoretic generalization of on-line learning and an application to boosting. J. Computer and System Sciences **55** (1997) 119–139
11. Garey, M.R., Johnson, D.S.: Computers and Intractability. Freeman, New York (1979)
12. Johnson, D.S.: Approximation algorithms for combinatorial problems. J. Computer and System Sciences **9** (1974) 256–278
13. Kearns, M.J., Vazirani, U.V.: An Introduction to Computational Learning Theory. The MIT Press, Cambridge, MA (1994)
14. Megiddo, N.: Linear programming in linear time when the dimension is fixed. SIAM J. Computing **12** (1983) 759–776
15. Motowani, R., Raghavan, P.: Randomized Algorithms. Cambridge University Press, New York (1994)
16. Radivojac, R., Obradoviv, Z., Brown, C.J., Dunker, A.K.: Improving sequence alignments for intrinsically disordered proteins. Pacific Symp. Biocomputing **7** (2002) 589–600

Dictionary Look-Up within Small Edit Distance

Abdullah N. Arslan and Ömer Eğecioğlu*

Department of Computer Science
University of California, Santa Barbara
Santa Barbara, CA 93106 USA
{arslan,omer}@cs.ucsb.edu

Abstract. Let $\mathcal{W}$ be a dictionary consisting of n binary strings of length m each, represented as a trie. The usual d-query asks if there exists a string in $\mathcal{W}$ within Hamming distance d of a given binary query string q. We present an algorithm to determine if there is a member in $\mathcal{W}$ within *edit distance* d of a given query string q of length m. The method takes time $O(dm^{d+1})$ in the RAM model, independent of n, and requires $O(dm)$ additional space.

1 Introduction

Let $\mathcal{W}$ be a dictionary consisting of n binary strings of length m each. A *d-query* asks if there exists a string in $\mathcal{W}$ within Hamming distance d of a given binary query string q. Algorithms for answering d-queries efficiently has been a topic of interest for some time, and have also been studied as the *approximate query* and the *approximate query retrieval* problems in the literature. The problem was originally posed by Minsky and Papert in 1969 [10] in which they asked if there is a data structure that supports fast *d-queries.*

The cases of small d and large d for this problem seem to require different techniques for their solutions. The case when d is small was studied by Yao and Yao [14] . Dolev et al. [5,6] and Greene et al. [7] have made some progress when d is relatively large. There are efficient algorithms only when $d = 1$; proposed by Brodal and Venkadesh [3], Yao and Yao [14], and Brodal and Gasieniec [2]. The small d case has applications in password security [9]. Searching biological sequence databases may also use the methods of answering d-queries.

Previous studies for the d-query problem have focused on minimizing the number of memory accesses for a d-query, assuming other computations are free, and used cell or bit probe models to express complexity. We assume a RAM model with constant memory access time and take into account all computations in the complexity analysis. Dolev et al. [6] presented bounds for the space and time complexity of the d-query problem under certain assumptions using various notions of proximity. In the model, $\mathcal{W}$ is stored in buckets, and preprocessing of $\mathcal{W}$ is allowed.

* Supported in part by NSF Grant No. EIA–9818320.

O.H. Ibarra and L. Zhang (Eds.): COCOON 2002, LNCS 2387, pp. 127–136, 2002.

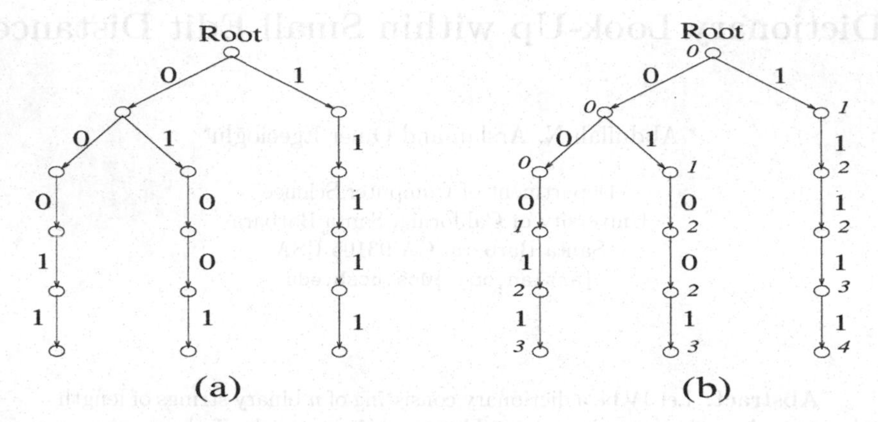

Fig. 1. a) An example trie with binary words 00011, 01001, and 11111 . b) The numbers in italic are the node weights computed with respect to query string 00100.

In this paper we consider answering d-queries efficiently without limiting ourselves to the construction of a new data structure parametrized by d. The variant of the original d-query problem that we consider is when the string-to-string *edit distance* is used as the distance measure instead of the ordinary case of Hamming distance. We assume that $\mathcal{W}$ is stored as a trie $\mathcal{T}_m$, and propose two algorithms for the d-query problem in this case. Our algorithms use the hybrid tree/dynamic programming approach [4]. The first one (Algorithm *LOOK-UP*$_{ed}$, Figure 4) requires $O(dm^{d+2})$ time in the worst case, and $O(dm^{d+1})$ space (in addition to the space requirements of the trie $\mathcal{T}_m$). This complexity is of interest for small values of d under investigation. The second algorithm (Algorithm *DFT-LOOK-UP*$_{ed}$, Figure 7) has time complexity $O(dm^{d+1})$, and additional space complexity of only $O(dm)$.

There is reason to believe that the average performance of both algorithms is much better when $\mathcal{W}$ is sparse.

2 Motivation: Hamming Distance Based Methods

Hamming distance between two binary strings is the number of positions they differ. A d-query asks if there is a member in a dictionary $\mathcal{W}$ whose Hamming distance is at most d from a given binary query string q.

We assume a trie representation $\mathcal{T}_m$ for $\mathcal{W}$, and assume for simplicity that $\mathcal{W}$ consists of binary words of length m each. A *trie* is a tree whose arcs are labeled by the symbols of alphabet Σ, in this case $\Sigma = \{0, 1\}$. The leaf nodes of $\mathcal{T}_m$ correspond to the words in $\mathcal{W}$, and when concatenated, the labels of arcs on a path from the root to a given intermediate node gives a prefix of at least one word in $\mathcal{W}$. Clearly, in the RAM model assumed, accessing a word in $\mathcal{W}$ takes $O(m)$ time. Figure 1 part (a) shows an example trie $\mathcal{T}_5$ representing a dictionary $\mathcal{W} = \{00011, 01001, 11111\}$.

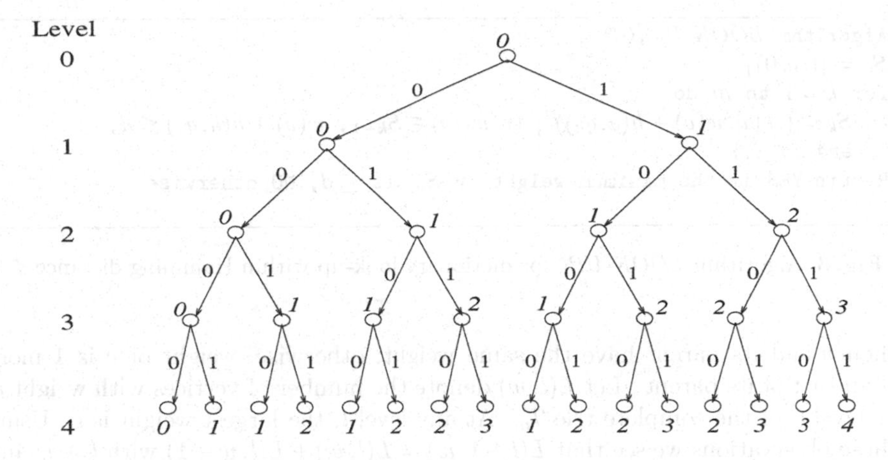

Fig. 2. In a complete binary trie of height 4, weights with respect to a given binary query string are shown in italic.

A naive method for answering a d-query is to generate the whole set of $\sum_{k=0}^{d}\binom{m}{k}$ strings differing from q in at most d positions, and with every string generated, perform a dictionary look-up in $\mathcal{T}_m$ for an exact member in $\mathcal{W}$. This naive *generate and test* algorithm takes $O(m^{d+1})$ time and $O(m)$ additional space to store a generated string at a time. Another naive method is to add all strings within Hamming distance d from any member in $\mathcal{W}$ to obtain a bigger dictionary $\mathcal{W}'$. Then any d-query can be answered in $O(m)$ time using the corresponding trie $\mathcal{T}'_m$ for an exact member. This latter method significantly increases the size of $\mathcal{W}$ by a number roughly $O(nm^d)$ m-bit members. Cost of constructing and maintaining $\mathcal{T}'_m$ may be extremely high.

For Hamming distance, we can improve the first naive algorithm above as follows. Let $s(v)$ denote the prefix corresponding to trie node v. Given a query string q, suppose that we assign weight w_h to each trie node v in $\mathcal{T}_m$ as

$$w_h(v) = h(s(v), q_{1\cdots|s(v)|}), \tag{1}$$

where h denotes Hamming distance. As an example, in Figure 1 (*b*) the weights of the nodes have been computed with respect to query string $q = 00100$. The idea is that we can prune the trie in our search for q at the nodes in $\mathcal{T}_m$ with $w_h(v) > d$.

Lemma 1. *Let N be the number of nodes in $\mathcal{T}_m$ with weight $\leq d$ as defined in (1). Then $N = O(m^{d+1})$.*

Proof. It is easy to see that N is maximized over all tries $\mathcal{T}_m$ when $\mathcal{T}_m$ is a complete trie over Σ, i.e. $\mathcal{T}_m$ contains all binary strings of length m. Figure 2 shows node weights of a complete trie with respect to q up to level 4 starting with the root at level 0. The root has weight 0. For any other vertex v at level l, if the arc from its parent to v has label q_l (the lth symbol of query string q)

```
Algorithm LOOK-UP_h(d)
S_0 = {(v_0, 0)}
for k = 1 to m do
    S_k = {(v[a], w(v) + h(a, q_k)) | (v, w(v)) ∈ S_{k-1}, w(v) + h(a, q_k) ≤ d,
    and a ∈ Σ}
Return YES if the minimum weight in S_m is ≤ d, NO otherwise
```

Fig. 3. Algorithm *LOOK-UP*$_h$ for dictionary look-up within Hamming distance d.

then v and its parent have the same weight; otherwise, weight of v is 1 more than that of its parent . Let $L(l, w)$ denote the number of vertices with weight w at level l of the complete trie $\mathcal{T}_m$. At any level l, the largest weight is l . Using these observations we see that $L(l+1, w) = L(l, w) + L(l, w-1)$ with $l \geq w$ and $L(l, 0) = 1$. Therefore $L(l, w)$ is the binomial coefficient $\binom{l}{w}$. Furthermore since the smallest level at which weight w appears in $\mathcal{T}_m$ is $l = w$, the total number of vertices with weight w in $\mathcal{T}_m$ is $\binom{w}{w} + \binom{w+1}{w} + \cdots + \binom{m}{w} = \binom{m+1}{w+1}$. Hence

$$N = \sum_{w=0}^{d} \binom{m+1}{w+1} = O(m^{d+1})$$

Based on the above lemma, Figure 3 outlines Algorithm *LOOK-UP*$_h$ for dictionary look-up within Hamming distance d. The algorithm explores all nodes v in $\mathcal{T}_m$ with weight $w_h(v) \leq d$, i.e. $s(v)$ is a prefix of a word in $\mathcal{W}$ whose Hamming distance from q is potentially within d. S_k stores the set of node-weight pairs $(v, w_h(v))$ for all nodes v at levels $\leq k$ with weight $w_h(v) \leq d$. The algorithm iteratively computes S_k from S_{k-1} by collecting all pairs $(v[a], w(v) + h(a, q_k))$ in S_k where $(v, w(v)) \in S_{k-1}$, $w(v) + h(a, q_k) \leq d$, and $a \in \Sigma = \{0, 1\}$. Clearly, if there is a member in $\mathcal{W}$ within Hamming distance d then it will be captured in S_m in which case the algorithm returns YES; otherwise it returns NO.

S_k contains $O(m^{d+1})$ node-weight pairs by Lemma 1. Therefore the time complexity in the assumed model is $O(m^{d+2})$. It also requires additional space to store $O(m^{d+1})$ trie nodes. The time complexity is no better in the worst case than that of the naive algorithm which generates and tries all possible strings within Hamming distance d from q. However for a sparse dictionary $\mathcal{W}$ Algorithm *LOOK-UP*$_h$ is bound to be much faster on the average.

3 Edit Distance Based *d*-Queries

Algorithm *LOOK-UP*$_h$ essentially computes the Hamming distance between q and a selected part of dictionary $\mathcal{W}$. Next we investigate the possibility of using a similar idea for d-queries defined with respect to the *edit distance*.

For the purposes of this paper, we use a simple type of edit distance. Given two strings $p = p_1 \cdots p_m$ and $q = q_1 \cdots q_m$, the edit distance $ed(p, q)$ is the minimum number of edit operations which transforms p into q. The edit operations are of three types: insert, delete, and substitute. Substituting a symbol by itself is called a *match*. A match operation is the only operation that does not contribute to the number of steps of the transformation. In terms of costs, all edit operations have cost 1 except for the match whose cost is 0. The usual framework for the analysis of edit distance is the *edit graph*. Edit Graph $G_{p,q}$ is a directed acyclic graph having $(m+1)^2$ lattice points (u,v) as vertices for $0 \leq u, v \leq m$. Horizontal and vertical arcs correspond to insert and delete operations respectively. The diagonal arcs correspond to substitutions. Each arc has a cost corresponding to the edit operation it represents. If we trace the arcs of a path from node $(0,0)$ to an intermediate node (i,j), and perform the indicated edit operations in the given order on $p_1 \cdots p_i$ then we obtain $q_1 \cdots q_j$. Edit distance between prefixes p_i and q_j is the cost of the minimum-cost path from $(0,0)$ to (i,j), and can be computed from the distances achieved at nodes $(i-1,j)$, $(i-1,j-1)$, and $(i,j-1)$. Hence it has a simple dynamic programming formulation [13]:

$$D_{i,j} = \min\{\ D_{i-1,j}+1,\ D_{i-1,j-1}+h(p_i,q_j),\ D_{i,j-1}+1\} \tag{2}$$

for $1 \leq i,j \leq m$, with $D_{i,j} = 0$ whenever $i = 0$ or $j = 0$.

3.1 Algorithm *LOOK-UP*$_{ed}$

With respect to a given binary query string q, we assign weight w_{ed} to any trie node v in $\mathcal{T}_m$ as

$$w_{ed}(v) = \min\{ed(s(v), r) \mid r \text{ is a prefix of } q\} \tag{3}$$

where *ed* denotes the edit distance.

Lemma 2. *Let N be the number of nodes in $\mathcal{T}_m$ with weight $\leq d$ as defined in (3). Then $N = O(m^{d+1})$.*

Proof. Proof is similar to the proof of Lemma 1 for the Hamming distance case. Analysis of N over a complete binary trie gives the maximum N. We omit the details. We remark however that in general $w_{ed}(v) \neq w_h(v)$, and for non-complete binary tries the distribution of weights over the nodes can differ significantly for Hamming and edit distances.

Algorithm *LOOK-UP*$_{ed}$ shown in Figure 4 extends the dynamic programming formulation (2) of the edit distance computation by considering all prefixes of all members in $\mathcal{W}$. $S_{i,j}$ stores all the node-weight pairs $(v, w_{ed}(v))$ where $s(v) = p_1 \cdots p_i$ for some $p \in \mathcal{W}$, and $w_{ed}(v) = ed(p_1 \cdots p_i, q_1 \cdots q_j)$.

The computations in *LOOK-UP*$_{ed}$ involve sets, as opposed to just scores of the ordinary edit distance computations. Edit operation and the trie nodes involved determine an action on the sets. Consider the operations resulting in at node (i,j) of the edit graph as shown in Figure 5. Let the operation be the

Algorithm $LOOK\text{-}UP_{ed}(d)$
$S_{i,-1} = \emptyset$ for all i, $1 \le i \le d$
$S_{-1,j} = \emptyset$ for all j, $1 \le j \le d$
$S_{0,0} = \{(v_0, 0)\}$
for $i = 0$ to m do
  for $j = \max\{0, i - \lceil d/2 \rceil\}$ to $\min\{m, i + \lfloor d/2 \rfloor\}$ do
  {
   if ($i = 0$ and $j = 0$) then continue with the next iteration
   else
   {
    $S'_{i-1,j} = \{(v[a], t+1) \mid t+1 \le d,\ (v,t) \in S_{i-1,j}$ and $a \in \Sigma\}$
    $S'_{i,j-1} = \{(v, t+1) \mid t+1 \le d,\ (v,t) \in S_{i-1,j}\}$
    $S'_{i-1,j-1} = \{(v[a], t + h(a, q_j) \mid t + h(a, q_j) \le d,\ (v,t) \in S_{i-1,j-1},\ a \in \Sigma\}$
    $S_{i,j} = \{(v,t) \mid t$ is the minimum weight
          paired with leaf v in $S'_{i-1,j} \cup S'_{i,j-1} \cup S'_{i-1,j-1}\}$
   }
  }
Return YES if the minimum weight in $S_{m,m}$ is $\le d$; otherwise NO

Fig. 4. Algorithm $LOOK\text{-}UP_{ed}$ for dictionary look-up within edit distance d.

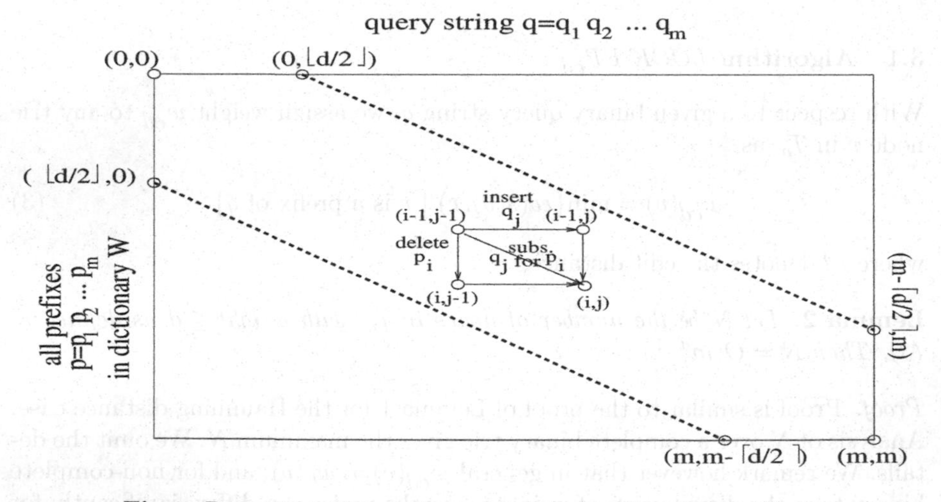

Fig. 5. Part of the edit graph explored during the computations.

deletion of symbol $p_i \in \Sigma$. For $(v,t) \in S_{i-1,j}$ if there is an arc from v to $v[a]$ with label $a \in \Sigma$ then the delete operation causes weight $t+1$ in $v[a]$. This potential weight assignment is reflected in set $S'_{i-1,j}$, and realized in set $S_{i,j}$ only if no weight smaller than $t+1$ is achieved by other edit operations resulting in $v[a]$. Now consider the insertion of q_j. For each $(v,t) \in S_{i,j-1}$, the pair $(v, t+1)$ is inserted into $S'_{i,j-1}$. Similarly for each $(v,t) \in S_{i-1,j-1}$, $(v[a], t + h(a, q_j))$ is inserted into $S'_{i-1,j-1}$. Subsequently, all node-weight pairs in sets $S'_{i-1,j}$, $S'_{i,j-1}$,

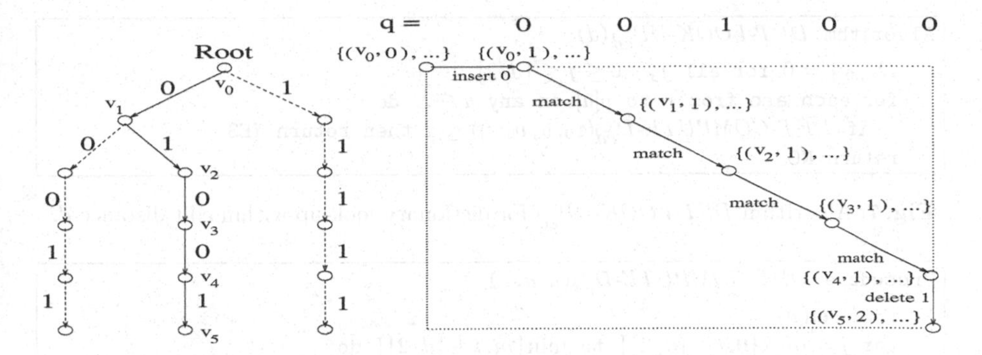

Fig. 6. Partial results on an edit path for a member within edit distance 2 of q.

and $S'_{i-1,j-1}$ are collected into set $S_{i,j}$ by including for each node at most one pair, namely the one with the minimum weight.

The computations on the edit graph can be restricted to a narrow diagonal band of the edit graph as shown in Figure 5, since any edit path with total weight at most d completely lies in this band.

We can easily show by induction that for all i, j with $0 \le i \le m$, and $\max\{0, i - \lceil d/2 \rceil\} \le j \le \min\{i, i + \lfloor d/2 \rfloor\}$, $(v, t) \in S_{i,j}$ iff there exists a trie node v such that $|s(v)| = i$ and $ed(s(v), q_1 \cdots q_j) = t$. This implies that $S_{m,m}$ includes a node-weight pair with weight $\le d$ for a leaf node iff there exists a member in $\mathcal{W}$ within edit distance d. Therefore the algorithm is correct.

Figure 6 shows an example edit path with partial results which identify member 01001 as within edit distance 2 of $q = 00100$.

By Lemma 2, computing each $S_{i,j}$ from $S_{i-1,j}$, $S_{i,j-1}$, $S_{i-1,j-1}$ takes $O(m^{d+1})$ time since the size of each of these sets is bounded by $O(m^{d+1})$. Therefore the time complexity of Algorithm *LOOK-UP*$_{ed}$ is $O(dm^{d+2})$, and it requires $O(dm^{d+1})$ additional space since it is enough to store the sets of the previous and current rows as the processing is done row by row.

3.2 Algorithm *DFT-LOOK-UP*$_{ed}$

Next we propose Algorithm *DFT-LOOK-UP*$_{ed}$ with which we improve the time complexity of the d-query problem with respect to edit distance to $O(dm^{d+1})$, and space complexity to $O(dm)$. The steps of the algorithm are shown in Figure 7. The algorithm is based on depth-first traversal (DFT) of trie $\mathcal{T}_m$, during which the entries of the dynamic programming matrix are computed row by row. For trie node v, $i = level(v)$ (Figure 9), and $\max\{0, i - \lceil d/2 \rceil\} \le j \le \min\{m, i + \lfloor d/2 \rfloor\}$ we define $D_{v,i,j}$ as

$$D_{v,i,j} = ed(s(v), q_1 \cdots q_j)$$

Algorithm *DFT-LOOK-UP*$_{ed}$ performs the initialization of scores for the first row, and invokes Procedure *DFT-COMPUTE-D*$_{ed}$ for each arc from root v_0 to

```
Algorithm DFT-LOOK-UP_ed(d)
    D_{v0,0,j} = 0 for all j,   0 ≤ j ≤ ⌊d/2⌋
    for each arc from v to v[a] on any a ∈ Σ do
        if DFT-COMPUTE-D_ed(v0, a, v0[a]) ≤ d then return YES
    return NO
```

Fig. 7. Algorithm *DFT-LOOK-UP*$_{ed}$ for dictionary look-up within edit distance d.

```
Procedure DFT-COMPUTE-D_ed(u, a, v)
    i = level(v)
    for j = max{0, i − ⌈d/2⌉} to min{m, i + ⌊d/2⌋} do
        D_{v,i,j} = min{ D_{u,i−1,j} + 1,  D_{u,i−1,j−1} + h(a, q_j),  D_{u,i,j−1} + 1}
    weight = min{D_{v,i,j} | max{0, i − ⌈d/2⌉} ≤ j ≤ min{m, i + ⌊d/2⌋}}
    if v is a leaf node or weight  > d then return weight
    return min{DFT-COMPUTE-D_ed(v, a, v[a]) | there is an arc
                incident from v to v[a] on a ∈ Σ}
```

Fig. 8. Procedure *DFT-COMPUTE-D*$_{ed}$ for computing the minimum edit distance achieved in subtrie rooted at v.

$v_0[a]$ with label a. If any of these invocations returns a value $\leq d$ then the algorithm returns YES; otherwise returns NO.

Given a parent node u, children node v, and symbol $a \in \Sigma$ of the arc connecting these two, Procedure *DFT-COMPUTE-D*$_{ed}(u, a, v)$ first computes the values in the row of node v using the values in the row of parent node u in the edit graph (Figure 9). If all computed entries in this row are $> d$, then the procedure returns the minimum of these numbers. Otherwise it traverses the subtrie rooted at v in depth-first manner, computes and returns the minimum edit distance achievable in the leaf nodes of this subtrie.

To show correctness, we claim that $D_{v,i,j}$ stores $ed(s(v), q_1 \cdots q_j)$ where $i = level(v)$. This can be shown by induction. Assume that before v is visited for the parent of v, and corresponding entries, the claim is true then following the computations for v we can easily see that the claim will be true for v after the processing is done for the entries of v. Another induction on the subtries of v reveals that the procedure call on v will return the minimum edit distance achieved in the leaves of the subtrie rooted at v. Therefore the algorithm returns YES iff there is a member in $\mathcal{W}$ within edit distance d of q.

Depth-first traversal visits $O(m^{d+1})$ trie nodes by Lemma 2. With every node visited $O(d)$ operations are performed. Therefore the time complexity of the algorithm is $O(dm^{d+1})$. Dept-first traversal requires that a branch of $O(m)$ trie nodes be stored. For each node $O(d)$ entries are maintained. Hence the space complexity of the algorithm is $O(dm)$.

We can adapt *DFT-LOOK-UP*$_{ed}$ (and *DFT-COMPUTE-D*$_{ed}$) to Hamming distance computations as well. In this case, we only need to consider the diagonal entries of the edit graph, and each entry is computed using only the value of the parent which is stored in the diagonal of the previous level. The result-

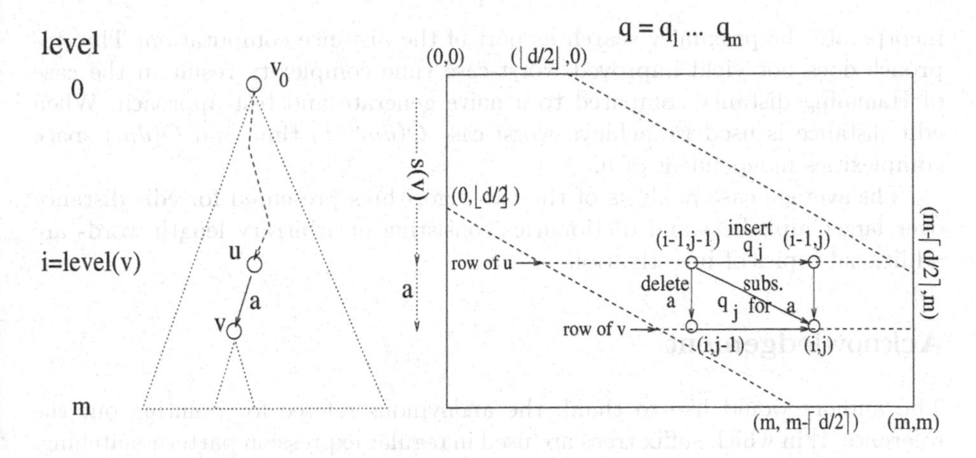

Fig. 9. Depth-first traversal on the trie, and the region of the dynamic programming matrix in which the computations are performed.

ing algorithm *DFT-LOOK-UP*$_h$ for Hamming distance based d-queries has time complexity $O(m^{d+1})$ and space $O(m)$.

4 Remarks

For clarity of presentation, we have assumed a dictionary of words of equal length. With some additional care our method can be generalized to the cases in which different word lengths, and larger alphabets are allowed for the dictionary.

Restricting the dynamic programming computation to a diagonal band of edit graph was used by Ukkonen [12]. We have essentially incorporated this idea in our method. For the purpose of developing new methods for d-queries, the idea of using sets to keep track of partial results may also be used in conjunction with suitable edit distance algorithms such as by Myers [11], and Kim et. al [8]. The algorithm in [8] is interesting in particular because it extends the definition of edit distance by allowing swaps.

As we remarked in section 2, the naive *generate and test* d-query algorithm for Hamming distance requires $O(m^{d+1})$ time and $O(m)$ space. If there exists an efficient algorithm to generate all binary strings within *edit distance* d of q then we can devise a similar *generate and test* d-query algorithm for the edit distance case. Similarly, the naive method for the Hamming distance in section 2 obtained by enlarging $\mathcal{W}$ can be adapted to the case of edit distance by efficient generation of words that are within edit distance d of the words in $\mathcal{W}$ if we agree to pay a very high cost for constructing, and maintaining the extended dictionary.

5 Conclusion

We have presented two algorithms *LOOK-UP* and *DFT-LOOK-UP* for answering d-queries in a dictionary of n binary words of length m. The algorithms incorporate the proximity search as part of the distance computation. This approach does not yield improved worst-case time complexity result in the case of Hamming distance compared to a naive generate and test approach. When edit distance is used we achieve worst-case $O(dm^{d+1})$ time and $O(dm)$ space complexities independent of n.

The average case analysis of the two algorithms presented for edit distance over larger alphabets and dictionaries consisting of arbitrary length words are additional topics of investigation.

Acknowledgement

The authors would like to thank the anonymous referee for pointing out the reference [1] in which suffix trees are used in regular expression pattern matching, and the hybrid tree/dynamic programming approach in [4], which is essentially the method presented here for a set of problems involving patterns in strings.

References

1. P. Bieganski. Genetic Sequence Data Retrieval and Manipulation based on Generalized Suffix Trees. *PhD thesis*, University of Minnesota, 1995.
2. G. S. Brodal and L. Gasieniec. Approximate dictionary queries, *in: Poc. 7th Combinatorial Pattern Matching, LNCS, Vol. 1075, Springer, Berlin*, 65–74, 1996.
3. G. S. Brodal and S. Velkatesh. Improved bounds for dictionary look-up with one error. *IPL*, 75, 57–59, 2000.
4. D. Gusfield. Algorithms on strings, trees, and sequences : computer science and computational biology. *Cambridge University Press*, 1997.
5. D. Dolev, Y. Harari, N. Linial, N. Nisan and M. Parnas. Neighborhood preserving hashing and approximate queries. *Proceedings of the Fifth ACM SODA*, 1994.
6. D. Dolev, Y. Harari and M. Parnas. Finding the neighborhood of a query in a dictionary. *Proceedings of the Second Israel Symposium on Theory of Computing and Systems*, 1993.
7. D. Greene, M. Parnas and F. Yao. Multi-index hashing for information retrieval. *Proceedings of 1994 IEEE FOCS*, pp. 722–731, November 1994.
8. D. K. Kim, J. S. Lee, K. Park and Y. Cho. Algorithms for approximate string matching with swaps. *J. of Complexity*, 15, 128–147, 1997.
9. U. Manber and S. Wu. An algorithm for approximate membership checking with applications to password security. *IPL*, 50, 191–197, 1994.
10. M. Minsky and S. Papert. Perceptrons. *MIT Press, Cambridge, MA*, 1969.
11. E. W. Myers. An O(ND) difference algorithm and its variations. *Algorithmica*, 1(2):251–266, 1986.
12. E. Ukkonen. Algorithms for Approximate String Matching. *Information and Control*, 64, 100-118, 1985.
13. R. A. Wagner and M. J. Fisher. The string-to-string correction problem. *JACM*, 21(1):168–173, January 1974.
14. A. C. Yao and F. F. Yao. Dictionary look-up with one error. *J. of Algorithms*, 25(1), 194–202, 1997.

Polynomial Interpolation of the Elliptic Curve and XTR Discrete Logarithm

Tanja Lange[1] and Arne Winterhof[2]

[1] Chair for Information Security, Ruhr-University Bochum, Universitätsstr. 150, 44780 Bochum, Germany
lange@exp-math.uni-essen.de
[2] Institute of Discrete Mathematics, Austrian Academy of Sciences, Sonnenfelsgasse 19, A-1010 Wien, Austria
arne.winterhof@oeaw.ac.at

Abstract. We prove lower bounds on the degree of polynomials interpolating the discrete logarithm in the group of points on an elliptic curve over a finite field and the XTR discrete logarithm, respectively.

1 Introduction

In cryptographic applications an important primitive used in the protocols is the discrete logarithm problem: Given a cyclic group $G =< \gamma >$ of order l and an element $\rho \in G$ determine the exponent x such that $\rho = \gamma^x$, $0 \leq x \leq l-1$. The integer x is called the *discrete logarithm of* ρ *to the base* γ. (For surveys on discrete logarithms see e. g. [2,5,6,7].) The first group suggested for use in practice was the multiplicative group of a finite field. In this particular case subexponential algorithms for solving the discrete logarithm problem are known which motivates considering other groups. An alternative used in practice is the group of points on an elliptic curve. Recently, Lenstra and Verheul [4] introduced a public key system called XTR which is based on a special method to represent elements of a multiplicative subgroup of a finite field. The security of this cryptosystem depends on the intractability of the *XTR discrete logarithm* (XTRDL) problem defined in Section 4.

For solving the discrete logarithm problem in the multiplicative group of a finite field (DL problem), the elliptic curve discrete logarithm (ECDL) problem, or the XTRDL problem it would be sufficient to have a polynomial over a finite field of low degree which reveals information on the DL, ECDL, or XTRDL, respectively, for a large set of elements. In the DL case it has been shown that such a polynomial does not exist [3,8,9]. In this paper we prove analogue results for the ECDL (Section 3) and the XTRDL (Section 4). Since the XTRDL problem is computationally equivalent to the discrete logarithm problem in a certain subgroup of a finite field we investigate this case (Section 5), as well. For the latter problem we also prove a lower bound on the *weight*, i. e. the number of nonzero coefficients, of a polynomial representing the discrete logarithm.

O.H. Ibarra and L. Zhang (Eds.): COCOON 2002, LNCS 2387, pp. 137–143, 2002.

2 Basic Notations and Preliminaries

We use the following ordering of the elements of the finite field $\mathbf{F}_q$ of order $q = p^r$ with a prime p and an integer $r \geq 1$. Let $\{\beta_0, \ldots, \beta_{r-1}\}$ be a basis of $\mathbf{F}_q$ over $\mathbf{F}_p$ and define ξ_k for $0 \leq k < q$ by

$$\xi_k = k_0\beta_0 + k_1\beta_1 + \ldots + k_{r-1}\beta_{r-1}$$

if

$$k = k_0 + k_1 p + \ldots + k_{r-1}p^{r-1} \qquad \text{with } 0 \leq k_i < p \text{ for } 0 \leq i < r.$$

The following result may be of independent interest (cf. [8, Lemma 3.3]).

Lemma 1. *Let $\gamma \in \mathbf{F}_q$ be an element of order l and $f(X) \in \mathbf{F}_q[X]$ a nonzero polynomial of degree at most $l-1$ with at least b zeros of the form γ^x with $0 \leq x \leq l-1$. Then for the weight of $f(X)$ we have*

$$\mathrm{w}(f) \geq \frac{l}{l-b}.$$

Proof. Put $t = \mathrm{w}(f)$, let $N \leq l-b$ be the number of $0 \leq x \leq l-1$ with $f(\gamma^x) \neq 0$ and T the number of pairs (y,i), $0 \leq y \leq l-1$, $0 \leq i \leq t-1$ with $f(\gamma^{y+i}) \neq 0$. Since $\gamma^x = \gamma^{x+l}$ we have $T = tN$. Using properties of Vandermonde matrices we can verify that for every $0 \leq y \leq l-1$ there exists an $0 \leq i \leq t-1$ with $f(\gamma^{y+i}) \neq 0$ and thus $t(l-b) \geq tN = T \geq l$. □

3 The Elliptic Curve Discrete Logarithm

In this section we restrict ourselves to odd characteristic. Let E be an elliptic curve defined over $\mathbf{F}_q$, $q \geq 7$, given by an equation of the form

$$E : Y^2 = X^3 + a_2X^2 + a_4X + a_6,\ a_i \in \mathbf{F}_q,\ i \in \{2,4,6\}.$$

(As usual we write the group law on elliptic curves additively.)
Let P be a point of order l on E and assume that we have a polynomial $f(X) \in \mathbf{F}_q[X]$ satisfying

$$f(x) = \xi_n,\ 1 \leq n \leq \lfloor l/2 \rfloor \iff \exists y \in \mathbf{F}_q : (x,y) = nP, \tag{1}$$

for $n \in S \subseteq \{N+1, \ldots, N+H\} \subseteq \{1, \ldots, \lfloor l/2 \rfloor\}$. (Thus from the result of $f(x)$ one determines the discrete logarithm of (x,y) to the base P for admissible values $(x,y) = \pm nP, n \in S$, in the sense that one is left with at most two choices for the index and can easily determine the correct one. Moreover, note that $\lfloor l/2 \rfloor < q$ by the Hasse-Weil bound.) Now we derive a lower bound on the degree of $f(X)$.

Theorem 1. *Let E be an elliptic curve over $\mathbf{F}_q$ $q \geq 7$ odd, P be a point on E of order l let $f(X)$ satisfy (1), and put $|S| = H - s$. Then we have*

$$\deg(f) \geq \left(1 - \frac{2}{p}\right)\frac{H-2}{3} - s - \frac{2}{3}.$$

Proof. Let $R \subseteq S$ consist of those $n \in S$ such that $\{n-1, n, n+1\} \subseteq S$, $n \notin \{1, N+H\}$, and such that $\xi_{n\pm1} = \xi_n \pm \xi_1$ (i. e. $n \not\equiv 0, p-1 \bmod p$). For the size of R we have $|R| \geq H - 3s - 2 - 2(\lfloor (N+H)/p \rfloor - \lfloor (N+1)/p \rfloor)$.
Put $m = \deg(f)$ and $f(X) = \sum_{i=0}^{m} f_i X^i$, $f_m \neq 0$. Put $nP = (x_n, y_n)$ for $n \in \{1, \ldots, \lfloor l/2 \rfloor\}$. We consider the following system of equations

$$\begin{aligned} f(x_{n+1}) - f(x_n) - \xi_1 &= \xi_{n+1} - \xi_n - \xi_1 = 0, \\ f(x_{n-1}) - f(x_n) + \xi_1 &= \xi_{n-1} - \xi_n + \xi_1 = 0, \end{aligned} \tag{2}$$

which holds for all $n \in R$.
By the usual addition formula on E we have

$$x_{n\pm1} = \frac{a(x_n) \mp 2y_n y_1}{(x_n - x_1)^2}$$

where

$$a(X) = x_1 X^2 + (x_1^2 + 2a_2 x_1 + a_4)X + a_4 x_1 + 2a_6.$$

Inserting this value into (2) leads to

$$\begin{aligned} f(x_{n\pm1}) - f(x_n) \mp \xi_1 &= \sum_{i=0}^{m} f_i \left(\frac{a(x_n) \mp 2y_n y_1}{(x_n - x_1)^2} \right)^i - f(x_n) \mp \xi_1 \\ &= \frac{u(x_n) \mp y_n v(x_n)}{(x_n - x_1)^{2m}} - f(x_n) \mp \xi_1 \\ &= 0 \end{aligned}$$

with uniquely determined polynomials $u(X)$ and $v(X)$ and thus

$$\frac{u(x_n)}{(x_n - x_1)^{2m}} - f(x_n) = 0,$$

where $\deg(u) \leq 2m$. Hence the polynomial

$$h(X) = (X - x_1)^{2m} f(X) - u(X)$$

of degree $3m$ has at least $|R|$ zeros and thus

$$3m \geq |R| \geq H - 3s - 2 - 2(\lfloor (N+H)/p \rfloor - \lfloor (N+1)/p \rfloor)$$

from which the result follows. □

Remarks:

1. For even $q \geq 8$ a similar result can be obtained. We define analogously a set R but with $n \equiv 2 \bmod 4$ instead of $n \not\equiv 0, p-1 \bmod p$. In this case we have $\xi_{n+1} = \xi_n + \xi_1$ and $\xi_{n-1} = \xi_n + \xi_3$ and $|R| \geq (H-5)/4 - s$. Using the known formula for adding points we obtain a polynomial $h(X)$ of degree $2m$ with at least $|R|$ zeroes and thus

$$\deg(f) \geq \frac{H-5}{8} - \frac{s}{2}.$$

2. Determining the weight of $f(X)$ is an interesting problem, however the method in the proofs of [8, Theorem 4.1] and [9, Theorem 1] does not carry through since the weight of $u(X)$ is not related to the weight of f.

4 The XTR Discrete Logarithm

Let p be a prime, $l > 3$ a prime divisor of $p^2 - p + 1$, and $\gamma \in \mathbf{F}_{p^6}$ an element of order l. For $\alpha \in \mathbf{F}_{p^6}$ we denote by $\mathrm{Tr}(\alpha) = \alpha + \alpha^{p^2} + \alpha^{p^4} \in \mathbf{F}_{p^2}$ the trace of α. Given $\xi \in \mathrm{Tr}(<\gamma>)$, the XTRDL problem is to find $0 \le x \le l-1$ such that $\xi = \mathrm{Tr}(\gamma^x)$. (Note that x is not unique but the least residues of p^2x and p^4x modulo l are the only other solutions of $\xi = \mathrm{Tr}(\gamma^x)$.)

For breaking XTR it would be sufficient to have a polynomial $f(X) \in \mathbf{F}_{p^2}[X]$ of low degree such that $f(\mathrm{Tr}(\gamma^x)) = \xi_x$ for a large number of $0 \le x \le l-1$, where as before $\xi_x \in \mathbf{F}_{p^2}$, $0 \le x \le l-1$, is defined by

$$\xi_x = x_0\beta_0 + x_1\beta_1 \quad \text{if} \quad x = x_0 + x_1 p, \quad 0 \le x_0, x_1 \le p-1,$$

where $\{\beta_0, \beta_1\}$ is a fixed basis of $\mathbf{F}_{p^2}$ over $\mathbf{F}_p$.

Theorem 2. *Let $f(X) \in \mathbf{F}_{p^2}[X]$ be a polynomial of degree d such that*

$$f(\mathrm{Tr}(\gamma^x)) = \xi_x \quad \text{for } x \in S$$

for a subset $S \subseteq \{0, 1, \ldots, l-1\}$. Then

$$d \ge \frac{|S|(|S|-1)}{5(l-1)(2p-1)}.$$

Proof. Since otherwise the result is trivial we may assume that $|S| \ge 2$ (hence $d > 0$). Consider

$$D = \{1 \le a \le l-1 \mid a \equiv y - x \bmod l, \ x, y \in S\}.$$

Obviously, there exists $a \in D$ such that there are at least

$$\frac{|S|(|S|-1)}{|D|} \ge \frac{|S|(|S|-1)}{l-1}$$

representations $a \equiv y - x \bmod l$, $x, y \in S$. Choose this a and put

$$R = \{x \in S \mid a + x \equiv y \bmod l \text{ with } y \in S\}.$$

We have $|R| \ge |S|(|S|-1)/(l-1)$. For $x \in R$ there are at most five different $\omega \in \mathbf{F}_{p^2}$, namely $\omega = \xi_a$, $\xi_a + \beta_1$, $\xi_a - \xi_l$, $\xi_a - \xi_l + \beta_1$, or $\xi_a - \xi_l - \beta_1$, such that

$$f(\mathrm{Tr}(\gamma^{a+x})) = \xi_x + \omega = f(\mathrm{Tr}(\gamma^x)) + \omega.$$

Therefore at least one of the five polynomials $h_\omega(X) \in \mathbf{F}_{p^6}[X]$,

$$h_\omega(X) = X^{dp}\left(f(\gamma^a X + (\gamma^a X)^{p-1} + (\gamma^a X)^{-p}) - f(X + X^{p-1} + X^{-p}) - \omega\right)$$

has at least $|R|/5$ zeros, where we used

$$\mathrm{Tr}(\gamma^x) = \gamma^x + \gamma^{p^2x} + \gamma^{p^4x} = \gamma^x + \gamma^{(p-1)x} + \gamma^{-px}.$$

The leading coefficient of $h_\omega(X)$ is $\gamma^{ad(p-1)} - 1$ times the leading coefficient of $f(X)$ and thus $d \geq l$ or

$$d(2p-1) = \deg(h_\omega) \geq \frac{|R|}{5} \geq \frac{|S|(|S|-1)}{5(l-1)}.$$

□

Remarks:

1. Since $\mathrm{Tr}(\gamma^x) = \mathrm{Tr}(\gamma^{p^2 x}) = \mathrm{Tr}(\gamma^{p^4 x})$ the largest possible S has cardinality $|S| = (l+2)/3 \leq (p^2 - p + 3)/3$.
2. It would be interesting to have a lower bound on the weight of a polynomial $f(X)$ with the properties of Theorem 2. Unfortunately, the method in the proofs of [8, Theorem 4.1] and [9, Theorem 1] fails.

5 The Discrete Logarithm in Multiplicative Subgroups of a Finite Field

Now we consider the case that the cyclic group is a multiplicative subgroup of $\mathbf{F}_q$ generated by some $\gamma \in \mathbf{F}_q$ of order l and prove bounds on the degree and weight of a polynomial interpolating the discrete logarithm in parts of this subgroup. This is of special interest in the setting of XTR.

Let $\mathrm{Tr}(\gamma^x) = g$, then $\gamma^x, \gamma^{p^2 x}$, and $\gamma^{p^4 x}$ are zeros of the polynomial $X^3 - gX^2 + g^p X - 1$. Since cubic equations over finite fields can be solved efficiently (see e. g. [1, Theorem 7.8.6]), we may consider the discrete logarithm problem in the subgroup $< \gamma >$ of $\mathbf{F}_{p^6}^*$ instead of the XTRDL problem.

Theorem 3. *Let $\gamma \in \mathbf{F}_q$ be an element of order l and $0 \leq N < N + H \leq l$. Let $f(X) \in \mathbf{F}_q[X]$ be a polynomial such that*

$$f(\gamma^x) = \xi_x \quad \textit{for } x \in S,$$

for a subset $S \subseteq \{N, \ldots, N+H-1\}$ of cardinality $|S| = H - s$. Then

$$\deg(f) > \left(1 - \frac{1}{p}\right)(H-2) - 2s$$

and for the weight of $f(X)$ we have

$$\mathrm{w}(f) \geq \frac{l}{l - (1-1/p)(H-2) + 2s}.$$

Proof. Let R be the set of $N \leq x \leq N + H - 1$ for which both $x \in S$ and $x + 1 \in S$. Then $|R| \geq H - 2s - 1$. We have $\xi_{x+1} = \xi_x + \xi_1$ if $x \not\equiv p - 1 \bmod p$. Hence

$$f(\gamma^{x+1}) = f(\gamma^x) + \xi_1$$

for $x \in R$ with $x \not\equiv p-1 \bmod p$. Therefore the polynomial

$$h(X) = f(\gamma X) - f(X) - \xi_1$$

has at least

$$|R| - \lfloor (N+H-1)/p \rfloor + \lfloor N/p \rfloor$$

zeros in $\mathbf{F}_q^*$ and is not identical to zero as $h(0) = -1$. Thus

$$\deg(f) \geq \deg(h) \geq |R| - \lfloor (N+H-1)/p \rfloor + \lfloor N/p \rfloor$$

and by Lemma 1 we have

$$\mathrm{w}(f) + 1 \geq \mathrm{w}(h) \geq \frac{l}{l - H + 2s + 1 + \lfloor (N+H-1)/p \rfloor - \lfloor N/p \rfloor}$$

from which the claim follows. □

Theorem 3 is only nontrivial if $|S| \geq \frac{H}{2}\left(1 + \frac{1}{p}\right) + 1 - \frac{1}{p}$. Now we extend the range of $|S|$ for nontrivial results on the degree of $f(X)$.

Theorem 4. *Let $\gamma \in \mathbf{F}_q$ be an element of order l. Let $f(X) \in \mathbf{F}_q[X]$ be a polynomial of degree d such that*

$$f(\gamma^x) = \xi_x \quad \text{for all } x \in S,$$

for a subset $S \subseteq \{0, \ldots, l-1\}$. Then

$$d > \frac{|S|(|S|-1)}{2 \cdot 3^{t-1}(l-1)},$$

where $t = \lceil \log_p(l) \rceil$.

Proof. Consider

$$D = \{1 \leq a \leq l-1 | a \equiv y - x \bmod l,\ x, y \in S\}.$$

Obviously, there exists $a \in D$ such that there are at least

$$\frac{|S|(|S|-1)}{|D|} \geq \frac{|S|(|S|-1)}{l-1}$$

representations $a \equiv y - x \bmod l$, $x, y \in S$. Choose this a and put

$$R = \{x \in S \mid a + x \equiv y \bmod l \text{ with } y \in S\}.$$

Due to the carries and reduction modulo l in the exponent, there are at most $2^{t-1} + 3^{t-1}$ possible elements $\omega \in \mathbf{F}_q$ such that

$$f(\gamma^{a+x}) = f(\gamma^x) + \omega,$$

namely the 2^{t-1} elements $\omega = \sum_{i\in I} \beta_i$ with $I \subseteq \{0,1,\ldots t-2\}$ in the case that $a+x < l$ and the 3^{t-1} elements $\omega = \sum_{i\in I}(\pm\beta_i)$ with $I \subseteq \{0,1,\ldots,t-2\}$ in the case that $a+x \geq l$. Therefore at least one of the polynomials

$$h_\omega(X) = f(\gamma^a X) - f(X) - \omega$$

has at least $|R|/(2^{t-1}+3^{t-1})$ zeros in $\mathbf{F}_q$. We choose this ω. The leading coefficient of $h_\omega(X)$ is $\gamma^{ad}-1$ times the leading coefficient of $f(X)$ and thus $d \geq l$ or $h_\omega(X)$ is not identical to zero. Thus

$$d = \deg(h_\omega) \geq \frac{|R|}{2^{t-1}+3^{t-1}} \geq \frac{|S|(|S|-1)}{(2^{t-1}+3^{t-1})(l-1)}.$$

□

Acknowledgment

Parts of this paper were written during a visit of the first author to the Austrian Academy of Sciences. She wishes to thank the Institute of Discrete Mathematics for hospitality and financial support.
We would also like to thank Igor Shparlinski for suggesting the XTRDL.

References

1. E. Bach and J.O. Shallit, Algorithmic number theory, Vol.1: Efficient algorithms. Cambridge: MIT Press 1996.
2. J. Buchmann and D. Weber, Discrete logarithms: recent progress. Coding theory, cryptography and related areas (Guanajuato, 1998), 42–56, Springer, Berlin, 2000.
3. D. Coppersmith and I.E. Shparlinski, On polynomial approximation of the discrete logarithm and the Diffie-Hellman mapping, J. Cryptology 13 (2000), 339–360.
4. A.K. Lenstra and E.R. Verheul, The XTR public key system, Advances in cryptology—CRYPTO 2000 (Santa Barbara, CA), 1–19, Lecture Notes in Comput. Sci., 1880, Springer, Berlin, 2000.
5. K.S. McCurley, The discrete logarithm problem. Cryptology and computational number theory (Boulder, CO, 1989), 49–74, Proc. Sympos. Appl. Math., 42, Amer. Math. Soc., Providence, RI, 1990.
6. A.J. Menezes, P.C. van Oorschot, and S.A. Vanstone, Handbook of applied cryptography. CRC Press Series on Discrete Mathematics and its Applications. CRC Press, Boca Raton, FL, 1997.
7. A.M. Odlyzko, Discrete logarithms and smooth polynomials. Finite fields: theory, applications, and algorithms (Las Vegas, NV, 1993), 269–278, Contemp. Math., 168, Amer. Math. Soc., Providence, RI, 1994.
8. I.E. Shparlinski, Number theoretic methods in cryptography. Complexity lower bounds. Progress in Computer Science and Applied Logic, 17. Birkhäuser Verlag, Basel, 1999.
9. A. Winterhof, Polynomial interpolation of the discrete logarithm, Des. Codes Cryptogr. 25 (2002), 63–72.

Co-orthogonal Codes
(Extended Abstract)

Vince Grolmusz

Department of Computer Science, Eötvös University, Budapest
Address: Pázmány P. stny. 1/C, H-1117, Budapest, Hungary
grolmusz@cs.elte.hu

Abstract. We define, construct and sketch possible applications of a new class of non-linear codes: co-orthogonal codes. The advantages of these codes are twofold: first, it is easy to decide whether two codewords form a unique pair (this can be used in decoding information or identifying users of some not-publicly-available or non-free service on the Internet or elsewhere), and the identification process of the unique pair can be distributed between entities, who perform easy tasks, and only the information, gathered from all of them would lead to the result of the identifying process: the entities, taking part in the process will not have enough information to decide or just to conjecture the outcome of the identification process.
Moreover, we describe a fast (and general) method for generating (non-linear) codes with prescribed dot-products with the help of multi-linear polynomials.

Keywords: non-linear codes, co-orthogonal codes, codes and set-systems, multi-linear polynomials, composite modulus

1 Introduction

In the present paper, we define, construct and sketch possible applications of a new class of non-linear codes: co-orthogonal codes. The advantages of these codes are twofold: first, it is easy to decide whether two codewords form a unique pair (this can be used in decoding information or identifying users of some not-publicly-available or non-free service on the Internet or elsewhere). Second, the identification process of the unique pair can be distributed between entities, who perform easy tasks, and only the information, gathered from all of them would lead to the result of the identifying process: the entities, taking part in the process will not have enough information to decide or just to conjecture the outcome of the identification process.

The identification is done by the following procedure: two sequences, $a = (a_1, a_2, \ldots, a_n)$ and $b = (b_1, b_2, \ldots, b_n)$ forms a *pair*, say, modulo 6, if 6 is a divisor of the sum

$$\sum_{i=1}^{n} a_i b_i.$$

O.H. Ibarra and L. Zhang (Eds.): COCOON 2002, LNCS 2387, pp. 144–152, 2002.

If, for every a there is exactly one b which forms a pair with a (and this fact can be decided easily), then, for example, keeping b secret would identify a. Our main result is that these codes (co-orthogonal codes) contain much more code-words, than the orthogonal codes, where the pairs are identified by the fact that in the sum above 6 is not a divisor of the sum. We also note, that taking 6 (or a non-prime power other small integer) is an important point here: we will get more code-words than in the case of primes!

1.1 Orthogonal Codes

Definition 1. *Let $n > 0$ and $r > 1$ be integers, and let Z_r denote modulo-r ring of integers. We call $A \subset Z_r^n$ an* orthogonal code, *if we can list the elements of A as $A = \{a^1, a^2, \ldots, a^\ell\} \cup \{b^1, b^2, \ldots, b^\ell\}$, such that for all $a^i \in A$: $a^i \cdot b^i \not\equiv 0 \pmod r$, but for all $i \neq j$: $a^i \cdot b^j \equiv 0 \pmod r$.*

In the definition above we allow $a^i = b^i$, and $u \cdot v$ denotes the dot-product (or scalar-product) of vectors u and v.

Example 1. Let $n = 2, r = 5$, and $A = \{a^1, a^2, b^1, b^2\}$, where $a^1 = (1,1), b^1 = (4,2), a^2 = (4,2), b^2 = (3,2)$. Then the pairwise dot-products can be given as a 2×2 matrix:

$$\begin{array}{c} \\ a^1 \\ a^2 \end{array} \begin{array}{c} \quad b^1 \qquad\quad b^2 \\ \begin{pmatrix} a^1 \cdot b^1 & a^1 \cdot b^2 \\ a^2 \cdot b^1 & a^2 \cdot b^2 \end{pmatrix} \end{array} = \begin{pmatrix} 6 & 5 \\ 20 & 16 \end{pmatrix} \equiv \begin{pmatrix} 1 & 0 \\ 0 & 1 \end{pmatrix} \pmod 5.$$

The Advantages of Orthogonal Codes. The orthogonal codes have very attractive properties:

- For each $a^i \in A$ it is easy to verify that a given vector $u \in A$ would serve as b^i: if $a^i \cdot u \not\equiv 0 \pmod r$ then $u = b^i$.
- The dot-product can be computed in a small memory: only modulo r computations (multiplications and additions) are needed.
- Suppose, that we have three players P_1, P_2, P_3, and the coordinates of the code-words of a^i and u are partitioned between them: suppose, that the first $n/3$ coordinates of the vectors are known for P_1, the second $n/3$ to P_2 and the third $n/3$ to P_3 (assume, that n is a multiple of 3). Then they can verify collectively whether $u = b^i$: for $j = 1, 2, 3$, P_j compute the dot-product of their known coordinates, and the mod r sum of their result gives the answer: if the value is not 0 modulo r, then $u = b^i$. Let us note, that the players *alone* tipically will not know the answer.
- An easy and fast parallel algorithm computes the dot-product of the two vectors, with n processors for the length-n vectors it can be done in $c \log n$ time.

Let us remark, that a trivial code with the highest possible rate has almost all properties mentioned above: indeed, consider code $B = \{0,1\}^n$, and let $a, b \in B$

form a pair if $a = b$. Since $a = b$ can be verified bitwise, it follows that the pair-verification can be done in parallel. However, as was noticed by Király [6], if $a \neq b$ then it will be witnessed by one or more processors or players, who knows only the bits (or sub-sequences) of a and b, so if a and b does not form a pair then it will be known for a player. This property can be fatally bad if our goal is to hide the outcome of the verification process.

The Disadvantage of Orthogonal Codes. It is easy to see, that if r is a prime, then the maximum size of an orthogonal code (that is, the cardinality of A, $|A|$) is at most $2n$, if the length of the code-words are n: (simply the matrix in Example 1 has full rank, because of the orthogonality property). That shows, that the rate of orthogonal codes are *extremely low* for r primes. It is not difficult to show, that the situation is not much better if r is a composite number: if r has ℓ prime divisors, then $|A|$ is at most $2\ell n$, which is still very small.

1.2 Co-orthogonal Codes

Definition 2. *We call* $A \subset Z_r^n$ *a* co-orthogonal code, *if we can list the elements of* A *as* $A = \{a^1, a^2, \ldots, a^\ell\} \cup \{b^1, b^2, \ldots, b^\ell\}$, *such that for all* $a^i \in A$: $a^i \cdot b^i \equiv 0 \pmod{r}$, *but for all* $i \neq j$: $a^i \cdot b^j \not\equiv 0 \pmod{r}$.

Example 2. Let $n = 2, r = 6$, and $A = \{a^1, a^2, a^3, a^4, a^5, a^6, b^1, b^2, b^3, b^4, b^5, b^6\}$, where $a^1 = (2,1), a^2 = (5,1), a^3 = (2,3), a^4 = (2,2), a^5 = (1,2), a^6 = (3,5)$, and $b^1 = (5,2), b^2 = (1,1), b^3 = (3,2), b^4 = (1,5), b^5 = (2,5), b^6 = (1,3)$. Then the pairwise dot-products can be given as a 6×6 matrix:

$$\begin{array}{c} \\ a^1 \\ a^2 \\ a^3 \\ a^4 \\ a^5 \\ a^6 \end{array}
\begin{array}{c} \begin{array}{cccccc} b^1 & b^2 & b^3 & b^4 & b^5 & b^6 \end{array} \\
\begin{pmatrix} 12 & 3 & 8 & 7 & 9 & 5 \\ 27 & 6 & 17 & 10 & 15 & 8 \\ 16 & 5 & 12 & 17 & 19 & 11 \\ 14 & 4 & 10 & 12 & 14 & 8 \\ 9 & 3 & 7 & 11 & 12 & 7 \\ 25 & 8 & 19 & 28 & 21 & 18 \end{pmatrix} \end{array}
\equiv \begin{pmatrix} 0 & 3 & 2 & 1 & 3 & 5 \\ 3 & 0 & 5 & 4 & 3 & 2 \\ 4 & 5 & 0 & 5 & 1 & 5 \\ 2 & 4 & 4 & 0 & 2 & 2 \\ 3 & 3 & 1 & 5 & 0 & 1 \\ 1 & 2 & 1 & 4 & 3 & 0 \end{pmatrix} \pmod{6}.$$

Note, that for code-length is 2 again, but we have $|A| = 12$.

We have shown in [2], that if r is a prime, then the rate of co-orthogonal codes are not much larger than the rate of orthogonal codes: there exist at most $O(n^{r-1})$ co-orthogonal code-words for any n . However, quite surprisingly, for non-prime-power, composite r's (e.g., $r = 6$), there are co-orthogonal codes of much larger rate (see Theorem 1).

The Advantages of Co-orthogonal Codes. It is obvious, that the co-orthogonal codes have the same advantages as the orthogonal codes: It is very easy to identify the matching code-word pairs; the computations can be done modulo a small number (in our example this number is 6); the identification process can be shared between players, not knowing the outcome of the identification. However, we can show, that the serious problem of the orthogonal codes, i.e., their very low rate does not appear here. We show this in the next section, with algorithmically fast constructions.

2 Constructing Co-orthogonal Codes Modulo a Composite Number

We give here some constructions for co-orthogonal codes. These constructions use techniques from papers [3], [2], and especially from [4]. The existence of (special) mod 6 co-orthogonal codes with high rate falsified old conjectures in extremal set theory (see [3] for details). This high rate of our codes facilitates the possible application of co-orthogonal codes. We note, that our codes here are binary (i.e., only 0 and 1 will appear in the code-words), but r, the modulus is a non-prime-power composite number.

Theorem 1. *Let m be a positive integer, and suppose that m has $r > 1$ different prime divisors: $m = p_1^{\alpha_1} p_2^{\alpha_2} \dots p_r^{\alpha_r}$. Then there exists $c = c(m) > 0$, such that for every integer $N > 0$, there exists an explicitly constructible binary co-orthogonal code H modulo m, of length N, such that $|H| \geq \exp\left(c\frac{(\log N)^r}{(\log\log N)^{r-1}}\right)$. Moreover, the minimum Hamming-distance between any two codewords of H is*

$$\frac{c' N \log\log N}{\log N}$$

for a positive c'.

Note, that $|H|$ grows faster even for $m = 6$ than any polynomial in n. Note also, that the code-generation is a fast polynomial algorithm (see Section 3.3).

2.1 k-Wise Co-orthogonal Codes

A generalization of the co-orthogonal codes is the *k-wise co-orthogonal codes.* While co-orthogonal codes are useful since their easy pair-identification property, k-wise co-orthogonal codes can be used for group-identification. Since dot-product (or scalar-product) can be defined only between only two vectors, we need a natural generalization here.

Definition 3. *Let $A = \{a_{ij}\}$ and $B = \{b_{ij}\}$ two $u \times v$ matrices over a ring R with unit element 1. Their* Hadamard-product *is an $u \times v$ matrix $C = \{c_{ij}\}$, denoted by $A \odot B$, and is defined as $c_{ij} = a_{ij} b_{ij}$, for $1 \leq i \leq u$, $1 \leq j \leq v$. Let $k \geq 2$. The k-wise dot product of vectors of length n, $a^1, a^2, \dots, a^k$ is computed as*

$$(a^1 \odot a^2 \odot \cdots \odot a^k) \cdot \mathbf{1},$$

where $\mathbf{1}$ denotes the length n all-1 vector.

Note, that if a^i is a characteristic vector of a subset A_i of an n-element ground-set (for $i=1,2,\ldots,k$), then $a^1 \odot a^2 \odot \cdots \odot a^k$ is a characteristic vector of $\bigcap_{i=1}^k A_i$, and the k-wise dot-product of $a^1, a^2, \ldots, a^k$ gives the *size* of this intersection.

Definition 4. *For a $k \geq 2$ we call $A \subset Z_r^n$ a* k-wise co-orthogonal code *modulo m, if the following holds:*

- *$\forall a^1 \in H$ there exist $a^2, a^3, \ldots, a^k \in H$ such that* $(a^1 \odot a^2 \odot \cdots \odot a^k) \cdot \mathbf{1} \equiv 0 \pmod{m}$,
- *If $\{a^2, a^3, \ldots, a^k\} \neq \{b^2, b^3, \ldots, b^k\}$, $b^i \in H$, $i = 2, 3, \ldots, k$, then*
$$(a^1 \odot b^2 \odot b^3 \odot \cdots \odot b^k) \cdot \mathbf{1} \not\equiv 0 \pmod{m}.$$

Now we can formulate the following result. The construction is – surprisingly – exactly the same as the construction for proving Theorem 1, and it is an improvement of a construction appeared in [5] for set-systems.

Theorem 2. *Let $n, t \geq 2$ integers, and let $p_1, p_2, \ldots, p_r$ be pairwise different primes, and let $m = p_1 p_2 \cdots p_r$. Then there exists a $c_m > 0$ and an explicitly constructible code H of length N, such that $|H| \geq \exp\left(\frac{c_m (\log N)^r}{(\log\log N)^{r-1}}\right)$, and H is k-wise co-orthogonal* for any $k \geq 2$.

The identification of a group is done with the easy computation of the k-wise dot-product of the codes of the members of the group. If this number is 0 modulo m, then the group is passed the identification, otherwise it failed.

3 Constructing Code $f(A)$ from f and A

Our construction is based on a method given in [4] and the BBR-polynomial [1]. In paper [4], we gave a general construction for hypergraphs with prescribed intersection sizes. Our construction described here can also be applied for constructing codes with prescribed dot-product matrices. For a detailed discussion of this problem, see [4].

Here we re-formulate this method in a form which is more suitable for our purposes in the present work.

Definition 5. *Let $f(x_1, x_2, \ldots, x_n) = \sum_{I \subset \{1,2,\ldots,n\}} \alpha_I x_I$ be a multi-linear polynomial, where $x_I = \prod_{i \in I} x_i$. Let $w(f) = |\{\alpha_I : \alpha_I \neq 0\}|$ and let $\mathrm{L}_1(f) = \sum_{I \subset \{1,2,\ldots,n\}} |\alpha_I|$.*

Definition 6. *Let $A = \{a^1, a^2, \ldots, a^\ell\} \subset \{0,1\}^n$ be a binary code. Then the matrix of A, denoted by $M(A)$, is an $n \times \ell$ 0-1 matrix, with column j equal to the code-word a^j, for $j = 1, 2, \ldots, \ell$.*

Definition 7. *Let $A = \{a^1, a^2, \ldots, a^\ell\} \subset \{0,1\}^n$ be a binary code, and let f be an n-variable multi-linear polynomial with positive integer coefficients. Then binary code $f(A) = \{c^1, c^2, \ldots, c^\ell\} \subset \{0,1\}^{\mathrm{L}_1(f)}$ is defined as follows: The rows of $M(f(A))$ correspond to monomials x_I's of f; there are α_I identical rows of $M(f(A))$, corresponding to the same x_I. The row, corresponding to x_I, is defined as the Hadamard-product of those rows i of $M(A)$, with $i \in I$.*

Example 3. Let $f(x_1,x_2,x_3,x_4) = x_1 + x_2 + 2x_3x_4$, and let the matrix $M(A)$ of code $A = \{a^1, a^2, a^3\}$ be

$$M(A) = \begin{array}{c} \\ 1 \\ 2 \\ 3 \\ 4 \end{array} \begin{array}{c} \begin{array}{ccc} a^1 & a^2 & a^3 \end{array} \\ \begin{pmatrix} 0 & 1 & 1 \\ 1 & 1 & 1 \\ 1 & 0 & 1 \\ 0 & 0 & 1 \end{pmatrix} \end{array}.$$

Then the matrix of code $f(A)$ is

$$M(f(A)) = \begin{array}{c} \\ x_1 \\ x_2 \\ x_3x_4 \\ x_3x_4 \end{array} \begin{array}{c} \begin{array}{ccc} c^1 & c^2 & c^3 \end{array} \\ \begin{pmatrix} 0 & 1 & 1 \\ 1 & 1 & 1 \\ 0 & 0 & 1 \\ 0 & 0 & 1 \end{pmatrix} \end{array}.$$

The most important property of code $f(A)$ is given in the following Theorem:

Theorem 3. *Let $t \geq 2$ and let $a^{i_\ell} \in A$ for $\ell = 1, 2, \dots, t$. Then*

$$f(a^{i_1} \odot a^{i_2} \odot \cdots \odot a^{i_t}) = (c^{i_1} \odot c^{i_2} \odot \cdots \odot c^{i_t}) \cdot \mathbf{1}. \tag{1}$$

An analogous theorem for set-systems appeared in [4]. We reproduce here its short proof.

Proof. Consider a monomial x_I of f, for some $I \subset \{1, 2, \dots, n\}$. Let us observe, that monomial x_I contributes one to the left hand side of equation (1) exactly when for all $j \in I$, the jth coordinate of every code-word $a^{i_1}, a^{i_2}, \dots, a^{i_t}$ are equal to 1. This happens exactly if the coordinate of $c^{i_1} \odot c^{i_2} \odot \cdots \odot c^{i_t}$, corresponding to monomial x_I, is 1, that means, that one is contributed to the right hand side of 3. □

3.1 Our Main Construction

Our binary co-orthogonal code will be constructed as $f(A)$, from a well-chosen polynomial f and a dense code A. For simplicity, in this preliminary version of this work, we prove Theorems 1 and 2 only for modulus $m = 6$.

Our f will be the BBR-polynomial. *Barrington, Beigel* and *Rudich* [1] showed, that for integers α and β, there exists an explicitly constructible, symmetric, n-variable, degree-$O(\min(2^\alpha, 3^\beta))$ polynomial f, (the BBR-polynomial), satisfying over $x = (x_1, x_2, \dots, x_n) \in \{0,1\}^n$:

$$f(x) \equiv 0 \pmod 6 \iff \sum_{i=1}^{n} x_i \equiv 0 \pmod{2^\alpha 3^\beta}. \tag{2}$$

Our A is defined as follows. Let A_0 denote all the weight-$2^\alpha 3^\beta - 1$ binary code-words of length $n-1$. Code A is got from code A_0 if we add a leading 1 for all codewords in A_0, consequently, each word in A is binary, has weight $2^\alpha 3^\beta$, and the dot-product of any two different words in A is non-zero, but less than $2^\alpha 3^\beta$. However, the dot-product of any $a \in A$ with itself is equal to $2^\alpha 3^\beta$.

Let us remark, that A itself is a co-orthogonal code modulo $2^\alpha 3^\beta$.

Moreover, for any $k \geq 2$, the dot-product of any k words (containing at least two different words) in A (see Definition 3) is also non-zero, and less than $2^\alpha 3^\beta$.

Now let α be the smallest integer that $n^{1/3} < 2^\alpha$, and let β be the smallest integer such that $n^{1/3} < 3^\beta$. Then the degree of f is $O(n^{1/3})$.

Let us consider now code $H = f(A)$. It contains at least $\binom{n}{n^{2/3}}$ code-words of length $N = \mathrm{L}_1(f) = O(n^{n^{1/3}})$, so

$$|H| = |f(A)| = \exp\left(c\frac{(\log N)^2}{(\log\log N)}\right).$$

And, from Theorem 3, for any $a \in H$ a forms a pair only with $b = a$, for any $b \in H$, modulo 6, this proves the first part of Theorem 1. The k-wise co-orthogonality follows also from Theorem 3.

For computing the minimum-distance of the code (for proving Theorem 1), we should note, that any two elements a^i and a^j of A differs in at least one bit. The weight of a^i is $2^\alpha 3^\beta = \Theta(n^{2/3})$, that means — because polynomial f is symmetric – that the corresponding code-word of $f(A)$, c^i has weight at least

$$\sum_{k=1}^{n^{1/3}} \binom{2^\alpha 3^\beta}{k} > \binom{n^{2/3}}{n^{1/3}}.$$

If we flip one 1-bit to zero, then at least

$$\binom{n^{2/3}}{n^{1/3}} - \binom{n^{2/3}-1}{n^{1/3}}$$

monomials of f become zero, that is, at least that many coordinates of c^i become zero by flipping any bit. Now the result follows.

3.2 Alternative Constructions

We would get more dense codes if we had a BBR polynomial with smaller degree, but, unfortunately, it is not known whether there exists such polynomial with lower degree. (For other applications of the BBR polynomial see [3] and [2]).

Alternatively, we can choose different codes A for the construction. For example, consider the following code A. Let vectors a^i be all the weight-$2^\alpha 3^\beta$ code-words of length n, and let b^i be the complement of a^i, for $i = 1, 2, \ldots, \binom{n}{2^\alpha 3^\beta}$. Then it is easy to see, that A is co-orthogonal code modulo $2^\alpha 3^\beta$. Then, from Theorem 3, with the BBR-polynomial f, code $f(A)$ is also a co-orthogonal code, but now modulo 6.

We list some further variants of codes A in Section 4.

3.3 Algorithmic Complexity of Computing $f(A)$

Let a^i be a code-word of A and let c^i the corresponding code-word of $f(A)$. Then the coordinates of c^i is computed as the values of monomials of f with substituting a^i: $f(a^i)$. The value of a degree-d monomial can be computed in $O(d)$ steps; so the length-N c^i can be computed in $N \log N$ steps.

4 Cryptographic Applications

Perhaps the most straightforward application is the following one: keep secret vector a^i, and accept vector x only if $a^i \cdot x \equiv 0 \pmod{m}$. From the co-orthogonal code, only $x = b^i$ satisfy this relation, but outside that code, many more x's may satisfy it; for example, $x = 0$ always satisfies this requirement.

Consequently, for any cryptographic application first we should verify whether x is in the code or not. We call this phase *membership testing.* If x fails the membership test, reject it. If x passes the membership test, then compute $y = a^i \cdot x \bmod m$ (even in a distributed way), and accept x iff y is 0, modulo m, and reject it otherwise.

Another problem with our main construction (Section 3.1) is the following: The pair of any $a \in f(A)$ is the same a itself!. That means, that if the verification process is distributed among players, any player who finds that a coordinate is different in a and b will know that they are not pairs. This problem can be avoided by applying some linear transformations for the codes, as described in Section 4.2.

4.1 Membership Testing

It is easy to see, that in any polynomial f, satisfying property (2), must contain monomials of degree one x_i $(i = 1, 2, \ldots, \ell)$ with a non-zero coefficient. That is also true for the BBR-polynomial f used in our construction.

Now, suppose that we need to verify whether $c \in f(A)$. We know, that which coordinates of a code-word c should correspond to monomials x_i $(i = 1, 2, \ldots, n)$, then, first we read only these (at most n) coordinates of c. The values of x_i should be equal to the coordinates of some $a^j \in A$: $x_i = a^j_i$, for $i = 1, 2, \ldots, n$. If this will not be satisfied, discard c, it is not in our code. Otherwise, compute from a^j and from f the code-word c^j. If $c = c^j$, accept c, otherwise reject.

This algorithm can be performed in $O(|c| \log(|c|)$ steps by the most straightforward implementation.

4.2 Non-binary, Non-self-paired Codes

We describe a quite general method to get non-binary from the binary co-orthogonal codes generated in Section 3.1. Note, that if the pair of a was a itself in the original construction, this property will disappear in the modified one:

Our idea comes from the following well-known identity:

$$x'A \cdot y = x' \cdot yA^T,$$

where x', y are length-n q-ary code-words, and A is an $n \times n$ matrix.

So, if m is a prime we can consider the m-element-field, and transform our code C into code C_A for any non-singular matrix A over the field as follows:

$$C_A = \{xA^{-1} : x \in C\} \cup \{yA^T : y \in C\}.$$

Clearly, if $x'A = x$, then $x \cdot y = x'A \cdot y = x' \cdot yA^T$, so if C was a co-orthogonal code, then C_A is also a co-orthogonal code over the field, and the pair of $x' = xA^{-1}$ is $xA^T = x'AA^T$, which tipically differs from x' (we should avoid unitary A's).

However, in our main construction m is composite, say $m = 6$. So, we should choose a non-singular $n \times n$ matrix A over the 2 element field, and another one, B, over the 3 element field. Suppose now, that C is a co-orthogonal code modulo 6. Then certainly

$$C_{A,B} = \{3xA^{-1} + 2xB^{-1} : x \in C\} \cup \{3yA^T + 2yB^T : y \in C\}$$

is a co-orthogonal code mod 6, and the pair of $3xA^{-1}+2xB^{-1}$ is exactly $3xA^T + 2xB^T$ (Note that A^{-1} is the inverse of A over GF_2, and B^{-1} is the inverse of B over the three element field.)

Acknowledgment

We are grateful to Zoltán Király and to Lajos Rónyai for discussions on this topic. The author acknowledges the partial support of János Bolyai Fellowship and research grants EU FP5 IST FET No. IST-2001-32012, OTKA T030059 and an ETIK grant.

References

1. David A. Mix Barrington, Richard Beigel, and Steven Rudich. Representing Boolean functions as polynomials modulo composite numbers. *Comput. Complexity*, 4:367–382, 1994. Appeared also in *Proc. 24th Ann. ACM Symp. Theor. Comput.*, 1992.
2. Vince Grolmusz. Low-rank co-diagonal matrices and Ramsey graphs. *The Electronic Journal of Combinatorics*, 7:R15, 2000. www.combinatorics.org.
3. Vince Grolmusz. Superpolynomial size set-systems with restricted intersections mod 6 and explicit Ramsey graphs. *Combinatorica*, 20:73–88, 2000.
4. Vince Grolmusz. Constructing set-systems with prescribed intersection sizes. Technical Report DIMACS TR 2001-03, DIMACS, January 2001. ftp://dimacs.rutgers.edu/pub/dimacs/TechnicalReports/TechReports/2001/2001-0 3.ps.gz.
5. Vince Grolmusz. Set-systems with restricted multiple intersections and explicit Ramsey hypergraphs. Technical Report DIMACS TR 2001-04, DIMACS, January 2001. ftp://dimacs.rutgers.edu/pub/dimacs/TechnicalReports/TechReports/2001/2001-0 4.ps.gz.
6. Zoltán Király. personal communication.

Efficient Power-Sum Systolic Architectures for Public-Key Cryptosystems in GF(2^m)

Nam-Yeun Kim, Won-Ho Lee, and Kee-Young Yoo

Department of Computer Engineering,
Kyungpook National University,
Deagu, Korea 702-701
knyeun@hanmail.net, purmi@purple.knu.ac.kr, yook@knu.ac.kr

Abstract. The current paper presents a new algorithm and two architectures for the power-sum operation ($AB^2 + C$) over GF(2^m) using a standard basis. The proposed algorithm is based on the MSB-first scheme and the proposed architectures have a low hardware complexity and small latency compared to conventional approaches. In particular, the hardware complexity and latency of the proposed parallel-in parallel-out array are about 19.8% and 25% lower, respectively, than Wei's. In addition, since the proposed architectures incorporate simplicity, regularity, modularity, and pipelinability, they are well suited to VLSI implementation and can be easily applied to inverse/division architecture.

1 Introduction

Finite or Galois fields GF(2^m) are widely used in many practical applications, such as error-correcting codes [1], public-key cryptography [2,3,4], digital signal processing [5], and so on. These applications usually require the computation of multiplication, power-sum, inverse/division, and exponentiation operations in GF(2^m). However, since these operations are quite time consuming, the design of a high speed and low-complexity arithmetic circuit with an efficient algorithm is needed. Among these operations, a power-sum is used as an efficient basic operation in decoding error-correcting codes. For example, the computation of $(S_1)^6 + S_3(S_1)^3 + S_5(S_1)^1 + (S_3)^2$ is required in a triple-error-correcting binary BCH step-by-step decoder, where S_1, S_3, and S_5 are the syndrome values calculated from the received words. The computation can be performed using only three power-sum operations and one multiplication; that is, $[(S_1)^3 + S_3]^2 + S_1[S_3(S_1)^2 + S_5]$ [6]. In addition, a power-sum is known as a basic operation for public-key cryptosystems, such as a Diffie-Hellman key exchange, ElGamal, RSA, and ECC [2,3,4] over GF(2^m). For example, when implementing an RSA cryptosystem, the modular exponentiation is a significant computational problem accomplished by performing iterations of the modular multiplication and squaring using large numbers (usually $> 500 bits$), $\beta^N = \beta^{N_0}[\beta^{N_1}[\beta^{N_2}[\cdots[\beta^{N_{m-2}}(\beta^{N_{m-1}})^2]^2]^2]^2]^2$. A simple computation algorithm for computing exponentiations in GF(2^m) is presented as [6]:

O.H. Ibarra and L. Zhang (Eds.): COCOON 2002, LNCS 2387, pp. 153–161, 2002.

$$\begin{aligned}
&\text{STEP 1}: \text{ If } N_{m-1} = 1 \text{ then } P = \beta \text{ else } P = \alpha^0 \\
&\text{STEP 2}: \text{ For } i = m-2 \text{ downto } 0 \\
&\text{STEP 3}: \quad \text{ If } N_i = 1 \text{ then } P = P^2\beta \text{ else } P = P^2\alpha^0
\end{aligned}$$

The final result is $P = \beta^N$. In this case, AB^2 operations can be used to compute the step 3 operations.

Systolic arrays for performing power-sum operations have already been proposed in [6,7,8] using a standard basis representation in GF(2^m). In [7], a systolic power-sum architecture is proposed along with a modified basic cell that can perform eight different types of computations by adding one MUX and one DEMUX. Wei [6] then presented architectures for an inverse and division in GF(2^m) based on the architecture in [7]. Yet these systolic array power-sum circuits still have certain shortcomings as regards cryptographic applications due to their high circuit complexity and long latency. As such, further research on an efficient power-sum circuit is needed.

In addition, many architectures over GF(2^m) have been developed using different bases; a normal, dual, and standard basis. A normal basis representation is very efficient for performing squaring and exponentiation or finding the inverse element, yet normal-basis multipliers require a basis conversion. A dual basis representation is also unsuitable for large finite fields in cryptography. Therefore, a standard basis offers the most regular and extensible features for hardware implementation and is easiest to use among the other representations.

Accordingly, the following concentrates on a power-sum operation for GF(2^m) using a standard basis. The proposed algorithm is then used as the basis for introducing a parallel-in parallel-out and serial-in serial-out systolic architecture. The hardware complexity and latency of the two architectures are lower than conventional architectures, plus they are well suited to VLSI implementation and can be easily applied to exponentiation architecture.

2 Algorithm

A finite field GF(2^m) has 2^m elements and it is assumed that all the $(2^m - 1)$ non-zero elements of GF(2^m) are represented using a standard basis. Let $A(x)$, $B(x)$, and $C(x)$ be three elements in GF(2^m) and $F(x)$ be the primitive polynomial, where,

$$\begin{aligned}
A(x) &= a_{m-1}x^{m-1} + a_{m-2}x^{m-2} + \cdots + a_1x + a_0 \\
B(x) &= b_{m-1}x^{m-1} + b_{m-2}x^{m-2} + \cdots + b_1x + b_0 \\
C(x) &= c_{m-1}x^{m-1} + c_{m-2}x^{m-2} + \cdots + c_1x + c_0 \\
F(x) &= f_{m-1}x^{m-1} + f_{m-2}x^{m-2} + \cdots + f_1x + f_0
\end{aligned}$$

The coefficients a_i, b_i, c_i, and f_i are the binary digits 0 and 1. If α is the root of $F(x)$, then $F(\alpha) = 0$, and $F(\alpha) \equiv \alpha^m = f_{m-1}\alpha^{m-1} + f_{m-2}\alpha^{m-2} + \cdots + f_1\alpha + f_0$, $F'(\alpha) \equiv \alpha^{m+1} = f'_{m-1}\alpha^{m-1} + f'_{m-2}\alpha^{m-2} + \cdots + f'_1\alpha + f'_0$ where f_i and $f'_i \in GF(2)(0 \le i \le m-1)$.

To compute an $AB^2 + C$ operation, the proposed algorithm commences with the following equation:

$$\begin{aligned} P &= AB^2 + C \bmod F(x) \\ &= A(b_{m-1}\alpha^{2m-2} + b_{m-2}\alpha^{2m-4} + \cdots + b_1\alpha^2 + b_0) + c_{m-1}\alpha^{m-1} + c_{m-2}\alpha^{m-2} \\ &\quad + \cdots + c_1\alpha + c_0 \bmod F(x) \\ &= (\cdots(\cdots((Ab_{m-1})\alpha^2 \bmod F(x) + Ab_{m-2})\alpha^2 \bmod F(x) + \cdots + Ab_{\lceil \frac{m}{2} \rceil})\alpha^2 \\ &\quad \bmod F(x) + \cdots + Ab_1 + c_3\alpha + c_2)\alpha^2 \bmod F(x) + Ab_0 + c_1\alpha + c_0) \end{aligned} \tag{1}$$

A recursive relation is derived that is suitable for efficient power-sum systolic-array implementation. The computation sequence of eq.(1) is written as the following algorithm:

Input : $A(x), B(x), C(x)$ and $F(x)$
Output : $P(x) = A(x)B(x^2) + C(x) \bmod F(x)$
STEP 1 : $P_1 = Ab_{m-1} \bmod F(x)$
STEP 2 : For $i = 2$ to $m - 1$
STEP 3 : $P_i = P_{i-1}\alpha^2 \bmod F(x) + Ab_{m-i} + c_{2(m-i)+1} + c_{2(m-i)}$

where the result P is equal to P_m. As shown in the upper computation sequence, the first term is very simple without any modulo reduction.

Beginning with the first term of eq.(1), Ab_{m-1}, the subsequent terms in the above equation are accumulated until reaching the end. The procedure of the new algorithm is as follows:

First,

$$\begin{aligned} P_1 &= Ab_{m-1} \\ &= \sum_{k=0}^{m-1} a_k b_{m-1} \alpha^k \\ &= \sum_{k=0}^{m-1} p_k^1 \alpha^k \end{aligned} \tag{2}$$

where, eq. (3) can be obtained;

$$p_k^1 = a_k b_{m-1} \tag{3}$$

In the general case,

$$\begin{aligned} P_i &= \left(P_{i-1}\alpha^2 + Ab_{m-i} + c_{2(m-i)+1} + c_{2(m-i)}\right) \bmod F(x) \\ &= \left(\sum_{k=0}^{m-1} p_k^{i-1}\alpha^k\alpha^2 + \sum_{k=0}^{m-1} a_k b_{m-i}\alpha^k + c_{2(m-i)+1} + c_{2(m-i)}\right) \bmod F(x) \end{aligned}$$

$$= \left(\sum_{k=0}^{m-1} p_k^{i-1} \alpha^{k+2} + \sum_{k=0}^{m-1} a_k b_{m-i} \alpha^k + c_{2(m-i)+1} + c_{2(m-i)} \right) \bmod F(x)$$

$$= \Big(p_{m-1}^{i-1} \alpha^{m+1} + p_{m-2}^{i-1} \alpha^m + \cdots + p_0^{i-1} \alpha^2 + a_{m-1} b_{m-i} \alpha^{m-1}$$

$$+ a_{m-2} b_{m-i} \alpha^{m-2} + \cdots + a_1 b_{m-i} \alpha + a_0 b_{m-i} + c_{2(m-i)+1}$$

$$+ c_{2(m-i)} \Big) \bmod F(x)$$

$$= \sum_{k=0}^{m-1} p_k^i \alpha^k \tag{4}$$

From the above procedures, eq. (5) can be derived.

$$p_k^i = p_{m-1}^{i-1} f_k' + p_{m-2}^{i-1} f_k + a_k b_{m-i} + c_k^i; \tag{5}$$

where,

$$y = \begin{cases} c_1^i = c_{2(m-i)+1} \\ c_0^i = c_{2(m-i)} \\ c_k^i = p_{k-2}^{i-1} \quad (\text{ for } 3 \le k \le m) \end{cases}$$

Thus the product P for $AB^2 + C$ in GF(2^m) can be computed efficiently using the above new recursive algorithm.

3 Systolic Architectures in GF(2^m)

3.1 Parallel-in Parallel-out

From the new power-sum algorithm, a corresponding parallel-in parallel-out systolic array architecture can be obtained by following the process in [10,11]. Fig.1 shows the proposed systolic power-sum circuit over GF(2^4). The inputs A, F, and F' enter the array in parallel from the top row, while B is from the leftmost column. The output P is transmitted from the bottom row of the array in parallel. Inputs A and B are both read into the system in the same order using the MSB-first,plus the output is also produced in the same order as the inputs. The MSB-first scheme supports pipelinability for a circuit that requires repeat operations, such as exponentiation and inverse/division, more easily than the LSB-first scheme.

In Fig.1, there is a traverse line in the (i, k) cell. This is required to pass the signal from the $(i-1, k-1)$ cell to the $(i, k+1)$ cell. When the cell is located in the first row; $k = m-1$, p_k^i must be connected to the p_{m-1}^i signal line, while p_k^{i-1} must be connected to the p_{m-2}^i signal line. Fig. 2 shows PE1 which represents the logic circuits of the top cells of the array, while Fig. 3 shows PE2 which represents the logic basic circuits. Fig 3 shows the basic cell for the general case where the circuit function is primarily governed by eq. (5), where the k-th bit (p_k^i) of P_i is the partial product. If the cell is located in the lowest two columns, p_{k-2}^{i-1} is zero. Note that the cells in the first row only need to calculate p_k^1, as

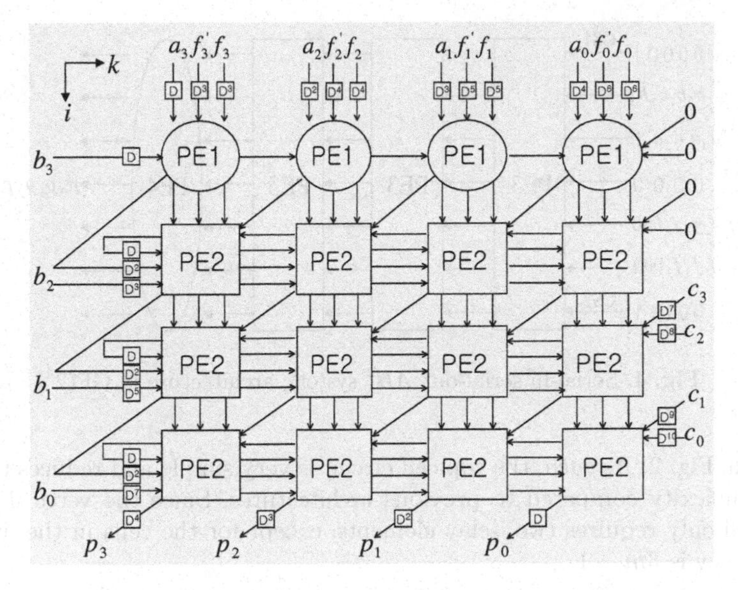

Fig. 1. Parallel-in parallel-out systolic architecture for $AB^2 + C$ in GF(2^4)

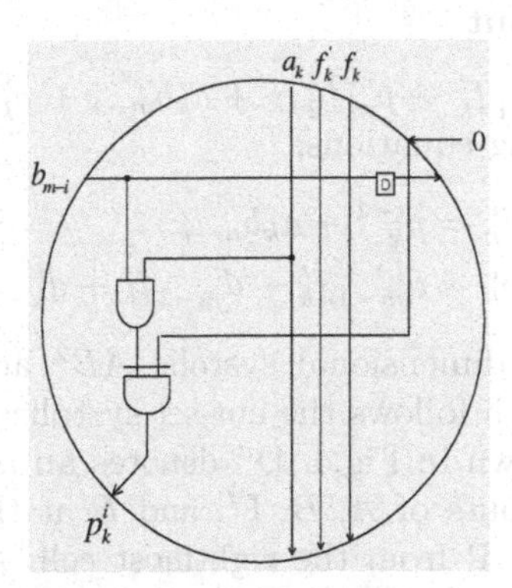

Fig. 2. Circuit for PE1(Processing Element 1)

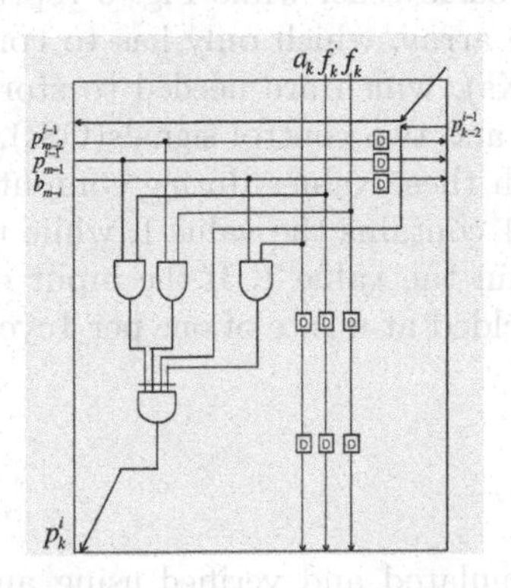

Fig. 3. Circuit for PE2

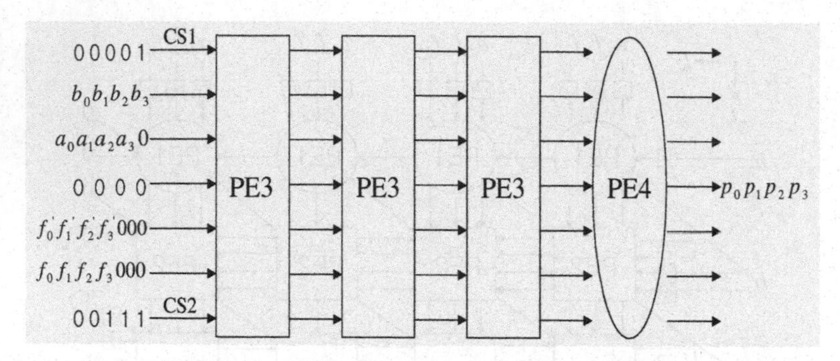

Fig. 4. Serial-in serial-out AB^2 systolic architecture in GF(2^4)

shown in Fig. 2. As such, the top cell circuit is very simple and reduces the total cell complexity compared to previous architectures. Since the vertical path of each cell only requires two delay elements, except for the cells in the first row, the latency is $3m - 1$.

3.2 Serial-in Serial-out

From eq. (5); $p_k^i = p_{m-1}^{i-1} f_k' + p_{m-2}^{i-1} f_k + a_k b_{m-i} + p_{k-2}^{i-1}$, two relations can be derived as in the following equations;

$$d_k^i = p_k^{i-1} + a_k b_{m-i};$$
$$p_k^i = d_{m-1}^i f_k' + d_{m-2}^i f_k + d_{k-2}^i \qquad (6)$$

Using eq. (6), a one-dimensional systolic AB^2 array is derived along the horizontal direction, which follows the cut-set systolization procedure described in [10]. The result is shown in Fig.4. D^n denotes an n-unit delay element. The system receives serial inputs of A, B, F$'$, and F at the leftmost cell and then generates a serial output P from the rightmost cell. All inputs, A, B, F$'$, and F, enter the system in the order of the most significant bit first. Fig. 5 shows the logic circuit for the basic cells, while Fig. 6 represents the logic circuit of the rightmost cell of the array, which only has to compute d_k^m . PE3 requires three multiplexers (MUXs), which are needed to store the values d_{m-1}^i, d_{m-2}^i and b_{m-i}in the i-th cell, and two control signals(CSs), CS1 and CS2, which are used to sustain or refresh these values during computations. Only the first bit among the m bits of CS1 contains the value 1, while the first three bits among the m bits of CS2 contain the value 1. If the input data arrives continuously, the output results are yielded at a rate of one per 1cycle after an initial delay of $2m - 1$.

4 Analysis

The new arrays were simulated and verified using an ALTERA MAX+PLUS simulator and FLEX10K device.

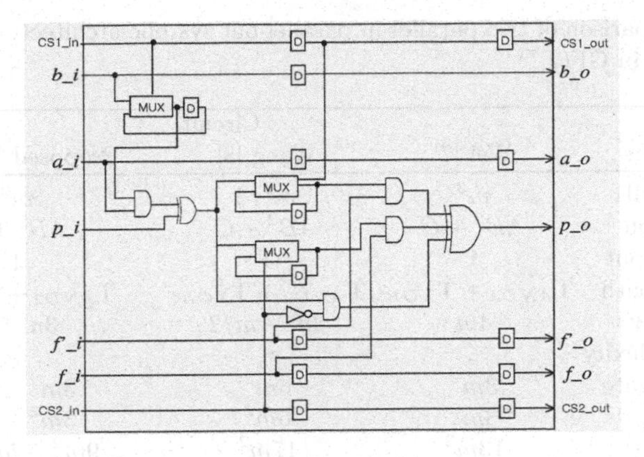

Fig. 5. The basic circuit for PE3

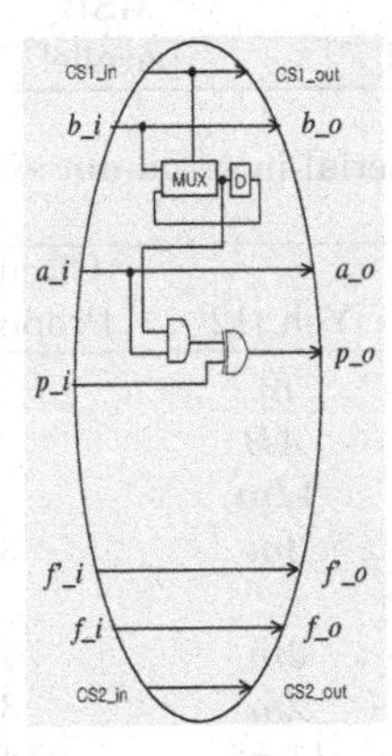

Fig. 6. The circuit for PE4

Table 1 shows a comparision between the proposed parallel architecture and the related circuit described in [7]. In [7], Wei proposed systolic arrays for performing a power-sum operation. Yet, as mentioned by Wang in [8], the systolic power-sum circuit proposed in [7] is inaccurate, therefore, to ensure its proper operation, three 1-bit latches need to be added to each cell of the circuit. Accordingly, Wang's assumption in [8] was followed for a proper comparison. The following assumptions were made in the comparison: 1) T_{AND2} and T_{XORi} denote the propagation delay through a 2-input AND gate and i-input XOR gate, respectively, 2) the 2-input AND gate, 2-input XOR gate, and 1-bit latch consisted of six, fourteen, and eight transistors, respectively [13], 3) the 2-input AND gate and 2-input XOR gate had 2.4ns and 4.2ns gate delays, respectively[13], and 4) the 3-input XOR gate and 4-input XOR gate were constructed using two and three 2-input XOR gates, respectively. Accordingly, the two arrays being compared had the same propagation delay of $T_{AND2} + 2T_{XOR2}$ through one cell [8]. The results clearly showed that the proposed array achieved the same

Table 1. Comparison of two parallel-in parallel-out systolic architectures for Computing power-sum in GF(2^m)

Item	Circuit Wei [7]	 Wang [8]	 Proposed [Fig. 1]
No.of cells	m^2	$m^2/2$	m^2
Function	AB^2+C	AB^2+C	AB^2+C
Throughput	1	1	1
Critical path	$T_{AND2}+T_{XOR3}$	$T_{AND2}+T_{XOR4}$	$T_{AND2}+T_{XOR4}$
Latency	$4m$	$2m-m/2$	$3m-1$
Cell Complexity			
1. AND gate	$3m^2$	$6m^2$	$3m^2-2m$
2. XOR gate	$3m^2$	$6m^2$	$3m^2-2m$
3. Latches	$13m^2$	$17m^2$	$9m^2-5m-3$
Data flow	Bi-directional	Unidirectional	Unidirectional
Algorithm	LSB	MSB	MSB
I/O format	parallel	parallel	parallel

Table 2. Comparison of two serial-in serial-out systolic architectures in GF(2^m)

Item	Circuit Yeh [12]	 Proposed [Fig. 4]
No.of cells	m	m
Function	AB	AB^2
Throughput	$1/m$	1
Latency	$3m$	$3m-1$
Cell Complexity		
1. AND gate	$3m$	$4m-3$
2. XOR gate	$2m$	$3m-2$
3. Latches	$11m^2$	$14m-13$
4. Mux	m	$3m-3$
I/O format	serial	serial
CS	2	2

throughput performance as that in [6, 7], yet exhibited advantages in terms of latency and hardware complexity [8]. The cell complexity of Wei's architecture was m^2(3AND+3XOR+13Latch), whereas that of the proposed parallel architecture was m^2(3AND+3XOR)-m(2AND+2XOR)+($9m^2-5m-3$)Latch. As such, the proposed architecture reduced the cell complexity by m(2AND+2XOR)+($4m^2+5m+3$)Latch. The latency of Wei's architecture[7] was $4m$, whereas that of the proposed architecture was $3m-1$. In particular, the hardware complexity and latency of the proposed parallel-in parallel-out array were about 19.8% and 25% lower, respectively, than Wei's over GF(2^m). The cell complexity of Wang's architecture was $m^2/2$(6AND+6XOR+17Latch). Therefore, it would seem that the proposed architecture reduced the cell complexity by m(2AND+2XOR)-($0.5m^2-5m-3$)Latch. The latency of the architecture of Wang [8] was $2m-m/2$ when the number of cells was $m^2/2$.

Table 2 shows a comparision between the proposed serial architecture and the related circuit described in [12]. The serial architecture had a hardware complexity of m(4AND+3XOR+14Latch)-(3AND+2XOR+13Latch) and latency of $3m-1$. In addition, the circuit had a throughput of 100%. Since Yeh [12] only computed an AB function, it is reasonable that the proposed serial AB^2 architecture exhibited a higher hardware complexity.

5 Conclusion

The current paper presented a new algorithm and a parallel-in parallel-out and a serial-in serial-out systolic architectures for performing the power-sum operation over GF(2^m) using a standard basis. The algorithm is based on the MSB-first scheme. Furthermore, since the proposed architectures have a low hardware complexity and small latency, they are efficient for computing exponentiation and inverse/division in GF(2^m) and well suited to VLSI implementation due to their simplicity, regularity, modularity, and pipelinability.

Acknowledgement

This work was supported by grant No. 2000-2-51200-001-2 from the Korea Science & Engineering Foundation

References

1. W.W. Peterson, E.J. Weldon: Error-correcting codes. MIT Press, MA (1972)
2. D.E.R. Denning: Cryptography and data security. Addison-Wesley, MA (1983)
3. A. Menezes : Elliptic Curve Public Key Cryptosystems, Kluwer Academic Publishers, Boston (1993)
4. R.L. Rivest, A. Shamir, and L. Adleman: A Method for Obtaining Digital Signatures and Public-key Cryptosystems. Comm. ACM. **21** (1978) 120–126
5. I.S. Reed and T.K. Truong: The use of finite fields to compute convolutions.IEEE Trans. Inform. Theory,**21** (1975) 208–213
6. S.W. Wei: VLSI architectures for computing exponentiations, multiplicative inverses, and divisions in GF(2^m). Proc. IEEE Trans. Circuits and Systems. **44** (1997) 847–855
7. S.W. Wei: A Systolic Power-Sum Circuit for GF(2^m). IEEE Trans. Computers. **43** (1994) 226–229
8. C.L. Wang and J.H. Guo: New systolic arrays for C+AB2, inversion, and division in GF(2^m). IEEE Trans. Computers **49** (2000) 1120–1125
9. C.W. Wu and M.K. Chang: Bit-Level Systolic Arrays for Finite-Field Multiplications. Journal of VLSI Signal Processing. **10** (1995) 85–92
10. S. Y. Kung: VLSI Array Processors. Prentice-Hall. **43** (1987)
11. K. Y. Yoo: A Systolic Array Design Methodology for Sequential Loop Algorithms., Ph.D. thesis, Rensselaer Polytechnic Institute, New York (1992)
12. C. S. Yeh, I. S. Reed, and T. K. Truong: Bit-Level Systolic Arrays for Finite-Field Multiplications. Journal of VLSI Signal Processing. **10** (1995) 85–92
13. Daniel D. Gajski: Principles of Digital Design. Prentice-Hall international, INC. (1997)

A Combinatorial Approach to Anonymous Membership Broadcast

Huaxiong Wang and Josef Pieprzyk

Centre for Advanced Computing – Algorithms and Cryptography
Department of Computing, Macquarie University, Australia
{hwang,josef}@comp.mq.edu.au

Abstract. A set system $(X, \mathcal{F})$ with $X = \{x_1, \ldots, x_m)$ and $\mathcal{F} = \{B_1, \ldots, B_n\}$, where $B_i \subseteq X$, is called an (n, m) cover-free set system (or CF set system) if for any $1 \leq i, j, k \leq n$ and $j \neq k$,

$$|B_i| \geq 2\left|B_j \bigcap B_k\right| + 1.$$

In this paper, we show that CF set systems can be used to construct anonymous membership broadcast schemes (or AMB schemes), allowing a center to broadcast a secret identity among a set of users in a such way that the users can verify whether or not the broadcast message contains their valid identity. Our goal is to construct (n, m) CF set systems in which for given m the value n is as large as possible. We give two constructions for CF set systems, the first one from error-correcting codes and the other from combinatorial designs. We link CF set systems to the concept of cover-free family studied by Erdös *et al* in early 80's to derive bounds on parameters of CF set systems. We also discuss some possible extensions of the current work, motivated by different application.

1 Introduction

Given a center and a set of users $\mathcal{U} = \{P_1, \ldots, P_n\}$. The center wishes to broadcast information to all users in such a way that (1) a single user from the group is sure that she is the intended recipient while (2) the other users are certain that the information is not intended for them. The security goal is that the users are not able to identify the intended recipient. Ideally, one would expect that the probability of guessing the right recipient by users is no better than $\frac{1}{n-1}$ (a non-intended user knows that the intended user must be someone different from him). Note that the probability of guessing the intended user by outsiders is expected to be no better than $1/n$. The above scheme is called *anonymous membership* scheme and have various applications in cryptographic protocols. The applications range from anonymous delegation (where one member of the group performs action on the behalf of the group), hiding the order of interactions (between the center and members of the group), cheating prevention (by hiding the order in which cryptographic operations are performed), etc.

A solution to the problem of anonymous membership could be based on public key cryptography in which the center encrypts some unambiguous text

O.H. Ibarra and L. Zhang (Eds.): COCOON 2002, LNCS 2387, pp. 162–170, 2002.

txt_i (also called secret identity) that indicates that the user P_i is the chosen one. The center broadcasts the cryptogram $E_{pk_i}(\mathrm{txt}_i)$, where pk_i is the public key of P_i. Since only P_i knows the private key corresponding to pk_i can decrypt the cryptogram and get the message txt_i, the secrecy of the anonymous membership is guaranteed. However, such a solution has two important drawbacks. The first the public-key encryption algorithm needs to be *key private* in the sense that an eavesdropper in possession of a ciphertext should not be able to tell which specific key out of the n known public keys is the one under which the ciphertext has been created. Obviously no all the public-key cryptosystems are key private. For example, in the RSA encryption, it is well known that the public modulus for different users must not be different thus the ciphertexts under different public keys may not be of the same length, so ciphertexts can be linked to the public key used. The second drawback is that all users in the group must perform expensive decryption no matter whether or not they are chosen. In many circumstances this may be unreasonable or unacceptable. The problem of key privacy in public-key encryption has been addressed in literature (see, for example, [1]).

This paper deals with unconditionally secure anonymous membership that typically offer a very efficient solution. Being more precise, our solutions require private communication to set up the system but once set up, users can very efficiently verify whether or not they are chosen.

Our work. We introduce anonymous membership broadcast schemes (AMB schemes for short) that allow a center to anonymously broadcast a secret identity to a group of users in a such way that only a single user learns that she has been chosen while the others know that they have not been. Our approach is combinatorial in nature and is secure against adversaries with unlimited computational power (unconditionally secure). We propose AMB schemes, based on a special set system, called *cover-free set system* (or *CF set system* for short). We give constructions of CF set systems from error-correcting codes, and certain combinatorial designs, such as μ-designs. The CF set systems give AMB schemes with significantly improved efficiency. The scheme based on error correcting codes deserves a special attention as it offers a logarithmic complexity for both communication and storage. This compares favorably with with trivial solutions that require linear complexity and storage. We link the CF set systems with extensively studied *cover-free families* introduced by Erdös *et al* in early 80's [5]. The established relation between AMB schemes and cover-free families allows us to derive bounds on various parameters for CF set systems and use them to assess the performance of AMB schemes. We also discuss possible extensions of the current work.

Related work. Anonymity is one of basic human rights that is guarded by the legal systems of all democratic countries. Anonymity has been studied in the context of secure electronic elections as their major security goal (for example, [4,9,10]). Although various aspects of anonymity have been studied, to our best knowledge, there is no efficient solution for the anonymous membership broadcast problem. There are, however, two works that are closely related to the AMB problem that we should mention. The first work by Fiat and Naor [6] considers broadcast encryption in which a center broadcasts a message to a *dynamically* changing subset of privileged users in such a way that only users in the subset learn the message. This is to say that the main security goals of broadcast encryption are the secrecy and integrity of the messages. The AMB problem addresses a different aspect of broadcast security, namely, the secrecy of the intended recipient's identity. Another concept relating to the AMB schemes is the secret set introduced by Molva and Tsudik [8]. A *secret set* is a subset of $\mathcal{U}$ that any user in $\mathcal{U}$ can test his/her membership in the subset but cannot determine neither the other users of the subset nor the cardinality of the subset. Obviously, the AMB problem can be seen as a special case of secret sets if only single users are considered as potential secret sets.

2 Anonymous Membership Broadcast Schemes

Assume that there is a trusted center and a universal set $\mathcal{U}$, consisting of n users $P_1, \ldots, P_n$, in an open communication network. The network is a collection of broadcast channels, i.e., the channels are publicly accessible and any information transmitted by the center will be received by every user in $\mathcal{U}$. During the setup of the system, the center generates and distributes secret information to each user through secure communication channels. Later on, the center wishes to broadcast an anonymous membership for a single user P_i in a such way that each user can verify his/her membership. In particular, P_i is able to make sure that she is the intended recipient while P_j $(j \neq i)$ is sure that he/she is not. We also assume that the secret identity is not known ahead of time. A cryptographic scheme (with the center and the group $\mathcal{U}$) that allows the center to broadcast secret identity that satisfies the two above-mentioned conditions, is called an *anonymous membership broadcast scheme* (or AMB schemes).

For the AMB schemes, we define the following three phases:

- *Initialization:* the center distributes secret information to each user in $\mathcal{U}$ (typically the center uses private communication channels existing between it and users).
- *Broadcast:* a secret identity of the chosen recipient P_i is broadcast to all users in $\mathcal{U}$.
- *Verification:* after seeing the broadcast message s, each user can verify his/her own membership, but gets no more information than he/she is supposed to known about the secret identity (or equivalently about the intended recipient).

Quality of an AMB scheme is measured by its security and by its efficiency.

- Security requires that each user except the intended one (whose secret identity has been broadcast) is not able to determine the secret identity of the recipient. More precisely, the probability that a user guesses the secret identity should be bounded by a fixed and pre-determined value. We call a AMB

scheme *perfect* if the probability is $1/(n-1)$ or each non-intended user knows that the intended recipient is somebody different from him/her.
- Efficiency is measured by the size of storage for secret information of the center and users, by the amount of communication overhead for broadcast and by computation performed by the center and the users. In this paper we are mainly concerned with the two first measures, i.e. storage and communication overhead.

Consider the following two trivial solutions for AMB schemes.
- *Solution 1.* The center randomly chooses a vector $\alpha = (a_1, \ldots, a_n) \in GF(2)^n$, and secretly gives a_i to P_i; $i = 1, \ldots, n$. For a secret identity of P_i, the center broadcasts $(b_1, \ldots, b_n) = (a_1, \ldots, a_{i-1}, a_i + 1, a_{i+1}, \ldots, a_n)$. Each user P_j can verify his/her membership by checking if $a_j = b_j$. The system is used once only.
- *Solution 2.* The center randomly chooses a permutation π on $\{1, \ldots, n\}$, and secretly gives $\pi(j)$ to user P_j; $1 \leq j \leq n$. For a secret identity of P_i, the center simply broadcasts $\pi(i)$. Each user P_j compares her secret information $\pi(j)$ and the broadcast $\pi(i)$ to verify her membership. The scheme is one-time.

We note that both the trivial solutions offer the perfect security, but have different communication and storage requirements. In Solution 1, the storages for the center and each user are n bits and 1 bit, respectively. The center needs to broadcast n bits. Solution 2 requires $n \log n$ and $\log n$ bits storage for the center and each user, respectively. The communication requires $\log n$ bits.

2.1 The Basic AMB Scheme

We observe that the cover-free set system can be used as a mathematical model of the AMB system. This observation gives us an advantage that security and efficiency parameters can be derived from parameters of the cover-free set system.

Definition 1. *Let $X = \{x_1, \ldots, x_m\}$ be a finite set and $\mathcal{F} = \{B_1, \ldots, B_n\}$ be a family of subsets of X. We call the set system $(X, \mathcal{F})$ an (n, m) cover-free set system (or CF set system) if the following condition is satisfied:*

$$|B_i| \geq 2\left|B_j \bigcap B_k\right| + 1, \quad \textit{for all } 1 \leq, i, j, k \leq n \textit{ and } j \neq k.$$

Given an (n, m) CF set system $(X, \mathcal{F})$ with $d = \min\{\lceil |B_i|/2 \rceil - 1 : 1 \leq i \leq n\}$. We construct an AMB scheme from $(X, \mathcal{F})$ as follows.

- *Initialization* – the center randomly chooses a vector $(a_1, \ldots, a_m) \in GF(2)^m$, and secretly sends a_i to user P_j if $x_i \in B_j$, for all $1 \leq i \leq m, 1 \leq j \leq m$.
- *Broadcast of identity of P_i* – the center randomly chooses a d-subset D from B_i and a d-subset C from $X \setminus B_i$. The center computes a vector $\beta = (b_1, \ldots, b_m)$ defined by

$$b_i = \begin{cases} a_i + 1 \text{ if } x_i \in B_i \setminus D, \text{ or } x_i \in C, \\ a_i \qquad \text{otherwise,} \end{cases}$$

and broadcasts $\beta = (b_1, \ldots, b_m)$ as the secret identity of P_i.

– *Verification* – each user P_j verifies his/her membership as follows. Assume that P_j has secret information $a_{j_1}, \ldots, a_{j_\ell}$ (i.e. $B_j = \{x_{j_1}, \ldots, x_{j_\ell}\}$), then P_j computes the vector $\gamma_j = (a_{j_1} + b_{j_1}, \ldots, a_{j_\ell} + b_{j_\ell})$, if the Hamming weight of γ_j is greater than d than P_j is the intended user, otherwise he/she is not.

Now we are ready to establish the relation between the CF set system and the AMB system.

Theorem 1. *Let $(X, \mathcal{F})$ be an (n, m) CF set system. The above construction from $(X, \mathcal{F})$ results in an AMB scheme with m bits broadcast, m bits storage for the center and $|B_i|$ bits storage for P_i for all $1 \leq i \leq n$.*

Proof. The parameters in the theorem are obvious. It is also clear that each user can verify his/her membership correctly. What we need is to prove the security of the resulting AMB scheme. A user P_j who is not the intended one, tries to guess the secret identity from the broadcast β. P_j's best strategy is to use his/her the secret information together with the public β to distinguish the identity of the chosen user P_i from the identities of the other users. The only information P_j can use to guess the membership of a user of P_i is the sub-vector of γ that corresponds to $B_j \cap B_i$, denoted by γ_{ij}. If the Hamming weight of γ_{ij} is greater than d, then P_i is the user of the secret identity. However, for each $i, i \neq j$, the Hamming weight of γ_{ij} runs between 0 and d, no matter whether P_i is the intended user or not. Thus, P_j will never be certain which one is the chosen one.

We note that the above construction does not provide perfect security for the resulting AMB scheme. Indeed, from P_j's point of view the probability that P_k is the intended user is higher than that of P_i if the Hamming weight of γ_{jk} is larger than γ_{ji}. So P_j's best strategy is to choose i as the secret identity if the Hamming weight of γ_{ji} is maximal. To provide sound security for the scheme, the probability of success of this attack must be bounded by some pre-designated value. For the applications for AMB schemes, from Theorem 1 we would like to have (n, m) CF set system with n as large as possible.

3 Constructions of Cover-Free Set Systems

Two constructions of (n, m) CF set systems are given. The first construction is based on error-correcting codes and the other uses combinatorial designs.

3.1 CF Set Systems from Error-Correcting Codes

Let Y be an alphabet of q elements. An (N, M, D, q) *code* is a set $\mathcal{C}$ of M vectors in Y^N such that the Hamming distance between any two distinct vectors in $\mathcal{C}$ is at least D.

Theorem 2. *If there is an (N, M, D, q) code, then there exists an (M, Nq) CF set system provided $N < 2D$.*

Proof. Let $\mathcal{C}$ be an (N, M, D, q) code. We write each codeword as $c_i = (c_{i1}, \ldots, c_{iN})$ with $c_{ij} \in Y$, where $1 \le i \le M, 1 \le j \le N$. Set $X = \{1, \ldots, N\} \times Y$ and $\mathcal{F} = \{B_i : 1 \le i \le M\}$, where for each $1 \le i \le M$ we define $B_i = \{(j, c_{ij}) : 1 \le j \le N\}$. It is easy to check that $|X| = Nq$, $|\mathcal{F}| = M$ and $|B_i| = N$. For each pair of i, k, we have $|B_i \cap B_k| = |\{(j, c_{ij}) : 1 \le j \le N\} \cap \{(j, c_{kj}) : 1 \le j \le N\}| = |\{j : c_{ij} = c_{kj}\}| \le N - D$. The condition that $N = |B_i| > 2|B_k \cap B_\ell|$ follows directly from the assumption that $N < 2D$. So $(X, \mathcal{F})$ is a CF set system.

The advantage of the construction from error-correcting codes is that the resulting AMB schemes have a very promising performance. For example, we know that there are algebraic-geometric codes breaking Gilbert-Varshamov bound. That is, for given q and $N < 2D$, we can construct (N, M, D, q) code with $N = O(\log M)$. Thus, we obtain the following corollary.

Corollary 1. *There exist (n, m) CF set systems with $m = O(\log n)$.*

3.2 CF Set Systems from Combinatorial Designs

Although constructions based on error-correcting codes produce CF set systems with very good performance. However, the performance is in general asymptotic, that is to say, the larger n gives better performance. Sometimes, when n is small, constructions from combinatorial designs can give better performance.

A $\mu - (v, r, \lambda)$ *design* is a set system $(X, \mathcal{F})$, where $|X| = v$, $|B| = r$ for every $B \in \mathcal{F}$, and every μ-subset of X occurs in *exactly* λ blocks in $\mathcal{B}$. We will only be interested in $\mu - (v, r, 1)$ designs. It is well known that in a $\mu - (v, r, 1)$ design, the number of blocks n is exactly $\binom{v}{\mu}/\binom{r}{\mu}$. Assume that there exists a $\mu - (v, r, 1)$ design $(X, \mathcal{B})$. Then for each pair $B_i, B_j \in \mathcal{B}$, we trivially have $|B_i \cap B_j| \le \mu - 1$. The following result is immediate.

Theorem 3. *If there exists a $\mu - (v, r, 1)$ design, then there exists a $(\binom{v}{\mu}/\binom{r}{\mu}, v)$ CF set system provided $r > 2(\mu - 1)$.*

There are many results on existence and constructions of $\mu - (v, r, 1)$ designs for $r = 2, 3$ [2]. On the other hand, no $\mu - (v, r, 1)$ design with $v > r > \mu$ is known to exist for $\mu \ge 6$. Furthermore, it is known that for $3 \le r \le 5$, a $2 - (v, r, 1)$ design exists if and only if $v \equiv 1$, or $r \bmod (r^2 - r)$. Taking $r = 3$, we obtain that there exist $(v(v-1)/6, v)$ CF set systems for all $v \equiv 1$, or $3 \bmod 6$. Using a similar argument, it is not difficult to show that other combinatorial designs, such as BIBDs, Steiner systems, orthogonal arrays, packing designs and many others can also be used to construct CF set systems for different parameters.

4 Bounds on CF Set Systems

In this section, we show that CF set systems are closely related to the well-known combinatorial objects of *cover-free families* studied by Erdös *et al* [5].

Definition 2. *A set system* $(X, \mathcal{F})$ *with* $X = \{x_1, \ldots, x_m\}$ *and* $\mathcal{F} = \{B_i \subseteq X \mid i = 1, \ldots, n\}$ *is called an* (n, m, t)-cover-free family *(or* (n, m, t)*-CFF) if for any subset* $\Delta \subseteq \{1, \ldots, n\}$ *with* $|\Delta| = t$ *and any* $i \notin \Delta$, $|B_i \setminus \bigcup_{j \in \Delta} B_j| \geq 1$.

Constructions and bounds for (n, m, t)-CFF were studied by numerous authors (see, for example, [3,5,12]). It is shown in [12] that for (n, m, t)-CFF with $t \geq 2$, $m \geq c \frac{t^2}{\log t} \log n$ for some constant c. On the other hand, Erdös *et al* [5] showed that for any $n > 0$, there exists an (n, m, t)-CFF with $m = O(t^2 \log n)$ and $|B_i| = O(t \log n)$. This result is, however, non-constructive. Although Kumar *et al* [7] gave a probabilistic construction of CFF that meets the bound, explicit constructions that can achieve or get close to Erdös *et al* bounds (see, for example, [7,12,13]), are still of high interest.

The following lemma establishes the link between CF set system and CFF.

Lemma 1. *An* (n, m) *CF set system is an* $(n, m, 2)$*-CFF.*

Proof. Let $(X, \mathcal{F})$ be an (n, m) CF set system. For any triple i, j, k, we have, from the definition of CF set system, $|B_i| > 2|B_i \cap B_j|$ and $|B_i| > 2|B_i \cap B_k|$, it follows that $|B_i \setminus (B_j \cup B_k)| \geq |B_i| - (|B_i \cap B_j| + |B_i \cap B_k|) \geq 1$. So $(X, \mathcal{F})$ is an $(n, m, 2)$-CFF.

It is easy to give examples to prove that the reverse implication to that in Lemma 1 is not true. So that not all $(n, m, 2)$-CFFs are (n, m) CF set systems. This indicates that there is a need for a further study of the CF set systems.

Let $M(n)$ denote the minimal value of m for which an (n, m) CF set system exists. From Lemma 1 and the bounds on CFF, we have the following result.

Theorem 4. $M(n) = \Theta(n)$.

5 Generalizations

AMB Systems Secure against Collusion Attacks. In the basic AMB systems, it is assumed that single users from $\mathcal{U}$ may try to attack the system. It is easy to see that the system does not provide protection against a collusion of the users in $\mathcal{U}$. We say an AMB scheme is *w-resilient* if any up to w users, even if they collude, cannot correctly guess the identity of the intended user, unless the intended user is one of the attackers. The basic AMB scheme based on CF set systems, can be converted into its w-resilient version in a straightforward manner. We only need to modify the condition $|B_i| > 2|B_j \cap B_k|$ in the CF set system to the following condition for the system $(X, \mathcal{F})$:

$$|B_i| \geq 2|B_j \cap (B_{j_1} \cup \cdots \cup B_{j_w})| + 1,$$

for any i and any $w + 1$ distinct elements $j, j_1, \ldots j_w$ in $\{1, \ldots, n\}$. We call a set system $(X, \mathcal{F})$ satisfying the above condition a w-resilient CF set system. Using a similar approach to that we have applied for the design of AMB schemes from CF set systems, we can construct w-resilient AMB-schemes from w-resilient CF set systems.

Constructions of Secret Sets. As we have already mentioned, the AMB schemes are relevant to the concept of secret sets introduced by Molva and Tsudik [8]. A set, more precisely a subset of a universal $\mathcal{U}$, is called a *secret set* if the following conditions are satisfied: (C1) any user (whether a set user or not) can verify his/her memberships in the set; (C2) no one, with the exception of the originator of the set, can verify another user's membership in the set; (C3) no one, with the exception of the originator of the set, can determine with certainty the number of users in the set.

The major difference between a AMB scheme and a secret set is that the latter has the additional security requirements of (C2) and (C3). It is obvious that a secret set can be constructed from an AMB scheme. Indeed, for an AMB scheme to achieve C2, it only needs to execute multiple, parallel AMB schemes, each independently broadcasts an anonymous membership. For C3, some *dummy* anonymous membership broadcasts are added to the system to raise the uncertainty of the cardinality of the anonymous users, that is to include AMB scheme, but no one is the anonymous membership. Such a construction is inefficient, as to achieve C3, it requires n single AMB schemes to be executed. Depending on the application in hand, if the secret set is defined by C1 and C2 only and the size of the set is small, then the constructions of secret set from AMB schemes should offer some advantages.

Computationally Secure Schemes. The basic AMB scheme is unconditionally secure if the key initialization is implemented through the privately secure channels. It is also one-time in the sense that every key bit can be used once only. The basic schemes, however, can be efficiently implemented for multiple usage in the conditionally secure setting. Note that in the initialization phases, the center needs to securely send a subset of bits to each user, and a secret bit may be shared by a subset of users. A natural question is: how can this be achieved in the conditionally secure setting? Each $x_i \in X, 1 \leq i \leq m$, is associated with a subset of users $\mathcal{U}_i = \{P_j \mid \text{if } x_i \in B_j\}$. Instead of having a common bit a_i distributed from the center, $\mathcal{U}_i$ and the center execute a conference key agreement protocol to obtain a common secret key $k_{\mathcal{U}_i}$. A common secret bit (or many secret bits) can be extracted from $k_{\mathcal{U}_i}$ using certain cryptographic techniques. For example, using $k_{\mathcal{U}_i}$ as the seed of a pseudorandom number generator results in many common secret bits for the users in $\mathcal{U}_i$. These secret bits can be later used as a_i's for the basic AMB schemes. Each user P_i needs to involve in $|B_i|$ different conference key agreement protocols.

Acknowledgment

The work was in part supported by Australian Research Council grant A00103078.

References

1. M. Bellare, A. Boldyreva, A. Desai and D. Pointcheval, Key-Privacy in Public-Key Encryption, *Advances in Cryptology – Asiacrypt'01*, LNCS, **2248**(2001), 566-582.

2. P. J. Cameron and J. H. Van Lint, Designs, Graphs, Codes, and their Links, Cambridge University Press, Cambridge 1991.
3. Y. Desmedt, R. Safavi-Naini, H. Wang, L. M. Batten, C. Charnes and J. Pieprzyk, *Broadcast Anti-jamming Systems*, Computer Networks, **35**(2001), 223-236.
4. D. Chaum, The dining cryptographers problem: unconditional sender and recipient untraceability, *J. of Cryptology*, **1** 1988, 65-75.
5. P. Erdös, P. Frankl, and Z. Furedi, Families of finite sets in which no set is covered by the union of r others, *Israel Journal of Mathematics*, **51**(1985), 79-89.
6. A. Fiat and M. Naor, Broadcast encryption, *Advances in Cryptology– Crypto '93*, LNCS, **773**(1994), 480-490.
7. R. Kumar, S. Rajagopalan and A. Sahai. Coding constructions for blacklisting problems without computational assumptions, *Advances in Cryptology – CRYPTO '99*, LNCS, **1666**(1999), 609-623.
8. R. Molva and G. Tsudik, Secret sets and applications, *Information Processing Letters*, **65**(1998), 47-55.
9. M. Reiter and A. Rubin, Crowds: Anonymity for Web transactions, *ACM Transactions on Information and System Security*, **1**(1998), 66-92.
10. A. Salomaa, Verifying and recasting secret ballots in computer networks, *New Results and New Trends in Computer Science*, LNCS, **555**(1991), 283-289.
11. D. R. Stinson. On Some Methods for Unconditionally Secure Key Distribution and Broadcast Encryption. *Designs, Codes and Cryptography* **12**(1997), 215-243.
12. D. S. Stinson, R. Wei and L. Zhu, Some new bounds for cover-free families, *Journal of Combinatorial Theory, A*, **90**(2000), 224-234.
13. H. Wang and C. Xing, Explicit constructions of perfect hash families from algebraic curves over finite fields, *Journal of Combinatorial Theory, A*, **93**(2001), 112-124.

Solving Constraint Satisfaction Problems with DNA Computing

Evgeny Dantsin and Alexander Wolpert

School of Computer Science, Roosevelt University
430 South Michigan Ave., Chicago, IL 60605, USA
{edantsin,awolpert}@roosevelt.edu

Abstract. We demonstrate how to solve constraint satisfaction problems (CSPs) with DNA computing. Assuming that DNA operations can be faulty, we estimate error probability of our algorithm. We show that for any k-CSP, there is a polynomial-time DNA algorithm with bounded probability of error. Thus, k-CSPs belong to a DNA analogue of **BPP**.

1 Introduction

After eight years of intensive research in DNA computing it is still not clear whether DNA computing can compete (or will be able to compete in the near future) with existing "silicon" computing. So far problem instances solved with DNA are much smaller than instances of the same problems cracked by electronic computers. Another question that should be addressed is the ability to control errors in DNA computation. Note that electronic computers are faulty too but it is possible to control their errors. In this paper we join the hunt for applications utilizing advantages of DNA computing; in particular we attack the question of error control.

The paper presents a DNA algorithm for solving constraint satisfaction problems (CSPs). Many particular problems that can be stated as CSPs have been already studied in publications on DNA computing. For example, Lipton [10] proposed a DNA algorithm for SAT, Bach et al [4] proposed a DNA algorithm for the 3-colorability problem, etc. We show how to satisfy general constraints with DNA. Our algorithm makes use of the JOIN operation (the name is inspired by "join" in databases, see also [15]). This operation is natural for CSPs as well as for DNA computing. For k-CSPs, the algorithm runs in polynomial time.

The JOIN operation is implemented using well known biochemical DNA manipulations such as EXTRACT, APPEND, MERGE and others, e.g. [5,13,15]. Some of them can introduce errors. For example, EXTRACT can have two kind of errors: false negative error (a strand containing a given substrand is not extracted) and false positive error (a strand not containing a given substrand is extracted). We analyze how such errors affect the result of our algorithm to estimate its probability of error.

To decrease error probabilities, we employ the technique proposed by Karp et al [9]. This method [9] makes EXTRACT error-resilient without a big sacrifice

O.H. Ibarra and L. Zhang (Eds.): COCOON 2002, LNCS 2387, pp. 171–180, 2002.

in the running time. More exactly, EXTRACT is converted into a DNA algorithm whose error probability δ is arbitrary small and whose running time is $O(\log^2 \delta)$. Using this construction, we convert our join-based algorithm into a DNA algorithm with error probability bounded by an arbitrary small constant. For k-CSPs, its running time is still polynomial. That is, any k-CSP can be solved by a polynomial-time DNA algorithm with bounded probability of error. Thus, any k-CSP belongs to a complexity class that can viewed as a DNA analogue of **BPP** (for **BPP** see e.g. [12]).

The paper is organized as folows. Section 2 contains definitions and notation related to CSPs. In Section 3 we define the JOIN operation and show how to solve CSPs with JOIN. Basic DNA operations used in our model of DNA computation are described in Section 4. Section 5 gives a DNA implementation of our algorithm for CSPs. In Sections 6 and 7 we modify our algorithm into an error-resilient algorithm.

2 Constraint Satisfaction Problems

In a constraint satisfaction problem (CSP) we are given: (i) a finite set of variables that range over a finite domain D, and (ii) constraints $C_1, \ldots, C_m$ on values of these variables. We need to determine whether there is a valuation of the variables that satisfies all $C_1, \ldots, C_m$. For example, the satisfiability problem (SAT) can be naturally restated as a CSP in which constraints are disjunctions of literals. Other examples are the graph colorability and solving equations over finite domains.

More formally, each CSP is specified by a finite domain D and a set of predicates defined on D. The predicates are required to be computable in polynomial time. Any atomic proposition $p(x_1, \ldots, x_k)$, where p is a predicate symbol and $x_1, \ldots, x_k$ are variables ranging over D, is called a *constraint.* A *CSP instance* is a finite set $\{C_1, \ldots, C_m\}$ of constraints.

An *assignment* for variables $x_1, \ldots, x_l$ is a valuation of these variables in D. We denote such an assignment by $\{x_1 \leftarrow d_1, \ \ldots, \ x_l \leftarrow d_l\}$ where $d_1, \ldots, d_r \in D$. An assignment for one variable is called a *unit* assignment. An assignment for all variables occurring in a CSP instance I is called a *full assignment* for I. A full assignment α for I is called a *solution* to I if α satisfies every constraint in I, i.e., every constraint evaluates to *true* under α. By a *CSP* we mean the following decision problem: given a CSP instance I, determine whether there a solution to I. Such a problem is called a *k-CSP* if each predicate has its arity at most k, i.e., each constraint contains at most k variables.

Let C be a constraint and α be an assignment for the variables occurring in C. We call α a *solution* to C if α satisfies C. Any constraint C can be identified with the set of all solutions to C.

For an assingnment α, the set of all variables occuring in α is denoted by $Var(\alpha)$. For a set S of assignments, we write $Var(S)$ to denote $\cup_{\alpha \in S} Var(\alpha)$.

3 Solving CSPs with Joins

Assignments α_1 and α_2 are called *consistent* if they agree on the common variables in $Var(\alpha_1) \cap Var(\alpha_2)$. We define the *join* of consistent assignments α_1 and α_2 to be the following assignment denoted by $\alpha_1 \bowtie \alpha_2$:

- $\alpha_1 \bowtie \alpha_2$ agrees with α_1 on all variables in $Var(\alpha_1) - Var(\alpha_2)$;
- $\alpha_1 \bowtie \alpha_2$ agrees with α_2 on all variables in $Var(\alpha_2) - Var(\alpha_1)$;
- $\alpha_1 \bowtie \alpha_2$ agrees with both α_1 and α_2 on all variables in $Var(\alpha_1) \cap Var(\alpha_2)$;
- $\alpha_1 \bowtie \alpha_2$ assigns values to only variables in $Var(\alpha_1) \cup Var(\alpha_2)$.

For example, the join of consistent assignments $\{x_1 \leftarrow 0,\ x_3 \leftarrow 1\}$ and $\{x_2 \leftarrow 1,\ x_3 \leftarrow 1\}$ is the assignment $\{x_1 \leftarrow 0,\ x_2 \leftarrow 1,\ x_3 \leftarrow 1\}$.

We also extend the join operation to sets of assignments. For sets S_1 and S_2 of assignments, we define the *join* operation as follows:

$$S_1 \bowtie S_2 \;=\; \{\alpha \bowtie \beta \mid \alpha \in S_1,\ \beta \in S_2,\ \alpha \text{ and } \beta \text{ are consistent}\}$$

In particular, if all pairs α, β are inconsistent, $S_1 \bowtie S_2$ is empty. Our operation is essentially the same as the natural join of relations in databases, e.g. [1]. A similar (but different) DNA operation was introduced in [15].

We present an algorithm that uses the join operation for step-by-step generation of solutions to a CSP instance. Given an instance $\{C_1, \ldots, C_m\}$, the algorithm runs in m steps. At step i the algorithm modifies the current set S of assignments (satisfying $C_1, \ldots, C_{i-1}$) in order to satisfy C_i. After m steps the algorithm returns the set S of all solutions to $\{C_1, \ldots, C_m\}$.

Algorithm 1 (Join-based algorithm for CSPs).

Input: Sets $S_1, \ldots, S_m$ of assignments that represent constraints $C_1, \ldots, C_m$, i.e., each S_i is the set of all solutions to C_i.

Output: The set S of all solutions to the input instance $\{C_1, \ldots, C_m\}$.

1. $S \leftarrow S_1$
2. **for** $i \leftarrow 2$ **to** m **do**
   (a) $S \;\leftarrow\; S \bowtie S_i$
   (b) **if** $S = \emptyset$ **then return** (*"no solution"*)
3. **return** (S)

The algorithm takes time $O(m \cdot t_{\bowtie})$ where $t_{\bowtie}$ is the maximum running time of the join operation. The space is $O(n \cdot |D|^n)$ where n is the number of variables occurring in $\{C_1, \ldots, C_m\}$ and $|D|$ is the number of elements in D. Our next step is to implememnt Algorithm 1 with DNA.

4 DNA Operations

Most papers on the comlexity of DNA algorithms use DNA computation models based on more or less close collections of DNA operations. In this paper, we use a collection of operations similar to the operations in [5].

To solve CSPs with DNA, assignments are encoded by *DNA strands.* For now, a *strand* can be thought of as a string over the alphabet $\Sigma = \{A, C, G, T\}$, see e.g. [2,13]. Given a CSP instance, we fix an encoding of unit assignments by strands (note that the set of all possible unit assignments for the instance is finite). For a unit assignment α, we denote a strand that encodes α by $E(\alpha)$. We assume that all unit assignments are encoded by strands of the same length. Encodings of unit assignments induce encodings of arbitrary assignments: if an assignmnet α consists of unit assignments $\alpha_1, \ldots, \alpha_r$, we encode α by a concatenation of strands $E(\alpha_1), \ldots, E(\alpha_r)$.

Strands are contained in *tubes.* We associate with each tube T the set $\mathcal{S}_T$ of all assignments encoded by the strands in T, i.e., $\mathcal{S}_T$ consists of all assignments α such that T contains $E(\alpha)$.

A DNA computation is a computation of a Turing machine augmented by operations on tubes. Operations take tubes and/or assignments (written on Turing machine tapes) as inputs and return tubes and/or assignments.

Extract. This operation takes a tube T and a unit assignment α as input and returns a tube with all strands that contain $E(\alpha)$ as a substrand. The extraction is a variation of the operation SEPARATE which separates the strands in a tube T into two tubes T_1 and T_2 such that T_1 consists of the strands containing a given string $s \in \Sigma^*$ and T_2 contains the rest of T. The extraction operation is denoted by EXTRACT(T, α).

Append. This operation takes a tube T and a unit assignment α as input and returns a tube that contains all strands of T concatenated with $E(\alpha)$. The operation is denoted by APPEND(T, α).

Merge. This operation takes tubes T_1 and T_2 as input and returns a tube that contains all strands of T_1 and T_2. The operation is denoted by MERGE(T_1, T_2, R).

Detect. This operation is a Boolean operation that returns *true* if an input tube T contains at least one strand and retuns *false* otherwise. The operation is denoted by DETECT(T).

Duplicate. This operation takes a tube T as input and returns another tube with the same set of strands as in T without destroying T. The duplication (as well as AMPLIFY) can be implemented with ANNEAL and POLYMERASE, see e.g. [6]. The operation is denoted by DUPLICATE(T).

Create. To start a DNA computation, we need to create the contents of an initial tube. The create operation takes a set S of assignments as input and returns a tube with strands that are encodings of the assignments in S. The operation is denoted by CREATE(S).

5 DNA Implementation of the Join-Based Algorithm

To implement Algorithm 1, we introduce a new DNA operation for computing joins. This operation is denoted by JOIN. We show how to implement JOIN using the basic DNA operations above.

Loosely speaking, the JOIN operation takes a tube T and a set S of assignments as input and returns a tube containing strands that encode $\mathcal{S}_T \bowtie S$. The input also contains additional information, namely the set $Var(\mathcal{S}_T)$ of variables. Of course this set is determined by T, however the computation of $\mathcal{S}_T$ from T with our DNA operations has high complexity. Instead, Algorithm 2 maintains the list $Var(\mathcal{S}_T)$ and passes it to the JOIN operation as part of input.

In the description of Procedure JOIN and Algorithm 2 below, we use the following notation. DNA operations are written using arrows, for example $T_1 \leftarrow$ EXTRACT(T, α). This means that T_1 contains the result of EXTRACT(T, α) even if the operation required several tubes and the result was first returned in a different tube. In particular, we write $T \leftarrow$ EXTRACT(T, α) to denote that the contents of T changes to the result returning by EXTRACT(T, α). We denote the operation of emptying a tube T by writing $T \leftarrow \emptyset$.

If an assignment α contains a unit assignment for a variable x then this unit assignment is denoted by $\alpha|_x$.

Procedure JOIN(T, S, V)

Input: A tube T with strands; a set S of assignments; the set V of variables occurring in $\mathcal{S}_T$.

Output: A tube T_1 with strands that encode the join of $\mathcal{S}_T$ and S, i.e., $\mathcal{S}_{T_1} = \mathcal{S}_T \bowtie S$.

1. $T_1 \leftarrow \emptyset$; $U \leftarrow V \cap Var(S)$; $W \leftarrow Var(S) - U$
2. **for each** assignment $\alpha \in S$ **do**
   (a) $T_2 \leftarrow$ DUPLICATE(T)
   (b) **for** each variable $x \in U$ **do** $T_2 \leftarrow$ EXTRACT$(T_2, \alpha|_x)$
   (c) **if** DETECT(T_2) **then for** each variable $y \in W$ **do** $T_2 \leftarrow$ APPEND$(T_2, \alpha|_y)$
   (d) $T_1 \leftarrow$ MERGE(T_1, T_2)
3. **return** (T_1)

Comment: Step 1 of the procedure divides the set $Var(S)$ of variables into two subsets: the set U of variables "occurring" in T and the set W of variables "not occurring" in T. Then for each assignment α in S, the procedure extracts all strands that "agree" with α on the common variables of U and expands these strands by adding the projection of α onto the "new" variables of W. All extracted and expanded strands are accumulated in a tube T_1 that eventually contains the join of $\mathcal{S}_T$ and S.

Procedure JOIN, we now implement Algorithm 1 with DNA.

Algorithm 2 (DNA implementation of the join-based algorithm).

Input: Sets $S_1, \ldots, S_m$ of assignments that represent constraints $C_1, \ldots, C_m$, i.e., each S_i is the set of all solutions to C_i.

Output: A tube T containing strands that encode all solution to $\{C_1, \ldots, C_m\}$, i.e., S_T is the set of all solutions to $\{C_1, \ldots, C_m\}$.

1. $T \leftarrow$ CREATE(S_1); $V \leftarrow Var(S_1)$
2. **for** $i \leftarrow 2$ **to** m **do**
   (a) $T \leftarrow$ JOIN(T, S_i, V); $V \leftarrow V \cup Var(S_i)$
   (b) **if not** DETECT(T) **then return** (*"no solution"*)
3. **return** (T)

The running time of DNA algorithms is measured by the number of applications of the basic DNA operations. The space complexity is called the *volume* and is measured by the maximum number of strands in all tubes, where the maximum is taken over all steps of the algorithm.

Proposition 1. *Given a k-CSP instance, Algorithm 2 makes $O(mk|D|^k)$ applications of DNA operations, where m is the number of constraints in the instance and $|D|$ is the cardinality of the domain. The volume is at most $|D|^n$ where n is the number of variable in the instance.*

Proof. The algorithm performs one CREATE operation plus $m - 1$ JOIN operations plus $m-1$ DETECT operations. When running JOIN(T, S, V), the number of applications of each of DUPLICATE, DETECT, and MERGE is at most $|S|$, which is not greater than $|D|^k$. The number of applications of EXTRACT and APPEND for one assignment $\alpha \in S$ is at most $|Var(S)|$, which is not greater than k. Therefore, the overall number of applications is $O(mk|D|^k)$. The volume is obviously bounded by the number of all possible assignments, i.e., by $|D|^n$.

6 Probabilistic Nature of DNA Operations

So far we have assumed that DNA computations are *error-free*, i.e., they work perfectly without any errors. However, in reality DNA computations can be faulty because some DNA operations can introduce errors. In particular, among the operations defined in Section 4, the operation APPEND, MERGE, and CREATE are error-free, while EXTRACT, DETECT, and DUPLICATE are faulty.

Extract. It is well known that EXTRACT is probabilistic in nature [3,9,7,11]. Recall that this operation takes a tube T and a unit assignment α as input and returns a tube T_1 with those strands of T that contain the substrand $E(\alpha)$. However, the following errors can occur:

1. A *false negative error:* a strand $s \in T$ containing $E(\alpha)$ does not end up in T_1. The probability of false negative error is denoted by ϵ and is estimated as $\epsilon \approx 10^{-1}$, see e.g. [11].
2. A *false positive error:* a strand $s \in T$ not containing $E(\alpha)$ ends up in T_1. The probability of false positive error is denoted by γ and is estimated as $\gamma \approx 10^{-6}$, see e.g. [11].

Single and double strands. To examine DUPLICATE and DETECT, we need to distinguish between different types of strands. By a *single strand* we mean a string over $\Sigma = \{A, C, G, T\}$ together with its *linear orientation*, either with orientation $5' \to 3'$ or with orientation $3' \to 5'$, see [14] for details. These two types of strands are denoted by $\uparrow s$ and $\downarrow s$ respectively. A *double strand* $\updownarrow s$ consists of a single strand $\uparrow s$ intertwined with its *Watson-Crick complement* $\downarrow \overline{s}$, see [14]. The operations DUPLICATE and DETECT can be implemented using the following three operations, see [8,6]:

1. Transformation of every pair of single strands $\uparrow s$ and $\downarrow s$ into the double strand $\updownarrow s$.
2. Denaturation of every double strand $\updownarrow s$ into its single strand components $\uparrow s$ and $\downarrow s$.
3. Shortening every strand by a sequence of length l, where l is the length of encodings of unit assignments.

Note that these operations can be regarded as error-free, see e.g. [11,9].

Duplicate. Let T be an input tube for DUPLICATE. Let t be a *tail tag*, i.e., a strand not occurring in encoding of unit assignments. We assume that t has the same length as encodings of unit assignments. To implement DUPLICATE, we append t to every strand in T (using APPEND). Then we apply the transformation of each strand $\uparrow s$ into $\updownarrow s$. Furthemore we apply the denaturation of double strands into their single strand components. Finally, we use EXTRACT to find strands containing the tag t and shorten them to eliminate t. Note that errors of DUPLICATE can appear only because of errors occurring in EXTRACT.

Detect. This operation can be implemented in different ways. In our method, we decide that a tube is not empty if its volume is greater than a threshold volume τ. We double the volume of T until its volume becomes greater than τ. If after $\lceil \log \tau \rceil$ doubles the volume of T is not greater than τ, we decide that T is empty. Since these doubles change the original contents of T, we start DETECT with duplicating of T. Therefore, DETECT can be faulty because of using DUPLICATE.

7 Error-Resilient Computation

Our purpose is to analyze errors of Algorithm 2 and to estimate their probabilities. Like the EXTRACT operation, Algorithm 2 can have two types of error:

1. A *false negative* error: the resulting tube T does not contain an encoding of some solution to the input CSP instance, i.e., a false negative error is the *loss of a solution.*
2. A *false positive* error: the resulting tube T contains an encoding of an assignment that is not a solution to the input CSP instance, i.e., a false positive error is the *acquisition of a pseudo-solution.*

7.1 Probabilities of False Negative Error and False Positive Error

Clearly, Algorithm 2 returns its result with a false negative error if at least one of the operations EXTRACT, DUPLICATE, or DETECT introduces a false negative error. Recall that the algorithm invokes JOIN(T, S_i, V) for $i = 2, \ldots, m$ and also invokes DETECT(T). In turn, each run of JOIN(T, S_i, V) includes:

- $|S_i|$ runs of DUPLICATE; let $\mathcal{E}^i_{dup}$ be the event that at least one of these runs has a false negative error;
- $|S_i|$ runs of DETECT; let $\mathcal{E}^i_{det}$ be the event that at least one of these runs has a false negative error;
- at most $|S_i| \cdot k$ runs of EXTRACT; let $\mathcal{E}^i_{ext}$ be the event that at least one of these runs has a false negative error.

Now we estimate probabilities $\mathbf{Pr}[\mathcal{E}^i_{dup}]$, $\mathbf{Pr}[\mathcal{E}^i_{det}]$, and $\mathbf{Pr}[\mathcal{E}^i_{ext}]$. Suppose that for each run of EXTRACT and for each strand s, the probability that s contains a given substrand but does not end up in the resulting tube is not greater than ϵ. Then we have

$$\mathbf{Pr}[\mathcal{E}^i_{dup}] \leq 2\,\epsilon\,|D|^n$$
$$\mathbf{Pr}[\mathcal{E}^i_{det}] \leq 2\,\epsilon\,|D|^n$$
$$\mathbf{Pr}[\mathcal{E}^i_{ext}] \leq k\,\epsilon\,|D|^n$$

The first inequality holds because DUPLICATE involves two EXTRACTs, and any test tube contains at most $|D|^n$ strands. The second inequality holds because DETECT invokes one DUPLICATE. The third inequality holds because $\mathcal{E}^i_{ext}$ happens if at least one of k consequitive EXTRACTs introduces a false negative error. Note that the number of strands in any tube does not exceed $|D|^n$. The probability that a false negative error occurs in JOIN(T, S_i, V) is not greater than

$$\mathbf{Pr}[\mathcal{E}^i_{dup} \cup \mathcal{E}^i_{det} \cup \mathcal{E}^i_{ext}] \;\leq\; (4 + k)\,\epsilon\,|D|^n$$

Finally, the probability that Algorithm 2 has a false negative error is the sum the probabilities of false negative error for all JOIN(T, S_i, V) where $i = 2, \ldots, m$ plus the probability of false negative error of the final DETECT. Thus we have for Algorithm 2:

$$\mathbf{Pr}[\text{Algorithm 2 has a false negative error}] \;\leq\; ((m-1)(4+k)+2)\,\epsilon\,|D|^n \quad (1)$$

The probability that Algorithm 2 has a false positive error can be estimated in a similar way. Assume that for each run of EXTRACT and for each strand s, the probability that s does not contain a given substrand but ends up in the resulting tube is not greater than γ. Then, repeating the arguments above, we have

$$\mathbf{Pr}[\text{Algorithm 2 has a positive error}] \;\leq\; (((m-1)(4+k)+2)\,\gamma\,|D|^n \quad (2)$$

These bounds $O(mk\epsilon|D|^n)$ and $O(mk\gamma|D|^n)$ use the upper bound $|D|^n$ on the number of strands in any tube. Instead, we could estimate the number of strands at each step i: when JOIN(T, S_i, V) is performed, the number of strands in a tube does not exceed $|D|^{ik}$. Then we would have the bounds $O(k\epsilon|D|^{mk})$ and $O(k\gamma|D|^{mk})$ respectively.

7.2 Bounded Probabilities of Error

The above bounds on probabilities of error depend on ϵ and γ. Is it possible to modify EXTRACT so that these probabilities decreases? The answer is yes. It is shown in [9] that for any given constant δ, the operation EXTRACT can be converted into a DNA algorithm such that its false negative and false positive errors are not greater than δ. The running time of this algorithm depends on ϵ, γ, and δ, namely the complexity of the algorithm is $\Theta(\lceil \log_\epsilon \delta \rceil \lceil \log_\gamma \delta \rceil)$. This result is proved in [9] for a slightly less general settings: strands are encodings of bit strings, and EXTRACT returns strands that encode strings containing a given bit in a given position. However, this case can be easily generalized for our settings.

Proposition 2 (Karp et al [9]). *There exists a DNA algorithm* ER-EXTRACT *(error-resilient extraction) with the following properties:*

1. *The algorithm runs on the following input: a tube T, a unit assignment α, and a number δ such that $0 < \delta < 1$. The algorithm returns a tube T_1.*
2. *For each strand $s \in T$ such that s contains $E(\alpha)$, the probability that s does not end up in T_1 is not greater than δ.*
3. *For each strand $s \in T$ such that s does not contain $E(\alpha)$, the probability that s ends up in T_1 is not greater than δ.*
4. *Assuming that the basic DNA operations* EXTRACT *and* MERGE *run in constant time, the algorithm* ER-EXTRACT *runs in time $\Theta(\lceil \log_\epsilon \delta \rceil \lceil \log_\gamma \delta \rceil)$, where ϵ and γ have the same meaning as above.*

Proof. Straighforward generalization of the proof of [9, Theorem 3.1].

Using Proposition 2, we can solve a k-CSP in polynomial time with bounded two-sided error probability. More exactly, the probabilities of false negative and false positive errors are both less than a constant $c < 1/2$. The constant c does not depend on the input size. Loosely speaking, we show that any k-CSP belongs to a DNA counterpart of **BPP** (for **BPP** see e.g. [12]).

Proposition 3. *For any k-CSP $\mathcal{P}$, there is a DNA algorithm $\mathcal{A}_\mathcal{P}$ that solves $\mathcal{P}$ in polynomial time with bounded error probability. Namely, the probabilities of false negative and false positive errors of $\mathcal{A}_\mathcal{P}$ are less than $1/4$.*

Proof. We obtain $\mathcal{A}_\mathcal{P}$ from Algorithm 2 by replacing every EXTRACT operation (including those in DUPLICATE and DETECT) by the ER-EXTRACT algorithm. Namely, we take ER-EXTRACT with

$$\delta \leq \frac{1}{4\left((m-1)(4+k)+2\right)|D|^n}. \tag{3}$$

Then, according to inequalities (1) and (2), we have

$$\mathbf{Pr}[\mathcal{A}_\mathcal{P} \text{ has a false negative error}] \leq ((m-1)(4+k)+2)\,\delta\,|D|^n \leq 1/4$$
$$\mathbf{Pr}[\mathcal{A}_\mathcal{P} \text{ has a false positive error}] \leq ((m-1)(4+k)+2)\,\delta\,|D|^n \leq 1/4$$

It remains to estimate the running time of $\mathcal{A}_{\mathcal{P}}$. It follows from Proposition 2 that ER-EXTRACT with δ satisfying (3) runs in time

$$\begin{aligned}\Theta(\lceil \log_\epsilon \delta \rceil \lceil \log_\gamma \delta \rceil) &= \Theta(\lceil \log_\epsilon (m^{-1}|D|^{-n}) \rceil \lceil \log_\gamma (m^{-1}|D|^{-n}) \rceil) \\ &= \Theta(n^2 \log^2 m).\end{aligned}$$

Since Algorithm 2 uses a polynomial number of applications of EXTRACT (Proposition 1), we have a polynomial bound on the running time of $\mathcal{A}_{\mathcal{P}}$.

References

1. S. Abiteboul, R. Hull, and V. Vianu. *Foundations of Databases.* Addison-Wesley, 1995.
2. L. M. Adleman. Molecular computation of solutions to combinatorial problems. *Science*, 266:1021–1024, November 1994.
3. L. M. Adleman. On constructing a molecular computer. In R. Lipton and E. Baum, editors, *DNA Based Computers*, volume 27 of *DIMACS Series in Discrete Mathematics and Theoretical Computer Science*, pages 1–21. American Mathematical Society, 1995.
4. E. Bach, A. Condon, E. Glaser, and C. Tanguay. DNA models and algorithms for NP-complete problems. In *Proceedings of the 11th Annual IEEE Conference on Computational Complexity*, pages 290–300, 1996.
5. R. Beigel and B. Fu. Solving intractable problems with DNA computing. In *Proceedings of the 13th Annual IEEE Conference on Computational Complexity*, pages 154–169, 1998.
6. D. Boneh, C. Dunworth, and R. J. Lipton. Breaking DES using a molecular computer. In R. Lipton and E. Baum, editors, *DNA Based Computers*, volume 27 of *DIMACS Series in Discrete Mathematics and Theoretical Computer Science*, pages 37–66. American Mathematical Society, 1995.
7. D. Boneh, C. Dunworth, R. J. Lipton, and J. Sgall. Making DNA computers error resistant. In *DNA Based Computers II*, volume 44 of *DIMACS Series in Discrete Mathematics and Theoretical Computer Science*, pages 163–170. American Mathematical Society, 1996.
8. K. Chen and V. Ramachandran. A space-efficient randomized DNA algorithm for k-SAT. In *Proceedings of the 6th International Workshop on DNA-Based Computers*, pages 199–208, 2000.
9. R. M. Karp, C. Kenyon, and O. Waarts. Error-resilient DNA computation. *Random Structures and Algorithms*, 15(3-4):450–466, 1999.
10. R. J. Lipton. DNA solutions of hard combinatorial problems. *Science*, 268:542–548, April 1995.
11. C. C. Maley. DNA computation: theory, practice, and prospects. *Evolutionary Computation*, 6(3):201–230, 1998.
12. R. Motwani and P. Raghavan. *Randomized Algorithms.* Cambridge University Press, 1995.
13. G. Păun, G. Rozenberg, and A. Salomaa. *DNA Computing: New Computing Paradigms.* Springer, 1998.
14. P. A. Pevzner. *Computational Molecular Biology.* MIT Press, 2000.
15. J. H. Reif. Parallel molecular computation: models and simulations. *Algorithmica*, 25(2):142–176, 1999.

New Architecture and Algorithms for Degradable VLSI/WSI Arrays

Wu Jigang[1], Heiko Schröder[2], and Srikanthan Thambipillai[1]

[1] Centre for High Performance Embedded Systems,
School of Computer Engineering, Nanyang Technological University,
Singapore,639798, Republic of Singapore
{asjgwu,astsrikan}@ntu.edu.sg
[2] School of Computer Science and Information Technology,
RMIT, Melbourne, Australia
heiko@cs.rmit.edu.au

Abstract. The problem of reconfiguring a two-dimensional degradable VLSI/WSI array under the row and column routing constraints is NP-complete. This paper aims to decrease gate delay and increase the harvest. A new architecture with six-port switches is proposed. New greedy rerouting algorithms and new compensation approaches are presented and used to reform the reconfiguration algorithm. Experimental results show that the new reconfiguration algorithm consistently outperforms the latest algorithm, both in terms of the percentages of harvest and that of degradation of VLSI/WSI array.

Keywords: Degradable VLSI/WSI array, reconfiguration, greedy algorithm, fault-tolerance, NP-completeness.

1 Introduction

The mesh-connected processor array has a regular and modular structure and allows fast implementation of most signal and image processing algorithms. With the advancement in VLSI (very large scale integration) and WSI (wafer scale integration) technologies, integrated systems for mesh-connected processor arrays can now be built on a single chip or wafer. As the density of VLSI/WSI arrays increases, probability of the occurrence of defects in the arrays during fabrication also increases. In addition, when the arrays are installed in space-flight instruments such as satellite, defects possibly occur due to harsh environments. These defects obviously affect the reliability of the whole system. Thus fault-tolerant technologies must be employed to enhance the yield and reliability of VLSI/WSI arrays.

Generally, two methods for reconfiguration, namely, redundancy approach and degradation approach, are used in fault tolerant technologies. In the redundancy approach, a system is built with some of its components called spare elements. These spare elements are used to replace faulty elements in the reconfigurable system. Various techniques for redundancy approach have been described in [1]-[9]. The disadvantage of this approach is the dimension of the

O.H. Ibarra and L. Zhang (Eds.): COCOON 2002, LNCS 2387, pp. 181–190, 2002.

arrays is fixed. If spare elements cannot replace all the faulty elements, the system is not reconfigurable and has to be discarded. In the degradation approach, all elements in a system are treated in uniform way, there are no spare element. This approach uses as many fault-free elements as possible to construct a target system. [10], [11] and [12] have studied the problem of two-dimensional degradable arrays under the following three different routing constraints, 1) *row and column bypass*, 2) *row bypass and column rerouting*, and 3) *row and column rerouting* on the four-port switch model with bypass link. They have shown that most problems that arise under these constraints are NP-complete. The problem turns out to be very difficult if rerouting in both row and column direction is considered at the same time.

In this paper, we consider the reconfiguration problem of two-dimension VLSI/WSI arrays under the constraint *row and column rerouting*. This problem has been proved to be NP-complete[10]. We propose a six-port switch model to replace the four-port switch model. For the column rerouting on the selected rows, we present a new greedy column rerouting algorithm based on a new idea called *local compensation*. The time complexity of the proposed reconfiguration algorithm is of the same order as that of Low's algorithm[12], denoted *RCRoute* in this paper, but the performance becomes significantly more powerful.

2 Definitions and Preliminaries

The original array after manufacturing is called a host array (degradable array) which contains faulty elements. Degradable subarray of a host array which contains no faulty element is called a target array (logical array). The rows (columns) in the host array and target array are called physical rows (columns) and logical rows (columns), respectively.

An element in the host array is represented by $e(i, j)$, where i is its row index and j is its column index, all switches and links in an array are assumed to be fault-free since they have very simple structure[10][11][12].

In a host array, if $e(i, j+1)$ is a faulty element, $e(i, j)$ can communicate with $e(i, j+2)$ directly and data will bypass $e(i, j+1)$. This scheme is called row bypass scheme. If the element $e(i, j)$ can connect directly to $e(i', j+1)$ with external switches, where $|i' - i| \leq d$, this scheme is called row rerouting scheme, d is called row compensation distance.

The *column bypass scheme* and the *column rerouting scheme* can be similarly defined. By limiting the compensation distance to 1, we essentially localize the movements of reconfiguration in order to avoid complex reconfiguration algorithm. In all figures of this paper, the shaded boxes stand for faulty elements and the white ones stand for the fault-free elements.

In this paper, $row(e)$ ($col(e)$) denotes the physical row (column) index of element e. H (S) denotes the host (logical) array. N denotes the number of fault-free elements in host array. R_i denotes the ith logical row. $e(i, j)$ ($er(i, j)$) denotes the element located in the ith row and in the jth column of host (logical) array.

3 New Architecture

Algorithm *RCRoute*[12] is based on four-port switch model shown by Fig. 1(a).

In Fig. 1(a), each element has four ports to connect its up, down, left and right neighbors, respectively. In the row bypass scheme, the data will bypass the faulty element through an internal bypass link from its left port to its right port without being processed, *i.e.*, a faulty element can be converted into a connecting element. No external switch is needed. But in the row rerouting scheme, each row have both bypass and rerouting capacities. External switches are necessary between two neighboring elements in two adjacent columns. The pros of this switch model is its simple construction. But it is due to its simple construction which provides less functions that the reported reconfiguration algorithms can not obtain high harvest. Assume the fault-free elements u and v are the neighbors located in same physical row, this switch model does not support reconfiguring u and v into same logical column.

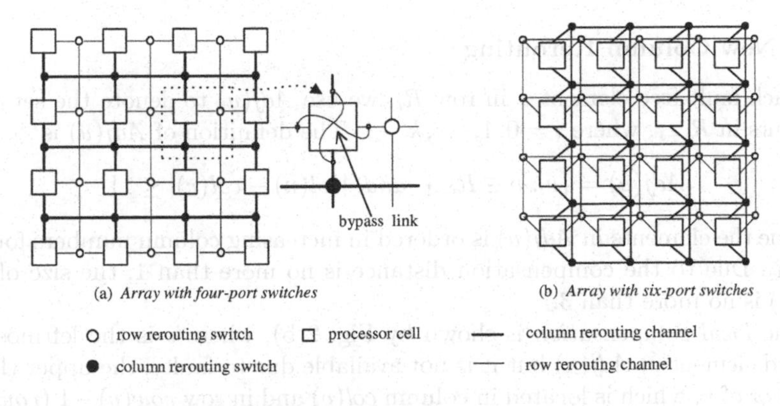

Fig. 1. Old architecture and new architecture

We design a new switch model shown in Fig. 1(b) to replace the four-port switch model. The new model combines one four-port switch and one bypass link into one six-port switch which consists of pass gates to establish all possible connection pair among the six input rails. In other words, a new switch has 6 ports, these ports can arbitrarily be connected pairwise. The only restriction is that no port can be connected to more than 1.

The new model overcomes the weakness of the old model in its switch function. In Fig. 1(a), there is a two-gate delay (for row and column rerouting) when one processor is faulty, and no column rerouting is supported for two neighboring processors in same physical row. But in Fig. 1(b), there is no gate delay when processor is faulty. Especially, the six-port model supports the rerouting of two neighboring processors in same physical row into same logical column. Fig. 2(a) shows the different cases for the rerouting of u, v and w into same logical column.

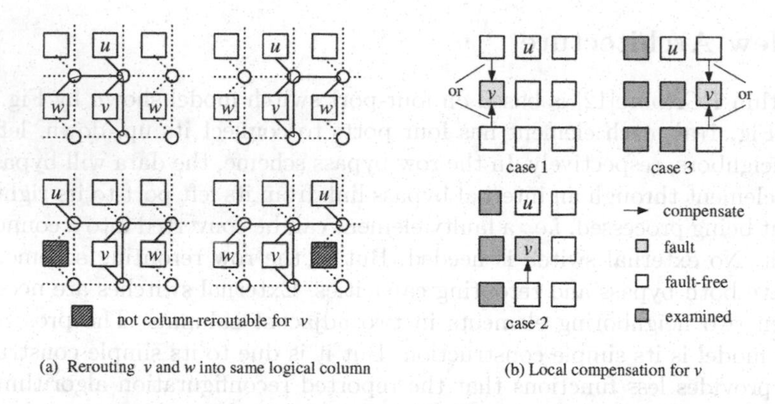

Fig. 2. Rerouting scheme and local compensation scheme

4 Algorithms

4.1 New Column Rerouting

For each fault-free element u in row R_i, we use *Adj*(u) to denote the set of the elements in R_{i+1}, where $i = 0, 1, \cdots, k-1$. The definition of *Adj*(u) is

$$Adj(u) = \{v : v \in R_{i+1} \ \ and \ \ |col(u) - col(v)| \leq 1\}.$$

Assume the elements in *Adj*(u) is ordered in increasing column numbers for each $u \in R_i$. Due to the compensation distance is no more than 1, the size of each *Adj*(u) is no more than 3.

The *local compensation* is shown by Fig. 2(b), where v is the leftmost unmarked element of *Adj*(u) but it is not available due to faulty, the upper (lower) neighbor of v, which is located in column $col(v)$ and in row $row(v) - 1$ ($row(v) + 1$), will be examined to compensate the element v whenever possible when v is the current element in the column rerouting. The upper neighbor of v is examined first. If the upper neighbor is not available, then the lower neighbor of v is examined. Connected on the six-port switch model, a target array is said to *contain all selected rows* if each logical column in this target array contains exactly one fault-free element from each of the selected rows under the local compensation. We use *Local_Comp* to stand for the procedure to short the paper.

In column rerouting algorithm, each step attempts to connect the current element u to the leftmost element of *Adj*(u) that has not been previously examined. In *Greedy_Column_Rerouting*[12] (*GCR* in short), if this step fails in doing so, no logical column that contains the current element u can be formed, the backtrack is doing next. But in *New_GCR* shown by Fig. 3, the local compensation will be experimented. If this local compensation fails either, the *New_GCR* has to backtrack to previous element, say p, that was connected to u and attempts to connect p to the leftmost element of *Adj*$(p) - \{u\}$ that has not been previous examined. For the sake of simplicity in column rerouting, it is not allowed rerouting v_{i+1} into the successor of v_i if $row(v_{i+1}) < row(v_i)$ in the *New_GCR*.

That means for the result logical column with elements $v_0, v_1, \cdots, v_{k-1}$, their index of the physical row satisfy $row(v_0) \le row(v_1) \le \cdots \le row(v_{k-1})$.

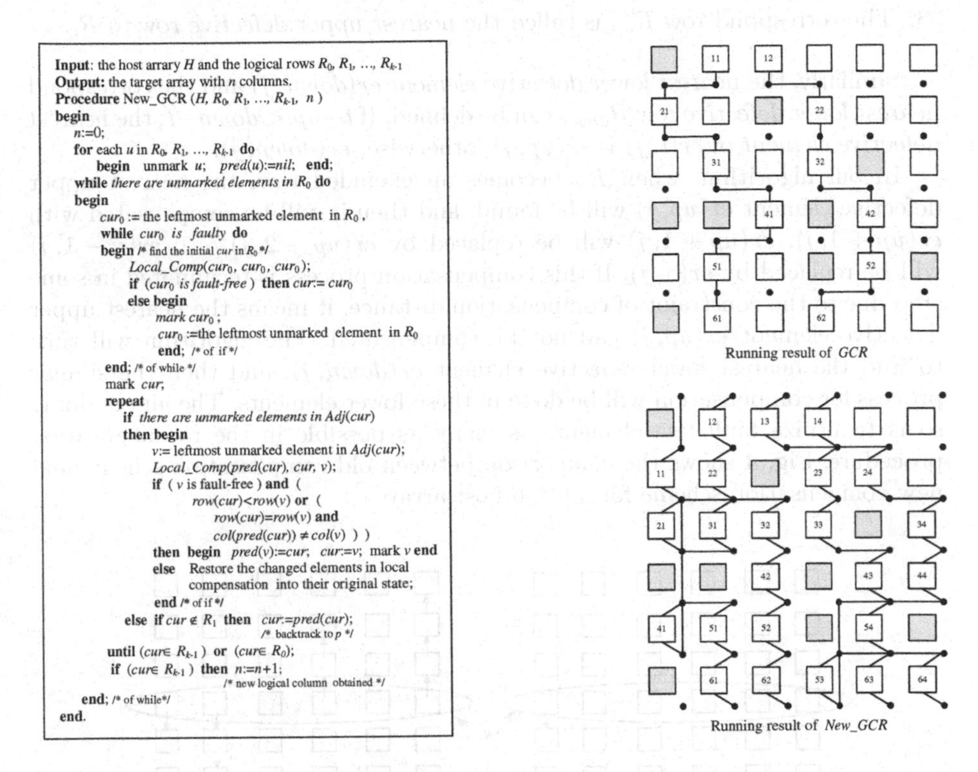

Input: the host arrary H and the logical rows R0, R1, ..., Rk-1
Output: the target array with n columns.
Procedure New_GCR (H, R0, R1, ..., Rk-1, n)
begin
n:=0;
for each u in R0, R1, ..., Rk-1 do
begin unmark u; pred(u):=nil; end;
while there are unmarked elements in R0 do
begin
cur0 := the leftmost unmarked element in R0 .
while cur0 is faulty do
begin /* find the initial cur in R0 */
Local_Comp(cur0, cur0, cur0);
if (cur0 is fault-free) then cur:= cur0
else begin
mark cur0 ;
cur0 :=the leftmost unmarked element in R0
end; /* of if */
end; /* of while */
mark cur;
repeat
if there are unmarked elements in Adj(cur)
then begin
v:= leftmost unmarked element in Adj(cur);
Local_Comp(pred(cur), cur, v);
if (v is fault-free) and (
row(cur)<row(v) or (
row(cur)=row(v) and
col(pred(cur)) ≠ col(v)))
then begin pred(v):=cur; cur:=v; mark v end
else Restore the changed elements in local compensation into their original state;
end /* of if */
else if cur ∉ R1 then cur:=pred(cur);
/* backtrack to p */
until (cur∈ Rk-1) or (cur∈ R0);
if (cur∈ Rk-1) then n:=n+1;
/* new logical column obtained */
end; /* of while*/
end.
Running result of GCR
Running result of New_GCR

Fig. 3. *New_GCR* and running example

The analysis of time complexity of *New_GCR* is similar to that of *GCR*. In detail, for each $u \in R_i$, at most 8 interconnections (from local compensation Fig. 2(b)) are examined at each step. Thus the number of valid interconnections in a host array with N fault-free elements is no more than $8N$. The time complexity of *New_GCR* is $O(N)$ since each valid interconnection is examined at most twice in *New_GCR*.

4.2 New Row-Compensation Strategy

Given the logical rows R_0, R_1, $\cdots$, R_s, assume the element $er(t, j)$ is a fault-free one in their jth column, where $0 < t \le s$. In order to express our idea shortly, we give the following definition.

1. The elements $er(1, j)$, $er(2, j)$,$\cdots$, $er(t-1, j)$ are called the upper elements of $er(t, j)$, the rows $R_0, R_1, \cdots, R_{t-1}$ are called *upper rows* of the row R_t.

2. The element $er(up, j)$ is called the *nearest upper defective element* of $er(t, j)$ if and only if $er(up, j)$ is defective and $er(up+1, j)$, $er(up+2, j)$, $\cdots$, $er(t-1, j)$ are fault-free elements, where $0 \leq up < t$.
3. The correspond row R_{up} is called the *nearest upper defective row* to R_t.

Similarly, the *nearest lower defective element* $er(down, j)$ and the correspond *nearest lower defective row* R_{down} can be defined. If $t - up < down - t$, the *nearest defective element* of $er(t, j)$ is $er(up, j)$, otherwise, $er(down, j)$.

In our algorithm, when R_γ becomes an excluded row, the nearest upper defective element $er(up, j)$ will be found, and then it will be compensated with $er(up+1, j)$, $er(up+1, j)$ will be replaced by $er(up+2, j)$, $\cdots$, $er(\gamma - 1, j)$ will be replaced by $er(\gamma, j)$. If this compensation process is terminated in some step due to the constraint of compensation distance, it means the nearest upper defective element $er(up, j)$ can not be compensated. The algorithm will turn to find the nearest lower defective element $er(down, j)$, and then the similar process for compensation will be done in these lower elements. The aim of doing so is to utilize fault-free elements as many as possible in the reconfiguration procedure. Fig. 4 shows the comparison between old compensation scheme and new compensation scheme for a 6×6 host array.

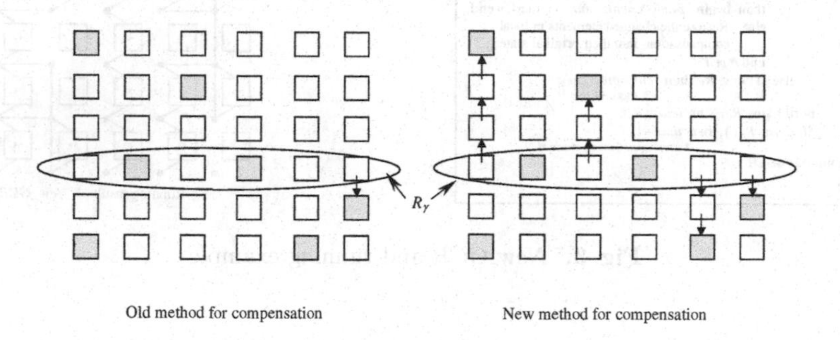

Fig. 4. Row-compensation schemes

The function *Up_Comp*(m', γ) shown in Fig. 5(a) is used to search for the nearest upper defective element $er(up, j)$ of the given fault-free element $er(\gamma, j)$ in jth column, then do compensations and replacements from $er(up+1, j)$ to $er(\gamma, j)$, one after the other. The worst case in step 1 is $up = 0$ and $\gamma = m' - 1$, *i.e.*, all interconnections on fault-free elements in jth column have been tested, thus at most $O(N_j)$ tests are spend, where N_j is the number of the fault-free elements in jth column. This is also the worst case of step 2.2 and the same number of tests are spend. Hence, the worst complexity of the function is $O(N_j)$.

Symmetrically, we can describe a function *Down_Comp*(m', γ) with the same worst complexity as that of the *Up_Comp*(m', γ), which is used to search for the lower nearest defective element $er(down, j)$ of $er(\gamma, j)$ and implement the compensation for it if necessary. *Up_Comp*(m', γ) and *Down_Comp*(m', γ) form

```
Input: the logical array {R_i},0≤i≤m' with R_γ.
Output: the value of integer variable flag.
Function Up_Comp( m', γ);
/* Find the nearest upper defective element and do compensations*/
begin
1 Search for the nearest upper defective element
    er(up, j) among the upper elements of the er(γ, j);
2 if ( up = -1) /* without fault elements */
  then flag := 0
  else /* do compensation and some replacements */
  begin
  2.1 k:=up+1;
  2.2 while (k ≤ γ) do
        switch ( row(er(k, j)) - row(er(k-1,j)) )
        begin
        case 1: /* for the neighbor row */
              if ( er(k,j) = 'y') /* successful */
              then {er(k-1,j):='d'; k:=k+1}
              else/* unsuccessful, jump over while-loop*/
                   k:= γ+1;
              break;
        case 2: /* for the next to neighbor row */
              if ( er(k,j) = 'u' ) /* successful */
              then {er(k-1,j):='d'; k:=k+1}
              else /*unsuccessful, jump over while-loop*/
                   k:= γ+1;
              break;
        default: /* over the compensation distance */
              k:= γ +1;
        end of switch & while;
  2.3 if ( k ≠ γ+1 ) /* successful compensation */
      then flag:=1
      else {flag:=0;
            Restore the changed elements to their
            original state;}
  end of if ;
3. return( flag );
end.
```

(a)

```
Input: the logical array {R_i},0≤i≤m' with R_γ.
Output: the compensated logical row;
Procedure Overall_Comp(m', γ)
/*Compentate a defective element with R_γ*/
begin
for j:= 0 to n do
  if ( m' > 2 ) then
  begin
  if er(γ,j) ≠ 'n' then
    begin
    flag := Up_Comp( m', γ);
    if flag = 0 then Down_Comp(m', γ);
    end of if;
  end of if;
end.
```

(b).(i)

```
Input: the host array H, the given constants r, c.
Output: a target array
Procedure New_Row_First(H, m, n, r, c)
begin
1.  Let S={ R_0, R_1, ..., R_m },
    New_GCR (H, S, n);
    /* Construct initial target arrays */
2. m' := m;
3. while (m' ≥ r) and (n < c) do
   begin
   3.1 Call approach[13] to select R_γ from S;
   3.2 Overall_Comp(m', γ) ; /* compensate other rows with R_γ*/
   3.3 Delete the row R_γ from S;
   3.4 New_GCR (H, S, n);
   3.5 m':=m'-1;
   end of while;
4.  if (m' ≥ r) and ( n' > c)
    then the target array is obtained
    else "algorithm failed";
end.
```

(b).(ii)

Fig. 5. Algorithms for compensation

the procedure *Overall_Comp* shown in Fig. 5(b).(i). It runs before R_γ is excluded and implements the compensations for whole logical array with all fault-free elements in R_γ. The complexity of *Overall_Comp* is $O(\sum_{0 \le j < n} N_j)$, *i.e.*, $O(N)$.

4.3 Reconfiguration Algorithm

New_Row_First shown in Fig. 5(b).(ii) is used to find a target array of maximum size based on row. Let m' be the number of logical rows and n' be the number of logical column of target array. The current logical array is $S = \{R_0, R_1, \cdots, R_{m'}\}$. Initially, all rows in the host array are selected for inclusion into the target array. Thus each logical row in S is also a physical row and each of them has only bypass capability. Its time complexity is $O((m-r)N)$. Similarly, we can describe a procedure *New_Column_First* to find a target array of maximum size based on column. Its time complexity is $O((n-c)N)$. We ends this subsection with the following description of main algorithm.

Algorithm New_RCRoute
begin
1. Call *New_Row_First* to find a maximal target arrays $m_1 \times n_1$ based on row.
2. Call *New_Column_First* to find a maximal target arrays $m_2 \times n_2$ based on column.
3. The result target arrays is $\max\{m_1 \times n_1, m_2 \times n_2\}$.
end.

The largest array derived from *New_Row_First* and *New_Column_First* is taken as the target array for H. The time complexity of algorithm *New_RCRoute* is $O(\max\{(m-r)N, (n-c)N\})$.

5 Experimental Results

In order to make a fair comparison between *New_RCRoute* and *RCRoute*, we have implemented them in C on a personal computer—Intel Pentium-II 233 MHZ. The *harvest* and *degradation*[12] for each target array are calculated for the same random host arrays in which the faults were generated by a uniform random generator. In our experiments, the size of each target array obtained by *New_RCRoute* is compared with 1) an upper bound on the size of target array and 2) the size of the target array obtained by *RCRoute*[12]. The upper bound of the target array size is calculated with the same method used by *RCRoute*. Tables 1-2 summarize the experimental results for the random host arrays with different size.

Table 1. The Comparison of Maximal Target Arrays with Theoretical Maximums

Host array			*RCRoute*			*New_RCRoute*		
Size $r \times c$	Fault (%)	Theorical Maximum	Target Array	Harvest (%)	Degrad. (%)	Target Array	Harvest (%)	Degrad. (%)
64×64	0.1	64×63	64×63	98.53	1.56	64×63	98.53	1.56
64×64	1.0	63×64	64×61	96.25	4.69	63×63	97.86	3.10
64×64	10.0	61×60	63×49	83.72	24.63	60×59	96.01	13.57
128×128	0.1	127×128	128×127	99.32	0.78	127×128	99.31	0.78
128×128	1.0	127×127	126×124	96.32	4.64	127×126	98.65	2.33
128×128	10.0	117×126	125×99	83.92	24.47	121×117	96.01	13.59
256×256	0.1	255×256	255×254	98.93	1.17	255×255	99.31	0.78
256×256	1.0	255×254	255×248	97.47	3.50	253×254	99.03	1.94
256×256	10.0	234×252	256×196	85.07	23.44	240×236	96.03	13.57
512×512	0.1	511×512	512×508	99.32	0.78	512×510	99.71	0.39
512×512	1.0	509×509	511×498	98.06	2.92	509×507	99.44	1.56
512×512	10.0	469×503	512×388	84.20	24.22	505×452	96.75	12.93

Table 1 shows the comparison in the sizes of the target arrays derived from *RCRoute*, *New_RCRoute* and the *Theorem_Upper_Bound*[12], respectively. For a same random instance, e.g., a host array of size 256×256 with 6553 faulty elements, we calculate the target array with *Theorem_Upper_Bound*[12], *RCRoute* and *New_RCRoute*, respectively. The size of its theoretical maximal target array is 234×252. The size of the target array derived from the *New_RCRoute*, 240×236, is more closer to the theoretical maximal size than that derived from *RCRoute*, 256×196. Furthermore, the *harvest* of *New_RCRoute* is 96.03%, which is greater than that of *RCRoute*, 85.07%. Meanwhile, the *degradation* of *New_RCRoute* is less than that of *RCRoute*.

Table 2. The Comparison of Maximal Square Arrays

Host array		*RCRoute*			*New_RCRoute*		
Size $r \times c$	Fault (%)	Target Array	Harvest (%)	Degrad. (%)	Target Array	Harvest (%)	Degrad. (%)
64×64	0.1	63×63	96.99	3.10	64×63	98.53	1.56
64×64	1.0	62×62	94.77	6.15	63×63	97.86	3.10
64×64	10.0	54×53	77.62	30.13	58×59	92.81	16.49
128×128	0.1	127×127	98.54	1.56	127×127	98.54	1.56
128×128	1.0	124×124	94.79	6.15	125×125	96.32	4.63
128×128	10.0	106×106	76.20	31.42	118×118	94.43	15.01
256×256	0.1	254×254	98.54	1.56	255×255	99.32	0.78
256×256	1.0	249×249	95.56	5.39	253×254	99.05	1.94
256×256	10.0	208×207	73.00	34.30	237×237	95.23	14.29
512×512	0.1	509×509	98.93	1.17	511×510	99.51	0.59
512×512	1.0	498×498	95.56	5.39	508×507	99.24	1.75
512×512	10.0	413×413	72.30	34.93	474×473	95.03	14.47

Table 2 shows the performance comparisons for the maximal square target array between *New_RCRoute* and *RCRoute*. For each random instance, we compare the maximal square target arrays derived from *RCRoute* with the maximal square target arrays derived from *New_RCRoute*. The size of square target array derived from *RCRoute* is far less than the size of square target array derived from *New_RCRoute* for the same random host array. For example, for a random host array with size 128×128 and 10% faulty elements, the size of square target array derived from *RCRoute* is only 106×106 which is far less than 118×118 derived from *New_RCRoute* for the same random host array. The *harvest* is increased from 76.20% derived from *RCRoute* into 94.43% derived from *New_RCRoute*. In addition, the related degradation is decreased.

6 Conclusions

We have discussed the reconfiguration algorithms for the degradable VLSI/WSI array using *row and column rerouting*. We have presented a new switch model and new greedy column rerouting algorithms for the selected rows included in target array. The improved algorithms have been implemented and experimental results demonstrate the efficiency of the proposed algorithms. The effect of multiple row exclusion and partial row exclusion in each iteration, on the efficiency of the algorithm will be an interesting topic for future investigation.

Acknowledgment

We are grateful to Mr. Ashish Panda for pointing out oversights in an earlier draft of this paper.

References

1. Mangir T. E. and Avizienis A.: Fault-tolerant design for VLSI: effect of interconnection requirements on yield improvement of VLSI design. IEEE Trans. Computers. **31** (1982) 609-615
2. Greene J. W. and Gamal A. E.: Configuration of VLSI array in the presence of defects. J. ACM. **31** (1984) 694-717
3. Lam C. W. H., Li H. F., Jakakumar R.: A study of two approaches for reconfiguring fault-tolerant systoric array. IEEE Trans. Computers. **38** (1989) 833-844
4. Koren I., Singh, A. D.: Fault tolerance in VLSI circuits. Computer. **23**(1990) 73-83
5. Chen Y. Y., Upadhyaya S. J., Cheng C. H.: A comprehensive reconfiguration scheme for fault-tolerant VLSI/WSI array processors. IEEE Trans. Computers. **46**(1997) 1363-1371
6. Horita T., Takanami I.: Fault-tolerant processor arrays based on the 1.5-track switches with flexible spare distributions. IEEE Trans. on Computers. **49** (2000) 542-552
7. Kuo S. Y., Fuchs W. K.: Efficient spare allocation for reconfigurable arrays. IEEE Design and Test. **4** (1987) 24-31
8. Wey C. L., Lombardi F.: On the repair of redundant RAM's. IEEE Trans. on CAD of Integrated Circuits and Systems. **6** (1987) 222-231
9. M. G. Sami and R. Stefabelli. Reconfigurable architectures for VLSI processing arrays. Proc. IEEE. **74** (1986) 712-722
10. Kuo S. Y., Chen I. Y.: Efficient reconfiguration algorithms for degradable VLSI/WSI arrays. IEEE Trans. Computer-Aided Design. **11** (1992) 1289-1300
11. Low C. P., Leong H. W.: On the reconfiguration of degradable VLSI/WSI arrays. IEEE Trans. Computer-Aided Design of integrated circuits and systems. **16** (1997) 1213-1221
12. Low C. P.: An efficient reconfiguration algorithm for degradable VLSI/WSI arrays. IEEE Trans. on Computers. **49** (2000) 553-559

Cluster: A Fast Tool to Identify Groups of Similar Programs

Casey Carter[1] and Nicholas Tran[2]

[1] Department of Computer Science, University of Illinois at Urbana-Champaign
Urbana, IL 61801, USA
ccarter@cs.uiuc.edu

[2] Department of Mathematics & Computer Science, Santa Clara University
Santa Clara, CA 95053-0290, USA
ntran@math.scu.edu

Abstract. `cluster` is a tool to partition a large pool of C programs into groups according to structural similarity. Its method involves calculating an alignment score for each program against a mosaic made of randomly selected code fragments of fixed size from the pool. The scores are then grouped togother so that the distance between two adjacent members of a group is at most some threshold value. `cluster` is effective in identifying tight clusters of similar programs and is capable of distributing its workload over a network of workstations to achieve very fast running times. As a tool, `cluster` is highly configurable: the user can adjust its alignment scoring scheme and clustering threshold as well as obtain visual alignments of programs suspected to be similar.

1 Introduction

The problem of detecting plagiarism in programming assignments can be divided into two subproblems: a) defining and measuring similarity between two programs; and b) grouping mutually similar programs together. A robust solution to the first subproblem is to use string alignment methods, which can be made to be resistant to systematic name changes, variations in white spaces and comments, and reordering of statements and functions [5,7]. The best alignment for two strings of length s and t can be found using dynamic programming in time $O(st)$. A software tool based on this approach has been implemented and shown to be effective against simple plagiarism techniques such as name changes, reordering of statements and functions, and adding/removing comments and white spaces [2].

For the second subproblem, a natural approach is to identify the desired clusters with connected components of a graph G whose vertices are the programs under consideration. The edges of G connect pairs of programs whose alignment scores exceed a given threshold value V. The running time of this algorithm is dominated by the time it takes to compute the best alignments for all $O(n^2)$ possible pairs of programs; finding the connected components takes only $O(n)$ time [1].

Thus a naive implementation of a plagiarism detector using the string alignment and connected-component clustering methods would have running time

O.H. Ibarra and L. Zhang (Eds.): COCOON 2002, LNCS 2387, pp. 191–199, 2002.

$\Theta(n^2l^2)$, which becomes impractical even for small values of n and average program size l. The bulk of this computation would be wasted however, since the number of plagiarism incidents found in a set of programs is usually small.

In this paper we describe the design and implementation of `cluster`, which is a randomized alignment-based plagiarism detector with running time $O(n^2l)$. `cluster` constructs a mosaic from randomly selected code fragments of fixed size taken from each program in the pool and finds the best alignment score between the mosaic and each of the programs. The list of n scores is then sorted and clustered according to a threshold value as before. `cluster` is effective because alignment scores against the mosaic for similar programs are approximately the same and very likely to be higher than the rest (especially if the group of similar programs is large.) It is fast because the code size of each contribution is kept to a fixed constant, i.e. the size of the mosaic is $O(n)$; the average time for one alignment is $O(nl)$, so the total running time for n alignments is $O(n^2l)$. Furthermore, since the n alignments can be performed independently, `cluster` can distribute its workload on a network of w workstations using the parallel virtual machine simulator `PVM` to achieve a speedup factor of w. Experimental data show that `cluster` is practical for all realistic values of n and l. For example, for $n = 200$ programs, average length $l = 25$ KBytes, and fragment size $f = 1.25$ KBytes $(= .05 * l)$, `cluster` takes about 45 minutes using a network of six 400-MHz Pentium PCs running Debian/GNU Linux 2.1.

`cluster` is highly configurable. The user can customize its scoring scheme for computing best alignments as well as its cluster threshold value. Effectiveness of values chosen for these parameters may be judged from `cluster`'s visual display of the clusters and of alignments of programs clustered together. `cluster` is a significantly improved extension of the work reported in [2], which is concerned mainly with the subproblem of aligning two programs. `cluster` is implemented in C++ and Tcl/Tk on a Linux platform, although it can be easily ported to Windows NT.

The rest of this paper is organized as follows. Section 2 and 3 explain the underlying string alignment and clustering algorithms. Section 4 describes the design and implementation of `cluster`, Section 5 presents experimental data obtained from running `cluster` on various data sets, and Section 6 discusses possible future improvements.

2 Alignment Algorithm

This section explains the string alignment algorithm used by `cluster` to measure similarity between two programs. An alignment of two strings s and t (of possibly different lengths) is obtained by inserting spaces in the strings so that their lengths become the same. Note that there are many possible alignments. For example, two alignments of the strings "mastery" and "stars" are

```
_masters        masters
sta___rs        __stars
```

Each column of an alignment is a pair of characters, which can either be a match (score m), a mismatch (score d), or a gap (score g). The score of a consecutive block of columns is simply the sum of the individual scores, and the score of an alignment is defined as that of its highest-scoring block(s) (computational biology literature calls this *local alignment.*) For the above example, if $m = 1$, $d = -1$, and $g = -2$, then "rs/rs" and "sters/stars" are the highest-scoring blocks in the first and second alignments respectively. The alignment scores are then 2 and 3. Alignment scores are related to edit distances and are used extensively in computational biology to detect relationships between DNA strands [4,6].

The optimal alignment score between two strings is the maximum score among all alignments. This value can be computed using dynamic programming. Formally, given two strings s and t, define $D(i,j)$ to be the optimal alignment score between the two substrings $s[1..i]$ and $t[1..j]$; $\max_{1\leq i\leq|s|,1\leq j\leq|t|} D(i,j)$ is the value we are looking for. Define $score(s[i],t[j]) = \begin{cases} m, \text{if } s[i] = t[j], \\ d, \text{otherwise.} \end{cases}$ The following recurrence relation gives us a method to compute the solution:

$$D(i,j) = \max \begin{cases} D(i-1,j-1) + score(s[i],t[j]), \\ D(i-1,j) + g, \\ D(i,j-1) + g, \\ 0 \end{cases}$$

Boundary conditions are given by $D(1,i) = i*g$ and $D(j,1) = j*g$. Elements of the matrix D can be computed by initializing the first row and column with the boundary conditions and then evaluating the elements from left to right and top to bottom. This is possible since $D(i,j)$ depends only on $D(i-1,j-1)$, $D(i-1,j)$, and $D(i,j-1)$. Running time of evaluating the best alignment is $O(|s||t|)$. Space requirement is $O(\min(|s|,|t|))$, since only two rows are needed by the computation at any time.

Finding a best alignment (i.e. the positions where spaces are inserted) as opposed to only the best alignment score can be done trivially in linear time with access to the full matrix D as follows. Starting at the position (r,c) of an element with the largest value in D, determine whether a space should be inserted at position r of the first string or at position c of the second string by examining $D(r-1,c-1)$, $D(r-1,c)$, and $D(r,c-1)$ (see the recurrence relation above.) Repeat this process for the selected neighbor until its value is 0. However, the space required to store D is $O(|s||t|)$, which becomes prohibitive even for medium-sized s and t, so this method is not practical. Instead a more sophisticated algorithm is used to obtain the alignment in $O(|s||t|)$ time and $O(|s|+|t|)$ space [3].

3 Clustering Algorithm

This section explains the method of clustering mutually similar programs by finding connected components of a graph. Given n programs, a graph G of n

vertices is constructed so that two vertices a and b of G are connected by an edge with their alignment score as weight, but only if the weight exceeds a user-selected threshold value th. Connected components of this graph can be naturally identified with the desired clusters. The connected components of G can be found in time $O(n)$ with a simple modification of any tree traversal algorithm. This running time is dominated by the time required to find the best alignment score for each pair of vertices, which is $O(n^2)$.

4 Design and Implementation

The heart of `cluster` was implemented in 4,600 lines of C++ and its graphical user interface in 3,200 lines of Tcl/Tk. It takes as input a set of programs, a scoring scheme consisting of weights for alignment, and a cluster threshold. First each program is parsed to generate a much smaller token stream. Next a mosaic is generated from these token streams and the best alignment scores between the mosaic and each token stream are computed. This is repeated a small number of times (default value is 5) to ensure that only similar programs will consistently earn high scores. The alignment scores for each program are added together and normalized. Finally, the scores are separated into clusters using the input threshold value. The algorithm of `cluster` appears in Figure 1.

4.1 Input/Output Formats

Upon startup, `cluster` asks the user to select a set of programs using a standard file selection dialog box. The programs are then clustered, and non-singleton groups are displayed in a linear list separated by markers. Highest-scoring programs appear at the top. The user can change the cluster threshold value to get an instant update of the grouping. Figure 2 shows the result output by `cluster` on a set of programs.

`cluster` also provides a visual alignment of any two programs belong to the same cluster upon request. The alignment displays the full text of the first program. Portions perfectly matched with the second program are displayed in green. Two matched tokens representing different strings (eg. string literals) are displayed in green and enclosed in a box. Two mismatched tokens are displayed in red and enclosed in a box. A gap is enclosed in a solid maroon box. Finally, portions of text which were not used in the alignment are displayed in black. An example appears in Figure 3.

4.2 `parse()`

The function `parse()` is a lexical analyzer generated by the Unix tool `flex` to convert a C program into a stream of tokens, each representing either an arithmetic or logical operation, a punctuation symbol, a C macro, a keyword, a numeric or string constant, or an identifier. Token numbers for keywords and special symbols are predefined. Those for identifiers are assigned dynamically

```
Input:  Programs p[0], ..., p[n-1], scoring scheme S, cluster threshold V
Output: Non-singleton clusters C1, C2, ..., Ck of similar programs

cluster(p[0], ..., p[n-1], S, V)
{
        for (i = 0; i < n; ++i)
                t[i] = parse(p[i]);
                score[i] = 0;
        endfor

        for (j = 0; j < iterations; ++j)
                tm = make_mosaic(t[0], ..., t[n-1]);
                for (i = 0; i < n; ++i)
                        score[i] += align(t[i], tm, S);
                endfor
        endfor

        ave = average(score[0], ..., score[n-1]);
        std = std_dev(score[0], ..., score[n-1]);
        for (i = 0; i < n; ++i)
                score[i] = (score[i] - ave) / std;
        endfor

        return group(score[0], ..., score[n-1], V);
}
```

Fig. 1. `cluster` algorithm.

with the use of a symbol table (shared by both programs), so that two occurrences of a variable name are replaced by two occurrences of some integer. White spaces and comments are discarded. The purpose of this tokenizing process is two-fold: to reduce a program to its parse tree, which is usually much shorter; and to remove inessential information before is performed. A token stream can be divided into a series of modules, which are defined as blocks of text beginning right after one top-level right brace and ending at the next top-level right brace. (Actually, if the preceding right brace belongs to a structure definition, then the new module begins after the semicolon ending the definition.)

4.3 `make_mosaic()`

The function `make_mosaic()` selects a random fragment from each program's token stream and concatenates them to form a new mosaic token stream. The size f of the fragments is a constant; its default value is 100 tokens, but this value can be set at run-time by the user. To select a fragment, a random starting position in the token stream is picked, which belongs to some module M in the original program. The next f tokens are returned, but only if they all belong to M; otherwise, the last f tokens of M are returned. If M has less than f tokens,

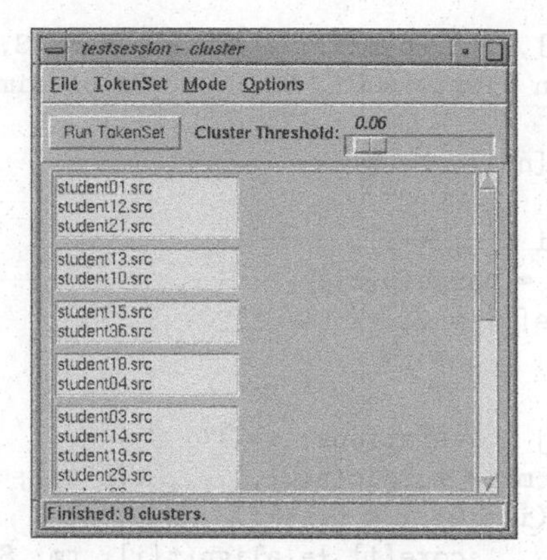

Fig. 2. Clustering of similar programs.

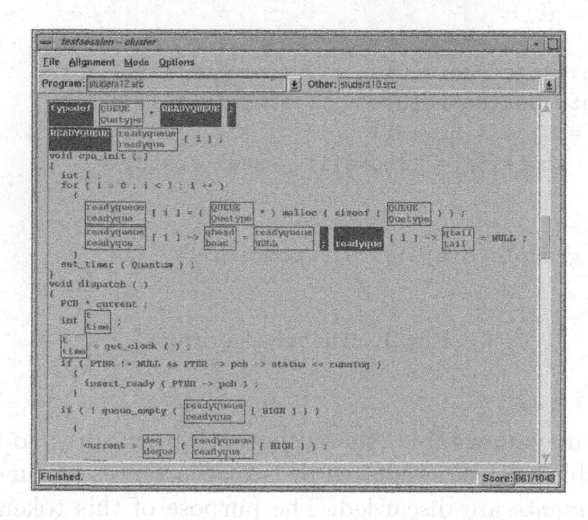

Fig. 3. Visual alignment of two similar programs.

then the whole module is returned, and another random fragment of size $f-|M|$ is selected.

4.4 `align()`

Given two token streams t_1 and t_2, `align()` finds a best alignment and its score between each module of t_1 and the whole stream t_2. This technique guarantees a high score in the case t_1 is simply a reordering of the modules of t_2. `align()` uses the following default scheme to compute the alignment score:

- matches of two identifier tokens are considered very significant and have weight 2; other matches have weight 1;
- mismatches of two identifiers may be due to systematic name changes and are disregarded, i.e. they have weight 0; other mismatches (eg. between an

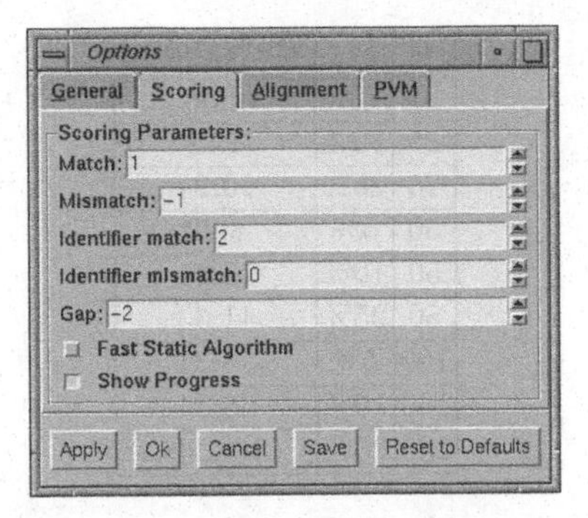

Fig. 4. Modifying weights used by `cluster`'s scoring scheme.

operator and an identifier) are considered more significant and have weight -1;
- gaps should be introduced with restraint and thus have weight -2.

Weights used by the scoring scheme are user-configurable as shown in Figure 4.

4.5 `group()`

The function `group()` takes a list of real numbers, sorts it, and then inserts a boundary between two consecutive elements if their values differ more than the cluster threshold. Only clusters with at least two members are displayed. Note that this method is much simpler than the connected-component algorithm; it is based on the assumption that two similar programs are likely to have similar alignment scores with the same mosaic.

4.6 Optional Features

Although `cluster` is designed as a stand-alone application, its computation can be distributed over a network of workstations to achieve even faster running time with the help of the parallel virtual machine simulator `PVM`. An on-going research project in distributed computing at the University of Tennessee, `PVM` is a software package that simulates a message-passing parallel machine with a heterogeneous network of computers running Unix and/or Windows NT. Programs for PVM can be written in C, C++, and Fortran. This software allows the user to distribute computations over a network of computers to solve much larger problems than possible with a single machine. PVM is very easy to install and use and can be downloaded from `http://www.epm.ornl.gov/pv`.

When used with PVM over a network of p processors, `cluster` achieves a perfect speedup factor of p, due to independence of the n alignment computations.

On the other hand, if quality of the clustering is more important than speed, the user can run `cluster` in the slow mode, which will perform all $O(n^2)$

n	f	Time (min.)
25	5%	2:06
25	10%	4:11
25	25%	10:24
50	5%	8:23
50	10%	16:41
50	25%	41:33
100	5%	33:31
100	10%	66:34
100	25%	166:29

Fig. 5. `cluster`'s running times on one processor.

n	f	Time (min.)	Speed up
25	5 %	0:23	5.479
25	10%	0:44	5.725
25	25%	1:47	5.831
50	5 %	1:27	5.792
50	10%	2:52	5.825
50	25%	7:05	5.865
100	5%	5:45	5.831
100	10%	11:33	5.764
100	25%	28:06	5.923

Fig. 6. `cluster`'s running times on six processors using PVM.

alignments and cluster the results using the connected-component algorithm described in Section 3.

5 Experimental Setup and Result

We ran `cluster` on a set of 24 real-life homework programs of average size 18 KBytes on one processor, with a contribution of 5% from each program to the mosaic. `cluster` identified correctly all groups of similar programs in this set in 80 seconds. The running time of the naive implementation of finding alignments for all pairs took 1503 seconds.

We tested `cluster` on a 400-MHz Pentium II PC running Debian/GNU Linux using programs of average size 13 KBytes and different values for n and f. The running times shown in Figure 5 agree with our theoretical analysis. Even on a set of 100 programs, with a contribution of 25% from each program to the mosaic, `cluster` took less than 3 hours. We reran `cluster` on the same dataset using PVM to distribute the computations across six identical 400-MHz processors. The obtained running times shown in Figure 6 show that PVM achieved an almost perfect speedup factor of 6.

6 Discussion

We have designed and implemented a fast tool to partition a large pool of C programs into groups according to structural similarity. Our tool uses string alignment methods to measure similarity between input programs and a mosaic consisting of fragments randomly selected from each input program. Clusters of mutually similar programs are deduced from these scores. The user can choose to display alignments between two programs belonging to a suspected cluster with similar blocks of code highlighted. Scoring scheme as well as similarity threshold values are easily configurable. Experimental data obtained by running `cluster` on real-life homework programs show that our tool is effective (see [2] for a real-life example) and achieves a significant speedup over the naive method of aligning all pairs of programs.

Future improvements include smarter handling of header files, detection of extraneous program fragments (dead code), and parsers for other languages such as C++ and Java.

It would be interesting to perform a rigorous analysis of the probability that `cluster` fails to detect similarity between programs. It should be pointed out, however, that a good theoretical performance guarantee may not measure up in practice, because plagiarism can be a subjective notion, and string alignment is but one method to measure similarity between two strings. In the end, detection plagiarism requires human intervention; the main utility of `cluster` is in making the job easier for the human operator by i) quickly narrowing the pool of suspects; ii) allowing customization of similarity criteria; and iii) providing visual evidence of similarity between two programs.

References

1. Cormen, T., Leiserson, C., and Rivest, R. *Introduction to Algorithms.* MIT Press and McGraw Hill, 1992.
2. Gitchell, D., and Tran, N. Sim: A utility for detecting similarity in computer programs. *SIGCSE Bulletin (ACM Special Interest Group on Computer Science Education) 31* (1999).
3. Hirschberg, D. A linear space algorithm for computing maximal commonsubsequences. *Communications of the ACM 18* (1975), 341–343.
4. Huang, X., Hardison, R. C., and Miller, W. A space-efficient algorithm for local similarities. *Computer Applications in the Biosciences 6*, 4 (1990), 373–381.
5. Hunt, J. W., and Szymanski, T. G. A fast algorithm for computing longest common subsequences. *Communications of the ACM 20*, 5 (May 1977), 350–353.
6. Myers, E. W., and Miller, W. Optimal alignments in linear space. *Computer Applications in the Biosciences 4*, 1 (1988), 11–17.
7. Smith, T. F., and Waterman, M. S. Identification of common molecular subsequences. *Journal of Molecular Biology 147* (1981), 195–197.

Broadcasting in Generalized de Bruijn Digraphs

(Extended Abstract)

Yosuke Kikuchi, Shingo Osawa, and Yukio Shibata

Department of Computer Science, Gunma University,
1-5-1 Tenjin-cho, Kiryu, Gunma, 376-8515 Japan
{kikuchi,shingo,shibata}@msc.cs.gunma-u.ac.jp

Abstract. This work deals with a broadcasting on generalized de Bruijn digraphs. Broadcasting on digraphs corresponds to one on networks with monodirection communication links in practice. The efficiency of broadcasting is affected by network topology. Generalized de Bruijn digraph is one of useful network models. We propose protocols for broadcasting on generalized de Bruijn digraphs using Kronecker product of graphs. The protocol presented in this paper constructs a k-ramified tree.

Keywords: generalized de Bruijn digraph, Imase Itoh digraph, broadcasting, Kronecker product, k-ramified tree.

1 Introduction

In communication networks including interconnection networks for parallel computers, it is important that each processor element(PE) has common information. Broadcasting is one of the such communication schemes. Broadcasting is used for loading copies of a single message originated by one PE, called the originator, to all other PEs in the network. Due to the importance of broadcasting, a great deal of researches have been devoted to obtaining efficient broadcasting algorithms. Whereas the topology of the communication network has an effect on the efficiency of broadcasting. On inquiring into broadcasting, there are some constraints concerned with ability of elements and links in the network. Furthermore, the network is synchronous, that is, PE's in the network are able to communicate simultaneously each other by using their ports. The required time of sending one message from one PE to another PE by one link is called the round. So, the constraints in the network are summarized as follows;

1. Each PE is able to send a message to at most one PE in a round;
2. Each PE is able to receive a message from at most one PE in a round;
3. Each PE is not able to send and receive messages simultaneously in a round;
4. Each PE is able to communicate to at most one adjacent PE.

The constraints 1. and 2. implicate that store and forward routing is used in the network. Given a network and a PE that is a message originator, there arises one question, "what is the minimum number of rounds required to complete broadcasting from the message originator?" Thus, given issue is the minimization of the number of rounds required to complete broadcasting from the message

O.H. Ibarra and L. Zhang (Eds.): COCOON 2002, LNCS 2387, pp. 200–209, 2002.

originator. To investigate broadcasting theoretically, a communication network is modeled by a graph or a digraph. A PE corresponds to a vertex of the graph or the digraph and a link corresponds to an edge or an arc. Monodirectional links are represented by arcs. Bidirectional half duplex links are represented by edges. Bidirectional full duplex links are represented by symmetric arcs. Digraphs, thus networks with monodirectional links, are only dealt with in this paper. It is preferred that a network has fewer links, for example in the design of VLSI systems, many links increase the design cost. Whereas it is desirable to complete broadcasting in fewer rounds. There is a trade off between the number of links and the number of rounds required to complete broadcasting. For completing broadcasting efficiently, it is important to choose an appropriate network topology. There are various studies on broadcasting in complete graphs[3], hypercubes [4,5,11], Cayley graphs [7] and de Bruijn digraphs [2] and there is a survey on broadcasting and gossiping [8]. One executing broadcasting, de Bruijn digraph has a good property as network topology. De Bruijn graph has more vertices than hypercube does, when the diameter and the maximum degree are fixed [1]. Generalized de Bruijn digraphs are independently introduced by Imase and Itoh [9,10], and Reddy, Pradhan and J. Kuhl [14], so called Imase Itoh digraphs. This generalization removes the restriction on the cardinality of vertex sets of these digraphs and makes these digraphs more valuable as network models. We deal with broadcasting in generalized de Bruijn digraphs. The rest of the paper is organized as follows. Section 2 gives definition and terminology. We introduce the k-ramified tree that plays an important role in the broadcasting protocol in Section 4. In Section 3 we state simple lower bounds on the number of rounds to complete broadcasting on symmetric complete digraphs in which the arcs of this symmetric complete digraphs do not represent full duplex links. Section 4 deals with protocols to execute broadcasting on generalized de Bruijn digraphs. For the design of our protocols, we use Kronecker product of digraphs. Section 5 concludes with discussions and further studies.

2 Definition and Terminology

We state some classes of digraphs needed in this paper and Kronecker product of graphs. $V(G)$ and $A(G)$ are the vertex set and the arc set of a digraph $G(V, A)$, respectively. There is an arc from u to v if $(u, v) \in A(G)$. The distance from a vertex u to a vertex v, denoted $d(u, v)$, is the length of the shortest path from u to v. We will use $ecc(u)$ to denote the eccentricity of the vertex u, that is, $\max_{v \in V(G)} d(u, v)$. The diameter of a digraph G, denoted $diam(G)$, is given by $diam(G) = \max_{u \in V(G)} \{ecc(u)\}$. If a digraph is not strongly connected, then $diam(G) = \infty$. The *generalized de Bruijn digraph* $G_B(n, d)$ is defined by congruence equations.

$$\begin{cases} V(G_B(n,d)) = \{0, 1, 2, \dots, n-1\}, \\ A(G_B(n,d)) = \{(x,y) | y \equiv dx + i \pmod{n}, 0 \le i < d\}. \end{cases}$$

If $n = d^D$, $G_B(n, d)$ is the de Bruijn digraph $B(d, D)$.

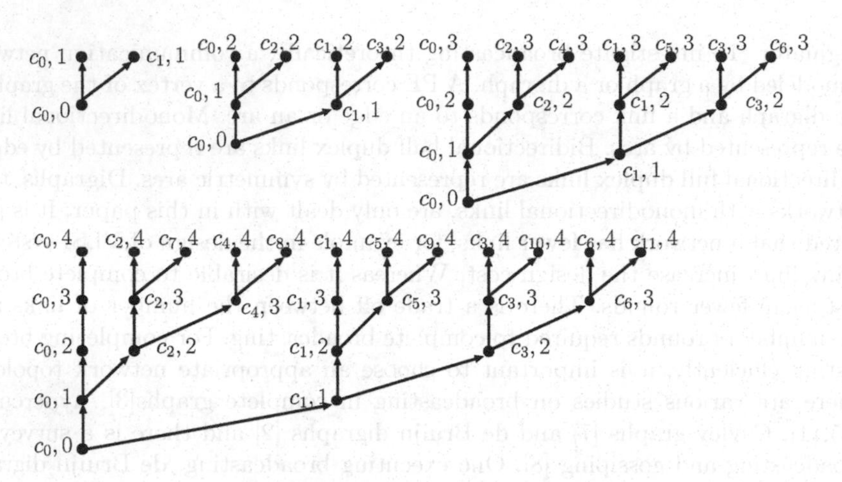

Fig. 1. The 2-ramified trees for $t = 1, 2, 3$ and 4.

The *path* P_n has n vertices and $n-1$ arcs. For $V(P_n) = \{v_1, v_2, \ldots, v_n\}$, the arc set $A(P_n) = \{(v_i, v_i + 1) | 1 \leq i \leq n-1\}$. P_n can be drawn so that all of its vertices and same directed arcs lie on a single line.

For a strongly connected digraph G, the shortest path spanning tree (breadth-fast search tree) with root r, denoted by $ST(r)$, is a spanning tree with root r and contains a shortest path between r and v in G for each vertex v in G. The shortest path spanning tree has an important role in studies for broadcasting and also in this paper. We consider the designing of the efficient protocol for broadcasting, applying the shortest path spanning tree.

We introduce the k-ramified tree. The k-ramified tree with depth t is denoted by $Tr(k,t)$, where k and t are positive integers and $k \geq 2$. Each vertex of $Tr(k,t)$ has a label $\langle c, l \rangle$, l means the depth in $Tr(k,t)$. A child of $\langle c, l \rangle$ has a label either $\langle c, l+1 \rangle$ or $\langle c', l+1 \rangle$. Children of $\langle c, l \rangle$ are classified into two types: the $\langle c, l+m \rangle$ ($m \geq 1$) is called the *lineal child*, then the arc is said to be the *lineal arc*. The $\langle c', l+1 \rangle$ is the *collateral child*, then the arc is called the *collateral arc*. In addition, arcs connecting vertices on depths k and $k+1$ are called *depth* $(k+1)$ *arcs*. For the vertex $\langle c, l \rangle$, the collateral child of $\langle c, l \rangle$ and the collateral children of lineal children of $\langle c, l \rangle$ are called collateral descendants of $\langle c, l \rangle$. There is an arc from the vertex $\langle c, l \rangle$ to its each child in $Tr(k,t)$. The definition of $Tr(k,t)$ is recursively done. The $Tr(k,0)$ is the trivial graph. $Tr(k,1)$ is the complete binary tree with depth 1. The root's label is $\langle c, 0 \rangle$. One of the children has the label $\langle c, 1 \rangle$ and another child has the label $\langle c', 1 \rangle$. The subgraph induced by all vertices whose depths are less than or equal to $t-1$ of $Tr(k,t)$ is $Tr(k,t-1)$. The vertex $\langle c, t-1 \rangle$ has children $\langle c, t \rangle$ and $\langle c't \rangle$ if the vertex $\langle c, t-k-1 \rangle$ does not exit in $Tr(k,t-1)$. The vertex $\langle c, t-1 \rangle$ has only one child $\langle c, t \rangle$ if the vertex $\langle c, t-k-1 \rangle$ exits in $Tr(k,t-1)$.

Figure 1 shows examples for the 2-ramified tree. Each vertex in $Tr(k,t)$ has at most two children. Thus $Tr(k,t)$ is a kind of binary tree and if $t \leq k$, then $Tr(k,t)$ is the complete binary tree whose depth is t.

The quasi k-ramified tree with depth t is denoted by $qu\text{-}Tr(k,t)$, where k and t are positive integers and $k \geq 2$. There is an arc from the vertex $\langle c,l\rangle$ to its each child in $Tr(k,t)$. The definition of $qu\text{-}Tr(k,t)$ is recursively given as well as $Tr(k,t)$. The $qu\text{-}Tr(k,0)$ is the trivial graph. $qu\text{-}Tr(k,1)$ is P_2. Then the label of the root is $\langle c,0\rangle$ and the label of child of the root is $\langle c,1\rangle$. The subgraph induced by all vertices whose depths are less than or equal to $t-1$ of $Tr(k,t)$ is $Tr(k,t-1)$. The vertex $\langle c,t-1\rangle$ has children $\langle c,t\rangle$ and $\langle c't\rangle$ if the vertex $\langle c,t-k-1\rangle$ does not exit in $Tr(k,t-1)$. The vertex $\langle c,t-1\rangle$ has only one child $\langle c,t\rangle$ if the vertex $\langle c,t-k-1\rangle$ exits in $qu\text{-}Tr(k,t-1)$. Each vertex in $qu\text{-}Tr(k,t)$ also has at most two children. Thus $qu\text{-}Tr(k,t)$ is a kind of binary tree.

Let $l_{k,s-1}$ (respectively, $qu\text{-}l_{k,s-1}$) be the number of vertices with depth $s-1$ in $Tr(k,t)$ (respectively, $qu\text{-}Tr(k,t)$). Then $qu\text{-}l_{2,s-1}$ is equal to the sth Fibonacci number $F(s)$, whereas $l_{k,s-1}$ is equal to $F(s+2)-1$. For a given integer n and a k-ramified tree, $f_k(n)$ denotes the integer r so that the maximum $l_{k,r}$ is the maximum value not exceeding n. For a given integer n and a quasi k-ramified tree, $qu\text{-}f_k(n)$ denotes the integer r so that the maximum $qu\text{-}l_{k,r}$ is the maximum value not exceeding n.

We use Kronecker product to design protocol to construct a spanning tree of generalized de Bruijn digraph. The *Kronecker product* (also known as the *tensor product*, *composition*, or *categorical product*) results in a graph $G = G_1 \times G_2$ whose vertex set is $V(G) = V(G_1) \times V(G_2)$ and arc set $A(G) = \{((u_1,u_2),(v_1,v_2))|u_1v_1 \in A(G_1) \text{ and } u_2v_2 \in A(G_2)\}$, where $V(G_i)$ is the vertex set of G_i and $A(G_i)$ is the arc set of G_i.

3 Broadcasting in the Network

We suppose that the length of one message can be equal to the amount of data sent per 1 round. For a broadcasting in the network, links used in the broadcasting induce a tree structure with originator at the root. That means this constructs a spanninig tree with originator at the root in the graph corresponding to the network. Then we design a protocol to obtain a suitable spanninig tree for the broadcasting, when given a root. For broadcasting with a rooted vertex u, the broadcasting time of u is the number of rounds to complete the broadcasting. For a digraph G and the vertex u, $b_G(u)$ denotes the minimum broadcasting time of u and $b(G)$ denotes the minimum broadcasting time of any broadcasting on G. For a digraph G such that $|V(G)| = n$ and the vertex u, it is clear that the lower bound of $b_G(u)$ is $\max\{ecc(u), \lceil \log_2 n \rceil\}$. Furthermore it is also obvious that the lower bound of $b(G)$ is $\max\{diam(G), \lceil \log_2 n \rceil\}$ for digraph G with $|V(G)| = n$. For an arbitrary graph G, determining $b(G)$ is known to be NP-complete [15]. The vertex which has received a message can send the message to a neighbor vertex. Thus if m vertices have received a message in a round, then $2m$ or fewer vertices will be able to receive the message at the next round. For a k-regular digraph G without loops, each vertex may send message to at most k vertices. In the k-ramified tree $Tr(k,t)$, the vertex $\langle c,l\rangle$ may have at most k collateral descendants. The rounds of broadcasting on k-regular digraph G without loops correspond to the depth of k-ramified tree. Then the number of vertices with depth s in the k-ramified tree is an upper bound of the number of vertices which could have received message up to s rounds. From this fact, we obtain the next statement.

Lemma 1. *If G is a k-regular digraph without loops and $|V(G)| = n$, then the lower bound of $b(G)$ is* $\max\{f_k(n)+1, diam(G)\}$

The symmetric complete digraph is the simple example which satisfies Lemma 1. Next we consider the case that G is a k-regular digraph and has at most one loop. The loop is unnecessary for broadcasting. If an originator has a loop, the originator has $k-1$ neighbor vertices in G.

Lemma 2. *If G is a k-regular digraph which has at most one loop and $|V(G)| = n$, then the lower bound of $b(G)$ is* $\max\{qu\text{-}f_k(n)+1, diam(G)\}$.

4 Broadcasting in the Generalized de Bruijn Digraphs

There always exist several loops in the generalized de Bruijn digraph as well as in the de Bruijn digraph. The generalized de Bruijn digraph $G_B(n,d)$ has exactly $\gcd(d-1,n)\lceil d/\gcd(d-1,n)\rceil$ loops [6,12]. Then we make a digraph from the generalized de Bruijn digraph $G_B(n,d)$ by deleting loops and this digraph is denoted $G_B^-(n.d)$. We consider the broadcasting by a shortest spanning tree on $G_B(n,d)$. The diameter of the generalized de Bruijn digraph $G_B(n,d)$ is $\lceil \log_d n \rceil$. Thus the depth of the shortest path spanning tree of $G_B(n,d)$ is at most $\lceil \log_d n \rceil$. We give two protocols for broadcasting on the generalized de Bruijn digraphs. One of the protocols is using the shortest path spanning tree, another is using a tree related to k-ramified tree. These two protocols are designed using Kronecker product of the generalized de Bruijn digraph $G_B^-(n,d)$ and the path P_2. One of the vertices of P_2 is labeled 0, another is labeled 1. The vertices whose second order label is 0 are called 0 level vertices and the vertices whose second order label is 1 are called 1 level vertices in $G_B^-(n,d) \times P_2$. We describe a protocol by constructing a shortest path spanning tree and giving a procedure of broadcasting using this tree simultaneously.

procedure *Make_SPS-tree_protocol*
{ u is an originator of $G_B(n,d)$.}
  Construct $G_B^-(n,d) \times P_2$;
  Let V_1 be a set of level 1 verteies in $G_B^-(n,d) \times P_2$;
  Assign 0 to the vertex $(u,0)$;
  $S := \{(u,0)\}$;
  Delete the vertex $(u,1)$ from the set V_1;
  repeat
    Set k up the minimum among numbers assigned to vertices in the set S;
    Choose the vertex $(v,0)$ which is assigned k in S;

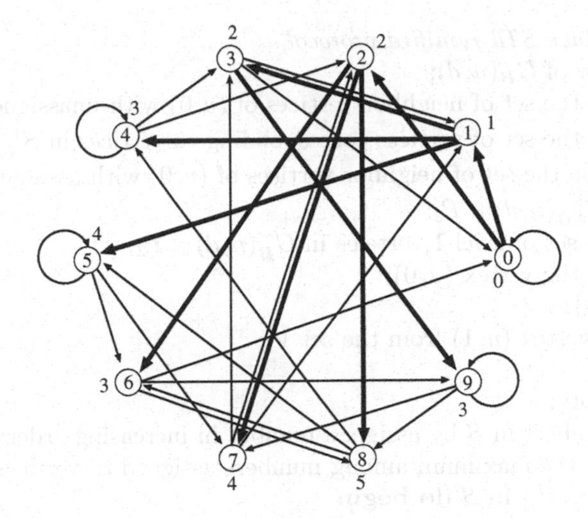

Fig. 2. Broadcasting using the shortest path spanning tree rooted at the vertex 0 obtained by SPS-tree protocol for $G_B(10, 3)$.

for $i := 0$ **to** $d - 1$ **do begin**
    $k := k + 1$;
    Assign k to $(dv + i, 1) \in V_1$ and $(dv + i, 0)$;
    Delete the vertex $(dv + i, 1)$ from the set V_1;
    $S := S \cup \{(dv + i, 0)\}$;
end
Delete the vertex $(v, 0)$ from the set S;
until V_1 is empty.

Lemma 3. *SPS-tree protocol generates a shortest path spanning tree of the generalized de Bruijn digraph.*

The thick arc is the arc used in the broadcasting from vertex 0 of $G_B(10, 3)$ in Figure 2. The numbers written inside the vertices are the labels of vertices of $G_B(10, 3)$. Each number which is assigned to the vertex indicates the round of receiving the message of the vertex 0 in the broadcasting. Figure 2 shows that the depth of the spanning tree obtained by SPS-tree protocol is 3. Due to the relation between the number of vertices and the out degree of each vertex, it is unable to construct a spanning tree with depth 2 for $G_B(10, 3)$. The vertex 8 receives the message at round 5 in Figure 2. However if we can construct a spanning tree using the arc from vertex 6 to vertex 8 instead of the arc from vertex 2 to vertex 8, then we obtain a broadcasting completed in 4 rounds.

Next we present a protocol constructing a spanning tree related to d-ramified tree and quasi q-ramified tree, and giving the procedure of broadcasting using this tree simultaneously for $G_B(n, d)$.

procedure *Make_STd-ramified_protocol*
{u is originator of $G_B(n,d)$;
Let $S^1_{N(v,0)}$ be the set of neighbor vertices of $(v,0)$ with unassigned numbers;
Let $S^0_{N(v,0)}$ be the set of vertices corresponding to vertices in $S^1_{N(v,0)}$;
Let $AS^1_{N(v,0)}$ be the set of neighbor vertices of $(v,0)$ with assigned numbers.}
  Construct $G^-_B(n,d) \times P_2$;
  Let V_1 be a set of level 1 verteies in $G^-_B(n,d) \times P_2$;
  Assign 0 to the vertex $(u,0)$;
  $S := \{(u,0)\}$;
  Delete the vertex $(u,1)$ from the set V_1;
  repeat
    A is empty;
    Sort members in S by assigned number in increasing order;
    Set k up the maximum among numbers assigned to vertices in the set S;
    foreach $(v,0)$ **in** S **do begin**
      Compute $S^1_{N(v,0)}$ and $S^0_{N(v,0)}$;
      Find the neighbor of $(v,0)$ such that
        $|S^1_{N(dv+i,0)}| - |AS^1_{N(dv+i,0)}|$ is the largest number;
      Assign $k+1$ to $(dv+i,0)$ and $(dv+i,1)$;
      Delete the vertex $(dv+i,1)$ from the set V_1;
      $A := A \cup \{(dv+i,0)\}$;
    end
    $S := S \cup A$;
  until V_1 is empty.

Obtained spanning trees are not necessarily the shortest path spanning trees. Each number which is assigned to the vertex indicates the round at which the message from u is received in the broadcasting.

Lemma 4. *STd-ramified protocol generates a spanning tree of the generalized de Bruijn digraph.*

In Figure 3, the thick arc is the arc used in the broadcasting from vertex 0 by applying STd-ramified protocol for $G_B(10,3)$. Figure 3 shows that the depth of the spanning tree obtained by STd-ramified protocol is 4. The vertex 8 receives the message at round 4 in Figure 3 and this broadcasting is completed in 4 rounds.

Lemma 5. *For any SPS-tree protocol for $G_B(n,d)$, there is a STd-ramified protocol such that it assigns the number less than the number assigned by the SPS-tree protocol for each vertex.*

We construct the following trees from the spanning trees of $G_B(n,d)$ obtained by above two protocols, then each label of the tree contains two elements. One of the elements is the label of $G_B(n,d)$, another is the round at which the vertex receives the message. If the vertex $u \in G_B(n,d)$ receives a message at round k, then the vertex corresponding to u has the label $\langle u,k \rangle$ in this tree. There is an

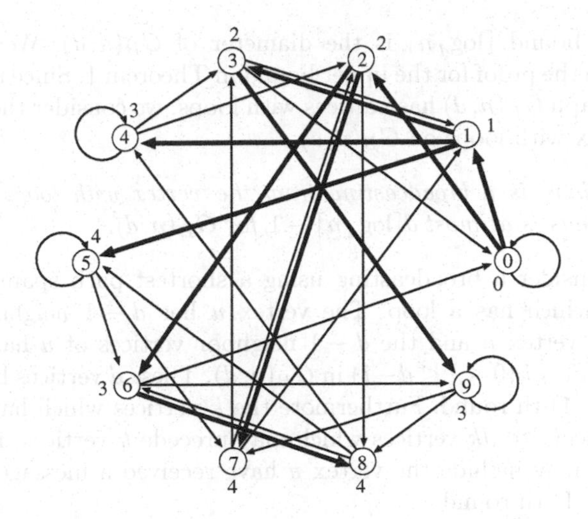

Fig. 3. Broadcasting using the spanning tree rooted at the vertex 0 obtained by STd-ramified protocol for $G_B(10, 3)$.

arc from the vertex u to the vertex u' in the spanning tree if and only if there is an arc from $\langle u, k\rangle$ to $\langle u', k+1\rangle$. We call the tree constructed by SPS-tree protocol the SPS tree and the tree constructed by STd-ramified protocol the STR tree, respectively. From Lemma 5, the next corollary holds.

Corollary 1. *The depth of STR tree is less than or equal to that of SPS tree.*

There is a following relation between the STR tree of $G_B(n, d)$ and the d-ramified tree.

Lemma 6. *Let the root of STR tree correspond to the vertex which has no loop in $G_B(n, d)$. If the depth of STR tree is t, then $t \geq f_k(n) + 1$ and the STR tree is a subgraph of $Tr(d, t)$.*

Lemma 7. *Let the root of STR tree correspond to the vertex which has a loop in $G_B(n, d)$. If the depth of STR tree is t, then $t \geq f_k(n) + 1$ and the STR tree is a subgraph of $qu - Tr(d, t)$.*

It is difficult to determine the number of rounds for broadcasting using STd-ramified protocol. Whereas we consider the generalized Fibonacci numbers involved in the number of rounds for broadcasting using STd-ramified protocol [3]. We give an upper bound for the number of rounds for broadcasting using STd-ramified protocol.

Theorem 1. *For the generalized de Bruijn digraph $G_B(n, d)$,*

$$\lceil \log_d n \rceil \leq b(G_B(n, d)) \leq d\lceil \log_d n \rceil.$$

The lower bound $\lceil \log_d n \rceil$ is the diameter of $G_B(n,d)$. We prepare some lemmas to give the proof for the upper bound in Theorem 1. Since the generalized de Bruijn digraph $G_B(n,d)$ has vertices with loops, we consider the broadcasting from the vertex with loops on $G_B(n,d)$.

Lemma 8. *There is a broadcasting from the vertex with loops such that the number of rounds is at most* $d\lceil \log_d n \rceil - 1$ *for* $G_B(n,d)$.

Proof. We consider a broadcasting using a shortest path spanning tree from the vertex u which has a loop. The vertex u has $d-1$ neighbor vertices in $G_B(n,d)$. The vertex u and the $d-1$ neighbor vertices of u have consecutive labels, that is, $du+i$ $(0 \leq i \leq d-1)$ in $G_B(n,d)$. These d vertices have a message from u by $(d-1)$-th round. Furthermore the k vertices which have consecutive labels are adjacent to dk vertices which may precede k vertices. Hence $d^{\lceil \log_d n \rceil}$ vertices which may include the vertex u have received a message from u down to $(d\lceil \log_d n \rceil - 1)$-th round. □

Next lemma states the broadcasting from the vertex which has no loop.

Lemma 9. *There is a broadcasting from the vertex with no loop such that the number of rounds is at most* $d\lceil \log_d n \rceil$ *for* $G_B(n,d)$.

Proof. We consider a broadcasting using shortest a path spanning tree from the vertex u which has no loop. The vertex u has d neighbor vertices in $G_B(n,d)$. There is no guarantee that the vertex u and d neighbor vertices of u have consecutive labels in $G_B(n,d)$. The d neighbor vertices have a message from u by d-th round. Furthermore the k vertices which have consecutive labels are adjacent to dk vertices which may include preceding k vertices. Hence $d^{\lceil \log_d n \rceil}$ vertices which may include the vertex u have received a message from u down to $d\lceil \log_d n \rceil$-th round. □

Since the vertex u and its neighbor vertices do not always have the consecutive labels in $G_B(n,d)$, the upper bound of Lemma 9 is not sharper than that of Lemma 8. Theorem 1 is the direct consequence of Lemma 8 and 9.

5 Conclusion and Further Studies

We propose protocols for broadcasting using Kronecker product of the generalized de Bruijn digraph and a path. We introduce k-ramified trees and quasi k-ramified trees concerned with an efficient broadcasting. We also estimate the minimum broadcasting time of the generalized de Bruijn digraphs.

The design of protocol for broadcasting using Kronecker product seems to be applicable to other classes of graphs and digraphs. The k-ramified tree and the quasi k-ramified tree will give an easy method to analyze the minimum broadcasting time of other classes of graphs and digraphs. The bounds of the minimum broadcasting time of the generalized de Bruijn digraphs will be made better by investigating the relation between the k-ramified trees (the quasi k-ramified trees) and the generalized de Bruijn digraphs.

Acknowledgement

This research is partly supported by Grant-in-Aid for Scientific Research (C) (No. 13680392) of Japan Society for the Promotion of Science.
This research has also been partly supported by the Kayamori Foundation of Informational Science Advancement.

References

1. J.-C. Bermond and C. Peyrat, De Bruijn and Kautz networks: a competitor for the hypercube?, *Hypercube and Distributed Computers*, (F. Andrè and J. P. Verjus, Eds.), Elsevier North-Hollamd, Amsterdam, 1989.
2. J.-C. Bermond and P. Fraigniaud, Broadcasting and gossiping in de Bruijn networks, *SIAM J. Comp.*, 23(1994) 212–225.
3. J. Bruck, R. Cypher and C-H Ho, Multiple message Broadcasting with generalized Fibonacci trees, 4th IEEE Symposium on parallel and Distributed Processing, (1992) 424–431.
4. G.-M. Chiu, A fault-tolerant broadcasting algorithm for hypercubes, *Info. Processing letters*, 66(1998) 93–99.
5. K. Diks, S. Dobrev, E. Kranakis, A. Pelc and P. Ružička, Broadcasting in unlabeled hypercubes with a linear number of messages, *Info. Processing letters*, 66(1998) 181–186.
6. D. Z. Du and F. K. Hwang, Generalized de Bruijn digraphs, *Networks*, 18(1988) 27–38.
7. C. GowriSankaran, Broadcasting on recursively decomposable Cayley graphs, *Discrete Appl. Math.*, 53(1994) 171–182.
8. S. T. Hedetniemi, S.Hedetniemi and A. Liestman, A survey of gossiping and broadcasting in communication networks, *Networks*, 18(1986) 319–349.
9. M. Imase and M. Itoh, Design to minimize diameter on buildin-block network, *IEEE Trans. Comp.*, **C-30** (1981) 439–442.
10. M. Imase and M. Itoh, A design for directed graphs with minimum diameter, *IEEE Trans. Comp.*, **C-32** (1983) 782–784.
11. S. Johnsson and C.-T. Ho, Optimum broadcasting and personalized communication in hypercubes, *IEEE Trans. Comp.*, **C-38** (1989) 1249–1268.
12. M. Mora, O. Serra and M. A. Fiol, General properties of *c*-circulant digraphs, *Ars Comb.*, **25C** (1998) 241–252.
13. F. T. Leighton, *Introduction to parallel algorithms and architectures: arrays · trees · hypercubes*, Morgan Kaufmann, San Mateo, 1992.
14. S. M. Reddy, D. K. Pradhan and J. Kuhl, Directed graphs with minimal diameter and maximum node connectivity, *School of Engineering Oakland Univ. Tech. Report*, 1980.
15. P. J. Slater, E.J. Cockayne and S.T. Hedetniemi, Information dissemination in trees, *SIAM J. Comp.*, 10 (1981) 692–701.

On the Connected Domination Number of Random Regular Graphs

William Duckworth and Bernard Mans

Department of Computing, Macquarie University
Sydney NSW 2109 Australia
{billy,bmans}@ics.mq.edu.au

Abstract. A connected dominating set (CDS) of a graph, G, is a set of vertices, $C \subseteq V(G)$, such that every vertex in $V(G) \setminus C$ is incident to at least one vertex of C in G and the subgraph induced by the vertices of C in G is connected. In this paper we consider a simple, yet efficient, randomised greedy algorithm for finding a small CDS of regular graphs. We analyse the average-case performance of this heuristic on random regular graphs using differential equations. In this way we prove an upper bound on the size of a minimum CDS of random regular graphs.

1 Introduction

A *dominating set* of a graph, G, is a set of vertices, $D \subseteq V(G)$, such that every vertex of G either belongs to D or is incident with a vertex of D in G. A *connected* dominating set, $\mathcal{C}$, of a graph, G, is a dominating set such that the subgraph induced by the vertices of $\mathcal{C}$ in G is connected. Define the minimum cardinality of all connected dominating sets of G as the *connected domination number* of G and denote this by $\gamma_c(G)$. This problem is of interest in many contexts, in particular, it has many applications in distributed and network environments.

For an arbitrary graph, G, the problem of determining $\gamma_c(G)$ is a well known NP-hard optimisation problem (see, for example, [10]) and is polynomially equivalent to the *maximum leaf spanning tree* problem. The non-leaf vertices of a spanning tree of a graph form a connected dominating set of the graph. Define $\lambda(G)$ to be the maximum number leaves in any spanning tree of G, so that for any n-vertex graph, G, $\lambda(G) = n - \gamma_c(G)$.

Solis-Oba [11] showed that the maximum leaf spanning tree problem is approximable within 2. Galbiati, Maffioli and Morzenti [7] showed that the same problem is not approximable within $1 + \epsilon$ for any $\epsilon > 0$ (unless P=NP).

A graph, G, is said to be d-regular if every vertex of G is incident with precisely d other vertices of G. When discussing any d-regular graph on n vertices, it is assumed that dn is even, d is constant and $d \geq 3$. We consider such graphs that are undirected, unweighted and contain no loops or multiple edges.

Storer [12] showed that for every connected cubic (i.e. 3-regular) graph, G, $\lambda(G) \geq \lceil (n/4) + 2 \rceil$. Griggs, Kleitman and Shastri [8] showed that for every cubic graph, G, that has no subgraph isomorphic to "$K_4 - e$" (K_4 with one edge removed) $\lambda(G) \geq \lceil (n/4) + (4/3) \rceil$.

O.H. Ibarra and L. Zhang (Eds.): COCOON 2002, LNCS 2387, pp. 210–219, 2002.

Griggs and Wu [9] showed that for every connected n-vertex graph, G, with minimum degree at least 4, $\lambda(G) \geq (2n+8)/5$ and for every connected n-vertex graph, G, with minimum degree at least 5, $\lambda(G) \geq (n+4)/2$. For a connected, n-vertex graph, G, with minimum degree $k \geq 6$, the exact value of $\lambda(G)$ remains unknown. The results of [9,12] are essentially the best possible since there exist infinite families of n-vertex, d-regular graphs such that $\lambda(G) \leq \lceil (d-2)n/(d+1) \rceil + 2$.

Duckworth and Wormald [6] gave a new derivation, at least to within an additive constant, of the main result of [12] and also showed that for every cubic graph, G, of girth at least 5, $\gamma_c(G) \leq 2n/3 + O(1)$. The linear programming technique that was used to analyse the performance of the algorithms that were presented, also demonstrated the existence of infinitely many cubic graphs for which the algorithms only achieve these bounds. An example was given of a family of cubic graphs of girth at least 5 for which $\gamma_c(G) \geq 4n/7 - O(1)$.

We focus our attention on regular graphs that are generated u.a.r. (uniformly at random). We use the notation $\mathbf{P}$ (probability), $\mathbf{E}$ (expectation) and say that a property, $\mathcal{B} = \mathcal{B}_n$, of a random regular graph on n vertices holds a.a.s. (asymptotically almost surely) if $\lim_{n\to\infty} \mathbf{P}(\mathcal{B}_n)$=1.

As far as the authors are aware, the only bound known on $\gamma_c(G)$ for random regular graphs is due to Duckworth [5] who showed that for a random cubic graph, G, $\gamma_c(G)$ is a.a.s. less than $0.5854n$. This bound was achieved by using differential equations to analyse the performance of a randomised algorithm.

For a d-regular graph on n vertices, G, a trivial lower bound on $\gamma_c(G)$ may be derived by considering the degrees of the vertices in the spanning tree that has a set of internal vertices of size $\gamma_c(G)$. Let $\mathcal{I}$ denote this set of internal vertices. Note that all vertices of $\mathcal{I}$ have degree at most d in the tree. All other vertices in the tree have degree 1 and there are $n-1$ edges in the tree. This implies $d\mathcal{I} + n - \mathcal{I} \geq 2(n-1)$ hence, $\gamma_c(G) \geq n/(d-1)$ (asymptotically).

Alon [1] proved by probabilistic methods that, for n sufficiently large, the size of a smallest *dominating set* of a n-vertex graph with minimum degree d is at least $(1+o_d(1))(1+\ln(d+1))n/(d+1)$. Therefore, clearly, for such a graph, G, the same bound also holds for $\gamma_c(G)$. In fact, Caro, West and Yuster [3] showed that, for n-vertex graphs of minimum degree d, the size of a minimum connected dominating set is essentially the same as the size of a minimum dominating set, for n and d sufficiently large.

In this paper we consider a simple, yet efficient, randomised greedy algorithm for finding a small connected dominating set of regular graphs. We analyse the average-case performance of this heuristic on random regular graphs using differential equations. In this way we prove an upper bound on the size of a minimum connected dominating set of random regular graphs.

The columns UB in Table 1 summarise our results and give a.a. sure upper bounds on $\gamma_c(G)$ when G is a random d-regular graph on n vertices; these are the main results of this paper. For each d we also include the values $n/(d-1)$ (as a comparison to the trivial lower bound) and the value $\ln(d+1)n/(d+1)$ (as a comparison to the results of Alon [1]).

Table 1. Bounds on $\gamma_c(G)$ for a random d-regular graph, G

d	UB	$n/(d-1)$	$\frac{\ln(d+1)n}{(d+1)}$	d	UB	$n/(d-1)$	$\frac{\ln(d+1)n}{(d+1)}$
3	0.5854n	0.5000n	0.3466n	20	0.1493n	0.0526n	0.1450n
4	0.4565n	0.3333n	0.3219n	30	0.1121n	0.0345n	0.1108n
5	0.3860n	0.2500n	0.2986n	40	0.0910n	0.0256n	0.0906n
10	0.2397n	0.1111n	0.2180n	50	0.0771n	0.0204n	0.0771n

2 Random Graphs and Differential Equations

In this section we introduce the model we use to generate a regular graph u.a.r. and give an overview of an established method of analysing the performance of randomised algorithms.

The model we use was first described in its simplest form by Bollobás [2]. For a d-regular graph on n vertices, first take dn points in n buckets labelled $1 \ldots n$ with d points in each bucket and then choose u.a.r. a disjoint pairing of the dn points. The buckets represent the vertices of the randomly generated graph and each pair represents an edge whose end-points are given by the buckets of the points in the pair. With probability bounded below by a positive constant, loops and multiple edges do not occur (see, for example, Wormald [13, Section 2.2]).

One method of analysing the performance of a randomised algorithm is to use a system of differential equations to express the expected changes in variables describing the state of the algorithm during its execution. Wormald [14] gives an exposition of this method and Duckworth [4] applies this method to several graph-theoretic optimisation problems. In order to analyse an algorithm using a system of differential equations, we incorporate the algorithm as part of a pairing process that generates a random regular graph. In this way, we generate the random graph in the order that the edges are examined by the algorithm.

We may consider the generation process as follows. Initially, all vertices have degree zero. Throughout the execution of the generation process vertices will increase in degree until all vertices have degree d. Once the degree of a vertex reaches d, the vertex is said to be *saturated* and the generation is complete when all vertices are saturated. During the generation process, we refer to the graph being generated as the *evolving graph*.

During the generation of a random regular graph we choose the pairs sequentially. The first point, p_i, of a pair may be chosen by any rule but in order to ensure that the regular graph is generated u.a.r., the second point, p_j, of that pair must be selected u.a.r. from all the remaining free (i.e. unpaired) points. The freedom of choice of p_i enables us to select it u.a.r. from the vertices of a particular degree in the evolving graph and we say that selecting p_j u.a.r. from all the remaining free points denotes selecting a *mate* for p_i. Using $B(p_k)$ to denote the bucket that the point p_k belongs to, we say that the edge from $B(p_i)$ to $B(p_j)$ is *exposed* and that the vertex represented by $B(p_j)$ is *hit* by this exposed edge.

In what follows, we denote the set of vertices of degree i of the evolving graph, at time t, by $V_i = V_i(t)$ and let $Y_i = Y_i(t)$ denote $|V_i|$. (For such variables, in the remainder of the paper, the parameter t will be omitted where no ambiguity arises.) We can express the state of the evolving graph at any point during the execution of the algorithm by considering the variables Y_i where $0 \leq i \leq d-1$. In order to analyse a randomised algorithm for finding a small connected dominating set, $\mathcal{C}$, of regular graphs, we calculate the expected change in this state over a predefined unit of time in relation to the expected change in the size of $\mathcal{C}$. Let $C = C(t)$ denote $|\mathcal{C}|$ at any stage of an algorithm (time t) and let $\mathbf{E}\Delta X$ denote the expected change in a random variable X conditional upon the history of the process. We then use the equations representing $\mathbf{E}\Delta Y_i$ and $\mathbf{E}\Delta C$ to derive a system of differential equations. The solutions to these differential equations describe functions which represent the behaviour of the variables Y_i. Wormald [14, Theorem 6.1] describes a general result which guarantees that the solutions of the differential equations almost surely approximate the variables Y_i and C with error $o(n)$. The expected size of the connected dominating set may be deduced from these results.

3 A Simple Heuristic

In order to analyse our algorithm for finding a small connected dominating set, $\mathcal{C}$, of a random, n-vertex, d-regular graph, we combine the algorithm with a pairing process that u.a.r. generates the graph as described in Section 2. Recall that V_i denotes the set of vertices of current degree i in the evolving graph, G. For a vertex, $v \in V(G)$, $N(v)$ denotes the set of vertices incident with v in G along exposed edges and we let $deg(v) = |N(v)|$. The combined algorithm and pairing process is given in Figure 1; a description is given below.

We say that the combined algorithm and pairing process proceeds in *operations*. Each operation represents the process of selecting a vertex, u, for possible inclusion in $\mathcal{C}$ along with the exposing of a number of edges incident with u.

The first operation is unique in the sense that it is the only operation in which u is selected from V_0. From this point, until the completion of the algorithm, there will always exist a non-saturated vertex in the evolving graph that has degree strictly greater than zero. This follows from the well-known fact that random regular graphs are a.a.s. connected (see, for example, [13]). The first vertex of $\mathcal{C}$, u, is selected u.a.r. from V_0 and all its incident edges are exposed. Note that after the first operation, Y_1 is non-zero.

The remainder of the algorithm is divided into two stages denoted by the two while loops. The first stage represents the period of time after the first operation up until the first time Y_1 reaches zero. The second stage represents the remainder of the process.

For each operation in the first stage, we select a vertex, u, u.a.r. from V_1. We then expose one edge incident with u to a vertex v (say). If v now has degree 1, we add u to $\mathcal{C}$ and expose its remaining incident edges. Otherwise, the operation terminates without increasing the size of $\mathcal{C}$.

```
Select u u.a.r. from V_0;
C ← {u};
Expose all remaining edges incident with u;
    while (Y_1 > 0) do
        Select u u.a.r. from V_1;
        Expose an edge incident with u;
        if ({N(u) ∩ V_1} ≠ ∅) C ← C ∪ u;
                        Expose all remaining edges incident with u;
    enddo
    while (Y_0 > 0) do
        k ← min_{v∈V(G)} {deg(v) > 1};
        if (Y_1 > 0) Select u u.a.r. from V_1;
                     Expose k − 1 edges incident with u;
        else         Select u u.a.r. from V_k;
                     Expose an edge incident with u;
        if ({N(u) ∩ V_1} ≠ ∅) C ← C ∪ u;
                        Expose all remaining edges incident with u;
    enddo
```

Fig. 1. Combined algorithm and pairing process

Throughout the second stage, we let k denote the current minimum degree of all vertices that have non-zero degree. If $k = 1$, we select a vertex, u, u.a.r. from V_1 and expose $k-1$ edges incident with u. Otherwise, we select a vertex, u, u.a.r. from V_k and expose one edge incident with u. If any of the new neighbours of u now have degree 1, u is added to $\mathcal{C}$ and all remaining edges incident with u are exposed. Otherwise, the operation terminates without increasing the size of $\mathcal{C}$.

For operations that select u u.a.r. from V_1 during the second stage, the rationale behind exposing $k - 1$ edges incident with u is an attempt to control the current minimum degree of the vertices that have non-zero degree. Notice then how the current minimum degree of all vertices that have non-zero degree is increasing throughout the process.

At the start of the process all vertices have degree zero and the first vertex of $\mathcal{C}$ is chosen u.a.r. from V_0. Each subsequent vertex, u, chosen for possible addition to $\mathcal{C}$, is selected u.a.r. from those vertices of a particular non-zero degree. This ensures that $\mathcal{C}$ remains connected in G. Once u has been chosen, one or more edges incident with u are exposed. We add u to $\mathcal{C}$ if one or more neighbours of u, along these exposed edges, now has degree 1. This ensures $\mathcal{C}$ is dominating in G at the end of the process.

Several variations of this algorithm were considered. For reasons of brevity, we do not give the full details here but the algorithm presented gave the best performance of all the algorithms that we analysed.

4 Algorithm Analysis

In what follows, we refer to the combined algorithm and pairing process as the algorithm. The actual algorithm itself, to be run on a random regular graph once it has been generated, may be extracted from this process.

For each operation after the first, a non-saturated vertex is selected for possible addition to $\mathcal{C}$. We define a Type k operation to be one in which in which the chosen vertex is selected u.a.r. from V_k.

Once the first vertex of $\mathcal{C}$ has been chosen, and all its incident edges have been exposed, we split the remainder of the analysis into $d-1$ distinct ordered phases; Phase 1, Phase 2, ... , Phase $d-1$. We informally define Phase k as the period of time from the first Type k operation up to but not including the first Type k' operation where $k' > k$. Phase k terminates as soon as Y_k reaches zero.

We define a *clutch* to be a series of operations in Phase k from a Type k operation up to but not including the next Type k operation. Note that in Phase 1, a clutch consists of just one Type 1 operation but a clutch in later phases may consist of more than one operation. In particular, a clutch in later phases will consist of one Type k operation and possibly some Type 1 operations.

We calculate the expected change in the variables Y_i, $0 \leq i \leq d-1$, and the expected change in the size of $\mathcal{C}$ for a clutch in each Phase. These equations are then used to form a system of differential equations.

For each edge exposed in the evolving graph two points are chosen. The first is chosen u.a.r. from a given set and the second is chosen u.a.r. from all the remaining free points. Let s denote the number of free points available in all buckets at a given stage (time t). Note that $s = s(t) = \sum_{i=0}^{d-1}(d-i)Y_i$. For our analysis it is convenient to assume that $s > \epsilon n$ for some arbitrarily small but fixed $\epsilon > 0$. Later, we discuss the last steps of the algorithm, where $s \leq \epsilon n$.

The probability that, when selecting a free point u.a.r. from all the remaining free points (at time t), the point belongs to a vertex of degree j is P_j where

$$\mathrm{P}_j = \mathrm{P}_j(t) = \frac{(d-j)Y_j}{s}, \qquad 0 \leq j \leq d-1.$$

The expected change in Y_i due to changing the degree of a vertex from i to $i+1$ by exposing an edge *to* it (at time t) is $\rho_i + o(1) = \rho_i(t) + o(1)$ where $\rho_0 = -\mathrm{P}_0$ and $\rho_i = \mathrm{P}_{i-1} - \mathrm{P}_i$, $1 \leq i \leq d-1$. To justify this, note that when the second point in a pair is selected u.a.r., the number of free points in the buckets corresponding to vertices of degree i is $(d-i)Y_i$. Selecting one of these points causes Y_i to decrease. However, the number of free points in the buckets corresponding to vertices of degree $i-1$ (when $i > 0$) is $(d+1-i)Y_{i-1}$ and selecting one of these points causes Y_i to increase. For $i > 0$, these two quantities are added because expectation is additive. The $o(1)$ term is due to the fact that the values of all these variables may change by a constant during the course of the operation being examined. Since $s > \epsilon n$ the error is in fact $O(1/n)$.

Note that Phase 1 of the algorithm is unique in the sense that it is the only phase in which all operations are certain to be of the same type. We therefore treat this phase separately and then go on to treat the later phases in general.

In Phase 1 we select a vertex, u, u.a.r. from V_1 and expose an incident edge to a vertex v (say). The decision whether to add u to $\mathcal{C}$ depends on the degree of v and further edges may then be exposed. The expected change in Y_i when performing an operation in Phase 1 (at time t) is $\mathbf{E}\Delta Y_i + o(1)$ where

$$\mathbf{E}\Delta Y_i = \mathbf{E}\Delta Y_i(t) = -\delta_{i1} + \rho_i + \mathrm{P}_0(d-2)\rho_i + (1-\mathrm{P}_0)\delta_{i2}, \quad 0 \leq i \leq d-1, \quad (1)$$

in which δ_{ij} denotes the Kronecker delta function.

The expected change in the size of the connected dominating set when performing an operation in Phase 1 (at time t) is $\mathbf{E}\Delta\mathcal{C} + o(1)$ where

$$\mathbf{E}\Delta\mathcal{C} = \mathbf{E}\Delta\mathcal{C}(t) = \mathrm{P}_0, \quad (2)$$

as we only add u to $\mathcal{C}$ if v had degree zero at the start of the operation.

In Phase k, $2 \leq k \leq d-1$, there are two types of operation. For simplicity, we treat operations of Type 1 first and then combine the equations given by these operations with those for operations of Type k. This enables us to calculate the necessary equations for a clutch of operations.

When performing an operation of Type 1 in Phase k we select a vertex, u, u.a.r. from V_1 and expose $k-1$ incident edges. The probability that we hit a vertex of degree zero when these edges are exposed is given by

$$1 - (1 - \mathrm{P}_0)^{k-1} + o(1).$$

The expected change in Y_i when performing an operation of Type 1 in Phase k, $2 \leq k \leq d-1$, (at time t) is $\psi_1 + o(1) = \psi_1(t) + o(1)$ where

$$\psi_1 = -\delta_{i1} + (k-1)\rho_i + (1-(1-\mathrm{P}_0)^{k-1})(d-k)\rho_i + (1-\mathrm{P}_0)^{k-1}\delta_{ik}, \quad 0 \leq i \leq d-1.$$

For an operation of Type k in Phase k, we first select a vertex, u, u.a.r. from V_k and expose one of its remaining incident edges to a vertex v (say). The decision whether to add u to $\mathcal{C}$ depends upon the degree of v. The expected change in Y_i when performing an operation of Type k in Phase k (at time t) is $\psi_k + o(1)$ where

$$\psi_k = \psi_k(t) = -\delta_{ik} + \rho_i + \mathrm{P}_0(d-k-1)\rho_i + (1-\mathrm{P}_0)\delta_{ik+1}, \quad 0 \leq i \leq d-1.$$

The expected change in the size of the connected dominating set when performing an operation of Type k in Phase k is the probability that we hit a vertex of degree zero when the first edge is exposed and this is

$$\mathrm{P}_0 + o(1).$$

We say that a *birth* denotes the generation of a vertex in V_1 by performing an operation in Phase k, $2 \leq k \leq d-1$. The expected number of births from performing an operation of Type k (at time t) is $\nu_k + o(1)$ where

$$\nu_k = \nu_k(t) = \mathrm{P}_0(1 + (d-k-1)\mathrm{P}_0).$$

With probability that the first edge is exposed to a vertex of degree zero we have one birth and, should this happen, we add the probability that other vertices of degree zero are hit when the remaining edges are exposed.

Similarly, the expected number of births from performing an operation of Type 1 (at time t) is $\nu_1 + o(1)$ where

$$\nu_1 = \nu_1(t) = (k-1)\mathrm{P}_0 + (1-(1-\mathrm{P}_0)^{k-1})(d-k)\mathrm{P}_0.$$

Consider the Type k operation at the start of the clutch to be the first generation of a *birth-death* process in which the individuals are the vertices in V_1, each giving birth to a number of children (essentially independent of the others) with expected number ν_1. Then, the expected number in the jth generation is $\nu_k {\nu_1}^{j-1}$ and the expected number of births in the clutch is $\nu_k/(1-\nu_1) + o(1)$.

For a clutch of operations in Phase k, $2 \le k \le d-1$, we have one Type k operation and for each birth we have a Type 1 operation (a.a.s.). The expected change in Y_i for a clutch in Phase k (at time t) is $\mathbf{E}\Delta Y_i + o(1)$ where

$$\mathbf{E}\Delta Y_i = \mathbf{E}\Delta Y_i(t) = \psi_k + \frac{\nu_k}{1-\nu_1}\psi_1, \qquad 0 \le i \le d-1. \tag{3}$$

The expected change in the size of the connected dominating set for a clutch in Phase k (at time t) is $\mathbf{E}\Delta C + o(1)$ where

$$\mathbf{E}\Delta C = \mathbf{E}\Delta C(t) = \mathrm{P}_0 + \frac{\nu_k}{1-\nu_1}(1-(1-\mathrm{P}_0)^{k-1}). \tag{4}$$

The Type k operation at the start of the clutch contributes an increase in the size of $\mathcal{C}$ if a vertex of degree zero is hit when the first edge is exposed. For each birth, the accompanying Type 1 operation contributes an increase in the size of $\mathcal{C}$ if a vertex of degree zero is hit when the first $k-1$ edges are exposed.

The combined algorithm and pairing process is analysed using differential equations and this way we prove the following theorem.

Theorem 1. *For each $d \ge 3$, where d remains constant, there exists a constant, c, such that for a random d-regular graph on n vertices, the size of a minimum connected dominating set is a.a.s. at most $cn + o(n)$.*

Proof. For Phase 1, Equation (1) representing the expected change in the variables Y_i, $0 \le i \le d-1$, for an operation in Phase 1 (since a clutch consists of just one operation in this phase) forms the basis of a differential equation. Write $Y_i(t) = nz_i(t/n)$, $s(t) = n\xi(t/n)$, $\mathrm{P}_j(t) = n\bar{\mathrm{P}}_j(t/n)$ and $\rho_i(t) = n\bar{\rho}_i(t/n)$. The differential equation suggested is

$$\frac{dz_i}{dx} = -\delta_{i1} + \bar{\rho}_i + \bar{\mathrm{P}}_0(d-2)\bar{\rho}_i + (1-\bar{\mathrm{P}}_0)\delta_{i2}, \qquad 0 \le i \le d-1 \tag{5}$$

where xn represents the number, t, of operations.

Using Equation (2) representing the expected increase in C for an operation in Phase 1 and writing $C(t) = nz(t/n)$ suggests the differential equation for z as

$$\frac{dz}{dx} = \bar{\mathrm{P}}_0. \tag{6}$$

For Phase k, $2 \leq k \leq d-1$, Equation (3) representing the expected change in the variables Y_i for a clutch of operations forms the basis of a differential equation. Write $\psi_k(t) = \bar{\psi}_k(t/n)$, $\psi_1(t) = \bar{\psi}_1(t/n)$, $\nu_k(t) = \bar{\nu}_k(t/n)$, $\nu_1(t) = \bar{\nu}_1(t/n)$, $\mathrm{P}_j(t) = n\bar{\mathrm{P}}_j(t/n)$ and $\rho_i(t) = n\bar{\rho}_i(t/n)$. The differential equation suggested is

$$\frac{dz_i}{dx} = \bar{\psi}_k + \frac{\bar{\nu}_k}{1-\bar{\nu}_1}\bar{\psi}_1, \qquad 0 \leq i \leq d-1, \tag{7}$$

where xn represents the number, t, of clutches.

Using Equation (4) representing the expected increase in C for a clutch in Phase k and writing $C(t) = nz(t/n)$ suggests the differential equation for z as

$$\frac{dz}{dx} = \bar{\mathrm{P}}_0 + \frac{\bar{\nu}_k}{1-\bar{\nu}_1}(1-(1-\bar{\mathrm{P}}_0)^{k-1}). \tag{8}$$

The solution to these systems of differential equations represent the cardinalities of the sets V_i and $\mathcal{C}$ (scaled by $1/n$) for given t. The initial conditions for Phase 1 are $z_0(0) = 1$, $z(0) = 0$ and $z_i(0) = 0$ where $1 \leq i \leq d-1$.

Wormald [14, Theorem 6.1] describes a general result that we may use to show that the functions representing the solutions to the differential equations almost surely approximate the variables Y_i/n and C/n with error $o(1)$.

For Phase 1 and for arbitrary small ϵ, define R to be the set of all (t, z_i, z) for which $t > -\epsilon$, $\xi > \epsilon$, $z_1 > \epsilon$, $z > -\epsilon$ and $z_i < 1+\epsilon$ where $0 \leq i \leq d-1$. Then, R defines a domain for the process so that [14, Theorem 6.1] may be applied.

It is simple to verify that parts (*i*), (*ii*) and (*iii*) of [14, Theorem 6.1] hold, however, note in particular that since $\xi > \epsilon$ inside R, the assumption that $s > \epsilon n$ used in deriving these equations is justified. The conclusion of [14, Theorem 6.1] therefore holds. This implies that, until a point arbitrarily close to where they leave R, the random variables Y_i/n and C/n a.a.s. remain within $o(1)$ of the corresponding deterministic solutions to the differential equations.

We compute the ratio dz_i/dz and we have

$$\frac{dz_i}{dz} = \frac{-\delta_{i1} + \bar{\rho}_i + \bar{\mathrm{P}}_0(d-2)\bar{\rho}_i + (1-\bar{\mathrm{P}}_0)\delta_{i2}}{\bar{\mathrm{P}}_0}, \qquad 0 \leq i \leq d-1$$

where differentiation is with respect to z and all functions may be taken as functions of z.

By solving this we find that the solution hits a boundary of the domain at $z_1 = \epsilon$. At this point, we formally define Phase 1 as the period of time from time t=0 to the time t_1 such that $z = t_1/n$ is the solution of $z_1 = \epsilon$.

From the point in Phase 1 after which [14, Theorem 6.1] does not apply until the start of Phase 2, the change in each variable per step is bounded by a constant, hence in $o(n)$ steps, the change in the variables is $o(n)$.

The initial conditions for Phase k, $2 \leq k \leq d-1$, are given by the final conditions for Phase $k-1$ and the equations are given by (7) and (8). We apply Theorem 6.1 from [14] to the process within Phase k. For arbitrary small ϵ, define R to be the set of all (t, z_i, z) for which $t > -\epsilon$, $\xi > \epsilon$, $z_k > \epsilon$, $z > -\epsilon$ and $z_i < 1+\epsilon$ where $0 \leq i \leq d-1$.

We compute the ratio dz_i/dz and we have

$$\frac{dz_i}{dz} = \frac{\bar{\psi}_k + \frac{\bar{\nu}_k}{1-\bar{\nu}_1}\bar{\psi}_1}{\bar{\mathrm{P}}_0 + \frac{\bar{\nu}_k}{1-\bar{\nu}_1}(1-(1-\bar{\mathrm{P}}_0)^{k-1})}, \qquad 0 \le i \le d-1.$$

By solving these differential equations, for each phase, we find that the solution hits a boundary of the domain at $z_k = \epsilon$. At this point, we formally define Phase k as the period of time from time t_{k-1} to the time t_k such that $z = t_{k-1}/n$ is the solution of $z_{k-1} = \epsilon$ and $z = t_k/n$ is the solution of $z_k = \epsilon$.

From the point in Phase $d-1$ after which [14, Theorem 6.1] does not apply until the end of the algorithm, the change in each variable per step is bounded by a constant, hence in $o(n)$ steps, the change in the variables is $o(n)$.

The equations were solved (numerically) using a Runge-Kutta method. The solution of $z_{d-1} = \epsilon$ in Phase $d-1$ corresponds to the size of the connected dominating set (scaled by $1/n$) when no vertices remain, thus proving the theorem.

References

1. Alon, N.: Transversal Numbers of Uniform Hypergraphs. Graphs and Combinatorics **6**(1) (1990) 1–4
2. Bollobás, B.: Random Graphs. Academic Press (1985)
3. Caro, T., West, D.B. and Yuster, R.: Connected Domination and Spanning Trees with Many Leaves. SIAM Journal on Discrete Mathematics **13**(2) (2000) 202–211
4. Duckworth, W.: Greedy Algorithms and Cubic Graphs. PhD Thesis, Department of Mathematics and Statistics, The University of Melbourne (2001)
5. Duckworth, W.: Minimum Connected Dominating Sets of Random Cubic Graphs. The Electronic Journal of Combinatorics, **9**(1) (2002) #R7
6. Duckworth, W. and Wormald, N.C.: Linear Programming and the Worst-Case Analysis of Greedy Algorithms on Cubic Graphs. *Submitted*
7. Galbiati, G., Maffioli, F. and Morzenti, A.: A Short Note on the Approximability of the Maximum Leaves Spanning Tree Problem. Information Processing Letters **52**(1) (1994) 45–49
8. Griggs, J.R., Kleitman, D.J. and Shastri, A.: Spanning Trees with Many Leaves in Cubic Graphs. Journal of Graph Theory **13**(6) (1989) 669–695
9. Griggs, J.R. and Wu, M.: Spanning Trees in Graphs of Minimum Degree 4 or 5. Discrete Mathematics **104**(2) (1992) 167–183
10. Haynes, T.W., Hedetniemi, S.T. and Slater, P.J.: Domination in Graphs: Advanced Topics. Marcel Dekker Inc., New York (1998)
11. Solis-Oba, R.: 2-Approximation Algorithm for Finding a Spanning Tree with Maximum Number of leaves. In: Proceedings of the 17th European Symposium on Algorithms, Springer (1998) 441–452
12. Storer, J.A.: Constructing Full Spanning Trees for Cubic Graphs. Information Processing Letters **13**(1) (1981) 8–11
13. Wormald, N.C.: Models of Random Regular Graphs. In: Surveys in Combinatorics, Cambridge University Press (1999) 239–298
14. Wormald, N.C.: The Differential Equation Method for Random Graph Processes and Greedy Algorithms. In: Lectures on Approximation and Randomized Algorithms, PWN (1999) 73–155

On the Number of Minimum Cuts in a Graph

L. Sunil Chandran* and L. Shankar Ram

Department of Computer Science and Automation, Indian Institute of Science, Bangalore, 560012, India
{sunil,shankar}@csa.iisc.ernet.in

Abstract. We relate the number of minimum cuts in a weighted undirected graph with various structural parameters of the graph. In particular, we upper-bound the number of minimum cuts in terms of the radius, diameter, minimum degree, maximum degree, chordality, expansion, girth etc. of the graph.

1 Introduction

Let $G = (V, E)$ be a graph or a multi-graph with positive weights on its edges. We denote $|V|$ by n and $|E|$ by m. By an unweighted graph, we mean that all the edges have unit weight. Let $(A, \overline{A})$ denote a cut of G, defined by the subsets $A \subset V$ and $\overline{A} = V - A$. We denote by $E(A, \overline{A})$, the set of edges in the cut, i.e., $E(A, \overline{A}) = \{(u, v) \in E : u \in A \textit{ and } v \in \overline{A}\}$. The weight of the cut $(A, \overline{A})$ is defined as the sum of weights on all the edges in $E(A, \overline{A})$, and will be denoted by $w(A, \overline{A})$. A minimum cut $(S, \overline{S})$ is one with the minimum weight over all cuts in G. (Some authors use the words *global minimum cuts* or *connectivity cuts* instead of minimum cuts). We will denote the weight of the minimum cut in G by $\lambda(G)$. Note that if G is unweighted, $\lambda(G)$ is same as the edge connectivity of the graph, i.e., the minimum number of edges whose removal disconnects the graph.

Note that the minimum cut in a graph may not be unique. We use $\Lambda(G)$ to denote the number of minimum cuts in G. The problem of counting the number of minimum cuts in a weighted undirected graph arises in various aspects of network reliability, like testing the super–λ–ness of a graph [2], estimating the probabilistic connectedness of a stochastic graph in which edges are subject to failure with probability p [4,5,6,16] and other areas [17]. For example, for a sufficiently small p, the probabilistic connectedness of G can be approximated as $P(G, p) \approx 1 - \Lambda(G)p^{\lambda(G)}(1-p)^{|E|-\lambda(G)}$, suggesting the importance of counting and bounding $\Lambda(G)$.

It is well known that for any weighted graph G, $\Lambda(G) \le \binom{n}{2}$ and this upper bound is achieved if G is a cycle C_n of n nodes with each edge having weight $\frac{\lambda(G)}{2}$ [7,1,11]. It is interesting to explore whether there exist tighter bounds for $\Lambda(G)$ when the graph satisfies various properties. For example, Bixby [1] studies $\Lambda(G)$ in terms of the weight of the minimum cuts $\lambda(G)$, in the special case where

* This research is supported in part by the Infosys Fellowship.

O.H. Ibarra and L. Zhang (Eds.): COCOON 2002, LNCS 2387, pp. 220–229, 2002.

all the edge weights are positive integers and $\lambda(G)$ is an odd integer. For this case, Bixby [1] shows that $\Lambda(G) \leq \lfloor \frac{3n}{2} \rfloor - 2$. In the case of unweighted simple graphs it is shown by Lehel, Maffray and Priessmann [12] that if $\lambda(G) = k$ where $k \geq 4$ is an even positive integer, then $\Lambda(G) \leq \frac{2n^2}{(k+1)^2} + \frac{(k-1)n}{k+1}$. When $k > 5$ is an odd integer, they show that $\Lambda(G) \leq (1 + \frac{4}{k+5})n$. The inherent structural difference between graphs with odd and even edge connectivity was pointed out by Kanevsky [13] also.

In this paper, we provide upper bounds for $\Lambda(G)$ in terms of many other important parameters of graphs. We assume weighted graphs. Multi-graphs, as far as the results here are concerned, can be considered as a special case of weighted graphs, since the multi-edges can be replaced by a single edge of appropriate weight without affecting the value of $\Lambda(G)$. Our only assumption about the weights is that they are positive. Note that, for the purposes of this paper, this assumption is equivalent to the assumption that the weights are at least 1, since multiplying the weights on every edge by the same constant will not change $\Lambda(G)$. While our upper bounds are valid for weighted undirected graphs and multi graphs, in most cases, the properties in terms of which the upper bounds are stated depend only on the structure of those graphs. In other words, the radius or minimum degree in terms of which we describe the upper bounds are those of the underlying unweighted simple graph and do not depend on the weights of the edges.

There is an abundance of literature regarding the determination of $\lambda(G)$ and finding a minimum cut in G. The problem of enumerating all the minimum cuts is considered by many authors [7,15,10,8], and various structures (for example, the cactus structure) are invented to efficiently represent all the minimum cuts in a graph. The fact that the performance of some of these algorithms depends on the number of minimum cuts in the graph also makes it interesting to look for tighter upper bounds for $\Lambda(G)$ when G satisfies certain properties. (For example a randomized algorithm due to Karger builds a data structure that represents all minimum cuts in $O(\Lambda(G) + n \log n)$ space). See [10] for a brief survey of results regarding the enumeration of all minimum cuts.

The slightly different question of upper–bounding the number of approximate minimum cuts– i.e., those cuts having weight at most $f\lambda(G)$, where $f > 1$ is a constant, is considered in [18,11,14,19] etc. For example, Karger [11], using probabilistic analysis shows that there are at most $O(n^{2f})$ cuts of the above kind in a graph of n nodes. Nagamochi et.al. [14] show that the number of cuts of weight at most $\frac{4}{3}\lambda(G)$ is upper–bounded by $\binom{n}{2}$. Williamson and Henzinger [19] show an upper bound of $O(n^2)$ for the number of cuts of weight at most $\frac{3}{2}\lambda(G)$, extending the arguments of [14].

1.1 Our Results

Radius and Diameter: If $G = (V, E)$ is a connected graph, the *eccentricity* of a node $v \in V$ is defined as $e(v) = \max distance(v, u)$ over all the nodes $u \in V$. The radius of the graph G, $r(G) = \min_{v \in V} e(v)$. A vertex v is a central node if $e(v) = r(G)$. Diameter of G, $d(G) = max_{v \in V} e(v)$. (Note that, in this paper, by "distance", we mean only the distances in the underlying unweighted graph. Thus radius, eccentricity, diameter etc. have nothing to do with the weights). We show that $\Lambda(G) \le (r+1)n - (2r+1) \le (d+1)n - (2d+1)$, where G is a weighted graph and r, d are the radius and diameter of G. As a special case, we observe that if there is a node which is a neighbour of every other node in the graph, i.e., if $r(G) = 1$, then $\Lambda(G) \le 2n - 3$. We illustrate the tightness of this bound by constructing a weighted clique $\mathcal{K}_n$ for which $\Lambda(\mathcal{K}_n) = 2n - 3$.

Minimum and Maximum Degree: Let the minimum degree and maximum degree of G be δ and Δ respectively. (Note that minimum and maximum degrees have nothing to do with the weights, i.e., $\delta = \min_{u \in V} |N(u)|$ and $\Delta = \max_{u \in V} |N(u)|$, $N(u)$ being the set of neighbours of the node u). We show that $\Lambda(G) \le (\frac{3n}{2(\delta+1)} + 1.5)n - (\frac{3n}{\delta+1} + 2)$ and $\Lambda(G) \le \frac{(n-\Delta+3)n}{2} - (n - \Delta + 2)$. Note that these bounds become significant when the concerned parameters are reasonably large. Also it is easy to get an upper bound involving both δ and Δ, by extending the techniques discussed in the paper.

Chordality: Let C be a simple cycle of a weighted undirected graph G. Any edge in the induced subgraph on the nodes of C, $G(C)$, other than the cycle edges themselves is called a chord of C. C is called an induced cycle (or chordless cycle), iff C does not have any chords. The length of the largest induced cycle in a graph G is called chordality of G. A graph G is called k–chordal iff the chordality of G is at most k. $\Lambda(G) \le \frac{(k+1)n}{2} - k$, where k is the chordality of the underlying unweighted (simple) graph, corresponding to G. We also show the tightness of the bound by exhibiting a k-chordal graph G for arbitrarily large n, such that $\Lambda(G) = \frac{(k+1)n}{2} - k$.

The word "chordality" originates from the well known subclass of perfect graphs, the chordal graphs. A graph G is chordal iff there is no induced cycle of length 4 or more in G. We define the chordality of a chordal graph to be 3. All graphs other than chordal graphs have chordality ≥ 4. Some other important classes of graphs with low chordality value are the cocomparability graphs, chordal bipartite graphs and weakly chordal graphs, all of which are known to be 4–chordal.

Note that C_n (the cycle on n nodes) is the graph with maximum chordality amongst all graphs on n nodes. Also, it is a graph which contains the maximum number of minimum cuts possible, namely $\binom{n}{2}$. (In fact our bound given above, shows that C_n with each edge having weight $\frac{\lambda}{2}$ is the *only* graph which contains $\binom{n}{2}$ minimum cuts, the weight of the minimum cut being λ). The fact that the maximum value of $\Lambda(G)$ is achieved by the graph of largest chordality motivates a study of the influence of chordality on $\Lambda(G)$.

Edge Expansion: Let $E(X, \overline{X})$ be the set of edges in the cut $(X, \overline{X})$. Then the edge expansion $\beta_e(G)$ is defined as $\beta_e(G) = \min \frac{|E(X,\overline{X})|}{|X|}$, over all subsets X of V with $|X| \le \frac{n}{2}$. We define the weighted edge expansion, $\beta_w(G) = \min \frac{w(X,\overline{X})}{|X|}$ over all subsets X of V, such that $|X| \le \frac{n}{2}$. A related concept is that of vertex expansion (or simply the expansion), $\beta_v(G)$, which is defined as $\beta_v(G) = min\frac{|N(X)|}{|X|}$ over all subsets X of V, such that $|X| \leq \frac{n}{2}$ where $N(X)$ denotes the set of neighbours of X, namely the set $\{u \in V - X : (u,v) \in E \; for \; some \; v \in X\}$.

We show that $\Lambda(G) \leq \frac{(\frac{\lambda}{\beta}+3)}{2}n - (\frac{\lambda}{\beta}+2)$ where $\beta = \beta_w(G)$. It is easy to note that (since we can assume without any loss of generality that all weights are at least 1) $\beta_w(G) \geq \beta_e(G) \geq \beta_v(G)$. Thus we can substitute $\beta_v(G)$ or $\beta_e(G)$ in the place of β in the above formula.

Expansion is a concept which has found application in areas ranging from complexity theory to coding theory. A further motivation for looking for a bound in terms of the expansion properties of the graph is the well known fact that *almost all graphs* are good expanders, i.e., if we randomly construct a graph by selecting each edge with certain probability p (greater than a threshold value), the probability that the expansion of the resulting graph being bounded below by a constant tends to 1 as $n \to \infty$.

Girth: Girth is the length of the smallest (induced) cycle in G. The intuition that expansion properties can control the number of minimum cuts in a graph, leads us to a bound in terms of girth also. We show that if G is an unweighted graph with girth g and minimum degree δ, the $\Lambda(G) < (\frac{n}{x+1}+1)n - (\frac{2n}{x+1}+1)$, where $x = \left(\frac{\delta-3}{2}\right)^{\frac{g-3}{2}}$. Note that this is in contrast with the bound in terms of chordality, the length of the largest induced cycle.

2 Preliminaries

Consider an undirected graph $G = (V,E)$ with a weight function $w : E \to \Re^+$. Let U and W be disjoint subsets of V. Let $E(U,W) = \{(u,v) \in E : u \in U, v \in W\}$, be the set of edges between the vertices in U and the vertices in W. Also, let $w(U,W)$ be the sum of the weights on the edges in $E(U,W)$. We denote the induced subgraph on the set of nodes U by $G(U)$. The set of neighbours of a node $u \in V$, will be denoted by $N(u)$, i.e., $N(u) = \{v \in V : (u,v) \in E\}$. Also $N(U) = \bigcup_{u \in U} N(u) - U$. As mentioned in the introduction, $\lambda(G)$ denotes the weight of a minimum cut and $\Lambda(G)$ denotes the number of minimum cuts in G. (The omitted proofs will be available in the full paper).

Lemma 1. *If $(S,\overline{S})$ is a minimum cut of a connected undirected graph G, then $G(S)$ and $G(\overline{S})$ are connected.*

Definition 1. *Let $(X,\overline{X})$ and $(Y,\overline{Y})$ be two cuts in a weighted undirected graph. $(X,\overline{X})$ and $(Y,\overline{Y})$ are said to cross each other iff all the four sets $X \cap Y$, $X \cap \overline{Y}$, $\overline{X} \cap Y$ and $\overline{X} \cap \overline{Y}$ are non empty. Then $(X,\overline{X})$ and $(Y,\overline{Y})$ are called a crossing pair of cuts.*

Lemma 2. *A pair of cuts $(S,\overline{S})$ and $(P,\overline{P})$ does not cross if and only if S(or $\overline{S}$) is a subset of P or $\overline{P}$. (i.e., $S \subseteq P$, $S \subseteq \overline{P}$, $\overline{S} \subseteq P$ or $\overline{S} \subseteq \overline{P}$)*

Proof. Follows from the definition of a crossing pair of cuts. ■

Lemma 3 (Bixby [1], Dinitz, Karzanov and Lomosonov [7]). *Let $(X, \overline{X})$ and $(Y, \overline{Y})$ be a crossing pair of minimum cuts in a weighted undirected graph G. Let $A = X \cap Y$, $B = \overline{X} \cap Y$, $C = X \cap \overline{Y}$ and $D = \overline{X} \cap \overline{Y}$. Then,*

1. $w(A, B) = w(B, D) = w(D, C) = w(C, A) = \frac{\lambda(G)}{2}$
2. $w(A, D) = w(B, C) = 0$. *That is,* $E(A, D) \cap E(B, C) = \emptyset$

Lemma 4. *If $(P, \overline{P})$ and $(S, \overline{S})$ are a crossing pair of minimum cuts, then $E(P, \overline{P}) \cap E(S, \overline{S}) = \emptyset$.*

Definition 2. *A circular partition $\mathcal{C} = (U_0, U_1, U_2, \cdots, U_{k-1})$ of the vertices of a graph G is a partition of the set of vertices V of G into disjoint non-empty subsets $U_0, U_1, \cdots, U_{k-1}$ such that*

1. $w(U_i, U_{i+1 \bmod k}) = \frac{\lambda(G)}{2}$, *for* $0 \leq i \leq k-1$.
2. $w(U_i, U_j) = 0$, *if* $i \neq j+1 \bmod k$ *or* $i \neq j-1 \bmod k$, *i.e.,* $E(U_i, U_j) = \emptyset$.
3. *For* $0 \leq i \leq k-1$, *the cut* $(U_i, \overline{U}_i)$ *– which is a minimum cut by conditions 1 and 2 – does not cross with any other minimum cut* $(A, \overline{A})$ *in* G.

Definition 3. *A cut $(A, \overline{A})$ is called a union cut with respect to a circular partition $\mathcal{C} = (U_0, U_1, \cdots, U_{k-1})$ iff $A = \bigcup_{j=i}^{i+b-1 \bmod k} U_j$ where $2 \leq b \leq k-2$. (Note that both A and $\overline{A}$ contain at least 2 subsets in $\mathcal{C}$).*

The cut $(A, \overline{A})$ is called a subset cut with respect to $\mathcal{C}$ iff $A \subseteq U_i$ or $\overline{A} \subseteq U_i$ for some i.

Lemma 5. *Let $\mathcal{C} = (U_0, U_1, \cdots, U_{k-1})$ be a circular partition of G. Then any minimum cut $(S, \overline{S})$ of G is either a union cut or a subset cut with respect to $\mathcal{C}$. Moreover, every union cut with respect to $\mathcal{C}$ is a minimum cut in G.*

Lemma 6. *Let G be a weighted undirected graph. Then G has a circular partition $\mathcal{C} = (U_0, U_1, \cdots, U_{k-1})$, where $k \geq 4$, if and only if there exists a crossing pair of minimum cuts in G.*

For a circular partition $\mathcal{C}$ of G, let the *partition number* $p(\mathcal{C})$ be defined as the number of subsets in $\mathcal{C}$. We define the *partition number of the graph G* as follows.

Definition 4. *The partition number $p(G)$ of a graph G is defined as $p(G) = 3$, if there is no circular partition for G. Otherwise, $p(G) = \max p(\mathcal{C})$, over all circular partitions $\mathcal{C}$ of G.*

Note that if there is a crossing pair of minimum cuts in G, then $p(G) \geq 4$, by Lemma 6. Otherwise, $p(G) = 3$.

Definition 5. *By contraction of a subset of vertices $X \subset V$, we mean replacing all the vertices in X by a single vertex x and adding the edges (y, x) for each $y \in N(X)$. The weight of the edge (y, x) (where $y \in N(X)$) is assigned to be $w(y, x) = \sum_{z \in X} w(y, z)$, where $(y, z) \in E(G)$. We denote the graph obtained after the contraction operation by G/X. We will refer to the operation of undoing the effect of a contraction (i.e., restoring G from G/X) by putting back X in the place of x as expanding the node x.*

Lemma 7. *If $(S, \overline{S})$ is a minimum cut in a weighted undirected graph G such that no other minimum cut $(A, \overline{A})$ crosses with $(S, \overline{S})$, then $\Lambda(G) = \Lambda(G/S) + \Lambda(G/\overline{S}) - 1$.*

Lemma 8. *If there are no crossing pairs of minimum cuts in G, then $\Lambda(G) \leq 2n - 3$. Moreover, there exists a graph on n nodes, G_n, (for every $n \geq 2$), such that $\Lambda(G_n) = 2n - 3$.*

3 The Partition Number and the Structural Parameters

3.1 Partition Number, $p(G)$

Lemma 9. *Let G be a weighted undirected graph. If $(X, \overline{X})$ is a minimum cut of G such that no other minimum cut crosses with $(X, \overline{X})$, then $p(G/X) \leq p(G)$.*

In the following Lemma, we upper–bound $\Lambda(G)$ in terms of the partition number. The tightness of the Lemma will be established in Theorem 8.

Lemma 10. *Let $G = (V, E)$ be a weighted undirected graph, where $|V| = n \geq 2$ and let the partition number $p(G) \leq p$. Then, $\Lambda(G) \leq \frac{(p+1)n}{2} - p$.*

In the rest of the paper, we show that various structural parameters of a graph can influence the partition number $p(G)$. Thus by means of Lemma 10, we relate the number of minimum cuts, $\Lambda(G)$, with many seemingly unrelated properties of the graph.
Remark: Please note that if $n \geq 2$ and $x \geq p$, then $\frac{(x+1)n}{2} - x \geq \frac{(p+1)n}{2} - p$. In most of the theorems below, we show that p is upper–bounded by a function $f(y)$ of y where y is some parameter of G, thereby showing that $\Lambda(G) \leq \frac{(f(y)+1)n}{2} - f(y)$.

3.2 Radius and Diameter

If $G = (V, E)$ is a connected graph, the *eccentricity* of a node $v \in V$ is defined as $e(v) = \max distance(v, u)$ over all the nodes $u \in V$. The radius of the graph G, $r(G) = \min_{v \in V} e(v)$. A vertex v is a central node if $e(v) = r(G)$. Diameter of G, $d(G) = max_{v \in V} e(v)$. (Note that, in this paper, by "distance", we mean only the distances in the underlying unweighted graph. Thus radius, eccentricity, diameter etc. have nothing to do with the weights).

Theorem 1. *If r is the radius of a weighted undirected graph G, then $\Lambda(G) \leq (r+1)n - (2r+1)$ (where $n \geq 2$).*

Proof. Suppose there are no crossing pairs of minimum cuts in G. If follows by Lemma 8 that $\Lambda(G) \leq 2n - 3$. Since the radius is at least 1, it is easy to verify that $\Lambda(G) \leq (r+1)n - (2r+1)$ in this case. Otherwise, by Lemma 6, there exists a circular partition $\mathcal{C} = (U_0, U_1, \cdots, U_{p-1})$ for G, where $p = p(G) \geq 4$. Let $x \in U_i$

be a central node of G. Let $y \in U_{i+\lfloor \frac{p}{2} \rfloor \bmod p}$. Clearly, $distance(x, y) \geq \lfloor \frac{p}{2} \rfloor$. That is, $r \geq \lfloor \frac{p}{2} \rfloor$ or $p \leq 2r+1$. Now, by Lemma 10, we get $\Lambda(G) \leq (r+1)n - (2r+1)$. ■

We note that the bound given by the above Theorem can be tight. For example, consider C_{2n+1}, the cycle on $2n+1$ nodes. Clearly, the radius of C_{2n+1} is n and the number of minimum cuts $= \binom{2n+1}{2} = (n+1)(2n+1) - (2n+1)$.

Observe that similar arguments as given for the case of radius hold good for the diameter also. Thus,

$$\Lambda(G) \leq (d+1)n - (2d+1)$$

This can also be verified from $\Lambda(G) \leq (r+1)n - (2r+1) \leq (d+1)n - (2d+1)$ by noting that $d \geq r$ and $n \geq 2$.

3.3 Politician Node

An interesting special case of Theorem 1 is when $radius(G) = 1$. Then, there exists a node which is adjacent to every other node of the graph. (Such a node is called a *politician* node). Thus, if there is a politician node in the graph, then $\Lambda(G) \leq 2n - 3$, by Theorem 1. In fact more is true.

Theorem 2. *If there is a politician node u in G, then there cannot be any crossing pairs of minimum cuts in G.*

Proof. If there is a crossing pair of minimum cuts, then by Lemma 6, there is a circular partition $\mathcal{C} = (U_0, U_1, \cdots, U_{k-1})$ $(k \geq 4)$. Without loss of generality let $u \in U_0$. Clearly, u cannot be adjacent to any node in U_2, by the definition of circular partition, contradicting the assumption that u is a politician node. ■

Note that in a complete graph, K_n, every node is a politician node. Thus, there are no crossing pairs of minimum cuts in a clique. Below, we show that the number of minimum cuts $\Lambda(K_n) = 2n - 3$, thus illustrating that the bound of Lemma 8 is tight. Moreover, since radius of a clique is 1, this is a tight example for Theorem 1 too. Since a complete graph is a chordal graph, the example below illustrates the tightness of Theorem 7 also.

Theorem 3. *For any $n \geq 2$ and $\lambda > 0$, there exists a weighted complete graph $\mathcal{K}_n$ such that $\lambda(\mathcal{K}_n) = \lambda$ and $\Lambda(\mathcal{K}_n) = 2n - 3$. Moreover every node x of $\mathcal{K}_n$ defines a minimum cut $(\{x\}, \overline{\{x\}})$ of $\mathcal{K}_n$.*

3.4 Maximum and Minimum Degree

The maximum degree $\Delta(G)$ (when it is reasonably high) can also constrain the number of minimum cuts $\Lambda(G)$.

Theorem 4. *If Δ is the maximum degree of a weighted undirected graph G, then $\Lambda(G) \leq \frac{(n-\Delta+3)n}{2} - (n - \Delta + 2)$, where $n \geq 2$.*

Proof. Suppose there is no crossing pairs of minimum cuts in G, then by Lemma 8, $\Lambda(G) \leq 2n-3 \leq \frac{(n-\Delta+3)n}{2} - (n-\Delta+2)$, which will be true if $0 \leq n^2-(\Delta+3)n+(2\Delta+2)$ or $0 \leq (n-\Delta-1)(n-2)$ which is true since $n \geq 2$ and $\Delta \leq n-1$. Now, if there is a crossing pair of minimum cuts in G, then by Lemma 6, there is a circular partition $\mathcal{C} = (U_0, U_1, \cdots, U_{p-1})$ ($p = p(G) \geq 4$). Without loss of generality , let the maximum degree node $u \in U_1$. Then, $|U_0 \cup U_1 \cup U_2| \geq \Delta + 1$ since every neighbour of u must be in U_0, U_1 or U_2. Thus, $p \leq 3 + (n - \Delta - 1) = n - \Delta + 2$ since each U_i of the circular partition must contain at least 1 node. By Lemma 10, $\Lambda(G) \leq \frac{(n-\Delta+3)n}{2} - (n - \Delta + 2)$. ■

Interestingly, the minimum degree of the graph can also control the number of minimum cuts.

Theorem 5. *If δ is the minimum degree of a weighted undirected graph G, then $\Lambda(G) \leq (\frac{3n}{2(\delta+1)} + 1.5)n - (\frac{3n}{\delta+1} + 2)$, where $n \geq 2$.*

Proof. If there are no crossing pairs of minimum cuts in G, it can be easily verified that $\Lambda(G) \leq 2n - 3 \leq (\frac{3n}{2(\delta+1)} + 1.5)n - (\frac{3n}{\delta+1} + 2)$ for $n \geq 2$. Otherwise consider a circular partition $\mathcal{C} = (U_0, U_1, \cdots, U_{p-1})$ ($p = p(G) \geq 4$). Group the subsets in $\mathcal{C}$ into $\lfloor \frac{p}{3} \rfloor$ triplets $(U_{3i}, U_{3i+1}, U_{3i+2})$ for $0 \leq i \leq \lfloor \frac{p}{3} \rfloor - 1$. $|U_{3i}| + |U_{3i+1}| + |U_{3i+2}| \geq \delta + 1$ since each neighbour of a node $u \in U_{3i+1}$ must be in one of the three sets in the corresponding triplet. Thus, $\lfloor \frac{p}{3} \rfloor (\delta + 1) \leq n$ and the result follows by Lemma 10. ■

3.5 Chordality

Recall that the chordality of a graph is the length of the longest induced cycle in the graph. We upper–bound $\Lambda(G)$ in terms of chordality in the following Theorem. Its tightness is established in Theorem 8.

Theorem 6. *If G is a weighted undirected graph with chordality k, then $\Lambda(G) \leq \frac{(k+1)n}{2} - k$, where $n \geq 2$.*

Proof. If there are no crossing pairs of minimum cuts in G, then by Lemma 8, $\Lambda(G) \leq 2n - 3 \leq \frac{(k+1)n}{2} - k$, since k is at least 3 by definition and $n \geq 2$. Otherwise, consider a circular partition $\mathcal{C}$ for G such that $p(\mathcal{C}) = p(G)$. If $p(\mathcal{C}) > k$, clearly there is an induced cycle in G with length $> k$, contradicting the k-chordality of G. It follows that $p(G) \leq k$. Therefore, by Lemma 10, $\Lambda(G) \leq \frac{(k+1)n}{2} - k$. ■

Theorem 7. *If G is a weighted chordal graph, then $\Lambda(G) \leq 2n-3$, where $n \geq 2$. Moreover there are no crossing pairs of minimum cuts in G. Also, there exists a weighted chordal graph G, for every $n \geq 2$, such that $\Lambda(G) = 2n - 3$.*

Proof. Since for chordal graphs $k = 3$ (by definition), $\Lambda(G) \leq 2n - 3$ follows from Theorem 6. If there is a crossing pair of minimum cuts in G, then there is a circular partition $\mathcal{C}$ for G with $p(\mathcal{C}) \geq 4$ by Lemma 6. This immediately implies an induced cycle of length ≥ 4, contradicting the fact that G is chordal. Finally since complete graphs are chordal graphs, the construction of Theorem 3, establishes the tightness of this bound. ■

3.6 A Tight Construction

We establish the tightness of Theorem 6 and Lemma 10, by the following construction.

Theorem 8. *For $k \geq 3$ and $\lambda > 0$, there exists a family $\mathcal{G}$ of weighted undirected k-chordal graphs such that for each graph $G_n \in \mathcal{G}$ with n nodes, $\Lambda(G_n) = \frac{(k+1)n}{2} - k$, weight of the minimum cut $= \lambda$ and $p(G_n) = k$. Moreover, every node u of G_n defines a minimum cut $(\{u\}, \overline{\{u\}})$.*

4 Edge Expansion and Girth

In this section we upper–bound $\Lambda(G)$ in terms of the ratio $\frac{\lambda}{\beta}$, where $\lambda = \lambda(G)$ and $\beta = \beta_w(G)$ is the weighted edge expansion of G. It can be easily verified that $1 \leq \frac{\lambda}{\beta} \leq \frac{n}{2}$. Our bound works for $\frac{\lambda}{\beta} \in [1, \frac{n}{3})$.

Lemma 11. *If G is a weighted undirected graph with weighted edge expansion $\beta_w(G) = \beta$, then for any minimum cut $(S, \overline{S})$, $|S| > \frac{n}{2}$ or $|S| \leq \left\lfloor \frac{\lambda}{\beta} \right\rfloor$.*

Proof. Suppose $|S| \leq \frac{n}{2}$. Then, $w(S, \overline{S}) = \lambda \geq \beta|S|$ by definition of weighted edge expansion. Since, $|S|$ is an integer, we have $|S| \leq \left\lfloor \frac{\lambda}{\beta} \right\rfloor$. ■

Theorem 9. *Let G be a weighted undirected graph with $\lambda(G) = \lambda$ and the weighted edge expansion $\beta_w(G) = \beta$. If $1 \leq \left\lfloor \frac{\lambda}{\beta} \right\rfloor < \frac{n}{3}$, then $\Lambda(G) \leq \frac{(\lfloor \frac{\lambda}{\beta} \rfloor + 3)}{2} n - (\left\lfloor \frac{\lambda}{\beta} \right\rfloor + 2)$.*

The observation that the edge expansion can control the number of minimum cuts allows us to obtain a upper bound for $\Lambda(G)$ in terms of the girth in the case of *unweighted graphs.* The following classical results are not very difficult to prove.

Lemma 12. *If $(S, \overline{S})$ is a minimum cut of an unweighted undirected graph G, then $|S| = 1$ or $|S| \geq \delta$, where δ is the minimum degree of G.*

Lemma 13 (See Bollabas [3], page 126.). *Let g be the girth of an unweighted undirected graph G, with $m \geq 2n$. Then, $g \leq 2 \left\lceil \log_{\lfloor \frac{\sigma}{2} \rfloor} n \right\rceil + 1$, where $\sigma = \frac{2m}{n}$, is the average degree. (m is the number of edges of G).*

Lemma 14 (See Harary [9]). *If δ is the minimum degree and $\lambda(G) = \lambda$ be the size of a minimum cut (i.e., edge connectivity) in an undirected unweighted graph G, then $\delta \geq \lambda$.*

Theorem 10. *If G is an undirected unweighted graph with minimum degree δ (at least 6) and girth g (at least 4), then $\Lambda(G) < (\frac{n}{x+1} + 1)n - (\frac{2n}{x+1} + 1)$, where $x = \left(\frac{\delta-3}{2}\right)^{\frac{g-3}{2}}$.*

Acknowledgements

We thank Dr. Ramesh Hariharan for reading the manuscript and for useful suggestions.

References

1. R.E. Bixby, *The Minimum Number of Edges and Vertices in a Graph with Edge Connectivity n and m n-bonds.* Networks, Vol.5,(1975) 253-298.
2. F.T. Boesch, *Synthesis Of Reliable Networks – A Survey.* IEEE Transactions On Reliability, vol R–35, (1986) 240-246.
3. B. Bollabos, *Extremal Graph Theory.* Academic Press, London, 1978.
4. M.O. Ball, J.S. Provan, *Calculating bounds on reachability and connectedness in stochastic networks.* Networks, vol 5, (1975) 253-298.
5. M.O. Ball, J.S. Provan, *The Complexity Of Counting Cuts And Of Computing The Probability That A Graph Is Connected.* SIAM Journal of Computing, 12, (1983) 777-788.
6. M.O. Ball, J.S. Provan, *Computing Network Reliability In Time Polynomial In The Number Of Cuts.* Operations Research, 32, (1984) 516-521.
7. E.A. Dinits, A.V. Karzanov, M.V. Lomosonov, *On the Structure of a Family of Minimal Weighted Cuts in a Graph.* Studies in Discrete Optimization [In Russian], A.A. Friedman (Ed), Nauka, Moscow (1976) 290-306.
8. H.N. Gabow, *A Matroid Approach To Finding Edge Connectivity And Packing Arborescences.* Proceedings Of 23rd Annual ACM-SIAM Symposium On Theory Of Computing, (1991) 112-122.
9. F. Harary, *Graph Theory.* Addison-Wesley Reading, MA, 1969.
10. Lisa Fleischer, *Building Chain And Cactus Representations Of All Minimum Cuts From Hao–Orlin In The Same Asymptotic Run Time.* Journal Of Algorithms, 33, (1999) 51-72.
11. D.R. Karger, *Random Sampling In Cut, Flow and Network Design Problems.* In Proceedings Of 6th Annual ACM-SIAM Symposium On Discrete Algorithms, (1995), 648-657.
12. Jeno Lehel, Frederic Maffray, Myriam Preissmann, *Graphs With Largest Number Of Minimum Cuts.* Discrete Applied Mathematics, 65, (1996) 387-407.
13. A. Kanevsky, *Graphs With Odd And Even Edge Connectivity Are Inherently Different.* Tech. report, TAMU-89-10, June 1989.
14. H. Nagamochi, K. Nishimura, T. Ibaraki, *Computing All Small Cuts In An Undirected Network.* SIAM Journal Of Discrete Math, 10(3), (1997) 469-481.
15. D. Naor, Vijay. V. Vazirani, *Representing and Enumerating Edge Connectivity Cuts in RNC* Workshop On Algorithms and Data structures (1991) LNCS 519 273-285.
16. J.S. Provan, *Bounds On The Reliability Of Networks.* IEEE Transactions On Reliability, R-35, (1986) 26-268.
17. J.C. Picard, M. Queyranne, *On The Structure Of All Minimum Cuts In A Network And Applications.* Mathematical programming Study, 13, (1980) 8-16.
18. V.V. Vazirani and M. Yannakakis, *Suboptimal Cuts: Their Enumeration, Weight, And Number.* Lecture Notes in Computer Science, 623, Springer-Verlag, (1992) 366-377.
19. M.R. Henzinger, D.P. Williamson, *On The Number Of Small Cuts.* Information Processing Letters, 59, (1996), 41-44.

On Crossing Numbers of 5-Regular Graphs

G.L. Chia[1] and C.S. Gan[2]

[1] Institute of Mathematical Sciences, University of Malaya, 50603 Kuala Lumpur, Malaysia
[2] Faculty of Engineering and Technology, Multimedia University, 75450 Malacca, Malaysia

Abstract. The paper attempts to classify 5-regular graphs according to their crossing numbers and with given number of vertices. In particular, it is shown that there exist no 5-regular graphs on 12 vertices with crossing number one. This together with a result in [2] imply that the minimum number of vertices in a 5-regular graph with girth three and crossing number one is 14.

Let G be a graph. The *crossing number* of G, denoted $cr(G)$, is the minimum number of pairwise intersections of its edges when G is drawn in the plane. Throughout this paper, we adopt the following notations. Let K_n and C_n denote the complete graph and the cycle respectively each on n vertices. Let $K_{m,n}$ denote the complete bipartite graph whose bipartite sets V_1 and V_2 are such that $|V_1| = m$ and $|V_2| = n$. Also, let $\overline{G}$ denote the complement of the graph G.

Let $\mathcal{G}(r, n, c)$ denote the set of all r-regular connected graphs on n vertices having crossing number c. Then n is even if r is odd. Obviously $\mathcal{G}(5, n, c)$ is empty if $n \leq 4$. Also, it is easy to see that $\mathcal{G}(5, 6, c)$ is empty if and only if $c \neq 3$. In fact, if $G \in \mathcal{G}(5, 6, c)$, then G is the complete graph on 6 vertices so that $cr(G) = 3$. Hence we see that $\mathcal{G}(5, 6, 3) = \{K_6\}$.

Let $G \in \mathcal{G}(5, 8, c)$. Then G is either $\overline{C_8}$, $\overline{C_3 \cup C_5}$ or $\overline{C_4 \cup C_4}$. In [2], it is shown that $cr(\overline{C_8}) = 2$. Since $\overline{C_r \cup C_s}$ contains $K_{r,s}$ as a subgraph, it follows that $cr(\overline{C_3 \cup C_5}) \geq 4$ and $cr(\overline{C_4 \cup C_4}) \geq 4$. Now Fig. 1 depicts a drawing of $\overline{C_3 \cup C_5}$ and $\overline{C_4 \cup C_4}$ each with 4 crossings. Hence we have $cr(\overline{C_3 \cup C_5}) = 4$ and $cr(\overline{C_4 \cup C_4}) = 4$.

We summarize these observations in the following proposition.

Proposition 1. *$\mathcal{G}(5, 8, c)$ is empty unless $c \in \{2, 4\}$. Moreover $\mathcal{G}(5, 8, 2) = \{\overline{C_8}\}$ and $\mathcal{G}(5, 8, 4) = \{\overline{C_3 \cup C_5}, \overline{C_4 \cup C_4}\}$.*

Let $G \in \mathcal{G}(5, 10, c)$. Then evidently G is non-planar. In [2], we have shown that $cr(G) \geq 2$.

Proposition 2. *$\mathcal{G}(5, 10, c)$ is empty unless $c \geq 2$.*

Question 1. *Is it true that $\mathcal{G}(5, 10, 2)$ is an empty set?*

We have not come across any 5-regular graphs on 10 vertices having crossing number 2.

Let $G \in \mathcal{G}(5, 12, c)$. Then it is well-known that $c = 0$ if and only if G is the icosahedron (see for example [1]).

O.H. Ibarra and L. Zhang (Eds.): COCOON 2002, LNCS 2387, pp. 230–237, 2002.

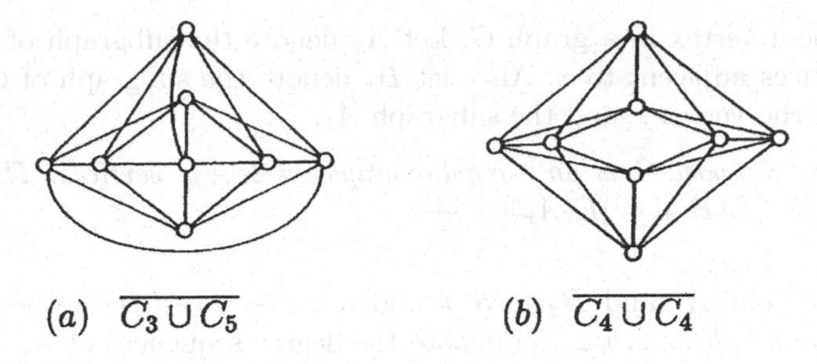

(a) $\overline{C_3 \cup C_5}$ (b) $\overline{C_4 \cup C_4}$

Fig. 1. Graphs with 4 crossings

Proposition 3. *$\mathcal{G}(5,12,0)$ consists of just the icosahedron.*

In this paper, we show that if G is a non-planar 5-regular graph on 12 vertices, then $cr(G) \geq 2$.

Proposition 4. *$\mathcal{G}(5,12,c)$ is empty if $c = 1$.*

The *girth* of a graph is the length of a smallest cycle in it. Let $f(r,g,c)$ denote the minimum number of vertices in an r-regular graph with girth g and crossing number c. In [2], we have shown that $12 \leq f(5,3,1) \leq 14$. Combining Proposition 4 and this result, we have the following theorem.

Theorem 1. *The minimum number of vertices in a 5-regular graph with girth 3 and crossing number 1 is given by $f(5,3,1) = 14$.*

The rest of the paper is to prove Proposition 4.

Lemma 1. *Let $G \in \mathcal{G}(5,12,1)$. Then there exists an edge e such that $G - e$ is a planar graph with only one quadrilateral.*

Proof. It is clear that there exists an edge e such that $G - e$ is planar.

Suppose X is a connected planar graph on n vertices and m edges. Let f and f_i denote the number of faces and i-faces in X respectively. Then

$$\begin{aligned} 2m &= \sum_{i \geq 3} i f_i \\ &= 3\sum_{i\geq 3} f_i + \sum_{i \geq 4}(i-3) f_i \\ &= 3f + \sum_{i \geq 4}(i-3) f_i \end{aligned}$$

Since $f = m + 2 - n$, we have

$$m = 3(n-2) - \sum_{i \geq 4}(i-3) f_i \tag{1}$$

As $G - e$ has 12 vertices and 29 edges, it follows from Equation (1) that $f_4 = 1$ and $f_i = 0$ for all $i \geq 5$. □

Let x be a vertex in a graph G. Let A_x denote the subgraph of G induced by the vertices adjacent to x. Also, let B_x denote the subgraph of G obtained by deleting the vertex x and the subgraph A_x.

Lemma 2. *Suppose G is an r-regular graph on $2r + s$ vertices. Then for any vertex $x \in G$, $|E(B_x)| = |E(A_x)| + \frac{rs}{2}$.*

Proof. Note that A_x and B_x have r and $r + s - 1$ vertices respectively. Let $(a_1, \dots, a_r)$ and $(b_1, \dots, b_{r+s-1})$ denote the degree sequences of A_x and B_x respectively.

Note that the number of edges in the subgraph $G - x$ is $\frac{r(2r+s-2)}{2}$ and that it is also equal to $|E(A_x)| + |E(B_x)|$ + number of edges from A_x to B_x.

Therefore,

$$\frac{\sum_{i=1}^{r} a_i}{2} + \frac{\sum_{i=1}^{r+s-1} b_i}{2} + \sum_{i=1}^{r}(r - 1 - a_i) = \frac{r(2r + s - 2)}{2}$$

which, on simplification, leads to $\frac{\sum_{i=1}^{r+s-1} b_i}{2} = \frac{\sum_{i=1}^{r} a_i}{2} + \frac{rs}{2}$. □

Let G be a graph. The *removal number* of G, denote $rem(G)$, is defined to be the minimum number of edges in G whose removal results in a planar graph. Obviously $cr(G) \geq rem(G)$. In the event that $rem(G) = 1$, then G contains an edge e such that $G - e$ is planar. Such an edge is called a *p-critical edge* of G.

Lemma 3.([2]) *Let G be a graph with a unique p-critical edge. Then $cr(G) \geq 2$.*

Let U be a proper subgraph of G. Assume that U is connected. Let $G(U)$ denote the graph obtained from G by contracting all edges in the subgraph U.

Suppose W is the subgraph $G - U$. Let $E(U, W) = \{(u, w) \in E(G) : u \in U, w \in W\}$.

Remark 1. *Let U be a connected proper induced subgraph of a graph G with $cr(G) = 1$. Let W be the subgraph $G - U$ and assume that W is connected.*

(a) Suppose U and W are both planar while $G(U)$ and $G(W)$ are both non-planar. Then clearly, any p-critical edge of G must come from $E(U, W)$.

(b) Suppose W is 2-edge connected and $G(W)$ is non-planar. Then W is planar.

To see that Remark 1(b) is true, assume that W is non-planar. Then any p-critical edge e of G must come from $E(W)$. Since W is 2-edge connected, $W - e$ is connected. This implies that $G(W - e)$ is non-planar since $G(W - e) \cong G(W)$. But this is a contradiction since $G - e$ is a planar graph.

Let G be a graph. If G is not a null graph, let G^- denote any graph obtained by deleting an edge from G. If G is not a complete graph, let G^+ denote any graph obtained by adding a new edge to G. Let W_n denote the graph obtained by joining a new vertex to every vertex of the cycle C_{n-1}, $n \geq 4$.

Lemma 4. *Suppose $G \in \mathcal{G}(5,12,1)$. Then K_5^-, $K_{3,3}$, $K_{2,4}$ and W_6^+ are forbidden subgraphs of G.*

Proof. Suppose $H \in \{K_5^-, K_{3,3}, K_{2,4}, W_6^+\}$ is a subgraph of G. Let U be the subgraph of G induced by the vertices of H. Let W be the subgraph $G - U$.

(a) Suppose H is K_5^-.

Then U is either K_5^- or K_5. Hence U (respectively W) has $9+\alpha$ (respectively $14+\alpha$) edges for some α such that $0 \leq \alpha \leq 1$.

If W is disconnected, then W has an isolated vertex x (since W has 7 vertices). In this case, x is adjacent to all the vertices of U giving a K_6^- or K_6 in G. But then $rem(G) \geq 2$, a contradiction.

If W is connected, then $G(W)$ is K_6^- or K_6. But then $rem(G) \geq 2$, a contradiction.

(b) Suppose H is $K_{3-r,3+r}$ where $r \in \{0,1\}$.

Then U (and so is W) has $9-r+\alpha$ edges for some $\alpha \geq 0$. As such if W is disconnected, then W has an isolated vertex x.

(i) Suppose $r = 0$.

If U has two or more vertices of degree 5, then U contains K_5^- as a subgraph. But this contradicts (a) and hence U has at most one vertex of degree 5.

If W is connected, then $G(W)$ contains $K_{3,4}$ as a subgraph. If W is disconnected, then the vertex x (which is isolated) in W is adjacent to 5 vertices of U giving $K_{3,4}$ as a subgraph in G. In either case, $cr(G) \geq 2$, a contradiction.

(ii) Suppose $r = 1$.

Let S be the partite set of $K_{2,4}$ having 4 vertices. If S contains a vertex of degree 5 in U, then there is a $K_{3,3}$ in U. But this is impossible by (i). Therefore every vertex in S is of degree at most 4 and hence is adjacent to some vertices in W.

Hence if W is connected, then $G(W)$ contains $K_{3,4}$ as a subgraph implying that $rem(G) \geq 2$, a contradiction. If W is disconnected, then the vertex x (which is isolated) in W is adjacent to at least three vertices of S giving $K_{3,3}$ as a subgraph in G, a contradiction to (i).

(c) Suppose H is W_6^+

Then U is also W_6^+ otherwise U contains K_5^- as a subgraph (which is impossible by (a)). It follows that W is a connected graph on 6 vertices and 11 edges.

Now, among all the nine graphs on 6 vertices and 11 edges, only the graph W_6^+ contains no forbidden subgraphs mentioned in (a) and (b). Hence W is also the graph W_6^+.

Now both U and W are planar while $G(U)$ and $G(W)$ are both non-planar. Since $G \in \mathcal{G}(5,12,1)$, there is an edge e in G such that $G - e$ is planar. By Remark 1(a), $e \in E(U,W)$. However, it is routine to check that for any edge $f \in E(U,W)$, the subgraph $G^* = G - f$ is such that $G^*(U)$ and $G^*(W)$ are both non-planar which means that G^* is non-planar, a contradiction and the proof is finished. □

Lemma 5. *Suppose $G \in \mathcal{G}(5,12,1)$. Then G does not contain a subgraph H such that $|V(H)| = 5$ and $|E(H)| = 8$.*

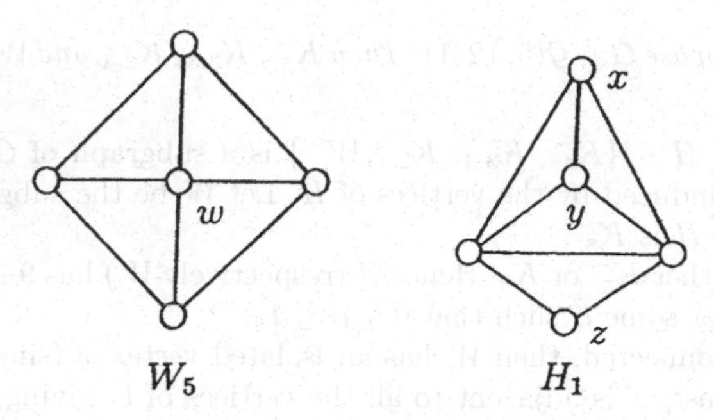

Fig. 2.

Proof. Suppose H is a subgraph of G on 5 vertices and 8 edges. Let U be the subgraph induced by the vertices of H. Then U is also H because otherwise U contains K_5^- or K_5 as subgraph and is impossible by Lemma 4.

It is easy to see that H is either W_5 or else the graph H_1 shown in Fig. 2.

Let W be the subgraph $G-U$. Then W has precisely 7 vertices and 13 edges.

If W contains a vertex v of degree ≤ 2, then v is adjacent to 3 or more vertices in U giving rise to some forbidden subgraph described in Lemma 4. Hence each vertex of W is of degree ≥ 3 and so W is connected. In fact, for any edge e in W, $W-e$ is connected. That is W is 2-edge connected.

Evidently, $G(W)$ is non-planar. As such, by Remark 1(b), W is a planar graph.

If W contains no vertices of degree 5, then $G(U)$ is a graph on 8 vertices having 20 edges so that $rem(G(U)) \geq 2$ implying $cr(G) \geq 2$, a contradiction. Hence there are only two possible degree sequences for W, namely $D_1 = (3,3,3,4,4,4,5)$ and $D_2 = (3,3,3,3,4,5,5)$.

If D_1 is the degree sequence of W, then $G(U)$ is a graph on 8 vertices having 19 edges. If D_2 is the degree sequence of W, then $G(U)$ contains $K_{3,3}$ as a subgraph because W contains $K_{2,3}$ as a subgraph. In either case, $G(U)$ is non-planar.

By Remark 1(a), any p-critical edge of G is contained in $E(U,W)$.

Let $e \in E(U,W)$. If U is the graph H_1, then $G^* = G-e$ is a non-planar graph because $G^*(W)$ contains $K_{3,3}$ as a subgraph (since the degree of the vertices x,y,z are ≤ 3). Hence U is the graph W_5. If e is not incident to the vertex w of W_5, then $G^* = G-e$ is non-planar becuase $G^*(W)$ contains $K_{3,3}$ as a subgraph. Hence e is incident to the vertex w and e is the unique p-critical edge of G. By Lemma 3, $cr(G) \geq 2$, a contradiction.

This completes the proof of the lemma. □

Lemma 6. *Suppose $G \in \mathcal{G}(5,12,1)$. Then K_4 is not a subgraph of G.*

Proof. Suppose K_4 is a subgraph of G.

Let U be the subgraph K_4. Then $W = G-U$ is the subgraph on 8 vertices having 16 edges.

Let v be a vertex in W. If $d_W(v) \leq 3$, then v is adjacent to at least two vertices of U giving a subgraph H_1 in G. This is impossible by Lemma 5. Hence W is a 4-regular graph. Consequently $G(U)$ is a graph on 9 vertices having 24 edges so that $rem(G(U)) \geq 3$ implying that $cr(G) \geq 3$, a contradiction. □

Definition 1. *A 4-cycle $a_1a_2a_3a_4$ is called a C^* cycle if it is chordless and there exists no vertex v such that v is adjacent to a_i for all $i = 1, \ldots, 4$.*

Lemma 7. *Suppose $G \in \mathcal{G}(5, 12, 1)$. Then for any vertex x in G, A_x does not contain an independent set Φ of three vertices such that $\sum_{v \in \Phi} d_{A_x}(v) \leq 3$.*

Proof. Let Φ be an independent set of three vertices in A_x. We observe that

(O1) if there is a vertex y in B_x having two neighbors u and v in Φ, then $xuyv$ is a C^* cycle in G.

If $xuyv$ is not a C^* cycle, then there exists a vertex w in G such that w is adjacent to x, u, y and v giving W_5 as a subgraph in G. But this is impossible by Lemma 5.

Assume that $\sum_{v \in \Phi} d_{A_x}(v) \leq 3$. Then there are at least 9 edges from Φ to B_x. We shall obtain a contradiction to Lemma 1 by showing that for any $e \in E(G)$, $G - e$ contains at least two C^* cycles.

Suppose B_x contains a vertex with 3 neighbors in Φ and another vertex with 2 (or more) neighbors in Φ. Then by observation (O1), this means that, for any edge e in G, $G - e$ has two or more C^* cycles.

Hence B_x contains 3 vertices each having 2 neighbors in Φ. But again, by observation (O1), this means that, for any edge e in G, $G - e$ has two or more C^* cycles.

This completes the proof. □

Let $G \in \mathcal{G}(5, 12, 1)$ and let x be a vertex in G. We shall now use the above lemmas to draw some conclusions that would narrow down the number of possibilities for the subgraph A_x of G.

(C1) $4 \leq |E(A_x)| \leq 7$.

If $|E(A_x)| \leq 3$, then A_x contains an independent set Φ of three vertices such that $\sum_{v \in \Phi} d_{A_x}(v) \leq 3$, a contradiction to Lemma 7. By Lemma 5, $|E(A_x)| \leq 7$.

(C2) It is not possible that $|E(A_x)| = 6 + k$, $k \in \{0, 1\}$.

Suppose $|E(A_x)| = 6 + k$, $k \in \{0, 1\}$. Then B_x is a graph on 6 vertices having $11 + k$ edges according to Lemma 2.

If every vertex of B_x has degree $\geq 4 + k$, then $|E(B_x)| \geq 12 + 3k$. Hence B_x contains a vertex z such that $d_{B_x}(z) \leq 3 + k$. But then $B_x - z$ is a subgraph on 5 vertices and having at least 8 edges. But this is a contradiction to Lemma 5.

(C3) A_x has no cycles of length 3 or 4.

This follows from Lemmas 5 and 6.

(C4) If $|E(A_x)| = 5$, then A_x is a cycle on 5 vertices.

If A_x is not a cycle on 5 vertices, then it contains either a triangle or a cycle of length 4. But this is impossible by (C3).

(C5) If $|E(A_x)| = 4$, then A_x is a path on 5 vertices.

If A_x is not a path on 5 vertices, then either A_x contains a 3-cycle or a 4-cycle (which is impossible by (C3)) or else A_x is a tree with a vertex of degree at least 3 and contains three vertices of degree 1. But this means that there is a forbidden independent set of three vertices satisfying the conditions of Lemma 7.

Hence we see that there are only two remaining cases to be considered for A_x.

(C6) A_x is either a path or a cycle on 5 vertices.

In the next two lemmas, we shall get rid of these two possibilities thereby proving Proposition 4.

Lemma 8. *Suppose $G \in \mathcal{G}(5, 12, 1)$. Then for any vertex x in G, A_x is not a path on 5 vertices.*

Proof. Suppose A_x is a path $v_1v_2 \ldots v_5$ on 5 vertices.

By Lemma 2, B_x is a graph on 6 vertices having 9 edges. If B_x contains a vertex y such that $d_{B_x}(y) \leq 1$, then $B_x - y$ is a graph on 5 vertices having at least 8 edges, a contradiction to Lemma 5. Hence each vertex of B_x has degree at least 2.

(O2) Let z be a vertex of degree 2 in B_x. Then z gives rise to at least two C^* cycles in G.

This is because z is adjacent to three vertices of A_x which do not form a subpath on 3 vertices in A_x (otherwise we have a W_5 in G which is impossible by Lemma 5).

We assert that there is at most one vertex of degree 2 in B_x.

Suppose y_1 and y_2 are two vertices of degree 2 in B_x. Then by observation (O2), each y_i gives rise to at least two C^* cycles in G. This in turn implies that $G - e$ contains at least two C^* cycles for any edge e in G. However this is a contradiction to Lemma 1. This proves the assertion.

Therefore there are only two possible degree sequences for B_x, namely $D_1 = (3, 3, \ldots, 3)$ and $D_2 = (2, 3, \ldots, 3, 4)$.

If B_x has D_1 as its degree sequence, then B_x is the graph $K_3 \times K_2$ since $K_{3,3}$ is a forbidden subgraph by Lemma 4. Clearly in this case, B_x contains three C^* cycles.

If B_x has D_2 as its degree sequence, then B_x is one of the three graphs J_1, J_2, and J_3 shown in Fig. 3. The graphs J_2 and J_3 each has two or more C^* cycles while J_1 has only one C^* cycle.

Notice that $\Phi = \{v_1, v_3, v_5\}$ is an independent set in A_x and that there are 8 edges joining Φ to B_x. Hence either (i) there is a vertex in B_x which has three neighbors in Φ giving rise to three C^* cycles in G or else (ii) there are two vertices in B_x each having at least 2 neighbors in Φ giving rise to at least two C^* cycles in G.

By Lemma 1, there is an edge e in G such that $G - e$ is planar. Let U be the subgraph induced by the vertices in $A_x \cup \{x\}$. Then evidently $G(U)$ is a non-planar graph. This implies that the edge e must come either from B_x or else from those that joins a vertex in A_x to a vertex in B_x. But then in either case, we see that $G - e$ contains two or more C^* cycles, a contradiction to Lemma 1.

This completes the proof of the lemma. □

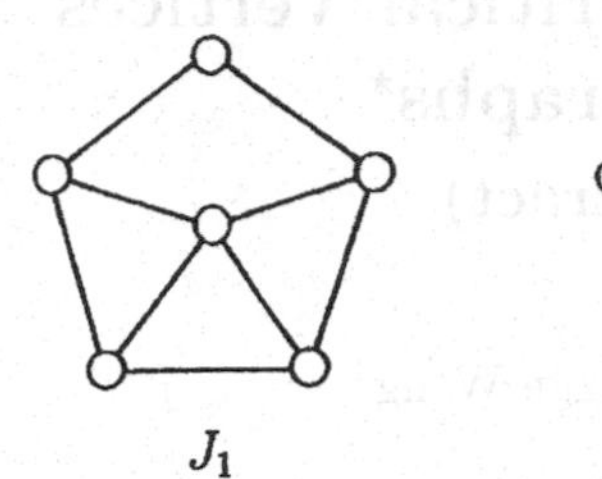
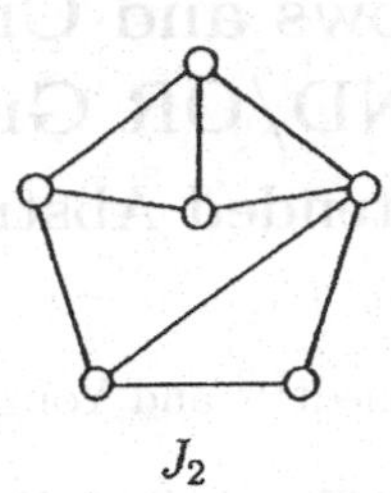
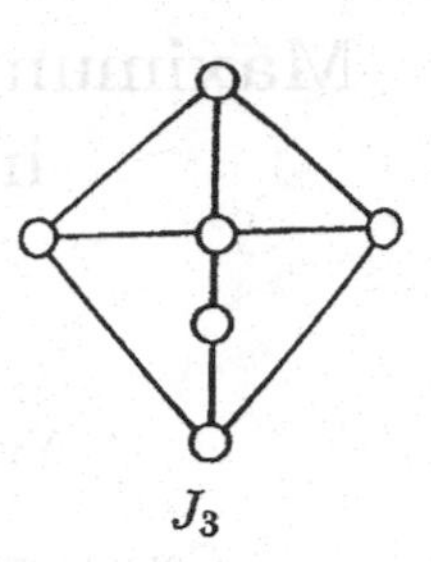

J_1 J_2 J_3

Fig. 3.

Lemma 9. *Suppose $G \in \mathcal{G}(5, 12, 1)$. Then for any vertex x in G, A_x is not a cycle on 5 vertices.*

Proof. Suppose A_x is a cycle $v_0v_1 \ldots v_4$ on 5 vertices.

In view of conclusion (C6) and Lemma 8, we see that

(O3) for any vertex u in G, A_u is a cycle on 5 vertices.

Hence it follows that B_x is a wheel W_6 on 6 vertices. Let the vertices of B_x be denoted $z, u_0, u_1, \ldots, u_4$ where $d_{B_x}(z) = 5$ and u_i is adjacent to u_{i+1}, $i = 0, 1, \ldots, 4$ (the operation on the subscripts is reduced modulo 5).

Now each vertex v_i, $i = 0, 1, \ldots, 4$ is adjacent to two vertices of B_x which must be adjacent in view of observation (O3). Likewise each vertex u_i, $i = 0, 1, \ldots, 4$ is adjacent to two adjacent vertices of A_x.

Without loss of generality, assume that v_i is adjacent to u_i and u_{i+1} for some i. Since A_{v_i} is a cycle, we may assume without loss of generality that v_{i-1} is adjacent to u_i for some i. This forces v_{i+1} to be adjacent to u_{i+1} by observation (O3). Applying (O3) to the vertex u_{i+1}, we see that v_{i+1} must be adjacent to u_{i+2}. Continue with same argument, eventually we have G isomorphic to the icosahedron. However this is a contradiction and the proof is finished. □

References

1. Chartrand, G., Lesniak, L.: Graphs & Digraphs. 3rd edn. Chapman & Hall, New York (1996)
2. Chia, G.L., Gan, C.S.: Minimal regular graphs with given girths and crossing numbers, (submitted)
3. Guy, R.K., Hill, A.: The crossing number of the complement of a circuit. Discrete Math. **5** (1973) 335–344
4. Kleitman, D.J.: The crossing number of $K_{5,n}$. J. Combinat. Theory. **9** (1970) 315–323
5. McQuillan, D., Richter, R.B.: On 3-regular graphs having crossing number at least 2. J. Graph Theory. **18** (1994) 831–839
6. Richter, B.: Cubic graphs with crossing number two. J. Graph Theory. **12** (1988) 363–374

Maximum Flows and Critical Vertices in AND/OR Graphs*
(Extended Abstract)

Yvo Desmedt[1,2] and Yongge Wang[3]

[1] Computer Science, Florida State University, Tallahassee
Florida FL 32306-4530, USA
desmedt@cs.fsu.edu

[2] Department of Mathematics, Royal Holloway, University of London, UK

[3] Department of Software and Information Systems,
University of North Carolina at Charlotte,
9201 University City Blvd, Charlotte, NC 28223
ywang@uncc.edu

Abstract. We will study this problem and present an algorithm for finding the minimum-time-cost solution graph in an AND/OR graph. We will also study the following problems which often appear in industry when using AND/OR graphs to model manufacturing processes or to model problem solving processes: finding maximum (additive and non-additive) flows and critical vertices in an AND/OR graph. Though there are well known polynomial time algorithms for the corresponding problems in the traditional graph theory, we will show that generally it is **NP**-hard to find a non-additive maximum flow in an AND/OR graph, and it is both **NP**-hard and co**NP**-hard to find a set of critical vertices in an AND/OR graph. We will also present a polynomial time algorithm for finding a maximum additive flow in an AND/OR graph, and discuss the relative complexity of these problems.

1 Introduction

Structures called AND/OR graphs are useful for depicting the activity of production systems (see, e.g., Nilsson [8]). Wang, Desmedt, and Burmester [10] used AND/OR graphs to make a critical analysis of the use of redundancy to achieve network survivability in the presence of malicious attacks. That is, they used AND/OR graphs to model redundant computation systems consisting of components which are based on computations with multiple inputs. Roughly speaking, an AND/OR graph is a directed graph with two types of vertices, labeled $\wedge$-*vertices* and $\vee$-*vertices*. The graph must have at least one input (source) vertex and one output (sink) vertex. In this case, processors which need all their inputs in order to operate could be represented by $\wedge$-vertices, whereas processors which can choose (using some kind of voting procedure) one of their "redundant"

* Research supported by DARPA F30602-97-1-0205.

O.H. Ibarra and L. Zhang (Eds.): COCOON 2002, LNCS 2387, pp. 238–248, 2002.

inputs could be represented by $\vee$-vertices. It should be noted that our following definition is different from the standard definitions in artificial intelligence (see, e.g., [8]). That is, the directions of the edges are opposite. The reason is that we want to use the AND/OR graphs to model redundant computation systems too.

Definition 1. *An* AND/OR graph $G(V_\wedge, V_\vee, INPUT, output; E)$ *is a graph with a set* $V_\wedge$ *of* $\wedge$-vertices, *a set* $V_\vee$ *of* $\vee$-vertices, *a set* $INPUT \subset V_\wedge$ *of* input vertices, *an* output vertex $output \in V_\vee$, *and a set of* directed edges E. *The input vertices have no incoming edges and the output vertex has no outgoing edges.*

Assume that an AND/OR graph is used to model a redundant computation system or a problem solving process. Then, information (for example, mobile codes) must flow from the input vertices to the output vertex. And a valid computation in an AND/OR graph can be described by a *solution graph.*

Definition 2. *Let* $G(V_\wedge, V_\vee, INPUT, output; E)$ *be an AND/OR graph. A* solution graph $P = (V_P, E_P)$ *is a minimum subgraph of* G *satisfying the following conditions: (1).* $output \in V_P$. *(2). For* $\wedge$*-vertex* $v \subset V_P$*, all incoming edges of* v *in* E *belong to* E_P. *(3). For* $\vee$*-vertex* $v \in V_P$*, there is exactly one incoming edge of* v *in* E_P. *(4). There is a sequence of vertices* $v_1, \ldots, v_n \in V_P$ *such that* $v_1 \in INPUT, v_n = output$*, and* $(v_i \rightarrow v_{i+1}) \in E_P$ *for each* $i < n$.

Wang, Desmedt, and Burmester [10] have studied the problem of finding vertex disjoint solution graphs in an AND/OR graph. Specifically, they have showed that it is **NP**-hard to find vertex disjoint solution graphs in an AND/OR graph. These problems are mainly related to that of achieving dependable computation using redundancy. Minimum-cost solution graphs have been studied extensively in artificial intelligence and many heuristic algorithms for finding minimum-cost solution graphs have been presented (see, e.g., [2,6,7,8]). When a problem solving process is modeled by an AND/OR graph, the minimum-cost solution graph can be used to attack the problem with the least resource. However, in many cases we want to solve the problem within the shortest time period and we assume that we have as many concurrent resources as we need to run all concurrent processes. Maximum-flow minimum-cut theorem (see, e.g., [3,5]) has played an important role in the study of networks. For example, it is used to direct the traffic in networks such as the Internet. However, this theorem is only for networks that have one kind of vertices: $\vee$-vertices (that is, directed graphs). And it is not applicable for networks which can only be modeled by AND/OR graphs. Indeed, in artificial intelligence and distributed computation systems, the realization of many important projects (such as the construction of a dam, of a shopping center, of a housing estate or of an aircraft; the carrying out of a sequence of manufacturing steps; the programming of a test flight of an aircraft; etc.) are dependent on the computations with multiple inputs and many problems in the realization are that of finding maximum flows in AND/OR graphs.

In this paper, we will consider the following problem: is there an equivalent theorem of maximum-flow minimum-cut theorem for AND/OR graphs? That is, does there exist a polynomial time heuristic algorithm for finding maximum flows in AND/OR graphs? We will show that this problem is **NP**-hard. We will also consider the problems of finding critical vertices in AND/OR graphs and show these problems are even “harder”, that is, they lie in the second level of the polynomial time hierarchy, which is believed to be harder than **NP**-complete problems. However, if we modify the flow structure and make it additive, as Martelli and Montanari [7] did for cost structures, then we will have a polynomial time (heuristic) algorithm for finding maximum additive flows in ADN/OR graphs.

The organization of the paper is as follows. We first present in Section 2 a polynomial time algorithm for finding a minimum-time-cost solution graph in an AND/OR graph. In Section 3 we discuss the problem of finding maximum flows in AND/OR graphs and prove the **NP**-hardness of several problems. Section 4 is devoted to the problem of finding critical vertices in an AND/OR graph. Several problems related to critical vertices are shown to be “harder”, that is, they lie in the second level of the polynomial time hierarchy. In Section 5 we present a polynomial time algorithm for finding maximum additive flows in AND/OR graphs.

We will use (fairly standard) notions of complexity theory. We refer the reader to [4] for definitions of these. Here we only give an informal description of a few notions. A *polynomial time many one reduction* (denoted by $\leq_m^p$) from a problem A to another problem B is a polynomial time computable function f with the property that $f(x) \in B$ if and only if $x \in A$ for all inputs x. A *polynomial time Turing reduction* (denoted by $\leq_T^p$) from a problem A to another problem B is a polynomial time oracle Turing machine M with the property that, for any input x, the Turing machine M with access to the oracle B will decide in polynomial time whether $x \in A$. For a complexity class $\mathcal{C}$ and a problem $A \in \mathcal{C}$, by saying that A is $\mathcal{C}$-complete we mean that every problem in $\mathcal{C}$ can be reduced to A by a polynomial time many one reduction.

Due to the space limit, we will not give proofs for our results, the details are referred to the full version of this paper.

2 Minimum-Time-Cost Solution Graphs and PERT Graphs

The shortest path problem and maximum flow problem are among the oldest problems in graph theory (see, e.g., [5]). They appear either directly, or as subsidiary problems, in many applications. Amongst others, we can mention the following: vehicle routing problems, some problems of investment and of stock control, many problems of dynamic programming with discrete states and discrete time, network optimization problems, and the problem of a continuous electrical current through a network of dipoles. However, as showed in [10], traditional graphs do not present a model for all problems in practice, e.g., it is always the case that a processor needs more than one type of inputs. AND/OR graphs seem to be a possible candidate for modeling these problems with multiple inputs. For a given AND/OR graph $G(V_\wedge, V_\vee, INPUT, output; E)$, if we associate with each edge $e \in E$ a rational number $l(e)$ called the length of the edge, then we can define a minimum-cost solution graph of G to be a solution graph $P(V_P, E_P)$ in G whose total length

$$l(P) = \sum_{e \in E_P} l(e)$$

is a minimum. In addition to its important applications in artificial intelligence (see, e.g., [8]), minimum-cost solution graphs have many practical applications in distributed computation systems with multiple inputs, because length $l(e)$ may equally well be interpreted as being a cost of transportation along e, the time through e, and so on. Chang and Slagle [2] have proposed a heuristic search algorithm for finding minimum-cost solutions in an AND/OR graph, but it was subsequently shown by Sahni [9] that with above definition of the cost, this problem is **NP**-hard. Thus their algorithm cannot be implemented efficiently in practice. By modifying the cost structure and making it additive, Martelli and Montanari [7] were able to formulate a polynomial time "marking" (with or without heuristic functions) algorithm AO* for AND/OR graphs (see also [6] for more discussions on the heuristic algorithm AO*). Roughly speaking, in the new cost structure, the cost of one edge may be counted as many times as it will be used in the unfolded AND/OR tree of the solution graph.

In the practice of distributed computation systems, many systems can be modeled by AND/OR graphs with only one $\vee$-vertex, that is, the output vertex is the only $\vee$-vertex. For such kind of AND/OR graphs, it is easy to find minimum-cost solutions in them with respect to the above "non-additive" cost definition.

Theorem 1. *There is a polynomial time algorithm to find a minimum-cost solution graph in an AND/OR graph with only one $\vee$-vertex.*

Proof. Given an AND/OR graph G with only one $\vee$-vertex, it is straightforward that there are at most k solution graphs in it, where k is the number of incoming edges of the $\vee$-vertex *output*. Whence it is easy to find a minimum-cost solution graph in it by an exhaustive search. Q.E.D.

For a problem solving process modeled by an AND/OR graph, the minimum-cost solution graph can be used to attack the problem with the least resource. However, in many cases we want to solve the problem within the shortest time period and we assume that we have as many concurrent resources as we need to run all concurrent processes. In the following, we present an algorithm for finding minimum-time-cost solution graphs in acyclic AND/OR graphs.

First we present the definition of *PERT digraphs* (**Program Evaluation and Review Technique**). A *PERT digraph* is an AND/OR graph $G(V_\wedge, V_\vee, INPUT, output; E)$ with the following properties: (1). $INPUT = \{in\}$ has only one element. (2). G has no directed circuits. (3). G has only one $\vee$-vertex *output* which has only one incoming edge. (4). Every vertex $v \in V_\wedge$ is on some directed path from *in* to *output*.

PERT digraphs have been used to model the central scheduling problems (see, e.g., [5]). A PERT digraph has the following interpretation. Every edge represents a process. All the processes which are represented by edges of in^+, can be started right away. For every vertex v, the processes represented by edges of v^+ can be started when all the processes represented by edges of v^- are completed. Note that we use v^- and v^+ to denote the incoming and outgoing edges of v respectively. For a given PERT digraph, we want to know how soon the whole project can be completed; that is, what is the shortest time, from the moment the processes represented by in^+ are started, until the process represented by $output^-$ is completed. We assume that the resources for running the processes are unlimited. For this problem to be well defined let us assume that each $e \in E$ has an assigned *length* $l(e)$, which specifies the time it takes to execute the process represented by e. The minimum completion time can be found by the following algorithm:

1. Assign in the label $O(\lambda(in) \leftarrow 0)$. All other vertices are "unlabeled".
2. Find a vertex, v, such that v is unlabeled and all edges of v^- emanate from labeled vertices. Assign
$$\lambda(v) \leftarrow \max_{e=(u \rightarrow v)} \{\lambda(u) + l(e)\}.$$
3. If $v = output$, halt; $\lambda(output)$ is the minimum completion time. Otherwise, go to Step 2.

For more discussions on the above algorithm it is referred to [5].

We define a *redundant PERT digraph* to be an AND/OR graph with the following properties: (1). $INPUT = \{in\}$ has only one element. (2). G has no directed circuits. (3). Every vertex $v \in V_\wedge \cup V_\vee$ is on some directed path from in to $output$.

As in a PERT digraph, every edge in a redundant PERT digraph represents a process. All the processes which are represented by edges of in^+, can be started right away. For every $\vee$-vertex v, the processes represented by edges of v^+ can be started when any one of the processes represented by edges of v^- is completed. And for every $\wedge$-vertex v, the processes represented by edges of v^+ can be started when all the processes represented by edges of v^- are completed.

Our problem deals with the question of finding a solution graph in a redundant PERT digraph such that the minimum completion time of the solution graph is a minimum. That is, if we have enough resources to run these processes concurrently, then we can solve the problem within the shortest time period. Such kind of minimum-time-cost solution graphs can be found by the following algorithm.

1. Assign in the label $O(\lambda(in) \leftarrow 0)$. All other vertices are "unlabeled".
2. Find a vertex, v, such that v is unlabeled and all edges of v^- emanate from labeled vertices. If v is an $\wedge$-vertex, then assign
$$\lambda(v) \leftarrow \max_{e=(u \rightarrow v)} \{\lambda(u) + l(e)\};$$
Otherwise assign
$$\lambda(v) \leftarrow \min_{e=(u \rightarrow v)} \{\lambda(u) + l(e)\};$$

3. If $v = output$, halt; $\lambda(output)$ is the minimum completion time of the minimum-time-cost solution graph. Otherwise, go to Step 2.

In Step 2, the existence of a vertex v, such that all the edges of v^- emanate from labeled vertices is guaranteed by Condition (2) and (3) of the definition for a redundant PERT digraph: If no unlabeled vertex satisfies the condition then for every unlabeled vertex, v, there is an incoming edge which emanates from another unlabeled vertex. By repeatedly tracing back these edges, one finds a directed circuits. Thus if such vertex is not found then we conclude that either Condition (2) or (3) does not hold.

It is easy to prove, by induction on the order of the labeling, that $\lambda(output)$ is the minimum completion time of the minimum-time-cost solution graph.

Once the algorithm terminates, by going back from *output* to *in*, via the edge which determined the label of the vertex, we can trace the minimum-time-cost solution graph. Clearly, they may be more than one minimum-time-cost solution graph.

3 The Maximum Flow Problem in an AND/OR Graph

Given an AND/OR graph $G(V_\wedge, V_\vee, INPUT, output; E)$, a *capacity function* c associated with G is a positive integral function defined on edges of G. A *flow* f in G is a positive integral function defined on edges of G with the property that for all $e \in E$, $0 \leq f(e) \leq c(e)$, and

$$\sum_{e \in v^-} f(e) = \sum_{e \in v^+} f(e) \qquad (1)$$

for all $v \in V_\vee$, where v^- is the set of incoming edges of v and v^+ is the set of the outgoing edges of v, and for all $v \in V_\wedge$,

$$\forall e_1, e_2 \in v^-(f(e_1) = f(e_2)) \text{ and } \forall e_1 \in v^- \forall e_2 \in v^+(f(e_2) \leq f(e_1)). \qquad (2)$$

For a vertex $v \in V_\vee \cup V_\wedge$, the *amount of flow* into v is defined to be the value $\sum_{e \in v^-} f(e)$. For a flow f in the AND/OR graph G, the *total flow* $F_f(G)$ is defined to be the amount of flow into the output vertex *output*. And we will use $F_c(G)$ to denote the maximum of $F_f(G)$ for all flows f in G, that is, $F_c(G) = \max\{F_f(G) : f \text{ is a flow in } G\}$.

Applications of the theory of flows in AND/OR graphs are extremely numerous and varied. For example, the optimal design and expansion of computation systems with multiple inputs, and the optimal design of a production manufacturing process. Though there are polynomial time algorithms for finding maximum flows in traditional graphs, we will show that the equivalent problem for AND/OR graphs is **NP**-hard. Specifically, we will show that the following problem MFAO (Maximum Flows for AND/OR Graphs) is **NP**-hard.

Instance: An AND/OR graph G, a capacity function c associated with G, and a positive integer k.

Question: Does there exist a flow f in G such that the total flow $F_f(G)$ is at least k?

Theorem 2. *MFAO is* **NP**-*complete for* $k = 1$ *(i.e., k is not a part of the input).*

Proof. It is clear that MFAO $\in$ **NP**. In order to prove that MFAO is **NP**-hard for $k = 1$, we can reduce the **NP**-complete problem $3SAT$ to MFAO. The reduction is similar to the reductions in [10]. The details of the proof could be found in the full version of this paper. Q.E.D.

In traditional graph theory, the problem of finding maximum flows in a graph is closely related to the problem of deciding the connectivity of the graph. However, Theorem 2 shows that there is a big difference between these two corresponding problems in AND/OR graphs. AND/OR graphs could have solution graphs but have no nonzero flow in them. AND/OR graphs could also have nonzero flows but do not have solution graphs in them. Such kind of examples could be found in the full version of this paper.

4 The Problem of Finding Critical Vertices in an AND/OR Graph

In this section, we assume the familiarity with the complexity classes within the Polynomial time Hierarchy like Σ_n^p and Π_n^p. For more details, it is referred to [4].

Let us consider the following scenarios: A redundant computation system (or a problem solving process) with multiple inputs (e.g., the electrical power distribution systems, the air traffic control system, etc.) is modeled by an AND/OR graph G with a capacity function c associated with it. And an adversary has the power to destroy k processors (that is, k vertices of the graph G) of the system. Then the adversary wants to know how to choose k vertices in the graph such that the destruction of these vertices results in the largest damage to the system. In another words, he wants to remove k vertices from the AND/OR graph such that the maximum flows of the resulting AND/OR graphs (in this case, the flows coming from the corrupted vertices are all 0) is a minimum. The designer of the system is also concerned with this scenario because he wants to know how robust his system is.

In order to state our problem more precisely, we first give two definitions. Given an AND/OR graph $G(V_\wedge, V_\vee, INPUT, output; E)$ with a capacity function c and a vertex set $U \subseteq (V_\wedge \cup V_\vee) \setminus (INPUT \cup \{output\})$, the capacity function c_U is defined by

$$c_U(e) = \begin{cases} 0 & \text{if } e \text{ is an outgoing edge of some vertex in } U, \\ c(e) & \text{otherwise.} \end{cases}$$

And for a number $k > 0$, a *set of critical vertices with respect to both c and k* is a vertex set $U \subseteq (V_\wedge \cup V_\vee) \setminus (INPUT \cup \{output\})$ with $|U| \leq k$ we have

– If $F_{c_U}(G)$ is the maximum (for the definition see the previous section) of all total flows in G with respect to the capacity function c_U then, for any other vertex set $U' \subseteq (V_\wedge \cup V_\vee) \setminus (INPUT \cup \{output\})$ with $|U'| = k$, $F_{c_U}(G) \leq F_{c_{U'}}(G)$, where $F_{c_{U'}}(G)$ is the maximum of all total flows in G with respect to the capacity function $c_{U'}$.

Whence the concise statement of our problem is the following: Given an AND/OR graph G with a capacity function c and a positive integer k, how can one find a set of critical vertices with respect to both c and k? We can show that this problem is **NP**-hard. Indeed, we can prove that the problem of deciding whether a given set of vertices is critical is in Π_2^p and is both **NP**-hard and co**NP**-hard. The CV (Critical Vertices) problem is defined as follows:

Instance: An AND/OR graph $G(V_\wedge, V_\vee, INPUT, output; E)$ with a capacity function c, and a vertex set $U \subseteq (V_\wedge \cup V_\vee) \setminus (INPUT \cup \{output\})$.
Question: Is U a set of critical vertices with respect to both c and $|U|$?

It is clear that U is a set of critical vertices if and only if for all flows f_U in G (with respect to the capacity function c_U) and for all $U' \subseteq (V_\wedge \cup V_\vee) \setminus (INPUT \cup \{output\})$ of size $|U|$, there is a flow $f_{U'}$ in G (with respect to the capacity function $c_{U'}$) such that $F_{f_U}(G) \leq F_{f_{U'}}(G)$. Whence CV is in $\Pi_2^p = \text{co}\Sigma_2^p$, that is, the second level of the polynomial time hierarchy.

Theorem 3. *CV is both* **NP***-hard and co***NP***-hard. Thus if* **NP**$\neq$co**NP***, then CV belongs neither to* **NP** *nor to co***NP***.*

Applications of critical vertices are varied. For example, in order to attack a computation system modeled by an AND/OR graph with the least resource, an adversary wants to choose a minimal set of critical vertices to corrupt (e.g., to bomb). And in order for a system designer to make the computation system dependable, he should pay more attention to the processors corresponding to the critical vertices.

It is often the case that a system designer wants to know how many faults a system can tolerate, that is, he is interested in the following problem CVB (Critical Vertices with a given Bound).

Instance: An AND/OR graph $G(V_\wedge, V_\vee, INPUT, output; E)$ with a capacity function c, two positive integers k and p.
Question: Does there exist a vertex set $U \subseteq (V_\wedge \cup V_\vee) \setminus (INPUT \cup \{output\})$ in G such that $|U| \leq k$ and $F_{c_U}(G) \leq p$?

It is clear that $F_{c_U}(G) \leq p$ if and only if for all flow f_U with respect to the capacity function c_U we have $F_{f_U}(G) \leq p$. Whence CVB belongs to the complexity class Σ_2^p. In the following we will show that CVB is both **NP**-hard and co**NP**-hard. We first introduce a restricted version SCV of CVB and show that the problem SCV is **NP**-complete. Given an AND/OR graph $G(V_\wedge, V_\vee, INPUT, output; E)$, a set $U \subseteq (V_\wedge \cup V_\vee) \setminus (INPUT \cup \{output\})$ is called a set of *strictly critical vertices* of G if, for any solution graph P in G, P passes through at least one vertex of U. Note that a set of strictly critical vertices is different from a vertex separator (though related) defined in [1]. The SCV (Strictly Critical Vertices) problem is defined as follows:

Instance: An AND/OR graph $G(V_\wedge, V_\vee, INPUT, output; E)$ and a positive integer $k \leq |(V_\wedge \cup V_\vee) \setminus (INPUT \cup \{output\})|$.
Question: Does there exist a size k set of strictly critical vertices?

Theorem 4. *1. SCV is* **NP**-*complete.*
2. CVB is **NP**-*hard.*
*3. CVB is co***NP**-*hard.*
4. If **NP**$\neq$*co***NP**, *then CVB belongs neither to* **NP** *nor to co***NP**.
5. $\overline{\text{CV}} \leq_T^p CVB$.

5 Maximum Additive Flows in an AND/OR Graph

We have mentioned in Section 2 that after Sahni [9] proved the **NP**-hardness of finding minimum-cost solution graphs in AND/OR graphs, Martelli and Montanari [7] modified the cost structure to be additive and got a (heuristic) polynomial time algorithm AO*. Based on the similar idea, we can make our definition of flows in an AND/OR graph "additive" and then get a polynomial time algorithm for finding maximum additive flows in AND/OR graphs. Like the definition of the additive cost of edges by Martelli and Montanari [7], we can unfold each $\vee$-vertex. That is, let the $\vee$-vertex have the copy function. More precisely, an *additive flow* f in an ADN/OR graph G is a flow function f defined on edges of G with the equation (1) replaced by the following equation (3):

$$f(e') \leq \sum_{e \in v^-} f(e) \tag{3}$$

for all $v \in V_\vee$ and $e' \in v^+$. For an additive flow f in the AND/OR graph G, the *total additive flow* $F'_f(G)$ is defined to be the amount of additive flow into the output vertex *output*. And we will use $F'_c(G)$ to denote the maximum of $F'_f(G)$ for all additive flow f in G, that is, $F'_c(G) = \max\{F'_f(G) : f \text{ is an additive flow in } G\}$.

Applications of the theory of additive flows in AND/OR graphs are varied. For example, data in computer systems are easy to copy. The additive flow may be interpreted that $\vee$-vertices in an AND/OR graph can "copy" data (note that in some production process, "hardwares" can not be easily "copied", whence can only be modeled by non-additive flows).

Theorem 5. *There is an efficient algorithm to compute the maximum additive flows in AND/OR graphs.*

Corollary 1. *For an AND/OR graph G with only one $\vee$-vertex and a capacity function c, there is a polynomial time algorithm for finding the maximum (non-additive) flow in it.*

As in Section 4, we can also define the set of critical vertices for additive flows and we denote by A-CV, A-SCV and A-CVB the corresponding versions of CV, SCV and CVB respectively. Since SCV = A-SCV and there is a polynomial time algorithm to compute the maximum additive flow in an AND/OR graph, we have the following theorem.

Theorem 6. *1. Both A-SCV and A-CVB are* **NP***-complete.*
2. A-CV $\in$ *co***NP** *and* $\overline{\text{A-CV}}$ *is* $\leq_T^p$*-complete for* **NP**.

Note that we still do not know whether $\overline{\text{A-CV}}$ is **NP**-complete, that is, whether $\overline{\text{A-CV}}$ is polynomial time many one complete for **NP**.

6 Comments and Open Problems

In this paper, we have discussed the problem of finding maximum flows and the problem of finding critical vertices in an AND/OR graph. As a summary, we list our main results in the following.

1. MFAO, SCV, A-SCV, A-CVB are **NP**-complete problems.
2. A-CV could be reduced to $\overline{3SAT}$ though we do not know whether A-CV is co**NP**-complete.
3. CVB lies between **NP**-complete and Σ_2^p-complete problems.
4. CV lies between co**NP**-complete and Π_2^p -complete problems.

The following interesting problems remain open yet.

1. Is CVB Σ_2^p-complete?
2. Can CVB be polynomial time Turing reduced to $\overline{\text{CV}}$?
3. Is CV Π_2^p-complete?
4. Is A-CV co**NP**-complete?
5. Can CVB be polynomial time Turing reduced to $3SAT$?

We conjecture that the answers to the above questions are all negative unless the polynomial time hierarchy collapses. In addition, it is interesting to show the exact relationship between $\overline{\text{CV}}$ and CVB?

References

1. M. Burmester, Y. Desmedt, and Y. Wang. Using approximation hardness to achieve dependable computation. In: *Proc. RANDOM '98*, pages 172–186. LNCS 1518, Springer Verlag, 1998.
2. C. L. Chang and J. R. Slagle. An admissible and optimal algorithm for searching AND/OR graphs. *Artificial Intelligence*, **2**:117–128, 1971.
3. L.R. Ford and D. R. Fulkerson. *Flows in Networks.* Princeton University Press, Princeton, NJ, 1962.
4. M. R. Garey and D. S. Johnson. *Computers and Intractability: A Guide to the Theory of* **NP***-Completeness.* W. H. Freeman and Company, San Francisco, 1979.

5. M. Gondran and M. Minoux. *Graphs and Algorithms.* John Wiley & Sons Ltd., New York, 1984.
6. D. Hvalica. Best first search algorithm in AND/OR graphs with cycles. *J. of Algorithms*, **21**:102–110, 1996.
7. A. Martelli and U. Montanari. Additive AND/OR graphs. In *Proceedings of the Third International Joint Conference on Artificial Intelligence,* pages 1–11, Morgan Kaufmann Publishers, Inc., 1973.
8. N. J. Nilsson. *Principles of Artificial Intelligence.* Tioga, 1980.
9. S. Sahni. Computationally related problems. *SIAM J. Comput.*, **3**:262–279, 1974.
10. Y. Wang, Y. Desmedt, and M. Burmester. **NP**-Hardness of dependable computation with multiple inputs. *Fundamenta Informaticae*, **42**(1):61–73, 2000.

New Energy-Efficient Permutation Routing Protocol for Single-Hop Radio Networks

Amitava Datta[1,*] and Albert Y. Zomaya[2]

[1] Department of Computer Science & Software Engineering
University of Western Australia
Perth, WA 6009, Australia
datta@cs.uwa.edu.au

[2] School of Information Technologies
University of Sydney
NSW 2006, Australia
zomaya@it.usyd.edu.au

Abstract. A radio network (RN) is a distributed system where each node is a small hand-held commodity device called a *station*, running on batteries. In a single-hop RN, every station is within the transmission range of every other station. Each station spends power while transmitting or receiving a message, even when it receives a message that is not destined for it. Since it is not possible to recharge batteries when the stations are on a mission, it is extremely important that the stations spend power only when it is necessary. In this paper, we are interested in designing an energy-efficient protocol for permutation routing which is one of the most fundamental problems in any distributed system. An instance of the *permutation routing* problem involves p stations of an RN, each storing $\frac{n}{p}$ items. Each item has a unique destination address which is the identity of the destination station to which the item should be sent. The goal is to route all the items to their destinations while spending as little energy as possible. We show that the permutation routing problem of n packets on an $RN(p,k)$ can be solved in $\frac{2n}{k} + (\frac{p}{k})^2 + p + 2k^2$ slots and each station needs to be awake for at most $\frac{6n}{p} + \frac{2p}{k} + 8k$ slots. When $k \ll p \ll n$, our protocol is more efficient both in terms of total number of slots and the number of slots each station is awake compared to the protocol by Nakano *et al.* [8].

1 Introduction

Wireless and mobile communication technologies have grown explosively in recent years. New demands for enhanced capabilities for these technologies will continue to grow in future. The communication in most cellular systems is based on robust infrastructures. However, radio networks should be rapidly deployable and self-organizing. Radio networks are useful for disaster relief, search-and-rescue, collaborative computing and interactive mission planning [5]. The

* This author's research is partially supported by Interactive Virtual Environments Centre (IVEC) and Australian Partnership in Advanced Computing (APAC).

O.H. Ibarra and L. Zhang (Eds.): COCOON 2002, LNCS 2387, pp. 249–258, 2002.

first radio network was the PRNET, a packet radio network, developed in the 1970s [2,5].

A radio network (RN) is a distributed system with no central arbiter, consisting of p radio transceivers called *stations*. The stations are usually small hand-held devices running on batteries and the batteries cannot be recharged while on a mission [3]. Hence, it is important that any protocol designed for an RN is power efficient, i.e., the stations spend as little power as possible. We assume that each of the p stations in the RN has a unique integer ID in the range $[1, p]$. Assigning unique IDs to the stations in an RN is a separate problem and has been solved by Nakano and Olariu [6]. They have shown that even if the stations do not have ID numbers initially, it is possible to devise a protocol which assigns a unique ID to each station and terminates in $O(\frac{p}{k})$ time slots with high probability, where k is the number of available channels.

In this paper, we are interested in low-mobility, single-hop RNs where every station is within the transmission range of every other station and the mobility of the stations is low or, at least much smaller compared to the time taken by a protocol to complete. An example of such a network is a collection of researchers in a conference room when they wish to exchange information among themselves without any preexisting infrastructure [8]. Designing routing protocols for single-hop RNs is important since routing in a multi-hop RN is often done by decomposing it into multiple single-hop RNs [2]. While the computational power of small hand-held devices is increasing at a rapid rate, the lifetime of batteries is not expected to improve significantly in the near future [3]. Moreover, recharging batteries may not be possible while on a mission. It is known that a significant amount of energy is spent by a station while transmitting or receiving packets. For example, the DEC Roamabout portable radio consumes about 5.76 watts during transmission, 2.88 watts during reception and 0.35 watts when it is idle [8,9]. A station spends power when it receives a packet which is not destined for it [3]. Hence, a station should not receive packets that are not meant for it if we want to make a protocol energy-efficient. However, for a single-hop RN, this is a nontrivial problem to solve as all stations are within the transmission range of each other.

Conflict resolution is one of the main issues in designing protocols for RNs. Carrier Sense Multiple Access (CSMA) is a simple and robust random access method for media access suitable for RNs [1]. However, a fraction of the available bandwidth is wasted for resolving random conflicts of messages [2]. Several conflict-free multiple access schemes have been proposed recently for radio networks. The most popular of these schemes is the Demand Assignment Multiple Access (DAMA) schemes proposed for transmission networks [4]. In the DAMA scheme, all the stations that want to transmit a message on a given channel are ordered in a logical ring according to which they are given transmission access to the channel [4]. From the viewpoint of designing energy-efficient protocols, the DAMA scheme is better since transmission contention resolution results in high energy consumption [10]. In a DAMA or reservation based protocol, collisions are avoided by reserving channels. Hence, there is no need for retransmission of packets lost due to collision. As in the paper by Nakano *et al.* [8], we are interested in designing a reservation based DAMA protocol for permutation routing on a radio network. To make such a protocol energy efficient, we need to measure its performance against two conflicting criteria, **i.** the overall number of time slots required for the protocol to terminate should be minimized and **ii.** the total number of slots for which an individual station is awake should be minimized.

The permutation routing problem is a useful abstraction for most routing problems in distributed systems. In the permutation routing problem, each station of a p-station RN stores $\frac{n}{p}$ packets and each packet has a unique destination. The task is to send the packets to their destination stations. The permutation routing problem has been explored for RNs recently in several papers. Nakano *et al.* [7] have designed a protocol that runs in $\frac{2n}{k} + k - 1$ time slots subject to k, the number of channels satisfying $k \leq \sqrt{\frac{p}{2}}$. The protocol by Nakano *et al.* [7] run extremely fast at the expense of high energy consumption, since every station must be awake for $\frac{2n}{k} + k - 1$ time slots. Note that since each station initially stores $\frac{n}{p}$ packets and receives $\frac{n}{p}$ packets, in the best case it is sufficient if a station is awake only for $\frac{2n}{p}$ slots.

Nakano *et al.* [8] have recently published a protocol for energy-efficient permutation routing in an RN. Their protocol routes n packets in a k-channel, p-station RN in at most $(2d + 2b + 1)\frac{n}{k} + k$ time slots with no station being awake for more than $(4d + 7b - 1)\frac{n}{p}$ time slots, where $d = \left\lceil \frac{\log \frac{p}{k}}{\log \frac{n}{p}} \right\rceil$, $b = \left\lceil \frac{\log k}{\log \frac{n}{p}} \right\rceil$, and $k \leq \frac{p}{2}$. Nakano *et al.* [8] argue that for most practical systems, the number of channels is much less than the number of stations and in turn, the number of stations is much less than the number of packets, i.e., $k \ll p \ll n$.

We show that the permutation routing problem of n packets on an $RN(p, k)$ can be solved in $\frac{2n}{k} + \left(\frac{p}{k}\right)^2 + p + 2k^2$ slots and each station needs to be awake for at most $\frac{6n}{p} + \frac{2p}{k} + 8k$ slots. Our protocol is more efficient both in terms of total number of slots and the number of slots each station is awake compared to the protocol by Nakano *et al.* [8] when $k \ll p \ll n$.

The rest of the paper is organized as follows. In Section 2, we present some preliminaries and give a brief overview of our protocol. We present our permutation routing protocol in Section 3.

2 Preliminaries

We consider a radio network $RN(p, k)$ with p stations and k channels. The i-th station is denoted by $S(i)$, $1 \leq i \leq p$. As mentioned before, we assume that the stations have been initialized by running an initializing protocol as in [6]. Hence, each station has a unique ID in the range $[1, p]$. Each station holds $\frac{n}{p}$ packets to be routed. Also, each station is the recipient of $\frac{n}{p}$ packets. Each packet has a unique destination address which is the address of one of the p stations. Station $S(i)$ knows the destinations of all the $\frac{n}{p}$ packets that it holds, however, $S(i)$ does not know which stations will send it packets. As in [6,8], we assume that each station has a local clock which keeps synchronous time by interfacing with a Global Positioning System (GPS). Time is divided into slots and all packet transmissions take place at slot boundaries. Each station has the computing capabilities of a laptop computer.

There are k transmission channels in the $RN(p,k)$. These k channels are denoted by $C(1), C(2), \ldots, C(k)$. One data packet can be transmitted in a time slot. In each time slot, a station can tune to one of the k channels and/or transmit a packet on one of the channels. These two channels may or may not be distinct. As in the paper by Nakano *et al.* [8], we assume that $k \leq \sqrt{\frac{p}{2}}$. This is a reasonable assumption, since in most real life situations, the number of channels is much less than the number of stations, i.e., $k \ll p$. If we want to design a protocol without collision, we need to ensure that only one station transmits a packet over a given channel in a time slot.

It is easy to design a simple protocol for permutation routing if energy efficiency is not an issue. Assume that there is only one channel, i.e., $k = 1$. Each station transmits its $\frac{n}{p}$ packets one by one by taking turn. Every other station receives a packet in every time slot and either accepts or rejects it depending on whether the packet is destined for it. It is easy to see that this protocol terminates in n slots, but every station should be awake for all of these n time slots. Hence, this protocol is not efficient in terms of power usage by each station.

Nakano *et al.* [8] have presented several protocols which are more energy efficient compared to the above protocol. The basic protocol in [8] runs in p rounds, in the j-th round all the packets destined for station $S(j)$ are routed to $S(j)$. In the j-th round, first a reservation protocol is run, when each station reserves the slots it requires for transmitting its packets destined for $S(j)$. Once the reservation protocol is complete, each station $S(i)$ knows exactly when it should wake up to send the packets destined for $S(j)$ among its $\frac{n}{p}$ packets. Hence, during the packet routing stage, each station needs to be awake for $\frac{n}{p}$ time slots for sending all of its $\frac{n}{p}$ packets and an additional $\frac{n}{p}$ time slots for receiving all the $\frac{n}{p}$ packets destined for it. Nakano *et al.* [8] have presented several efficient variations of this basic protocol.

The basic reservation protocol in [8] is the following. Consider for the time being that we have an $RN(p,1)$, i.e., a single channel RN populated by p stations. Station $S(i)$ has n_i items to transmit to station $S(j)$ for $1 \leq i \leq p$, $\sum_{i=1}^{p} n_i = \frac{n}{p}$, as station $S(j)$ receives $\frac{n}{p}$ items. The reservation protocol takes $p-1$ slots. In the first slot $S(1)$ sends n_1 to $S(2)$, in the second slot $S(2)$ sends $n_1 + n_2$ to $S(3)$ and in general in the i-th time slot, $1 \leq i \leq p-1$ station $S(i)$ sends $n_1 + n_2 + \ldots + n_i$ to station $S(i+1)$. Note that each station is awake only for two time slots, one slot for receiving a packet from the station just before it and one slot for sending a packet to the next station. At the end of the protocol, each station $S(i)$ knows $n_1 + n_2 + \ldots + n_{i-1}$. Hence, during the packet routing stage station $S(i)$ will wake up at time slot $n_1 + n_2 + \ldots + n_{i-1} + 1$ and start transmitting its n_i packets to station $S(j)$. After $n_1 + n_2 + \ldots + n_i$ time slots station $S(i)$ will go to sleep again. Note that this is a prefix sum computation over the integers n_i.

2.1 An Overview of Our Protocol

The key idea behind our protocol is the following. We divide the p stations $S(i), 1 \leq i \leq p$ into k groups $G(1), G(2), \ldots, G(k)$ with $\frac{p}{k}$ stations in each group. Note that initially each group $G(j)$ holds $\frac{n}{p} \times \frac{p}{k} = \frac{n}{k}$ packets and also each group $G(j)$ is the recipient of $\frac{n}{k}$ packets.

Our protocol consists of two stages. In the first stage, stations in each group $G(j), 1 \leq j \leq k$, receive all the $\frac{n}{k}$ packets that have destination addresses in $G(j)$. We perform this packet transmission by utilizing all the k channels in the following way. We assign channel $c(m), 1 \leq m \leq k$ exclusively to the stations in group $G(m)$. Note that the stations in $G(m)$ potentially hold packets whose destinations are in all the groups $G(1), G(2), \ldots, G(k)$. The $\frac{p}{k}$ stations in $G(m)$ use the channel $C(m)$ to transfer all these packets to their respective destination groups. However, at this point the packets are only sent to some station in their destination group, not necessarily to the correct destination station.

Once the first stage is complete, each group $G(m), 1 \leq m \leq k$ has received all the $\frac{n}{k}$ packets whose destinations are stations in $G(m)$. In the second stage, we again assign channel $C(m)$ to $G(m)$ and route the packets to their correct destination stations.

3 Permutation Routing on an $RN(p,k)$

3.1 First Stage

We first explain some notations. The packets in station $S(i)$ with destination address in group $G(j)$ are denoted by $S(i,j)$, $1 \leq i \leq p, 1 \leq j \leq k$. Consider group $G(m)$ and its $\frac{p}{k}$ stations $S((m-1)\frac{p}{k}+1), S((m-1)\frac{p}{k}+2), \ldots, S(m\frac{p}{k})$. We use the notation $SUM(m,j) = S((m-1)\frac{p}{k}+1,j) + S((m-1)\frac{p}{k}+2,j) + \ldots + S(m\frac{p}{k},j)$ and the notation $sum(m,j,r) = S((m-1)\frac{p}{k}+1,j) + S((m-1)\frac{p}{k}+2,j) + \ldots + S((m-1)\frac{p}{k}+r,j), 1 \leq r \leq \frac{p}{k}$. In other words, $SUM(m,j)$ is the number of packets in the stations in $G(m)$ with destinations in group $G(j)$ and $sum(m,j,r)$ is the prefix sum of the number of packets in stations $S((m-1)\frac{p}{k}+1), \ldots, S((m-1)\frac{p}{k}+r)$ with destinations in group $G(j)$. In particular, $SUM(m,j)$ and $sum(m,j,\frac{p}{k})$ are the same integer. Note that we can compute $SUM(m,j)$ as well as $sum(m,j,r), 1 \leq r \leq \frac{p}{k}$ in $G(m)$ by running the basic reservation protocol by Nakano *et al.* [8] as described in Section 2. We will henceforth refer to the computations of $SUM(m,j)$ and $sum(m,j,r)$ as prefix sum computations. We need the following lemma for the computation in this phase.

Consider a group $G(j)$ and its $\frac{p}{k}$ stations. The $\frac{p}{k}$ stations are destinations for $\frac{n}{k}$ packets overall. In the first stage, we want to send all the packets with destinations in $G(j)$ to the stations in $G(j)$. Moreover, for each group $G(m)$, we want to send the $SUM(m,j)$ packets to a distinct set of stations in $G(j)$. In other words, each station in $G(j)$ will receive packets from stations in only one group $G(m)$, $1 \leq m \leq k$.

Lemma 1. *We can send all the $\frac{n}{k}$ packets which are destined for $G(j)$ to the stations in $G(j)$, with the restriction that each station in $G(j)$ receives packets from only one group, if the following conditions are satisfied:*
– *Each station in $G(j)$ will receive at most $\frac{2n}{p}$ packets, and*
– $k \leq \sqrt{\frac{p}{2}}$

Proof. In the worst case, $k-1$ groups will have one packet each with destination in $G(j)$ and the remaining $\frac{n}{k}-k+1$ packets with destination in $G(j)$ will come from a single group, say, $G(s), s \leq k$. Since each station in $G(j)$ receives packets from only one group, $k-1$ distinct stations will receive one packet each from the $k-1$ groups. Hence, the remaining $\frac{p}{k}-k+1$ stations in $G(j)$ should receive the remaining $\frac{n}{k}-k+1$ packets. In other words, each of the $\frac{p}{k}-k+1$ stations will receive $\frac{\frac{n}{k}-k+1}{\frac{p}{k}-k+1} = \frac{n-k^2+k}{p-k^2+k}$ packets. For the lemma to hold, $\frac{n-k^2+k}{p-k^2+k} \leq \frac{2n}{p}$. Simplifying, we get the inequality,

$$k^2 \leq \frac{np}{2n-p}+k$$

If we replace the maximum value $\sqrt{\frac{p}{2}}$ for k, this inequality reduces to

$$\frac{p}{2} \leq \frac{np}{2n-p}+\sqrt{\frac{p}{2}}$$

Now, $\frac{np}{2n-p} > \frac{np}{2n} = \frac{p}{2}$, and hence $\frac{p}{2} < \frac{np}{2n-p}+\sqrt{\frac{p}{2}}$. Since this inequality is correct, the statement of the lemma holds.

Step 1.
In the first step, we assign channel $C(m)$ to group $G(m)$, $1 \leq m \leq k$ and do the following computations in parallel in all the groups. First in k rounds of prefix sum computations, we compute $SUM(m,j)$ and $sum(m,j,r)$, for $1 \leq j \leq k$ and $1 \leq r \leq \frac{p}{k}$. Each round of this computation requires $(\frac{p}{k}-1)$ slots and hence overall $(p-k)$ slots. Since in each round each station is awake for 2 slots, overall each station is awake for $2k$ slots.

After this computation, the last station in $G(m)$ i.e., $S(m\frac{p}{k})$ will hold the k integers $SUM(m,j)$, $1 \leq j \leq k$. The station $S(m\frac{p}{k})$ now computes the k indices $index(m,j) = \lceil SUM(m,j)/\frac{2n}{p} \rceil$, for $1 \leq j \leq k$. The purpose of computing $index(m,j)$ is to determine how many stations in $G(j)$ will receive packets from $G(m)$. Recall that from Lemma 1, we need to send upto $\frac{2n}{p}$ packets to each station. This is clearly a local computation step and does not require any slot.

Step 2.
Since the stations in group $G(j), 1 \leq j \leq k$ will receive packets from different groups over different channels, we need to inform each station in $G(j)$ at which time slot and over which channel its packets will arrive. We use the integers $index(m,j)$ computed in Step 1 for this computation.

We use a single channel for the computation in this step. The computation is done in k rounds. Consider the s-th round, $1 \leq s \leq k$. We do a prefix sum computation of the integers $index(1, s), index(2, s), \ldots, index(k, s)$ for the k groups. This prefix sum computation is again done by the basic protocol of Nakano *et al.* [8], as discussed in Section 2 and the last stations from each group participate in this computation. We denote $pre\_index(r, s) = index(1, s) + index(2, s) + \ldots, index(r-1, s)$. $pre\_index(r, s) + 1$ is the index of the first station in $G(s)$ that will receive packets from stations in $G(r), 1 \leq r \leq k$. The computation in each round requires $k - 1$ slots and hence the overall slot requirement is $k^2 - k$. Further, the last station from each group needs to be awake for 2 slots in each round and hence overall $2k$ slots over k rounds.

Note that the last station in each group now can inform all the stations in the group the integer $pre\_index(r, s)$. We can now use a different channel in each group and broadcast these k integers within the group in k slots. Each station needs to be awake for k slots for recieving these k integers. Hence, the overall slot requirement for this step is $k^2 - k + k = k^2$ and each station remains awake for $2k + k = 3k$ slots.

Step 3.

When we send the $\frac{n}{k}$ packets from group $G(m)$ to stations in all the other groups, we plan to use only $\frac{n}{k}$ slots. Hence, we need to inform each station in $G(m)$ when it should transmit its packets. Recall that the stations in $G(m)$ will use the channel $C(m)$ for transmitting the packets. We plan to transmit the packets from stations in $G(m)$ in the following way. First, the stations in $G(m)$ will transmit packets with destinations in $G(1)$ one by one, starting from the first station in $G(m)$ and until the last station. Then the packets with destinations in $G(2)$ will be transmitted and so on until the packets for $G(k)$ are transmitted. Hence, each station in $G(m)$ should know when it should start transmitting its packets to the k different groups. To simplify notations, we denote the j-th station in group $G(m)$ by $G(m, j), 1 \leq j \leq \frac{p}{k}$. Each station $G(m, j)$ will wake up k times to transmit its share of k batches of packets to the k groups. By $start\_time(m, j, q), 1 \leq m, q \leq k$, we denote the slot number when station $G(m, j)$ will start transmitting to a station in group $G(q)$. Our aim in this step is to compute all these starting slots. We use a different channel for each group and the computation is done in parallel in each group. We discuss the computation in $G(m)$.

Recall that the last station $G(m, \frac{p}{k})$ holds the k integers $SUM(m, j), 1 \leq m \leq q$ at the end of the computation in Step 1. Also, each station $G(m, j)$ holds the k prefix sums $sum(m, j, r), 1 \leq m, r \leq k$. Station $G(m, j)$ should send the packets to station $G(s)$ only after all the packets with destinations in stations $G(1), G(2), \ldots, G(s-1)$ as well as packets with destinations in $G(s)$ from all the stations $G(m, 1), G(m, 2), \ldots, G(m, j-1)$ have been sent. Note that the prefix sums $sum(m, j, s)$ computed in Step 1 already holds the latter integer. We need only to compute the quantity $SUM(m, 1) + SUM(m, 2) + \ldots + SUM(m, s-1)$. This is again a prefix sum computation of k quantities. The last station $G(m, \frac{p}{k})$ can compute this and in k slots broadcast the k prefix sums to all the stations in $G(m)$. Each station needs to be awake for k slots. After this each station in $G(m)$ can compute $start\_time(m, j, q)$ for $1 \leq j \leq \frac{p}{k}$ and $1 \leq q \leq k$.

Step 4.

After Step 3, each station knows the slot when it should start transmitting its packets. Each station also knows the address of the station where its packets will go. However, each station still does not know when to wake up to receive the packets. In other words, we need to inform each station when it will start receiving the packets and over which channel.

Recall that in Step 2, we have computed the integers $pre\_index(r, s)$ in the last stations of each group. In particular, $pre\_index(r, s) + 1, 1 \le s \le k$ is the index of the first station in $G(s)$ that will receive packets from stations in $G(r), 1 \le r \le k$. Consider the last station $G(r, \frac{p}{k})$ in group $G(r)$. After Step 2, $G(r, \frac{p}{k})$ has k such integers $pre\_index(r, s), 1 \le s \le k$. Also, in Step 1 we have computed k integers $SUM(r, s), 1 \le s \le k$ in $G(r, \frac{p}{k})$. If we broadcast both $SUM(r, s)$ and $pre\_index(r, s)$ to all the stations in $G(s)$, each station $G(s, t), 1 \le t \le \frac{p}{k}$ in $G(s)$ can identify the following :

- whether the index t of $G(s, t)$ within $G(s)$ is the same as the integer $pre\_index(r, s) + 1$,
- whether $G(s, t)$ will receive any packet from the stations in $G(r)$. Recall that each station will receive at most $\frac{2n}{p}$ packets. If $t \ne pre\_index(r, s)+1$, $G(s, t)$ will receive a packet if and only if $(pre\_index(r, s) + 1) + \left\lceil \frac{SUM(r,s)}{\frac{2n}{p}} \right\rceil = t$. In other words, $G(s, t)$ may not be the first station to receive packets from $G(r)$, but still it may receive some packets from $G(r)$.

Assume that $G(s, t_1)$ is the first station in $G(s)$ to receive packets from $G(r)$ and $G(s, t_2), t_2 > t_1$, is another station that receives packets from $G(r)$. Then $G(s, t_2)$ starts receiving its packets at time slot $\frac{2n}{p} \times (t_2 - t_1) + 1$ as the stations starting from $G(s, t_1)$ and until $G(s, t_2 - 1)$ each receives $\frac{2n}{p}$ packets due to Lemma 1. Hence, each station can decide exactly at which time slot it will start receiving packets.

The computation in this step is done in k rounds with k slots in each round. In the r-th slot of the i-th round, $1 \le i, r \le k$, the last station from $G(r)$ broadcasts the integers $pre\_index(r, i), 1 \le r \le k$ and $SUM(r, i)$ to all the stations in $G(i)$. Note that all the stations in $G(i)$ need to be awake for receiving these k broadcasts in the i-th round. Hence, overall this step takes k^2 slots. The last station in each group has to broadcast k items and has to receive k items. Hence, each station needs to be awake for at most $2k$ time slots.

Step 5.

Finally, in this step each group $G(i), 1 \le i \le k$ transmits its $\frac{n}{k}$ packets to all the other groups in $\frac{n}{k}$ slots. Since each station has to transmit $\frac{n}{p}$ packets and receive at most $\frac{2n}{p}$ packets, each station should be awake for at most $\frac{3n}{p}$ slots.

The total number of slots required for completion of Stage 1 is $\frac{n}{k} + p + 2k^2$ and each station needs to be awake for at most $\frac{3n}{p} + 8k$ slots. This is calculated by adding the slot requirements and awake times for Steps 1-5 in Stage 1.

3.2 Second Stage

At the end of Stage 1, stations in $G(i), 1 \leq i \leq k$ hold all the $\frac{n}{k}$ packets with destination in $G(i)$. Our task in this stage is to route all these $\frac{n}{k}$ packets within each group to their correct destinations. This routing is done in parallel in all the groups using channel $C(i), 1 \leq i \leq k$ for $G(i)$ and we discuss the routing only in $G(i)$.

First, each station sorts all its packets according to their destination indices and there are $\frac{p}{k}$ destination stations in $G(i)$. Note that, this sorting can be done locally in each station and does not require any slot. We denote the j-th station in $G(i)$ by $G(i,j), 1 \leq j \leq \frac{p}{k}$. Further, the number of packets in $G(i,j)$ with their destination station $G(i,m)$ is denoted by $l(m,j), 1 \leq m \leq \frac{p}{k}$. For example, the integer $l(2,3)$ denotes the number of packets in station $G(i,3)$ with destination station $G(i,2)$. All the $l(m,j)$ values can be computed locally in each station after the sorting.

Now the routing proceeds in $\frac{p}{k}$ stages. In the m-th stage, the stations in $G(i)$ route all the packets with destination station $G(i,m)$. First, a prefix sum of the integers $l(m,j), 1 \leq j \leq \frac{p}{k}$ is computed by the basic protocol of Nakano *et al.* [8] in $\frac{p}{k}$ slots. Each station needs to be awake for 2 slots during this prefix computation. Next, station $G(i,m)$ keeps awake for $\frac{n}{p}$ slots and all the other stations transmit their packets one after another. The slot when a station will transmit its packets is determined by the prefix sum that it has got.

Hence, after $\frac{p}{k}$ rounds of the above routing, all the stations receive their packets. The total slot requirement for computing the prefix sums in $\frac{p}{k}$ rounds is $(\frac{p}{k})^2$. The total slot requirement for routing the packets is $\frac{n}{k}$ as this is the total number of packets routed. Each station needs to be awake for $\frac{2p}{k}$ slots during the prefix sum computations and at most $\frac{2n}{p}$ slots for transmitting its packets. Also, each station needs to be awake for $\frac{n}{p}$ slots for receiving its packets. Hence, each station needs to be awake for $\frac{3n}{p} + \frac{2p}{k}$ slots overall in this stage.

Theorem 1. *The permutation routing problem of n packets on an* $RN(p,k)$ *can be solved in* $\frac{2n}{k} + (\frac{p}{k})^2 + p + 2k^2$ *slots and each station needs to be awake for at most* $\frac{6n}{p} + \frac{2p}{k} + 8k$ *slots.*

In Figure 1, we compare our protocol and the protocol by Nakano *et al.* [8]. It is clear that our protocol is more efficient both for T_t and T_a when $k \ll p \ll n$. Our protocol is worse than the protocol in [8] when n is not very large compared to p, or the number of channels k is relatively smaller. However, the improvements in both T_t and T_a are noticeable when n is much larger than p. For example, for $p = 100, n = 100,000$ and $k = 16$, our protocol terminates in 13,148 slots compared to the 71,432 slots required by the protocol in [8]. For the same case, in our protocol each station remains awake for 6,140 slots compared to 10,000 slots in [8]. In most real-life situations the number of packets n is usually much larger than the number of stations and we expect that our protocol is more suitable for such situations.

p =			100			1,000		
n =			1,000	10,000	100,000	10,000	100,000	1,000,000
k=2	[13]	T_t	3,502	25,002	250,002	45,002	350,002	2,500,002
		T_a	140	1,000	10,000	180	1,400	10,000
	our	T_t	3,608	12,608	102,608	261,008	351,008	1,251,008
		T_a	176	716	6,116	1,076	1,616	7,016
k=4	[13]	T_t	1,754	12,504	125,004	22,504	175,004	1,250,004
		T_a	140	1,000	10,000	180	1,400	10,000
	our	T_t	1,257	5,757	50,757	68,532	113,532	563,532
		T_a	142	682	6,082	592	1,132	6,532
k=8	[13]	T_t	1,001	7,147	71,432	11,258	87,508	625,008
		T_a	140	1,000	10,000	180	1,400	10,000
	our	T_t	622	2,872	25,372	19,253	41,753	266,753
		T_a	149	689	6,089	374	914	6,314
k=16	[13]	T_t	1,001	7,147	71,432	5,641	31,266	312,516
		T_a	140	1,000	10,000	210	1,000	10,000
	our	T_t	773	1,898	13,148	6,606	17,856	130,356
		T_a	200	740	6,140	313	853	6,253

Fig. 1. A comparison between our protocol and that by Nakano *et al.* [8]. T_t and T_a respectively denote the number of slots for completion and the maximum number of slots each station needs to be awake.

References

1. N. Abramson, "Multiple access in wireless digital networks", *Proc. IEEE*, Vol. 82, pp. 1360-1370, 1994.
2. D. Bertzekas and R. Gallager, *Data Networks*, 2nd Edition, Prentice Hall, 1992.
3. N. Bambos and J. M. Rulnick, "Mobile power management for wireless communication networks", *Wireless Networks*, Vol. 3, pp. 3-14, 1997.
4. M. Fine and F. A. Tobagi, "Demand assignment multiple access schemes in broadcast bus local area networks", *IEEE Trans. Computers*, Vol. 33, pp. 1130-1159, 1984.
5. M. Gerla and T. C. Tsai, "Multicaster, mobile, multimedia radio network", *Wireless Networks*, Vol. 1, pp. 255-265, 1995.
6. K. Nakano and S. Olariu, "Randomized initialization protocols for radio networks", *IEEE Trans. Parallel and Distributed Systems*, Vol. 11, pp. 749-759, 2000.
7. K. Nakano, S. Olariu and J. L. Schwing, "Broadcast-efficient protocols for mobile radio networks", *IEEE Trans. Parallel and Distributed Systems*, Vol. 10, pp. 1276-1289, 1999.
8. K. Nakano, S. Olariu and A. Y. Zomaya, "Energy-efficient permutation routing in radio networks", *IEEE Trans. Parallel and Distributed Systems*, vol. 12, No. 6, pp. 544-557, 2001.
9. S. Singh and C. S. Raghavendra, "PAMAS – Power aware multi-access protocol with signalling for ad-hoc networks", *ACM Computer Comm. Review*, Vol. 28, pp. 5-26, 1998.
10. K. Sivalingam, M. B. Srivastava and P. Agarwal, "Low power link and access protocols for wireless multimedia networks", *Proc. IEEE Vehicular Technology Conference (VTC '97)*, 1997.

Simple Mutual Exclusion Algorithms Based on Bounded Tickets on the Asynchronous Shared Memory Model

Masataka Takamura and Yoshihide Igarashi

Department of Computer Science, Gunma University, Kiryu, Japan 376-8515
{takamura,igarashi}@comp.cs.gunma-u.ac.jp

Abstract. We propose two simple algorithms based on bounded tickets for the mutual exclusion problem. These are modifications of the Bakery algorithm. An unattractive property of the Bakery algorithm is that the shared memory size is unbounded. The first algorithm based on bounded tickets uses one extra process that does not correspond to any user. It is lockout-free and mutual exclusion on the asynchronous single-writer/multi-reader shared memory model. We then modify it to reduce the shared memory size with the cost of another extra process. The maximum waiting time using each of them is bounded by $(n-1)c + O(nl)$, where n is the number of users, l is an upper bound on the time between two successive atomic steps, and c is an upper bound on the time that any user spends using the resource. The shared memory size needed by the first algorithm and the second algorithm are $(n+1)(1+\lceil\log(2n+1)\rceil)$ bits and $n(2+\lceil\log n\rceil)+2$ bits, respectively.

1 Introduction

Mutual exclusion is a problem of managing access to a single indivisible resource that can only support one user at a time. An early algorithm for the mutual exclusion problem proposed by Dijkstra [8] guarantees mutual exclusion, but it does not guarantee lockout-freedom. Subsequent algorithms are improvements on Dijkstra's algorithm by guaranteeing fairness to the different users [2,11,12,14,15,16,18,19] and by weakening the type of shared memory [4,5,6,7,10,14,15,16].

The Bakery algorithm for the mutual exclusion problem is due to Lamport [14]. It works in a way like a queue of customers in a bakery, where customers draw tickets. It only uses single-writer/multi-reader shared variables, and satisfies the first-in first-served property. These are attractive features. An unattractive property of the Bakery algorithm is that it uses unbounded size shared variables. The problem of bounding the size of shared variables is important and has been much studied [1,2,9,13,20]. One of the solutions to this problem is an algorithm using a general technique, bounded time-stamping [9]. The algorithms given in [1,13] are also solutions to this problem. The shared memory size of the bounded Bakery algorithm given in [1] is $O(n^2)$, where n is the number of the

O.H. Ibarra and L. Zhang (Eds.): COCOON 2002, LNCS 2387, pp. 259–268, 2002.

users in the system. The algorithm given in [13] uses single-writer shared memory of size $O(n \log n)$ bits together with a multi-writer shared variable although any concurrency does not occur concerning the multi-writer shared variable.

In this paper we also propose simple algorithms based on bounded tickets for the mutual exclusion problem on the asynchronous single-writer/multi-reader shared memory model. Initially we modify the Bakery algorithm so that it requires only bounded size single-writer/multi-reader shared variables. This provisional version guarantees mutual exclusion under the condition that there is always a user trying to use the resource. In order to remove this condition we use an extra process in our first algorithm called *n-bmexcl*1. The algorithm using the extra process guarantees lockout-freedom and mutual exclusion. The existence of an extra process may be unattractive feature, but the shared memory size for our algorithm is smaller than that of the algorithm given in [1]. We modify our algorithm to make a further reduction of the shared memory size with the penalty of the number of extra processes. For each of the algorithms proposed in this paper, the time from when a particular user tries to use the resource until it is actually granted to use the resource is bounded by $(n-1)c+O(nl)$, where n is the number of the users, l is an upper bound on the time between two successive atomic steps, and c is an upper bound on the time that any user is granted to use the resource. The waiting time for obtaining the grant to use the resource by the algorithms given in [1,13] is also $(c-1)c+O(nl)$. The waiting time by the original Bakery algorithm is bounded by $(n-1)c+O(n^2l)$ [17]. The shared memory size needed by our first algorithm and second algorithm are $(n+1)(1+\lceil \log(2n+1) \rceil)$ bits and $n(2+\lceil \log n \rceil)+2$ bits, respectively. The algorithms proposed in this paper is an improvement of the size of shared variables on the original Bakery algorithm and the algorithm given in [1].

2 Preliminaries

The computation model used here is the asynchronous single-writer/multi-reader shared memory model. It is a collection of processes and shared variables. The mutual exclusion problem is to devise a protocol of how to allocate a single indivisible and nonsharable resource among n users $U_1, \ldots, U_n$. Interactions between a process and its corresponding user are by input actions from the user to the process and by output actions from the process to the user. Each process is considered to be a state machine with signals entering and leaving the process, representing its input and output actions. All communication among the processes is via the shared memory. Such a state machine is called an I/O automaton [17]. We assume that the order of actions by processes can be serialized so that the serialization is consistent with the behavior of the whole system.

A user with access to the resource is modeled as being in the critical region. When a user is not involved in the resource, it is said to be in the remainder region. In order to gain admittance to the critical region, a process executes a trying protocol. The duration from the start of executing the trying protocol to the entrance of the critical region is called the trying region. After the end of the use of the resource by a user, the corresponding process executes an exit protocol. The duration of executing the exit protocol is called the exit region. These procedures can be repeated in cyclic order, from the remainder region to the trying region, to the critical region, to the exit region, and then back again to the remainder region.

We assume that the n processes are numbered $1, \dots, n$. Each process i corresponds to user U_i ($1 \leq i \leq n$). An extra process is introduced in our first algorithm named *n-bmexcl*1, and two extra processes are introduced in our second algorithm named *n-bmexcl*2. The extra process used in *n-bmexcl*1, does not correspond to any user, but each extra process used in *n-bmexcl*2 corresponds to a user. For *n-bmexcl*1 the inputs to process i from user U_i are try_i, which means a request by user U_i for access to the resource, and $exit_i$, which means an announcement of the end of the use of the resource by U_i. For *n-bmexcl*1 the outputs from process i to user U_i are $crit_i$, which means the grant of the resource to U_i, and rem_i, which tells U_i that it can continue with the rest of its work. For users U_{n-1} and U_n of *n-bmexcl*2, input and output correspondence between these users and processes are modified. A system solving the mutual exclusion problem should satisfy the following conditions.

(1) There is no reachable system state in which more than one user is in the critical region.
(2) If at least one user is in the trying region and no user is in the critical region, then at some later point some user enters the critical region.
(3) If a user is in the exit region, then at some later point the user enters the remainder region.
(4) If all users always return the resource, then any user wishing to enter the critical region eventually does so.

Conditions (1), (2), (3) and (4) above are called *mutual exclusion, progress for the trying region, progress for the exit region*, and *lockout-freedom*, respectively.

The following procedure *n-Bakery* is the Bakery algorithm quoted from [3,17], $(a, b) < (a', b')$ at line 07 means that $a < a'$, or $a = a'$ and $b < b'$. The entry section of the Bakery algorithm begins with a part called the doorway (from line 02 to line 04) where processes in the trying region obtain their tickets. Then processes with their tickets proceed to execute the major part of the algorithm. If process i completes executing the doorway before process j begins the doorway, process i enters the critical region before process j does so. Hence, the Bakery algorithm satisfies the first-in first-served property.

procedure *n-Bakery*
shared variables
for every $i \in \{1, \dots, n\}$:
choosing$(i) \in \{0, 1\}$, initially 0, writable by i and readable by all $j \neq i$;
ticket$(i) \in N$, initially 0, writable by i and readable by all $j \neq i$;

process i
input actions {inputs to process i from user U_i}: try_i, $exit_i$;

```
output actions {outputs from process i to user U_i}: crit_i, rem_i;
** Remainder region **
01: try_i:
02: choosing(i) := 1;
03: ticket(i) := 1 + max_{j≠i} ticket(j);
04: choosing(i) := 0;
05: for each j ≠ i do begin
06:     waitfor choosing(j) = 0;
07:     waitfor ticket(j) = 0 or (ticket(i), i) < (ticket(j), j) end;
08: crit_i;
** Critical region **
09: exit_i:
10: ticket(i) := 0;
11: rem_i;
```

The Bakery algorithm uses only single-writer/multi-reader shared variables, but for each i $(1 \leq i \leq n)$, the size of $ticket(i)$ is unbounded. The running time for the trying region by *n-Bakery* is bounded by $(n-1)c + O(n^2 l)$. The computation of the **for** statement at line 05 takes $O(n^2 l)$ time. The running time of this part can also be reduced to $O(nl)$ time by replacing the statements from line 05 to line 07 in *n-Bakery* with the statements from line 05 to line 10 in procedure *n-provision* in Section 3.

3 A Provisional Algorithm

The algorithm given in this section is a provisional one. It is a simple modification of the Bakery algorithm. For a technical reason, the domain of ticket numbers used by the algorithm is $\{-1, 0, \ldots, 2n-2\}$. A ticket with any number of $\{0, \ldots, 2n-2\}$ is called a regular ticket and the ticket with -1 is called the invalid ticket. The algorithm works correctly as a lockout-free mutual exclusion protocol under the condition that there always exists a process with a regular ticket after a process first gets a regular ticket. The order of the regular tickets is cyclic in *modulo* $2n-1$. It is not necessarily that regular ticket numbers appearing in the trying region at a time point in an execution are cyclically consecutive. We will show such a scenario in Example 1. Here a gap means a cyclically non-consecutive part in the set of regular tickets. For example, if $n = 5$ and the set of regular tickets appearing at a time point is $\{8, 0, 1, 4\}$, then from 2 to 3 and from 5 to 7 are gaps. Furthermore, there may be some processes with the same ticket number in the trying region. We define the order among the tickets in the trying region as follows: The first number just after the end of the *largest gap* is the smallest one. Starting from this number, the order is decided cyclically in *modulo* $2n-1$. For the set of regular tickets, $\{8, 0, 1, 4\}$, ticket 8 is the smallest one, and the order of these tickets are 8,0,1,4, since gap from 5 to 7 is larger than gap from 2 to 3. If two or more processes have the same ticket number, the order of these processes is the order of process identifiers.

We first define our terminology. Function *scanticket*() scans all shared variables $ticket(j)$ $(1 \leq j \leq n)$ and returns the set of pairs of regular tickets and process identifiers holding the regular tickets. Function $rmax(S)$ returns the largest ticket number in set S, if S is not the empty set, and returns -1 otherwise. Function $prev_i(S)$ returns the identifier of the largest process that is smaller than process i in the order of pairs of ticket numbers and process identifiers if process i is not the smallest one in S, and otherwise it returns an arbitrary process identifier except i itself.

```
procedure n-provision
shared variables
  for every i ∈ {1,...,n}:
    choosing(i) ∈ {0,1}, initially 0, writable by i and readable by all
        j ≠ i;
    ticket(i) ∈ {−1,0,...,2n − 2}, initially −1, writable by i and
        readable by all j ≠ i;
process i
  input/output actions: the same as in n-Bakery
  ** Remainder region **
 01:  try_i:
 02:  choosing(i) := 1;
 03:  ticket(i) := (1 + rmax(scanticket())) modulo 2n − 1;
 04:  choosing(i) := 0;
 05:  index := {1,2,...,n};
 06:  while index ≠ ∅ do
 07:     for each j ∈ index do
 08:        if choosing(j) = 0 then index := index − {j};
 09:  j := prev_i(scanticket());
 10:  waitfor ticket(j) = −1 or (ticket(i), i) < (ticket(j), j);
 11:  crit_i;
  ** Critical region **
 12:  exit_i:
 13:  ticket(i) := −1;
 14:  rem_i;
```

Example 1. Let $n = 5$. The following scenario is possible in a fair execution of *n-provision*. Assume $ticket(1) = 0, ticket(2) = 1, ticket(3) = 2, ticket(4) = 3$, and $ticket(5) = -1$ at a time point. Then process 5 enters the trying region. During the execution of $scanticket()$ by process 5, it observes $ticket(1) = 0$ but before observing $ticket(2)$ and $ticket(3)$, process 2 and process 3 quickly move to the critical region and exit the critical region in this order. Then process 5 observes $ticket(2) = -1$, $ticket(3) = -1$ and $ticket(4) = 3$, and obtains $ticket(5) = 4$ since the gaps observed by process 5 are from 1 to 2 and from 4 to 8 (in this case the largest gap is observed to be from 4 to 8). This example shows that process may observe more than one gap in a certain situation.

Observation 1. *If after a process first gets a regular ticket, $scanticket()$ never becomes empty at any time when it is called by any process, then the latest ticket number in the trying region is the first number of the largest gap in the set* *scanticket() called by the latest process. Note that this observation is not always correct if the size of the domain of regular tickets is less than $2n-1$. This is the reason why we take $\{-1, 0, \ldots, 2n-2\}$ as the domain of regular tickets.*

Theorem 1. *For an execution by procedure n-provision, if scanticket() never becomes empty after a process first gets a regular ticket then both mutual exclusion and lockout-freedom are satisfied in the execution.*

Proof Sketch. Assume that after a process first gets a regular ticket, *scanticket*() is always a non-empty set. From Observation 1, a new ticket number added at line 03 is cyclically larger by one than the largest ticket number among the ticket numbers in *scanticket*(). Hence, regular ticket numbers held by processes in the trying region are well ordered. Thus the order of ticket numbers in *scanticket*() can be correctly decided. Then the process, say process i, with the smallest ticket can send output signal $crit_i$ at line 11 to tell the corresponding user that the resource is now available. The smallest ticket is reset after the end of use of the resource at line 13 in the exit region. Therefore, mutual exclusion is satisfied in any fair execution by *n-provision* under the condition given in the theorem. Every step in any fair execution by *n-provision* is progressive if the condition given in the theorem is satisfied. Hence, the theorem holds. □

Theorem 2. *For any execution by n-provision such that scanticket() never becomes empty after a process first gets a regular ticket, the time from when a particular process enters its trying region until the process enters its critical region is bounded by $(n-1)c + O(nl)$. The total size of shared memory for n-provision is $n(2 + \lceil \log n \rceil)$ bits.*

Proof. The part from line 02 to line 04 of the program is called the doorway. The running time of the doorway is $O(nl)$ since each statement at lines 01, 02 and 04 can be executed in $O(l)$ time and the statement at line 03 can be executed in $O(nl)$ time. The **while** statement at lines 06 to 08 can be executed in $O(nl)$ time, since for any j such that $choosing(j) = 1$ in the doorway, $choosing(j)$ is set to 0 in $O(nl)$ time. The statement at line 09 can be executed in $O(nl)$ time. The statement at line 10 is completed in $(n-1)c + O(l)$ time, since the number of processes with smaller tickets at the beginning of executing the **waitfor** statement is at most $n-1$. The statement at line 11 is executed in $O(l)$ time. Hence, the running time in the trying region is $(n-1)c + O(nl)$. The algorithm needs n 1-bit shared variables, $choosing(i), 1 \leq i \leq n$, and n shared variables, $ticket(i)$ $(1 \leq i \leq n)$, each of which stores one of $2n$ distinct values, $-1, 0, 1, \ldots, 2n-2$. Hence, the total size of the shared memory is $n(2 + \lceil \log n \rceil)$ bits. □

In the case where *scanticket*() becomes empty at a time point during an execution of *n-provision*, it may not run properly as suggested in the next example.

Example 2. Let $n = 5$. The following scenario is possible in a fair execution of *n-provision*. Assume that $ticket(1) = 4$ and all other tickets are invalid at a time point. Then process 2 enters the trying region and observes $scanticket() = \{4\}$. Just after that observation by process 2, process 1 enters the critical region and quickly exits the critical region. Before process 2 sets $ticket(2) = 5$, process 3 enters the trying region and observes $scanticket() = \emptyset$. Then process 3 sets $ticket(3) = 0$. Immediately after that, process 4 enters the trying region and observes the two gaps, and then it sets its ticket to be 1. Consequently two gaps, $\{2,3,4\}, \{6,7,8\}$ are created. In this situation, a new comer to the trying region cannot decide which is the largest gap. That is, the new comer cannot decide which is the largest ticket, 0 or 5.

4 Mutual Exclusion Algorithms

We give two mutual exclusion algorithms that are modifications of *n-provision*. The first algorithm, called *n-bmexcl1*, uses an extra process that does not correspond to any user. It is first-in first-served as the Bakery algorithm. The extra process prevents the set of regular tickets from being empty. The order of tickets is defined as the order used in *n-provision*, but we take *modulo* $2n$ instead of *modulo* $2n-1$. The second algorithm, *n-bmexcl2*, is a marginal improvement on the first algorithm in the shared memory size.

```
procedure n-bmexcl1
shared variables
  for every i ∈ {1,2,...,n+1}:
    choosing(i) ∈ {0,1}, initially 0, writable by i and readable by all
      j ≠ i;
    ticket(i) ∈ {−1,0,...,2n−1}, initially −1 for 1 ≤ i ≤ n and
      0 for i = n+1, writable by i and readable by all j ≠ i;

process i  for 1 ≤ i ≤ n
  input/output actions: the same as in n-Bakery;
  ** Remainder region **
01: try_i:
02: choosing(i) := 1;
03: ticket(i) := (1 + rmax(scanticket())) modulo 2n;
04: choosing(i) := 0;
05: index := {1,2,...,n+1};
06: while index ≠ ∅ do
07:    for each j ∈ index do
08:       if choosing(j) = 0 then index := index − {j};
09: j := prev_i(scanticket());
10: waitfor ticket(j) = −1 or (ticket(i),i) < (ticket(j),j);
11: crit_i;
  ** Critical region **
12: exit_i:
13: ticket(i) := −1;
14: rem_i;

process n+1
  input/output actions:   none
```

01: **repeat**
02: $choosing(n+1) := 1$;
03: $ticket(n+1) := (1 + rmax(scanticket() - \{ticket(n+1)\}))$ $modulo\ 2n$;
04: $choosing(n+1) := 0$;
05: $index := \{1, 2, \ldots, n, n+1\}$;
06: **while** $index \neq \emptyset$ **do**
07: **for each** $j \in index$ **do**
08: **if** $choosing(j) = 0$ **then** $index := index - \{j\}$;
09: $j := prev_{n+1}(scanticket())$;
10: **waitfor** $ticket(j) = -1$ or $(ticket(n+1), n+1) < (ticket(j), j)$;
11: **waitfor** $|scanticket()| \geq 2$;
12: **forever**;

A ticket controlled by process $n+1$ always exists in the trying region. Hence, for *scanticket*() called by any process at any time point in a fair execution by *n-bmexcl*1, there is a unique largest gap in *scanticket*(). Any behavior of this extra process does not cause any harm. The mechanism of granting a process to enter the critical region in an execution by *n-bmexcl*1 is the same as in an execution by *n-provision*. The existence of the extra process affects the running time in the trying region of any other process by $O(nl)$ time. The next theorem is straightforward.

Theorem 3. *For any execution by n-bmexcl*1*, it satisfies lockout-freedom and mutual exclusion, and the time from when any user requests to use the resource until it is allowed to use the resource is bounded by* $(n-1)c + O(nl)$*. The total size of shared memory needed by n-bmexcl*1 *is* $(n+1)(1 + \lceil \log(2n+1) \rceil)$ *bits.*

The next algorithm, *n-bmexcl*2, is a variation of *n-bmexcl*1. For $n-1 \leq i \leq n$, the i-th user corresponds to process i and process $i+2$ in the next procedure, where inputs try_i and $exit_i$ are sent to process $i+2$, output $crit_i$ is sent from process i, and output rem_i is sent from process $i+2$.

procedure *n-bmexcl2*
shared variables
$flag(n-1) \in \{0, 1\}$, initially 0, writable by $n+1$ and readable by $n-1$;
$flag(n) \in \{0, 1\}$, initially 0, writable by $n+2$ and readable by n;
for every $i \in \{1, 2, \ldots, n\}$:
$choosing(i) \in \{0, 1\}$, initially 0, writable by i and readable by all $j \neq i$;
$ticket(i) \in \{-1, 0, 1, \ldots, 2n-2\}$, initially -1 for $1 \leq i \leq n-2$ and 0 for $n-1 \leq i \leq n$, writable by i and readable by all $j \neq i$;

process i for $1 \leq i \leq n-2$
input/output actions: the same as in *n-Bakery*;
** Remainder region **
01: try_i:
02: $choosing(i) := 1$;
03: $ticket(i) := (1 + rmax(scanticket()))\ modulo\ 2n-1$;

04: $choosing(i) := 0$;
05: $index := \{1, 2, \ldots, n\}$;
06 to 11: the same as lines 06 to 11 in *n-bmexcl*1;
 ** Critical region **
12 to 14: the same as lines 12 to 14 in *n-bmexcl*1;

process i for $n-1 \le i \le n$
 input actions: none;
 output action: $crit_i$;
01: **repeat**
02: $choosing(i) := 1$;
03: $ticket(i) := (1{+}rmax(scanticket() - \{ticket(i)\}))$ *modulo* $2n-1$;
04: $choosing(i) := 0$;
05 to 10: the same as lines 05 to 10 for process i $(1 \le i \le n-2)$;
11: $flag_i := flag(i)$;
12: **if** $flag_i = 1$ **then** $crit_i$;
13: **waitfor** $|scanticket()| \ge 2$;
14: **forever**;

process i for $n+1 \le i \le n+2$
 input actions: try_{i-2}, $exit_{i-2}$;
 output action: rem_{i-2};
 ** Remainder region **
01: try_{i-2}:
02: $flag(i-2) := 1$;
 ** Critical region **
03: $exit_{i-2}$:
04: $flag(i-2) := 0$;
05: rem_{i-2};

The running time for the trying region of *n-bmexcl*2 is the same as the running time for the trying region of *n-bmexcl*1 within a constant factor. In *n-bmexcl*2, there are n one-bit shared variables, $choosing(i)$, $1 \le i \le n$, n shared variables $ticket(i)$, $1 \le i \le n$ of $\lceil \log 2n \rceil$ bits, and two one-bit shared variables, $flag(n-1)$ and $flag(n)$. We, therefore, have the next theorem.

Theorem 4. *For any execution by n-bmexcl2, it satisfies lockout-freedom and mutual exclusion, and the time from when any user requests to use the resource until it is allowed to use the resource is bounded by* $(n-1)c + O(nl)$. *The total size of shared memory for n-bmexcl2 is* $n(2 + \lceil \log n \rceil) + 2$ *bits.*

5 Concluding Remarks

The algorithms proposed in this paper are improvements on the Bakery algorithm, but we use an extra process in *n-bmexcl*1 and two extra processes in *n-bmexcl*2. The shared memory size of each of our algorithms is smaller than the shared memory size of the algorithm given in [1], the latter does not use any extra process. We are interested in a problem whether further modifications are possible to remove the extra process(es) either in *n-bmexcl*1 or in *n-bmexcl*2 without any penalty.

References

1. U.Abraham, "Bakery Algorithms", Technical Report, Dept. of Mathematics, Ben Gurion University, Beer-Sheva, Israel, May 2001.
2. J.H.Anderson, "Lamport on mutual exclusion: 27 years of planting seeds", *Proceedings of the 27th Annual ACM Symposium on Principles of Distributed Computing*, Newport, Rhode Island, pp.3–12, 2001.
3. H.Attiya and J.Welch, "Distributed Computing: Fundamentals, Simulations and Advanced Topics", McGraw-Hill, New York, 1998.
4. J.E.Burns, "Mutual exclusion with linear waiting using binary shared variables", *ACM SIGACT News*, vol.10, pp.42–47, 1978.
5. J.E.Burns, P.Jackson, N.A.Lynch, M.J.Fischer, and G.L.Peterson, "Data requirements for implementation of N-process mutual exclusion using a single shared variable", *J. of the ACM*, vol.29, pp.183–205, 1982.
6. J.E.Burns, and N.A.Lynch, "Bounds on shared memory for mutual exclusion", *Information and Computation*, vol.107, pp.171-184, 1993.
7. A.B.Cremers and T.N.Hibbard, "Mutual exclusion of N processors using an $O(N)$-valued message variable", *5th International Colloquium on Automata, Languages and Programming*, Udine, Italy, *Lecture Notes in Computer Science*, vol.62, pp.165–176, 1978.
8. E.W.Dijkstra, "Solution of a problem in concurrent programming control", *Communications of the ACM*, vol.8, p.569, 1965.
9. D.Dolev and N.Shavit, "Bounded concurrent time-stamping", *SIAM J. on Computing*, vol.26, pp.418–455, 1997.
10. M.J.Fischer, N.A.Lynch, J.E.Burns, and A.Borodin, "Distributed FIFO allocation of identical resources using small shared space", *ACM Transactions on Programming Languages and Systems*, vol.11, pp.90–114, 1989.
11. Y.Igarashi, H.Kurumazaki, and Y.Nishitani, "Some modifications of the tournament algorithm for the mutual exclusion problem", *IEICE Transactions on Information and Systems*, vol.E.82-D, pp.368–375, 1999.
12. Y.Igarashi and Y.Nishitani, "Speedup of the n-process mutual exclusion algorithm", *Parallel Processing Letters*, vol.9, pp.475–485, 1999.
13. P.Jayanti, K.Tan, G.Friedland, and A.Katz, "Bounded Lamport's bakery algorithm", *Proceedings of SOFTSEM'2001*, *Lecture Notes in Computer Science*, vol.2234, Springer-Verlag, Berlin, pp.261–270, November 2001.
14. L.Lamport, "A new solution of Dijkstra's concurrent programming problem", *Communications of ACM*, vol.17, pp.453–455, 1974.
15. L.Lamport, "The mutual exclusion problem. Part II : Statement and solutions", *J. of the ACM*, vol.33, pp.327–348, 1986.
16. L.Lamport, "A fast mutual exclusion algorithm", *ACM Transactions on Computer Systems*, vol.5, pp.1-11, 1987.
17. N.A.Lynch, "Distributed Algorithms", Morgan Kaufmann, San Francisco, California, 1996.
18. G.L.Peterson, "Myths about the mutual exclusion problem", *Information Processing Letters*, vol.12, pp.115–116, 1981.
19. G.L.Peterson and M.J.Fischer, "Economical solutions for the critical section problem in a distributed system", *Proceedings of the 9th Annual ACM Symposium on Theory of Computing*, Boulder, Colorado, pp.91–97, 1977.
20. M.Takamura and Y.Igarashi, "A simplification of the Bakery algorithm based on bounded tickets for the mutual exclusion problem", Technical Report of IEICE, vol.101, no.376, COMP2001-45, pp.61–68, October 2001.

Time and Energy Optimal List Ranking Algorithms on the k-Channel Broadcast Communication Model

Koji Nakano

School of Information Science
Japan Advanced Insitute of Science and Technology
Tatsunokuchi, Ishikawa 923-1292, Japan
knakano@jaist.ac.jp

Abstract. A Broadcast Communication Model (BCM, for short) is a distributed system with no central arbiter populated by n processing units referred to as stations. The stations can communicate by broadcasting/receiving a data packet to one of k distinct communication channels. The main contribution of this paper is to present time and energy optimal list ranking algorithms on the BCM. We first show that the rank of every node in an n-node linked list can be determined in $O(n)$ time slots with no station being awake for more than $O(1)$ time slots on the single-channel n-station BCM. We then extend this algorithm to run on the k-channel BCM. For any small fixed $\epsilon > 0$, our list ranking algorithm runs in $O(\frac{n}{k})$ time slots with no station being awake for more than $O(1)$ time slots, provided that $k \leq n^{1-\epsilon}$. Clearly, $\Omega(\frac{n}{k})$ time is necessary to solve the list ranking problem for an n-node linked list on the k-channel BCM. Therefore, our list ranking algorithm on the k-channel BCM is time and energy optimal.

1 Introduction

A *Broadcast Communication Model* (BCM, for short) is a distributed system with no central arbiter populated by n processing units referred to as *stations* $S(1), S(2), \ldots, S(n)$. The fundamental characteristic of the model is the broadcast nature of communications. A data packet broadcast on a channel can be received by every station that have tuned to the channel. The nature of end units is immaterial: they can be processors in a parallel computing environment or radio transceivers in a wireless network. Likewise, the nature of the transmission channel is immaterial: it could be a global bus in a multiprocessor system or a radio frequency channel in a radio network. It is important to note that the BCM model provides a common generalization of bus-based parallel architectures, cluster computing environment, local area networks, and single-hop radio networks. Although the BCM is assumed to be operate in synchronous mode, we do not prescribe a particular synchronization mechanism. We feel that this is best left to the particular application. For example, in radio networks, synchronization may be provided by an interface to a commercially-available Global Positioning System [9].

O.H. Ibarra and L. Zhang (Eds.): COCOON 2002, LNCS 2387, pp. 269–278, 2002.

For simplicity, we assume that the BCM has the following collision detection capability [11]: In the BCM, the status of a channel is: *NULL:* if no station broadcasts on the channel in the current time slot; *SINGLE:* if exactly one station broadcasts on the channel in the current time slot; and *COLLISION:* if two or more stations broadcast on the channel in the current time slot. The status of a channel can be detected by stations that tune to it. Although our algorithms presented in this paper use the collision detection capability, it is not difficult to modify them such that they do not use this capability.

We assume that broadcasting/receiving data packets in the BCM is very costly. If the stations run on batteries and, therefore, saving battery power is exceedingly important, as recharging batteries may not be possible while on mission. It is well known that a station in the radio network expends power while its transceiver is active that is, while transmitting or receiving a packet [5,12]. Consequently, we are interested in developing algorithms that allow stations to power their transceiver off (i.e. go to sleep) to the largest extent possible. Accordingly, we judge the goodness of a algorithm by the following two yardsticks: *running time slots*: the overall number of time slots required by the algorithm to terminate *awake time slots*: for each individual station the total number of time slots when it has to be *awake* in order to broadcast/receive a data packet. As we are going to show later, the goals of optimizing these parameters are, of course, conflicting. It is relatively straightforward to minimize overall completion time at the expense of energy consumption.

A linked list is a basic data structure frequently used in many processing tasks. *A linked list* L of n nodes is specified by an array p such that $p(i)$ contains a pointer to the node following node i in the list of L. Figure 1 illustrates an example of a linked list. A node i is *the end* of the list if $p(i) = i$. Further, if there exists no node j such that $p(i) = j$, then node i is *the top* of the list. The list ranking problem asks to determine *the rank* of every node i $(1 \leq i \leq n)$, which is the distance to the end of the list. Nodes 2 and 8 in Figure 1 are the top and the end nodes of the list, respectively.

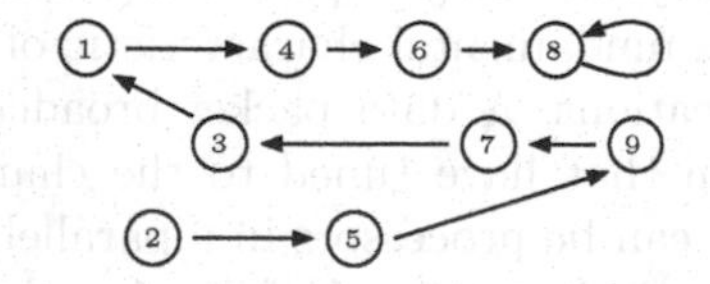

Fig. 1. An example of a linked list

The list ranking problem has been solved in several contexts [4,7,13]. It is well known that the list ranking problem can be solved in $O(\log n)$ time using n processors on the (EREW) PRAM [6,8]. This parallel list ranking algorithm uses *the pointer jumping technique* [1,8], which repeats changing each pointer such that a new pointer is the successor of the successor. The pointer jumping is repeated until all pointer points at the end of nodes. Further, it is known that the number of processors can be reduced to $\frac{n}{\log n}$ [4,8].

One of the straightforward strategies to design an energy-efficient algorithm on the BCM is to simulate a known PRAM algorithm. We are going to show that an energy-efficient list raking algorithm on the BCM can be obtained by simulating a known PRAM list ranking algorithm. First, it should be clear that any single step communication performed on the n-processor $O(n)$-memory-cell PRAM can be simulated by the n-station n-channel BCM in $O(1)$ time slots. This can be done by assigning $O(1)$ memory cells to each station. Then, read/write operations on the PRAM can be simulated using communication channels on the BCM in obvious way. Hence, any algorithm running in $O(\log n)$ time using n processors and $O(n)$ memory cells on the PRAM can be simulated by the n-station n-channel BCM in $O(\log n)$ time. Further, communication using n channels can be simulated on the k-channel BCM ($k \leq n$) in $O(\frac{n}{k})$ time slots. It is known that the list ranking problem can be solved in $O(\log n)$ time using n processors on the PRAM [6,8]. Thus, the list ranking problem can be solved in $O(\frac{n \log n}{k})$ time slots with each station being awake for at most $O(\log n)$ time slots. However this algorithm is not time and energy optimal.

The main contribution of this paper is to present time and energy optimal ranking algorithms on the BCM. Surprisingly, stations are awake for only $O(1)$ time slots in our list ranking algorithms. We first show that the rank of every node in an n-node linked list can be done in $O(n)$ time slots with no station being awake for more than $O(1)$ time slots on the single-channel n-station BCM. We then extend this algorithm to run on the k-channel BCM. For every small fixed $\epsilon > 0$, our algorithm runs in $O(\frac{n}{k})$ time slots with no station being awake for more than $O(1)$ time slots, provided that $k \leq n^{1-\epsilon}$. Clearly, every $p(i)$ must be broadcast at least once. Hence, $\Omega(\frac{n}{k})$ time is necessary to solve the list ranking problem for an n-node linked list on the k-channel BCM. Therefore, our algorithm is time and energy optimal.

2 List Ranking Using List Shrink

The main purpose of this section is to show fundamental techniques used in our time and energy-optimal list ranking algorithm.

We use a fundamental technique for solving the list ranking problem as follows: This technique uses two arrays of variables $q[i]$ and $r[i]$ for every i ($1 \leq i \leq n$). Initially, $q[i]$ is storing pointer $p(i)$ for every i, and $r[i] = 0$ if node i is the end of the list, and $r[i] = 1$ otherwise. During the execution of list ranking algorithms, $r[i]$ is always storing the distance from node i to $q[i]$. When the list ranking algorithm terminates, $q[i]$ is storing the pointer to the end of the list for every i. Thus, each $r[i]$ is storing the rank of node i. This technique used in the pointer jumping [8], which repeats operations $q[i] \leftarrow q[q[i]]$ and $r[i] \leftarrow r[i] + r[q[i]]$ for every i ($1 \leq i \leq n$) in parallel. After $\log n$ iterations, every $q[i]$ is storing the pointer to the end of the list, thus, $r[i]$ is storing the rank of node i. The reader should have no difficulty to confirm the correctness of the pointer jumping.

Our list ranking algorithm also uses the two arrays q and r. For a current linked list stored in array q, *a left sublist* is a sequence $\langle i_1, i_2, \ldots, i_m\rangle$ of nodes such that $i_1 > i_2 > \cdots > i_m$ and $i_{j+1} = q[i_j]$ for every j $(1 \leq j \leq m-1)$. A left sublist is *a maximal left sublist* if no other left sublist contains it. We say that nodes i_1 and i_m are *the head* and *the tail* of the maximal left sublist $\langle i_1, i_2, \ldots, i_m\rangle$. Similarly, we can define *a right sublist*, *a maximal right sublist*, and their *head* and *tail nodes*. In Figure 1, $\langle 9, 7, 3, 1\rangle$ is a maximal left sublist, and both $\langle 1, 4, 6, 8\rangle$ and $\langle 2, 5, 9\rangle$ are maximal right sublists. Further, node 1 is the tail node of $\langle 9, 7, 3, 1\rangle$ as well as the head of $\langle 1, 4, 6, 8\rangle$.

Our list ranking algorithm repeats shrinking maximal left and right sublists. Further, every leaf node that has no predecessor is eliminated. More precisely, our algorithm repeats `list-shrink` described as follows:

`list-shrink`
Step 1: shrink left sublists by procedure `left-shrink`.
Step 2: eliminate leaf nodes by procedure `leaf-elimination`.
Step 3: shrink right sublists by procedure `right-shrink`.
Step 4: eliminate leaf nodes by procedure `leaf-elimination`.

Figure 2 illustrates each step of `list-shrink` executed for the linked list in Figure 1. Somewhat surprisingly, `list-shrink` can be done in $O(n)$ time slots with no station being awake for more than $O(1)$ time slots. Further, `list-shrink` eliminates at least half of the nodes.

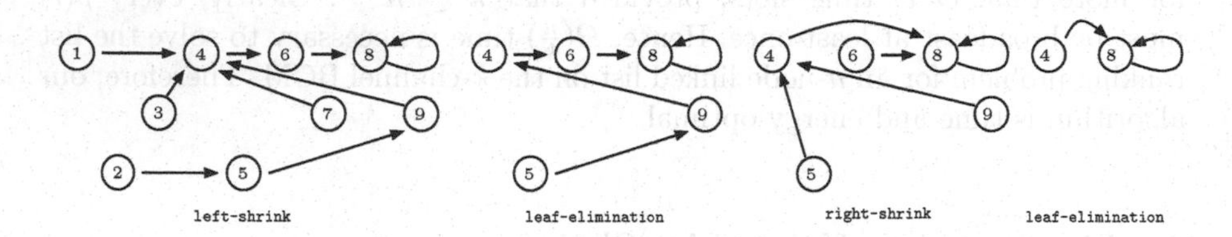

Fig. 2. Each step of `list-shrink`

We will show the details of each step of `list-shrink`. Let q and r be arrays storing pointers of a linked list and the distance as explained above. Procedure `left-shrink` is described as follows.

`left-shrink`
for $i \leftarrow 1$ to n do
  $S(i)$ broadcasts $q[i]$ and $r[i]$ on the channel.
  $S(j)$ satisfying $q[j] = i < j$ receives them and
    sets $q[j] \leftarrow q[i]$ and $r[j] \leftarrow r[j] + r[i]$.

Clearly, `left-shrink` takes n time slots. Further, each $S(i)$ is awake at time slots i and $p[i]$ and is asleep for the other time slots. Suppose that `left-shrink` is executed on the linked list in Figure 1. Note that $r[8] = 0$, and $r[i] = 1$ for all i

Table 1. The values of local variables for the list in Figure 1

node	1		2		3		4		5		6		7		8		9	
	q	r	q	r	q	r	q	r	q	r	q	r	q	r	q	r	q	r
initial input	4	1	5	1	1	1	6	1	9	1	8	1	3	1	8	0	7	1
`left-shrink`	4	1	5	1	4	2	6	1	9	1	8	1	4	3	8	0	4	4
`leaf-elimination`	$\underline{4}$	$\underline{1}$	$\underline{5}$	$\underline{1}$	$\underline{4}$	$\underline{2}$	6	1	9	1	8	1	$\underline{4}$	$\underline{3}$	8	0	4	4
`right-shrink`							8	2	4	5	8	1			8	0	4	4
`leaf-elimination`							8	2	$\underline{4}$	$\underline{5}$	$\underline{8}$	$\underline{1}$			8	0	$\underline{4}$	$\underline{4}$
`left-shrink`							8	2							8	0		
`leaf elimination`							$\underline{8}$	$\underline{2}$							8	0		
`right-shrink`															8	0		
`leaf-elimination`															8	0		
`rewind` ($T = 4$)															8	0		
`rewind` ($T = 3$)							$\underline{8}$	$\underline{2}$							8	0		
`rewind` ($T = 2$)							8	2	$\underline{8}$	$\underline{7}$	$\underline{8}$	$\underline{1}$			8	0	$\underline{8}$	$\underline{6}$
`rewind` ($T = 1$)	$\underline{8}$	$\underline{3}$	$\underline{8}$	$\underline{8}$	$\underline{8}$	$\underline{4}$	8	2	8	7	8	1	$\underline{8}$	$\underline{5}$	8	0	8	6

($i \neq 8$). It is easy to see that, after executing `left-shrink`, $q[i_1] = q[i_2] = \cdots = q[i_m]$ holds for each left sublist $\langle i_1, i_2, \ldots, i_m \rangle$, Table 1 illustrates for the values of q and r after each step of `list-shrink`. In the figure, the values of q and r are with underlines when the corresponding nodes are eliminated. Further, they are blank if the corresponding station is asleep.

After `left-shrink`, the graph may have several leaves, which are nodes having no predecessor. Clearly, we obtain a new shrunk list by removing the leaves. The following procedure `leaf-elimination` finds all leaves.

```
leaf-elimination
for i ← 1 to n do
    S(j) broadcasts j if j = p[i].
    S(i) monitors the channel.
        If the status of the channel is NULL then node i is a leaf.
```

In `leaf-elimination`, station $S(i)$ is awake at time slots i and $p[i]$. Hence, no station is awake for more than two time slots.

Procedure `right-shrink` performs the same operation in the opposite order as `left-shrink`. Since each step of `list-shrink` can be done in n time slots with every station being awake for two time slots, we have,

Lemma 1. *Procedure* `list-shrink` *takes* $4n$ *time slots with no station being awake for more than* 8 *time slots.*

We are going to prove that no more than half nodes are in the list after executing `list-shrink`. Suppose that a list has s maximal right sublists $L_1, L_2, \ldots, L_s$ in this order. For example, in Figure 1, $s = 2$ and $L_1 = \langle 1, 4, 6, 8 \rangle$ and $L_2 = \langle 2, 5, 9 \rangle$. For simplicity, we assume that L_1 and L_s contain the top and the end nodes of the whole list, respectively. We can show the proof similarly when this is not the case. From the definition, each maximal right sublist L_i

has at least two nodes. Further, no node is contained in two or more maximal right sublists. Hence, we have $2s \le n$. After Step 2, every node that are not in maximal right sublists are removed. For example, in Figure 2, nodes 3 and 7 are removed. Further, the head node (nodes 1 and 2 in Figure 2) of each maximal right sublist are removed. Clearly, the list obtained after Step 2 has at most s maximal right sublists. Note that the lists may have less than s maximal right sublists, because two adjacent sublists may be merged into one. Since every left sublist has two nodes, every node is in one of the maximal right sublists. Let $L'_1, L'_2, \ldots, L'_{s'}$ ($s' \le s$) denote the maximal sublists after Step 2. We can evaluate the number of nodes in the list obtained after Step 4 as follows. The obtained list contains at most two nodes in $L'_{s'}$. The two nodes are the end of the whole list and the head of $L'_{s'}$. The list also contains the tail of each L'_i ($2 \le i \le s'$). All nodes in L'_1 are eliminated in Step 4. Thus, the list thus obtained has at most s' ($\le s \le \frac{n}{2}$) nodes. Therefore, we have the following lemma.

Lemma 2. *After executing* `list-shrink` *on a list of n nodes, the resulting list has no more than $\frac{n}{2}$ nodes.*

Lemma 2 implies that all nodes but the end of the list are eliminated by repeating `list-shrink` for $\log n$ times. After that, the rank of every node can be computed by rewinding $2 \log n$ iterations of `leaf-elimination` as follows. For each node i but the end of the list, let $t(i)$ denote an integer such that node i has eliminated in $t(i)$-th ($1 \le i \le 2\log n$) `leaf-elimination`. For convenience, let $t(j) = 2\log n$ for the end node j of the whole list. Suppose that node $q[i]$ is the end of the list and $r[i]$ is storing the rank of node i. Then, for every node j satisfying $q[j] = i$, its rank is the sum of $r[j]$ and $r[i]$. Using this fact, the rank of node i is stored in $r[i]$ by the following procedure.

```
rewind
for T ← 2 log n downto 1 do
    for i ← 1 to n do
        if t(i) ≤ T then S(i) broadcasts q[i] and r[i]
        every S(j) satisfying t(j) = T and q[j] = i receives q[i] and r[i] and
            sets q[i] ← q[j] and r[j] ← r[j] + r[i].
```

Table 1 also shows the values of q and r during the execution of `rewind`. They have underlines, when the corresponding station is awake and change its q and r. Clearly, station i is awake for $O(t(i))$ time slots for list ranking. From Lemma 2, the number of nodes i satisfying $t(i) \ge 2A$ is at most $\frac{n}{2^A}$ for every $1 \le A \le \log n$. Therefore, we have,

Lemma 3. *The list ranking problem can be solved in $O(n \log n)$ time slots with at most $\frac{n}{2^A}$ ($1 \le A \le \log n$) station being awake for $O(A)$ time slots.*

3 Time and Energy Optimal List Ranking

Recall that, in each `list-shrink`, at most half of the nodes remain in the list. By renumbering the remaining nodes after each `leaf-elimination`, we can reduce the running time slots. More precisely, we give a unique number in the range $[1, n']$ to each remaining node, where n' is the number of remaining nodes.

For this purpose, we use an energy-optimal prefix-sums algorithm described next. Suppose that we have an array a of n numbers. Each $a(i)$ $(1 \leq i \leq n)$ is stored in $S(i)$. *The prefix-sums problem* asks to compute the i-th prefix-sum $prefix(i) = a(1) + a(2) + \cdots + a(i)$ for every i. The prefix sums problem can be solved in $n - 1$ time slots with every station being awake for at most two time slots. The details of the algorithm are spelled out as follows.

```
prefix-sums
S(1) sets prefix(1) ← a(1).
for i ← 1 to n − 1 do
    S(i) broadcasts prefix(i)
    S(i + 1) receives prefix(i) and sets prefix(i + 1) ← prefix(i) + a(i).
```

It is easy to see that every $S(i)$ leans $prefix(i)$ when `prefix-sums` terminates. Further each station $S(i)$ is awake for at time slots i and $i - 1$ and is asleep for the other time slots.

The remaining nodes can be renumbered using `prefix-sums`. Suppose that `list-shrink` is executed on a list of n nodes. Let $a(i) = 1$, if node i is remaining, and $a(i) = 0$ if node i is eliminated. By computing the prefix-sums of a, each remaining node i is assigned new ID $prefix(i)$. Then, every remaining node has a unique ID in the range $[1, n_1]$, where n_1 is the number of remaining nodes.

After assigning new IDs to the remaining nodes, we arrange them in stations $S(1), S(2), \ldots S(n_1)$ such that each $S(i)$ $(1 \leq i \leq n_1)$ is storing remaining node with new ID i. After that, pointers are changed according to the new ID using `node-transfer` as follows:

```
node-transfer
for i ← 1 to n/2 do
    if node j is remaining and prefix(j) = i then S(j) broadcasts q[j] and r[j].
    S(i) receives them. Let q'(i) denote the value of q[j].
for i ← 1 to n do
    S(i) broadcasts prefix(i).
    S(j) such that q'(j) = i receives and store it in q[j].
```

After executing `list-shrink` on the list in Figure 1, two nodes 4 and 8 are remaining. By `prefix-sums`, nodes 4 and 8 receives new IDs $prefix(4) = 1$ and $prefix(8) = 2$, respectively.

From Lemma 2, $n_1 \leq \frac{n}{2}$ holds. Thus, `node-transfer` correctly moves node i to $S(prefix(i))$ and runs in $\frac{3}{2}n$ time slots. If $S(i)$ has remaining node i, it is awake at time slots $prefix(i)$ and $\frac{n}{2} + i$. Further, for every $S(i)$ $(1 \leq i \leq n_1)$, it is awake at time slots $\frac{n}{2} + i$ and $\frac{n}{2} + q'(i)$. Thus no station is awake for more than four time slots.

After `node-transfer`, we execute `list-shrink` on the new list with n_1 nodes. Suppose that we have n_2 nodes after executing the second `list-shrink`. We use `prefix-sums` and `node-transfer` to move the n_2 nodes to n_2 stations

$S(\frac{n}{2}+1), S(\frac{n}{2}+2), \ldots, S(\frac{n}{2}+n_2)$. Continuing similarly, the list ranking problem can be solved. After i-th `list-shrink` ($1 \leq i \leq \log n - 1$), the n_i remaining nodes are moved to n_i stations $S(n-\frac{n}{2^{i-1}}+1), S(n-\frac{n}{2^{i-1}}+2), \ldots S(n-\frac{n}{2^{i-1}}+n_i)$. Using the n_i stations, `list-shrink` is executed on the new list of n_i nodes. This takes $O(\frac{n}{2^i})$ time slots and stations storing remaining nodes are awake for $O(1)$ time slots. From Lemma 2, $n_i \leq \frac{n}{2^i}$ holds for every i ($1 \leq i \leq \log n$). Hence, no station is working for more than two iterations after the first iteration of `list-shrink`. Thus, the $\log n$ iterations of `list-shrink`, `prefix-sums`, and `node-transfer` can be done in $O(n+\frac{n}{2}+\frac{n}{4}+\cdots+1) = O(n)$ time slots with each station being awake for $O(1)$ time slots. We can modify `rewind` according to new position of nodes. Finally, we have the following important theorem.

Theorem 1. *The list ranking of an n-node linked list given to n stations can be done in $O(n)$ time slots with no station being awake for more than $O(1)$ time slots on the single-channel BCM.*

4 List Ranking on the *k*-Channel BCM

This section is devoted to show that the list ranking can be done in $O(\frac{n}{k})$ time slots on the k-channel BCM with no station being awake for $O(1)$ time slots. Our idea is to simulate the single-channel list ranking algorithm on the k-channel BCM. We first show a list ranking algorithm on the BCM which has exactly $\sqrt{n}$ channels. We then go on to generalize this algorithm to run on the k-channel BCM. Due to the stringent page limitation, we only demonstrate how we simulate `left-shrink` and `leaf-elimination` on the $\sqrt{n}$-channel BCM. The other procedures can be simulated similarly.

Imagine that nodes are partitioned into $\sqrt{n}$ groups such that i-th ($1 \leq i \leq \sqrt{n}$) group consists of nodes in the range $[(i-1)\sqrt{n}+1, i\sqrt{n}]$. Each maximal sublist is partitioned into segments so that a segment consists of nodes in the same group. Procedure `left-shrink` is simulated by two subprocedures `left-shrink1` and `left-shrink2` that we describe next. In `left-shrink1`, each segment is shrunk. After removing leaf nodes, `left-shrink2` shrinks each maximal left sublist.

In `list-shrink1`, segments consist of nodes in the range $[(i-1)\sqrt{n}+1, i\sqrt{n}]$ are shrunk using channel i ($1 \leq i \leq \sqrt{n}$). For any pair i, j ($1 \leq i, j \leq \sqrt{n}$), let $|i,j|$ denote $(i-1)\sqrt{n}+j$. The details of `left-shrink1` are spelled out as follows:

`left-shrink1`
for $i \leftarrow 1$ to $\sqrt{n}$ do in parallel
  for $j \leftarrow 1$ to $\sqrt{n}$ do
    $S(|i,j|)$ broadcasts $q[|i,j|]$ and $r[|i,j|]$ on channel i.
    $S(|i,j'|)$ satisfying $q[|i,j'|] = |i,j| < |i,j'|$ receives them from channel i
      and sets $q[|i,j'|] \leftarrow q[|i,j|]$ and $r[|i,j'|] \leftarrow r[|i,j'|] + r[|i,j|]$.

Clearly, `left-shrink1` runs in $\sqrt{n}$ time slots with each station being awake for at most two time slots. After executing `left-shrink1`, $q[|i,j|]$ is storing a new pointer, which is the successor of the tail node of the segment. After that, leaves are eliminated using the $\sqrt{n}$ channel similarly as follows:

leaf-elimination
for $i \leftarrow 1$ to $\sqrt{n}$ in parallel
  for $j \leftarrow 1$ to $\sqrt{n}$ do
  $S(|i',j'|)$ broadcasts $|i',j'|$ on channel i if $|i,j| = p[|i',j'|]$.
  $S(|i,j|)$ monitors channel i.
    If the status of the channel is NULL then node $|i,j|$ is a leaf.

Clearly, all nodes in the maximal left sublists but the tails of the segments are removed by `leaf-elimination`. Note that the remaining tails in the same maximal left sublist are in distinctive groups. Using this fact, `left-shrink2` shrinks maximal left sublists. Recall that, in `left-shrink1`, nodes $|i,1|, |i,2|, \ldots, |i,\sqrt{n}|$ are broadcast on channel i in this order. On the other hand, in `left-shrink2`, nodes $|1,j|, |2,j|, \ldots, |\sqrt{n},j|$ are broadcast on channel j. This broadcast enables us to shrink maximal left sublists. The details are spelled out as follows:

left-shrink2
for $j \leftarrow 1$ to $\sqrt{n}$ in parallel
  for $i \leftarrow 1$ to $\sqrt{n}$ do
    $S(|i,j|)$ broadcasts $q[|i,j|]$ and $r[|i,j|]$ on channel j.
    $S(|i',j|)$ satisfying $q[|i',j|] = |i,j| < |i',j|$ receives them from channel j and sets $q[i',j] \leftarrow q[i,j]$ and $r[i',j] \leftarrow r[i',j] + r[i,j]$.

Again, `left-shrink2` runs in $\sqrt{n}$ time slots with each station being awake for at most two time slots. It is easy to see that, after executing `left-elimination` again, all nodes but one in each maximal sublist are removed. Similarly, we can simulate `right-shrink` and `rewind`, `prefix-sums`, `node-transfer` on the k-channel BCM.

After executing `list-shrink`, `prefix-sums`, and `node-transfer` on an n-node list, we obtain the shrunk list with less than n nodes. Let $n_1, n_2, \ldots, n_{\log n}$ be the number of nodes such that n_i is the number of remaining n nodes after i-th iteration of `list-shrink` and `prefix-sums` on the k-channel BCM. We use $\frac{n}{2^i}$ $(\geq n_i)$ processors and $\sqrt{\frac{n}{2^i}}$ channels to perform i-th iteration, which takes $O(\sqrt{\frac{n}{2^i}})$ time slots. It follows that $\log n$ iterations take at most $O(\sqrt{n} + \sqrt{\frac{n}{2^1}} + \sqrt{\frac{n}{2^2}} + \cdots + \sqrt{1}) = O(\sqrt{n})$ time slots. Thus, the list ranking can be done in $O(\sqrt{n})$ time slots with no station being awake for $O(1)$ time slots on the $\sqrt{n}$-channel BCM.

Next, let us consider the case when the BCM has less than $\sqrt{n}$ channels. Let k $(\leq \sqrt{n})$ be the number of available channels. Communication using $\sqrt{n}$ channels can be simulated in $\frac{\sqrt{n}}{k}$ time slots in obvious way. Hence, the list ranking problem can be solved in $\frac{\sqrt{n}}{k} \times O(\sqrt{n}) = O(\frac{n}{k})$ time slots. Thus, we have,

Lemma 4. *The list ranking of an n-node linked list given to n stations can be done in $O(\frac{n}{k})$ time slots with no station being awake for more than $O(1)$ time slots on the k-channel BCM provided that $k \leq \sqrt{n}$.*

It is not difficult to generalize our list ranking algorithm on the $\sqrt{n}$-channel BCM to run on the $n^{1-\frac{1}{c}}$-channel BCM for any fixed $c \geq 2$. For example, procedure `left-shrink` can be simulated by executing the $n^{1-\frac{1}{c}}$-channel version of `left-shrink` for c times, each of which runs in $O(n^c)$ time slots. Consequently, we have the following theorem:

Theorem 2. *For every $c \geq 2$, the list ranking of an n-node linked list given to n stations can be done in $O(cn^{\frac{1}{c}})$ time slots with no station being awake for more than $O(c)$ time slots on the $n^{1-\frac{1}{c}}$-channel BCM.*

Let $\epsilon = \frac{1}{c}$ be a small fixed real number. From above theorem, we have the following important corollary.

Corollary 1. *For any small fixed $\epsilon > 0$, the list ranking of an n-node linked list can be done in $O(\frac{n}{k})$ time slots with no station being awake for more than $O(1)$ time slots on the k-channel n-station BCM provided that $k \leq n^{1-\epsilon}$.*

References

1. S. G. Akl, *Parallel Computation: Models and Methods* Prentice Hall, 1997.
2. J. L. Bordim, J. Cui, T. Hayashi, K. Nakano, and S. Olariu, Energy-efficient initialization protocols for ad-hoc radio network, *IEICE Trans. on Fundamentals*, E83-A, 9, pp.1796-1803, 2000.
3. T. H. Cormen, C. E. Leiserson, R. L. Rivest, *Introduction to algorithms*, MIT Press, 1994.
4. R. Cole and U. Vishkin, Approximate parallel scheduling. Part I: The basic technique with applications to optimal parallel list ranking in logarithmic time, *SIAM J. Computing*, 17, 1, 128–142, 1988
5. K. Feher, *Wireless Digital Communications*, Prentice-Hall, Upper Saddle River, NJ, 1995.
6. A. Gibbons and W. Rytter, *Efficient parallel algorithms*, Cambridge University Press, 1988.
7. T. Hayashi, K. Nakano, and S. Olariu, Efficient List Ranking on the Reconfigurable Mesh, with Applications, *Theory of Computing Systems*, 31, 593–611, 1998.
8. J. JáJá, *An introduction to parallel algorithms*, Addison-Wesley, 1992.
9. E. D. Kaplan, *Understanding GPS: principles and applications*, Artech House, Boston, 1996.
10. F. T. Leighton, *Introduction to parallel algorithms and architectures*, Morgan Kaufmann, 1992.
11. K. Nakano and S. Olariu, Randomized initialization protocols for ad-hoc networks, *IEEE Trans. on Parallel and Distributed Systems*, 11, 7, 749–759, 2000.
12. R.A. Powers, Batteries for low-power electronics, *Proc. IEEE*, 83, pp.687–693, 1995.
13. M. Reid-Miller, *List Ranking and List Scan on the CRAY C-90*, Journal of Computer and System Sciences", (1996), 53, 3.

Energy-Efficient Size Approximation of Radio Networks with No Collision Detection*

Tomasz Jurdziński[1,2], Mirosław Kutyłowski[3], and Jan Zatopiański[2]

[1] Institute of Computer Science, Technical University Chemnitz
[2] Institute of Computer Science, Wrocław University
[3] Institute of Mathematics, Wrocław University of Technology

Abstract. Algorithms for radio networks are studied in two scenarios: (a) the number of active stations is known (or approximately known) (b) the number of active stations is unknown. In the second (more realistic) case it is much harder to design efficient algorithms. For this reason, we design an efficient randomized algorithm for a single-hop radio network that approximately counts the number of its active stations. With probability higher than $1 - \frac{1}{n}$, this approximation is within a constant factor, the algorithm runs in poly-logarithmic time and its energy cost is $o(\log\log n)$. This improves the previous $O(\log n)$ bound for energy. In particular, our algorithm can be applied to improve energy cost of known leader election and initialization protocols (without loss of time efficiency).

1 Introduction

Background. In recent years mobile and wireless communication focuses a lot of attention. New technologies provide communication means, quite exact positioning and global time via GPS systems, and finally, quite reasonable computing resources in mobile devices. This provides grounds for new applications such as law enforcement, logistics, disaster-relief, and so on.

This development redefines the demands on communication algorithms. Communication through radio channels enables easy broadcasting and delivery of messages in a distributed system. However, new problems arise: collision of messages sent simultaneously produces noise, sending and receiving periods have to be minimized due to energy consumption (pocket radio devices run on batteries). Moreover, we cannot control which stations are in use, even cannot say how many of them are switched on. This chaotic behavior has also one advantage: once we are able to run an algorithm on such a network, then it is very robust.

The model. A *radio network* considered in this paper (see e.g. [1,3,4,9,16,19]), or *RN* for short, consists of an unknown number of computing devices called here *stations* communicating through a shared channel. This model corresponds to

* partially supported by KBN, grant 8T11C 04419 and DFG, grant GO 493/1-1, and AXIT Polska

O.H. Ibarra and L. Zhang (Eds.): COCOON 2002, LNCS 2387, pp. 279–289, 2002.

hand-held devices running on batteries using a radio channel for communication. The devices are bulk-produced so we assume, as many authors, that the stations of a RN have no ID's or serial numbers.

We assume that the RN stations have local clocks synchronized to global time (this is realistic due to GPS technology). Communication is possible in time slots determined by global time, called *steps*. During a single step any station may send and/or receive a message. If exactly one station sends a message, then it is available to all stations listening at this moment. Thus, we consider here *single-hop* RN's. Many authors consider also the case in which several independent communication channels are available. Other important generalization considered is that a network is defined by a graph, where nodes denote stations and a station v is reachable from the station u if and only if there is an edge (u, v) in the graph (so-called *multi-hop RN*, [1,4,5,8]).

If more than one station sends during a step, then a *collision* occurs, the messages are scrambled so that the stations receive noise. We consider here a weak no-collision-detection model, no-CD for short, in which the stations cannot even recognize that a collision has occurred. It is quite often assumed that the station sending a message may simultaneously hear and thereby recognize that a collision has occurred [2,3,7,10,15,16,17,19]. This feature is quite strong algorithmically, but unavailable in some technologies (see [11] for results concerning the weaker model). Let us remark that the algorithm presented in this paper can be adopted to such a weak model.

The stations of a RN that are switched on are called *active*. An active station is *awake*, when it is sending a message or listening, otherwise it is *asleep*.

Complexity measures. Most important complexity measures for RN's are *time* and *energy cost* [15,16,17]: time is the number of steps required for executing an algorithm, energy cost is the maximal number of steps at which a station of a RN is awake. This is motivated by the fact that sending and receiving messages are main sources of energy consumption ([6,18]). For the sake of simplicity we neglect here differences between energy consumption of sending and listening.

RN's intensively use randomization. This is absolutely necessary, for instance, for symmetry breaking. In the case of a randomized algorithm energy cost e means that with high probability no station is awake for more than e steps. This is motivated by the fact that an accidental burst of energy consumption at a single station may exhaust its batteries and prevent the station to follow the algorithm. For this reason, a lot of recent research is focused on energy efficient RN algorithms (see e.g. [2,5,15,16,17]).

Known versus unknown number of stations. The algorithms for RN's are designed under the following assumptions:

scenario I: the number of active stations is known;
scenario II: the number of active stations is unknown, but can be approximated within a constant factor;
scenario III: there is no information on the number of active stations.

Since many algorithms for RN are designed for known number of active stations, it is desirable to design an efficient procedure for converting scenario III into scenario I or at least II. Therefore we consider the following *size approximation problem:*

> given a RN with unknown number n of active stations, find a number n' such that $\frac{1}{c}n \leq n' \leq c \cdot n$ for a certain constant c. At the end of the protocol each active station should be aware of n'.

In [2], a solution with logarithmic energy cost and $O(\log^2 n)$ time is given.

It is much harder to design algorithms in scenario I than in scenario III. For instance, an energy efficient solution to initialization problem in scenario I is proposed in in [16]: with probability at least $1 - \frac{1}{n}$ its energy cost is $O(\log \log n)$ and execution time is $O(n)$. It can be easily generalized to the case of scenario II; finding an energy efficient solution for scenario III was left as open problem.

For another important problem of leader election, energy efficient algorithms have been designed in [15]. The authors present a randomized algorithm that for n stations elects a leader in scenario I in time $O(\log f)$ and energy $O(\log \log f + \frac{\log f}{\log n})$ with probability $1 - 1/f$ for any $f \geq 1$. Moreover, they present algorithms that work in scenario III and elect a leader within $O(\log n)$ energy cost and $O(\log^2 n)$ time with probability $1 - \frac{1}{n}$.

A deterministic RN algorithm determining the number of active stations (for stations with distinct ID's from the set $\{1, \ldots, n\}$) is presented in [12, Theorem 2] (formulated as a solution for leader election):

Lemma 1 ([12]). *Consider a RN with at most n active stations with labels in the range $[1, n]$, each $t < n$ assigned to at most one active station. There is a deterministic protocol finding the station with the largest label and counting the number of active stations which runs in time $O(n)$ and has energy cost $O((\log n)^\varepsilon)$ for any $\varepsilon > 0$.*

2 New Results

Our main result is an energy efficient solution to size approximation problem:

Theorem 1. *There is a randomized algorithm for weak no-CD RN such that with probability at least $1 - \frac{1}{n}$ a number n_0 is found such that $\frac{1}{c}n_0 \leq n \leq cn_0$ (for some constant $c \geq 1$) within time $O(\log^{2+\varepsilon} n)$ with energy cost $O((\log \log n)^\varepsilon)$ for any constant $\varepsilon > 0$.*

An example for application of this result is that together with [16, Theorem 6.1] it solves an open problem from [16]:

Corollary 1. *There is an initialization protocol for a no-CD RN with n stations in scenario III that runs in time $O(n)$ and has energy cost $O(\log \log n)$ with probability at least $1 - \frac{1}{n}$.*

Corollary 1 also yields an algorithm for leader election in scenario III with poly-logarithmic time and energy cost $O(\log\log n)$ (with probability $1-\frac{1}{n}$). This improves exponentially energy cost upon the algorithms from [15], without loss of time efficiency, when time achieved with high probability (that is, $1-1/n$) is considered.

3 Size Approximation Algorithm

3.1 Basic Algorithm

Let us consider the following experiment. During a step each of n active stations sends a message with probability p. If exactly one station chooses to send, then we say that the step is *successful* and that the station that have sent a message *succeeds*. Note that the probability p_S that a step is successful, equals $np(1-p)^{n-1}$. The largest value of this expression is obtained for $p=1/n$; then $p_S \approx 1/e$. If we repeat the experiment independently l times, then the expected number of successful steps is approximately l/e. By Chernoff Bounds [14] we may bound the probability that the number of successes is far from its expectation.

Using these observations we construct the first, energy inefficient, algorithm. Let d be a sufficiently large constant.

Basic Algorithm

for $k=1,2,\ldots$ *run phase k*:
  repeat $d\cdot k$ times:
    each station sends with probability $1/2^k$ and
      listens all the time counting the number of successful
    steps
  if the number of successful steps is close to $d\cdot k/e$ then $n_0 \leftarrow 2^k$, halt

Using Chernoff Bounds one can show that the value of n_0 found by the algorithm is not far from the number of active stations. However, this simple algorithm has large energy cost: about $\log n$ phases are to be executed, during each phase every active station is listening all the time. So each station is awake for $\Theta(\log^2 n)$ steps.

3.2 Improvements Idea

The first change is that a station listens only when it sends a message. This reduces the energy cost of each station to the number of steps in which it sends. However, then no station knows the number of successful steps, but only the number of the successful steps during which it has been sending. To solve this problem we apply the algorithm from Lemma 1. For this algorithm each station has a "temporary" ID – the first i such that it has succeeded in step i during this phase. In this way, all successful stations learn the total number of successful steps (a slight adjustment is necessary since the same station may succeed more than once). In an extra step the station with the smallest "temporary" ID informs all stations about this number.

Still, every station has to listen at the end of each phase, and about $\log n$ phases are to be executed making energy cost $\Omega(\log n)$. A relatively simple remedy is to make such a step "all have to listen" only for the phases l such that $l = g(j)$ for $j \in \mathbb{N}$ where g is a function satisfying $g(j+1) \geq \lceil g(j)^{1+\epsilon} \rceil$ for some $\varepsilon > 0$ (again, we apply Lemma 1, for determining the first successful phase among $g(j-1)+1, \ldots, g(j)$). This does not postpone getting the final result very much: we may expect that the number of phases is at most $m^{1+\varepsilon}$, where m is such that Basic Algorithm finishes its work at phase m, i.e. $m = \Theta(\log n)$. Now the number of the "obligatory listening steps" is reduced to the minimal l such that $g(l) \geq \log n$.

Even with these changes, we are unable to guarantee with high probability that every station is awake for $o(\log \log n)$ steps. Namely, we have to guarantee low energy cost due to sending messages. For technical reasons, we split the protocol in two parts. During the first part, we execute the simple algorithm described above for $k = 1, 2, \ldots, k_0$, where k_0 is some constant. For the second part, we make a crucial change that every station sends at most once in the loop taken from Basic Algorithm. (However, some additional activities will be necessary for informing all stations about the number n found.) Then, it becomes "passive" and only listens to the results. The idea is that this modification does not change the number of successful steps at each phase substantially, because the number of stations "eliminated" in this way is very small with respect to n. However, we are faced with nasty technical details due to the fact that the steps are stochastically dependent. So, a careful argument is required to estimate the number of successes.

3.3 Description of the Algorithm

Let $g(1) = 2$, $g(i+1) = \lceil (g(i))^{1+\varepsilon} \rceil$ for a constant $\varepsilon > 0$; let d and k_0 be large enough constants.

Algorithm ApproxSize(ε):

(01) run k_0 phases of Basic Algorithm
(02) each station sets *status* $\leftarrow$ *fresh* and *temp2ID* $\leftarrow 0$
(03) `for` $k = k_0+1, k_0+2\ldots$ `do`
(04) each station sets *tempID* $\leftarrow 0$
(05) `for` $j = 1$ `to` $d \cdot k$ `do`
(06) each *fresh* station sends and listens with probability $1/2^k$
(07) each *fresh* station that sends a message sets *status* $\leftarrow$ *used*
(08) `if` a station have sent successfully `then` it sets *tempID* $\leftarrow j$
(09) using Lemma 1 for $n = dk$ and ε the stations with *tempID* $\neq 0$:
(10) compute N, the number of stations with *tempID* $\neq 0$ and
(11) find a station with the smallest *tempID*
(12) `if` $N \geq dk/20$ `then`
(13) the station with the smallest *tempID* sets its *temp2ID* $\leftarrow k$

(14) **if** $k = g(l)$ for an $l \in \mathbb{N}$ **then**
(15) all stations with *temp2ID* $\neq 0$ elect a leader (with the smallest *temp2ID*) using Lemma 1 for $n = k$ and ε
(16) a leader (if elected) sends its *temp2ID* and all other stations listen
(17) **if** not(noise/silence) **then**
(18) every station sets $n_0 \leftarrow 2^p$ after receiving p and halts

Let the loop consisting of lines (05)-(08) for $k = i$ be called *phase* i. Let *trial* j *of phase* i denote the jth execution of line (06) during phase i.

The following properties follow directly from the construction: if ApproxSize halts with $k = k''$, then the first value k' of k, for which the condition from line (12) has been satisfied, fulfills $\lceil (k')^{1+\varepsilon} \rceil \geq k''$. Also, after halting all active stations hold the same number $n_0 = 2^{k'}$ assigned in line (18).

4 Analysis of the Algorithm

4.1 Complexity Analysis

Let n be the number of active stations. We say that the protocol succeeds for a number k', if the condition in (12) is satisfied for $k = k'$. Let k_s be the smallest number k for which the condition in (12) is satisfied.

Lemma 2 (Main Lemma). *For a k such that $k \leq \log n - 6$ algorithm ApproxSize succeeds with probability $O(\frac{1}{n^2})$. For $k = \lceil \log n \rceil$ it does not succeed with probability $O(\frac{1}{n^2})$.*

We postpone the proof of this result and yield corollaries which establish Theorem 1.

Corollary 2. *ApproxSize running on n active stations halts and outputs n_0 such that $n_0 = \Theta(n)$ with probability $1 - q$, where $q = O(\log n / n^2)$.*

Proof. It follows from the fact that with high probability $\log n - 6 \leq k_s \leq \log n + 1$ and $n_0 = 2^{k_s}$. □

Corollary 3. *ApproxSize(ε) running on n active stations halts within $O(\log^{2+2\varepsilon} n)$ steps for any $\varepsilon \geq 0$ with probability $1 - q$, where $q = O(1/n^2)$.*

Proof. Assume that $\log n > k_0$. Our algorithm needs a constant time for steps (01)-(02) and with high probability it finishes the loop (03)-(18) for $k = k_f$ for some $k_f \leq (\log n)^{1+\varepsilon}$ (by Lemma 2). One execution of (05)-(08) takes time $O((\log n)^{1+\varepsilon})$. By Lemma 1, steps (09)-(11) take time $O((\log n)^{1+\varepsilon})$. The same applies to the line (15). So the loop (05)-(18) takes $O(\log^{1+\varepsilon} n)$ steps, and the whole protocol $O(\log^{2+2\varepsilon} n)$ steps. □

Corollary 4. *ApproxSize*(ε) *running on* n *active stations has energy cost* $O((\log\log n)^\epsilon)$ *with probability* $1-q$, *where* $q = O(1/n^2)$.

Proof. Assume that $\log n > k_0$. Every station needs a constant energy for lines (01)-(02) and with high probability it finishes the loop (03)-(18) for $k = k_f$ for some $k_f \le (\log n)^{1+\varepsilon}$. In lines (05)-(08) each station sends (and listens) at most once (after the step in which the station was awake for sending and listening, it becomes *used*). Every station takes part in lines (09)-(13) at most once (in iteration in which it becomes *used*) and is awake $O((\log\log n)^\varepsilon)$ times. Similarly, every station executes line (15) at most once (also $O((\log\log n)^\varepsilon)$ energy). Finally, each station listens in line (16), but with high probability, this line is executed $O(\log^{-1}(1+\varepsilon)\cdot\log\log(\log^{1+\varepsilon} n))$ times. □

Remark. If we put $g(l+1) = 2\cdot g(l)$, then we obtain energy bound $O(\log\log n)$ and time $O(\log^2 n)$ with probability bigger than $1-1/n$ (the same time bound as in the algorithm for leader election from [15], which requires logarithmic energy).

4.2 Proof of Main Lemma

Probability that a station becomes *used* is upper bounded by the expression

$$\sum_{k=k_0+1}^{\infty} dk/2^k \le c,$$

for some small constant c. Hence the expected number of used stations does not exceed cn. By choosing k_0 sufficiently large we get $c \le \frac{1}{4e}$. If we imagine that stations continue the lines (03)-(08) infinitely,, then the probabilities of being *used* are independent. Using Chernoff Bound on sums of independent random variables [14] we get:

Corollary 5. *Probability that the number of used stations exceeds* $n/2$ *is at most* $2^{-n/2}$.

One may be tempted to apply the results on Poisson trials in order to estimate the expected value of N in line (10) of ApproxSize and deviations from the expected value. The main technical problem that arises here is that our modification (making the station *used* after sending) makes different iterations of the loop in the lines (06)-(08) stochastically dependent. Moreover, one success may improve success probabilities in subsequent steps, even if the increase is not substantial. Note that nice properties of sums of Poisson random variables are based on the fact that success probabilities do not depend on the history of computation!

This section is organized as follows: First we analyze success probabilities in line (06) provided that less than $n/2$ stations are *used*. We analyze separately the stages for $k \le \log n - 6$ and for $k = \lceil \log n \rceil$. The core of the proof is showing a relationship between the number of successes in a line (06) during a phase and the sum of independent random variables.

Success probabilities. A simple calculation shows the following properties:

Proposition 1. *Assume that the number of unused stations before a step is at least $n/2$. Then for $k = \lceil \log n \rceil$, the probability that exactly one station sends at line (06) is at least* 0.1.

Proposition 2. *Assume that the number of unused stations before a step is at least $n/2$. Then for $k \leq \log n - 6$ and some constant c', the probability that line (06) is successful is not greater than $p_k = \frac{n}{2^{k-1}}\left(1 - \frac{1}{2^k}\right)^{n/2}$.*

Sums of independent random variables. Let $\mathbf{P}[A]$ denote probability of an event A. Let $p_k = \frac{n}{2^{k-1}}\left(1 - \frac{1}{2^k}\right)^{n/2}$ for $k \leq \log n - 6$, and $p_k = 0.1$ for $k = \lceil \log n \rceil$. We consider independent random variables $x_1, \ldots, x_{dk}$, where $x_i \in \{0,1\}$, $\mathbf{P}[x_i = 1] = p_k$ for $i \leq dk$ (from the context it will be clear which k do we mean.)

Lemma 3. *Let $X = \sum_{i=1}^{dk} x_i$. Then for sufficiently large n:*

(a) $\mathbf{P}\left[X > \frac{dk}{20}\right] < \frac{1}{n^2}$ *for* $k \leq \log n - 6$,
(b) $\mathbf{P}\left[X > \frac{dk}{20}\right] \geq 1 - \frac{1}{n^2}$ *for* $k = \lceil \log n \rceil$.

The proof of this lemma, based on Chernoff Bounds, is tedious, but straightforward so we omit it here.

Estimating the number of successes. Let $\mathcal{S}_i^j$ denote the event "the number of used stations is less than $n/2$ immediately before trial j of phase i". Recall that by Corollary 5, $\mathbf{P}\left[\mathcal{S}_i^j\right] \geq 1 - 2^{-n/2}$. First, we examine the number of successes during phase k, for a k such that $k \leq \log n - 6$, under assumption that event $\mathcal{S}_k^1$ holds. Let w_i be a random variable, such that $w_i = 1$ if the trial i at phase k is successful, and $w_i = 0$ otherwise. Let x_i be random variables defined as above. Note that, for every $U \subseteq S_k^i$, $\mathbf{P}[w_i = 1 \,|\, U] \leq \mathbf{P}[x_i = 1]$ if $k \leq \log n - 6$ (by Proposition 2) and $\mathbf{P}[w_i = 1 \,|\, U] \geq \mathbf{P}[x_i = 1]$ if $k = \lceil \log n \rceil$ (by Proposition 1).

Lemma 4. *Let $M = 2^{n/2}$, $k \leq \log n - 6$. For each $c > 0$ holds:*

$$\mathbf{P}\left[\sum_{i=1}^{dk} w_i > c \,\middle|\, \mathcal{S}_k^1\right] \leq \mathbf{P}\left[\sum_{i=1}^{dk} x_i > c\right] + \frac{2^{dk}-1}{M \cdot \mathbf{P}[\mathcal{S}_k^1]}\,.$$

Proof. We prove the lemma in a slightly more general form by inverse induction on j: let U be any event such that $U \subseteq \mathcal{S}_k^j$, then

$$\mathbf{P}\left[\sum_{i=j}^{dk} w_i > c \,\middle|\, U\right] \leq \mathbf{P}\left[\sum_{i=j}^{dk} x_i > c\right] + \frac{2^{(dk-j)}-1}{M \cdot \mathbf{P}[U]}\,. \tag{1}$$

The case $j = dk$ is given by Proposition 2. Now, for $j < dk$ we have

$$\mathbf{P}\left[\sum_{i=j}^{dk} w_i > c \,\middle|\, U\right] = \mathbf{P}\left[\sum_{i=j}^{dk} w_i > c \wedge \mathcal{S}_k^{j+1} \,\middle|\, U\right] + \mathbf{P}\left[\sum_{i=j}^{dk} w_i > c \wedge \neg\mathcal{S}_k^{j+1} \,\middle|\, U\right].$$

Since the second term is bounded from above by

$$\mathbf{P}\left[\neg\mathcal{S}_k^{j+1}\,|U\right] \le \frac{\mathbf{P}[\neg\mathcal{S}_k^{j+1}]}{\mathbf{P}[U]} \le \frac{1}{M\cdot\mathbf{P}[U]}\,,$$

we get

$$\mathbf{P}\left[\sum_{i=j}^{dk} w_i > c\,|\,U\right] \le \mathbf{P}\left[\sum_{i=j}^{dk} w_i > c \wedge \mathcal{S}_k^{j+1}\,|\,U\right] + \frac{1}{M\cdot\mathbf{P}[U]}\,.$$

The first term on the left hand side of the last inequality equals

$$\mathbf{P}\left[\sum_{i=j+1}^{dk} w_i > c \wedge w_j = 0 \wedge \mathcal{S}_k^{j+1}\,|\,U\right]$$
$$+\mathbf{P}\left[\sum_{i=j+1}^{dk} w_i > c-1 \wedge w_j = 1 \wedge \mathcal{S}_k^{j+1}\,|\,U\right].$$

Let the probabilities above be denoted by H_0 and H_1, respectively. We estimate H_0. Let W_0 denote the event $\mathcal{S}_k^{j+1} \wedge w_j = 0 \wedge U$. Then , by induction hypothesis:

$$\begin{aligned}
H_0 = \mathbf{P}\left[\sum_{i=j+1}^{dk} w_i > c\,|\,W_0\right] \cdot \mathbf{P}\left[W_0\,|\,U\right] \le\\
\left(\mathbf{P}\left[\sum_{i=j+1}^{dk} x_i > c\right] + \frac{2^{dk-j-1}-1}{M\cdot\mathbf{P}[W_0]}\right) \cdot \frac{\mathbf{P}[W_0]}{\mathbf{P}[U]} =\\
\left(\mathbf{P}\left[\sum_{i=j+1}^{dk} x_i > c\right] \cdot \frac{\mathbf{P}[W_0]}{\mathbf{P}[U]}\right) + \frac{2^{dk-j-1}-1}{M\cdot\mathbf{P}[U]} =\\
\left(\mathbf{P}\left[\sum_{i=j+1}^{dk} x_i > c\right] \cdot \mathbf{P}\left[\mathcal{S}_k^{j+1} \wedge w_j = 0\,|\,U\right]\right) + \frac{2^{dk-j-1}-1}{M\cdot\mathbf{P}[U]} \le\\
\left(\mathbf{P}\left[\sum_{i=j+1}^{dk} x_i > c\right] \cdot \mathbf{P}\left[w_j = 0\,|\,U\right]\right) + \frac{2^{dk-j-1}-1}{M\cdot\mathbf{P}[U]}\,.
\end{aligned}$$

Similarly we obtain

$$H_1 \le \left(\mathbf{P}\left[\sum_{i=j+1}^{dk} x_i > c-1\right] \cdot \mathbf{P}\left[w_j = 1\,|\,U\right]\right) + \frac{2^{dk-j-1}-1}{M\cdot\mathbf{P}[U]}\,.$$

So we get

$$\begin{aligned}
\mathbf{P}\left[\sum_{i=j}^{dk} w_i > c|U\right] \le \left(\mathbf{P}\left[\sum_{i=j+1}^{dk} x_i > c\right] \cdot \mathbf{P}\left[w_j = 0|U\right]\right) +\\
\left(\mathbf{P}\left[\sum_{i=j+1}^{dk} x_i > c-1\right] \cdot \mathbf{P}\left[w_j = 1|U\right]\right) + \frac{2(2^{dk-j-1}-1)+1}{M\cdot\mathbf{P}[U]}\,.
\end{aligned}$$

The last term equals $\frac{2^{dk-j}-1}{M\cdot\mathbf{P}[U]}$. In order to estimate the sum of the first two terms note that by replacing $\mathbf{P}\left[w_j = 1|U\right]$ by any number $p \ge \mathbf{P}\left[w_j = 1|U\right]$ and $\mathbf{P}\left[w_j = 0|U\right]$ by $1-p$ we do not decrease the sum. Indeed, it follows from the fact that

$$\mathbf{P}\left[\sum_{i=j+1}^{dk} x_i > c-1\right] \ge \mathbf{P}\left[\sum_{i=j+1}^{dk} x_i > c\right].$$

By taking $p = \mathbf{P}\left[x_1 = 1\right]$, we get

$$\begin{aligned}
\mathbf{P}\left[\sum_{i=j}^{dk} w_i > c|U\right] \le \left(\mathbf{P}\left[\sum_{i=j+1}^{dk} x_i > c\right] \cdot \mathbf{P}\left[x_1 = 0\right]\right) +\\
\left(\mathbf{P}\left[\sum_{i=j+1}^{dk} x_i > c-1\right] \cdot \mathbf{P}\left[x_1 = 1\right]\right) + \frac{2^{dk-j}-1}{M\cdot\mathbf{P}[U]} =\\
\mathbf{P}\left[\sum_{i=j}^{dk} x_i > c\right] + \frac{2^{dk-j}-1}{M\cdot\mathbf{P}[U]}.
\end{aligned}$$

This concludes the proof of inductive step for inequality (1). □

It follows from Lemma 4 that if there are at least $n/2$ unused stations at the beginning of phase k, $k \le \log n - 6$, then the probability that $\sum_{i=1}^{dk} w_i > c$ is bounded by

$$\mathbf{P}\left[\sum_{i=1}^{dk} x_i > c\right] + \frac{2^{dk}-1}{M \cdot \mathbf{P}\left[\mathcal{S}_k^1\right]} \le \mathbf{P}\left[\sum_{i=1}^{dk} x_i > c\right] + \frac{2^{dk}-1}{M-1}$$

(we have used the fact that $\mathbf{P}\left[\mathcal{S}_k^1\right] \ge 1 - 1/M$). For $c = dk/20$, we have $\mathbf{P}\left[\sum_{i=1}^{dk} x_i > c\right] \le n^{-2}$, so

$$\mathbf{P}\left[\sum_{i=1}^{dk} w_i > \tfrac{dk}{20}\right] = \mathbf{P}\left[\sum_{i=1}^{dk} w_i > \tfrac{dk}{20} \wedge \mathcal{S}_k^1\right] + \mathbf{P}\left[\sum_{i=1}^{dk} w_i > \tfrac{dk}{20} \wedge \neg\mathcal{S}_k^1\right] \le$$
$$\mathbf{P}\left[\sum_{i=1}^{dk} w_i > \tfrac{dk}{20} \mid \mathcal{S}_k^1\right] \cdot \mathbf{P}\left[\mathcal{S}_k^1\right] + 2^{-n/2} \le \frac{1}{n^2} + \frac{2^{d(\log n - 6)}-1}{2^{n/2}} + 2^{-n/2} = O\left(\frac{1}{n^2}\right).$$

For $\kappa = \lceil \log n \rceil$, a similar estimation from below of $\mathbf{P}\left[\sum_{i=1}^{\kappa} w_i \le c\right]$ can be obtained by considering the variables $v_i = 1 - w_i$ and $y_i = 1 - x_i$. Exactly as before we show that

$$\mathbf{P}\left[\sum_{i=1}^{\kappa} w_i \le c\right] = \mathbf{P}\left[\sum_{i=1}^{\kappa} v_i > d\kappa - c\right] \le$$
$$\mathbf{P}\left[\sum_{i=1}^{\kappa} y_i > d\kappa - c\right] + \frac{2^{d\kappa}-1}{M-1} = \mathbf{P}\left[\sum_{i=1}^{\kappa} x_i \le c\right] + \frac{2^{d\kappa}-1}{M-1}.$$

Since $\mathbf{P}\left[\sum_{i=1}^{dk} x_i \le d\kappa/20\right] \le n^{-2}$, so we get in a similar way that $\mathbf{P}\left[\sum_{i=1}^{dk} w_i < d\kappa/20\right] = O\left(\frac{1}{n^2}\right)$.

Notes and Comments. Our complexity analysis of the algorithm ApproxSize convince that it has desirable *asymptotic* behavior. Although we did not care about values of constants a more detailed probabilistic analysis and fine tuning for small n should show that the algorithm is quite practical.

References

1. Bar-Yehuda, R., Goldreich, O., Itai, A.: On the Time-Complexity of Broadcast in Multi-hop Radio Networks: An Exponential Gap Between Determinism and Randomization. Journal of Computer Systems Sciences 45(1) (1992), 104-126
2. Bordim, J.L., Cui, J., Hayashi, T., Nakano, K., Olariu, S.: Energy-Efficient Initialization Protocols for Ad-hoc Radio Networks. ISAAC'99, LNCS 1741, Springer-Verlag, 1999, 215–224
3. Chlebus, B.S.: Randomized Communication in Radio Networks. A chapter in ,,Handbook on Randomized Computing" P. M. Pardalos, S. Rajasekaran, J. H. Reif, J. D. P. Rolim, (Eds.), Kluwer Academic Publishers, to appear
4. Chlamtac, I., Kutten, S.: On Broadcasting in Radio Networks – Problem Analysis and Protocol Design. IEEE Trans. on Commun. 33 (1985), 1240–1246

5. Dessmark, A., Pelc, A.: Deterministic Radio Broadcasting at Low Cost. STACS'2001, LNCS 2010, Springer-Verlag, 158–169
6. Fifer, W.C., Bruno, F.J.: Low Cost Packet Radio. Proc. of the IEEE 75 (1987), 33–42
7. Gąsieniec, L., Pelc, A., Peleg, D.: The Wakeup Problem in Synchronous Broadcast Systems. SIAM Journal on Discrete Math. 14(2) (2001), 207–222
8. Gąsieniec, L., Lingas, A.: On Adaptive Deterministic Gossiping in Ad Hoc Radio Networks. ACM-SIAM SODA '2002
9. Gitman, I., Van Slyke, R.M., Frank, H.: Routing in Packet-Switching Broadcast Radio Networks. IEEE Trans. on Commun. COM-24 (1976), 926–930
10. Hayashi, T., Nakano, K., Olariu, S.: Randomized Initialization Protocols for Packet Radio Networks. IPPS'1999, IEEE 1999, 544–548
11. Jurdziński, T., Kutyłowski, M., Zatopiański, J.: Weak Communication in Radio Networks. Euro-Par'2002, LNCS , Springer-Verlag (accepted paper)
12. Jurdziński, T., Kutyłowski, M., Zatopiański, J.: Efficient Algorithms for Leader Election in Radio Networks. ACM PODC'2002, (accepted paper)
13. Jurdziński, T., Kutyłowski, M., Zatopiański, J.: Energy-Efficient Size Approximation for Radio Networks with no Collision Detection. Technical Report CSR-02-02, Technische Universität Chemnitz, Fakultät für Informatik, `http://www.tu-chemnitz.de/informatik/`
14. Motwani, R., Raghavan, P.: Randomized Algorithms. Cambridge University Press, 1995
15. Nakano, K., Olariu, S.: Randomized Leader Election Protocols in Radio Networks with No Collision Detection. ISAAC'2000, LNCS 1969, Springer-Verlag, 362–373
16. Nakano, K., Olariu, S.: Energy Efficient Initialization Protocols for Radio Networks with no Collision Detection. ICPP'2000, IEEE 2000, 263–270
17. Nakano, K., Olariu, S.: Energy-Efficient Randomized Routing in Radio Networks. Proc. of 4th Workshop on Discrete Algorithms and Methods for Mobile Computing and Communications (DIALM), 35–44
18. Salkintzis, A.K., Chamzas, C.: An In-band Power-saving Protocol for Mobile Data Networks. IEEE Trans. on Communication, COM-46 (1998), 1194–1205
19. Willard, D.E.: Log-logarithmic Selection Resolution Protocols in Multiple Access Channel. SIAM Journal on Computing 15 (1986), 468-477

A New Class of Symbolic Abstract Neural Nets: Tissue P Systems

C. Martín-Vide[1], J. Pazos[2], G. Păun[1], and A. Rodríguez-Patón[2]

[1] Research Group on Mathematical Linguistics, Rovira i Virgili University
Pl. Imperial Tàrraco 1, 43005 Tarragona, Spain
{cmv,gp}@astor.urv.es

[2] Faculty of Computer Science, Polytechnical University of Madrid
Campus de Montegancedo, Boadilla del Monte 28660, Madrid, Spain
{jpazos,arpaton}@fi.upm.es

Abstract. Starting from the way the inter-cellular communication takes place by means of protein channels and also from the standard knowledge about neuron functioning, we propose a computing model called a *tissue P system*, which processes symbols in a multiset rewriting sense, in a net of cells similar to a neural net. Each cell has a finite state memory, processes multisets of symbol-impulses, and can send impulses ("excitations") to the neighboring cells. Such cell nets are shown to be rather powerful: they can simulate a Turing machine even when using a small number of cells, each of them having a small number of states. Moreover, in the case when each cell works in the maximal manner and it can excite all the cells to which it can send impulses, then one can easily solve the Hamiltonian Path Problem in linear time. A new characterization of the Parikh images of ET0L languages are also obtained in this framework.

1 Introduction

This paper can be seen at the same time as a contribution to neural networks (of a symbolic type), to membrane computing (with cells arranged in "tissues"), to finite automata networks (working not with strings, but with multisets of symbols), to multiset processing, to (distributed) automata and language theory. The motivation is two-fold: the inter-cellular communication (of chemicals, energy, information) by means of complex networks of protein channels (see, e.g., [1], [11]), and the way the neurons co-operate, processing impulses in the complex net established by synapses (see, e.g., [1], [2]).

The common mathematical model of these two kinds of symbol-processing mechanisms is the net of finite state devices, and this is the type of computing mechanisms we are going to consider: networks of finite-automata-like processors, dealing with symbols, according to local states (available in a finite number for each "cell"), communicating through these symbols, along channels ("axons") specified in advance. Note that the neuron modelling was the starting point of the theory of finite automata ([13], [10]), that symbol processing neural networks have a rich (and controversial) history (see [5] and its references), and that

O.H. Ibarra and L. Zhang (Eds.): COCOON 2002, LNCS 2387, pp. 290–299, 2002.

networks of string-processing finite automata have appeared in many contexts ([6], [9], [12], etc), but our models are different in many respects from all these previous models.

Having in mind the bio-chemical reality we refer to, a basic problem concerns the organization of the bunch of symbols available in each node, and the easiest and most natural answer is: no organization. Formally, this means that we have to consider *multisets* of symbols, sets with multiplicities associated with their elements. In this way, we need a kind of finite automata dealing with multisets of symbols, a topic which falls into an area of (theoretical) computer science not very much developed, although some recent (see, e.g., [7]), or not so recent (see, e.g., [4]) approaches can be found in the literature. Actually, most of the vivid area of membrane computing (P systems) [15] is devoted to multiset processing (details at `http://bioinformatics.bio.disco.unimib.it/psystems`).

The computing models we propose here, under the name of *tissue P systems*, in short, *tP systems*, consist of several *cells*, related by protein channels. In order to preserve also the neural intuition, we will use the suggestive name of *synapses* for these channels. Each cell has a state from a given finite set and can process multisets of *objects*, represented by symbols from a given alphabet. The standard rules are of the form $sM \to s'M'$, where s, s' are states and M, M' are multisets of symbols. Some of the elements of M' may be marked with the indication "go", and this means that they have to immediately leave the cell and pass to the cells to which we have direct links through synapses. This communication (transfer of symbol-objects) can be done in a replicative manner (the same symbol is sent to all adjacent cells), or in a non-replicative manner; in the second case we can send all the symbols to only one adjacent cell, or we can distribute them, non-deterministically. One more choice appears in using the rules $sM \to s'M'$: we can apply such a rule only to one occurrence of M (that is, in a sequential, *minimal* way), or to all possible occurrences of M (a *parallel* way), or, moreover, we can apply a maximal package of rules of the form $sM_i \to s'M_i', 1 \leq i \leq k$, that is, involving the same states s, s', which can be applied to the current multiset (the *maximal* mode). By the combination of the three modes of processing objects and the three modes of communication among cells, we get nine possible behaviors of our machinery.

A way to use such a computing device is to start from a given initial configuration (that is, initial states of cells and initial multisets of symbol-objects placed in them) and to let the system proceed until reaching a halting configuration, where no further rule can be applied, and to associate a result with this configuration. Because of the nondeterminism, starting from one given initial configuration we can reach arbitrarily many different halting configurations, hence we can get arbitrarily many outputs. Another possibility is to also provide *inputs*, at various times of a computation, and to look for the outputs related to them. Here we will consider only the first possibility, of *generative* tP systems, and the output will be defined by sending symbols out of the system. To this aim, one cell will be designated as the output one, and in its rules $sM \to s'M'$ we will also allow that symbols from M' are marked with the indication "out";

such a symbol will immediately leave the system, contributing to the result of the computation.

At the first sight, such a machinery (a finite net of finite state devices) seems not to be very powerful, e.g., as compared with Turing machines. Thus, it is rather surprising to find that tP systems with a small number of cells (two or four), each of them using a small number of states (resp., at most five or four) can simulate any Turing machine, even in the non-cooperative case, that is, only using rules of the form $sM \to s'M'$ with M being a singleton multiset; moreover, this is true for all modes of communication for the minimal mode of using the rules, and, in the cooperative case, also when using the parallel or the maximal mode of processing objects. When the rules are non-cooperative and we use them in the maximal mode, a characterization of Parikh images of ET0L languages is obtained, which completes the study of the computing power of our devices (showing that in the *parallel* and *maximal* cases we dot not get computational universality).

The above mentioned results indicate that our cells are "very powerful"; as their power lies in using states, hence in remembering their previous work, a natural idea is to consider tP systems with a low bound on the number of states in each cell. In view of the previously mentioned results, tP systems with at most 1, 2, 3, or 4 states per cell are of interest. We only briefly consider this question here, and we show that even reduced tP systems as those which use only one state in each cell can be useful: using such a net we can solve the Hamiltonian Path Problem in linear time (this is a direct consequence of the structure of a tP system, of the maximal mode of processing objects, and of the power of replicating the objects sent to all adjacent cells); remember that HPP is an NP-complete problem.

The power of tP systems with a reduced number of states per component remains to be further investigated. Actually, many other natural research topics can be considered, with motivations from automata and language theory (variants, power, normal forms), neural networks (learning, dynamic sets of neurons, dynamic synapses), computability (other NP-complete problems treated in this framework), dynamic systems (reachable configurations), etc.

2 Some Mathematical Prerequisites

The computability notions we use here are standard and can be found in many books, so we specify only some notations.

A *multiset* over a set X is a mapping $M : X \longrightarrow \mathbf{N}$; for $a \in X$, we say that $M(a)$ is the *multiplicity* of a in M. Here we work only with multisets over finite sets X. For two multisets M_1, M_2 over some set X we write $M_1 \subseteq M_2$ if and only if $M_1(a) \leq M_2(a)$ for all $a \in X$ (we say that M_1 is *included* in M_2). The *union* of M_1, M_2 is the multiset $M_1 \cup M_2 : X \longrightarrow \mathbf{N}$ defined by $(M_1 \cup M_2)(a) = M_1(a) + M_2(a)$, for all $a \in X$. If $M_1 \subseteq M_2$, then we also define the *difference* multiset $M_2 - M_1 : X \longrightarrow \mathbf{N}$ by $(M_2 - M_1)(a) = M_2(a) - M_1(a)$, for all $a \in X$. For $Y \subseteq X$ and M a multiset over X, we define the *projection* on Y by $pr_Y(M)(a) = \begin{cases} M(a), & \text{if } a \in Y, \\ 0, & \text{otherwise} \end{cases}$.

For a given alphabet V, V^* is the language of all strings over V, including the empty string, denoted by λ. The *Parikh mapping* associated with V is denoted by Ψ_V. A multiset M over an alphabet V can be represented by a string $w \in V^*$ such that $\Psi_V(w)$ gives the multiplicities in M of the symbols from V; obviously, all permutations of w are representations of the same multiset. For a family FA of languages, we denote by $PsFA$ the family of Parikh images of languages in FA. By CF, CS, RE we denote the families of context-free, context-sensitive, and recursively enumerable languages, respectively.

3 Tissue P Systems

We now pass to the definition of our variant of membrane (P) systems, which can also be considered as a model of a symbolic neural net. We introduce it in the general form, then we will consider variants of a restricted type.

A *tissue P system*, in short, a tP system, of *degree* $m \geq 1$, is a construct

$$\Pi = (E, \sigma_1, \ldots, \sigma_m, syn, i_{out}), \text{ where}$$

1. E is a finite non-empty alphabet (of *chemical objects*, but we also call them *excitations/impulses*);
2. $syn \subseteq \{1, 2, \ldots, m\} \times \{1, 2, \ldots, m\}$ (*synapses* among cells);
3. $i_{out} \in \{1, 2, \ldots, m\}$ indicates the *output cell*;
4. $\sigma_1, \ldots, \sigma_m$ are *cells*, of the form $\sigma_i = (Q_i, s_{i,0}, w_{i,0}, P_i), 1 \leq i \leq m$, where:
   (a) Q_i is a finite set (of *states*);
   (b) $s_{i,0} \in Q_i$ is the *initial state*;
   (c) $w_{i,0} \in E^*$ is the *initial multiset* of impulses;
   (d) P_i is a finite set of *rules* of the form $sw \to s'xy_{go}z_{out}$, where $s, s' \in Q_i$, $w, x \in E^*$, $y_{go} \in (E \times \{go\})^*$ and $z_{out} \in (E \times \{out\})^*$, with the restriction that $z_{out} = \lambda$ for all $i \in \{1, 2, \ldots, m\}$ different from i_{out}.

A tP system as above is said to be *cooperative* if it contains at least a rule $sw \to s'w'$ such that $|w| > 1$, and *non-cooperative* in the opposite case.

Any m-tuple of the form $(s_1w_1, \ldots, s_mw_m)$, with $s_i \in Q_i$ and $w_i \in E^*$, for all $1 \leq i \leq m$, is called a *configuration* of Π; $(s_{1,0}w_{1,0}, \ldots, s_{m,0}w_{m,0})$ is the *initial configuration* of Π.

Using the rules from the sets $P_i, 1 \leq i \leq m$, we can define *transitions* among configurations. To this aim, we first consider three *modes of processing the stimuli* and three *modes of transmitting excitations* from a cell to another one. Let us denote $E_{go} = \{(a, go) \mid a \in E\}$, $E_{out} = \{(a, out) \mid a \in E\}$, and $E_{tot} = E \cup E_{go} \cup E_{out}$. For $s, s' \in Q_i, x \in E^*, y \in E^*_{tot}$, we write

$$sx \Longrightarrow_{min} s'y \text{ iff } sw \to s'w' \in P_i, w \subseteq x, \text{ and } y = (x - w) \cup w',$$
$$sx \Longrightarrow_{par} s'y \text{ iff } sw \to s'w' \in P_i, w^k \subseteq x, w^{k+1} \not\subseteq x,$$

$$\text{for some } k \geq 1, \text{ and } y = (x - w^k) \cup w'^k,$$

$$sx \Longrightarrow_{max} s'y \text{ iff } sw_1 \to s'w'_1, \ldots, sw_k \to s'w'_k \in P_i, k \geq 1,$$
$$\text{such that } w_1 \ldots w_k \subseteq x, y = (x - w_1 \ldots w_k) \cup w'_1 \ldots w'_k,$$
$$\text{and there is no } sw \to s'w' \in P_i \text{ such that } w_1 \ldots w_k w \subseteq x.$$

In the first case, only one occurrence of the multiset from the left hand side of a rule is processed (replaced by the multiset from the right hand of the rule, at the same time changing the state of the cell), in the second case a maximal change is performed with respect to a chosen rule, in the sense that as many as possible copies of the multiset from the left hand side of the rule are replaced by the corresponding number of copies of the multiset from the right hand side, while in the third case a maximal change is performed with respect to all rules which use the current state of the cell and introduce the same new state after processing the impulses.

We also write $sx \to_\alpha sx$, for $s \in Q_i, x \in E^*$, and $\alpha \in \{min, par, max\}$, if there is no rule $sw \to s'w'$ in P_i such that $w \subseteq x$. This encodes the case when a cell cannot process the current impulses in a given state (it can be "unblocked" after receiving new impulses from its ancestors).

The multiset w' from a rule $sw \to s'w'$ contains symbols from E, but also symbols of the form (a, go) (or, in the case of cell i_{out}, of the form (a, out)). Such symbols will be sent to the cells related by synapses to cell σ_i where the rule $sw \to s'w'$ is applied, according to the following modes:

- *repl*: each symbol a, for (a, go) appearing in w', is sent to each of the cells σ_j such that $(i, j) \in syn$;
- *one*: all symbols a appearing in w' in the form (a, go) are sent to one of the cells σ_j such that $(i, j) \in syn$, nondeterministically chosen; more exactly, in the case of modes *par* and *max* of using the rules, we first perform all applications of rules, and after that we send all obtained symbols to a unique descendant of the cell (that is, we do not treat separately the impulses introduced by each rule, but all of them in a package);
- *spread*: the symbols a appearing in w' in the form (a, go) are non-deterministically distributed among the cells σ_j such that $(i, j) \in syn$.

In order to formally define the transition among the configurations of Π we need some further notations. For a multiset w over E_{tot}, we denote by $go(w)$ the multiset of symbols $a \in E$ appearing in w in the form (a, go), and by $out(w)$ the multiset of symbols $a \in E$, appearing in w in the form (a, out). Clearly, $go(w)(a) = w((a, go))$ and $out(w)(a) = w((a, out)), a \in E$. Moreover, for a node i in the graph defined by syn we denote $ant(i) = \{j \mid (j, i) \in syn\}$ and $succ(i) = \{j \mid (i, j) \in syn\}$ (the ancestors and the successors of node i, respectively).

Now, for two configurations $C_1 = (s_1w_1, \ldots, s_mw_m), C_2 = (s'_1w''_1, \ldots, s'_mw''_m)$ we write $C_1 \Longrightarrow_{\alpha,\beta} C_2$, for $\alpha \in \{min, par, max\}, \beta \in \{repl, one, spread\}$, if there are $w'_1, \ldots, w'_m$ in E^*_{tot} such that $s_iw_i \Longrightarrow_\alpha s'_iw'_i$, $1 \leq i \leq m$, and

- for $\beta = repl$ we have $w''_i = pr_E(w'_i) \cup \bigcup_{j \in ant(i)} go(w'_j)$;

– for $\beta = one$ we have $w_i'' = pr_E(w_i') \cup \bigcup_{j \in I_i} go(w_j')$, where I_i is a subset of $ant(i)$ such that the set $ant(i)$ was partitioned into $I_1, \ldots, I_m$; at this transition, all non-empty sets of impulses of the form $\bigcup_{j \in I_k} go(w_j')$, $1 \leq k \leq m$, should be sent to receiving cells (added to multisets w_l'', $1 \leq l \leq m$);
– for $\beta = spread$ we have $w_i'' = pr_E(w_i') - go(w_i') \cup z_i$, where z_i is a submultiset of the multiset $\bigcup_{j \in ant(i)} go(w_j')$ such that $z_1, \ldots, z_m$ are multisets with the property $\bigcup_{j=1}^{m} z_j = \bigcup_{j \in ant(i)} go(w_j')$, and such that all $z_1, \ldots, z_m$ are sent to receiving cells (added to multisets w_l'', $1 \leq l \leq m$).

Note that in the case of the cell $\sigma_{i_{out}}$ we also remove all symbols $a \in E$ appearing in $w_{i_{out}}'$ in the form (a, out).

During any transition, some cells can do nothing: if no rule is applicable to the available multiset of impulses in the current state, then a cell waits until new impulses are sent to it from its ancestor cells.

A sequence of transitions among configurations of the tP system Π is called a *computation* of Π. A computation which ends in a configuration where no rule in no cell can be used, is called a *halting* computation. Assume that during a halting computation the tP system Π sends out, through the cell $\sigma_{i_{out}}$, the multiset z. We say that the vector $\Psi_E(z)$, representing the multiplicities of impulses from z, is *computed* (or *generated*) by Π. We denote by $N_{\alpha,\beta}(\Pi)$, $\alpha \in \{min, par, max\}$, $\beta \in \{repl, one, spread\}$, the set of all vectors of natural numbers generated by a tP system Π, in the mode (α, β). The family of all sets $N_{\alpha,\beta}(\Pi)$, generated by all cooperative tP systems with at most $m \geq 1$ cells, each of them using at most $r \geq 1$ states, is denoted by $NtP_{m,r}(Coo, \alpha, \beta)$; when non-cooperative tP systems are used, we write $NtP_{m,r}(nCoo, \alpha, \beta)$ for the corresponding family of vector sets. When one (or both) of the parameters m, r are not bounded, then we replace it (them) with $*$, thus obtaining families of the form $NtP_{m,*}(\gamma, \alpha, \beta)$, $NtP_{*,r}(\gamma, \alpha, \beta)$, etc.

We have 18 families of the form $NtP_{*,*}(\gamma, \alpha, \beta)$, but, as we will see below, not all of them are different.

4 An Example

Before investigating the power and the properties of tP systems, let us examine an example, in order to clarify and illustrate the previous definitions. Consider the rather simple tP system:

$$\begin{aligned}
\Pi_1 &= (\{a\}, \sigma_1, \sigma_2, \sigma_3, syn, 1),\\
\sigma_1 &= (\{s\}, s, a, \{sa \to s(a, go),\ sa \to s(a, out)\}),\\
\sigma_2 &= (\{s\}, s, \lambda, \{sa \to s(a, go)\}),\\
\sigma_3 &= (\{s\}, s, \lambda, \{sa \to s(a, go)\}),\\
syn &= \{(1,2), (1,3), (2,1), (3,1)\}.
\end{aligned}$$

The reader can easily check that we have:

$$N_{\alpha,repl}(\Pi_1) = \{(n) \mid n \geq 1\}, \text{ for } \alpha \in \{min, max\},$$

$$N_{par,repl}(\Pi_1) = \{(2^n) \mid n \geq 0\},$$
$$N_{\alpha,\beta}(\Pi_1) = \{(1)\}, \text{ for } \alpha \in \{min, par, max\}, \beta \in \{one, spread\}.$$

Indeed, in the non-replicative mode of communication, no further symbol is produced, hence we only generate the vector (1). In the replicative case, the symbols produced by the rule $sa \to s(a, go)$ from cell 1 are doubled by communication. When the rules are used in the parallel mode, then all symbols are processed at the same time by the same rule, which means that all symbols present in the system are doubled from a step to the next one, therefore, the powers of 2 are obtained. When the rules are used in the minimal mode, the symbols are processed or sent out one by one, hence all natural numbers can be obtained. In the maximal mode, we can send copies of a at the same time to cells 2 and 3, and outside the system, hence again any number of symbols can be sent out.

5 The Power of tP systems

The following relations are direct consequences of the definitions.

Lemma 1. (i) *For all* $1 \leq m \leq m', 1 \leq r \leq r', \gamma \in \{Coo, nCoo\}, \alpha \in \{min, par, max\}$, *and* $\beta \in \{repl, one, spread\}$, *we have:*

$$NtP_{m,r}(\gamma, \alpha, \beta) \subseteq NtP_{m',r'}(\gamma, \alpha, \beta) \subseteq NtP_{*,*}(\gamma, \alpha, \beta) \subseteq PsRE,$$
$$NtP_{m,r}(nCoo, \alpha, \beta) \subseteq NtP_{m,r}(Coo, \alpha, \beta).$$

(ii) *For all tP systems* Π, *cooperating or not, where each cell has at most one successor, and for all* $\alpha \in \{min, par, max\}$ *we have*

$$N_{\alpha,repl}(\Pi) = N_{\alpha,one}(\Pi) = N_{\alpha,spread}(\Pi).$$

As it is standard when considering a new computing device, we compare the power of tP systems with that of Turing machines and restricted variants of them. Refined classifications of the power of such machines are provided by the Chomsky and the Lindenmayer hierarchies. We start by considering the minimal mode of using the rules in a tP system, and this turns out to be computationally universal, a fact which makes natural the comparison with (Parikh images of) Chomsky families, in particular, $PsRE$. In a subsequent section we will consider the parallel and the maximal modes of using the rules, and this will make necessary the comparison with (Parikh images of) Lindenmayer families.

5.1 Comparison with Chomsky Families

Rather surprising, if we take into consideration the apparently weak ingredients of our models, when using the mode *min* of applying the rules, even the non-cooperative tP systems turn out to be computationally universal. (As expected, the same result holds true also when using cooperative rules, in all modes *min, par, max*.) In proving such results we try to keep as reduced as possible both the number of cells and the maximal number of states used by the cells.

Theorem 1. $PsRE = NtP_{2,5}(\gamma, min, \beta)$ *for all* $\gamma \in \{Coo, nCoo\}, \beta \in \{repl, one, spread\}$.

At the price of using two more cells, we can decrease the number of used states (the proof is omitted).

Theorem 2. $PsRE = NtP_{4,4}(\gamma, min, \beta)$ *for all* $\gamma \in \{Coo, nCoo\}, \beta \in \{one, spread\}$.

If we use cooperative rules, then we can further decrease both the number of cells and of states. Moreover, we can characterize $PsRE$ for all modes *min, par, max* of processing the impulses, and this completes the study of the cooperative case.

Theorem 3. $PsRE = NtP_{2,2}(Coo, \alpha, \beta)$ *for all* $\alpha \in \{min, par\}, \beta \in \{repl, one, spread\}$.

We do not know whether or not the results in Theorems 1, 2, and 3 are optimal in the number of cells and of states.

5.2 Comparison with Lindenmayer Families

The maximal mode of using the rules in a tP system resembles the parallel mode of rewriting the strings in an L system, and this makes the following results expected.

Theorem 4. (i) $PsE0L \subseteq NtP_{1,2}(nCoo, max, \beta)$ *for all* $\beta \in \{repl, one, spread\}$. (ii) $PsET0L \subseteq NtP_{1,3}(nCoo, max, \beta)$ *for all* $\beta \in \{repl, one, spread\}$.

For tP systems working in the *min* mode, we need further additional cells (and states) in order to simulate E0L and ET0L systems.

Theorem 5. $PsE0L \subseteq NtP_{2,3}(nCoo, min, \beta)$ *for all* $\beta \in \{repl, one, spread\}$.

In the case of ET0L systems we needed one more cell and one more state (but we do not know whether or not this result can be improved).

Theorem 6. $PsET0L \subseteq NtP_{3,4}(nCoo, min, \beta)$ *for all* $\beta \in \{repl, one, spread\}$.

Interestingly enough, the converse of assertion (ii) from Theorem 4 is also true, even in the following more general form (and this settles the study of modes *par* and *max*: they do not lead to computational universality).

Theorem 7. $NtP_{*,*}(nCoo, \alpha, \beta) \subseteq PsET0L$, *for all* $\alpha \in \{par, max\}$ *and* $\beta \in \{repl, one, spread\}$.

Together with assertion (ii) from Theorem 4 we get the following characterization of $PsET0L$, which precisely describes the power of the mode *max* in the non-cooperative case.

Theorem 8. *$NtP_{1,1}(nCoo, max, \beta) \subseteq NtP_{1,2}(nCoo, max, \beta) \subseteq NtP_{1,3}(nCoo, max, \beta) = NtP_{m,r}(nCoo, max, \beta) = NtP_{*,*}(nCoo, max, \beta) = PsET0L$, for all $m \geq 1, r \geq 3$.*

A more precise characterization of families $NtP_{m,r}(nCoo, par, \beta)$, $\beta \in \{repl, one, spread\}$, remains to be found (but we already know that such systems only generate Parikh images of ET0L languages).

6 Solving HPP in Linear Time

The architecture of tP systems and their way of working (especially the fact that in the maximal mode of using the rules we can process all impulses which may be processed in such a way that the same next state is obtained, irrespective which rules are used, and the fact that in the replicative mode one can send the same impulses to all successors of a cell) have an intrinsic computational power. More precisely, problems related to paths in a (directed) graph can be easily solved by a tP system, just by constructing a net with the synapses graph identical to the graph we deal with, constructing all paths in the graph with certain properties by making use of the maximal mode of applicating the rules and of the replicative communication, and checking the existence of a path with a desired property.

We illustrate this power of tP systems with the Hamiltonian Path Problem (HPP), which asks whether or not in a given directed graph $G = (V, U)$ (where $V = \{a_1, \ldots, a_m\}$ is the set of vertices, and $U \subseteq V \times V$ is the set of edges) there is a path starting in some vertex a_{in}, ending in some vertex a_{out}, and visiting all vertices exactly once. For simplicity, in what follows we assume that $a_{in} = a_1$ and $a_{out} = a_m$. It is know that the HPP is a NP-complete problem, hence it is one of the problems considered as intractable for the sequential computers (for the Turing machines).

Having a graph $G = (V, U)$ as above, we construct the tP system $\Pi = (E, \sigma_1, \ldots, \sigma_m, U, m)$, with

$$\begin{aligned}
E &= \{[z; k] \mid z \in V^*, 0 \leq |z| \leq m, 0 \leq k \leq m\},\\
\sigma_1 &= (\{s\}, s, [\lambda; 0], \{s[\lambda; 0] \rightarrow s([1; 1], go)\}),\\
\sigma_i &= (\{s\}, s, \lambda, \{s[z; k] \rightarrow s([zi; k+1], go) \mid z \in V^*, 1 \leq |z| \leq m-2,\\
&\qquad |z|_i = 0, 1 \leq k \leq m-2\}), \text{ for each } i = 2, 3, \ldots, m-1, \text{ and}\\
\sigma_m &= (\{s\}, s, \lambda, \{s[z; m-1] \rightarrow s([zm; m], out) \mid z \in V^*, |z| = m-1\}).
\end{aligned}$$

It is easy to see that $N_{max,repl}(\Pi) \neq \emptyset$ if and only if HPP has a solution for the graph G: the paths in G grow simultaneously in all cells of Π, because of the *max* mode of using the rules (each cell has only one state, hence all rules can be used at the same time). Moreover, the cell σ_m can work only after $m-1$ steps and a symbol is sent out of the net at the step m. Thus, it is enough to watch the tP system at step m and if any symbol is sent out, then HPP has a solution, otherwise we know that such a solution does not exist. (Note that the symbol sent out describes a Hamiltonian path in G.)

References

1. B. Alberts et al., *Essential Cell Biology. An Introduction to the Molecular Biology of the Cell,* Garland Publ. Inc., New York, London, 1998.
2. M.A. Arbib, *Brains, Machines, and Mathematics,* second ed., Springer-Verlag, Berlin, 1987.
3. M.A. Arbib, *The Methaphorical Brain: An Introduction to Schemes and Brain Theory,* Wiley Interscience, 1988.
4. J.P. Banatre, D. LeMetayer, Gamma and chemical reaction model: ten years after, in vol. *Coordination Programming: Mechanisms, Models, and Semantics* (C. Hankin, ed.), Imperial College Press, 1996, 3–41.
5. D.S. Blank *et al* (24 co-authors), Connectionist symbol processing: Dead or alive?, *Neural Computing Surveys,* 2 (1999), 1–40.
6. C. Choffrut, ed., *Automata Networks, Lecture Notes in Computer Science,* 316, Springer-Verlag, Berlin, 1988.
7. E. Csuhaj-Varju, C. Martin-Vide, V. Mitrana, Multiset automata, *Multiset Processing* (C.S. Calude, Gh. Paun, G. Rozenberg, A. Salomaa, eds), *Lecture Notes in Computer Science,* 2235, Springer-Verlag, 2001.
8. A. Dovier, A. Policriti, G. Rossi, A uniform axiomatic view of lists, multisets, and sets, and the relevant unification algorithms, *Fundamenta Informaticae,* 36, 2-3 (1998), 201–234.
9. F. Gecseg, *Products of Automata,* Springer-Verlag, Berlin, 1986.
10. S.C. Kleene, Representation of events in nerve nets and finite automata, *Automata Studies,* Princeton Univ. Press, Princeton, N.J., 1956, 2–42.
11. W.R. Loewenstein, *The Touchstone of Life. Molecular Information, Cell Communication, and the Foundations of Life,* Oxford Univ. Press, 1999.
12. A. Mateescu, V. Mitrana, Parallel finite automata systems communicating by states, *Intern. J. Found. Computer Sci.,* to appear.
13. W.S. McCulloch, W.H. Pitts, A logical calculus of the ideas immanent in nervous activity, *Bull. Math. Biophys.,* 5 (1943), 115–133.
14. M. Minsky, *Computation: Finite and Infinite Machines,* Prentice-Hall, 1967.
15. Gh. Păun, Computing with membranes, *Journal of Computer and System Sciences,* 61, 1 (2000), 108–143.

Transducers with Set Output

Jurek Czyzowicz[1], Wojciech Fraczak[2], and Andrzej Pelc[1]

[1] Dept. d'informatique, UQAH, CP 1250, succ. B, Hull PQ J8X 3X7, Canada
[2] Solidum Systems Corp, 1575 Carling Av., Ottawa ON K1Z 7M3, Canada

Abstract. We consider transducers with set output, i.e., finite state machines which produce a set of output symbols upon reading any input symbol. When a word consisting of input symbols is read, the union of corresponding output sets is produced. Such transducers are instrumental in some important data classification tasks, such as multi-field packet classification. Two transducers are called *equivalent* if they produce equal output upon reading any input word. In practical data classification applications, it is important to store in memory only one transducer of every equivalence class, in order to save memory space. This yields the need of finding, in any equivalence class, one transducer, called *canonical* which is easy to compute, given any transducer from this class. One of the results of this paper is the construction of an algorithm which completes this task. Assuming that the input and output alphabets are of bounded size, for a given n-state transducer T, our algorithm finds the canonical transducer $\Psi(T)$ equivalent to T in time $O(n \log n)$.

1 Introduction

Data classification is among crucial problems in network information processing. Packets arriving from a communication channel have to be divided into several classes, based on their content. Among different ways of classifying data, used in modern technology, the following approaches are the most significant: the network processor approach, used by such companies as Intel, Motorola, and Vitesse, the content addressable memory (CAM) approach [1], used, e.g., by Netlogic, IDT and Kawasaki, and the dedicated chip approach based on finite state automata, used, e.g., by Agere, Raqia, and Solidum Systems Corp. [2]. Among the three, the approach using a dedicated chip is by far the fastest: it enables classifying data packets at wire speed, i.e., it permits to complete the classification as soon as the last bit of the packet is read.

In particular, the mechanism of data classification used by Solidum Systems Corp. (see `http://www.solidum.com`) uses the concept of a transducer which is, roughly speaking, a finite state machine producing some output after reading every input symbol. The entire output produced after reading an input string, called the *tag* of this string, can be considered as the name of the class to which the processed packet is assigned.

Classic transducers [3,4,5,6,7] output strings of symbols after reading every input symbol, and concatenate these strings as the input word is read. However, for some important tasks, such as multi-field packet classification [8], also

O.H. Ibarra and L. Zhang (Eds.): COCOON 2002, LNCS 2387, pp. 300–309, 2002.

called multi-dimensional range matching [9], transducers producing strings of symbols are not a good tool. Indeed, in this context, the order in which output symbols are produced, is not important, and once a symbol has been produced during the classification process, producing this symbol again at a later stage does not change the result. Hence, in this important application of multi-field packet classification, it appears more natural to define tags as *sets* of output symbols, rather than *strings* of such symbols. More precisely, upon reading an input symbol, a (possibly empty) set of output symbols is produced, and the union of such consecutive sets is produced on every classification path. Transducers of this new type are called *set transducers* and are used as a theoretical tool forming the base of data classification technology in the above mentioned context. It turns out that classic transducers would be, indeed, entirely inappropriate to solve the multi-field packet classification problem: it can be shown that using classic transducers instead of set transducers in this context may result in an exponential growth of the number of states.

Transducers used in practical applications are very large: they are composed of hundreds of thousands of states. Hence it is necessary to decompose them into smaller building blocks, and store these blocks, reusing the same blocks for the construction of many different complex transducers. From the point of view of efficient memory use, it is of crucial importance to store in the memory only one such building block of given functionality, avoiding simultaneous storage of many equivalent transducers. (Two transducers are called *equivalent* if they produce equal output upon reading any input word: intuitively, they classify all packets in the same way.) To this end we need to choose, in every class of equivalent transducers, one representative, called the *canonical transducer* of this class. The choice of the canonical transducer must satisfy the property that, given any transducer, the canonical transducer equivalent to it can be produced efficiently. If this is the case, it becomes easy to avoid duplication of equivalent transducers stored in the memory: if a given transducer is needed as a building block, it is enough to compute the canonical transducer of its class and check if it is already stored in the memory. We add it to the memory only if it is not yet there.

Hence the problem of efficient finding of the canonical transducer equivalent to a given one is of significant practical importance. One of the results of this paper is the construction of an algorithm which completes this task. Assuming that we are given an n-state transducer T and that the alphabets are of bounded size, our algorithm finds the canonical transducer equivalent to T in time $O(n \log n)$.

Related work. Classic transducers (those with string output) have been extensively studied in the literature [3,4,5,6,7]. In this context, the particular problem of efficient finding of the canonical transducer equivalent to a given one has been considered in [7]. In our case this problem seems to be different because the idea from [7] directly using minimization cannot work: minimal set transducers are not unique.

As for the multi-field packet classification problem, which was our main reason for introducing and studying the concept of set transducers, its solutions in [8,9] were based on a combination of hardware and software tools, as opposed to our approach which uses a general purpose transducer entirely implemented in hardware, and thus permitting much faster classification.

2 Set Transducers

We consider transducers which output sets. They differ from usual sequential transducers in that they define partial mappings from words over an input alphabet into subsets of an output alphabet.

Let Σ and Δ be two disjoint finite sets of *input* and *output symbols*, respectively. A *deterministic finite automaton (dfa)* $A = (Q, s, \delta, F)$ consists of a finite set Q of states, a partial transition function $\delta : Q \times \Sigma \mapsto Q$, an initial state $s \in Q$, and a set of final states $F \subseteq Q$. We call *transition* of A any pair (x, a) of state and input symbol for which transition function $\delta(x, a)$ is defined.

A *labeling* L of an *dfa* $A = (Q, s, \delta, F)$ is a pair (I, λ) consisting of an *initial output* $I \subseteq \Delta$ and a partial *labeling function* $\lambda : Q \times \Sigma \mapsto 2^{\Delta}$ with the same domain as δ.

A *set transducer* $T = (A, L)$ is a pair consisting of a *dfa* A and labeling L of A. The transition function δ and labeling function λ naturally extend into $\hat{\delta} : Q \times \Sigma^* \mapsto Q$ and $\hat{\lambda} : Q \times \Sigma^* \mapsto 2^{\Delta}$ by:

$$\hat{\delta}(x, \varepsilon) \stackrel{\text{def}}{=} x, \quad \hat{\delta}(x, aw) \stackrel{\text{def}}{=} \hat{\delta}(\delta(x, a), w)$$
$$\hat{\lambda}(x, \varepsilon) \stackrel{\text{def}}{=} \emptyset, \quad \hat{\lambda}(x, aw) \stackrel{\text{def}}{=} \lambda(x, a) \cup \hat{\lambda}(\delta(x, a), w)$$

for $a \in \Sigma$, $w \in \Sigma^*$, and $x \in Q$, where $\varepsilon \in \Sigma^*$ denotes the empty word. Note that $\hat{\delta}$ and $\hat{\lambda}$ are partial functions: $\hat{\delta}(x, w)$ is defined if and only if $\hat{\lambda}(x, w)$ is.

From now on, we will use δ and λ instead of $\hat{\delta}$ and $\hat{\lambda}$, respectively. We will also assume that the underlying *dfa* A is always trimmed by eliminating non-essential states, i.e., for every state $x \in Q$ there exist $w, w' \in \Sigma^*$ such that $\delta(s, w) = x$ and $\delta(x, w') \in F$.

Every set transducer T defines a partial mapping $|T| : \Sigma^* \mapsto 2^{\Delta}$, from the set of words over input symbols into sets of output symbols. The mapping $|T|$ is defined for an input word w if the word leads from the initial state to a final state. The value of $|T|(w)$ is the union of all output sets produced on the path including the initial output:

$$|T|(w) \stackrel{\text{def}}{=} \begin{cases} I \cup \lambda(s, w) & \text{if } \delta(s, w) \in F \\ \text{not defined} & \text{otherwise} \end{cases} \tag{1}$$

Transducers T_1 and T_2 are called *equivalent* if $|T_1| = |T_2|$.

Consider transducers in Fig. 1. They define the same partial mapping, T, from words over alphabet $\{a, b\}$ into subsets of $\{\alpha, \beta\}$:

$$T = \{aa \mapsto \{\alpha\}, ab \mapsto \{\alpha, \beta\}, ba \mapsto \{\alpha, \beta\}, bb \mapsto \{\beta\}\}.$$

In the sequel, we will define a canonical transducer in every equivalence class. More precisely, we will describe a transformation Ψ on the class of all

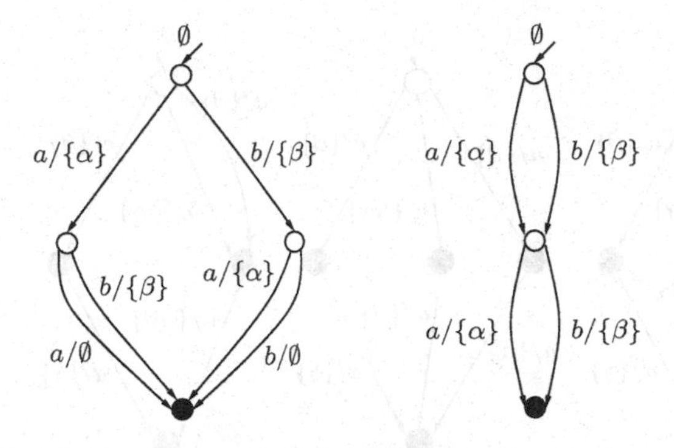

Fig. 1. Two equivalent set transducers

transducers such that $\Psi(T)$ is equivalent to T; and if T_1 is equivalent to T_2 then $\Psi(T_1) = \Psi(T_2)$. Our main goal will then be efficient computation of $\Psi(T)$, given a transducer T.

For any state x of a finite deterministic automaton $A = (Q, s, \delta, F)$ and any labeling $L = (I, \lambda)$ of A we define:

$$D(x, L) \stackrel{\text{def}}{=} \{\alpha \in \Delta \mid \forall w \in \Sigma^* \left((\delta(x, w) \in F) \Rightarrow (\alpha \in \lambda(x, w))\right)\} \tag{2}$$

$$U(x, L) \stackrel{\text{def}}{=} I \cup \{\alpha \in \Delta \mid \forall w \in \Sigma^* \left((\delta(s, w) = x) \Rightarrow (\alpha \in \lambda(s, w))\right)\} \tag{3}$$

Intuitively, $D(x, L)$ and $U(x, L)$ denote the intersection of output sets generated on all paths starting at, respectively ending in, state x.

We define the following transformation Φ on the class of all labelings of an automaton. It will be used later in the construction of transformation Ψ.

Definition 1. *Let $A = (Q, s, \delta, F)$ be an automaton with labeling $L = (I, \lambda)$. The labeling $\Phi(L) = (I', \lambda')$ of A is defined as follows:*

- $I' \stackrel{\text{def}}{=} I \cup D(s, L)$;
- $\lambda'(x, a) \stackrel{\text{def}}{=} (\lambda(x, a) \cup D(\delta(x, a), L)) \setminus (U(x, L) \cup D(x, L))$, *for every state $x \in Q$ and input symbol $a \in \Sigma$.*

We extend Φ on transducers, i.e., if $T = (A, L)$ then $\Phi(T) \stackrel{\text{def}}{=} (A, \Phi(L))$. Output symbols in labels of transducers can be sometimes moved along directed paths keeping the resulting transducer equivalent. Intuitively $\Phi(T)$ is the result of moving those symbols as close to the initial state as possible.

We will prove below that transformation Φ preserves the semantics of the transducer.

Theorem 1. *For any transducer T we have $|T| = |\Phi(T)|$.*

Proof. Let $T = (A, L)$, $A = (Q, s, \delta, F)$, $L = (I, \lambda)$, and $\Phi(L) = (I', \lambda')$.

We prove that $\forall w \in \Sigma^* \forall \alpha \in \Delta (\alpha \in |T|(w) \Leftrightarrow \alpha \in |\Phi(T)|(w))$.

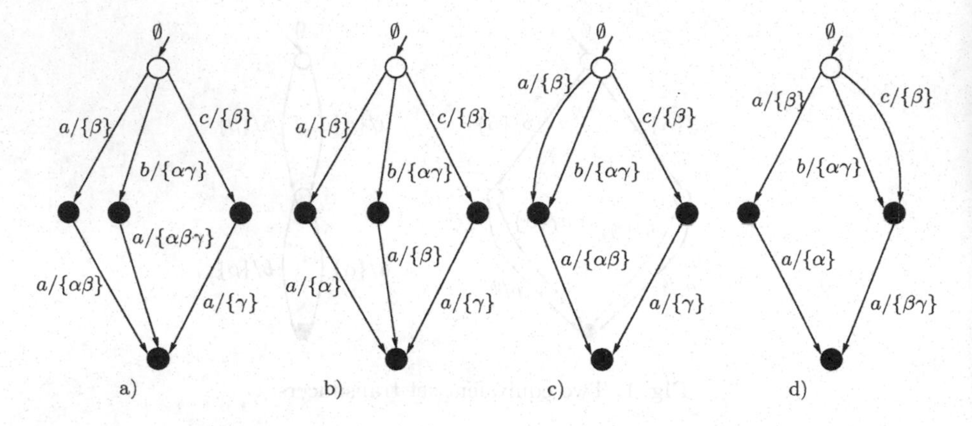

Fig. 2. Equivalent transducers

1. $\forall w \in \Sigma^* \forall \alpha \in \Delta(\alpha \in |\Phi(T)|(w) \Rightarrow \alpha \in |T|(w))$
   Let us take the shortest prefix w' of w, such that $\alpha \in I' \cup \lambda'(s, w')$.
   If w' is the empty word then $\alpha \in I \cup D(s, L)$ and thus $\alpha \in |T|(w)$.
   If w' is not the empty word, i.e., $w' = w''a$ for some input word w'' and an input symbol a, then $\alpha \in \lambda'(\delta(s, w''), a)$, i.e., $\alpha \in \lambda(\delta(s, w''), a)$ or $\alpha \in D(\delta(s, w'), L)$. In both cases this implies $\alpha \in \lambda(s, w)$ and thus $\alpha \in |T|(w)$.
2. $\forall w \in \Sigma^* \forall \alpha \in \Delta(\alpha \in |T|(w) \Rightarrow \alpha \in |\Phi(T)|(w))$
   Let us take the shortest prefix w' of w, such that $\alpha \in I \cup \lambda(s, w')$.
   If w' is the empty word then $\alpha \in I$. Since $I \subseteq I'$ we have $\alpha \in |\Phi(T)|(w)$.
   If w' is not the empty word, i.e., $w' = w''a$ for some input word w'' and an input symbol a, then $\alpha \in \lambda(\delta(s, w''), a)$ and $\alpha \notin U(\delta(s, w''), L)$.
   Two cases are possible:
   (a) $\alpha \in \lambda'(\delta(s, w''), a)$
   In this case α is obviously present in $\lambda'(s, w)$, i.e., $\alpha \in |\Phi(T)|(w)$.
   (b) $\alpha \notin \lambda'(\delta(s, w''), a)$
   By definition of $\Phi(L)$ we have $\alpha \in D(\delta(s, w''), L) \cup U(\delta(s, w''), L)$, thus $\alpha \in D(\delta(s, w''), L)$.
   If for all prefixes z of w'' we have $\alpha \in D(\delta(s, z), L)$ then α will be in I'. Otherwise, we find a non-empty prefix vb ($v \in \Sigma^*$, $b \in \Sigma$) of w'' such that $\alpha \notin D(\delta(s, v), L)$ and $\alpha \in D(\delta(s, vb), L)$ (notice that such a prefix vb must exist). By definition $\lambda'(\delta(s, v), b) = \lambda(\delta(s, v), b)) \cup D(\delta(s, vb), L) \setminus (D(\delta(s, v), L) \cup U(\delta(s, v), L))$. Thus $\alpha \in \lambda'(\delta(s, v), b)$ which implies $\alpha \in \lambda'(s, w)$, i.e., $\alpha \in |\Phi(T)|(w)$. □

A naive idea of finding the canonical transducer equivalent to a given transducer T, would be to minimize $\Phi(T)$, hoping that such a minimal transducer is unique. However, it turns out that this is not the case.

In Fig. 2 transducer (b) is the result of applying transformation Φ to transducer (a). Transducer (a) can be further minimized by collapsing states $\delta(s, a)$ and $\delta(s, b)$ yielding transducer (c), or by collapsing states $\delta(s, b)$ and $\delta(s, c)$ yielding transducer (d). This example shows that minimization of $\Phi(T)$ may, in case of an arbitrary transducer T, lead to many different transducers. Hence we have to proceed in a more subtle way. In the sequel, we define a class of transducers called *simple.* For any transducer T we construct a simple transducer T' equivalent to it. Then we show that $\Phi(T')$ is also a simple transducer, and finally we prove that there exists exactly one minimal simple transducer equivalent to $\Phi(T')$. This unique transducer will then be taken as the canonical transducer $\Psi(T)$.

3 Simple Transducers

In this section we introduce a class of transducers which, on one input word, never produce the same output symbol twice. This will enable us to produce efficiently a canonical (unique) representative of each equivalence class.

Definition 2. *A labeling $L = (I, \lambda)$ of a dfa $A = (Q, s, \delta, F)$ is* simple *if for any word ww' accepted by A, i.e., $\delta(s, ww') \in F$, we have $(I \cup \lambda(s, w)) \cap \lambda(\delta(s, w), w') = \emptyset$. A transducer is simple if its labeling is simple.*

In Fig. 2 transducer (b) is simple however (a), (c), and (d) are not.

We now describe a transformation Ξ which, given a transducer $T = (A, L)$, produces a transducer $\Xi(T)$ which is simple and equivalent to T.

Definition 3. *Let $T = (A, L)$ be a transducer with $A = (Q, s, \delta, F)$ and $L = (I, \lambda)$. We define $\Xi(T) = (A', L')$ with $A' = (Q', s', \delta', F')$ and $L' = (I, \lambda')$ in the following way:*

- *$Q' \subseteq Q \times 2^{\Delta}$, i.e., states are pairs of original states with subsets of output symbols;*
- *$s' = (s, I)$;*
- *$\delta'((x, O), a) = (\delta(x, a), O \cup \lambda(x, a))$;*
- *$(x, O) \in F'$ whenever $x \in F$;*
- *$I' = I$;*
- *$\lambda'((x, O), a) = \lambda(x, a) \setminus O$.*

The following lemma shows that Ξ transformation of the transducer preserves its semantics. The next lemma will prove that the resulting transducer is simple.

Lemma 1. *For every input word $w \in \Sigma^*$ we have:*

1. *$\delta'(s', w) = (\delta(s, w), \lambda(s, w) \cup I)$;*
2. *$I \cup \lambda(s, w) = I' \cup \lambda'(s', w)$.*

Proof. The property can be proved by induction on the length of w. It holds for w being empty word. Let suppose that the property holds for any w of length k. Consider a word w' of length $k + 1$, $w' = wa$.

1. $\delta'(s', wa) = \delta'(\delta'(s', w), a)$. By the induction hypothesis we have, $\delta'(s', wa) = \delta'((\delta(s, w), O), a) = (\delta(\delta(s, w), a), O')$, for $O = I \cup \lambda(s, w)$ and $O' = (O \cup \lambda(\delta(s, w), a))$. Finally, $O' = O \cup \lambda(\delta(s, w), a) = I \cup \lambda(s, w) \cup \lambda(\delta(s, w), a) = I \cup \lambda(s, wa)$. I.e., $\delta'(s', wa) = (\delta(s, wa), I \cup \lambda(s, wa))$.

2. $\lambda'(s', wa) = \lambda'(s', w) \cup \lambda'(\delta'(s', w), a)$ and $\delta'(s', w) = (\delta(s, w), I \cup \lambda(s, w))$. By the induction hypothesis and the definition of λ', $\lambda'(s', wa) = \lambda(s, w) \cup (\lambda(\delta(s, w), a) \setminus I \cup \lambda(s, w))$. Finally, $I' \cup \lambda'(s', wa) = I \cup \lambda(s, w) \cup \lambda(\delta(s, w), a) = I \cup \lambda(s, wa)$. □

Lemma 2. *Transducer $\Xi(T)$ is simple.*

Proof. Let $T = (A, L)$, $A = (Q, s, \delta, F)$, $L = (I, \lambda)$, $\Xi(T) = (A', L')$, $A' = (Q', s', \delta', F')$, and $L' = (I', \lambda')$. For every word ww' accepted by A (and thus, by Lemma 1, also by A') we have $(I' \cup \lambda'(s', w)) \cap \lambda'(\delta'(s', w), w') = (I \cup \lambda(s, w)) \cap (\lambda(\delta(s, w), w') \setminus (I \cup \lambda(s, w)) = \emptyset$. □

Lemmas 1 and 2 imply the following theorem.

Theorem 2. *For any transducer T there exists a simple transducer T' such that $|T| = |T'|$.*

The next lemma shows that the transformation Φ preserves simplicity.

Lemma 3. *If L is a simple labeling then so is $\Phi(L)$.*

Proof. Let $L = (I, \lambda)$, $\Phi(L) = (I', \lambda')$ be labelings of *dfa* $A = (Q, s, \delta, F)$.

Suppose that $\Phi(L)$ is not simple. Then there is a word ww' accepted by A such that $(I' \cup \lambda'(s, w)) \cap \lambda'(\delta(s, w), w') \neq \emptyset$. Since $I' \subseteq U(x, L) \cup D(x, L)$ for every state $x \in Q$, we have $I' \cap \lambda'(\delta(s, w), w') = \emptyset$ and thus $\lambda'(s, w) \cap \lambda'(\delta(s, w), w') \neq \emptyset$.

Without loss of generality we suppose that $w = w_1 a w_2$, $w' = b w_3$, and there exists an input symbol α such that $\alpha \in \lambda'(\delta(s, w_1), a) \cap \lambda'(\delta(s, w_1 a w_2), b)$. Hence

$$\alpha \in \lambda(\delta(s, w_1), a) \cup D(\delta(s, w_1 a), L) \quad (4)$$

$$\alpha \notin D(\delta(s, w_1 a w_2), L) \quad (5)$$

$$\alpha \in \lambda(\delta(s, w_1 a w_2), b) \cup D(\delta(s, w_1 a w_2 b), L) \quad (6)$$

(4) and (5) imply $\alpha \in \lambda(\delta(s, w_1), a w_2)$, and (6) implies $\alpha \in \lambda(\delta(s, w_1 a w_2), b w_3)$. Hence $\alpha \in \lambda(s, w) \cap \lambda(\delta(s, w), w')$, which contradicts L being simple. □

A transducer may be viewed as a finite state automaton accepting words over both (input and output) alphabets. The canonical form of the transducer will be obtained as the minimization of such automaton.

Definition 4. *Let $A = (Q, s, \delta, F)$ be a dfa and $L = (I, \lambda)$ its labeling. We call* in-out word *of the transducer $T = (A, L)$ any finite list of pairs $\omega = (a_1, o_1) \ldots (a_k, o_k)$ with $a_i \in \Sigma$, $o_i = \lambda(\delta(s, a_1 \ldots a_{i-1}), a_i)$, for $i \in [1, k]$, and such that $\delta(s, w) \in F$, for $w = a_1 a_2 \ldots a_k$.*

Theorem 3. *Let $T_1 = (A_1, \Phi(L_1))$, $T_2 = (A_2, \Phi(L_2))$ be equivalent simple transducers with labelings $\Phi(L_1) = (I_1, \lambda_1)$ and $\Phi(L_2) = (I_2, \lambda_2)$, respectively. Then $I_1 = I_2$ and the sets of in-out words of T_1 and T_2 are equal.*

Proof. $I_1 = I_2$ because this is the intersection of all possible results which for both transducers are the same if they are equivalent.

Let $A_1 = (Q_1, s_1, \delta_1, F_1)$ and $A_2 = (Q_2, s_2, \delta_2, F_2)$.

Take an input word w accepted by the transducers. By equivalence of T_1 and T_2 we have $|T_1|(w) = |T_2|(w)$, i.e., $I_1 \cup \lambda_1(s_1, w) = I_2 \cup \lambda_2(s_2, w)$. Since $I_1 = I_2$ and the labelings are simple thus $I_i \cap \lambda_i(s_i, w) = \emptyset$, for $i \in \{1, 2\}$, we have $\lambda_1(s_1, w) = \lambda_2(s_2, w)$.

Take a letter $\alpha \in \lambda_1(s_1, w)$. Let $w_1 a_1, w_2 a_2$ be unique prefixes of w, with $w_i \in \Sigma^*$ and $a_i \in \Sigma$, for $i \in \{1, 2\}$, such that $\alpha \in \lambda_1(\delta_1(s_1, w_1), a_1) \cap \lambda_2(\delta_2(s_2, w_2), a_2)$.

We prove that in-out words in T_1 and T_2 are equal by showing that w_1 and w_2 must be equal.

Suppose not. Without loss of generality, let w_1 be a proper prefix of w_2. By simplicity and definition of $\Phi(L_2)$, $\alpha \notin \lambda_2(s_2, w_1 a)$ and $\alpha \notin D(\delta_2(s_2, w_1 a), L_2)$. Thus there is a word w' such that $\alpha \notin \lambda_2(\delta_2(s_2, w_1 a), w')$ and $\delta_2(s_2, w_1 w') \in F_2$. Hence $\alpha \notin |T_2|(w_1 w')$ whereas $\alpha \in |T_1|(w_1 w')$, which contradicts the equivalence of T_1 and T_2. □

4 The Algorithm for Finding the Canonical Transducer

In this section we present an algorithm which, given a transducer T of size n, finds the canonical transducer $\Psi(T)$ equivalent to T, in time $O(n \log n)$.

Algorithm Canonical (T).

1. Compute $T' = \Xi(T)$ (a simple transducer equivalent to T).
2. Compute $\Phi(T')$.
3. Compute $\Psi(T)$, i.e., the transducer which is the result of minimization of $\Phi(T')$ treated as a *dfa* on in-out words.

The following result shows that, given any transducer T, Algorithm Canonical (T) correctly computes the canonical transducer equivalent to T.

Theorem 4. *For any equivalent transducers T_1, T_2, we have $\Psi(T_1) = \Psi(T_2)$.*

Proof. Let T_1 and T_2 be equivalent transducers. By Theorem 2 and transitivity of transducer equivalence, transducers $\Xi(T_1)$ and $\Xi(T_2)$ are equivalent. By Lemma 3 and Theorem 1, transducers $\Phi(\Xi(T_1))$ and $\Phi(\Xi(T_2))$ are simple equivalent transducers. By Theorem 3, these transducers, viewed as automata, accept the same sets of in-out words. Hence their minimization (as automata over in-out words) yields the same result. □

It remains to estimate the complexity of Algorithm Canonical. To this end, we need the following results.

Definition 5. *Let $A = (Q, s, \delta, F)$ be a dfa. A transition (x, a), where $x \in Q$ and $a \in \Sigma$ is called* cyclic transition *if there is a word $w \in \Sigma^*$ such that $\delta(x, aw) = x$. A labeling $L = (I, \lambda)$ of A is called* acyclic labeling *if every cyclic transition (x, a) in A is labeled by the empty set, $\lambda(x, a) = \emptyset$.*

Lemma 4. *Let $L = (I, \lambda)$ be a labeling of dfa $A = (Q, s, \delta, F)$, $x \in Q$ a state, and $w \in \Sigma^*$ an input word. If $\lambda(x, w) = \emptyset$ then $D(x, L) \subseteq D(\delta(x, w), L)$.*

Proof. From the definition of $D(x, L)$ we have:

$$D(x, L) \subseteq \{\alpha \in \Delta \mid \forall v \in w\Sigma^* \, ((\delta(x, v) \in F) \Rightarrow (\alpha \in \lambda(x, v)))\}$$

where $w\Sigma^*$ ($\subseteq \Sigma^*$) denotes all input words starting by w. Thus $D(x, L) \subseteq \lambda(x, w) \cup D(\delta(x, w), L)$, i.e., $D(x, L) \subseteq D(\delta(x, w), L)$ while $\lambda(x, w) = \emptyset$. □

Corollary 1. *If x and y are in an empty-labeled cycle, i.e., there is a word ww' such that $\delta(x, w) = y$, $\delta(y, w') = x$, and $\lambda(x, ww') = \emptyset$, then $D(x, L) = D(y, L)$.*

Lemma 5. *Given a simple transducer T of size n, the transducer $\Phi(T)$ can be computed in time $O(n)$.*

Proof. It follows from Definition 1 that it is enough to compute the values of $D(x, L)$ and $U(x, L)$, for all nodes x of T, in linear time. To this end, we construct the following multi-graph G^* (with labeled arcs), called the *component graph* of the graph G underlying T [10]. Nodes of G^* are strongly connected components of G, and arcs are defined as follows. If there is a transition in T from v to w with label a, then there is an arc from the component of v to the component of w, with the same label a.

Let $\mathcal{F}$ be the set of nodes X of G^*, such that at least one element of X is a finite state of T. For any node X of G^*, let $\Pi_d(X)$ be the set of all directed paths from X to a node in $\mathcal{F}$. For any path π, let $L(\pi)$ be the union of labels on π. Define $D^*(X) = \bigcap\{L(\pi) \mid \pi \in \Pi_d(X)\}$. Similarly, let J be the node of G^* containing the initial node of T. For any node X of G^*, let $\Pi_u(X)$ be the set of all directed paths from J to X. Define $U^*(X) = \bigcap\{L(\pi) \mid \pi \in \Pi_u(X)\} \cup I$, where I is the initial output of T.

By simplicity of T, all self-loops in G^* have empty labels. This implies that $D(x, L) = D^*(X)$ and $U(x, L) = U^*(X)$, where X is the strongly connected component of x. On the other hand, observe that, since all self-loops in G^* have empty labels, removing them from G^* does not affect the values of $D^*(X)$ and $U^*(X)$. Denote by $\overline{G}$ the multi-graph resulting from G^* after removing all self-loops. This is a directed acyclic multi-graph.

Given transducer T, the multigraph $\overline{G}$ can be computed in linear time, because strongly connected components of a graph can be obtained in time linear in its size [11]. On the other hand, all sets $D^*(X)$ and $U^*(X)$ can be computed in linear time, using topological sorting. Given those sets, and hence all sets $D(x, L)$ and $U(x, L)$, the value of $\Phi(T)$ can be computed in linear time, using Definition 1. □

Theorem 5. *Let T be an arbitrary transducer of size n, over bounded size alphabets. Algorithm Canonical (T) computes $\Psi(T)$ in time $O(n \log n)$.*

Proof. Since alphabets are of bounded size, the size of $\Xi(T)$ has the same order of magnitude as that of T, i.e., $O(n)$. By the construction from Definition 3, step 1 of Algorithm Canonical (T) (computing transducer $\Xi(T)$) takes time $O(n)$. By Lemma 5, step 2 of Algorithm Canonical (T) takes time $O(n)$. Finally, minimization of $\Phi(\Xi(T))$ takes time $O(n \log n)$ [3]. □

Remark. Notice that steps 1 and 2 of Algorithm Canonical (T) are executed in linear time. Time $O(n \log n)$ is needed only because of the minimization procedure in step 3, [3]. However, if the underlying graph of the transducer is a directed acyclic graph, then minimization can be carried out in linear time [12]. This implies that, in the case of such transducers, Algorithm Canonical (T) computes $\Psi(T)$ in linear time.

References

1. Azgani, S.: Using content-addressable memory for networking applications. Communications Systems Design **5** (1999)
2. Jenkins, C.: Speed and throughput of programable state machines for classification of OC192 data. In: Network Processors Conference, San Jose, California (2000) 6–24
3. Hopcroft, J., Ullman, J.: Introduction to Automata Theory, Languages, and Computation. Addison-Wesley (1979)
4. Eilenberg, S.: Automata, Languages, and Machines. Volume A. Academic Press (1974)
5. Berstel, J.: Transductions and Context-Free Languages. Teubner (1979)
6. Mohri, M.: Finite-state transducers in language and speech processing. Computational Linguistics **23** (1997) 269–311
7. Mohri, M.: Minimization algorithms for sequential transducers. Theoretical Computer Science **234** (2000) 177–201
8. Gupta, P., McKeown, N.: Packet classification on multiple fields. In: SIGCOMM. (1999) 147–160
9. Lakshman, T.V., Stiliadis, D.: High-speed policy-based packet forwarding using efficient multi-dimensional range matching. In: SIGCOMM. (1998) 203–214
10. Tarjan, R.E.: Depth first search and linear graph algorithms. SIAM Journal on Computing **1** (1972) 146–160
11. Tarjan, R.E.: Finding dominators in directed graphs. SIAM Journal on Computing **3** (1974) 62–89
12. Revuz, D.: Dictionnaires et lexiques: méthodes et algorithmes. PhD thesis, Institut Blaise Pascal, Paris, France (1991) LITP 91.44.

Self-assembling Finite Automata

Andreas Klein[1] and Martin Kutrib[2]

[1] Institut für Mathematik, Universität Kassel
Heinrich Plett Straße 40, D-34132 Kassel, Germany
klein@mathematik.uni-kassel.de
[2] Institut für Informatik, Universität Giessen
Arndtstraße 2, D-35392 Giessen, Germany
kutrib@informatik.uni-giessen.de

Abstract. We investigate a model of self-assembling finite automata. An automaton is assembled on demand during its computation from copies out of a finite set of items. The items are pieces of a finite automaton which are connected to the already existing automaton by overlaying states. Depending on the allowed number of such interface states, the degree, infinite hierarchies of properly included language families are shown. The presented model is a natural and unified generalization of regular and context-free languages since degrees one and two are characterizing the finite and pushdown automata, respectively. Moreover, by means of different closure properties nondeterministic and deterministic language families are separated.

1 Introduction

Self-assembly appears in nature in several ways. One of the simplest mechanisms is the merging of drops of water when placed close together. The process is directed by minimization of potential energy and, thus, an example for uncoded self-assembly. On the other extreme in complexity protein molecules inside biological cells self-assemble to reproduce cells each time they divide. In this example the assembly instructions are built in the components and, therefore, it is coded self-assembly [4]. Originally, the study of self-assembly was motivated by biologists. A well-studied example is the assembly of bacteriophages, a type of virus which infects bacterial cells [1]. Formal investigations in this field are accompanied by the development of corresponding computational models which are also of great interest from an engineering point of view. An introduction can be found in [8] where an automaton model of self-assembling systems is presented.

Here, in some sense, we adapt self-assembly to the theory of automata and formal languages. Basically, the idea is to assemble an automaton during its computation. Therefore, we provide a finite set of items, the so-called *modules*. The automata are assembled from module copies on demand. The assembling rules are encoded by the state transition function. Starting with one piece of a finite automaton during the computation so-called *assembling transitions* are

O.H. Ibarra and L. Zhang (Eds.): COCOON 2002, LNCS 2387, pp. 310–319, 2002.

traversed that direct the assembling of another copy of some item in a well-specified manner.

Each of the modules has a set of entry and a set of return states which together are called *interface states.* An assembling transition specifies how the new copy of the module fits to the already existing part of the automaton. The connection is made by overlaying the interface states by existing states. So the result of the self-assembly is a finite automaton, but the number of its states may depend on the input. It will turn out that the generative capacity of such models depend on their degree, i.e. the number of interface states of the modules.

Related to the work of the present paper are the so-called self-modifying finite automata [5,6,7,9]. In this model modifications of the automaton are allowed during transitions. The modifications include adding and deleting states and transitions. A weak form of self-modifying automata has been shown to accept the metalinear languages as well as some other families of context-free and non-context-free languages. Less restricted variants can accept arbitrarily hard languages, even non-recursive ones.

Since assembling modules sounds like calling subroutines another related paper is [3], where finite automata are considered that have a stack for storing return addresses (states). Every time a final state is entered the computation continues in the state at the top of the stack. Depending on the number of states which may be stored during one transition an infinite hierarchy in between the regular and context-free languages is shown.

Here by means of self-assembling finite automata of degree k we obtain a natural and unified generalization of finite automata and pushdown automata. In particular, infinite hierarchies depending on the degree are shown. For degree one and two the regular and context-free languages are characterized, respectively. Moreover, some closure properties are proved which lead to a separation result between nondeterministic and deterministic computations.

2 Self-Assembling Finite Automata

In order to introduce the model under consideration in more detail at first we define the basic items, the modules, which are used in the assembling process more formally:

Definition 1. *Let $u, v \in \mathbb{N}_0$ be constants. A* (nondeterministic) module with u entries and v exits *(u:v-module) is a system $\langle Q, I, O, A, \delta, F\rangle$, where*

1. *Q is the finite set of* inner states,
2. *$I = \{r_1, \ldots, r_u\}$ is the ordered set of u* entry states *such that $I \cap Q = \emptyset$,*
3. *$O = \{r_{u+1}, \ldots, r_{u+v}\}$ is the ordered set of v* return states *such that $O \cap (Q \cup I) = \emptyset$,*
4. *A is the finite set of* input symbols,
5. *The* module transition function *δ maps $Q \times A$ to the finite subsets of $Q \cup O \cup (\mathbb{N} \times Q^+ \times (Q \cup O)^+)$ and $I \times A$ to the subsets of Q,*
6. *$F \subseteq Q \cup I \cup O$ is the set of* accepting (or final) states.

So, the nondeterministic transition function may map states to states in which case we have state changes without assembling new items as usual.

In the second case δ requires to assemble a new copy of a module which is identified by an index from $\mathbb{N}$. The interface states I' and O' of the new module are overlayed by the states specified by $Q^+ \times (Q \cup O)^+$. From this point of view the restrictions of δ are convenient and natural: A return state is for exit purposes and, therefore, δ is not defined for states in O. Otherwise, a return state would at the same time be an entry state. Conversely, an entry state cannot be reached from inside the module. Otherwise it also would be a return state. Finally, after assembling a new module the computation should enter the module for at least one time step without assembling further modules, i.e., $I \times A$ is mapped to subsets of Q only.

Since modules are the basic items from which k-self-assembling finite automata are assembled, for their definition we need to ensure that only pieces are connected that fit together.

Definition 2. *Let $k \in \mathbb{N}_0$ be a constant. A* nondeterministic self-assembling finite automaton $\mathcal{M}$ of degree k (k-NFA) *is an ordered set $\langle M_0, \dots, M_m \rangle$ of modules over a common input alphabet A, where for all $0 \le i \le m$ the module $M_i = \langle Q, I, O, A, \delta, F \rangle$*

1. has at most k interface states, i.e. $|I| + |O| \le k$,
2. for all $(s, a) \in (Q \times A)$ the assembling transition
$(j, (p_1, \dots, p_u), (p_{u+1}, \dots, p_{u+v})) \in \delta(s, a)$ *implies*
(a) $j \le m$ and M_j is a u:v-module,
(b) $\{p_1, \dots, p_{u+v}\}$ are different and $s \in \{p_1, \dots, p_u\}$,
3. M_0 is a 0:0-module with a designated starting state s_0.

Condition 2b ensures that at most two states are overlayed and, moreover, an assembling transition transfers the computation into the new module.

The general behavior of a k-NFA is best described by configurations and their successor configurations.

A *configuration* c_t of $\mathcal{M}$ at some time $t \ge 0$ is a description of its global state which is a set of existing states S_t, transition and assembling rules given by a mapping δ_t from $S_t \times A$ to the finite subsets of $S_t \cup (\mathbb{N} \times S_t^+ \times S_t^+)$, the currently active state s_t, the current set of final states F_t and the remaining input word w_t. Thus, a configuration is a 5-tuple $c_t = (S_t, \delta_t, s_t, F_t, w_t)$.

The initial configuration $c_0 = (Q_0, \delta_0, s_0, F_0, w)$ at time 0 is defined by the input word $w = a_0 \cdots a_{n-1} \in A^*$ and the components of module $M_0 = \langle Q_0, \emptyset, \emptyset, A, \delta_0, F_0 \rangle$, where s_0 is the designated starting state from Q_0.

Let $c_t = (S_t, \delta_t, s_t, F_t, a_t \cdots a_{n-1})$ be a configuration, then for each element from $\delta_t(s_t, a_t)$ successor configurations $(S_{t+1}, \delta_{t+1}, s_{t+1}, F_{t+1}, a_{t+1} \cdots a_{n-1})$ are defined as follows.

During an ordinary state transition, as usual for finite automata, only the active state changes.

During an assembling transition a copy of the new module M_j has to be created, the active state has to be computed and the interface states have to be overlayed, what includes the appropriate update of the rules.

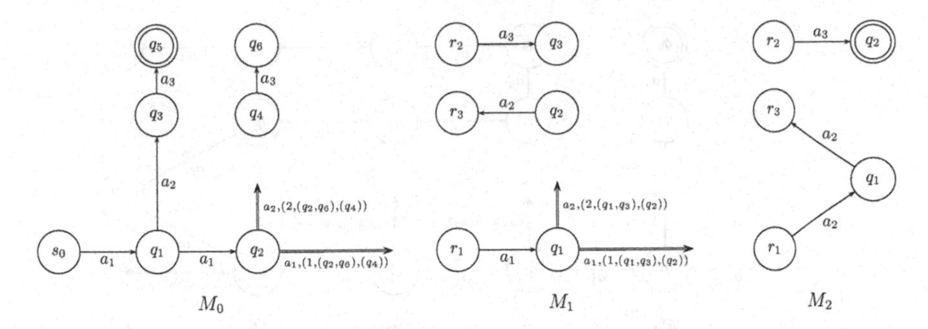

Fig. 1. Modules of a 3-DFA accepting L_3.

Let $(j,(p_1,\dots,p_u),(p_{u+1},\dots,p_{u+v})) \in (\mathbb{N} \times S_t^+ \times S_t^+)$ be an element from $\delta_t(s_t,a_t)$.

The copy $\bar{M}_j$ is created by renaming all states and correspondingly the transition rules such that they are different from the states in S_t. Set $S_{t+1} = S_t \cup \bar{Q}$ and identify the entry states of $\bar{M}_j$ by $p_1,\dots,p_u$ and the return states by $p_{u+1},\dots,p_{u+v}$. Accordingly, F_{t+1} is the union of all present final states.

In order to define s_{t+1} we first observe that by definition state s_t has to belong to $\{p_1,\dots,p_u\}$, say $s_t = p_l$. Next s_{t+1} is chosen from the set of possible successor states (under input a_t) of the unique state which is overlayed by p_l, i.e., the computation enters the newly assembled module.

It remains to join the mappings δ_t and $\bar{\delta}$ to δ_{t+1}. This is done by taking the mapping $\bar{\delta}$ of $\bar{M}_j$ and textually rename each occurrence of interface states by their overlaying states. Finally, the traversed assembling transition is replaced by an ordinary state transition.

An input word $a_0 \cdots a_{n-1}$ is *accepted* by a k-NFA iff the set of possible configurations at time n (i.e., after processing the whole input) is not empty and contains at least one configuration whose active state s_n belongs to the set of accepting states F_n.

A k-NFA is deterministic (k-DFA) iff for any input all configurations have deterministic mappings, i.e., $\delta_t : S_t \times A \to S_t \cup (\mathbb{N} \times S_t^+ \times S_t^+)$ is a partial function.

The *family of all languages that are acceptable* by some k-NFA (k-DFA) is denoted by $\mathscr{L}(k\text{-NFA})$ ($\mathscr{L}(k\text{-DFA})$).

The following example illustrates self-assembling finite automata more figurative. It becomes important for proving hierarchies in later sections.

Example 3. For any constant $k \geq 1$ let $A_k = \{a_1,\dots,a_k\}$ be an alphabet and $L_k = \{a_1^n \cdots a_k^n \mid n \in \mathbb{N}\}$.

In order to show that L_k is accepted by a k-DFA we present constructions for $k = 3$ which can easily be generalized to arbitrary k.

Figure 1 shows the graphical representation of a 3-DFA $\mathcal{M} = \langle M_0, M_1, M_2 \rangle$ that accepts L_3. Assembling transitions are indicated by double arrows. The assembled 3-DFA after accepting the input $a_1^4 a_2^4 a_3^4$ is depicted in Figure 2.

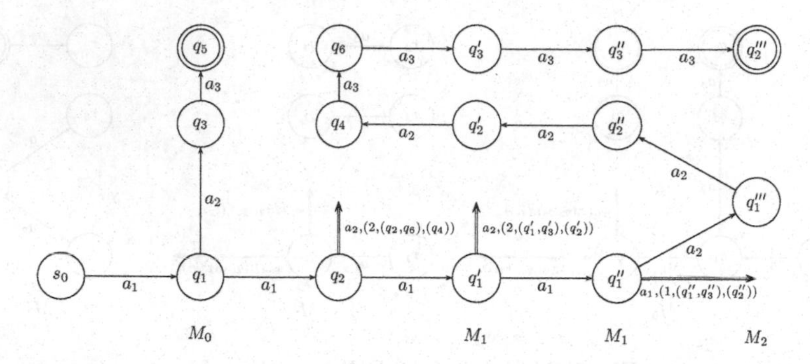

Fig. 2. Structure of a 3-DFA accepting L_3 after processing $a_1^4 a_2^4 a_3^4$.

The self-assembling finite automata in Example 3 have an important and interesting property, they are *loop-free*. Exactly this restriction of the model yields a natural and unified generalization of finite and pushdown automata. In particular, the generative capacity of self-assembling finite automata depends on their degree. Later on we are going to show infinite hierarchies of properly included language families for loop-free and unrestricted variants. Here we present the results concerning degrees 1 and 2. We prove that loop-free 1-DFAs accept exactly the regular and loop-free 2-NFAs exactly the context-free languages.

Definition 4. *Let $k \geq 1$ be a constant. A* computation *of a self-assembling finite automaton $\mathcal{M}$ of degree k is* loop-free *if $\mathcal{M}$ enters each of its existing states at most once. An* automaton *$\mathcal{M}$ is* loop-free *if all of its computations are loop-free.*

In order to distinguish loop-free languages we denote the family of all languages that are acceptable by some loop-free k-NFA (k-DFA) by $\mathscr{L}_{lf}(k\text{-NFA})$ ($\mathscr{L}_{lf}(k\text{-DFA})$).

Theorem 5. *Every context-free language is accepted by some loop-free* 2-NFA.

Proof. It is well known that for every context-free language not containing λ there exists a grammar in Greibach normal form. I.e., every production is of the form $X \to a\gamma$, where X is a variable, a a terminal and γ a possibly empty word of variables. In the following a loop-free 2-NFA is constructed that computes leftmost derivations of such a grammar $\mathcal{G}$. Subsequently, the empty word can be included simply by making the starting state final.

For each production $X \to aY_1 \cdots Y_n$ in $\mathcal{G}$ whose right-hand side has at least one variable a module $M_{Y_1 \cdots Y_n}$ is constructed as follows: $Q = \{Y_1, \ldots, Y_n\}$, $I = \{r_1\}$, $O = \{r_2\}$, $F = \emptyset$, $\delta(r_1, a) = \{Y_1\}$. For all $x \in A$:

$$Y_{i+1} \in \delta(Y_i, x) \text{ iff there exists the production } Y_i \to x \text{ in } \mathcal{G},$$
$$r_2 \in \delta(Y_n, x) \text{ iff there exists the production } Y_n \to x \text{ in } \mathcal{G},$$
$$(j, (Y_i), (Y_{i+1})) \in \delta(Y_i, x) \text{ iff there exists a production } Y_i \to xZ_1 \cdots Z_l \text{ in } \mathcal{G} \text{ and } M_j = M_{Z_1 \cdots Z_l}$$
$$(j, (Y_n), (r_2)) \in \delta(Y_n, x) \text{ iff there exists a production } Y_n \to xZ_1 \cdots Z_l \text{ in } \mathcal{G} \text{ and } M_j = M_{Z_1 \cdots Z_l}$$

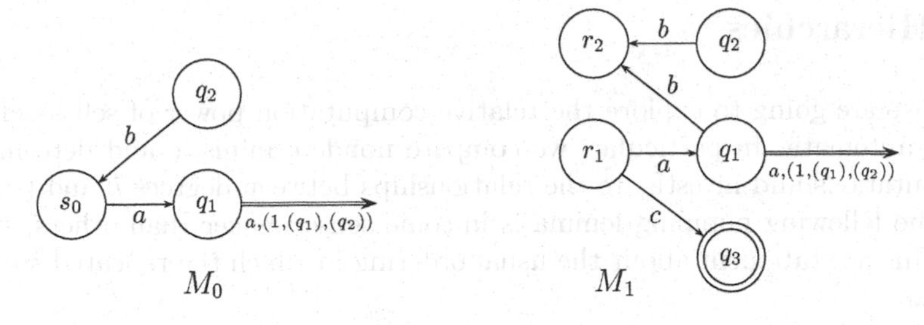

Fig. 3. Modules of a 2-DFA accepting L of Example 10.

Therefore, with input symbol x, the computation process assembles a module M_γ iff a leftmost derivation step of $\mathcal{G}$ generates $x\gamma$. The process returns from M_γ iff the variables γ have been completely replaced by terminals.

In order to complete the construction module M_0 is defined as the others, with the exception that r_1 is omitted, r_2 is now an inner state, the axiom of $\mathcal{G}$ is the second inner state, $F = \{r_2\}$ and the starting state is the axiom.

Altogether, the 2-NFA starts in a state that corresponds to the axiom of $\mathcal{G}$, simulates leftmost derivations and returns to the unique final state only if all variables that appear during the derivation of the input could be replaced. □

In order to complete the characterization we need also the converse of Theorem 5. A proof can be found in [2].

Theorem 6. *The family $\mathscr{L}_{lf}$(2-NFA) is equivalent to the context-free languages.*

Now we climb down the Chomsky-hierarchy and consider regular languages.

Theorem 7. *[2] Every* 1-NFA *language is regular.*

Theorem 8. *[2] Every regular language is accepted by some loop-free* 1-DFA.

Corollary 9. *The nondeterministic families $\mathscr{L}$(1-NFA), $\mathscr{L}_{lf}$(1-NFA) and the deterministic families $\mathscr{L}$(1-DFA), $\mathscr{L}_{lf}$(1-DFA) are equivalent to the regular languages.*

These results immediately raise the question for the power and limitations of loop-freeness in connection with self-assembling finite automata. Are all loop-free automata with a given degree equivalent to unrestricted automata with the same degree? Quite the contrary, there is a difference for all $k \geq 2$. The next example proves the claim for $k = 2$. In the next section it is extended to arbitrary degrees.

Example 10. The 2-DFA whose modules are depicted in Figure 3 accepts the language $L = \{a^{n_1}b^{n_1} \cdots a^{n_j}b^{n_j}a^m c \mid j \in \mathbb{N}, n_i \geq 2 \text{ for } 1 \leq i \leq j \text{ and } m < \max\{n_1, \ldots, n_j\}\}$. Since L is not context-free it does not belong to $\mathscr{L}_{lf}$(2-NFA).

In order to determine where the generative power of k-NFAs ends up here we state that for any $k \in \mathbb{N}$ the family $\mathscr{L}$(k-NFA) is a proper subfamily of the context-sensitive languages.

3 Hierarchies

Now we are going to explore the relative computation power of self-assembling finite automata. In particular, we compare nondeterministic and deterministic computations and investigate the relationships between degrees k and $(k+1)$.

The following pumping lemma is in some sense weaker than others, since it contains no statement about the usual ordering in which the repeated subwords appear.

Lemma 11. *Let $k \in \mathbb{N}$ be a constant and $\mathcal{M}$ be a loop-free k-NFA accepting a language L. Then there exists a constant $n \in \mathbb{N}$ such that every $w \in L$ with $|w| \geq n$ may be written as $x_0 y_1 x_1 y_2 \cdots y_k x_k$, where $|y_1| + |y_2| + \cdots + |y_k| \geq 1$, and for all $i \in \mathbb{N}$ there exists a word $w' \in L$ such that w' is in some order a concatenation of the (sub)words $x_0, x_1, \ldots, x_k$ and i times y_j for each $1 \leq j \leq k$.*

Proof. $\mathcal{M}$ consists of finitely many modules each having finitely many interface states. For long enough words from L there exists an accepting computation such that a module is assembled at least twice whereby the ordering of passed through interface states is identical. Obviously, the necessary input length can be calculated. It depends on $\mathcal{M}$ only and defines the constant n.

Now let w be an accepted input with $|w| \geq n$. We denote the first instance of the module by $\hat{M}$ and the second one by $\tilde{M}$. The input symbols consumed until $\hat{M}$ is assembled define the subword x_0. Thus, after processing x_0 an interface state of $\hat{M}$ and $\tilde{M}$ appears for the first time. Next we consider the sequence of input symbols until the next interface state (of $\hat{M}$ or $\tilde{M}$) is entered and define it to be y_1. We continue as follows (cf. Figure 4): Input sequences connecting return states of $\hat{M}$ with entry states of $\tilde{M}$, or entry states of $\tilde{M}$ with return states of $\hat{M}$ form the subwords x_j. Input sequences connecting each other pair of interface states from $\hat{M}$ or $\tilde{M}$ form the subwords y_j. The input sequence after entering an interface state for the last time forms the subword x_k.

Since there must exist at least one path from an entry state of $\hat{M}$ to an entry state of $\tilde{M}$, at least one subword y_i is not empty. On the other hand, since $\mathcal{M}$ is loop-free and its modules have at most k interface states, there exist at most k paths defining subwords y_j.

From the given accepting computation we derive another one by using a third copy $\bar{M}$ of the module and placing it in between the paths connecting $\hat{M}$ and $\tilde{M}$. The idea is as follows:

Since the interface states appearing in the same ordering $\bar{M}$ behaves like $\hat{M}$ when the computation enters one of its entry states. Thus, the connections between $\bar{M}$ and $\tilde{M}$ are identical to the connections between $\hat{M}$ and $\tilde{M}$. On the other hand, $\bar{M}$ behaves like $\tilde{M}$ when the computation enters one of its return states. Thus, the connections between $\hat{M}$ and $\bar{M}$ are identical to the connections between $\hat{M}$ and $\tilde{M}$. But the connections are exactly the subwords y_j. We conclude that the new accepting computation is for an input that is a concatenation of $x_0, \ldots, x_k$ and 2 times y_j for each $1 \leq j \leq k$. Trivially, we can insert i copies of $\bar{M}$. □

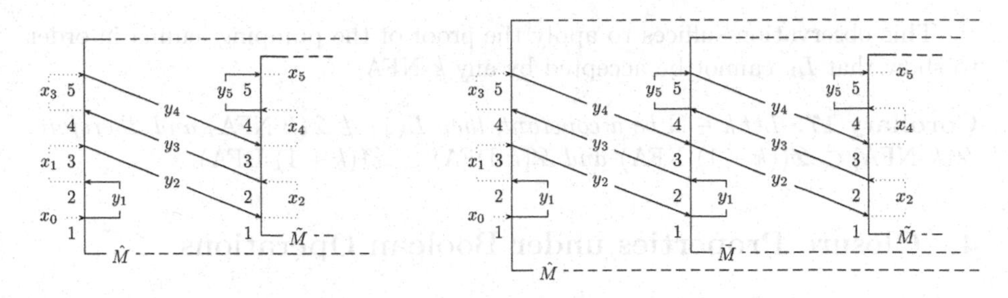

Fig. 4. Accepting computations for $x_0y_1x_1y_2x_2y_3x_3y_4x_4y_5x_5$ (left) and $x_0y_1x_1y_2y_1y_3x_3y_4y_2x_2y_3y_5y_4x_4y_5x_5$ (right).

We apply the pumping lemma to the language $L_k = \{a_1^n \cdots a_k^n \mid n \in \mathbb{N}\}$ of Example 3.

Lemma 12. *Let $k \in \mathbb{N}$ be a constant, then L_{k+1} does not belong to $\mathscr{L}_{lf}(k\text{-NFA})$.*

Since the constructions in Example 3 are deterministic and loop-free, as an immediate corollary we obtain hierarchies of loop-free self-assembling automata.

Corollary 13. *Let $k \in \mathbb{N}$ be a constant, then $\mathscr{L}_{lf}(k\text{-DFA}) \subset \mathscr{L}_{lf}((k+1)\text{-DFA})$ and $\mathscr{L}_{lf}(k\text{-NFA}) \subset \mathscr{L}_{lf}((k+1)\text{-NFA})$.*

Now we return to the question concerning the limitations of loop-freeness. The answer has been given for the cases $k = 1, 2$. For $k > 2$ the question is answered by the next result. It proves also that the pumping lemma does not hold for unrestricted self-assembling finite automata.

Lemma 14. *Let $k \in \mathbb{N}$ be a constant. There exists a language $L \in \mathscr{L}(3\text{-DFA})$ which does not belong to $\mathscr{L}_{lf}(k\text{-NFA})$.*

Proof. The witness for the assertion is the language $L = \{a_1^n(a_2^n a_3^n)^+ \mid n \in \mathbb{N}\}$.

A 3-DFA accepting L is a simple modification of the 3-DFA accepting the language L_3 given in Example 3. Simply insert a transition $\delta(q_5, a_2) = q_3$ in module M_0 and a transition $\delta(q_2, a_2) = q_1$ in module M_2.

By the pumping lemma it is easy to see that L cannot be accepted by any loop-free k-NFA for any $k \in \mathbb{N}$. □

Corollary 15. *Let $k \geq 2$ be a constant, then $\mathscr{L}_{lf}(k\text{-DFA}) \subset \mathscr{L}(k\text{-DFA})$ and $\mathscr{L}_{lf}(k\text{-NFA}) \subset \mathscr{L}(k\text{-NFA})$.*

The hierarchy result for loop-free self-assembling finite automata can be adapted to the unrestricted case. Though the pumping argument requires loop-freeness, the acceptors for the languages L_k may not contain any useful loop even in the unrestricted case.

Lemma 16. *Let $k, k' \in \mathbb{N}$ be two constants and $\mathcal{M}$ be a k'-NFA accepting the language L_k. Then during all accepting computations $\mathcal{M}$ does not enter any existing state more than once.*

This observation suffices to apply the proof of the pumping lemma in order to show that L_k cannot be accepted by any k-NFA.

Corollary 17. *Let $k \in \mathbb{N}$ be a constant, then $L_{k+1} \notin \mathscr{L}(k\text{-NFA})$ and, therefore, $\mathscr{L}(k\text{-NFA}) \subset \mathscr{L}((k+1)\text{-NFA})$ and $\mathscr{L}(k\text{-DFA}) \subset \mathscr{L}((k+1)\text{-DFA})$.*

4 Closure Properties under Boolean Operations

It turns out that the loop-free k-DFA languages are properly contained in the loop-free k-NFA languages. Thus, nondeterministic and deterministic language families are separated by different closure properties.

We start the investigation by showing that none of the families is closed under intersection.

Theorem 18. *Let $k \geq 2$ be a constant. Then there exist languages $L, L' \in \mathscr{L}_{lf}(k\text{-DFA})$ such that $L \cap L' \notin \mathscr{L}(k\text{-NFA})$. In particular, none of the families $\mathscr{L}_{lf}(k\text{-DFA})$, $\mathscr{L}_{lf}(k\text{-NFA})$, $\mathscr{L}(k\text{-DFA})$ and $\mathscr{L}(k\text{-NFA})$ is closed under intersection.*

Proof. Let $L = \{a_1^n \cdots a_k^n a_{k+1}^m \mid m, n \in \mathbb{N}\}$ and $L' = \{a_1^m a_2^n \cdots a_{k+1}^n \mid m, n \in \mathbb{N}\}$.

In order to construct a loop-free k-DFA for L and L' we only need minor modifications of the loop-free k-DFAs given in Example 3. But $L \cap L' = \{a_1^n \cdots a_{k+1}^n \mid n \in \mathbb{N}\} = L_{k+1}$ which by Corollary 17 cannot be accepted by any k-NFA. □

In order to disprove the nondeterministic closure under complement we prove the closure under union. The following technical lemma prepares for the construction. A proof can be found in [2].

Lemma 19. *Let $k \in \mathbb{N}$ be a constant. For any (loop-free) k-NFA $\mathcal{M}$ there exists an equivalent (loop-free) k-NFA $\mathcal{M}'$ such that there is no assembling transition from the starting state.*

Theorem 20. *Let $k \geq 2$ be a constant, then $\mathscr{L}_{lf}(k\text{-NFA})$ and $\mathscr{L}(k\text{-NFA})$ are closed under union.*

Proof. Let L and L' be two k-NFA-languages. As usual for nondeterministic devices the construction for $L \cup L'$ relies on the idea that an acceptor can initially guess whether the input belongs to L or L'. Technical problems are solved by applying Lemma 19. The details are presented in [2]. □

Corollary 21. *Let $k \geq 2$ be a constant, then $\mathscr{L}_{lf}(k\text{-NFA})$ and $\mathscr{L}(k\text{-NFA})$ are not closed under complement.*

The corollary follows from the closure under union by de Morgan's law: The closure under complement would imply the closure under intersection. Now we consider deterministic devices. One might expect their closure under complement by exchanging final and non-final states. But to this end the transition functions need to be total mappings.

In order to make them total we have to cope with the situation that states will eventually be overlayed by entry states of successor modules. Therefore, in general more than one transition function must be considered. On the other hand, some computations may enter these states without assembling the successor modules which implies that only one transition function must be total. The problem can be solved for loop-free k-DFAs [2].

Lemma 22. *Let $k \in \mathbb{N}$ be a constant and $\mathcal{M}$ be a loop-free k-DFA. Then there exists an equivalent loop-free k-DFA $\mathcal{M}'$ such that for any state in any reachable configuration the local transition function is totally defined.*

Theorem 23. *Let $k \in \mathbb{N}$ be a constant, then $\mathscr{L}_{lf}(k\text{-DFA})$ is closed under complement.*

Since $\mathscr{L}_{lf}(k\text{-DFA})$ has been shown not to be closed under intersection but to be closed under complement we obtain immediately:

Corollary 24. *Let $k \geq 2$ be a constant, then $\mathscr{L}_{lf}(k\text{-DFA})$ is not closed under union.*

From the different closure properties and for structural reasons the separation of nondeterministic and deterministic loop-free languages follows.

Theorem 25. *Let $k \geq 2$ be a constant, then $\mathscr{L}_{lf}(k\text{-DFA}) \subset \mathscr{L}_{lf}(k\text{-NFA})$.*

References

1. Casjens, S. and King, J. *Virus assembly*. Annual Review of Biochemistry 44 (1975), 555–604.
2. Klein, A. and Kutrib, M. *Self-assembling finite automata*. IFIG Research Report 0201, Institute of Informatics, University of Giessen, 2002. `http://www.informatik.uni-giessen.de/staff/kutrib/papers.html`
3. Nebel, M. E. *On the power of subroutines for finite state machines*. J. Aut., Lang. and Comb. 6 (2001), 51–74.
4. Penrose, L. S. *Self-reproducing machines*. Scientific American 200 (1959), 105–113.
5. Rubinstein, R. S. and Shutt, J. N. *Self-modifying finite automata*. Proc. of the IFIP 13th World Computer Congress. Vol. 1 : Technology and Foundations, 1994, pp. 493–498.
6. Rubinstein, R. S. and Shutt, J. N. *Self-modifying finite automata – power and limitations*. Technical Report WPI-CS-TR-95-4, Computer Science Department, Worcester Polytechnic Institute, 1995.
7. Rubinstein, R. S. and Shutt, J. N. *Self-modifying finite automata: An introduction*. Inform. Process. Lett. 56 (1995), 185–190.
8. Saitou, K. and Jakiela, M. J. *On classes of one-dimensional self-assembling automata*. Complex Systems 10 (1996), 391–416.
9. Wang, Y., Inoue, K., Ito, A., and Okazaki, T. *A note on self-modifying finite automata*. Inform. Process. Lett. 72 (1999), 19–24.

Repetition Complexity of Words

Lucian Ilie[*,**], Sheng Yu[***], and Kaizhong Zhang[†]

Department of Computer Science, University of Western Ontario
N6A 5B7, London, Ontario, Canada
{ilie,syu,kzhang}@csd.uwo.ca

Abstract. With ideas from data compression and combinatorics on words, we introduce a complexity measure for words, called *repetition complexity*, which quantifies the amount of repetition in a word. The repetition complexity of w, $\mathrm{R}(w)$, is defined as the smallest amount of space needed to store w when reduced by repeatedly applying the following procedure: n consecutive occurrences $uu \ldots u$ of the same subword u of w are stored as (u, n). The repetition complexity has interesting relations with well-known complexity measures, such as *subword complexity*, SUB, and *Lempel-Ziv complexity*, LZ. We have always $\mathrm{R}(w) \geq \mathrm{LZ}(w)$ and could even be that the former is linear while the latter is only logarithmic; e.g., this happens for prefixes of certain infinite words obtained by iterated morphisms. An infinite word α being ultimately periodic is equivalent to: (i) $\mathrm{SUB}(\mathrm{pref}_n(\alpha)) = \mathcal{O}(n)$, (ii) $\mathrm{LZ}(\mathrm{pref}_n(\alpha)) = \mathcal{O}(1)$, and (iii) $\mathrm{R}(\mathrm{pref}_n(\alpha)) = \lg n + \mathcal{O}(1)$. De Bruijn words, well known for their high subword complexity are shown to have almost highest repetition complexity; the precise complexity remains open. $\mathrm{R}(w)$ can be computed in time $\mathcal{O}(n^3(\log n)^2)$ and it is open, and probably very difficult, to find very fast algorithms.

Keywords: words, repetition complexity, Lempel-Ziv complexity, subword complexity, infinite words, iterated morphisms, ultimate periodicity, de Bruijn words, algorithms

1 Introduction

The repetitions in words constitute one of the oldest and most important properties of words. The study of repetitions dates back to the pioneering work of Thue [Th06,Th12] at the beginning of the last century. He was concerned with infinite repetition-free words. Ever since, various aspects of repetitions in words were quite extensively investigated, see, e.g., [Lo83], [ChKa97], [Lo02] and the references therein.

[*] corresponding author
[**] Research partially supported by NSERC grant R3143A01.
[***] Research partially supported by NSERC grant OGP0041630.
[†] Research partially supported by NSERC grant OGP0046373.

O.H. Ibarra and L. Zhang (Eds.): COCOON 2002, LNCS 2387, pp. 320–329, 2002.

In the present paper we investigate repetition in words from a new perspective: word complexity. Several measures of the complexity of words were proposed in the literature. The complexity of a word can be considered from different points of view: the shortest program to generate it (Kolmogorov [Ko65], see also [Ma66,Ch74]), the shortest compressed form (Lempel and Ziv [LeZi76]), the number of subwords ([Lo83,dB46]), the number of maximal repetitions ([KoKu99]), the highest order of repetitions ([De72]). The new complexity measure we introduce concerns the amount of repetition in a word; we call it *repetition complexity*. The basic idea is that the more the repetitions, the less the complexity. However, measures related with classical properties of words, such as the number of repetitions or highest order of repetitions, seem to be less appropriate.

Our complexity measure takes ideas from both data compression theory and combinatorics on words. Essentially, from a repetition of a word we remember only the base and the exponent; that is, we replace n consecutive occurrences of the same word $uu \cdots u$ by (u, n). The complexity is the minimum size to which a word is reduced by iteratively applying this procedure. As we shall see, the problem of doing optimally such reductions can be very intricate.

We investigate the repetition complexity from several points of view. It turns out that, aside from introducing challenging combinatorial problems, it is closely connected with well-known complexity measures, such as Lempel-Ziv complexity and subword complexity.

Due to the optimal compression it produces, [ZiLe77,HPS92], the Lempel-Ziv complexity turns out to be always lower than the repetition complexity. Moreover, there are arbitrarily long words for which Lempel-Ziv complexity is much smaller. To prove this, we use prefixes of infinite words obtained by iterating exponential prolongable morphisms. The general result we use here, interesting in its own, says that prefixes of such infinite words have logarithmic Lempel-Ziv complexity.

Next, we give a result which relates all three complexities: subword, Lempel-Ziv, and repetition. Certain orders of these complexities for prefixies of infinite words turn out to be equivalent with ultimate periodicity. Precisely, an infinite word being ultimately periodic is equivalent to any of the following three properties, where n states for the length of the prefixes:

(i) the subword complexity of its prefixes is linear in n,
(ii) the Lempel-Ziv complexity of prefixes is constant, and
(iii) the repetition complexity of prefixes is $\lg n$ plus a constant.

In particular, these provide new characterizations of ultimate periodicity of infinite words.

Another connection with the subword complexity is done via de Bruijn words. These are well known as having very high repetition complexity. We give a lower bound on their repetition complexity which is close to linear, that is, highest. We believe they have actually linear repetition complexity. However, this remains open.

Finally, we present an algorithm for computing the repetition complexity in time $\mathcal{O}(n^3(\log n)^2)$. Although this might seem slow, we give clear insight on why it seems a very difficult problem to find very fast algorithms to compute the repetition complexity. Other open questions are proposed at the end.

Due to limited space, all proofs are omitted.

2 Repetition Complexity

We give here the basic definitions and notations we need in the paper. For basic results in combinatorics on words we refer to [Lo83,ChKa97,Lo02].

Let A be a finite alphabet and A^* the free monoid generated by A with the identity ε; $A^+ = A^* - \{\varepsilon\}$. For $u, v \in A^*$, we say that v is a *subword* of u if $u = u'vu''$, for some $u', u'' \in A^*$; v is a *prefix* (resp., *suffix*) of u if $u' = \varepsilon$ (resp., $u'' = \varepsilon$). The prefix relation is denoted $\leq_{\mathrm{pref}}$ and the prefix of length n of w is denoted $\mathrm{pref}_n(w)$.

For any word $w \in A^*$, the *length* of w will be denoted by $|w|$. If $w = w_1 w_2 \cdots w_{|w|}$, where $w_i \in A$, then any integer $i, 1 \leq i \leq |w|$ is called a *position* in w; for $1 \leq i \leq |w|, i \leq j \leq |w|$, $w[i,j]$ denotes the subword $w_i w_{i+1} \cdots w_j$ of w; it has length $j-i+1$. For $n \geq 0$, the nth power of w, denoted w^n, is defined inductively by $w^0 = \varepsilon$, $w^n = w^{n-1}w$. w is *primitive* if there is no $n \geq 2$ such that $w = u^n$, for some word u. The *primitive root* of w, denoted $\rho(w)$, is the unique primitive word u such that $w \in u^+$. The *order* of w is $\mathrm{ord}(w) = \frac{|w|}{|\rho(w)|}$; we have $w = \rho(w)^{\mathrm{ord}(w)}$. A *period* of w is p such that $w_i = w_{i+p}$, for any $1 \leq i \leq |w| - p$.

A *repetition* in w is a subword of w of the form u^n for some nonempty word u and integer $n \geq 2$; n is the *order* and $|u|$ is the *period* of the repetition. For technical reasons, we formally define a repetition in w as a triple of positive integers (i, p, e) such that the word $(w[i, i+p-1])^e$ is a subword of w starting at position i; that is, we have at position i a repetition of order e and period p.

We use repetitions to reduce the representation of a word, that is, we iteratively replace n consecutive occurrences of the same word $uu \cdots u$ by u and n. While the former takes $n|u|$ units of space to represent, we assume that the latter needs only $|u| + \lceil \lg(n+1) \rceil$. We shall assume decimal representation for exponents but the results hold essentially unchanged for any base greater than two. Notice that, if n is in decimal then u^n is shorter than n consecutive u's, for any word u, as soon as $n \geq 2$; this helps avoiding special irrelevant cases in our reasoning.

We shall call this procedure a *reduction*; u^n is a reduced form of $uu \cdots u$. A word w can be iteratively reduced using the above procedure for repetitions inside w. However, some repetitions cannot be reduced simultaneously (because of overlapping), while further reductions can be applied inside already reduced repetitions. We formally define the repetition complexity below.

Let $D = \{0, 1, \ldots, 9\}$ be the set of decimal digits, $D \cap A = \emptyset$, and let $\langle, \rangle, {}^\wedge$ be three new letters; put $T = A \cup D \cup \{\langle, \rangle, {}^\wedge\}$. For a positive integer n, $\mathrm{dec}(n) \in D^*$ is the decimal representation of n. (For a word w, $|\mathrm{dec}(|w|)| = \lceil \lg(|w|+1) \rceil$ is the length of the decimal representation of the length of w.) Define the binary relation $\Rightarrow \subseteq T^* \times T^*$

$$u \Rightarrow v \quad \text{iff} \quad u = u_1 x^n u_2, v = u_1 \langle x \rangle^\wedge \langle \mathrm{dec}(n) \rangle u_2,$$
$$\text{for some } u_1, u_2 \in T^*, x \in A^+, n \geq 2.$$

Let $\Rightarrow^*$ be the reflexive and transitive closure of $\Rightarrow$; if $u \Rightarrow^* v$, then v is a reduced form of u. Define also a morphism $h : T^* \to (A \cup D)^*$ which simply erases all letters from $\{\langle, \rangle, \hat{}\}$ and keeps those in $A \cup D$ unchanged. The *repetition complexity* of a word $w \in A^*$, denoted $\mathrm{R}(w)$, is formally defined as

$$\mathrm{R}(w) = \min_{w \Rightarrow^* u} |h(u)|.$$

Such an u with $\mathrm{R}(w) = |h(u)|$ is called a *shortest reduced form* of w and $w \Rightarrow^* u$ is an *optimal reduction* of w.

We notice that our reduction relation $\Rightarrow$ is not confluent. For instance, if $w = \mathsf{ababcbc}$, then we have two reductions which cannot be continued any further: $w \Rightarrow \langle\mathsf{ab}\rangle\hat{}\langle 2\rangle\mathsf{cbc}$ and $w \Rightarrow \mathsf{aba}\langle\mathsf{bc}\rangle\hat{}\langle 2\rangle$. Actually, both are optimal reductions.

Example 1 Consider the word $w = \mathsf{ababaababababbbabb}$. Several possible reductions for w are shown below (the first is optimal and so $\mathrm{R}(w) = 10$):

$w \Rightarrow \langle\mathsf{ababa}\rangle\hat{}\langle 2\rangle\mathsf{bbbabb} \Rightarrow \langle\mathsf{ababa}\rangle\hat{}\langle 2\rangle\langle\mathsf{b}\rangle\hat{}\langle 3\rangle\mathsf{abb} \Rightarrow \langle\langle\mathsf{ab}\rangle\hat{}\langle 2\rangle\mathsf{a}\rangle\hat{}\langle 2\rangle\langle\mathsf{b}\rangle\hat{}\langle 3\rangle\mathsf{abb}$
$w \Rightarrow^* \langle\mathsf{ab}\rangle\hat{}\langle 2\rangle\mathsf{aaba}\langle\mathsf{babb}\rangle\hat{}\langle 2\rangle$
$w \Rightarrow^* \mathsf{a}\langle\mathsf{ba}\rangle\hat{}\langle 2\rangle\langle\mathsf{ab}\rangle\hat{}\langle 2\rangle\mathsf{a}\langle\mathsf{b}\rangle\hat{}\langle 3\rangle\mathsf{a}\langle\mathsf{b}\rangle\hat{}\langle 2\rangle$

The next lemma gives the bounds for the R-complexity.

Lemma 2 *For any $w \in A^*$ with $|w| \geq 2$, $1 + |\operatorname{dec}(|w|)| \leq \mathrm{R}(w) \leq |w|$.*

The next result concerns words with highest repetition complexity. Using results from combinatorics of words we show that there are many of such words.

Theorem 3 *The number of words over three (or more) letters of maximum repetition complexity is exponential in terms of the length.*

A property expected from a complexity measure is subadditivity. The complexity we introduced has it.

Lemma 4 *For any u, v, $\mathrm{R}(uv) \leq \mathrm{R}(u) + \mathrm{R}(v)$.*

3 The Definition of Repetition Complexity

We discuss here our choice of defining the repetition complexity. Another choice could have been the highest order of a repetition. This is a local property which does not necessarily affect the whole word. If the highest order is very low, then it becomes more relevant. For instance, if it is less than 2, then we obtain our highest complexity, but for higher values we can have totally different words. For instance, the word $((\cdots(\mathsf{a}_1^2\mathsf{a}_2)^2\cdots\mathsf{a}_n)^2$ has highest order of repetition 2 but it clearly contains a lot more repetition than a prefix of a 2^--free word (see [ChKa97]). The difference with respect to the highest order of repetition is smallest but the repetition complexities are quite different: logarithmic for the former and linear for the latter.

The number of repetitions is another candidate. A good example here is the Fibonacci word defined by $f = \lim_{n\to\infty} f^n(\mathsf{a})$, where $f(\mathsf{a}) = \mathsf{ab}$, $f(\mathsf{b}) = \mathsf{a}$. By [Cr81], any prefix of length n of f has $\Theta(n \lg n)$ maximal repetitions (i.e., repetitions which cannot be extended). A much less complex word, a^n has only one maximal repetition. For further results concerning the number of repetitions in words, see [KoKu99] and the references therein.

Of course, the repetition in a^n is very long. We should therefore take into account both the number of repetitions and their lengths. But our complexity does it. Moreover, it takes implicitly into account overlappings among the repetitions by the fact that overlapping repetitions cannot be reduced simultaneously.

Finally, one could argue that the exponents should be counted as one unit of space each, as in a RAM model. But then, an infinite word like *aaa*... would have all prefixes reduced to size two which is unreasonable.

4 Subword and Lempel-Ziv Complexities

We recall two basic complexity measures of words to which we compare the repetition complexity. These are the subword complexity and Lempel-Ziv complexity.

For a word w, the *subwords complexity* of w is the number of subwords of w, denoted by $\text{SUB}(w)$. The next lemma gives the optimal range for the subword complexity.

Lemma 5 *For any w, we have* $|w| + 1 \le \text{SUB}(w) \le 1 + \frac{1}{2}(|w|^2 + |w|)$.

Essential for us in the above lemma is the fact that, on fixed alphabets, $\text{SUB}(w)$ is at least linear and at most quadratic; e.g., for $w = \mathsf{a}^n\mathsf{b}^n$, $\text{SUB}(w) = \Theta(|w|^2)$.

One of the most famous complexity measure is the one introduced by Lempel and Ziv [LeZi76] in connection with their algorithm for data compression, see [ZiLe77,ZiLe78,CrRy94].

For a word w, we define the *e-decomposition*[1] of w as the (unique) decomposition $w = w_1w_2\cdots w_k$ such that, for any $1 \le i \le k$ (with the possible exception of $i = k$), w_i is the shortest prefix of $w_iw_{i+1}\cdots w_k$ which does not occur before in w; that is, w_i does not occur as a subword of $\pi(w_1w_2\cdots w_i)$, where the application π removes the last letter of its argument.

The complexity measure introduced by Lempel and Ziv represents the *number of subwords in the e-decomposition* of w; we denote it by $\text{LZ}(w)$.

Example 6 Consider the word $w = \mathsf{aababbabbabb}$. The e-decomposition of w is $w = \mathsf{a.ab.abb.abbabb}$, where the subwords are marked by dots. Therefore, $\text{LZ}(w) = 4$.

[1] 'e' comes from 'exhaustive'; Lempel and Ziv [LeZi76] called this decomposition the *exhaustive production history* of w; it is called *f-factorization* by [CrRy94] and *s-factorization* by [Ma89].

We notice that the e-decomposition can be defined in the same way for right-infinite words; at each step we take the longest prefix of the remaining infinite suffix which does not appear before; this prefix may be the remaining suffix of the infinite word, in which case it is the last element of the decomposition.

The next lemma is a weak form of a result of [LeZi76] which states the bounds for the LZ-complexity.

Lemma 7 $\text{LZ}(w) = \mathcal{O}\Big(\frac{|w|}{\lg |w|}\Big)$.

5 Relation with Lempel-Ziv Complexity

We shall compare in this section our complexity with the Lempel-Ziv complexity. We start by investigating closer the R-complexity. As defined above, $\text{R}(w)$ is the size of $h(v)$ for an optimal reduction $u \Rightarrow^* v$. At each step in this reduction, we use a repetition (i, p, e) in w to decrease the size; denote the space saved by reducing w according to this repetition by $\text{red}(w, i, p, e) = (e-1)p - |\operatorname{dec}(e)|$. When w is understood, we write simply $\text{red}(i, p, e)$. Of course, the saving in space does not depend on the position of the repetition in w, but we still keep i as an argument in order to be able to identify precisely what repetition we are talking about.

The next lemma shows how an optimal reduction is obtained.

Lemma 8 *For any word w, there is an ordered sequence of $m \geq 0$ repetitions in w*

$$(i_1, p_1, e_1), (i_2, p_2, e_2), \ldots, (i_m, p_m, e_m) \tag{1}$$

such that

$$\text{R}(w) = |w| - \sum_{k=1}^{m} \text{red}(i_k, p_k, e_k), \tag{2}$$

and any two repetitions (i_k, p_k, e_k) and (i_l, p_l, e_l), $1 \leq k < l \leq m$, are

(i) either disjoint, i.e., $[i_k, i_k + p_k e_k - 1] \cap [i_l, i_l + p_l e_l - 1] = \emptyset$,

(ii) or the one appearing later in (1) is contained in the first period of the other, i.e., $i_k \leq i_l$ and $i_l + p_l e_l \leq i_k + p_k$.

We give next an example of an application of Lemma 8.

Example 9 For the word

$$w = \mathsf{aabbbabbbbaaababbbabbbbaaabbababab},$$

an ordered sequence (1) can be $(2, 13, 2), (2, 4, 2), (3, 1, 3), (11, 1, 3), (29, 2, 3)$. The space saved by each of them is, in order, 12, 3, 1, 1, 3. Finally, $\text{R}(w) = |w| - (12 + 3 + 1 + 1 + 3) = 34 - 20 = 14$; w can be written as $w = \mathsf{a}((\mathsf{ab}^3)^2\mathsf{ba}^3\mathsf{b})^2\mathsf{b}(\mathsf{ab})^3$.

We establish next the connection in one direction with the Lempel-Ziv complexity and give some non-trivial examples showing the optimality of the result.

Theorem 10 *For any word w, $\mathrm{R}(w) \geq \mathrm{LZ}(w)$.*

Example 11 Consider $n \geq 1$ and n different letters $\mathsf{a}_i, 1 \leq i \leq n$, and construct the word

$$w_n = ((\cdots(\mathsf{a}_1^9\mathsf{a}_2)^9\mathsf{a}_3)^9 \cdots \mathsf{a}_{n-1})^9\mathsf{a}_n.$$

We have $|w_n| = \frac{9^n-1}{8}$ and $\mathrm{R}(w_n) = 2n-1 = \Theta(\lg |w_n|)$. Denoting $x_n = w_{n-1}^8\mathsf{a}_n$, we have $w_n = w_{n-1}x_n$. The e-decomposition of w_n is $w_n = \mathsf{a}_1.x_2.x_3.x_4.\cdots.x_n$ and so $\mathrm{LZ}(w_n) = n = \Theta(\lg |w_n|)$. Therefore the result in Theorem 10 cannot be improved by more than a constant.

We now consider the relation in the opposite direction. We show that we have the opposite case: there are words of high R-complexity but low LZ-complexity. We shall need a result about infinite words which is interesting in itself. We denote by A^ω the set of right-infinite words over A. For $w \in A^*$, we denote $w^\omega = www\cdots$.

A morphism $\varphi : A^* \to A^*$ is called *prolongable* on $a \in A$ if $\varphi(a) \in aA^+$. If φ is prolongable on a, then $\lim_{n\to\infty} \varphi^n(a) \in A^\omega$ exists and we denote it by $\varphi^\infty(a)$. φ is called *exponential* if there are an integer $n \geq 1$ and a real $c > 1$ such that, for any $w \in A^*$, $|\varphi^n(w)| \geq c^n|w|$.

Lemma 12 *Let $\varphi : A^* \to A^*$ be an exponential morphism prolongable on $a \in A$ and denote $\alpha = \varphi^\infty(a)$. Then, $\mathrm{LZ}(\mathrm{pref}_n(\alpha)) = \mathcal{O}(\lg n)$.*

Remark 13 We notice that the condition on φ being exponential in Lemma 12 is essential as shown by the following example. Take $\varphi : \{\mathsf{a},\mathsf{b},\mathsf{c}\}^* \to \{\mathsf{a},\mathsf{b},\mathsf{c}\}^*$, given by $\varphi(\mathsf{a}) = \mathsf{a}$, $\varphi(\mathsf{b}) = \mathsf{ba}$, $\varphi(\mathsf{c}) = \mathsf{cba}$. φ is prolongable on c and we have the e-decomposition of $\varphi^\infty(\mathsf{c})$

$$\varphi^\infty(\mathsf{c}) = \mathsf{c.b.a.baa.baaa.baaaa.baaaaa.baaaaaa}\cdots$$

Hence $\mathrm{LZ}(\mathrm{pref}_n(\varphi^\infty(\mathsf{c}))) = \Theta(\sqrt{n})$.

Lemma 12 gives a relation in the other direction between R and LZ.

Theorem 14 *There are arbitrarily long words w for which $\mathrm{R}(w) = |w|$ and $\mathrm{LZ}(w) = \mathcal{O}(\lg |w|)$.*

6 Periodic Infinite Words and Complexity of Prefixes

We show in this section a strong connection between all low R-, LZ-, and SUB-complexities for prefixes of infinite words. This gives us also new characterizations of ultimately periodic infinite words. (The characterization (ii) resembles the famous one by Coven and Hedlund [CoHe73].)

Theorem 15 *For any infinite word α, the following assertions are equivalent:*
(i) α is ultimately periodic,
(ii) $\mathrm{SUB}(\mathrm{pref}_n(\alpha)) = \mathcal{O}(n)$,
(iii) $\mathrm{LZ}(\mathrm{pref}_n(\alpha)) = \mathcal{O}(1)$,
(iv) $\mathrm{R}(\mathrm{pref}_n(\alpha)) = \lg n + \mathcal{O}(1)$.

Remark 16 We notice that in Theorem 15 we have at (ii) and (iii) the order of the lower bound for the respective complexity from Lemmas 5 and 7 while at (iv) we have a stronger condition: the lower bound in Lemma 2 plus a constant instead of $\mathcal{O}(\lg n)$. In fact, $\mathcal{O}(\lg n)$ is not good as shown by the following example. Consider the word $w_k = ((\cdots(\mathsf{bab})^{10})\mathsf{ba}^2\mathsf{b})^{10^2}\mathsf{ba}^3\mathsf{b})^{10^3}\cdots\mathsf{ba}^k\mathsf{b})^{10^k}$. We have $w_k \leq_{\text{pref}} w_{k+1}$, so we can construct the infinite word $\boldsymbol{w} = \lim_{k\to\infty} w_k$. It can be shown that $\text{R}(\text{pref}_n(\boldsymbol{w})) = \mathcal{O}(\lg n)$. But $\boldsymbol{w}$ is not ultimately periodic. Therefore, we have from the R-complexity a slightly finer characterization of ultimately periodic words.

7 De Bruijn Words and Subword Complexity

We next investigate the case of words with high subword complexity. We consider *de Bruijn* words $\boldsymbol{b}_k$ (see [dB46]) which have very high SUB-complexity. For $k \geq 1$, a de Bruijn word $\boldsymbol{b}_k \in A^*$ has the properties $|\boldsymbol{b}_k| = \text{card}(A)^k + k - 1$ and $\text{SUB}(\boldsymbol{b}_k) \cap A^k = A^k$; that is, $\boldsymbol{b}_k$ has as subwords all words of length k and any two subwords of length k of $\boldsymbol{b}_k$ starting from different positions are different. (There are many such words but our result holds for all of them.)

If $\text{card}(A) = l$, then the number of all subwords of $\boldsymbol{b}_k$ is $\text{SUB}(\boldsymbol{b}_k) = \frac{l^k-1}{l-1} + \frac{l^k(l^k+1)}{2}$. So, not only that $\text{SUB}(\boldsymbol{b}_k)$ is of the order of maximal subword complexity in Lemma 5, but also the difference between the upper bound in Lemma 5 and $\text{SUB}(\boldsymbol{b}_k)$ is of strictly lower order: $\mathcal{O}(|\boldsymbol{b}_k| \lg |\boldsymbol{b}_k|)$. We show that de Bruijn words have also high repetition complexity.

Theorem 17 $\text{R}(\boldsymbol{b}_k) = \Omega\left(\frac{|\boldsymbol{b}_k| \lg\lg |\boldsymbol{b}_k|}{\lg |\boldsymbol{b}_k|}\right)$.

8 Computing the Repetition Complexity

We show in this section that the repetition complexity can be computed in time $\mathcal{O}(n^3(\log n)^2)$. Due to the very intricate nature of repetitions, this problem is by no means trivial. (A good example of how complex can be the repetitions in a word are the Fibonacci words.) In fact, it can be seen as a restricted case of the optimal data compression which is NP-complete; see [StSz78,GaJo79].

We give next another example which, although simple from algorithmic point of view, shows that a word can have exponentially many optimal reductions; which again makes the problem hard. Consider the Morse-Hedlund infinite word $\boldsymbol{m}$, [MoHe44], defined as $\boldsymbol{m} = m^\infty(\mathsf{a})$, where $m : \{\mathsf{a},\mathsf{b},\mathsf{c}\}^* \to \{\mathsf{a},\mathsf{b},\mathsf{c}\}^*$, $m(\mathsf{a}) = \mathsf{abc}$, $m(\mathsf{b}) = \mathsf{ac}$, $m(\mathsf{c}) = \mathsf{b}$. For the morphism φ given by $\varphi(\mathsf{a}) = \mathsf{ababa}$, $\varphi(\mathsf{b}) = \mathsf{a'b'a'b'a'}$, $\varphi(\mathsf{c}) = \mathsf{a''b''a''b''a''}$, we have that $\varphi(\text{pref}_n(\boldsymbol{m}))$ has length $5n$ and 2^n optimal reductions.

We present next our method to compute $\text{R}(w)$. The following observation is the basis of our algorithm. For any non-empty word w, we have

$$\text{R}(w) = \min\left(\min_{w=uv}\left(\text{R}(u) + \text{R}(v)\right), \min_{k|\text{ord}(w)}\left(\text{R}(\rho(w)^{\frac{\text{ord}(w)}{k}}) + |\text{dec}(k)|\right)\right). \quad (3)$$

Based on (3), we use then dynamic programming to compute the repetition complexities of all subwords of w.

Theorem 18 *The repetition complexity of w, for $|w| = n$, can be computed in time $\mathcal{O}(n^3(\log n)^2)$.*

Here is where $\mathcal{O}(n^3(\log n)^2)$ comes from. We compute the repetition complexity for all subwords, which are n^2. For each, we investigate linearly many possiblities and therefore we get n^3. The logarithms come from arithmetic operations and comparisons. The preprocessing for computing the primitive roots and the orders for all subwords of w can be done in time $\mathcal{O}(n^2(\log n)^2)$.

9 Conclusions and Further Research

We invetigate the repetitions in words from the point of view of complexity of words. Our work is related to the study of repetitions in words in general, see, e.g., [KoKu99], but our goals are different. We want to measure the complexity of words from this perspective. We introduce the notion of repetition complexity of a word and discuss its appropriateness by comparison with other potential candidates.

We give results which relate our complexity to well-known complexity measures like subword or Lempel-Ziv complexity. These turn out to give interesting results about infinite words. We mention here several problems which deserve further investigation.

The algorithm we gave for computing the repetition complexity is, of course, not very fast (compared to usual algorithms dealing with repetitions in words, e.g., [Cr81,ApPr83,MaLo84,KoKu99]), but it seems very difficult to give very fast algorithms. Notice that we used dynamic programming so, using this idea we cannot find algorithms with sub-quadratic time. Completely new ideas and properties of words are needed for fast algorithms.

Another problem is that the algorithm is not of much use if we try to compute (or only approximate) the repetition complexity of some families of words, say all prefixes of the Fibonacci infinite word. Some different tools for lower bounds are needed.

We showed in Theorem 17 that de Bruijn words have high repetition complexity. We believe they have in fact linear complexity, that is, $\mathrm{R}(\boldsymbol{b}_k) = \Theta(|\boldsymbol{b}_k|)$.

A related complexity which we did not discuss here can be naturally defined using rational repetitions. For instance, consider the words $\mathsf{abcdabc}$ and $\mathsf{abcdefg}$. Both have R-complexity 7 as none contains any integer repetitions, although the former contains clearly more repetition than the latter. Using rational powers, we may write $\mathsf{abcdabc} = (\mathsf{abcd})^{3/4}$ which takes only 6 units of space.

Finally, a problem which we have not approached concerns the connection between our complexity and randomness. It should be investigated how much random are the words with high repetition complexity, in particular the square-free words.

References

ApPr83. Apostolico, A., and Preparata, F., Optimal off-line detection of repetitions in a string, *Theoret. Comput. Sci.* **22** (1983) 297 – 315.

dB46. de Bruijn, N.G., A combinatorial problem, *Proc. Kon. Ned. Akad. Wetensch.* **49** (1946) 758–764.

Ch74. Chaitin, G.J, Information-theoretic limitations of formal systems, *J. Assoc. Comput. Mach.* **21** (1974) 403 – 424.

ChKa97. Choffrut, C., and Karhumäki, J., Combinatorics of Words, in: G. Rozenberg, A. Salomaa, eds., *Handbook of Formal Languages, Vol. I*, Springer-Verlag, Berlin, 1997, 329 – 438.

CoHe73. Coven, E.M., and Hedlund, G., Sequences with minimal block growth, *Math. Sytems Theory* **7** (1973) 138 – 153.

Cr81. Crochemore, M., An optimal algorithm for computing the repetitions in a word, *Inform. Proc. Lett.* **12** (5) (1981) 244 – 250.

CrRy94. Crochemore, M., and Rytter, W., *Text Algorithms*, Oxford Univ. Press, 1994.

CrRy95. Crochemore, M., and Rytter, W., Squares, cubes, and time-space efficient string matching, *Algorithmica* **13** (1995) 405 – 425. Oxford Univ. Press, 1994.

De72. Dejean, F., Sur un théorème de Thue, *J. Combin. Theory, Ser. A* **13** (1972) 90–99.

GaJo79. Garey, M.R., Johnson, D.S., *Computers and Intractability. A Guide to the Theory of NP-completeness*, W.H. Freeman and Co., San Francisco, 1979.

HPS92. Hansel, G., Perrin, D., and Simon, I., Compression and entropy, *Proc. of STACS'92*, LNCS 577, Springer-Verlag, 1992, 515 – 528.

Ko65. Kolmogorov, A.N., Three approaches to the quantitative definition of information, *Probl. Inform. Transmission* **1** (1965) 1 – 7.

KoKu99. Kolpakov, R., and Kucherov, G., Finding maximal repetitions in a word in linear time, *Proc. of FOCS'99*, 596 – 604.

LeZi76. Lempel, A., and Ziv, J., On the complexity of finite sequences *IEEE Trans. Information Theory* **22**(1) (1976) 75–81.

Lo83. Lothaire, M., *Combinatorics on Words*, Addison-Wesley, Reading, MA, 1983.

Lo02. Lothaire, M., *Algebraic Combinatorics on Words*, Cambridge Univ. Press, 2002.

MaLo84. Main, M., and Lorentz, R., An $\mathcal{O}(n \lg n)$ algorithm for finding all repetitions in a string, *J. Algorithms* **5** (1984) 422 – 432.

Ma89. Main, M., Detecting leftmost maximal periodicities, *Discrete Appl. Math.* **25** (1989) 145 – 153.

Ma66. Martin-Löf, P., The definition of random sequences, *Inform. and Control* **9** (1966) 602 – 619.

MoHe44. Morse, M., and Hedlund, G., Unending chess, symbolic dynamics and a problem in semigroups, *Duke Math. J.* **11** (1944) 1 – 7.

StSz78. Storer, J.A., Szymanski, T.G., The macro model for data compression, *Proc. of 10th STOC*, 1978, 30 – 39.

Th06. Thue, A., Uber unendliche Zeichenreihen, *Norske Vid. Selsk. Skr. Mat.-Nat. Kl. (Kristiania)* **7** (1906) 1 – 22.

Th12. Thue, A., Uber die gegenseitige Lage gleicher Teile gewisser Zeichenreihen, *Norske Vid. Selsk. Skr. Mat.-Nat. Kl. (Kristiania)* **5** (1912) 1 – 67.

ZiLe77. Ziv, J., and Lempel, A., A universal algorithm for sequential data compression, *IEEE Trans. Information Theory* **23** (3) (1977) 337 – 343.

ZiLe78. Ziv, J., and Lempel, A., Compression of individual sequences via variable length encoding, *IEEE Trans. Information Theory* **24** (5) (1978) 530 – 536.

Using PageRank to Characterize Web Structure

Gopal Pandurangan[1], Prabhakar Raghavan[2], and Eli Upfal[1]

[1] Computer Science Department, Brown University
Box 1910, Providence, RI 02912-1910, USA
{gopal,eli}@cs.brown.edu*

[2] Verity Inc., 892 Ross Drive, Sunnyvale, CA 94089, USA
pragh@verity.com

Abstract. Recent work on modeling the Web graph has dwelt on capturing the degree distributions observed on the Web. Pointing out that this represents a heavy reliance on "local" properties of the Web graph, we study the distribution of PageRank values (used in the Google search engine) on the Web. This distribution is of independent interest in optimizing search indices and storage. We show that PageRank values on the Web follow a power law. We then develop detailed models for the Web graph that explain this observation, and moreover remain faithful to previously studied degree distributions. We analyze these models, and compare the analyses to both snapshots from the Web and to graphs generated by simulations on the new models. To our knowledge this represents the first modeling of the Web that goes beyond fitting degree distributions on the Web.

1 Introduction

There has been considerable recent work on developing increasingly sophisticated models of the structure of the Web [1, 3, 4, 9, 13, 14]. The primary drivers for such modeling include developing an understanding of the evolution of the Web, better tools for optimizing Web-scale algorithms, mining communities and other structures on the Web, and studying the behavior of content creators on the Web. Prior modeling has dwelt on fitting models to the observed degree distribution of the Web. While this represents a significant step (both empirically and analytically), a weakness of this approach is the heavy reliance on a single set of parameters – the degree distribution. Moreover, the degree distribution is a very "local" property of graphs, something that is well recognized from at least two distinct viewpoints: (1) as a ranking mechanism, ordering the Web pages in search results by in-degree (popularity of linkage) is very easy to spam and thus not reliable; (2) from a graph-theoretic standpoint, it is easy to exhibit "very different" graphs that conform to the same degree distribution. Indeed, the first of these reasons led to the PageRank function [8] used in the Google engine.

In this paper we present a more detailed approach to modeling, to explain the distributions of *PageRank* values on the Web. Our model augments the de-

* Supported in part by NSF grant CCR-9731477 and NSF ITR grant CCR-0121154.

O.H. Ibarra and L. Zhang (Eds.): COCOON 2002, LNCS 2387, pp. 330–339, 2002.

gree distribution approach, so that as a by-product we achieve previous models' success in explaining degree distributions.

Our study of PageRank distributions is also of independent interest for Web search and ranking pages. For search engines employing PageRank and associated ranking schemes, it is important to understand whether, for instance, 99% of the total PageRank is concentrated in (say) 10% of the pages. This (especially in conjunction with query distribution logs) has implications for compressing inverted indices and optimizing the available storage.

2 Background and Related Work

The Web as a graph. View the Web as a *directed* graph whose nodes are html pages. Each hyperlink is a directed edge in the natural manner. The *in-degree* of a node is the number of edges (hyperlinks) into it; a simplistic interpretation of the in-degree of a page is as a popularity count. The *out-degree* of a node is the number of links out of it; this is simply the number of `href` tags on the page. The *degree distribution* of a graph is the function of the non-negative integers that specifies, for each $k \geq 0$, what fraction of the pages have degree k; there are naturally two degree distributions for a directed graph, the in-degree distribution and the out-degree distribution.

These distributions have been the objects of considerable prior study [1, 3, 4, 9, 13, 14], on various snapshots of the Web ranging from the Web pages at a particular university to various commercial crawls of the Web. Despite the varying natures of these studies, the in-degree distribution appears to be very well approximated by the function $c/k^{2.1}$ where c is the appropriate normalization constant (so that the fractions add to one). Likewise, the out-degree distributions seem to be very well approximated by the function $c_o/k^{2.7}$. Such distributions are known as *power law* distributions.

Recent work of Dill et al. [10] provides some explanation for this "self-similar" behavior: that many properties of the Web graph are reflected in sub-domains and other smaller snapshots of the Web. Indeed, this will provide the basis for some of our experiments, in which we derive an understanding of certain properties of the Web by studying a crawl of the `brown.edu` domain. (This methodology was pioneered by Barabasi et al. [3,4], who extrapolated from the `nd.edu` domain of Notre Dame University. They made a prediction on the diameter of the undirected version of the Web graph, in which one ignores link directions.)

Other properties of the Web graph that have been studied (analytically or empirically) include connectivity [9], clique distributions [13] and diameter [7].

PageRank. The *PageRank* function was presented in [8,17] and is reportedly used as a ranking mechanism in the commercial search engine Google [12]. It assigns to each Web page a positive real value called its PageRank. In the simplest use of the PageRank values, the documents matching a search query are presented in decreasing order of PageRank.

The original intuition underlying PageRank was to visualize a random surfer who browsed the Web from page to page. Given the current location (page) q of the surfer, the successor location is a page reached by following a hyperlink out of page q uniformly at random. Thus each hyperlink is followed with probability proportional to the out-degree of q. In this setting, the PageRank of each page is the frequency with which, in the steady state, the page q is visited by such a surfer. Intuitively, the surfer frequently visits "important" pages such as `yahoo.com` because many pages hyperlink to it. Moreover, by calculations from elementary probability theory, the PageRank of a page q is increased if those pages that hyperlink to q have high PageRank themselves. An immediate difficulty with this notion: some pages, or an (internally) connected cluster of pages may have no hyperlinks out of them, so that the random surfer may get stuck. To address this, Brin and Page [8] introduced a *decay* parameter p: at each step, with probability p the surfer proceeds with the random walk, and with probability $1-p$, the surfer "teleports" to a completely random Web page, independent of the hyperlinks out of the current page. We refer to [8, 17] for details on the mathematics of PageRank and its practical implementation using the decay parameter.

3 Web Graph Models

The classical random graph models of *Erdös-Renyi* [5] do not explain the power law properties of the degree distribution nor the the superabundance of clique-like structures [14] in the Web graph. Thus, it is clear that the Web graph does not conform to the Erdös-Renyi model. One of the first models to explain the power law property was proposed by Barabasi *et al.* which has two key features: (1) nodes and edges are added to the graph one at a time (*uniform growth*) and (2) each incoming node chooses to connect to a node node q in proportion to the current in-degree of q (*preferential attachment*). This model yields Web graphs whose in-degree distributions have been shown to converge to the distribution $\approx 1/k^2$ [3, 4].

However, as noted earlier, empirical studies have shown that in-degrees are in fact distributed as $\approx 1/k^{2.1}$ (rather than $1/k^2$). To help explain the exponent of 2.1, Kumar *et al.* [15] introduced the following more detailed process by which each edge chooses the node to point to. Some fraction of the time (a parameter they call $\alpha \in [0, 1]$) the edge points to a node chosen uniformly at random. The rest of the time (a fraction $1 - \alpha$), the edge picks an intermediate node v at random, and *copies* the destination of a random edge out of v. In other words, the new edge points to the destination of an edge e, chosen at random from the outgoing edges of a random node v. Kumar *et al.* offer the following behavioral explanation for this process: some fraction of the time a content creator creating a page refers to a random new topic and thus creates a link (edge) to a random destination. The remainder of the time, the content creator copies a hyperlink off an existing page (in this case v), having decided that this is an interesting link. They then explain a number of empirical observations on the Web graph including the in-degree exponent of 2.1 and the large number of clique-like structures observed by [14]. Their model can be viewed as a generalization of the models of Barabasi and others, parameterized by α. We will henceforth refer to this model as the *degree-based selection model.* Could it be that this model would also explain the PageRank distributions we observe on the Web?

Before we address this question, we next introduce a new model inspired by the α model above. Suppose that each edge chose its destination at random a fraction $\beta \in [0, 1]$ of the time, and the rest of the time chose a destination in proportion to its *PageRank.* Following the behavioral motivation of Kumar *et al.*, this can be thought of as a content-creator who chooses to link to random pages some fraction of the time, and to pages highly rated by a PageRank-based engine such as Google the remainder of the time. In other words, content creators are more likely to link to pages that score high on PageRank-based search results, because these pages are easy to discover and link to. This is not implausible from the behavioral standpoint, and could help capture the PageRank distributions we observe (just as in-degree based linking helped explain in-degree distributions in prior work). We will call this the *PageRank-based selection model.*

However, this now raises the question: if we could develop a model that explained observed PageRank distributions, could it be that we lose the ability to capture observed degree distributions? To address this, we now present the most general model we will study. There are two parameters $a, b \in [0, 1]$ such that $a + b \leq 1$. With probability a an edge points to a page in proportion to its in-degree. With probability b it points to a page in proportion to its PageRank. With the remaining probability $1 - a - b$, it points to a page chosen uniformly at random from all pages. We thus have a family of models; using these 2-parameter models we can hope to simultaneously capture the two distributions we investigate – the PageRank distribution (representing global properties of the graph), and the in-degree distribution (representing local properties of the graph). We will call this the *hybrid selection model.*

4 Experiments

To set the context for exploring the models in Section 3, we study the distribution of PageRanks (as well as of the in- and out-degrees) on several snapshots of the Web.

Brown University domain. Our first set of experiments was on the Web graph underlying the Brown University domain (`*.brown.edu`). Our approach is motivated by recent results on the "self-similar" nature of the Web (e.g., [10]): a thematically unified region (like a large subdomain) displays the same characteristics as the Web at large. The Brown Web consisted of a little over 100,000 pages (and nearly 700,000 hyperlinks) with an average in-degree (and thus out-degree) of around 7. This is very close to the average in-degree reported in large crawls of the Web [14]. Our crawl started at the Brown University homepage (`www.brown.edu` – "root" page) and proceeded in breadth-first fashion; any

URL outside the `*.brown.edu` domain was ignored. We did prune our crawl – for example, URL's with /cgi-bin/ were not explored.

Our experiments show that the in-degree and out-degree distribution follows a power law with exponent 2.1 and 2.7 respectively. The plots are strikingly similar to the ones reported on far larger crawls of the Web (see [9, 14]). For example [9] report exactly the same power law exponents on a crawl of over 200 million pages and 1.5 billion hyperlinks.

However, the most interesting result of our study was that of the PageRank distribution. We first describe our PageRank computation. As in [17], we first pre-process pages which do not have any hyperlinks out of them (i.e., pages with out-degree 0): we assume that these have links back to the pages that point to them [2]. This is intuitively more justifiable than just dropping these pages: we expect surfers to trace back their trail when they reach a dead end. In our PageRank computation we set the decay parameter to 0.9; this is a typical value reportedly used in practice (e.g., [8] uses 0.85), and the convergence is fast (under 20 iterations). Similar fast convergence is reported in [8, 17]. However, varying the decay parameter does not significantly change our results, as long as the parameter is fairly close to 1. In particular, we get essentially the same results for decay parameter values down to 0.8.

The main result of our PageRank distribution plot (Figure 1) is that a large majority of pages (except those with very small PageRank) follow a power law with an exponent close to 2.1. That is, the fraction of nodes having PageRank r is proportional to $1/r^{2.1}$. This appears to be the same as the in-degree exponent; more on this later. In Section 5 we will give an analysis suggesting this PageRank distribution, based on various models from Section 3.

WT10g data. We repeated our experiments on the WT10g corpus [18], a recently released, 1.69 million document testbed for conducting Web experiments. The results are almost identical to those on the Brown Web; the in-degree, out-degree, and PageRank distributions follow power laws with exponent close to 2.1, 2.7 and 2.1 respectively. Figure 1 shows the plot of PageRank distribution of the wt10g corpus. The power law here appears much sharper than in the Brown Web. Also, unlike the Brown Web, the plot has slope 2.1 across almost the entire spectrum of PageRank values, except for those with very low PageRank values; a possible explanation is that unlike the Brown domain, the WT10g corpus is constructed by a careful selection of Web pages so as to characterize the *whole Web* [18].

5 Fitting the Models: Analysis and Simulations

In this section we address some of the modeling questions raised in Section 3. Having obtained the empirical distributions in Section 4, we first give analytical predictions of the shape of the PageRank distributions for the degree-based and PageRank-based selection models of Section 3. The intent is to infer what choices of these model parameters would give rise to the distributions observed in our experiments. Finally, in Section 5.3 we generate random graphs according to

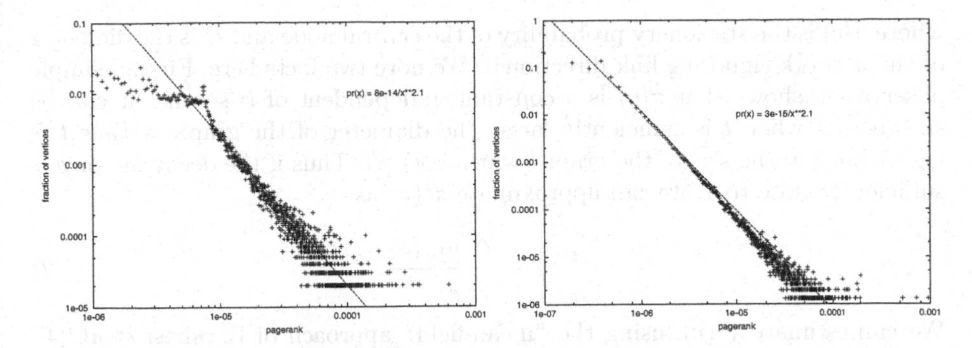

Fig. 1. Log-log plot of the PageRank distribution of the Brown domain (left) and the WT10g (right). A vast majority of the pages (except those with very low PageRank) follow a power law with exponent close to 2.1.

these fitted models, to see if in fact they give rise to graphs that match the distributions observed on the Web.

5.1 Degree-Based Selection

Consider a graph evolving in a sequence of *time steps* – as noted in Section 3 such evolution is not only realistic in the context of the Web, it is also a feature of all Web graph models. A single node with r outgoing edges is added at every time step. (We assume that we start with a single node with a self-loop at time 0 [6].) Each edge chooses its destination node independently with probability proportional to 1+in-degree[1] of each possible destination node. This model is essentially the one analyzed by Barabasi *et al.* and is a special case of the α model (where $\alpha = 0$) of Kumar *et al.*

Let $\pi^t(v)$ represent the PageRank of v at time step t. We can interpret the PageRank as the stationary probability of a random walk on the underlying graph, with the teleport operation (Section 2) being modeled by a "central" node c. At each step, the surfer either decides to continue his random walk with probability p or chooses to return to the central node with probability $1-p$; from the central node he jumps to a random node in the graph. To write an expression for $\pi^t(v)$ it is useful to define $f^t(v)$, the "span" of v at time t: the *sum* of the in-degrees of all nodes in the network (including v itself) that have a path to v that does not use the central node (we also refer to the nodes contributing to the span as "span nodes"). Since each edge contributes a $1/r$ fraction of the stationary probability of its source node (using the standard stationary equations (see [16])), we can bound $\pi^t(v)$ for the above random walk as follows:

$$\frac{f^t(v)\pi(c)p^D}{rt} \le \pi^t(v) \le \frac{f^t(v)\pi(c)}{rt} \tag{1}$$

[1] We assume that each incoming node has "weight" 1, otherwise there won't be any non-trivial growth.

where $\pi(c)$ is the stationary probability of the central node and D is the diameter of the network (ignoring link directions). We note two facts here. First, a simple observation shows that $\pi(c)$ is a constant, independent of t; second, it can be shown that when t is sufficiently large, the diameter of the graph at time t is logarithmic in the size of the graph (which is t) [7]. Thus if the decay factor p is sufficiently close to 1, we can approximate $\pi^t(v)$ as

$$\pi^t(v) \approx \frac{f^t(v)\pi(c)}{rt}. \tag{2}$$

We can estimate $f^t(v)$, using the "mean-field" approach of Barabasi *et al.* [4]. Treating $f^t(v)$ as continuous, we can write the differential equation for the rate of change of $f^t(v)$ with time: $\frac{d(f^t(v))}{dt} = \frac{f^t(v)}{t}$, where the right hand side denotes the probability that an incoming edge connects to one of the span nodes of v. The solution to the above equation with the initial condition that node v was added at time t_v is $f^t(v) = t/t_v$. Using this in equation (2), and assuming that nodes are added at equal time intervals, we can show that the probability density function F for $\pi^t(v)$ is: $F(\phi) \approx \pi(c)/rt\phi^2$, implying that the PageRank follows a power law with exponent 2, independent of r and t. Simulations of this model (shown in Figure 2) agree well with this prediction.

As already mentioned in Section 3, the in-degree distribution of this model follows a power law with exponent 2, the same as the PageRank distribution derived above. However, the empirically observed power laws of both PageRank and in-degree have exponents of 2.1; thus the degree-based selection model does not quite match the in-degree and PageRank exponents observed in practice. Now a natural question is whether we can make it match both the distributions by changing α, i.e., by incorporating a random selection component in choosing nodes. The answer is surprisingly[2] yes; more on this in Section 5.3.

5.2 PageRank-Based Selection

We show that power law emerges for the PageRank and degree distributions in this model (we assume $\beta = 0$, i.e, the node selection is based only on Pagerank), but the exponents are different from the degree-based model.

Using the same argument as before, we can show that Equation (2) holds. However, $f^t(v)$ here follows a different differential equation than the one in the previous analysis: $\frac{d(f^t(v))}{dt} \approx \frac{f^t(v)r}{2rt}$. The reasoning is as follows. The probability that $f^t(v)$ increases by one is the probability that the incoming node chooses any one of the nodes in the span to connect to, which is proportional to the sum of the PageRanks of all the span nodes of v. To calculate this probability, we see that each directed edge contributes nearly *twice* to the sum (if p is sufficiently

[2] Surprising because, it is not the case that PageRank and in-degree distributions are related – as suggested by the the similarity of the power law exponents of the two distributions. It follows from our analysis above, that even when nodes are selected uniformly at random (i.e., $\alpha = 1$), a power law (with a small exponent) emerges for the PageRank; but the degree distribution is Poisson.

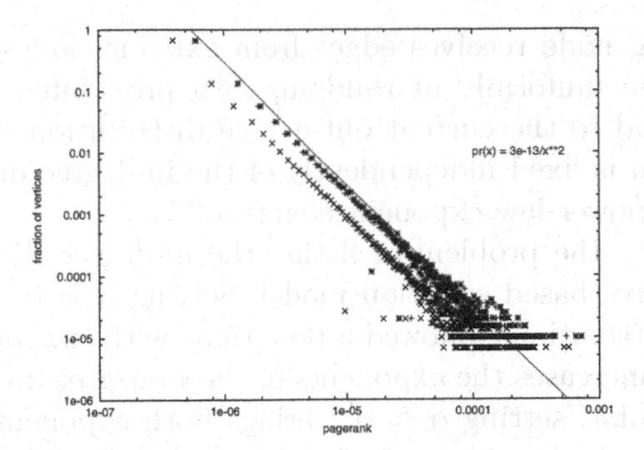

Fig. 2. Log-log plot of degree-based selection with $\alpha = 0$. The number of nodes shown is 300,000 (+), 200,000 (*) and 100,000 (x). It clearly shows that the slope is 2, confirming the power law predicted by analysis.

large) and the total PageRank is thus proportional to the sum of the degrees which is $2rt$.

Plugging the solution of the above differential equation in Equation (2), we can show that the probability density function F for $\pi^t(v)$ in this model is: $F(\phi) \approx (\pi(c))^2/r^2t^2\phi^3$, i.e., predicting that the PageRank follows a power law with exponent 3. Analogously, we can show that the degree also a follows a power law with exponent 3. Simulations also agree quite well with this prediction.

Thus, the PageRank-based selection model with $\beta = 0$ does not match the empirically observed in-degree and PageRank exponents. Can we hope to match the observations by varying β? Unlike the degree-based selection model, the answer is no; increasing β will only increase the power law exponent (above 3) for the in-degree distribution. This can be verified by experiments. We are thus left with the degree-based selection model and the hybrid selection model of Section 3 as candidates for explaining the observations.

5.3 Simulations of the Generative Models

An accurate model of the Web graph must conform with the experimentally observed in-degree, out-degree, and PageRank distributions. We simulated the degree-based and hybrid selection models defined in section 3 under various parameters to find settings that generate the observed empirical distributions. We simulated graphs of size up to 300,000 nodes, and we varied the average number of new edges generated per new node generation (time step). In particular, to be "close" to the real Web's average out-degree (and in-degree), we focused on the range in which the average number of edges added per new node is around 7. We obtained essentially the same results for the power laws, irrespective of the size (from 10,000 nodes onwards) or the number of outgoing edges.

Our first step was fitting the out-degree distribution. Following Kumar *et al.*, we use the degree-based copying model with a suitable value of β to fit the out-degree distribution to a power law with exponent 2.7. At each time step, the incoming node receives edges from existing nodes. With probability β a node is chosen uniformly at random, with probability $1-\beta$ the node is chosen proportional to the current out-degree distribution. Note that the out-degree distribution is fixed independently of the in-degree distribution. We use $\beta = 0.45$ to get a power law exponent equal to 2.7.

We turn now to the problem of fitting the in-degree distribution. We first simulated the degree-based selection model. Setting $\alpha = 0$, both the in-degree and PageRank distributions followed a power law with exponent 2. We observed that increasing α increases the exponents in the in-degree and PageRank distributions. In particular, setting $\alpha \approx 0.2$ brings both exponents to the empirical value of 2.1. This value is unique; by increasing or decreasing α we lose the fit. Thus, we found a setting of the parameters for which the degree-based selection model simultaneously fits all the three distributions.

Since degree-based selection model fits the empirical data, a natural question is whether PageRank-based selection is irrelevant in modeling the Web graph. To answer this, we experimented with the 2-parameter hybrid selection model proposed in Section 3. Surprisingly when $a = b \approx 0.33$, we could again simultaneously fit all three distributions. Thus we have an alternative model, with a substantial PageRank-based selection component, that fits the Web empirical data. As mentioned in Section 3, this model is plausible from the behavioral standpoint.

6 Conclusion

We present experimental and analytical studies of PageRank distribution on the Web graph, and use it to develop more accurate generative models for the evolution of the Web graph. We consider three possible models: degree-based selection, PageRank-based selection, and a hybrid model. Our analysis shows that the PageRank-based selection model cannot fit the empirical data. For the two other models we found settings of parameters under which the model fits simultaneously the in-degree and out-degree distributions and the PageRank distribution. A natural question for further study is whether one of these models describes the Web better than the other. Another interesting question is investigating the relationship between PageRank and in-degree which may shed new insight into Web structure.

Acknowledgments

We are very grateful to Joel Young for providing us with his Web crawler and for many hours of help.

References

1. L. Adamic and B. Huberman. Power Law distribution of the World Wide Web, Technical Comment on [3], *Science*, **287**, 2000, 2115a.

2. Arvind Arasu, Junghoo Cho, Hector Garcia-Molina, Andreas Paepcke, Sriram Raghavan. Searching the Web. *ACM Transactions on Internet Technology*, **1**(1), 2001, 2-43.
3. A. Barabasi and R. Albert. Emergence of Scaling in Random Networks. *Science* , **286**(509), 1999.
4. A. Barabasi, R.Albert and H. Jeong. Mean-field theory for scale-free random graphs. *Physica A*, **272**, 1999, 173-187.
5. B. Bollobas. *Random Graphs.* Academic Press, 1990.
6. B. Bollobas, O. Riordan, J. Spencer, and G. Tusnady. The degree sequence of a scale-free random graph process. *Random Structures and Algorithms*, **18**(3), 2001, 279-290.
7. B.Bollobas and O. Riordan. The diameter of a scale-free random graph. *preprint*, 2001.
8. S. Brin and L. Page. The anatomy of a large-scale hypertexual Web search engine. In *Proceedings of the 7th WWW conference*, 1998.
9. A. Broder, R. Kumar, F. Maghoul, P. Raghavan, S. Rajagopalan, R. Stata, Andrew Tomkins, J. Weiner. Graph Structure in the Web. In *Proceedings of the 9th WWW Conference*, 2000.
10. S. Dill, R. Kumar, K. McCurley, S. Rajagopalan, D. Sivakumar, and A. Tomkins. Self-Similarity in the Web. In *Proceedings of the 27th International Conference on Very Large Databases (VLDB)*, 2001.
11. D. Gibson, J.M. Kleinberg and P. Raghavan. Inferring Web communities from link topology. In *Proceedings of the ACM Symposium on Hypertext and Hypermedia*, 1998.
12. Google Inc. `http://www.google.com`
13. J. Kleinberg, S. Ravi Kumar, P. Raghavan, S. Rajagopalan and A. Tomkins. The Web as a graph: measurements, models and methods. In *Proceedings of the 5th Annual International Computing and Combinatorics Conference (COCOON)*, 1999.
14. R. Kumar, P. Raghavan, S. Rajagopalan, and A. Tomkins. Trawling the Web for Emerging Cyber-Communities. In *Proceedings of the 8th WWW Conference*, 1999, 403-416.
15. R. Kumar, P. Raghavan, S. Rajagopalan, D. Sivakumar, A. Tomkins, and E. Upfal. Stochastic Models for the Web. In *Proceedings of the 41st Annual Symposium on the Foundations of Computer Science (FOCS)*, 2000.
16. R. Motwani and P. Raghavan. *Randomized Algorithms*, Cambridge University Press, 1995.
17. L. Page, S. Brin, R. Motwani, and T. Winograd. The PageRank Citation Ranking: Bringing order to the Web, *Technical Report*, Computer Science Department, Stanford University, 1998.
18. WT10g collection draft paper.
`http://www.ted.cmis.csiro.au/TRECWeb/wt10ginfo.ps.gz`

On Randomized Broadcasting and Gossiping in Radio Networks

Ding Liu and Manoj Prabhakaran

Department of Computer Science, Princeton University, Princeton, NJ 08544, USA
{dingliu,mp}@cs.princeton.edu

Abstract. This paper has two parts. In the first part we give an alternative (and much simpler) proof for the best known lower bound of $\Omega(D \log(N/D))$ time-steps for randomized broadcasting in radio networks with unknown topology. In the second part we give an $O(N \log^3 N)$-time randomized algorithm for gossiping in such radio networks. This is an improvement over the fastest previously known algorithm that works in time $O(N \log^4 N)$.

1 Introduction

We consider two classical problems of distributing information in radio networks: *broadcasting* and *gossiping*. In broadcasting, the goal is to distribute a message from a distinguished source node to all other nodes in the network. In gossiping each node in the network holds a message, and the goal is to distribute each message to all nodes in the network. In both problems we want to use as less time as possible to finish the task. The radio network is an abstraction of communication networks with minimal assumptions and features and it can model many situations. Communication in radio networks and variants thereof have been widely studied for a long time [1, 4, 7–11, 15–18].

A radio network [4, 8] is modeled as a directed graph $G(V, E)$ where $|V| = N$. The nodes of the graph represent processors in the network and they are assigned different identifiers from the set $\{1, 2, \ldots, N\}$. A directed edge from node u to node v means that u can send messages to v, and we say that v is an *out-neighbor* of u and u is an *in-neighbor* of v. Time is divided into discrete time steps. All nodes have access to a global clock and work synchronously. Two prominent features of the radio network model are: (1) Processors have no knowledge of the network topology. They only know the size of the network N [1]. (2) A node v receives the message from its in-neighbor u in a step if and only if u is the *only* in-neighbor of v that is transmitting in that time step. If two or more in-neighbors of v transmit then a *collision* occurs and none of the messages is received by v. Furthermore v cannot distinguish such collisions from the situation where none of its neighbors is transmitting. See [3, 4, 8, 14, 19] for discussions on this and related models.

[1] In our lower bound we let them know the diameter of the network D also.

O.H. Ibarra and L. Zhang (Eds.): COCOON 2002, LNCS 2387, pp. 340–349, 2002.

This model is suited for channels with high noise and unreliable *collision detection*. Further the topology of the network is considered unknown to the processors, which makes it suitable to model mobile networks or networks with dynamically configured topology or unreliable links[2]. Thus algorithms for radio networks have potentially wide applications.

On the other hand one is also interested in lower bounds in this model, because along with efficient algorithms in less stricter models (for instance see [12, 13]) it demonstrates the importance of the various components in the communication model. Our model allows unlimited computational power and unbounded message size. This makes the lower bounds strong, whereas the algorithms typically do not exploit such unreasonable assumptions.

Previous results. Chlebus [6] surveys the state of the art in randomized protocols for radio networks. For deterministic broadcasting in unknown radio networks, the best currently known upper bound is due to Chrobak, Gąsieniec and Rytter [9]. Their algorithm runs in time $O(N \log^2 N)$ in a network with N nodes. Their algorithm is non-constructive in the sense that they show the existence of such algorithms without explicitly constructing one. Recently Indyk [16] gave a constructive solution with similar bounds. The best known lower bound is $\Omega(N \log N)$ [5, 7]. Bar-Yehuda, Goldreich and Itai [4] gave a randomized algorithm that achieves broadcast in expected time $O(D \log N + \log^2 N)$. For lower bounds we let the algorithm know the diameter of the network, D also. Kushilevitz and Mansour [17] established a lower bound of $\Omega(D \log (N/D))$. For deterministic gossiping Chrobak, Gąsieniec and Rytter [9] presented the first sub-quadratic algorithm whose running time is $O(N^{3/2} \log^2 N)$. Again their algorithm is non-constructive and a constructive solution with similar time bound was recently provided by Indyk [16]. Chrobak, Gąsieniec and Rytter also gave a randomized $O(N \log^4 N)$-time algorithm for gossiping in radio networks with unknown topology [10].

Our results. In section 3 we give an alternative proof for the $\Omega(D \log (N/D))$ lower bound on randomized broadcasting algorithms. Our proof is essentially different from, and much simpler than the previous one [17], which is complicated and involves a reduction from the general case to a special "uniform" case. We hope that our proof will give a fresh view of the lower bounds for the problem, and may help in bridging the gap between the known upper and lower bounds. In section 4 we give a randomized $O(N \log^3 N)$-time algorithm for gossiping in radio networks with unknown topology. Our basic algorithm is Monte Carlo type and it easily yields a Las Vegas algorithm with expected running time $O(N \log^3 N)$. Our algorithm follows the one in [10], and the essential difference is that we replace a deterministic procedure in that algorithm by a new randomized procedure. Finally in section 5 we note that a linear time deterministic *gossiping* algorithm exists for symmetric networks, which is asymptotically optimal.

[2] For gossiping to be feasible, we require that the underlying network is strongly connected. For broadcasting to be feasible, we require that all nodes are reachable from the source. These are the only assumptions on the network topology.

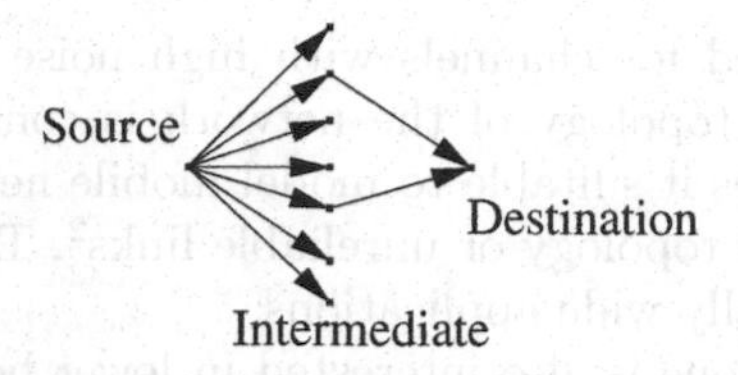

Fig. 1. The graph family $\mathcal{D}_m$.

2 Preliminaries

An algorithm in the radio network is a distributed protocol that for each node v and each time step t, specifies the action (either Transmit or Receive) of v at step t, possibly based on all past messages received by v. The model allows *unlimited computational power* and all computations are carried out between time steps. If v transmits at time step t the algorithm also specifies the message, which could be arbitrarily long. The running time of a Monte Carlo broadcasting algorithm is the smallest t such that for any connected network topology[3], and for any assignment of identifiers to the nodes, the probability of completing the broadcast no later than at step t is $\Omega(1)$. For a Las Vegas algorithm, the running time is defined as the expected time of completion. The running time for a gossiping algorithm is defined similar to that of broadcasting algorithm. Both of our upper and lower bounds are based on some observations about a special family of networks denoted by $\mathcal{D}_m$ which we define next.

Definition 1. *The network family* $\mathcal{D}_m$ *is the set of all* $2^m - 1$ *networks of the following type. There are totally* $m+2$ *nodes: one source, one destination and* m *intermediate nodes (Fig. 1). There are edges from the source to all intermediate nodes, and edges from some intermediate nodes (at least one) to the destination.*

We consider broadcasting on family $\mathcal{D}_m$. Initially only the source holds the message and it transmits first. After that every intermediate node gets the message. Now the goal is to send the message to the destination. It is easy to show that any deterministic algorithm for $\mathcal{D}_m$ must use m steps in the worst case. On the other hand, there exists a randomized broadcasting algorithm that runs in $O(\log m)$ time. We give one below; a similar algorithm appears in [4]. Note that the algorithm is executed in parallel by all the nodes.

Procedure DECAYINGBROADCAST (m)
for $i \leftarrow 1$ to $\log m + 1$ **do**
  with probability $1/2^{i-1}$ Transmit

DECAYINGBROADCAST runs in $O(\log m)$ time and achieves a constant success probability[4]. To see this, note that if there are k edges connecting intermediate nodes to the destination, then in the i-th round the success probability is

[3] By *connected* we mean each node is reachable from the source.
[4] "success" means the destination receives the message.

$k/2^{i-1}(1 - 1/2^{i-1})^{k-1}$. So in round $i = \lceil \log k \rceil + 1$ this probability is lower bounded by a constant, at least 1/8. In the next section we shall see that the running time of this algorithm is asymptotically optimal.

3 The $\Omega(D \log (N/D))$ Lower Bound for Randomized Radio Broadcast: A Simple Proof

In [17] an $\Omega(D \log (N/D))$ lower bound for randomized (Las Vegas type) broadcasting algorithm was established. Here we give an alternative proof of the same result. We think our proof is interesting not only because it is much simpler than the previous one, but also it uses a completely different approach. We hope that our proof will open a new line of attack for proving lower bounds on similar problems.

Theorem 1. *For any Monte Carlo broadcasting algorithm (i.e., randomized broadcasting algorithm that succeeds with probability $\Omega(1)$) there is a network with N nodes and diameter D, in which the algorithm takes $\Omega(D \log (N/D))$ expected time.*

The proof of Theorem 1 is based on the well-known Yao's minimax principle (see [20], Theorem 3). This principle reduces the task of proving randomized lower bound to that of proving deterministic lower bound. In order to prove Theorem 1, we pick a probability distribution over a suitable family of networks with N nodes and diameter D, such that any deterministic algorithm that succeeds with probability $\Omega(1)$ for this distribution has expected running time $\Omega(D \log (N/D))$. In view of the minimax principle this will prove Theorem 1. On the other hand, the proof of [17] is based on direct analysis of randomized algorithms. We first prove two preliminary lemmas.

Lemma 1. *There exists a probability distribution $\mathcal{P}$ over $\mathcal{D}_m$ such that the probability for the destination to get the message in any* one *deterministic step is $O(1/\log m)$.*

Proof. We partition $\mathcal{D}_m$ into m subfamilies $\bigcup_{k=1}^{m} \mathcal{D}_m^k$, where $\mathcal{D}_m^k$ is the set of networks with exactly k intermediate nodes connected to the destination. We pick $\mathcal{P}$ in two steps. First for each k $(1 \leq k \leq m)$ we assign weight $c/(k \log m)$ to subfamily $\mathcal{D}_m^k$, where c is a normalization factor such that the weights add up to 1. When m is large c is close to $\ln 2$. Then for each $\mathcal{D}_m^k$ we evenly distribute its total weight to all $\binom{m}{k}$ networks belonging to it. In other words, each network with exactly k intermediate nodes connected to the destination is assigned weight $c/(k\binom{m}{k} \log m)$.

Now look at any deterministic step and suppose that j $(1 \leq j \leq m)$ intermediate nodes are transmitting in that step. Over the distribution $\mathcal{P}$ we picked, the success probability is easily seen to be

$$c \cdot \sum_{k=1}^{m} \frac{j\binom{m-j}{k-1}}{k\binom{m}{k} \log m} = \frac{c}{\log m} \cdot \frac{j}{m} \cdot \sum_{k=1}^{m-j+1} \frac{\binom{m-j}{k-1}}{\binom{m-1}{k-1}}$$

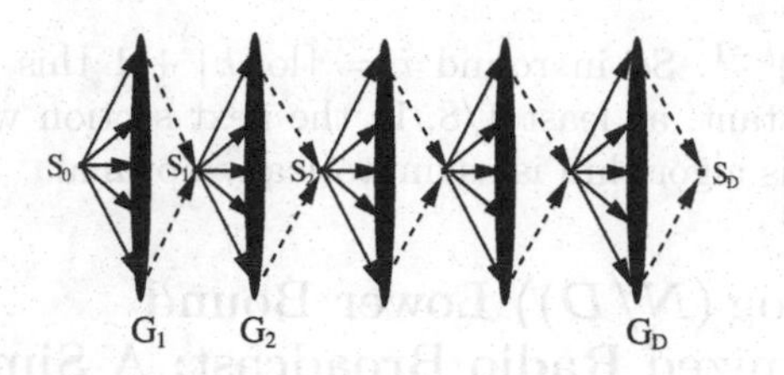

Fig. 2. Structure of the network family $\mathcal{F}_{N,D}$.

Note that we have used the fact $k\binom{m}{k} = m\binom{m-1}{k-1}$. In order to show that this probability is $O(1/\log m)$ we only need to show that for *any* j, $\sum_{k=1}^{m-j+1} \binom{m-j}{k-1}/\binom{m-1}{k-1}$ is $O(m/j)$. When $j = 1$ this is obvious. So we assume that $j > 1$. Let $A_k = \binom{m-j}{k-1}/\binom{m-1}{k-1}$; then for $k \geq 1$, $A_{k+1}/A_k = (m-j-k+1)/(m-k) < (m-j+1)/m$. Also $A_1 = 1$, so $\sum_{k=1}^{m-j+1} \binom{m-j}{k-1}/\binom{m-1}{k-1}$ is bounded by the sum of an infinite geometric series with initial value 1 and decreasing ratio $(m-j+1)/m$. This series sums to $m/(j-1) \leq 2m/j$. □

We define the running time of a deterministic algorithm as the time taken till the broadcast succeeds or till the algorithm terminates, whichever is earlier. Note that the algorithm may succeed before it terminates (as the nodes may not realize that all nodes received the message). In the following lemma it is this definition of running time that is used[5].

Lemma 2. *Over the probability distribution $\mathcal{P}$ in Lemma 1, the expected running time of any deterministic broadcasting algorithm for $\mathcal{D}_m$ that succeeds with probability $\Omega(1)$ is $\Omega(\log m)$.*

Proof. Let $p = \Omega(1)$ be the success probability of the algorithm over $\mathcal{P}$. Let t be the smallest number such that the algorithm succeeds for $p/2$ fraction (according to $\mathcal{P}$) of the inputs in at most t steps. Thus for at least $p/2$ fraction, the algorithm runs for at least t steps. So the expected running time is at least $tp/2$. But by Lemma 1, succeeding in $p/2$ fraction of $\mathcal{P}$ requires $t = \Omega(\log(m)p/2) = \Omega(\log m)$ steps. □

By Yao's minimax principle Lemma 2 implies a $\Omega(\log m)$ lower bound for Monte Carlo algorithms for $\mathcal{D}_m$, and thus DECAYINGBROADCAST given in section 2 (with an initial transmission by the source) is asymptotically optimal.

To prove the general lower bound of $\Omega(D \log (N/D))$ we construct a family $\mathcal{F}_{N,D}$ of networks, pick a probability distribution $\mathcal{P}^*$ over $\mathcal{F}_{N,D}$, and prove the same lower bound on the expected running time (over $\mathcal{P}^*$) of any deterministic algorithm that succeeds with probability $\Omega(1)$. Theorem 1 then follows from Yao's minimax principle.

Figure 2 illustrates the structure of $\mathcal{F}_{N,D}$. It consists of D layers G_1 through G_D and each G_i is a network in $\mathcal{D}_m$ where $m = \lfloor N/D \rfloor$. Note that the graph

[5] which makes the lower bound stronger.

is directed from left to right, and there are no edges within each layer. Every network in $\mathcal{F}_{N,D}$ has $\Theta(N)$ nodes and diameter $\Theta(D)$. The probability distribution $\mathcal{P}^*$ is picked by letting each G_i $(1 \leq i \leq D)$ *independently* comply to the probability distribution $\mathcal{P}$ specified in the proof of Lemma 1. The intuition is that on each layer the algorithm is expected to run $\Omega(\log(N/D))$ steps and so the total time is $\Omega(D \log(N/D))$. Following is the proof of Theorem 1.

Proof. Let t_i $(1 \leq i \leq D)$ be the number of steps a deterministic algorithm A spends on layer G_i (i.e., the duration for which the message does not reach S_i after reaching S_{i-1}). They are random variables. If the message never reaches S_{i-1} then t_i is 0. The expected running time of A is $\mathbf{E}(\sum_{i=1}^{D} t_i) = \sum_{i=1}^{D} \mathbf{E}(t_i)$. We would like to use Lemma 2 to prove that $\mathbf{E}(t_i) = \Omega(\log(N/D))$. But there are a few subtleties. (i) $\mathbf{E}(t_i)$ is over $\mathcal{P}^*$ but Lemma 2 is about $\mathcal{P}$; (ii) we should take into account the possibility that the algorithm can "learn" from *history* and behave differently on networks with identical G_i; in other words, t_i not only depends on the topology of G_i but also those of G_1 through G_{i-1}. Below we take care of both these subtleties.

Fix a layer i and consider the success probability on this layer: this is no less than the overall success probability and hence is $\Omega(1)$. Now we partition the input space of A into finer subspaces such that: (i) within each subspace layers G_1 through G_{i-1} are all fixed; (ii) each subspace has equal weight (for this we may subdivide some instances having the same layers G_1 through G_{i-1} into finer subspaces). Since in $\mathcal{P}^*$, G_i is independent of the earlier layers, for each subspace restricted to G_i, the distribution is $\mathcal{P}$. Within each subspace the algorithm behaves identically on G_i. For each subspace consider the probability of success at the layer G_i, conditional to that subspace (i.e., for fixed G_1 through G_{i-1}). Then, in at least $\Omega(1)$ fraction of the subspaces the algorithm must have $\Omega(1)$ success probability. These subspaces are called *good*. We apply Lemma 2 to each *good* subspace to get a $\Omega(\log(N/D))$ lower bound on $\mathbf{E}_{\mathcal{P}}(t_i)$ [6], where the expectation is over the subspace (with distribution $\mathcal{P}$ in G_i). Taking the expectation over distribution $\mathcal{P}^*$ amounts to averaging over all subspaces. Since there are $\Omega(1)$ fraction good subspaces the expectation $\mathbf{E}_{\mathcal{P}^*}(t_i)$ is also $\Omega(\log(N/D))$.

We do this for each i. Linearity of expectation gives the lower bound. □

4 The $O(N \log^3 N)$ Randomized Algorithm for Gossiping

In this section we give an $O(N \log^3 N)$-time randomized algorithm for gossiping. The algorithm is an improvement over the recent work in [10], which gives an $O(N \log^4 N)$-time randomized algorithm.

[10] describes their algorithm in terms of *Distributed Coupon Collection*. There a simple randomized and distributed procedure DistCouponColl is

---

[6] W.l.o.g we assume that the nodes do not start the algorithm until they receive the first message (because we may let each node know the topology of the *previous* layers; it can then simulate messages it should have received, until it receives the actual message). Thus t_i is indeed the running time of the algorithm in that layer.

described, in which each node is a *bin* and each message a *coupon*. In a time step, each bin can be opened or left closed; if at some time step exactly one bin is opened all the coupons in that bin are *collected*. There may be many copies of a coupon in the network. The aim of DISTCOUPONCOLL(s) is to collect all coupons (i.e., at least one copy of each message), and for that each node repeats for s times the following: with probability $1/N$ open itself. Lemma 3 proved in [10], tells us how large an s we need for a good probability of collecting all coupons.

Lemma 3. [10] *If we have N bins and N coupons and each coupon has at least K copies (each copy belonging to a different bin), then for any constant ε, $0 < \varepsilon < 1$, if we run* DISTCOUPONCOLL(s) *with $s = (4N/K)\ln(N/\varepsilon)$, with probability at least $1 - \varepsilon$ all coupons will be collected.*

The overall algorithm in [10] is as follows: there are $\log N$ stages, and in each stage (with high probability) the number of copies of *each* message in the network is doubled; when stage i begins $K = 2^i$ copies of each message should be present in the network. For this, in stage i the DISTCOUPONCOLL is performed $(4N/K)\ln(N/\varepsilon)$ times so that at the end of the stage each message would have got *collected*. When a node is opened, it does a limited broadcast to double the number of copies of its coupons. Using a deterministic procedure called LTDBROADCAST, this takes time $O(K\log^2 N)$. Thus each stage takes $O(N\log^2 N\log(N/\varepsilon))$ time and has an error probability of ε. Since there are $\log N$ stages, with $\varepsilon = \epsilon/\log N$, the overall error probability is bounded by ϵ and the total time is $O(N\log^3 N\log(N/\epsilon))$.

The New Algorithm

Our new algorithm runs in two phases: in the first phase it does $\log N$ round-robins. After this each message has at least $\log N$ copies in the network. The second phase is identical to the old algorithm, except that the LTDBROADCAST is replaced by a new procedure RANDLTDBROADCAST given here. As we shall see, this allows us to save an asymptotic factor of $\log N$ in the running time.

Recall that in section 2 we give a randomized procedure DECAYINGBROADCAST that finishes broadcasting in $\mathcal{D}_m$ with constant probability. Here we use it to send a broadcast message to a new node with constant probability. In other words, if some nodes have not got the message yet, then there exists one such node v such that at least one in-neighbor of v has the message. Regarding v as the *destination* node (Fig. 1), one round of DECAYINGBROADCAST gives a constant probability of sending the message to v. In RANDLTDBROADCAST we use repeated DECAYINGBROADCAST for carrying out limited broadcast. In fact, if we repeat it $O(N)$ times we get a simple $O(N\log N)$ algorithm for broadcast whose error probability can be easily bounded by a constant by using Markov inequality. But in order to use this as a module in our final algorithm we will need tighter Chernoff-type bounds.

RANDLTDBROADCAST$_v$ is executed in parallel by all the nodes. Each node has a local Boolean flag *active*$_v$; initially this variable is set to True for some nodes (the "open" bins) and set to False for all other nodes. Note that if during some round of RANDLTDBROADCAST$_v$ node v receives a message, $active_v$ becomes True in the next round.

Procedure RANDLTDBROADCAST$_v$ (N, K)
for $i \leftarrow 1$ to cK **do** { c is an absolute constant to be determined later}
  Round i:
  if $active_v$ **then**
    DECAYINGBROADCAST$_v$ (N)
  else
    Receive for $\log N + 1$ steps
    if received a new message **then**
      $active_v \leftarrow True$

Analysis of RANDLTDBROADCAST Time taken is clearly $O(K \log N)$. We define Boolean random variables X_i, for $1 \leq i \leq cK$, as follows. $X_i = 1$ if a new node receives the message or all nodes have already received the message at round i. Otherwise $X_i = 0$. It is clear that if $\sum_{i=1}^{cK} X_i \geq K$ then the limited broadcast has succeeded in getting the message to at least K new nodes. Since X_i's are *not independent* of each other, we cannot directly use the Chernoff Bounds. But they have the following property: For all 2^{i-1} settings of $(x_1, \ldots, x_{i-1})$, $Pr(X_i = 1 | X_1 = x_1, \ldots, X_{i-1} = x_{i-1}) \geq 1/8$, as guaranteed by DECAYING-BROADCAST. This allows us to use Chernoff-Bound-like argument to bound the error probability as summarized in the following lemma. The proof, omitted here, follows essentially along the same line as that of the Chernoff bound.

Lemma 4. *When the network is initialized with a single node as active, in* $O(K \log N)$ *time,* RANDLTDBROADCAST(N, K) *succeeds in broadcasting the messages in that node to at least* K *nodes with probability at least* $1 - \exp(-\alpha K)$, *where* α *can be made arbitrarily large by choosing a sufficiently large* c.

Now we are ready to analyze our final algorithm. The first phase takes time $O(N \log N)$. The second phase consists of $O(\log N)$ stages. In the i-th stage there are $O(N/K \ln(N/\varepsilon))$ calls to RANDLTDBROADCAST, with each call takes $O(K \log N)$ time. Hence each stage takes $O(N \log N \ln(N/\varepsilon))$ time and the whole phase takes $O(N \log^2 N \log(N/\varepsilon))$ time. The second phase dominates the overall running time.

The first phase is deterministic. In the second phase, there are two sources of error: the RANDLTDBROADCAST and the DISTCOUPONCOLL. First we analyze the error probability due to RANDLTDBROADCAST in the i-th stage. Recall that $K \geq \log N$. Error probability for each invocation of RANDLTDBROADCAST(N, K) is $\leq \exp(-\alpha K) \leq 1/N^2$, by choosing a sufficiently large c, and for $s_i = O(\frac{N}{K} \log \frac{N}{\varepsilon})$ invocations it is $O(\log(N/\varepsilon)/N)$. Thus (as long as ε is not exponentially small in N) the error probability due to RANDLTDBROADCAST is comfortably $o(1)$. In fact, by choosing c sufficiently large this error probability can be driven down to any inverse polynomial in N. Note that for $K = o(\log N)$,

$(N/K)\exp(-\alpha K)$ is not $O(1)$. This is the reason for having a separate first phase, so that before the second phase starts, the number of copies per message is large enough.

For DISTCOUPONCOLL in each stage, by Lemma 3 we bound the error probability by ε. Since we have $\varepsilon = \epsilon/\log N$, and there are less than $\log N$ stages, the error probability due to DISTCOUPONCOLL is bounded by ϵ. This dominates the error probability of the overall algorithm.

Theorem 2. *For any given constant ϵ, $0 < \epsilon < 1$,* RANDGOSSIP*(N, ϵ) run on an N node radio network completes gossiping in time $O(N \log^2 N \log(N/\epsilon))$ with probability at least $1 - \epsilon$.*

Finally like in [10], this Monte Carlo algorithm can be converted into a Las Vegas algorithm with expected running time $O(N \log^3 N)$.

5 Gossiping in Symmetric Radio Networks

A radio network is *symmetric* if for any two nodes u and v, whenever there is an edge from u to v, there is also an edge from v to u. For symmetric networks broadcasting can be done deterministically in linear time [7]. Here we note that the same is true for gossiping.

Theorem 3. *There exists a deterministic algorithm that finishes gossiping in $5N - 4$ rounds in any unknown symmetric radio networks with N nodes.*

The algorithm has three phases and it is an extension of the broadcast algorithm in [7]. The first phase is a round-robin, after which each node knows all its neighbors. The second and the third phases are Depth-First-Search (DFS) initiated by a pre-determined root node. Though the nodes have no knowledge on the topology of the network, they can carry out DFS in a distributed way in linear time [2]. At the end of the second phase the root has all the messages. In the third phase this message is sent to all the nodes in the network using DFS. We omit the details.

Acknowledgements

We would like to thank Andy Yao for introducing us to radio networks. Also we thank Amit Chakrabarti for useful discussions.

References

1. Alon, N., Bar-Noy, A., Linial, N., Peleg, D.: A lower bound for radio broadcast, Journal of Computer and System Sciences 43 (1991), 290–298.
2. Awerbuch, B.: A New Distributed Depth-First-Search Algorithm, Information Processing Letters, 20, (1985), 147–150.

3. Bar-Yehuda, R., Goldreich, O., Itai, A.: Efficient emulation of single-hop radio network with collision detection on multi-hop radio network with no collision detection, Distributed Computing 5 (1991), 67–71.
4. Bar-Yehuda, R., Goldreich, O., Itai, A.: On the time-complexity of broadcast in multi-hop radio networks: An exponential gap between determinism and randomization, Journal of Computer and System Sciences 45 (1992), 104–126.
5. Bruschi, D., Del Pinto, M.: Lower bounds for the broadcast problem in mobile radio networks, Distributed Computing 10 (1997), 129–135.
6. Chlebus, B.S.: Randomized communication in radio networks, A chapter in *Handbook on Randomized Computing*, eds. Pardalos, P.M., Rajasekaran, S., Reif, J., Rolim, J.D.P., Kluwer Academic Publishers, to be published.
7. Chlebus, B.S., Gąsieniec, L., Gibbons, A.M., Pelc, A., Rytter, W.: Deterministic broadcasting in unknown radio networks, Proc. 11th Annual ACM-SIAM Symp. on Discrete Algorithms, SODA (2000), 861–870.
8. Chlebus, B.S., Gąsieniec, L., Östlin, A., Robson, J.M.: Deterministic radio broadcasting, Proc. 27th International Colloquium on Automata, Languages and Programming, ICALP (2000).
9. Chrobak, M., Gąsieniec, L., Rytter, W.: Fast broadcasting and gossiping in radio networks, Proc. 41st Annual IEEE Conference on Foundations of Computer Science, FOCS (2000), 575–581.
10. Chrobak, M., Gąsieniec, L., Rytter, W.: A randomized algorithm for gossiping in radio networks, Proceedings of 7th Annual International Computing and Combinatorics Conference, COCOON (2001), 483-492.
11. Clementi, A.E.F., Monti, A., Silvestri, R.: Selective families, superimposed codes, and broadcasting in unknown radio networks, Proc. 12th Annual ACM-SIAM Symp. on Discrete Algorithms, SODA (2001), 709–718.
12. Diks, K., Kranakis, E., Krizanc, D., Pelc, A.: The impact of knowledge on broadcasting time in radio networks, Proc. 7th European Symp. on Algorithms, ESA (1999), 41–52.
13. Gaber, I., Mansour, Y.: Broadcast in radio networks, Proc. 6th Annual ACM-SIAM Symp. on Discrete Algorithms, SODA (1995), 577–585.
14. Gallager, R.: A perspective on multiaccess channels, IEEE Transactions on Information Theory, 31 (1985), 124–142.
15. Hedetniemi, S.M., Hedetniemi, S.T., Liestman, A.L.: A survey of gossiping and broadcasting in communication networks, Networks 18(1988), 319-359.
16. Indyk, P.: Explicit constructions of selectors with applications, Proc. 13th Annual ACM-SIAM Symp. on Discrete Algorithms, SODA (2002), 697–704.
17. Kushilevitz, E., Mansour, Y.: An $\Omega(D \log (N/D))$ lower bound for broadcast in radio networks, SIAM Journal on Computing 27 (1998), 702–712.
18. Kushilevitz, E., Mansour, Y.: Computation in noisy radio networks, Proc. 9th Annual ACM-SIAM Symp. on Discrete Algorithms, SODA (1998), 236–243.
19. Tanenbaum, A.S.: Computer Networks, Prentice-Hall, Englewood Cliffs, NJ, 1981.
20. Yao, A. C-C.: Probabilistic computations: Towards a unified measure of complexity, Proc. 18th Annual IEEE Conference on Foundations of Computer Science, FOCS (1977), 222–227.

Fast and Dependable Communication in Hyper-rings

Tom Altman[1], Yoshihide Igarashi[2], and Kazuhiro Motegi[2]

[1] Department of Computer Science, University of Colorado at Denver, Denver, CO 80217, USA
taltman@carbon.cudenver.edu

[2] Department of Computer Science, Gunma University, Kiryu, Japan 376-8515
igarashi@comp.cs.gunma-u.ac.jp

Abstract. A graph $G = (V, E)$ is called a hyper-ring with N nodes (N-HR for short) if $V = \{0, ..., N-1\}$ and $E = \{\{u, v\} | v - u$ modulo N is a power of $2\}$. The following results are shown. We prove that the node-connectivity κ of an N-HR is equal to its degree, say δ, by presenting an algorithm for the explicit construction of δ node-disjoint paths connecting nodes s and t. The length of these paths is bounded by $\lceil \log D \rceil + 3$, where D is the positional distance between s and t. Finally, we show a node-to-node communication scheme for HRs that requires only $\lceil \log D \rceil + 3$ rounds, even in the presence of up to $\delta - 1$ node failures.

Keywords: Hyper-ring, connectivity, broadcasting, network, reliability

1 Introduction

Various topologies have been proposed for interconnecting processors in large scaled parallel and/or distributed systems. Hyper-rings (HRs) [2] and their variations have appeared in the literature under several names, including optimal broadcasting scheme [1] and binary jumping networks [8]. Detailed discussions on the structure, properties, and advantages of this family of networks and their fault tolerance were presented in [2,7,8].

Node-connectivity of a network is fundamental in the analysis of network reliability and/or security. Menger's Theorem relates the connectivity of a graph G, denoted by $\kappa(G)$, to the size of the smallest set (among the maximal sets) of pair-wise internally node-disjoint paths between any pair of nodes s and t [12]. Note that in any G, $\kappa \le \delta$, where δ denotes the minimal nodal degree of G. Hence, in a network, multiple copies of a message may be sent through a number of disjoint paths and fault tolerance can be achieved in this manner [10].

In addition to increased reliability, disjoint paths provide an excellent mechanism for transmitting secret messages through a network. Using Rabin's Information Dispersal Algorithm [11], or partitioning a secret message into submessages, where every i-th packet is transmitted via the i-th mod κ path, may prevent (or at least make it more difficult) for adversaries to intercept and decipher it.

Han et al. [8], have extended the results obtained in [7] addressing the problem of broadcasting in faulty binary jumping networks, a directed version of

O.H. Ibarra and L. Zhang (Eds.): COCOON 2002, LNCS 2387, pp. 350–359, 2002.

HRs with roughly half the number of edges. However, their work was limited to the general broadcasting problem (one source sending the same message to all nodes in the network), and not to the more interesting problem of an explicit construction of the κ node-to-node disjoint paths, investigated here.

It is relatively easy to show that for any N-HR, $\delta - 2 \leq \kappa \leq \delta$, by applying the results of [4] to a *reduced* HR in which the maximal positional distance between any connected nodes is $2^{\lfloor \log N \rfloor - 1}$ and not the standard $2^{\lfloor \log N \rfloor}$. The reduced HR would be a circulant with the property of *convexity*, which is sufficient (but not a necessary) condition for $\kappa = \delta$ [4,5].

The major contribution of this paper is showing that HRs' connectivity is maximal by presenting an algorithm that generates the δ node-disjoint paths between any source node s and destination node t. Moreover, the length of all of these paths is bounded from above by $\lceil \log D \rceil + 3$, where $D \leq \lfloor N/2 \rfloor$ is the positional distance between s and t.

The rest of the paper is organized as follows. In Section 2, we present two fundamental methods of HR construction. There, we also introduce Semi Hyper-rings (SHRs) and enriched SHRs, discuss their connectivities, and node-disjoint path construction for SHRs. These will play a fundamental role in our node-to-node disjoint-path construction and communication schemes of Section 3.

2 Preliminaries

2.1 HR Construction

Let us begin with a formal definition of a hyper-ring.

Definition 1. A graph $G = (V, E)$ is called a *hyper-ring* with N nodes (N-HR for short) if $V = \{0, ..., N-1\}$ and $E = \{\{u, v\} | [v - u]_N$ is a power of 2 $\}$, where $[m]_r$ is m modulo r.

Definition 2. The *positional distance* between node s and node t of N-HR is defined to be $min\{[t - s]_N, [s - t]_N\}$.

The number of edges between nodes in an HR is roughly twice that of a hypercube (HC) with the same number of nodes, but the proposed organization possesses a number of advantages over the HC. In particular, for any N we can construct an N node HR, whereas a hypercube must contain exactly 2^k nodes for some k. An example of an 11-HR is shown in Figure 1. Note that what appears to be a connection of positional distance 3 is really a (counterclockwise) connection of positional distance 8. This class of edges, connecting nodes $2^{\lfloor \log N \rfloor}$ away from each other, will be addressed in more detail in the path construction procedures of this section.

The HCs have a natural and an elegant recursive structure that does not seem to extend to HRs in an obvious way. In particular, the number of edges in HRs does not grow monotonically with the number of nodes [2]. It turns out, however, that HRs do possess an interesting recursive construction that encompasses HCs

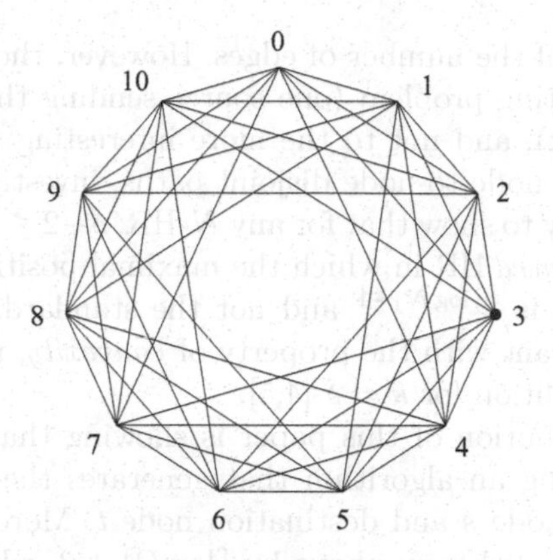

Fig. 1. An example of an 11-HR.

in a sense as a special case. examine First, let us observe a simple (nonrecursive) construction of N-HRs. The following procedure is straightforward and needs no further explanation.

for $i := 0$ **to** $N - 1$ **do**
  for $k := 0$ **to** $\lfloor \log N \rfloor$ **do**
    if there is no edge between i and $[i + 2^k]_N$
      then connect node i to nodes labeled $[i \pm 2^k]_N$

Observe that in HRs, the two edges usually connecting nodes that are either $2^{\lfloor \log N \rfloor}$ or $2^{\lfloor \log N \rfloor - 1}$ positional distance away from each other either do not exist as such (examine the $(2^i + 2^j)$-HR family), may merge into one (2^k-HR family), or crisscross (all of the remaining HRs). Again, see edges (0,8) and (0,$[-8]_{11} = 3$) in Figure 1.

Assume that an N-HR has already been constructed. Suppose we wish to construct a $2N$-HR from it. The following procedure [2], takes as input an N-HR_0 and returns a $2N$-HR.

make a duplicate copy of N-HR_0, call it N-HR_1
for $i := 0$ **to** $N - 1$ **do**
  relabel node i in N-HR_0 to $2i$
  relabel node i in N-HR_1 to $2i + 1$
for $i := 0$ **to** $2N - 1$ **do**
  connect node i and node $[i + 1]_{2N}$

Figure 2 shows the connections for one node of the HR constructed using the doubling construction procedure. It shows that if δ is the degree of an N-HR, $N > 1$, then the degree of $2N$-HR is $\delta + 2$. Previously, it was shown that the doubling of an HR increases its connectivity from κ to $\kappa + 2$ [3]. The essence of the proof is captured by the two paths containing nodes α and β in Figure 2, however, it will not be discussed here.

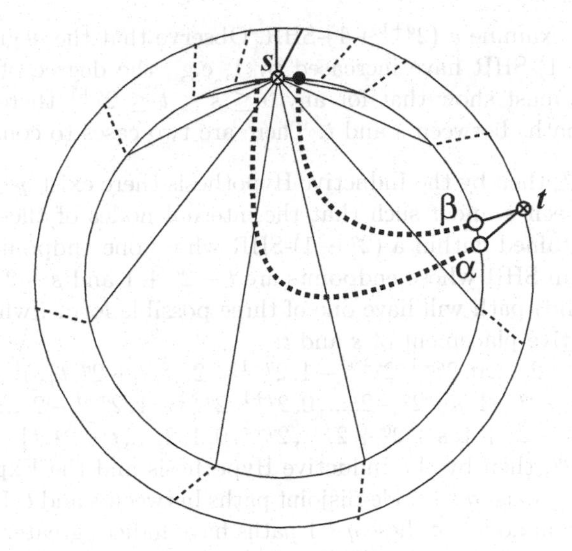

Fig. 2. An HR constructed using the doubling procedure and the two additional node-disjoint paths via α and β.

2.2 Semi Hyper-rings

Before we can proceed with the node-disjoint path construction algorithm for HRs, let us introduce a related structure and examine its connectivity and path algorithms.

Definition 3. A *Semi Hyper-ring* is a graph corresponding to any segment of P consecutive integers, (P-SHR for short) in which any two nodes (integers) are connected iff they are a power of 2 positional distance away from each other.

Lemma 1. The node-connectivity of a P-SHR is equal to $\lceil \log P \rceil$, which is also the degree of its first and last nodes.

Proof. It is obvious that the degree of the first and last nodes of any P-SHR is $\lceil \log P \rceil$. Observe that for all $r > 0$, the SHRs of size $2^r + 1$ through 2^{r+1} will have the same connectivity. This is due to the Expansion Lemma (see, e.g., [12]) which states that if G is a k-connected graph, and G' is obtained from G by adding a new vertex y with at least k neighbors in G, then G' is k-connected. This allows us to focus on SHRs of size $P = 2^q + 1$, e.g., the integers $0, 1, ..., 2^q$.

We show, by induction on q, that P-SHRs have connectivity of $\lceil \log P \rceil$. For $q = 1$, the 3-SHR is obviously 2-connected.

Inductive Hypothesis: Assume that any $(2^q + 1)$-SHR is $q + 1$-connected, i.e., between any two nodes s and t, there exist $q+1$ node-disjoint paths. Equivalently, if S_1 and S_2 are two disjoint sets of $q+1$ nodes each, all located within the same $(2^q + 1)$-SHR, then there exist $q + 1$ node-disjoint paths within it that pairwise connect the nodes of both sets.

Let us now examine a $(2^{q+1}+1)$-SHR. Observe that the degrees of all nodes in the $(2^{q+1}+1)$-SHR have increased by 1, e.g., the degree of node 0 is now $q+2$, etc. We must show that for any $0 \le s < t \le 2^{q+1}$ there are now $q+2$ node-disjoint paths between s and t. There are two cases to consider:

1. If $t-s \le 2^q$, then by the Inductive Hypothesis there exist $q+1$ node-disjoint paths between s and t such that the internal nodes of these paths will be totally contained within a (2^q+1)-SHR whose one endpoint is either node s, or t, or an SHR whose endpoints are $t-2^q+1$ and $s+2^q-1$.
   The $(q+2\text{nd})$ path will have one of three possible forms, which will depend on the relative placement of s and t:
   $\{s, s-1, s-2, ..., 0, 2^{q+1}, 2^{q+1}-1, 2^{q+1}-2, ..., t+2^q, t\}$, if $s, t \le 2^q$;
   $\{s, s-2^q, s-2^q-1, s-2^q-2, ..., 0, 2^{q+1}, 2^{q+1}-1, 2^{q+1}-2, ..., t\}$, if $s, t \ge 2^q$;
   $\{s, s+2^q, s+2^q+1, s+2^q+2, ..., 2^{q+1}, 0, 1, 2, ..., t-2^q, t\}$, if $s < 2^q < t$.
2. If $t-s > 2^q$, then by the Inductive Hypothesis and the Expansion Lemma there already exist $q+1$ node-disjoint paths between s and t. Furthermore, all of the internal nodes of these $q+1$ paths have indices greater than s and less than t. The $(q+2\text{nd})$ path will have the form: $\{s, s-1, s-2, ..., 0, 2^{q+1}, 2^{q+1}-1, 2^{q+1}-2, ..., t\}$. □

2.3 Path Construction for SHRs

For ease of presentation and clarity, let us visualize the problem of SHR/HR node-disjoint path construction as a simple (one player) blue/green pebble game, which has only three rules:

1. The only pebble movements (hops) allowed are via the existing edges.
2. A pebble may not be placed on a node previously visited by any other pebble.
3. If node i is occupied by a pebble, the player may *capture* it by hopping a different color pebble to i and removing both pebbles from the game.

The object of the game is to maximize the number of captured pebbles, given some initial pebble configuration. It should be clear that the traces of the pebble hops uniquely define a set of node-disjoint paths and that a capture indicates a completion of a path between the nodes where the pebbles were originally placed.

Let us now present procedure *CompressSHR* that given t as input generates the $\lceil \log t \rceil$ node-disjoint paths between 0 and t within the $(t+1)$-SHR. We will describe the steps of this procedure in terms of pebble movements.

procedure *CompressSHR(t)*
Initialize: place $\lceil \log t \rceil$ blue and $\lceil \log t \rceil$ green pebbles in locations $\{2^0, 2^1, ..., 2^{\lfloor \log t \rfloor}\}$ and $\{t-2^0, t-2^1, ..., t-2^{\lfloor \log t \rfloor}\}$, respectively.

0. Check for any green and blue pebbles placed on the same node, (including blue on t and green on 0) and if found, capture them.

while any pebbles are left **do**
1. Identify the green pebble g that is closest and to the right of the rightmost blue pebble b.
2. For all of the remaining green pebbles to the right of b use an appropriate *single-hop* to move them to the left of b.
   If any land on another blue pebble, consider it a capture.
3. Construct the appropriate sequence of hops for g to capture b.

It is clear that the green pebble movements represent the node t to node s paths and that they are node-disjoint (see Figure 3). Observe that at each loop iteration at least one capture is made, limiting the number of loops to $\lceil \log t \rceil$, making the maximal number of hops made by any pebble to be no more than $\lceil \log t \rceil$. Furthermore, if for some reason (e.g., an incorrect initial placement of one of the blue pebbles) step 1 fails, then let g be the green pebble closest to the rightmost blue pebble and continue with step 3, for that particular while-loop iteration. Later, we will see how that may become necessary during the HR path construction.

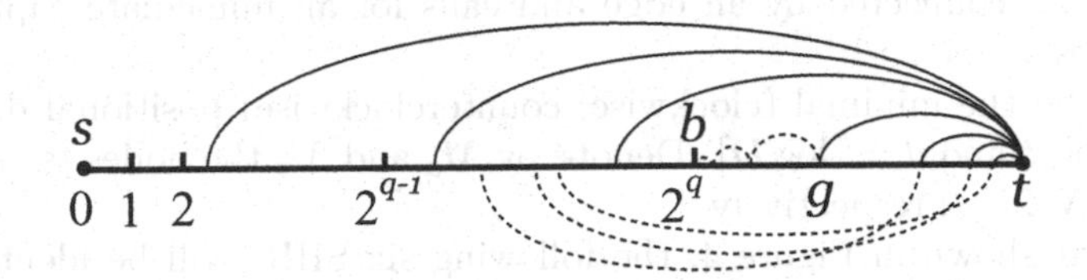

Fig. 3. Compression of an SHR and node-disjoint path construction.

Definition 4. A P-SHR is said to be *enriched* if it is extended by allowing additional connections between all pairs of nodes that are of k positional distance away from each other, for a fixed k, $3 \leq k < P$, which is not a power of 2.

Hence, the degree of the first and last nodes in any enriched P-SHR is equal to $\lceil \log P \rceil + 1$. Observe that procedure *CompressSHR*(t) can be naturally extended to include enriched SHRs by initially placing additional blue and green pebbles in locations k and $t - k$, respectively.

Observation 1 *Procedure CompressSHR will capture all of the pebbles in any (enriched) P-SHR.*

Let $s < t$ be two nodes in an N-HR that are $D \leq \lfloor N/2 \rfloor$ positions away from each other. We may construct a (possibly enriched) D-SHR$_{(s,t)}$ that spans s and t by removing from the N-HR all of the nodes (and their edges) outside the line segment (s, t), (i.e., outside $\{s, s+1, ..., t-1, t\}$).

Let us focus on the family of N-HRs that possess the crisscrossing edges connecting nodes of $2^{\lfloor \log N \rfloor}$ positional distance.

Observation 2 *If $D \geq N - 2^{\lfloor \log N \rfloor}$ then the degree of nodes s and t as well as the number of node-disjoint paths between s and t in the D-SHR$_{(s,t)}$ is equal to* $\lceil \log D \rceil + 1$, *otherwise it is* $\lceil \log D \rceil$.

The next result will allow us to obtain the disjoint-path length bounds for not only SHRs, but eventually, HRs.

Lemma 2. *For any (enriched) D-SHR, the length of the node-disjoint paths constructed by procedure CompressSHR is bounded by* $\lceil \log D \rceil + 2$.

3 Path Construction for HRs

We now present our algorithm for the construction of node-disjoint paths in HRs. First, given N, s, and t we will make the initial pebble placement by putting the blue (and green) pebbles on nodes which are exactly $\pm 2^i, i = 0, ..., \lfloor \log N \rfloor$, positions away from s (and t), respectively (making any immediate capture(s), if any blue/green pair of pebbles has been found on the same node). Observe that a placement of a blue pebble on t and a green one on s indicates that s and t are directly connected by an edge and calls for an immediate capture of the two pebbles.

Let D be the minimal (clockwise, counterclockwise) positional distance between s and t and $d = \lceil \log D \rceil$. Denote by M_s and M_t the nodes $[s + \lfloor N/2 \rfloor]_N$ and $[t - \lfloor N/2 \rfloor]_N$, respectively.

Next, as shown in Figure 4, the following six SHRs will be identified. Note that, in certain extreme cases, (e.g., $2K$-HRs and $D = K$), SHR$_B$, SHR$_{B'}$, SHR$_C$, and SHR$_{C'}$ may contain no nodes, and s and t may serve as end nodes for both SHR$_A$ and SHR$_E$.

1. SHR$_A$ is the segment (s, t) of size $2 \leq D \leq \lfloor N/2 \rfloor$.
2. SHR$_B$ is bounded by $s - 1$ and the closer of two nodes $(s - 2^d, M_t + 1)$. SHR$_{B'}$ is bounded by $t + 1$ and the closer of two nodes $(t + 2^d, M_s - 1)$.
3. SHR$_C$ is bounded by closer of two nodes $(s - 2^d - 1, M_t + 1)$, and $M_t + 1$. SHR$_{C'}$ is bounded by closer of two nodes $(t + 2^d + 1, M_s - 1)$, and $M_s - 1$.
4. SHR$_E$ is bounded by M_t and M_s.

We can easily capture all of the pebbles in SHR$_A$ by *CompressSHR*.

Next, we will capture the pebbles from the nodes of SHR$_B$ and SHR$_{B'}$. We single-hop the blue pebbles from SHR$_B$ to SHR$_{B'}$ by taking their 2^{d+1} (clockwise) connections. While it is clear that all of the blue pebbles would wind up in SHR$_{B'}$, there are two potential problems.

1. One blue pebble may already be in SHR$_{B'}$ via the 2^d-connection. To address this problem, one of the blue pebbles should be left behind in SHR$_B$, the question of which one is answered below.
2. One blue pebble may already be in SHR$_{B'}$ via the $2^{\lfloor \log N \rfloor}$-connection. If so, a second blue pebble would have to be left behind in SHR$_B$.

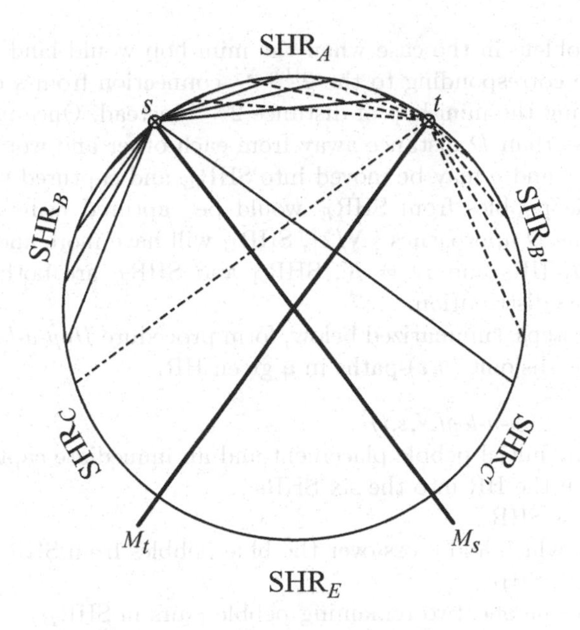

Fig. 4. Partitioning of an N-HR.

A quick check would determine if any of the SHR$_B$ blue pebbles could land on the above-mentioned (blue) problem pebbles in SHR$_{B'}$ (and those would be the ones that would stay behind in SHR$_B$. If no such pebble(s) exist in SHR$_B$, then we pick them randomly. In any case, at most $d+1$ blue (and the same number of green) pebbles would be present in SHR$_{B'}$ after this 2^{d+1} cross-over. Observe that the one/two blue pebble(s) left behind will now be used within SHR$_B$ to join the one/two green pebbles that (by symmetry) would have been in SHR$_B$ from the beginning. At this time *CompressSHR* would be used on SHR$_{B'}$.

We now turn our attention to SHR$_C$ and SHR$_{C'}$. Observe that if $D \neq 2^d$, the pebbles in SHR$_C$ are in locations (blue) $2^{d+1}, 2^{d+2}, ...$ and (green) $2^{d+1} - D, 2^{d+2} - D, ...$, positions away from s. The blue/green pebble pairs in SHR$_C$ from locations $(2^{d+j}, 2^{d+j} - D)$ away from s will be appropriately moved and captured. The positional distance between the blue and green pebbles within these pairs is bounded by D. The possible presence of the pebbles (via the $2^{\lfloor \log N \rfloor}$-connection) is not a significant problem here, since we are joining at most two pairs of pebbles within at least a 2D-sized subsegments of SHR$_C$. The blue/green pebble pairs in SHR$_{C'}$ would be captured in a symmetrical fashion.

If $D = 2^d$, then we would proceed as above, except there could be two *lonely* pebbles left: a blue pebble, b, located 2^{d+1} positions away from s in SHR$_C$ and a green pebble, g, located 2^{d+1} positions away from t in SHR$_{C'}$. These will be paired-up by allowing b to cross-over into SHR$_{C'}$ via a 2^{d+2} connection. Of course a direct hop from b's original position is not possible since it would collide with another blue pebble's location (the 2^{d+1} edge of s). Therefore, b could take a mini-hop of distance 2^{d-1} toward s and then take the cross-over edge 2^{d+2}. The (potential) problem in the case when the mini-hop would land b on a location of blue pebble corresponding to the $2^{\lfloor \log N \rfloor}$ connection from s can be resolved easily by making the mini-hop of distance 2^{d-2} instead. Once in $\mathrm{SHR}_{C'}$, b and g would be less than D distance away from each other and would be captured. Alternatively, b and g may be moved into SHR_E and captured there.

Finally, the pebbles from SHR_E would be captured using *CompressSHR*. Observe that as D approaches $\lfloor N/2 \rfloor$, SHR_E will have more and more pebbles. In fact, for $2K$-HRs and $D = K$, SHR_A and SHR_E are both K-SHRs with identical pebble distribution.

The above steps, summarized below, form procedure *Hop-a-log* that will construct the node-disjoint (s,t)-paths in a given HR.

procedure *Hop-a-log(N,s,t)*
Make the initial pebble placement and an immediate capture check
Partition the HR into the six SHRs
Compress SHR_A
Identify which and cross-over the blue pebbles from SHR_B to $\mathrm{SHR}_{B'}$
Compress $\mathrm{SHR}_{B'}$
Capture the one/two remaining pebble pairs in SHR_B
Capture the $(2^{d+j}, 2^{d+j} - D)$ pairs in SHR_C and $\mathrm{SHR}_{C'}$
If they exist, capture the two *lonely* pebbles using the 2^{d+2} hop
Compress SHR_E.

Lemma 3. *For any HR the* Hop-a-log *procedure will generate $\kappa = \delta$ node-disjoint paths between any source node s and destination node t. Moreover, the length of each such path is bounded by $\lceil \log D \rceil + 3$.*

Proof. Follows from proof of Lemma 1, Observation 2, and Lemma 2, and the fact that the sizes of all of the SHRs on which *CompressSHR* was used, were bounded by D. □

Theorem 1. *A node-to-node communication scheme using procedure* Hop-a-log *will send a message from s to t within $\lceil \log D \rceil + 3$ rounds, even in the presence of $\kappa - 1$ node failures.*

4 Concluding Remarks

Besides facilitating a fast and reliable node-to-node communication, HRs provide an excellent means for network broadcasting. Assuming a multi-port broadcasting model in which a node can send a message to all of its immediate neighbors in one round, broadcasting in HRs may be carried out in only $\lceil \log N \rceil$ rounds to send information from any source to all destination nodes, if no nodes have failed. The bound on the path lengths in Lemma 3, however, guarantees that an HR broadcasting scheme using this procedure will send a message from any source to all destinations within $\lceil \log N \rceil + 2$ rounds, even in the presence of $\kappa - 1$ node failures.

Hyper-rings appear to be the first log-sparse connection architecture that allows for dependable communication in time of $\lceil \log D \rceil + 3$, where D is the positional distance between the source and destination nodes. This *speed of locality* is especially important for fast and dependable none-to-node communication protocols and distributed computing. Examples of such applications would include distributed pattern recognition algorithms, battlefield communication management, monitoring systems, etc. HRs are certainly worthy of further studies.

References

1. N. Alon, A. Barak, and U. Mauber, On disseminating information reliably without broadcasting, *7th Int. Conference on Distributed Computer Systems* (1987) 74-81.
2. T. Altman, Y. Igarashi, and K. Obokata, Hyper-ring connection machines, *Parallel Computing* 21 (1995) 1327-1338.
3. T. Altman, Reliable communication schemes for hyper-rings, *28th Int. Southeastern Conference on Combinatorics, Graph Theory, and Computing* (1997).
4. F. Boesch and A. Felzer, A general class of invulnerable graphs, *Networks* 2 (1972) 261-283.
5. F. Boesch and R. Tindell, Circulants and their connectivities, *J. of Graph Theory* 8 (1984) 487-499.
6. E. van Doorn, Connectivity of circulant digraphs, *J. of Graph Theory* 10 (1986) 9-14.
7. Y. Han, R. Finkel, An optimal scheme for disseminating information, *The 22nd Int. Conference on Parallel Processing* (1988) 198-203.
8. Y. Han, Y. Igarashi, K. Kanai, and K. Miura, Broadcasting in faulty binary jumping networks, *J. of Parallel and Distributed Computing* 23 (1994) 462-467.
9. W. Knödel, New gossips and telephones, *Discrete Mathematics* 13 (1975) 95.
10. A. Pelc, Fault-tolerant broadcasting and gossiping in communication networks, *Networks* 28 (1996) 143-156.
11. M. O. Rabin, Efficient dispersal of information for security, load balancing, and fault tolerance, *J. of ACM* 36 (1989) 335-348.
12. D. West, *Introduction to Graph Theory,* Prentice Hall (1996).

The On-Line Heilbronn's Triangle Problem in Three and Four Dimensions

Gill Barequet

The Technion–Israel Institute of Technology, Haifa 32000, Israel
barequet@cs.technion.ac.il
http://www.cs.technion.ac.il/~barequet

Abstract. In this paper we show lower bounds for the on-line version of Heilbronn's triangle problem in three and four dimensions. Specifically, we provide incremental constructions for positioning n points in the 3-dimensional (resp., 4-dimensional) unit cube, for which every tetrahedron (resp., pentahedron) defined by four (resp., five) of these points has volume $\Omega(\frac{1}{n^{3.333...}})$ (resp., $\Omega(\frac{1}{n^{5.292...}})$).

1 Introduction

Heilbronn posed his famous (off-line) triangle problem [5] about 50 years ago:

> Given n points in the unit square, what is $\mathcal{H}_2(n)$, the maximum possible area of the *smallest* triangle defined by some three of these points?

There is a large gap between the best currently-known lower and upper bounds for $\mathcal{H}_2(n)$, $\Omega(\log n/n^2)$ [3] and $O(1/n^{8/7-\varepsilon})$ (for any $\varepsilon > 0$) [2]. A comprehensive survey of the history of this problem (excluding the results of Komlós et al.) is given by Roth in [6].

In [1] we presented a generalization of the triangle problem to d dimensions:

> Given n points in the d-dimensional unit cube, what is $\mathcal{H}_d(n)$, the maximum possible volume of the *smallest* simplex defined by some $d+1$ of these points?

It was shown in [1] that $\mathcal{H}_d(n) = \Omega(\frac{1}{n^d})$. This lower bound was achieved by a specific example and by a probabilistic argument. Lefmann [4] slightly improved this bound, showing by using uncrowded hypergraphs that $\mathcal{H}_d(n) = \Omega(\frac{\log n}{n^d})$. In particular, $\mathcal{H}_3(n) = \Omega(\frac{\log n}{n^3})$ and $\mathcal{H}_4(n) = \Omega(\frac{\log n}{n^4})$.

The on-line variant of the triangle problem is harder than the off-line variant because the value of n is not specified in advance. In other words, the points are positioned one after the other in a d-dimensional unit cube, while n is incremented by one after every point-positioning step. The procedure can be stopped at any time, and the already-positioned points must have the property that every subset of $d+1$ points define a polytope whose volume is at least some quantity, where the goal is to maximize it. It was shown in [1] that this quantity is $\Omega(\frac{1}{n^4})$ in three dimensions.

O.H. Ibarra and L. Zhang (Eds.): COCOON 2002, LNCS 2387, pp. 360–369, 2002.

In this paper we use nested packing arguments of balls to provide a sharper lower bound in three dimensions, and use a much more complex version of this method to give a lower bound in four dimensions. Specifically, we provide incremental procedures for positioning n points (one by one) in a 3-dimensional (resp., 4-dimensional) unit cube so that all the tetrahedra (resp., pentahedra) defined by quadruples (resp., 5-tuples) of the points have volume $\Omega(\frac{1}{n^{10/3}}) = \Omega(\frac{1}{n^{3.333...}})$ (resp., $\Omega(\frac{1}{n^{127/24}}) = \Omega(\frac{1}{n^{5.292...}})$). Obviously, these results do not match the known lower bounds for the off-line variant of the problem.

In a sense our method is a nontrivial generalization of Schmidt's incremental construction [7] that shows that $\mathcal{H}_2(n) = \Omega(\frac{1}{n^2})$. The planar construction of Schmidt does not use, however, nested packing arguments as do we in the current work.

2 The Construction in Three Dimensions

We first develop the construction in three dimensions.

2.1 Notation and Plan

We use the following notation: Let p_1, p_2, p_3, p_4 be any four points in $\Re^3$. Then, $|p_1, p_2|$ denotes the distance between two points p_1, p_2; $|p_1, p_2, p_3|$ denotes the area of the triangle $p_1p_2p_3$; and $|p_1, p_2, p_3, p_4|$ denotes the volume of the tetrahedron $p_1p_2p_3p_4$. We denote by C^3 the unit cube. The line defined by the pair of points p_i, p_j is denoted by ℓ_{ij}.

We want to construct a set S of n points in C^3 such that

(i) $|p_ip_j| \geq \frac{a}{n^{1/3}}$, for any pair of distinct points p_i, $p_j \in S$ and for some constant $a > 0$.
(ii) $|p_ip_jp_k| \geq \frac{b}{n}$, for any triple of distinct points p_i, p_j, $p_k \in S$ and for some constant $b > 0$.
(iii) $|p_ip_jp_kp_l| \geq \frac{c}{n^{10/3}}$, for any quadruple of distinct points p_i, p_j, p_k, $p_l \in S$ and for some constant $c > 0$.

The goal is to construct S incrementally. That is, assume that we have already constructed a subset S_v of v points, for $v < n$, that satisfies conditions (i)–(iii) above. We want to show that there exists a new point $p \in C^3$ that satisfies

(i') $|pp_i| \geq \frac{a}{n^{1/3}}$, for each point $p_i \in S_v$.
(ii') $|pp_ip_j| \geq \frac{b}{n}$, for any pair of distinct points p_i, $p_j \in S_v$.
(iii') $|pp_ip_jp_k| \geq \frac{c}{n^{10/3}}$, for any triple of distinct points p_i, p_j, $p_k \in S_v$.

We will show this by summing up the volumes of the 'forbidden' portions of C^3 where one of the inequalities (i')–(iii') is violated, and by showing that the sum of these volumes is less than 1, implying the existence of the desired point p, which we then add to S_v to form S_{v+1}, and continue in this manner until the entire set S is constructed.

2.2 The Construction

The forbidden regions where one of the inequalities (i') is violated are v balls of radius $a/n^{1/3}$. Their overall volume (within C^3) is at most $v \cdot \frac{4\pi a^3}{3n} = O(\frac{a^3 v}{n}) = O(a^3)$.

The forbidden regions where one of the inequalities (ii') is violated are $\binom{v}{2}$ cylinders Q_{ij}, for $1 \le i < j \le v$, where the cylinder Q_{ij} has radius $2b/(n|p_i p_j|)$. Their overall volume (within C^3) is at most $\sum_{1 \le i < j \le v} \frac{4\sqrt{3}\pi b^2}{n^2 |p_i p_j|^2}$. To bound this sum, we fix p_i and sum over p_j. We use a spherical packing argument that exploits the fact that S_v satisfies (i). Specifically, we have

$$\sum_{j \ne i} \frac{1}{|p_i p_j|^2} \le \sum_{t=1}^{O(n^{1/3}/a)} \frac{M_t n^{2/3}}{a^2 t^2},$$

where M_t is the number of points of S_v that lie in the spherical shell centered at p_i with inner radius $at/n^{1/3}$ and outer radius $a(t+1)/n^{1/3}$. There are $O(n^{1/3}/a)$ such spherical shells. (Further shells would be outside C^3). Because of (i), the number of such points is $M_t = O(t^2)$. This follows by an argument of packing spheres of volume a^3/n within a shell whose volume is $a^3 t^2/n$. Hence the above sum is $O(n/a^3)$. Repeating this for each $p_i \in S_v$, we see that the total volume of the forbidden cylinders is $O(b^2 v/(a^3 n)) = O(b^2/a^3)$.

Finally, the forbidden regions where one of the inequalities (iii') is violated are $\binom{v}{3}$ slabs σ_{ijk}, for $1 \le i < j < k \le v$, where the slab σ_{ijk} has width $12c/(n^{10/3}|p_i p_j p_k|)$. Their overall volume (within C^3) is at most

$$\sum_{1 \le i < j < k \le v} \frac{36c}{n^{10/3}|p_i p_j p_k|}. \tag{1}$$

To bound this sum, we fix p_i, p_j and sum over p_k. We use a cylindrical packing argument that exploits the fact that S_v satisfies (i) and (ii). Specifically, we have

$$\sum_{k \ne i,j} \frac{1}{|p_i p_j p_k|} \le \frac{N_0 n}{b} + \sum_{t=1}^{O(n^{1/3}/a)} \frac{2 N_t n^{1/3}}{at|p_i p_j|}, \tag{2}$$

where N_0 is the number of points of S_v that lie in the inner-most cylinder of the packing, and N_t is the number of points of S_v that lie in the cylindrical shell centered at ℓ_{ij} with inner radius $at/n^{1/3}$ and outer radius $a(t+1)/n^{1/3}$. There are $O(n^{1/3}/a)$ such cylindrical shells. Obviously $N_0 = O(n^{1/3}/a)$, since the volume of the inner-most cylinder is $a^2/n^{2/3}$ and because of (i). Also, because of (i), $N_t = O(tn^{1/3}/a)$. This follows by an argument of packing spheres of volume a^3/n within a shell whose volume is $a^2 t/n^{2/3}$. Hence the above sum is $O(n^{4/3}/(ab) + n/(a^3|p_i p_j|))$. We now substitute this bound in (1) and sum over all p_i, p_j to obtain the upper bound

$$O\left(\sum_{1 \le i < j \le v} c\left(\frac{1}{abn^2} + \frac{1}{a^3 n^{7/3}|p_i p_j|}\right) \right).$$

The sum of the first summand is $O(cv^2/(abn^2)) = O(c/(ab))$. We bound the sum of the second summand as above, fixing p_i and using a spherical packing argument within spherical shells centered at p_i. Arguing as above, we obtain

$$\sum_{j\neq i} \frac{c}{a^3 n^{7/3}|p_i p_j|} \leq \frac{c}{a^3 n^{7/3}} \sum_{t=1}^{O(n^{1/3}/a)} \frac{M_t n^{1/3}}{a^4 t} = \frac{c}{n^2} \sum_{t=1}^{O(n^{1/3}/a)} \frac{O(t^2)}{at^4} = O(\frac{c}{a^6 n^{4/3}}).$$

And summing this over all p_i, we obtain a final bound of $O(cv/(a^6 n^{4/3})) = o(1)$.

In summary, the overall total volume of the forbidden regions is $O(1)$, which can be made smaller than 1 by an appropriate choice of the constants a, b, c. More precisely, the forbidden volume is at most $C_1 a^3 + C_2 b^2/a^3 + C_3 c/(ab)$, where C_1, C_2, C_3 are real constants that do not depend on a, b, c. It is easy to make this term at most 1 by first choosing any value for a such that $0 < a \leq \sqrt[3]{1/(3C_1)}$, then choosing any value for b such that $0 < b \leq \sqrt{a^3/(3C_2)}$, and finally choosing any value for c such that $0 < c \leq ab/(3C_3)$. This completes the construction.

3 The Construction in Four Dimensions

We are now ready to use the same ideas for the more-involved four-dimensional construction. For clarity of exposition we omit the tuning constants from the $O(\cdot)$ and $\Omega(\cdot)$ notations.

3.1 Notation and Plan

Our current notation and plan are similar to those of the 3-dimensional construction. Let p_1, p_2, p_3, p_4, p_5 be any five points in $\Re^4$. As in the previous section, $|p_1, p_2|$ denotes the distance between two points p_1, p_2; $|p_1, p_2, p_3|$ denotes the area of the triangle $p_1 p_2 p_3$; $|p_1, p_2, p_3, p_4|$ denotes the (3-dimensional) volume of the tetrahedron $p_1 p_2 p_3 p_4$; and $|p_1, p_2, p_3, p_4, p_5|$ denotes the (4-dimensional) volume of the pentahedron $p_1 p_2 p_3 p_4 p_5$. We denote by C^4 the 4-dimensional unit cube. Again, the line defined by the pair of points p_i, p_j is denoted by ℓ_{ij}.

We want to construct a set S of n points in C^4 such that

(i) $|p_i p_j| \geq \frac{a}{n^{1/4}}$, for any pair of distinct points p_i, $p_j \in S$ and for some constant $a > 0$.
(ii) $|p_i p_j p_k| \geq \frac{b}{n^{2/3}}$, for any triple of distinct points p_i, p_j, $p_k \in S$ and for some constant $b > 0$.
(iii) $|p_i p_j p_k p_l| \geq \frac{c}{n^{43/24}}$, for any quadruple of distinct points p_i, p_j, p_k, $p_l \in S$ and for some constant $c > 0$.
(iv) $|p_i p_j p_k p_l p_m| \geq \frac{d}{n^{127/24}}$, for any 5-tuple of distinct points p_i, p_j, p_k, p_l, $p_m \in S$ and for some constant $d > 0$.

The goal is again to construct S incrementally. That is, assume that we have already constructed a subset S_v of v points, for $v < n$, that satisfies conditions (i)–(iv) above. We want to show that there exists a new point $p \in C^4$ that satisfies

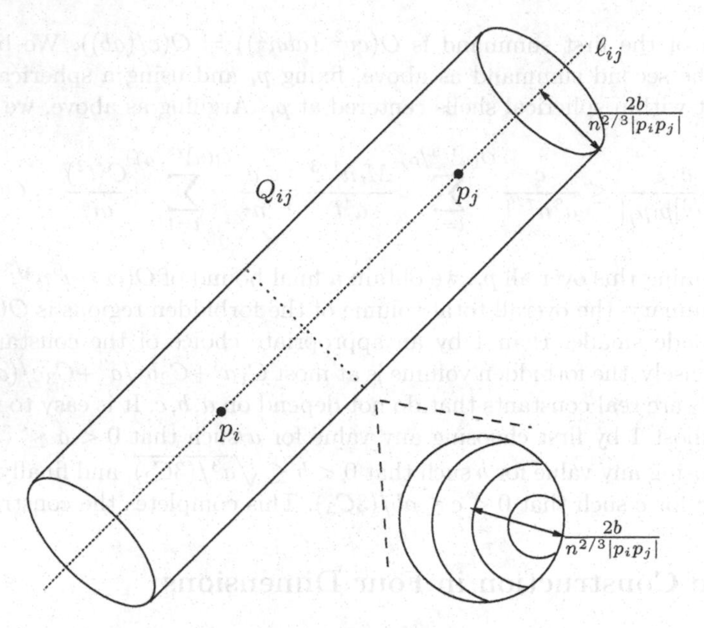

Fig. 1. A cylinder in $\Re^4$

(i') $|pp_i| \geq \frac{a}{n^{1/4}}$, for each point $p_i \in S_v$.
(ii') $|pp_ip_j| \geq \frac{b}{n^{2/3}}$, for any pair of distinct points p_i, $p_j \in S_v$.
(iii') $|pp_ip_jp_k| \geq \frac{c}{n^{43/24}}$, for any triple of distinct points p_i, p_j, $p_k \in S_v$.
(iv') $|pp_ip_jp_kp_l| \geq \frac{d}{n^{127/24}}$, for any quadruple of distinct points p_i, p_j, p_k, $p_l \in S_v$.

We will show this by summing up the volumes of the 'forbidden' portions of C^4 where one of the inequalities (i')–(iv') is violated, and by showing that the sum of these volumes is less than 1, implying the existence of the desired point p, which we then add to S_v to form S_{v+1}, and continue in this manner until the entire set S is constructed.

3.2 Forbidden Balls

The forbidden regions where one of the inequalities (i') is violated are v 4-dimensional balls of radius $a/n^{1/4}$. Their overall volume (within C^4) is at most $v \cdot O(\frac{1}{n}) = O(\frac{v}{n}) = O(1)$.

3.3 Forbidden Cylinders

The forbidden regions where one of the inequalities (ii') is violated are $\binom{v}{2}$ 4-dimensional "cylinders" Q_{ij}, for $1 \leq i < j \leq v$. The cylinder Q_{ij} is centered at ℓ_{ij}, its length is at most $\sqrt{4} = 2$, and its cross-section perpendicular to ℓ_{ij} is a 3-dimensional sphere of radius $2b/(n^{2/3}|p_ip_j|)$ (see Figure 1). The overall volume of the cylinders (within C^4) is at most

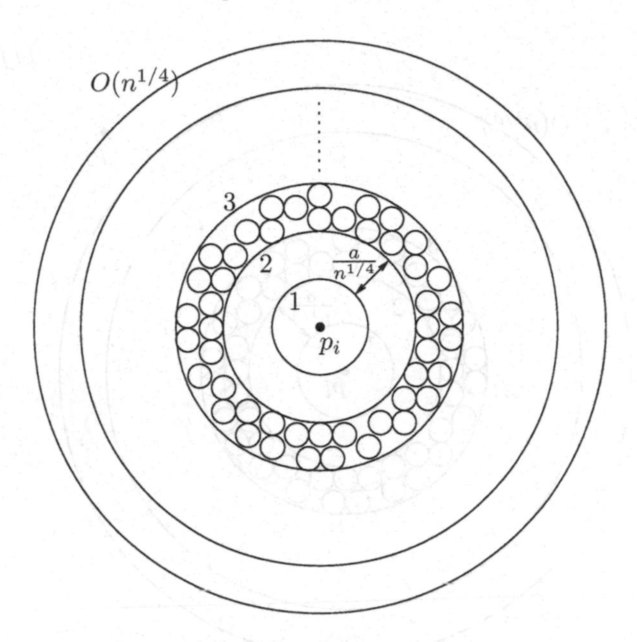

Fig. 2. A spherical packing of balls in $\Re^4$

$$\sum_{1 \le i < j \le v} \frac{64\pi b^3}{3n^2|p_ip_j|^3}. \tag{3}$$

To bound this sum, we fix p_i and sum over p_j. We use a 4-dimensional spherical packing argument that exploits the fact that S_v satisfies (i). Specifically, we have

$$\sum_{j \ne i} \frac{1}{|p_ip_j|^3} \le \sum_{t=1}^{O(n^{1/4})} \frac{M_t n^{3/4}}{a^3 t^3}, \tag{4}$$

where M_t is the number of points of S_v that lie in the 4-dimensional spherical shell centered at p_i with inner radius $at/n^{1/4}$ and outer radius $a(t+1)/n^{1/4}$; see Figure 2. There are $O(n^{1/4})$ such spherical shells. Because of (i), the number of such points is $M_t = O(t^3)$. This follows by an argument of packing spheres of volume $\Omega(1/n)$ within a shell whose volume is $O(t^3/n)$.

Hence the sum in Equation (4) is $O(n)$. Summing this over all p_i, we obtain a final bound of $O(vn)$. Substituting this in Equation (3), we see that the total volume of the forbidden cylinders is $O(v/n) = O(1)$.

3.4 Forbidden Prisms

The forbidden regions where one of the inequalities (iii') is violated are $\binom{v}{3}$ 4-dimensional "prisms" ϕ_{ijk}, for $1 \le i < j < k \le v$. The base area of ϕ_{ijk} is at most 4, and its "height" (in the 3rd and 4th dimensions) is a 2-dimensional circle

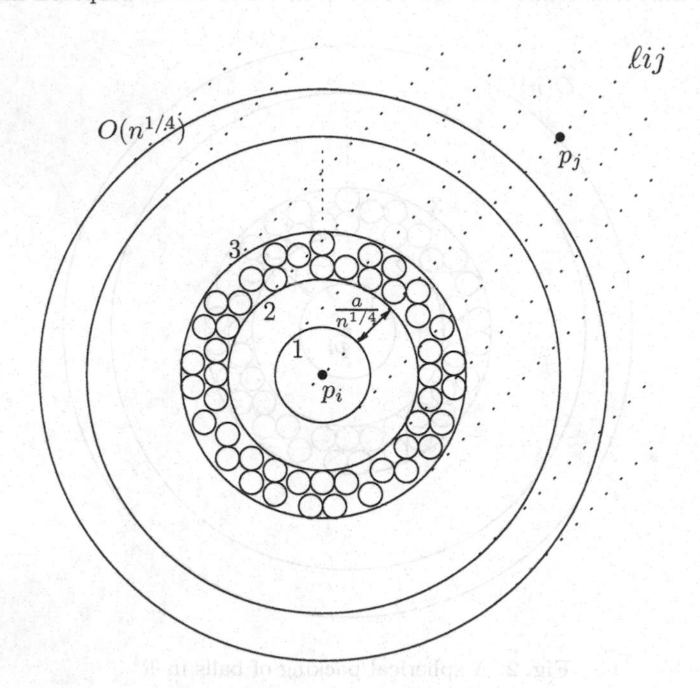

Fig. 3. A 4-D cylindrical packing (an extruded 3-D spherical packing) of balls in $\Re^4$

of radius $O(1/(n^{43/24}|p_ip_jp_k|))$. The overall volume of the prisms (within C^4) is at most

$$\sum_{1 \le i < j < k \le v} O(\frac{1}{n^{43/12}|p_ip_jp_k|^2}). \tag{5}$$

To bound this sum, we fix p_i, p_j and sum over p_k. We use a 4-dimensional cylindrical packing argument that exploits the fact that S_v satisfies (i) and (ii). The cylinders are centered at ℓ_{ij}. (See Figure 3; the line ℓ_{ij} is emanating from p_i toward p_j through the 4th dimension.) Specifically, we have

$$\sum_{k \ne i,j} \frac{1}{|p_ip_jp_k|^2} \le \frac{N_0 n^{4/3}}{b^2} + \sum_{t=1}^{O(n^{1/4})} \frac{4N_t n^{1/2}}{a^2t^2|p_ip_j|^2}, \tag{6}$$

where N_0 is the number of points of S_v that lie in the inner-most 4-dimensional cylinder of the packing (centered at ℓ_{ij} and of radius $a/n^{1/4}$), and N_t is the number of points of S_v that lie in the cylindrical shell centered at ℓ_{ij} with inner radius $at/n^{1/4}$ and outer radius $a(t+1)/n^{1/4}$.

Obviously $N_0 = O(n^{1/4})$, since the volume of the 3-dimensional cross-sectional sphere of the inner-most cylinder is $O(1/n^{3/4})$ and because of (i). Also, we have $N_t = O(t^2 n^{1/4})$. This follows by an argument of packing spheres of volume $\Omega(1/n)$ within a shell whose volume is $O(t^2/n^{3/4})$.

Hence the sum in Equation (6) is $O(n^{19/12} + \frac{n}{|p_i p_j|^2})$. Substituting this in Eq. (5), we obtain the upper bound on the total volume of the forbidden prisms

$$O\left(\sum_{1\le i<j\le v}\left(\frac{1}{n^2}+\frac{1}{n^{31/12}|p_ip_j|^2}\right)\right)=O\left(\frac{v^2}{n^2}+\frac{1}{n^{31/12}}\sum_{1\le i<j\le v}\frac{1}{|p_ip_j|^2}\right). \quad (7)$$

We bound the sum in the second summand similarly to our bounding of the term in Equation (4) (in Section 3.3). We fix p_i and use a 4-dimensional spherical packing argument within spherical shells centered at p_i. Arguing as above, we obtain

$$\sum_{j\ne i}\frac{1}{|p_ip_j|^2}\le\sum_{t=1}^{O(n^{1/4})}\frac{M_t n^{1/2}}{a^2t^2}=\sum_{t=1}^{O(n^{1/4})}\frac{O(t^3)n^{1/2}}{a^2t^2}=O(n).$$

Summing this over all p_i, we obtain a final bound of $O(vn)$. Substituting this in Equation (7), we see that the total volume of the forbidden prisms is $O(\frac{v^2}{n^2}+\frac{v}{n^{19/12}})=O(1)$.

3.5 Forbidden Slabs

The forbidden regions where one of the inequalities (iv') is violated are $\binom{v}{4}$ 4-dimensional slabs σ_{ijkl}, for $1\le i<j<k<l\le v$, whose "base" is a 3-dimensional tetrahedron with volume at most 8. The height (in the 4th dimension) of the slab σ_{ijkl} is $O(1/(n^{127/24}|p_ip_jp_kp_l|))$. The overall volume of the slabs (within C^4) is at most

$$\sum_{1\le i<j<k<l\le v}O\left(\frac{1}{n^{127/24}|p_ip_jp_kp_l|}\right). \quad (8)$$

To bound this sum, we fix p_i, p_j, p_k and sum over p_l. We use a 2-dimensional quasi-circular packing argument that exploits the fact that S_v satisfies (i), (ii), and (iii). The packing consists of the Cartesian products of the plane $\pi_{ijk}\in\Re^4$ that passes through p_i, p_j, and p_k, and circles that lie in the 2-space orthogonal to π_{ijk} and whose centers belong to π_{ijk} (see Figure 4). Specifically, we have

$$\sum_{l\ne i,j,k}\frac{1}{|p_ip_jp_kp_l|}\le\frac{L_0 n^{43/24}}{c}+\sum_{t=1}^{O(n^{1/4})}\frac{L_t n^{1/4}}{at|p_ip_jp_k|}, \quad (9)$$

where L_0 is the number of points of S_v that lie in the inner-most shape of the packing (centered at π_{ijk}) and of radius $a/n^{1/4}$, and L_t is the number of points of S_v that lie in the shell centered at π_{ijk} with inner radius $at/n^{1/4}$ and outer radius $a(t+1)/n^{1/4}$.

Obviously $L_0=O(n^{1/2})$, since the volume of the inner-most shape is $O(1/n^{1/2})$ and because of (i). Also, we have $L_t=O(tn^{1/2})$. This follows by an argument of packing spheres of volume $\Omega(1/n)$ within a shell whose volume is $O(t/n^{1/2})$.

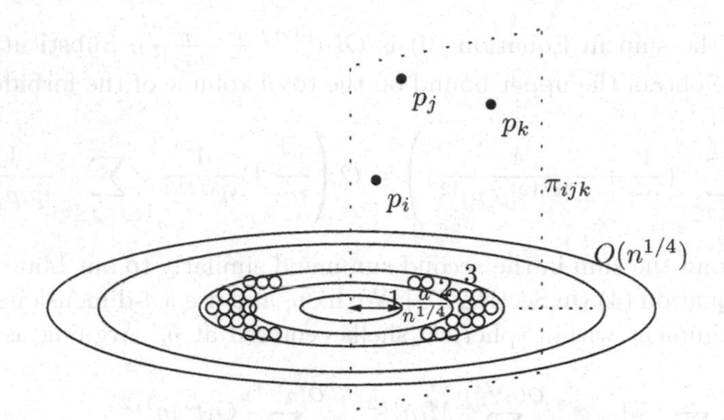

Fig. 4. A 2-D quasi-circular packing of balls in $\Re^4$

Hence the sum in Equation (9) is $O(n^{55/24} + \frac{n}{|p_i p_j p_k|})$. Substituting this in Eq. (8), we obtain the upper bound on the total volume of the forbidden slabs

$$O\left(\sum_{1 \le i < j < k \le v} (\frac{1}{n^3} + \frac{1}{n^{103/24}|p_i p_j p_k|})\right)$$
$$= O\left(\frac{v^3}{n^3} + \frac{1}{n^{103/24}} \sum_{1 \le i < j < k \le v} \frac{1}{|p_i p_j p_k|}\right). \quad (10)$$

We bound the sum in the second summand similarly to our bounding of the term in Equation (6) (in Section 3.4). We fix p_i, p_j and use a 4-dimensional cylindrical packing argument within cylindrical shells centered at ℓ_{ij}. Arguing as above, we obtain

$$\sum_{k \ne i,j} \frac{1}{|p_i p_j p_k|} \le \frac{N_0 n^{2/3}}{b} + \sum_{t=1}^{O(n^{1/4})} \frac{2N_t n^{1/4}}{at|p_i p_j|} = O(n^{11/12} + \frac{n}{|p_i p_j|}).$$

Summing this over all p_i, p_j, we obtain a bound of

$$O\left(v^2 n^{11/12} + n \sum_{1 \le i < j \le v} \frac{1}{|p_i p_j|}\right).$$

Substituting this in Equation (10), we see that the total volume of the forbidden slabs is

$$O\left(\frac{v^3}{n^3} + \frac{v^2}{n^{27/8}} + \frac{1}{n^{79/24}} \sum_{1 \le i < j \le v} \frac{1}{|p_i p_j|}\right). \quad (11)$$

We bound the sum in the third summand similarly to our bounding of the term in Equation (4) (in Section 3.3). We fix p_i and use a 4-dimensional spherical packing argument within spherical shells centered at p_i. Arguing as above, we obtain

$$\sum_{j \neq i} \frac{1}{|p_i p_j|} \leq \sum_{t=1}^{O(n^{1/4})} \frac{M_t n^{1/4}}{at} = \sum_{t=1}^{O(n^{1/4})} \frac{O(t^3) n^{1/4}}{at} = O(n).$$

Summing this over all p_i, we obtain a final bound of $O(vn)$.

Substituting this in Equation (11), we see that the total volume of the forbidden slabs is $O(\frac{v^3}{n^3} + \frac{v^2}{n^{27/8}} + \frac{v}{n^{55/24}}) = O(1)$.

3.6 Epilogue

In summary, the total volume of the forbidden regions is $O(1)$, which can be made smaller than 1 by an appropriate choice of the constants a, b, c, d. Careful calculation shows that the forbidden volume is at most $C_1 a^4 + C_2 P_b(a, b) + C_3 P_c(a, b, c) + C_4 P_d(a, b, c, d)$, where C_1, C_2, C_3, C_4 are real constants independent of a, b, c, d, and $P_b(a, b)$, $P_c(a, b, c)$, $P_d(a, b, c, d)$ are rational polynomials. We simply set a so that $C_1 a^4 \leq 1/4$. Then, given a value of a, is it easy to choose b so as to make $C_2 P_b(a, b) \leq 1/4$. Similarly, the values of c and d are chosen so that the third and fourth summands, respectively, are at most $1/4$.

4 Conclusion

We have presented constructions that solve the on-line version of Heilbronn's triangle problem in three and four dimensions. The same technique can be applied in higher dimensions (with increasing complexity of computations).

Acknowledgment

The author wishes to thank Micha Sharir for helpful discussions on the triangle problem and on nested packing arguments.

References

1. G. Barequet, A lower bound for Heilbronn's triangle problem in d dimensions, *SIAM J. on Discrete Mathematics*, 14 (2001), 230–236.
2. J. Komlós, J. Pintz, and E. Szemerédi, On Heilbronn's triangle problem, *J. London Mathematical Society (2)*, 24 (1981), 385–396.
3. J. Komlós, J. Pintz, and E. Szemerédi, A lower bound for Heilbronn's problem, *J. London Mathematical Society (2)*, 25 (1982), 13–24.
4. H. Lefmann, On Heilbronn's problem in higher dimension, *Proc. 11th Ann. ACM-SIAM Symp. on Discrete Algorithms*, San Francisco, CA, 60–64, 2000.
5. K.F. Roth, On a problem of Heilbronn, *Proc. London Mathematical Society*, 26 (1951), 198–204.
6. K.F. Roth, Developments in Heilbronn's triangle problem, *Advances in Mathematics*, 22 (1976), 364–385.
7. W.M. Schmidt, On a problem of Heilbronn, *J. London Mathematical Society (2)*, 4 (1971), 545–550.

Algorithms for Normal Curves and Surfaces

Marcus Schaefer[1], Eric Sedgwick[2], and Daniel Štefankovič[3]

[1] DePaul University (mschaefer@cs.depaul.edu)
[2] DePaul University (esedgwick@cs.depaul.edu)
[3] University of Chicago (stefanko@cs.uchicago.edu)

Abstract. We derive several algorithms for curves and surfaces represented using normal coordinates. The normal coordinate representation is a very succinct representation of curves and surfaces. For embedded curves, for example, its size is logarithmically smaller than a representation by edge intersections in a triangulation. Consequently, fast algorithms for normal representations can be exponentially faster than algorithms working on the edge intersection representation. Normal representations have been essential in establishing bounds on the complexity of recognizing the unknot [Hak61, HLP99, AHT02], and string graphs [SSŠ02]. In this paper we present efficient algorithms for counting the number of connected components of curves and surfaces, deciding whether two curves are isotopic, and computing the algebraic intersection numbers of two curves. Our main tool are equations over monoids, also known as word equations.

1 Introduction

Computational topology is a recent area in computational geometry that investigates the complexity of determining properties of topological objects such as curves and surfaces [BE$^+$99, DEG99]. For example, it is known that we can decide whether two curves on a surface are homotopic in linear time if the surface is represented by a triangulation, and the curves as sequences of intersections with the triangulation [DG99].

In 1930 Kneser [Kne30] introduced a representation for curves and surfaces in which these objects are described by their *normal coordinates*. This led to the theory of normal surfaces which was used by Haken in 1961 to show that the unknot could be recognized by an algorithm (which, much later, was shown to run in exponential time). Haken's approach was pushed further by Hass, Lagarias, and Pippenger who showed that the unknot could be recognized in **NP** [HLP99]. To this end they had to verify in polynomial time that a special type of normal surface was an essential disk. The result of [HLP99] was recently extended by Agol, Hass, and Thurston [AHT02]. The main contribution of [AHT02] was a polynomial time algorithm for computing the number of connected components of a normal surface. This immediately implies polynomial time algorithms for checking whether a normal surface is connected, and whether it is orientable.

The theory of normal curves is much simpler than the theory of normal surfaces. Nevertheless, it was one of three essential ingredients in the proof that

O.H. Ibarra and L. Zhang (Eds.): COCOON 2002, LNCS 2387, pp. 370–380, 2002.

string graphs can be recognized in **NP**, a problem that had only recently been shown to be decidable at all [SSŠ02].

With this paper we attempt to initiate a more systematic study of algorithms for normal curves and surfaces. We believe that the examples mentioned earlier show that this is a worthwhile attempt. There is also a more theoretical justification of this approach: we get a clearer picture of the computational complexity of a problem if we ensure that the object representations are *succinct*, meaning that they are not compressible. It is easy to show that the representation of an embedded curve as a sequence of intersections with a triangulation is never succinct: if we consider it as a word over an alphabet made up of the edges of the triangulation, it can always be compressed to polylogarithmic size (see Section 4.1). For embedded curves this means that the normal coordinates are a more succinct and natural way of representation.

Among other things we will give efficient algorithms to count components of curves and surfaces, decide whether two curves are isotopic, and compute the algebraic intersection number of a pair of curves.

A final word about technique. We extend the ideas from [SSŠ02] where we based our algorithms on word equations. We believe the connection between word equations and normal curves and surfaces is a very strong one, and, as far as we can tell, has not been observed before. There have, of course, been algorithms based on words in the fundamental group of a manifold, but words in this context are always words in a group. The original observation here is that the recent, and powerful, results about computational aspects of monoids can be applied to embeddings and immersions of curves and surfaces [PR98, GKPR96, Ryt99, DM01]. For example, the algorithm for counting components suggested by Agol, Hass, and Thurston is quite involved, whereas our algorithm follows quite naturally from the decidability of word equations with prescribed lengths by using some topology. A second example is the recognition problem for string graphs. Currently this problem can only be solved with the help of algorithms for trace monoids [DM01, SSŠ02].

2 Normal Coordinates

In the following M will always be a compact orientable 2-manifold with boundary ∂M, unless stated otherwise. Let b be the number of the boundary components of M and g the genus of M. The Euler characteristic of M is $\chi = 2 - 2g - b$.

A *simple arc* is a homeomorphic image of the interval $[0,1]$. A simple arc $\gamma : [0,1] \to M$ such that both its endpoints $\gamma(0), \gamma(1)$ are on the boundary ∂M and the internal points $\gamma(x), 0 < x < 1$ are in the interior of M is called a *properly embedded arc.* A simple closed arc $\gamma : [0,1] \to M$, $\gamma(0) = \gamma(1)$ such that all the points $\gamma(x), 0 \leq x \leq 1$ are in the interior of M is called a *properly embedded circle.* A *curve* is an embedded collection of properly embedded arcs and properly embedded circles. Note that a curve cannot be self intersecting, since it is embedded.

Two curves γ_1, γ_2 are *isotopic rel boundary* ($\gamma_1 \sim \gamma_2$) if there is a continuous deformation of γ_1 to γ_2 which does not move the points on the boundary ∂M and during the deformation the curve stays embedded. From now on by isotopy we mean isotopy rel boundary.

Let $T = (V_T, E_T)$ where V_T is a set of points in M and E_T is a embedded set of embedded arcs in M with both endpoints in V_T. T is a *triangulation* of M if every connected component C of $M - E_T$ is homeomorphic to an open disc and in a closed walk along the boundary of C we meet 3 points from V_T (not necessarily distinct). If the manifold M has non-empty boundary and $V_T \subseteq \partial M$ then we call the triangulation T *minimal*.

Let T be a triangulation of M. Let γ be a curve. We say that γ is *normal* w.r.t. T if all the intersections with T are transversal and if γ enters a triangle $t \in T$ via an edge e then it leaves t via an edge different from e (i.e no two consecutive intersections of γ with T belong to the same edge $e \in T$).

Let γ be a curve in M. Let $c \sim \gamma$ minimize the number of intersections with T. If c enters and leaves some $t \in T$ through the same edge e then the number of intersections of c with T can be reduced by pulling the curve c, a contradiction. Hence for each non-trivial curve γ in M there is $c \sim \gamma$ which is normal w.r.t. T. If γ is an arc we can fix one of its endpoints as the initial point of γ, and number the intersection points of γ with T in order (starting with 0). We call this number the *index* of an intersection point along γ.

An isotopy which for each edge $e \in T$ maps points in e to points in e is called a *normal isotopy*. If the triangulation T is minimal then every isotopy is a normal isotopy.

Given a curve γ in normal position w.r.t. T we can write on each edge of the triangulation how many times γ intersects it. Such a representation is called a *representation using normal coordinates*. In each triangle the numbers written on its sides determine the behavior of the curve inside the triangle up to isotopy (for this we need that the curve is properly embedded and in normal position w.r.t. T). Two curves which have the same numbering are normally isotopic (assuming that the positions of the points on the boundary agree).

The size of the representation using normal coordinates is the total bitlength of the labels. The number of intersections of γ with $e \in T$ will be denoted $\gamma(e)$.

The segments into which an edge $e \in T$ is cut by γ are called *ports*. The ports are representatives of points in $M - \gamma$. A port is given by $(u, v) \in T$ on which it occurs and a number from $\{0, \ldots, \gamma(e)\}$ encoding its order on (u, v). Similarly an intersection point of γ with T is given by $(u, v) \in T$ on which it occurs and a number from $\{0, \ldots, \gamma(e) - 1\}$.

3 Results

The goal of this paper is to prove the following results.

Theorem 1. *Let M be a compact 2-manifold. Let T be a triangulation of M. Let γ, δ be curves in M given by normal coordinates w.r.t. T. Let p, q be ports of γ in T. Let r be an intersection point of γ with T. The following problems can be solved in polynomial time.*

(a) *Find the normal coordinates of the connected component* γ *containing* r.
(b) *Count the number of connected components of* γ.
(c) *List all the non-isotopic connected curves* $\gamma_1, \ldots, \gamma_n$ *which occur in* γ. *For each* γ_i *find the number of occurrences of* γ_i *in* γ.
(d) *Decide whether* γ *and* δ *are isotopic (assuming* γ, δ *are normal isotopy disjoint).*
(e) *Decide whether ports* p, q *are in the same connected component of* $M - \gamma$. *If they are in the same connected component find an embedded arc in* $M - \gamma$ *connecting them.*
(f) *Compute the algebraic intersection number of* γ, δ.
(g) *If* γ *is a properly embedded arc, we can for each intersection point with* T *compute its index along* γ *(arbitrarily declaring one of* γ*'s endpoints the first intersection).*

Theorem 2. *Let* M *be a compact 2-manifold. Let* T *be a triangulation of* M. *Let* γ, δ *be curves in* M *given by normal coordinates w.r.t.* T. *The following problems can be solved by a Las Vegas algorithm (an algorithm solving the problem in expected polynomial time with zero probability of error).*

(a) *Decide whether* γ *and* δ *are isotopic (assuming* $\partial M \neq \emptyset$*).*
(b) *Locate the intersection point with index* n *of* γ *with* T.

Problem (*b*), and therefore (*a*) (by reduction), can be solved in polynomial time, but the proof is more complicated, and we do not include it here. Part (*b*) of Theorem 1 can be extended to normal surfaces; that is, we can count the number of connected components of a surface given in normal coordinates in polynomial time. This result was first shown in [AHT02], however, our proof is simpler, and is based on ideas developed independently in [SSŠ02]. For lack of space, we do not include the proof in this version.

4 Word Equations

Let Σ be an *alphabet.* Words in Σ^* are represented by *straight line programs* (SLP), a special type of context free grammar. A *straight line program* P of length n is a sequence of assignments $x_i =$ expr for $1 \leq i \leq n$ where expr is either a symbol from Σ or $x_j x_k$, $1 \leq j, k < i$. For a word w given by SLP we denote the length n of the program by $|SLP(w)|$. Note that there are strings (e.g. a^m) for which $|SLP(w)|$ is exponentially smaller than $|w|$.

Lemma 1 ([GKPR96]). *Let* p, t *be words represented by SLP's. The following two problems can be solved in time polynomial in* $|SLP(p)|$ *and* $|SLP(t)|$. *How many times does* p *occur in* t*? What is the position of the first occurrence of* p *in* t*?*

Lemma 2. *Let* t *be a word given by an SLP. The following problem can be solved in time polynomial in* $|SLP(t)|$. *Given two positions* $i \leq j$ *in* t *find an SLP for the substring* $t[i \ldots j]$ *of* t.

Let Θ be an alphabet of variables disjoint from Σ. A *word equation* $u = v$ is a pair of words $(u, v) \in (\Sigma \cup \Theta)^* \times (\Sigma \cup \Theta)^*$. The *size of the equation* $u = v$ is $|u| + |v|$. A *solution of the word equation* $u = v$ is a morphism $h : (\Sigma \cup \Theta)^* \to \Sigma^*$ such that $h(a) = a$ for all $a \in \Sigma$ and $h(u) = h(v)$ (h being a morphism means that $h(wz) = h(w)h(z)$ for any $w, z \in (\Sigma \cup \Theta)^*$). The *length of the solution* h is $\sum_{x \in \Theta} |h(x)|$. A *word equation with specified lengths* is a word equation $u = v$ and a function $f : \Theta \to \mathbb{N}$. The solution h has to respect the lengths, i.e. we require $|h(x)| = f(x)$ for all $x \in \Theta$. Let $h : (\Sigma \cup \Theta)^* \to \Sigma^*$ be a solution of an equation $u = v$. The SLP encoding of h is the sequence of SLP encodings of $h(x)$ for all $x \in \Theta$. The size of the encoding is $|SPL(h)| = \sum_{x \in \Theta} |SLP(h(x))|$. The usefulness of SLP encoding for word equations is demonstrated by following result.

Theorem 3 ([PR98]). *Let $u = v$ be a word equation with lengths specified by a function f. Assume that $u = v$ has a solution respecting f. The $SLP(h)$ of the lexicographically least solution h can be found in polynomial time in the size of the equation and the size of the binary encoding of f.*

4.1 Curve Coloring Equations

In this section we construct a system of word equations with given lengths which will allow us to color connected components of a curve γ on M normal w.r.t. a triangulation T. The alphabet used in the word equations encodes the colors. Let $t \in T$ be a triangle with vertices u, v, w. We add the following twelve variables to the system.

$$x_{t,(u,v)}, x_{t,(v,u)}, x_{t,(u,w)}, x_{t,(w,u)}, x_{t,(v,w)}, x_{t,(w,v)}, y_{t,u}, y_{t,v}, y_{t,w}, y_{u,t}, y_{v,t}, y_{w,t} \tag{1}$$

The variable $x_{t,(u,v)}$ encodes the order in which the colors occur on (u, v). We specify $|x_{t,(u,v)}| = \gamma(u, v)$. The variable $y_{t,u}$ encodes the colors of the (directed) segment of $x_{t,(u,v)}$ whose edges pass from (w, u) to (v, u). See Figure 1.

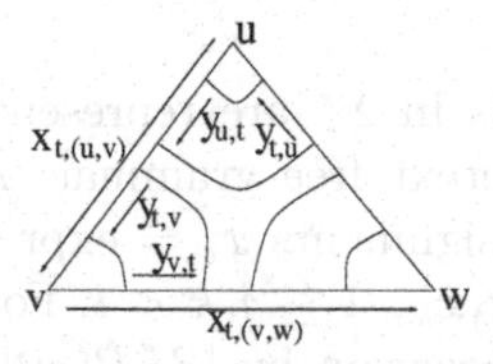

Fig. 1. Some of the variables for triangle t.

The following equations are called *triangle constraints*.

$$\begin{array}{lll} x_{t,(u,v)} = y_{u,t}y_{t,v} & x_{t,(v,u)} = y_{v,t}y_{t,u} & x_{t,(v,w)} = y_{v,t}y_{t,w} \\ x_{t,(w,v)} = y_{w,t}y_{t,v} & x_{t,(u,w)} = y_{u,t}y_{y,w} & x_{t,(w,u)} = y_{w,t}y_{t,u} \end{array} \tag{2}$$

Note that the lengths of the x variables determine the lengths of the y variables, for example $|y_{u,t}| = (|x_{t,(u,v)}| + |x_{t,(u,w)}| - |x_{t,(v,w)}|)/2$. For each edge (u, v) which is contained in two triangles $s, t \in T$ we add the following *edge constraint*.

$$x_{s,(u,v)} = x_{t,(u,v)} \tag{3}$$

Without additional equations the system has a solution in which all words consist of a's only. We can add additional equations specifying colors of some intersection points of γ with T. If the constraints are consistent, the components of γ in the resulting coloring will be monochromatic, otherwise there will not be a solution. (This will be useful when counting connected components later.)

Now we can prove part (a) of Theorem 1. Find the lexicographically smallest solution of the curve coloring equation with r colored by color b over alphabet $\Sigma = \{a, b\}$. Assigning to each edge $(u, v) \in t \in T$ the number of b's in $x_{t,(u,v)}$ yields a normal coordinate representation of γ_r.

Let γ be a properly embedded arc. There is a triangulation T' such that γ is an edge of T and T' is in normal position w.r.t. T. Set up curve coloring equations for the edges of T in the triangulation T' and color each edge of T with different color. In the solution the word x_γ written on γ is the edge intersection representation of γ in T. By Lemma 3 x_γ can be compressed to size polynomial in $\log|\gamma|$ and size of the triangulations T' and T. Hence the edge intersection representation is compressible.

4.2 Region Coloring Equations

We can modify the curve coloring equations (1), (2), (3) to color the connected components of $M - \gamma$. The variable $x_{t,(u,v)}$ will now encode the colors of the ports of γ on (u, v), hence we specify $|x_{t,(u,v)}| = \gamma(u, v) + 1$. The variable $y_{t,u}$ will encode colors of regions extending from (w, u) to (v, u). For each $t \in T$ we add a variable z_t which encodes the color of the center region (the region which has ports on all edges of t). The triangle constraints become

$$\begin{array}{lll} x_{t,(u,v)} = y_{u,t} z_t y_{t,v} & x_{t,(v,u)} = y_{v,t} z_t y_{t,u} & x_{t(v,w)} = y_{v,t} z_t y_{t,w} \\ x_{t,(w,v)} = y_{w,t} z_t y_{t,v} & x_{t(u,w)} = y_{u,t} z_t y_{y,w} & x_{t,(w,u)} = y_{w,t} z_t y_{t,u} \end{array} \qquad (4)$$

The edge constraints remain unchanged. Without additional constraints the system has a solution with all regions colored by the same color a. We can add additional equations specifying colors of some ports. In the resulting coloring (if it exists) each region will be monochromatic.

Let $p \in (u, v)$ be a port of γ in T, where (u, v) belongs to triangles $s, t \in T$. We would like to modify the equations to color the connected components of $M - \gamma - p$. It is enough to modify the edge constraints of (u, v) and (v, u). We replace the equation $x_{s,(u,v)} = x_{t,(u,v)}$ by the equations

$$\begin{array}{c} x_{s,(u,v)} = w_1 w_2 w_3, \; w_1 w_4 w_3 = x_{t,(u,v)} \\ |w_1| = p, \; |w_2| = |w_4| = 1, \; |w_3| = \gamma(u, v) - p. \end{array}$$

Similarly we can set up equations for coloring components of $M - \gamma$ with polynomially many ports of γ in T removed.

4.3 Intersection Counting Equations

If we are given an edge e, and a number that specifies an intersection of γ along e, we want to compute the index of that intersection.

Again, this problem can be used solving word equations. In a first step we need to determine for each edge (u,v) belonging to a triangle t how many of the intersections of γ along (u,v) enter t, and how many leave t (we think of traversing γ starting at the starting point we fixed). Call these two numbers $\gamma^i(u,v,t)$ and $\gamma^o(u,v,t)$. We know that $\gamma^i(u,v,t)+\gamma^o(u,v,t)=\gamma(u,v)$ for edges (u,v) belonging to t.

We can determine the numbers $\gamma^i(u,v,t)$ and $\gamma^o(u,v,t)$ by setting up the triangle and edge constraints shown in (2) and (3), and specifying that $|x_{t,(u,v)}| = 4\gamma(u,v)$. We then force the starting point of γ to be equal to the string *cabc*. The solution for each edge will be a concatenation of *cabc*'s and *cbac*'s, depending on whether the edge enters or leaves the triangle. Because of Lemma 1 we can count the occurrence of these substrings in polynomial time, and therefore we can compute $\gamma^i(u,v,t)$ and $\gamma^o(u,v,t)$ in polynomial time.

With this information we set up a new set of equations. Let $m=|\gamma|+1$. For each triangle t we take two copies of the variables in (1), one for intersections coming into t, and one for outgoing intersections. Distinguish the two set of variables by superindexing them by i or o. For every edge (u,v) that belongs to triangles s and t we set up the following equations:

$$\begin{array}{lll} x^o_{s,(u,v)}a = ax^i_{t,(u,v)} & ax^o_{s,(v,u)} = x^i_{t,(v,u)}a & x^i_{t,(u,v)} = y^i_{u,t}y^i_{t,v} \\ x^i_{t,(v,u)} = y^i_{v,t}y^i_{t,u} & x^o_{t,(u,v)} = y^o_{u,t}y^o_{t,v} & x^i_{t,(v,u)} = y^o_{v,t}y^o_{t,u} \end{array} \quad (5)$$

We also specify that $|x^i_{t,(u,v)}| = |x^i_{t,(u,v)}| = m\cdot\gamma^i(u,v,t)$, and $|x^o_{t,(u,v)}| = |x^o_{t,(u,v)}| = m\cdot\gamma^o(u,v,t)$, and set the startpoint of γ to be ba^{m-1}. Equations (5) ensure that at every intersection point the b is moved by one position.

Suppose now we are given a coordinate k along an edge (u,v) of the triangulation. Since we can determine the direction of the intersection, we know the triangle t that γ enters along that point. Using a set of equations like the one we used earlier to count the number of intersections in each direction, we can determine the coordinate k' of the intersection crossing into t. We can then retrieve the string $x^i_{t,(u,v)}[k'*m\ldots(k'*m+m-1)]$, and determine the position of the single b in that string which immediately gives us the index we sought. This proves part (g) of Theorem 1.

5 The Algorithms

5.1 Counting Connected Components

Let T be a triangulation of M, where τ is the number of triangles in T. Let γ be a curve in M given by normal coordinates w.r.t. T. The number of components k of γ can be exponential in the input size. However it is known that the components fall into few normal isotopy classes [Kne30].

Lemma 3. *The components of γ fall into at most 6τ normal isotopy classes.*

We omit the proof. We can now prove parts (b) and (c) of Theorem 1. The algorithm for computing the number of components of γ works as follows. Pick an edge (u, v) in T which is intersected by γ. Let r be the intersection of γ and (u, v) closest to u. Compute the normal coordinates of the connected component γ_r of γ which contains r. Using binary search find the intersection point ℓ of γ with (u, v) whose connected component γ_ℓ is normally isotopic to γ_r. The components of points between r and ℓ are exactly the components of γ which are normally isotopic to γ_r. Hence we can increase the component count by $(\ell - r + 1)$ and run the algorithm for the curve $\gamma - (\ell - r + 1)\gamma_r$. Lemma 3 implies that the number of repetitions of the algorithm is bounded by 6τ.

The described algorithm also finds non-normally-isotopic $\gamma_1, \ldots, \gamma_n$ which occur in γ and their counts. Using part (d) of Theorem 1 we can merge the counts for isotopic γ_i.

5.2 Deciding Isotopy I

Let R be a union of connected components of $M - \gamma$ given as a solution of the region coloring equations, where R is colored with color b and $M - R$ is colored with color a. We want to compute the Euler characteristic of R. We will cut R by γ and T into regions and use the formula $\chi = V - E + F$. To compute V, E, F we just need to compute these quantities for each triangle $t \in T$ with appropriate weights and then sum them together. The edges in $\partial t - \partial M$ get weight $1/2$ and the vertices which are also vertices of T get weight one over the number of triangles they occur in. We will only need to compute the numbers of occurrences of b, ab, bb, ba in each $x_{t,(u,v)}$ which can be done using Lemma 1.

Now we can prove part (d) of Theorem 1. Because of part (c) it is enough to consider the case when γ is connected. It is known (see [FM97]) that two *disjoint* properly embedded arcs are isotopic iff they bound a disc in M. Let $\alpha = \gamma + \delta$, the curve whose normal coordinates is the sum of the coordinates of γ and δ. Since γ and δ are normal isotopy disjoint the curve α contains both γ, δ as components. We color the port between starting points of γ and δ with b and check that the resulting region has Euler characteristic 1. The case where γ and δ are properly embedded circles is similar, we only have to check whether they co-bound an annulus.

5.3 Connecting Two Points

In this section we prove part (e) of Theorem 1. First we want to decide whether ports p and q are in the same connected component of $M - \gamma$. Add an equation coloring p with b to the region coloring equations. The ports p, q are in the same connected component of $M - \gamma$ iff port q has color b in the lexicographically smallest solution of the equations.

Suppose that p and q are in the same connected component. We want to find a curve connecting them. Let $t \in T$ and let $r_{t,1}, r_{t,2}, r_{t,3}$ be the ports of the center region of t. A shortest path connecting p and q does not enter the center region of t twice and hence one of the ports $r_{t,1}, r_{t,2}, r_{t,3}$ can be removed while keeping p, q in the same component. We find the port which can be removed, remove it, and move on to another triangle which does not have a port removed. After removing a port of the center region of each triangle the components of

$M - \gamma - \{r_{t,?}; t \in T\}$ are either discs or annuli. Finally we remove both p, q and color just one side of p with color b. For one of the sides of p the coloring will reach q. Let $x_?$ be the solution of this system of equations. The colored component which reached from p to q is is a sequence of rectangular regions. The curve connecting p, q which runs in the middle of the colored component intersects edge $(u, v) \in T$, $\#_b x_{t,(u,v)}$ times.

5.4 Deciding Isotopy II

Let T be a triangulation of M and let γ be a curve in M given by normal coordinates w.r.t. T. We will show how to compute normal coordinates of $\gamma' \sim \gamma$ w.r.t. a triangulation T' which is a modification of T. The modifications we will consider are called bistellar moves (flip, add, drop) shown on the Figure 2. The moves can be applied only when all the triangles shown in Figure 2 are distinct.

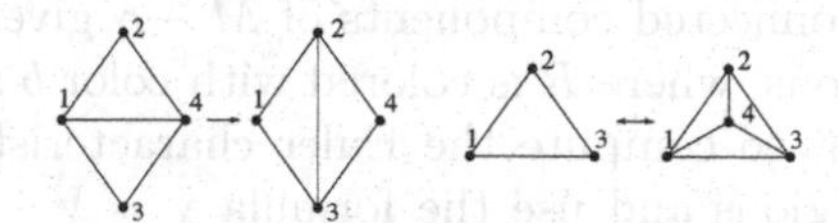

Fig. 2. The bistellar moves.

The new coordinates after the add move can be chosen to be $\gamma(14) = 0$, $\gamma(24) = \gamma(12)$, $\gamma(34) = \gamma(13)$. For the flip move $\gamma(23) = (\gamma(12) + \gamma(24) + \gamma(13) + \gamma(34))/2 + |\gamma(12) - \gamma(24) + \gamma(34) - \gamma(13)|/2 - \gamma(14)$.

For the drop move we can w.l.o.g. assume that the curve γ inside the triangle 123 looks as the one at the Figure 3. The e curves around the point 4 can be removed because they are homotopic to a point. We also need to pull the f curves since they are not in normal position w.r.t. T'.

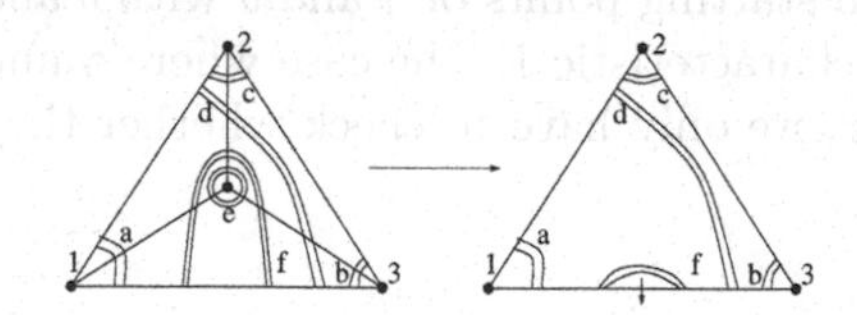

Fig. 3. The drop move.

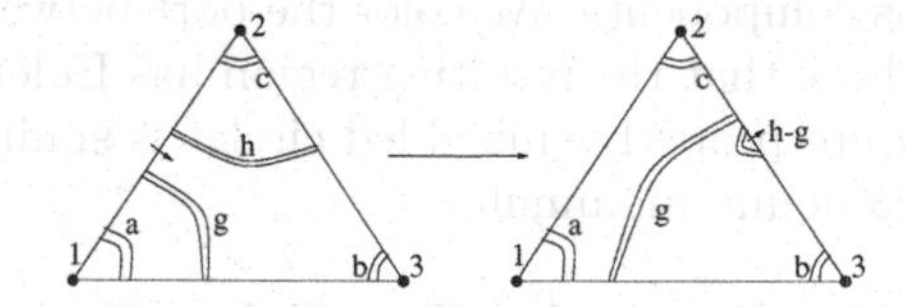

Fig. 4. Pulling the curve to a normal position (w.l.o.g. $h \geq g$).

The pulling might propagate to neighboring triangle as shown in Figure 4. We cannot follow the propagation directly because it can happen exponentially many times.

When both g and h from Figure 4 are non-zero we say that the pulled curves split. For each triangle $t \in T$ at most once split occurs in t (otherwise the curve γ would have to self-intersect).

The algorithm for pulling the curves will work as follows. Using word equations we will find the segment of the pulled curves until the first split. We pull the curves, perform the split and repeat. The number of splits and hence repetitions is bounded by the number of triangles in T.

To find the first split we extract the outermost of the f pulled curves, call it δ. Then we cut δ in t. If δ is still connected we cut it once more at its midpoint (using part (b) of Theorem 2). We obtain two embedded arcs α, β starting in t. The first split occurs when α, β choose different route inside a triangle t. The split can be easily found using binary search.

Lemma 4. *Let M be a 2-manifold with non-empty boundary. Let T be a triangulation of M. Let γ be a curve in M given by normal coordinates w.r.t. T. We can find a minimal triangulation T' of M and normal coordinates of $\gamma' \sim \gamma$ w.r.t. T'*

Proof. Using bistellar moves and by possibly adding new triangles we can remove vertices of T which are not on the boundary ∂M of M. Let $v \in T$ be a vertex in the interior of M. If deg $v = 3$ then there are three distinct triangles neighboring v and we can apply the drop move to eliminate v. If deg $v > 3$ then we can apply a sequence of flip moves to decrease the degree of v to 3 and then eliminate it with a drop. If deg $v = 2$, then we have two triangles t_1, t_2 glued together along two adjacent edges. If either t_1 or t_2 is glued to another triangle, we can apply a flip to increase the degree to 3 and then flip. If not, then M is a disk consisting of two triangles, and we will need to add another triangle to the boundary, and proceed as in the following case. The last case is when deg $v = 1$. This implies that a single triangle is glued to itself. If it is also attached to another triangle, then we can apply a flip to increase the degree of v to 2. If it is not, then M is a disk consisting of a single triangle and we must attach an additional triangle, and then apply the flip.

Two curves given by normal coordinates w.r.t. a minimal triangulation are isotopic iff the coordinates are equal (and the points on ∂M agree). Hence as a corollary of Lemma 4 we obtained a proof of part (a) of Theorem 2.

5.5 Finding the n-th Intersection Point

There is a simple randomized (Las Vegas) algorithm to find the n-th intersection point of an oriented embedded arc γ with the triangulation T. The intersection points are numbered by $0, \ldots, |\gamma|$.

Pick a random intersection point r of γ with T. Using the curve coloring equations color one side with b's and the other side with a's. We obtain two embedded curves γ_1, γ_2 which when glued together at r yield γ. If $|\gamma_1| < n$ then find the $n - |\gamma_1|$th intersection point of γ_2 with T, otherwise find nth intersection point of γ_1 with T. The expected number of repetitions is $O(\log |\gamma|)$.

5.6 Computing the Algebraic Intersection Number

The *algebraic intersection number* of two oriented curves γ, δ in an orientable surface is defined as follows. For each intersection of γ and δ in which δ crosses from left to right add $+1$, if it crosses from right to left add -1. The algebraic intersection number is invariant under isotopy.

Since the algebraic intersection number of γ and δ is invariant under isotopy, we can fix a drawing of the curves. For each edge (u, v) of the triangulation we choose which of the curves will intersect (u, v) in the half closer to u, the other one will intersect (u, v) in the half closer to v. Now we draw the curves so that the segments of γ, δ in each triangle are geodesics.

For a triangle $t \in T$ with vertices u, v, w we need to compute the number of segments of γ oriented from (u, v) to (u, w). Take the curve coloring equations for the curve 4γ. Using the same idea we saw earlier, color the copies *cabc* (in that order) at one endpoint. The occurrences of *cabc*, and *cbac* show the orientation of γ along an edge. Counting the number of occurrences of *cabc*, and *cbac* then gives us the result.

References

AHT02. I. Agol, J. Hass, and W. Thurston. 3-manifold knot genus is NP-complete. In *Proceedings of the 33th Annual ACM Symposium on Theory of Computing (STOC-2002)*, 2002.

BE+99. M. Bern, D. Eppstein, et al. Emerging challenges in computational topology. ACM Computing Research Repository, September 1999.

DEG99. T. Dey, H. Edelsbrunner, and S. Guha. Computational topology. In B. Chazelle, J.E. Goodman, and R. Pollack, editors, *Advances in Discrete and Computational Geometry*, volume 223 of Contemporary Mathematics. American Mathematical Society, 1999.

DG99. T. Dey and S. Guha. Transforming curves on surfaces. *JCSS: Journal of Computer and System Sciences*, 58(2):297–325, 1999.

DM01. V. Diekert and A. Muscholl. Solvability of equations in free partially commutative groups is decidable. In *ICALP 2001*, pages 543–554, 2001.

FM97. A. Fomenko and S. Matveev. *Algorithmic and computer methods for three-manifolds.* Kluwer, 1997.

GKPR96. L. Gąsieniec, M. Karpinski, W. Plandowski, and W. Rytter. Efficient algorithms for Lempel-Ziv encoding. *in Proceedings of SWAT'96, LNCS 1097*, pages 392–403, 1996.

Hak61. W. Haken. Theorie der Normalflächen. *Acta Mathematica*, 105:245–375, 1961.

HLP99. J. Hass, J. Lagarias, and N. Pippenger. The computational complexity of knot and link problems. *Journal of ACM*, 46(2):185–211, 1999.

Kne30. H. Kneser. Geschlossene Flächen in dreidimensionalen Mannigfaltigkeiten. *Jahresbericht der Deutschen Mathematikver-Vereinigung*, pages 248–260, 1930.

PR98. W. Plandowski and W. Rytter. Application of Lempel-Ziv encodings to the solution of words equations. In *Automata, Languages and Programming*, pages 731–742, 1998.

Ryt99. W. Rytter. Algorithms on compressed strings and arrays. In *Proceedings of 26th Annual Conference on Current Trends in Theory and Practice of Infomatics.*, 1999.

SSŠ02. M. Schaefer, E. Sedgwick, and D. Štefankovič. Recognizing string graphs in np. In *Proceedings of the 33th Annual ACM Symposium on Theory of Computing (STOC-2002)*, 2002.

Terrain Polygon Decomposition, with Application to Layered Manufacturing*

Ivaylo Ilinkin[1], Ravi Janardan[1], and Michiel Smid[2]

[1] Dept. of Computer Science & Engineering and Army High Performance Computing Research Center, University of Minnesota, Minneapolis, MN 55455, USA
{ilinkin,janardan}@cs.umn.edu

[2] School of Computer Science, Carleton University, Ottawa, Canada, K1S 5B6
michiel@scs.carleton.ca

Abstract. Efficient algorithms are given for decomposing a simple polygon into two special polygons, each with the property that every boundary and interior point can be connected to a single edge by a perpendicular line segment interior to the polygon. This allows efficient construction of certain classes of 3D parts via Layered Manufacturing.

1 Introduction

Let $\mathcal{P}$ be a simple, n-vertex polygon [12]. $\mathcal{P}$ is a *terrain polygon* (*terrain*, for short) if it has an edge, e, called a *base*, to which every point in $\mathcal{P}$ can be joined by a perpendicular line segment interior to $\mathcal{P}$ [3] (Fig. 1(a)). We consider two problems:

1. Decide if $\mathcal{P}$ is a terrain and, if it is, then find a base. We give a simple algorithm that runs in $O(n)$ time.
2. If $\mathcal{P}$ is not a terrain, then decide if it can be decomposed by a line into two terrains. If so, then compute a decomposing line and a base for each terrain. We give an algorithm which runs in $O(n \log n)$ time, if the terrains have a common base (on the decomposing line), and in $O(n^2 \log n)$ time, otherwise. (Fig. 1(b) and 1(c))

This problem arises in Layered Manufacturing (LM), where physical prototypes of 3D solids are created directly from their digital models ([8]). The 3D digital model (a polyhedron) is oriented suitably and sliced by a plane into parallel 2D layers. The layers (polygons) are "printed" successively on a fabrication device, each layer on top of the previous one, so that the 3D model is realized as a stack of 2D layers.

* Research of II and RJ supported, in part, by NSF grant CCR–9712226. This effort is also sponsored, in part, by the Army High Performance Computing Research Center under the auspices of the Department of the Army, Army Research Laboratory cooperative agreement number DAAD19-01-2-0014, the content of which does not necessarily reflect the position or the policy of the government, and no official endorsement should be inferred. Research of MS supported by NSERC.

O.H. Ibarra and L. Zhang (Eds.): COCOON 2002, LNCS 2387, pp. 381–390, 2002.

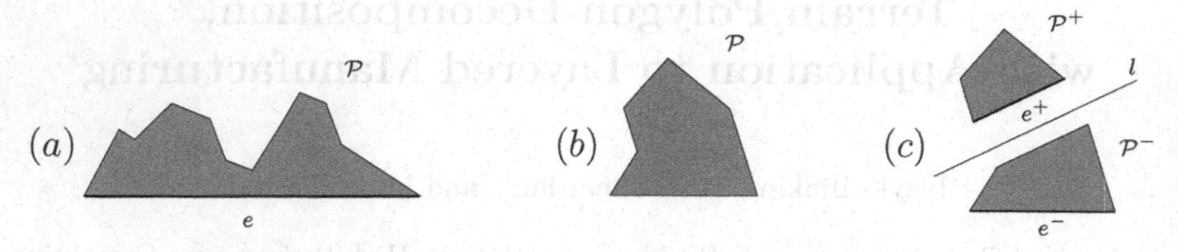

Fig. 1. (*a*) a terrain polygon; (*b*) a polygon which is not a terrain; (*c*) decomposition into terrains, with bases e^+ and e^-

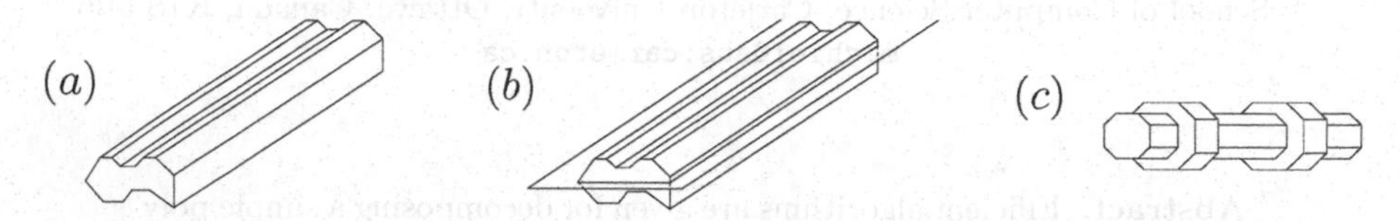

Fig. 2. (*a*) a long, slender generic object of uniform cross-section; (*b*) decomposition with a common base; (*c*) an object whose cross-sections are scaled versions of each other

A key process-planning step in LM is the analysis of the model to determine the need for *supports*. These are temporary structures, generated as the model is built, to prop-up portions of the model that do not have previously-built layers under them. Supports affect adversely the efficiency of LM and can be reduced by orienting the model suitably [1,9,10,11].

The terrain decomposition problem comes up quite naturally in this setting. Consider building a long and slender part of uniform cross-section, e.g. a piston rod, a drive shaft, or a gun barrel (Fig. 2(*a*)). No supports are needed if it is built along the long axis, but the process will be slow owing to the large number of layers (a foot-long part could have a thousand layers). Furthermore, the stacking of so many layers and the high height-to-width ratio will lead to instabilities and inaccuracies in the part.

However, if the cross-section of the part is decomposable into two terrains, then the part can be built without supports, while avoiding the disadvantages mentioned above. We divide the 3D model into two pieces using a plane which contains the decomposing line and the long axis of the model, build each piece on the facet containing the base edge, and then glue the pieces back together. No supports are needed, since the cross-section of each piece is a terrain (Fig. 2(*b*)). This approach also works if the cross-sections are not uniform but are scaled versions of one another (Fig. 2(*c*)).

1.1 Related Work

We are aware of three closely related results. Asberg *et. al.* [1] identify the class of objects that can be built without supports using Stereolithography, and give a result similar to ours for problem 1. In the context of casting, Rosenbloom and Rappaport [13] give an efficient algorithm to decompose a simple polygon into two terrains with a common base and give a somewhat different solution from ours, with the same bounds. Also for casting, Bose *et. al.* [2] show how to decompose a polyhedron by a plane into two 3D terrains with a common base. This algorithm could be used to solve the 2D problem, but not as efficiently as in [13] or here. We describe our results for these problems in some detail, since our solutions for the general problem, where the bases are different, relies on these results. (We are not aware of any prior results for the general problem.)

In [3], Fekéte and Mitchell have proven that it is NP-complete to decide if a polygon with holes or a polyhedron of genus zero can be decomposed into $k > 1$ terrains. In related work, [6], we have shown how to decompose a polyhedron into two connected components by a plane normal to a fixed direction $\mathbf{d}$, so that the supports are minimized when the components are built in directions $\pm\mathbf{d}$.

Due to space constraints, we omit several proofs and details; these can be found in [7].

2 Recognizing Terrains

The following lemma characterizes a terrain polygon. We assume that all edge normals of the polygon $\mathcal{P}$ have been translated to the origin. Furthermore, by "$\mathcal{P}(e)$ is a terrain" we mean that $\mathcal{P}$ is a terrain with base e.

Lemma 1. *A simple polygon $\mathcal{P}$ is a terrain if and only if there is an edge $e \in \mathcal{P}$ such that, for any other edge $e' \in \mathcal{P}$, $\mathbf{n}_e \cdot \mathbf{n}_{e'} \leq 0$, where $\mathbf{n}_e$ and $\mathbf{n}_{e'}$ are the outward-directed unit-normals of e and e', respectively.*

2.1 The Recognition Algorithm

Let $\mathcal{S}^1$ be the unit-circle. Each edge normal $\mathbf{n}_e$ yields a *normal-point* n_e on $\mathcal{S}^1$. (Equal normals are represented by different normal-points.) Let N be the set of normal-points.

Lemma 2. *W.l.o.g. assume that no normal-point of N coincides with $(-1,0)$ or $(1,0)$; otherwise, rotate N suitably. Let n_a (resp. n_b) be the first (resp. last) clockwise point of N on the upper half of $\mathcal{S}^1$. Let n_c (resp. n_d) be the first (resp. last) clockwise point of N on the lower half of $\mathcal{S}^1$. Then $\mathcal{P}$ is a terrain if and only if at least one of the edges a, b, c, or d is a base.*

Proof. The "if" part is obvious. For the "only if" part, assume that none of a, b, c, or d is a base. Let e be any other edge of $\mathcal{P}$. Then n_e lies between n_a and n_b or between n_c and n_d, w.l.o.g. n_a and n_b. Since the angle between n_a and n_b is less than π, the angle between at least one of n_e and n_a or n_e and n_b is less than $\pi/2$. Thus, by Lemma 1, e is not a base, so $\mathcal{P}$ is not a terrain. □

We can identify n_a, n_b, n_c, and n_d in $O(n)$ time and use Lemma 1 to test if any is a base in $O(n)$ time.

Theorem 1. *It is possible to decide in $O(n)$ time and $O(n)$ space whether an n-vertex simple polygon is a terrain. If it is, then a base can be identified within these bounds.*

The following lemma and corollary will be used in Section 3.

Lemma 3. *Let $\mathcal{P}$ be a simple polygon and let $\mathcal{Q}$ be its convex hull. If $\mathcal{P}(e)$ is a terrain, then e is an edge of $\mathcal{Q}$ and $\mathcal{Q}(e)$ is a terrain.*

Corollary 1. *If $\mathcal{P}(e)$ is a terrain, then the interior angles at e are at most $\pi/2$.*

3 Decomposition into Two Terrains with a Common Base

We assume that successive collinear edges of $\mathcal{P}$ have been merged into a single edge. Let l be a line such that $l \cap \mathcal{P}$ is a connected segment $\overline{uv}$. Let C^+ (resp. C^-) be the boundary of $\mathcal{P}$ from u to v (resp. v to u) in counter-clockwise order. Let $\mathcal{P}^+$ (resp. $\mathcal{P}^-$) be the polygon bounded by $C^+ \cup \overline{uv}$ (resp. $C^- \cup \overline{uv}$).

Lemma 4. *Let l be a line decomposing $\mathcal{P}$ into simple polygons $\mathcal{P}^+$ and $\mathcal{P}^-$, as above, such that $\mathcal{P}^+(\overline{uv})$ and $\mathcal{P}^-(\overline{uv})$ are terrains. Then (i) u and v are both vertices of $\mathcal{P}$; or (ii) u is a vertex of $\mathcal{P}$ and v is the point where l intersects perpendicularly the interior of an edge, e, of $\mathcal{P}$; or (iii) u and v are points where l intersects perpendicularly the interiors of edges, say e_u and e_v, of $\mathcal{P}$, respectively.*

Proof. If u and v are both vertices of $\mathcal{P}$, then case (i) holds. If u is a vertex and v is in the interior of an edge e, then by Corollary 1, the interior angle at v in $\mathcal{P}^+$ is at most $\pi/2$. Similarly, the interior angle at v in $\mathcal{P}^-$ is also at most $\pi/2$. Since the sum of these angles is π, the two interior angles at v are both $\pi/2$. Thus, case (ii) holds. Case (iii) is similar. □

Lemma 5. *If $\mathcal{P}^+(\overline{uv})$ and $\mathcal{P}^-(\overline{uv})$ are terrains, then the element-pair (u, v) or (u, e) or (e_u, e_v) in Lemma 4 determining the decomposing line l is an antipodal pair of the convex hull, $\mathcal{Q}$, of $\mathcal{P}$.*

Proof. Let (x, y) be the element-pair in question and let $V_{\overline{uv}}$ be the strip erected perpendicularly to $\overline{uv}$. Clearly, $\mathcal{P}$ is contained in $V_{\overline{uv}}$. Since x is either a vertex or an edge of $\mathcal{P}$ perpendicular to l, x lies on a bounding line of $V_{\overline{uv}}$. Since $\mathcal{P}$ is completely on one side of this bounding line, $x \in \mathcal{Q}$. Similarly, $y \in \mathcal{Q}$. Since $\mathcal{P}$ is contained between the parallel bounding lines of $V_{\overline{uv}}$, it follows that (x, y) is an antipodal pair of $\mathcal{Q}$. □

Thus, it suffices to consider only the antipodal pairs of $\mathcal{Q}$ in computing l. There are three types of antipodal pairs: $(vertex, vertex)$-pair, $(vertex, edge)$-pair, and $(edge, edge)$-pair. Since the first two types are handled similarly, we discuss only the first and third types, abbreviated as VV-pair and EE-pair, respectively. The following theorem summarizes the result.

Theorem 2. *It is possible to determine in $O(n \log n)$ time and $O(n)$ space whether an n-vertex simple polygon $\mathcal{P}$ can be decomposed by a line l into two terrain polygons, $\mathcal{P}^+(e)$ and $\mathcal{P}^-(e)$, where $e = l \cap \mathcal{P}$. If l exists, then it can be computed within the same bounds.*

3.1 Handling Antipodal VV-Pairs

Let $p_1, p_2, \ldots, p_n$ and $q_1, q_2, \ldots, q_m$ be the vertices of $\mathcal{P}$ and $\mathcal{Q}$, respectively, in counter-clockwise order. Consider the set, L, of all lines, l_{ij}, defined by the antipodal VV-pairs (q_i, q_j). For each l_{ij} we can test in $O(n)$ time whether it decomposes $\mathcal{P}$ into two simple polygons, $\mathcal{P}_{ij}^+$ and $\mathcal{P}_{ij}^-$ and whether $\mathcal{P}_{ij}^+(e_{ij})$ and $\mathcal{P}_{ij}^-(e_{ij})$ are terrains, where $e_{ij} = l_{ij} \cap \mathcal{P}$. Since there are $O(n)$ antipodal VV-pairs the running time is $O(n^2)$. We show how to improve this to $O(n \log n)$.

Let q_i be a vertex of $\mathcal{Q}$. The vertices antipodal to q_i form a chain, $C(q_i)$, on $\mathcal{Q}$. Furthermore, if q_i' and q_i'' are the first and last vertices of $C(q_i)$ in counter-clockwise order, then q_i' is the last vertex of $C(q_{i-1})$, and q_i'' is the first vertex of $C(q_{i+1})$ [12, page 173]. $C(q_{i+1})$ comes immediately after $C(q_i)$ and the two share q_i'' (see Fig. 3(a)).

The antipodal pairs of $\mathcal{Q}$ can be enumerated in $O(n)$ time using the *rotating calipers approach* [12,14]. The decomposition lines determined by antipodal VV-pairs can be computed during the enumeration, as follows: We use data structures that can answer the following types of query given a polygon $\mathcal{P}$: (1) determine whether a line, l, through two vertices of $\mathcal{P}$ decomposes $\mathcal{P}$ into two simple polygons; and (2) determine whether $\mathcal{P}(e)$ is a terrain, where e is an edge of $\mathcal{P}$.

For the first type of query we use the ray-shooting data structure of [5]. If the ray directed from u to v meets the boundary of $\mathcal{P}$ at v, then l_{uv} decomposes $\mathcal{P}$ into two simple polygons.

The second type of query can be answered efficiently using a leaf-oriented, balanced, binary-search tree, $\mathcal{T}$, whose leaves store the normal-points of the edges of $\mathcal{P}$ in the order in which they appear around the unit-circle $\mathcal{S}^1$. Given an edge e of $\mathcal{P}$, by Lemma 1 it suffices to locate the neighbors of n_e in $\mathcal{T}$ and check that they are each at distance at least $\pi/2$ from n_e.

We use separate trees for $\mathcal{P}^+$ and $\mathcal{P}^-$, denoted $\mathcal{T}^+$ and $\mathcal{T}^-$, respectively, which maintain the invariant that whenever a line l from L decomposes $\mathcal{P}$ into two simple polygons, $\mathcal{T}^+$ and $\mathcal{T}^-$ contain the normal-points of the edges of $\mathcal{P}^+$ and $\mathcal{P}^-$, respectively, except for $e = l \cap \mathcal{P}$. What remains to be shown is how to maintain $\mathcal{T}^+$ and $\mathcal{T}^-$ efficiently as we consider different antipodal VV-pairs (u, v).

Let $\mathcal{Q}^+$ be the polygon defined by the chain of edges along the boundary of $\mathcal{Q}$ in counter-clockwise order from u to v, together with the edge $\overline{uv}$. Define $\mathcal{Q}^-$ similarly. Since $\mathcal{Q}$ is a convex polygon, the line l_{uv} containing $\overline{uv}$ will always decompose $\mathcal{Q}$ into two simple polygons. Whenever l_{uv} also decomposes $\mathcal{P}$ into two simple polygons, we let $\mathcal{P}^+$ be the polygon defined by the chain of edges along the boundary of $\mathcal{P}$ in counter-clockwise order from u to v, together with the edge $\overline{uv}$. $\mathcal{P}^-$ is defined similarly.

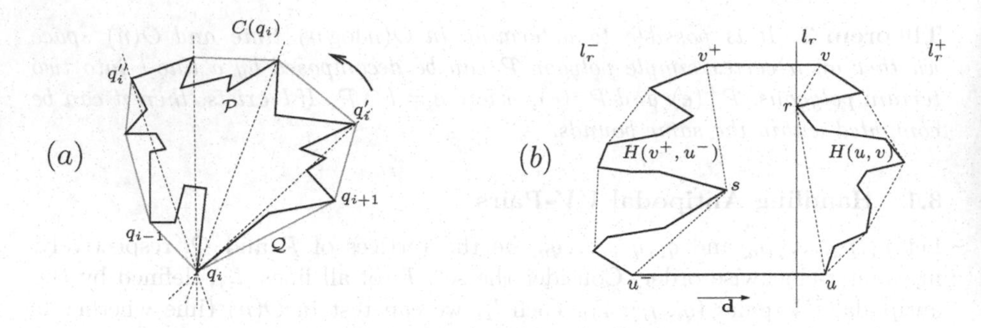

Fig. 3. Illustrating (a) the algorithm of Section 3.1 for antipodal VV-pairs; and (b) the algorithm of Section 3.2 for antipodal EE-pairs

Let $\overline{q_k q_{k+1}}$ be an edge of $\mathcal{Q}$ and let $H(q_k, q_{k+1})$ denote the chain of edges along the boundary of $\mathcal{P}$ from q_k to q_{k+1} in counter-clockwise order. Clearly, if a line l from L decomposes $\mathcal{P}$ into two simple polygons, then $\overline{q_k q_{k+1}}$ is an edge of $\mathcal{Q}^+$ if and only if $H(q_k, q_{k+1})$ belongs to $\mathcal{P}^+$. Therefore, it is sufficient to keep track of the edges of $H(q_k, q_{k+1})$ only when $\overline{q_k q_{k+1}}$ changes from $\mathcal{Q}^+$ to $\mathcal{Q}^-$ and vice versa. This offers an efficient way to maintain $\mathcal{T}^+$ and $\mathcal{T}^-$. Each edge $\overline{q_k q_{k+1}}$ of $\mathcal{Q}$ changes exactly twice between $\mathcal{Q}^+$ and $\mathcal{Q}^-$, and this happens precisely when either u or v is advanced from q_k to q_{k+1}. Whenever $\overline{q_k q_{k+1}}$ changes from $\mathcal{Q}^+$ to $\mathcal{Q}^-$, the normal-points of the edges of $H(q_k, q_{k+1})$ are deleted from $\mathcal{T}^+$ and inserted in $\mathcal{T}^-$. Since, for any two edges, $\overline{q_k q_{k+1}}$ and $\overline{q_j q_{j+1}}$, $H(q_k, q_{k+1})$ and $H(q_j, q_{j+1})$ have no edges in common, each edge of $\mathcal{P}$ will be inserted and deleted exactly twice from $\mathcal{T}^+$ and $\mathcal{T}^-$.

As shown in [7], this algorithm runs in $O(n \log n)$ time and $O(n)$ space.

3.2 Handling Antipodal *EE*-Pairs

The antipodal EE-pairs present new challenges, since unlike the VV-pairs (and VE-pairs), where the decomposition line is determined uniquely, here only the direction of the decomposition line is known. However, it is still possible to determine efficiently a suitable decomposition line (if it exists), as we show below.

We would like to identify all lines, l, determined by EE-pairs, (e', e''), of $\mathcal{P}$ that satisfy case (iii) of Lemma 4. By Lemma 5, it follows that (e', e'') is an antipodal EE-pair of $\mathcal{Q}$, and e' and e'' are edges of both $\mathcal{P}$ and $\mathcal{Q}$.

The antipodal EE-pairs can be enumerated using the following observation: Let (u, v) be an antipodal VV-pair, let u^- be the clockwise neighbor of u, and let v^+ be the counter-clockwise neighbor of v. If both u and u^- are antipodal to $\overline{vv^+}$, then $(\overline{u^-u}, \overline{vv^+})$ is an antipodal EE-pair. Therefore, the antipodal EE-pairs can be enumerated while enumerating the antipodal VV-pairs.

The antipodal EE-pair $(\overline{u^-u}, \overline{vv^+})$ determines a family of parallel lines perpendicular to the lines through $\overline{u^-u}$ and $\overline{vv^+}$. Let $\mathbf{d}$ be the unit-normal that is directed from u^- to u. Let $H(u, v)$ and $H(v^+, u^-)$ be the counter-clockwise chains of vertices from u to v and v^+ to u^-, respectively. Let r be the extreme vertex of $H(u,v)$ in direction $-\mathbf{d}$ and let s be the extreme vertex of $H(v^+,u^-)$ in direction $\mathbf{d}$. Let l_r be the line through r that is perpendicular to $\mathbf{d}$. Clearly, the vertices of $H(u,v)$ are all in exactly one of the closed half-planes of l_r, which we denote as l_r^+. Then l_r decomposes $\mathcal{P}$ into two simple polygons if and only if s is not in l_r^+ (see Fig. 3(*b*)). If l_s is the line through s that is perpendicular to $\mathbf{d}$, then any line, l, in the open strip determined by l_r and l_s is a candidate decomposition line.

To find the extreme vertex of the simple path $H(u,v)$ in a given direction it is sufficient to consider its convex hull. To answer an extreme-vertex query efficiently, we need a data structure for dynamic maintenance of the convex hull of $H(u,v)$. Since the changes to $H(u,v)$ occur only at its ends, we can use the data structure of [4], which has amortized query and update time of $O(\log n)$ for a sequence of n operations. Denote this structure by $\mathcal{H}^+$. Similarly, let $\mathcal{H}^-$ be the data structure for maintaining the convex hull of $H(v^+,u^-)$.

Suppose that l decomposes $\mathcal{P}$ into two simple polygons, $\mathcal{P}^+$ and $\mathcal{P}^-$, as determined by querying $\mathcal{H}^+$ and $\mathcal{H}^-$. We need to test whether $\mathcal{P}^+(e_r)$ and $\mathcal{P}^-(e_r)$ are terrains, where $e_r = l_r \cap \mathcal{P}$. We update $\mathcal{T}^+$ and $\mathcal{T}^-$ to reflect the fact that portions of $\overline{u^-u}$ and $\overline{vv^+}$ are in $\mathcal{P}^+$ and $\mathcal{P}^-$ by inserting the corresponding normal-points. We query $\mathcal{T}^+$ with the normal-point of $-\mathbf{d}$ to determine if $\mathcal{P}^+(e_r)$ is a terrain. Similarly, we query $\mathcal{T}^-$ with the normal-point for $\mathbf{d}$ to determine if $\mathcal{P}^-(e_r)$ is a terrain.

The overall algorithm takes $O(n\log n)$ time and $O(n)$ space (see [7]).

4 Decomposition into Two Terrains without a Common Base

We first define the notion of a *cusp*. A vertex v of $\mathcal{P}$ is a *cusp* w.r.t. a line l containing v if both of v's neighbors in $\mathcal{P}$ are in the same closed half-plane of l and the interior angle at v is strictly greater than π. An edge e of $\mathcal{P}$ is a *cusp* w.r.t. a line l containing e if both of e's endpoints are cusps. Note that if a line l contains a vertex or an edge of $\mathcal{P}$ that is a cusp w.r.t. l, then l decomposes $\mathcal{P}$ into more than two simple polygons.

Lemma 6. *Let L be a non-empty family of lines that decompose $\mathcal{P}$ into two non-empty terrains that do not have a common base. Then there is a line, l, in L which intersects the boundary of $\mathcal{P}$ at points u and v, such that (i) u and v are vertices of $\mathcal{P}$ and neither u nor v is a cusp w.r.t. l; or (ii) the segment joining u and v contains an edge of $\mathcal{P}$ that is not a cusp w.r.t. l; or (iii) u is a vertex of $\mathcal{P}$ that is not a cusp with respect to l, v is in the interior of an edge of $\mathcal{P}$, and l is perpendicular to the line containing some edge of $\mathcal{P}$.*

Proof. Let l be a line in L. Since l decomposes $\mathcal{P}$ into two non-empty simple polygons, $l \cap \mathcal{P}$ contains exactly one line segment, whose endpoints are denoted u and v. If u and v are vertices of $\mathcal{P}$, then neither can be a cusp with respect to l, since otherwise l would decompose $\mathcal{P}$ into more than two simple polygons.

Similarly, if $\overline{uv}$ contains an edge of $\mathcal{P}$, that edge cannot be a cusp. Therefore, case (*i*) or case (*ii*) holds.

If u and v are not vertices of $\mathcal{P}$, let $\mathcal{P}^{+}_{\overline{uv}}$ be the terrain for which $\overline{uv}$ is not a base. Let r be the vertex of $\mathcal{P}^{+}_{\overline{uv}}$ that is closest to l, and let l_r be the line l translated to r. We first show that r is not a cusp with respect to l_r. Assume for a contradiction that r is a cusp. Let f and g be the edges of $\mathcal{P}^{+}_{\overline{uv}}$ incident to r and let ϕ and ψ be the portions of the interior angle at r between l_r and each of f and g, respectively (note that $\phi + \psi < \pi$). Consider the outward unit-normals of $\overline{uv}$, f, and g translated to r. The normal-points $n_{\overline{uv}}$, n_f, and n_g, partition the unit-circle into three disjoint arcs of lengths $\phi + \psi$, $\pi - \phi$, and $\pi - \psi$ (Fig. 4). Since each arc is of length less than π, no point on the unit-circle is at distance at least $\pi/2$ from each of $n_{\overline{uv}}$, n_f, and n_g. However, $\mathcal{P}^{+}_{\overline{uv}}(e)$ is a terrain, for some edge e of $\mathcal{P}^{+}_{\overline{uv}}$, and, furthermore, e is distinct from $\overline{uv}$, f, and g (the interior angle at r is greater than π, and, therefore, neither f nor g can be a base for $\mathcal{P}^{+}_{\overline{uv}}$). Therefore, n_e must be at distance at least $\pi/2$ from each of $n_{\overline{uv}}$, n_f, and n_g, a contradiction. (This proof can also be adapted to show that l_r does not go through an edge of $\mathcal{P}$ which is a cusp [7].)

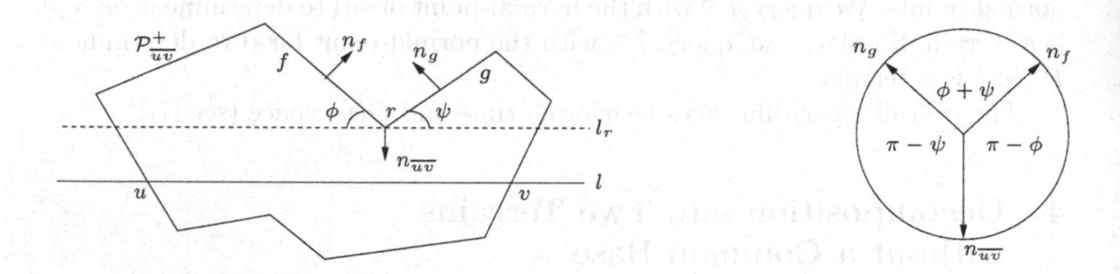

Fig. 4. Illustrating, by contradiction, that vertex r cannot be a cusp

Since r is not a cusp w.r.t. l_r and since no vertex is strictly crossed during the translation of l to r, l_r also decomposes $\mathcal{P}$ into two simple polygons. Furthermore, since the sets of normal-points associated with the two polygons in the decomposition remains the same during the translation, l_r also decomposes $\mathcal{P}$ into two terrains; thus $l_r \in L$.

If l_r goes through an edge of $\mathcal{P}$, then case (*ii*) holds. Otherwise, let $\overline{qr} = l_r \cap int(\mathcal{P})$ and w.l.o.g. let $\mathcal{P}^{+}_{\overline{qr}}$ be the terrain for which $\overline{qr}$ is not a base. Let l'_r be the line l_r rotated by an angle β about r in the half-plane containing $\mathcal{P}^{+}_{\overline{qr}}$, where β is the smallest angle such that l'_r either goes through a vertex r' of $\mathcal{P}$ or becomes perpendicular to the line containing an edge of $\mathcal{P}$; note that l'_r is in L. If the latter case applies, then case (*iii*) holds. Otherwise, similar to the earlier discussion, we can show that l'_r does not go through a vertex or an edge that is a cusp of $\mathcal{P}$ with respect to l'_r, and therefore, either case (*i*) or case (*ii*) holds. □

Using Lemma 6, we can decide if $\mathcal{P}$ can be decomposed into two terrains: Enumerate all lines that go through two vertices of $\mathcal{P}$ (cases (*i*) and (*ii*)), or go through a vertex of $\mathcal{P}$ and are perpendicular to the line containing an edge of $\mathcal{P}$ (case (*iii*)). For each line test whether it decomposes $\mathcal{P}$ into two polygons $\mathcal{P}^+$ and $\mathcal{P}^-$, and whether $\mathcal{P}^+$ and $\mathcal{P}^-$ are terrains. Clearly, the problem can be solved in time $O(n^3)$. We present an improved algorithm, whose performance is summarized below.

Theorem 3. *It is possible to determine in $O(n^2 \log n)$ time and $O(n)$ space if there is a line which decomposes a polygon $\mathcal{P}$ into two terrains. If it exists, it can be computed within these bounds.*

4.1 The Algorithm

Case (*a*)**:** *Handling the lines determined by two vertices of $\mathcal{P}$.*

For each vertex u of $\mathcal{P}$ we enumerate the lines through u by visiting the vertices of $\mathcal{P}$ from u^+ to u^- in counter-clockwise order. Let v be a vertex of $\mathcal{P}$ and let C^+ and C^- be the boundary of $\mathcal{P}$ from u to v, and v to u, respectively, taken in counter-clockwise order. During the walk from u^+ to u^- we maintain two leaf-oriented, balanced, binary-search trees, $\mathcal{T}^+$ and $\mathcal{T}^-$, whose leaves store the normal-points of the edges of $C^+ \cup \overline{uv}$ and $C^- \cup \overline{uv}$, respectively, in the order in which they appear around $\mathcal{S}^1$. Whenever the line l_{uv} through u and v decomposes $\mathcal{P}$ into two simple polygons, $\mathcal{P}^+$ and $\mathcal{P}^-$, the normal-points of their edges are stored in $\mathcal{T}^+$ and $\mathcal{T}^-$, respectively.

Initially $\mathcal{T}^+$ contains only the normal-point of $\overline{uu^+}$, $\mathcal{T}^-$ contains the normal-points of the edges of $\mathcal{P}$, and v is the vertex u^+. During the transition from v to v^+ we perform the following steps:

First, we update $\mathcal{T}^+$ and $\mathcal{T}^-$, to reflect the fact that the edge $\overline{vv^+}$ switches from C^- to C^+, and that the edge $\overline{uv}$ is replaced by $\overline{uv^+}$. The normal-point of $\overline{vv^+}$ is removed from $\mathcal{T}^-$ and inserted in $\mathcal{T}^+$, and the normal-points of $\overline{uv}$ in $\mathcal{T}^+$ and $\mathcal{T}^-$ are replaced by the appropriate normal-points of $\overline{uv^+}$.

Next, we test whether l_{uv} decomposes $\mathcal{P}$ into two simple polygons. We do this by using the algorithm in [5] to test if the ray originating from a point at infinity along l_{uv} and directed towards u intersects the boundary of $\mathcal{P}$ exactly twice.

Finally, if l_{uv} decomposes $\mathcal{P}$ into two simple polygons, $\mathcal{P}^+$ and $\mathcal{P}^-$, we test whether each polygon is a terrain. To test $\mathcal{P}^+$ we find in $\mathcal{T}^+$ the neighbors, n_a and n_d, of the normal-point $(-1, 0)$, and the neighbors, n_b and n_c, of the normal-point $(1, 0)$. As shown in the proof of Lemma 2, these are the only candidates for bases of $\mathcal{P}^+$. For each candidate normal-point it is sufficient to test whether its neighbors in $\mathcal{T}^+$ are at distance at least $\pi/2$. Similarly, we use $\mathcal{T}^-$ to test $\mathcal{P}^-$.

Case (*b*)**:** *Handling the lines that go through a vertex of $\mathcal{P}$ and are perpendicular to the line containing an edge of $\mathcal{P}$.*

Let e be any edge of $\mathcal{P}$ and let l_e be the line containing e. We sweep a line, l, perpendicular to l_e which visits the vertices of $\mathcal{P}$ in sorted order. We define l^+ to be the closed half-plane in the direction of the sweep, and l^- be the closed half-plane in the opposite direction. During the sweep we maintain data structures

$\mathcal{T}^+$ and $\mathcal{T}^-$ similar to part (a). Here $\mathcal{T}^+$ and $\mathcal{T}^-$ contain the normal-points of the edges of $\mathcal{P}$ (or portions thereof) that are in l^+ and l^-, respectively, and the normal-points of the directions normal to l and pointing into l^- and l^+, respectively. Whenever l decomposes $\mathcal{P}$ into two simple polygons, $\mathcal{P}^+$ and $\mathcal{P}^-$, the normal-points of their edges are stored in $\mathcal{T}^+$ and $\mathcal{T}^-$, respectively.

Initially $\mathcal{T}^-$ is empty and $\mathcal{T}^+$ contains the normal-points of the edges of $\mathcal{P}$. At each vertex we consider the incident edges. The normal-points of the edges that lie in l^- are removed from $\mathcal{T}^+$. Next, we test whether l decomposes $\mathcal{P}$ into two terrains as described in part (a). Finally, the normal-points of the edges that lie in l^+ are inserted in $\mathcal{T}^-$.

The overall algorithm takes $O(n^2 \log n)$ time and $O(n)$ space (see [7]).

References

1. B. Asberg, G. Blanco, P. Bose, J. Garcia-Lopez, M. Overmars, G. Toussaint, G. Wilfong, and B. Zhu. Feasibility of design in stereolithography. *Algorithmica*, 19:61–83, 1997.
2. P. Bose, D. Bremner, and M. van Kreveld. Determining the castability of simple polyhedra. *Algorithmica*, 19(1–2):84–113, Sept. 1997.
3. S. P. Fekéte and J. S. B. Mitchell. Terrain decomposition and layered manufacturing. *Intl. J. Comput. Geom. & Appl.*, 11:647–668, 2001.
4. J. Friedman, J. Hershberger, and J. Snoeyink. Efficiently planning compliant motion in the plane. *SIAM J. Comput.*, 25:562–599, 1996.
5. J. Hershberger and S. Suri. A pedestrian approach to ray shooting: Shoot a ray, take a walk. *J. Algorithms*, 18:403–431, 1995.
6. I. Ilinkin, R. Janardan, J. Majhi, J. Schwerdt, M. Smid, and R. Sriram. A decomposition-based approach to layered manufacturing. In *Proc. 7th WADS*, volume 2125 of *LNCS*, pages 389–400. Springer-Verlag, 2001. To appear in *Comput. Geom. Theory Appl.*
7. I. Ilinkin, R. Janardan, and M. Smid. Terrain polygon decomposition, with application to Layered Manufacturing. `http://www.cs.umn.edu/~janardan/terrain.pdf`.
8. C. C. Kai and L. K. Fai. *Rapid Prototyping: Principles and Applications in Manufacturing.* Wiley and Sons, Inc., 1998.
9. J. Majhi, R. Janardan, J. Schwerdt, and M. Smid. Multi-criteria optimization algorithms for layered manufacturing. In *Proc. 14th Ann. ACM Symp. Comput. Geometry*, pages 19–28, 1998. To appear in *Intl. J. Math. Algorithms.*
10. J. Majhi, R. Janardan, J. Schwerdt, M. Smid, and P. Gupta. Minimizing support structures and trapped area in two-dimensional layered manufacturing. *Comput. Geom. Theory Appl.*, 12:241–267, 1999.
11. J. Majhi, R. Janardan, M. Smid, and P. Gupta. On some geometric optimization problems in layered manufacturing. *Comput. Geom. Theory Appl.*, 12:219–239, 1999.
12. F. Preparata and M. Shamos. *Computational Geometry: An Introduction.* Springer-Verlag, New York, NY, 1985.
13. D. Rappaport and A. Rosenbloom. Moldable and castable polygons. *Comput. Geom. Theory Appl.*, 4:219–233, 1994.
14. G. Toussaint. Solving geometric problems with the rotating calipers. In *Proc. IEEE MELECON '83*, pages A10.02/1–4, 1983.

Supertrees by Flipping*

D. Chen[1], O. Eulenstein[1], David Fernández-Baca[1], and M. Sanderson[2]

[1] Department of Computer Science, Iowa State University, Ames, IA 50011, USA
{jackie,oeulenst,fernande}@cs.iastate.edu
[2] Section of Evolution and Ecology, University of California, Davis, CA 95616, USA
mjsanderson@ucdavis.edu

Abstract. The input to a supertree problem is a collection of phylogenetic trees that intersect pairwise in their leaf sets; the goal is to construct a single tree that retains as much as possible of the information in the input. This task is complicated by inconsistencies due to errors. We consider the case where the source trees are rooted and are represented by the clusters they exhibit. The problem is to find the minimum number of *flips* needed to resolve all inconsistencies, where each flip moves a taxon into or out of a cluster. We prove that the minimum flip problem is $\mathcal{NP}$-complete, but show that it is fixed-parameter tractable and give an approximation algorithm for a special case.

1 Introduction

All species of living organisms are thought to be related to each other by a largely tree-like pattern of ancestry and descent — a phylogeny. The tree of life contains millions of species, but phylogenetic analyses typically include on the order of only 10 – 100 species at a time. This stems partly from limitations of data, such as the small number of sequences typically available for homologous genes, and partly from limitations associated with the computational complexity of optimization-based tree-building methods. Thus, it is unlikely that conventional tree-building methods that use comparative data, such as sequences, as their inputs will scale well to data sets involving very large collections of species.

An alternative strategy for combining phylogenetic information is to use a collection of smaller *source trees* as the input and construct directly a larger tree. Such a *supertree* includes all or most of the taxa (labels) from the collection of source trees while preserving the phylogenetic information from those trees [26]. Ideally supertrees also make statements about relationships that were not possible from any single source tree alone. Biologists have been constructing such supertrees informally for some time [8, 22], but the development of formal algorithms to solve problems associated with supertree construction has permitted unprecedented analyses of large phylogenetic groups in the last few years [3, 20, 24, 29].

* Research of D. Chen, O. Eulenstein, and M. Sanderson supported in part by the National Science Foundation (NSF) under grant no. 0075319. D. Fernández-Baca was supported in part by NSF under grant CCR-9520946.

O.H. Ibarra and L. Zhang (Eds.): COCOON 2002, LNCS 2387, pp. 391–400, 2002.

Previous Results: All supertree methods have as their input a collection of phylogenies and as output a supertree or collection of supertrees [13, 27, 28]. Most versions of the supertree problem apply only to phylogenies with the same taxa set; in this context, they are called *consensus tree methods* [5, 16, 18, 19]. One of the few supertree methods that consider phylogenies that do not necessarily share all taxa is *matrix representation using parsimony* (MRP). This approach seeks the most parsimonious (MP) tree(s) [11] from a data set of incomplete binary characters in which each character represents a cluster (clade) from one of the source trees [2, 4, 24, 25]. Taxa present in the cluster are scored 1, those absent in cluster are scored 0, and those not sampled on that source tree are scored by a ?. The wide availability of algorithms for MP [12] have made this approach for building supertrees the choice of phylogeneticists working with real data sets. However, MP is $\mathcal{NP}$-complete [14].

Other supertree problems have algorithms with polynomial running times. Aho et al. [1, 17] devised an efficient algorithm for building a supertree when the input trees are compatible. Semple and Steel [27] modified this procedure to handle incompatibility. Their approach, called the *MinCutSupertree algorithm* (MC), is guided by a local optimization criterion, although some global properties can be shown for the output tree. In any event, the notion of supertree is sufficiently broad and ill-defined that it is not obvious which problem should be solved. Hence, much work remains in deciding which method can best be justified on biological and computational grounds.

Contributions of This Paper: All empirical studies attempting to construct supertrees have found that source trees conflict with one another — errors are present. This paper describes a new way to construct supertrees motivated by a notion of *error correction.* Any tree can be represented by set of clusters, each cluster comprising a set of taxa (labels). One notion of *error* in such a cluster system is the presence of an incorrect label in a cluster or the absence of one that should be present. In a matrix representation of a tree, such errors correspond to *flips* from $0 \to 1$ or $1 \to 0$. When the matrix representations of conflicting source trees are combined in a single matrix, the matrix no longer represents perfectly any phylogenetic tree. A natural optimization problem is to find the minimum number of flips that converts the matrix into one that represents a phylogenetic tree; we call this the *flip problem.*

We show that the flip problem and *simplified* versions of it are $\mathcal{NP}$-complete. For input trees with a same taxa set, we prove that the flip-problem is fixed parameter tractable and fixed ratio approximable. In a computational study [7] that is omitted for brevity, we show that the flip problem outperforms MRP and MC, if solved exactly. The supertree web page [6] provides an exact (exponential time) algorithm to solve the flip problem.

The flip problem is related to *fractional character compatibility* (FCC) [19]. While FCC is not intended as a supertree method, it can be viewed as a flip problem for unrooted phylogenies over the same taxa set.

2 Definitions and Notation

Throughout the paper M is a finite set, $\wp(N)$ denotes the power set of a set N, and $A \Delta B$ denotes the symmetric difference between the sets A and B.

Depending on the proof, we use either set-theoretic or graph-theoretic representations of characters; each leads to corresponding representations of flipping.

2.1 Set-Theoretic Definitions and Notations

Definition 1 (characters). *A* binary character *over M is a bipartition $C = \{C_\circ, C_\bullet\}$ of $M' \subseteq M$. $C_\circ$ and $C_\bullet$ are, respectively, the* 0-state *and the* 1-state *of C. The ordered pair $(C_\circ, C_\bullet)$ is a* directed binary character *over M. The set of all directed binary characters over M is denoted $\mathcal{P}_M$. Unless stated otherwise a directed binary character over M is said to be a* character. *If $C \in \mathcal{P}_M$ is a bipartition of M, it is called* complete. *$\mathcal{C}_M$ denotes the subset of all complete characters in $\mathcal{P}_M$. A* completion *of $P \in \mathcal{P}_M$ is a complete character $C \in \mathcal{C}_M$, where $P_\circ \subseteq C_\circ$ and $P_\bullet \subseteq C_\bullet$.*

Definition 2 (compatibility and phylogenetic trees). *A* phylogenetic tree *T over $M' \subseteq M$ is a rooted tree with leaf set M'. The set of all leaves that descend from the same non-root node of T is a* cluster *in T. Given $A \subseteq M$ the minimal subtree in T connecting A is denoted by $T_{|A}$. Character C has* convex states *in T, iff $T_{|C_\circ}$ and $T_{|C_\bullet}$ do not intersect in their sets of internal nodes. A set $\mathcal{C}$ of characters is* compatible *iff there exists a phylogenetic tree $T_\mathcal{C}$ in which every character $C' \in \mathcal{C}$ has convex states. In this case $T_\mathcal{C}$ and $\mathcal{C}$ are said to be* consistent.

Theorem 1 ([10, 15]). *A set $\mathcal{C} \subseteq \mathcal{C}_M$ of complete characters is compatible iff for every pair $\{C, C'\} \subseteq \mathcal{C}$, $C_\bullet \cap C'_\bullet \in \{\emptyset,\ C_\bullet,\ C'_\bullet\}$.*

The compatibility of a set of complete or incomplete characters can be tested in polynomial time [23].

Definition 3 (flips). *For $C \in \mathcal{P}_M$ and a set $F \in \wp(M)$, called a* flip set *for C, we define the* flip-operation *$\triangleleft$ as follows,*

$$C' = C \triangleleft F :\Longleftrightarrow C'_\circ = C_\circ \Delta F' \wedge C'_\bullet = C_\bullet \Delta F'$$

where $F' = F \cap (C_\circ \cup C_\bullet)$. We call F simply a flip *when C is obvious. F is a* d-flip *if $F \subseteq C_\bullet$, and an* i-flip *if $F \subseteq C_\circ$. Flip operations are generalized to character tuples $\mathcal{C} = (C_i)_{i=1}^r \in \mathcal{P}_M^r$ and flip tuples $\mathcal{F} = (F_i)_{i=1}^r \in \wp(M)^r$ by $\mathcal{C} \triangleleft \mathcal{F} := (C_i \triangleleft F_i)_{i=1}^r$. $s(\mathcal{F}) := \Sigma_{i=1}^r |F_i|$ is the* size *of $\mathcal{F}$.*

Each cluster in a phylogenetic tree T over $M' \subseteq M$ can be represented by a character C where $C_\bullet$ contains all taxa in the cluster and $C_\circ = M' - C_\bullet$. A tuple $\mathcal{T} = (T_1, \ldots, T_l)$ of input trees for the supertree problem is represented by a tuple $\mathcal{C} = (C_{11}, \ldots, C_{1m_1}, \ldots, C_{l1}, \ldots, C_{lm_l})$ of characters, where $(C_{j1}, \ldots, C_{jm_j})$ represents all clusters of T_j, for each $j \in \{1, \ldots, l\}$.

Definition 4. *The* flip problem *(*FP*) is defined as follows.*

Given *A character tuple* $\mathcal{C} \in \mathcal{P}_M^r$ *(representing the clusters in a tuple of phylogenetic trees),* $r \in \mathbb{N}$*, and a number* $k \in \mathbb{N}$*.*
Question *Does there exist a flip tuple* $\mathcal{F} \in \wp(M)^r$ *where* $s(\mathcal{F}) \leq k$*, such that* $\mathcal{C} \triangleleft \mathcal{F}$ *is compatible?*

When $\mathcal{F}$ *is required to be a d-flip (i-flip) tuple the problem is called the* D-Flip Problem (DFP) *(*I-Flip Problem (IFP)*).*

2.2 Graph-Theoretic Definitions

Definition 5 (Character graph). *Let* $\mathcal{C} = (C_1, \ldots, C_r) \in \mathcal{C}_M^r$*. The* character graph *for* $\mathcal{C}$ *is the bipartite graph* $G = (X, Y, E)$*, where* $X = \{1, \ldots, r\}$*,* $Y = M$*, and* $E = \{\{x, y\} \mid x \in X, y \in M, y \in C_{x\bullet}\}$*.*

The flip operation on sets (Definition 3) corresponds to editing the character graph. In particular, flipping a taxon in a character is equivalent to deleting or inserting an edge in the character graph.

Definition 6. *The* Π Edit Problem (EP(π)) *is defined as follows.*

Given *A graph property* Π*, a bipartite graph* $G = (X, Y, E)$*, and* $k \in \mathbb{N}$*.*
Question *Does there exist a set* $F \subseteq X \times Y$*, where* $|F| \leq k$*, such that the graph* $G' = (V, E \Delta F)$ *is a* Π *graph?*

When we place the restriction that $F \subseteq E$ *(*$F \subseteq (A - E)$*) the problem is called the* Π Deletion Problem (DP(π)) *(*Π Insertion Problem (IP(π))*).*

Definition 7 ($\mathcal{M}$-free graph property). *An* $\mathcal{M}$-graph *is a cycle-free path of length* 4*. The property* $\mathcal{M}$-free *is the set of all bipartite graphs* $G = (X, Y, E)$ *that do not have an induced* $\mathcal{M}$*-graph whose degree-*1 *nodes are in* Y*.*

The following proposition, whose proof is omitted, relates the set-theoretic and graph-theoretic definitions of flipping. Its first part expresses Theorem 1 in graph-theoretic terms.

Proposition 1.

1. $\mathcal{C} \in \mathcal{C}_M^r$ *is compatible iff its character graph is* $\mathcal{M}$*-free.*
2. *Problems (i)* EP($\mathcal{M}$*-free*) *and* FP*, (ii)* DP($\mathcal{M}$*-free*) *and* DFP*, and (iii)* IP($\mathcal{M}$*-free*) *and* IFP*, are polynomially equivalent.*

3 $\mathcal{NP}$-Completeness Results

We now show that FP, DFP, and IFP are $\mathcal{NP}$-complete when the input characters are complete. The $\mathcal{NP}$-completeness for (partial) characters follows directly.

For brevity the proofs of Lemmas 1 and 2 are omitted, but can be found in [7].

Lemma 1. *Let $\mathcal{C} = (C_1, C_2) \in \mathcal{C}_M^2$, such that $|C_{1\bullet}| = |C_{2\bullet}| = 3$ and $|C_{1\bullet} \cap C_{2\bullet}| \leq 1$. If $\mathcal{F} = (F_1, F_2) \in \wp(M)^2$ is a flip-tuple, such that $|F_1| = 0$, $|F_2| = 1$, and $\mathcal{C} \triangleleft \mathcal{F} = (C'_1, C'_2)$ is compatible, then $C'_{1\bullet} \cap C'_{2\bullet} = \emptyset$.*

Lemma 2. *Let $\mathcal{C} = (C_i)_{i=1}^r \in \mathcal{C}_M^r$ such that $|C_{i\bullet}| = 3$, and $|C_{i\bullet} \cap C_{j\bullet}| \leq 1$ for $i \neq j$. If there exists a flip tuple $\mathcal{F} = (F_i)_{i=1}^r \in \wp(M)^r$, such that $|F_i| = 1$ for each i and $(C'_1, \ldots, C'_r) = \mathcal{C} \triangleleft \mathcal{F}$ is compatible, then $2r \leq |\bigcup_{i \in \{1,\ldots,r\}} C'_{i\bullet}|$.*

The version of exact 3-cover defined below is known to be $\mathcal{NP}$-complete [19].

Definition 8 (Constrained Exact 3-Cover (CX3C)).

Given *A set X with $|X| = 3q$ for some $q \in \mathbb{N}$, and a collection $\mathcal{S}$ of 3-element subsets of X, such that for any set $\{S, S'\} \subseteq \mathcal{S}$, $|S \cap S'| \leq 1$.*

Question *Does there exist a set $\mathcal{S}' \subseteq \mathcal{S}$ that covers X? ($\mathcal{S}'$ covers X, iff $X = \bigcup_{S \in \mathcal{S}'} S$ and for any $\{S, S'\} \subseteq \mathcal{S}'$, $S \cap S' = \emptyset$.)*

Theorem 2. FP *is $\mathcal{NP}$-complete.*

Proof. Clearly, FP $\in \mathcal{NP}$. The reduction CX3C $\leq_p^m$ FP is shown by a modification of a proof given in [19].

Construction: Let a possible instance for CX3C be a set X, where $|X| = 3bq$ for some $q \in \mathbb{N}$, and $\mathcal{S} = \{S_1, \ldots, S_r\}$ be a collection of 3-element subsets of X. From this instance a possible instance for FP is constructed as follows: For every $i \in \{1, \ldots, r\}$ a character C_i is constructed, such that $C_{i\bullet} = S_i$ and $C_{i\circ} = X - S_i$. From the resulting characters the character r-tuple $\mathcal{C} = (C_1, \ldots, C_r)$ is constructed and the number $k = 2(r - q)$ is calculated.

Obviously, $\mathcal{C}$ and k can be calculated in polynomial time. Thus, the $\mathcal{NP}$-completeness of FP derives from the following statement.

Claim: $\mathcal{S}$ contains a 3-cover for X iff there exists a flip-tuple $\mathcal{F}$ where $s(\mathcal{F}) \leq k$ and $\mathcal{C} \triangleleft \mathcal{F}$ is compatible.

Proof of claim:
"$\Longrightarrow$": Let $\mathcal{S}' = \{S_1, \ldots, S_q\}$ be a subset of $\mathcal{S}$ that is a 3-cover for X. Let $\mathcal{F} = (F_1, \ldots, F_q, F_{q+1}, \ldots F_r)$ be a flip tuple where $F_i = \emptyset$ for any $i \in \{1, \ldots, q\}$ and $F_i \subset C_{i\bullet}$, such that $|F_i| = 2$ for any $i \in \{q+1, \ldots, r\}$. Thus, $s(\mathcal{F}) = k$. Now we show in two steps that $\mathcal{C} \triangleleft \mathcal{F}$ is compatible.
First we argue that $C_i - F_i$ is compatible with any character in $\mathcal{C}_M$ for any $i \in \{q+1, \ldots, r\}$. We have $C_i \triangleleft F_i = C_i - F_i$ where $F_i \subset C_i$, since F_i is a d-flip for C_i. Hence, $|C_i \triangleleft F_i| = |C_i - F_i| = 1$, since $|F_i| = 2$ and $|C_i| = 3$. By Theorem 1, a character $C \in \mathcal{C}_M$ where $|C_\bullet| = 1$, is compatible with any character in $\mathcal{C}_M$. Thus, $C_i \triangleleft F_i$ is compatible with any character in $\mathcal{C}_M$.
Second we argue that $C'_i = C_i \triangleleft F_i$ and $C'_j = C_j \triangleleft F_j$ are compatible for different $i, j \in \{1, \ldots, q\}$. It is $C'_i = C_i$ and $C'_j = C_j$, since $F_i = F_j = \emptyset$. Next we have $C_{i\bullet} \cap C_{j\bullet} = \emptyset$, since by our construction $C_{i\bullet} = S_i$, $C_{j\bullet} = S_j$, and S_i, S_j are elements of the 3-cover for X. Thus, $C'_{i\bullet} \cap C'_{j\bullet} = \emptyset$. From this and Theorem 1 it follows that C'_i and C'_j are compatible.
From the first and second part above, it follows directly that $\mathcal{C} \triangleleft \mathcal{F}$ is compatible.

"$\Longleftarrow$": Suppose that there exists a flip-tuple $\mathcal{F}$ such that $s(\mathcal{F}) \leq k$ and $\mathcal{C} \triangleleft \mathcal{F} = (C'_1, \ldots, C'_r)$ is compatible. W.l.o.g. let $\mathcal{F} = (F_0, \ldots, F_i, F_{i+1}, \ldots, F_j, F_{j+1}, \ldots F_r)$ such that $|F_k| = 0$ for $k \in \{1, \ldots, i\}$, $|F_k| = 1$ for $k \in \{i+1, \ldots, j\}$, and $|F_k| > 1$ for $k \in \{j+1, \ldots, r\}$. Let $f_0 = i$, $f_1 = j - i$, and $f_{2\leq} = r - j$. We show $3/2(f_0 - q) \geq f_1 \geq 2(f_0 - q)$. Hence, $q = f_0$ and $\{S_1, \ldots, S_q\}$ is a 3-cover for X.

It holds that $k = 2(n - q) \geq f_1 + 2f_{2\leq}$ and replacing n by $f_0 + f_1 + f_{2\leq}$ yields $f_1 \geq 2(f_0 - q)$. By construction, we have that $3q \geq |\bigcup_{l\in\{1,\ldots j\}} C'_l|$. By Lemma 1 we have $|\bigcup_{l\in\{1,\ldots j\}} C'_l| = 3f_0 + |\bigcup_{l\in\{i+1,\ldots j\}} C'_l|$. Lemma 2 states $|\bigcup_{l\in\{i+1,\ldots j\}} C'_l| \geq 2f_1$. Thus $3q \geq 3f_0 + 2f_1$; that is, $3/2(f_0 - q) \geq f_1$. □

The proof of the next theorem is similar to that of Theorem 2.

Theorem 3. DFP *is* $\mathcal{NP}$*-complete.*

The following theorem can be proved by a reduction from the $\mathcal{NP}$-complete *chaingraph insertion problem* [30], using Proposition 1 .

Theorem 4. IP($\mathcal{M}$*-free) is* $\mathcal{NP}$*-complete.*

4 Fixed Ratio Approximation Algorithm

Theorem 5. *Let* $G = (X, Y, E)$ *be a character graph such that the nodes in* X *have an upper degree bound* d*. The optimization versions of* EP($\mathcal{M}$*-free) and* DP($\mathcal{M}$*-free) can be approximated within a factor of* $2d$*.*

Proof. Our argument is based on a general result for node deletion and editing problems [21]. Define $\mathcal{M}(G)$ to be the set of induced $\mathcal{M}$–graphs on a bipartite graph G, and $Inc(v)$ to be the set of edges that are incident on a node v.

```
Approx(G)
1  F_A = ∅
2  while A = (X, Y, E − F_A) contains an induced M-graph
3      do let path ⟨a, b, c, d, e⟩ be an induced M-graph in A
4          F_A = F_A ∪ Inc(b) ∪ Inc(d)
5  return F_A
```

Let $F_{A_0} = F_A$, $A_0 = A$, and $n \in \mathbb{N}_0$ be the number of executions of line 4 for some input. At the ith execution of line 4, let A_i denote the graph A, F_{A_i} the edge deletion set F_A, and $\mathcal{M}_i$ the induced $\mathcal{M}$-graph $\mathcal{M}$.

Correctness: From Claim 1 it follows that $\emptyset \subseteq \mathcal{M}(A_i) \subset \mathcal{M}(A_{i-1}) \subseteq \mathcal{M}(G)$ for any $i \in \{1, \ldots, n\}$. Thus the algorithm terminates and $\mathcal{M}(A_n) = \emptyset$.

Approximation ratio: Let F_O be an optimal solution for either EP($\mathcal{M}$-free) or DP($\mathcal{M}$-free). We show that any $\mathcal{M}_i$ is *deleted* by an edge $e_i \in F_O$ (existence) that does not delete $\mathcal{M}_j$ for $j > i$ (uniqueness). The algorithm deletes any $\mathcal{M}_i$ by at most $2d$ edge deletions. It follows that $|F_A| \leq |F_O|\, 2d$.

Existence: From Claim 1 below, it follows that $\mathcal{M}(A_i) \subset \mathcal{M}(G)$, $i \in \{0, \ldots, n\}$. Thus $\mathcal{M}_i$ must be *deleted* by some $e_i \in F_O$ that is incident on the nodes of M_i. Uniqueness: From Claim 2 below, it follows that $\mathcal{M}(A_{i+1}) \subseteq \mathcal{M}(O)$ for $O = (X, Y, (E - A_i)\Delta\{e_i\})$. Thus there exists no induced $\mathcal{M}$-graph in A_{i+1} that is *deleted* by e_i, but not by the algorithm.

Claim 1: Let $B = (X, Y, E)$ and $A = (X, Y, E - Inc(x))$ for some $x \in X$, then $\mathcal{M}(A) \subseteq \mathcal{M}(B)$.

Proof: Suppose that there exists $\mathcal{M} \in \mathcal{M}(B) - \mathcal{M}(A)$. Thus the cycle-free path $\mathcal{M}$ in B is *created* from A by removing some edge in $Inc(x)$. Hence x must be a node on the path $\mathcal{M}$ in $\mathcal{M}(B)$. As a contradiction x is not connected to any other node in B and thus can not be on the path $\mathcal{M}$. □

Claim 2: Let $B = (X, Y, E)$, and $\mathcal{M} \in \mathcal{M}(B)$, described by the path $\langle a, b, c, d, e \rangle$. Further let $A = (X, Y, E - F_A)$ for $F_A = Inc(b) \cup Inc(d)$, and $O = (X, Y, E\Delta F_O)$, for some $F_O \subseteq \{\{x, y\} \mid x \in \{b, d\},\ y \in \{a, c, d\}\}$. Then $\mathcal{M}(A) \subseteq \mathcal{M}(O)$.

Proof of claim : Suppose that there exists $\mathcal{M}' \in \mathcal{M}(A) - \mathcal{M}(O)$. From $\mathcal{M}' \in \mathcal{M}(A)$ we have that (i) $\mathcal{M}'$ does not contain the nodes b, c, since their edge degree in B is 0; (ii) $\mathcal{M}' \in \mathcal{M}(G)$, follows from Claim 1. From (i) and (ii) follows that the path $\mathcal{M}'$ in G does not contain the nodes b and d. We now show as a contradiction that $\mathcal{M}'$ in G contains either b or d. From $\mathcal{M}' \in \mathcal{M}(G)$ and $\mathcal{M}' \notin \mathcal{M}(O)$ follows that some edge modifications in F_O *removed* $\mathcal{M}'$ in O. Since these edges *remove* $\mathcal{M}'$ from G, they are incident on nodes that are on the path $\mathcal{M}'$ in G. Edges in F_O are incident on b or d. It follows b or d is on the path $\mathcal{M}'$ in G. □

5 Fixed Parameter Tractability

Definition 9. *Let $k \in \mathbb{N}$. The* $\mathcal{M}_k$-free edit problem (EP($\mathcal{M}_k$-free)) *is defined as follows.*

Given *A bipartite graph $G = (X, Y, E)$, a partition $\{E_\bullet, E_\circ, E_?\}$.*
Question *Does there exist a set $F \subseteq X \times Y$ where $|F| \leq k$, such that $G' = (V, E\Delta F)$ is $\mathcal{M}$-free?*

When we place the restriction that $F \subseteq E$ ($F \subseteq (A - E)$), the problem is called the $\mathcal{M}_k$-free deletion problem (DP($\mathcal{M}_k$-free)) *(*$\mathcal{M}_k$-free insertion problem (IP($\mathcal{M}_k$-free))*).*

A decision problem parameterized by k is *fixed parameter tractable* ($\mathcal{FPT}$), if there is an algorithm that correctly decides the problem in $O(f(k)p(n))$, for an input of size n. $p(n)$ is a polynomial function and f an arbitrary function [9].

Theorem 6. *Let $k \in \mathbb{N}$, then* EP($\mathcal{M}_k$*-free*)) *is in $\mathcal{FPT}$.*

Proof. Let $G = (X, Y, E)$ be the given character graph. We construct a search tree T for G of height at most $k + 1$, where each node v in the tree T is labeled by a set of edge modifications $m(v)$, and $g(v) = (V, E\Delta m(v))$. For any internal node v in T it holds that the graph $g(v)$ is not $\mathcal{M}$-free, and v has exactly 6 children that represent all possible edge modifications to eliminate a particular induced $\mathcal{M}$-graph in $g(v)$.

Construction: Create the root r of the search tree T with labels $g(r) = G$ and $m(r) = \emptyset$. For any leaf l in the search tree of height less then $k+1$, search for an induced $\mathcal{M}$-graph in $g(l)$. If an induced $\mathcal{M}$-graph $\mathcal{M}$ exists, then there are exactly 6 different single edge modifications to eliminate $\mathcal{M}$ (4 edge deletions and 2 edge insertions). For each edge modification e create a new node v. Label v with $m(n) = m(l) \cup \{e\}$ and $g(v) = (V, E\Delta m(v))$. Make v a child of l. Note: $g(v)$ is $\mathcal{M}$-free, if v is a leaf of depth smaller then $k+1$. Also the character graph $g(v)$ is not $\mathcal{M}$-free, if v is an internal node of T.

Let F be a minimal set such that $G_F = (X, Y, E\Delta F)$ is $\mathcal{M}$-free.

Claim 1: There exists a leaf l in the search tree T, such that $m(l) \subseteq F$.

Proof: For the root r of T it holds that $m(r) = \emptyset$. Thus, if r is a leaf the claim holds. Otherwise r is an internal node in T such that $m(v) \subseteq F$. In general, let v be an internal node in T such that $m(v) \subseteq F$. The character graph $g(v)$ is not $\mathcal{M}$-free, since it is an internal node in T. Hence there exists an induced M-graph M in $g(v)$. Since $m(v) \subseteq F$ and G_F is $\mathcal{M}$-free, there exists an edge $e \in F - m(v)$ that eliminates M. By the construction of T there exists a child u of v, such that $m(u) = m(v) \cup \{e\}$. Hence $m(u) \subseteq F$. Since T is of finite height, there exists a leaf l in T such that $m(l) \subseteq F$. □

Claim 2: T is a complete tree of height $k+1$, iff $|F| > k$.

Proof: *Case* $|F| \leq k$: From Claim 1 it follows that there exists a leaf l in T such that $m(l) \subseteq F$. Thus, $|m(l)| \leq k$ and further the depth of l is smaller or equal k. It follows that T is not a complete tree of height $k+1$. *Case* $|F| > k$: T has no leaf at a depth smaller then $k+1$. Otherwise F would not be a minimal set of edge modifications. By the construction of T, its height is at most $k+1$. Thus, T is a complete tree of height $k+1$. □

Clearly, the search tree can be constructed and tested for a possible completion of height $k+1$ in $O(6^k n^5)$ time. Thus, $\mathcal{M}$-free (k) EP is in $\mathcal{FPT}$. □

Corollary 1. *Let* $k \in \mathbb{N}$. DP($\mathcal{M}_k$*-free*) *and* IP($\mathcal{M}_k$*-free*) *are in* $\mathcal{FPT}$.

6 Discussion and Open Problems

Experimental results [6, 7] suggest that solving the minimum-flip problem exactly reconstructs supertrees more accurately than Semple and Steel's MC algorithm or MRP. Since the minimum-flip problem is $\mathcal{NP}$-complete, our studies were limited to relatively small trees ($\leq$ 16 taxa)[6, 7]. We have developed a heuristic algorithm for the minimal flip problem that allows us to handle larger trees; experiments to be reported elsewhere indicate that the results are close to solutions provided by MRP heuristics, but outperform MC.

An important task is to determine if there exists a *polynomial time approximation scheme* for the minimum flip problem. In fact no positive results for solving the minimum flip problem for (partial) characters are known.

References

1. A. V. Aho, Y. Sagiv, T. G. Szymanski, and J. D. Ullman, *Inferring a tree from lowest common ancestors with an application to the optimization of relational expressions*, SIAM Journal on Computing **10** (1981), no. 3, 405–421.
2. B. R. Baum, *Combining trees as a way of combining data sets for phylogenetic inference, and the desirability of combining gene trees*, Taxon **41** (1992), 3–10.
3. O. R. P. Bininda-Emonds, J. L. Gittleman, and A. Purvis, *Building large trees by combining phylogenetic information: a complete phylogeny of the extant Carnivora (Mammalia)*, Biol. Rev. **74** (1999), 143–175.
4. D. R. Brooks, *Hennig's parasitological method: a proposed solution*, Syst. Zool. **30** (1981), 325,331.
5. David Bryant, John Tsang, Paul E. Kearney, and Ming Li, *Computing the quartet distance between evolutionary trees*, Symposium on Discrete Algorithms, 2000, pp. 285–286.
6. D. Chen, O. Eulenstein, D.Fernández-Baca, and M. Sanderson, http://genome.cs.iastate.edu/supertree.
7. ______, *Supertrees by flipping*, Tech. Report TR02-01, Iowa State University, Dept. of Computer Science, Iowa State University, Department of Computer Science, 226 Atanasoff Hall, Ames, IA 50011-1040 USA, January 2002.
8. M. J. Donoghue, *Phylogenies and the analysis of evolutionary sequences, with examples from seed plants*, Evolution **43** (1989), 1137–1156.
9. R. G. Downey and M. R. Fellows, *Parameterized compllexity*, Springer, 1997.
10. G. F. Estabrook, C. Johnson, and F. R. McMorris, *An idealized concept of the true cladistic character?*, Mathematical Bioscience **23** (1975), 263–272.
11. J. S. Farris, *On comparing the shapes of taxonomic trees*, Systematic Zoology **22** (1976), 50–54.
12. J. Felsenstein, *PHYLIP homepage*, `http://evolution.genetics.washington.edu/phylip.html`.
13. A. D. Gordon, *Consensus supertrees: the synthesis of rooted trees containing overlapping sets of labelled leaves*, J. Classif. **9** (1986), 335–348.
14. R. L. Graham and L. R. Foulds, *Unlikelihood that minimal phylogenies for a realistic biological study can be constructed in reasonable computation time*, Math. Biosci. **60** (1982), 133–142.
15. D. Gusfield, *Algorithms on strings, trees, and sequences*, Cambridge University Press, 1979.
16. Henzinger, King, and Warnow, *Constructing a tree from homeomorphic subtrees, with applications to computational evolutionary biology*, SODA: ACM-SIAM Symposium on Discrete Algorithms (A Conference on Theoretical and Experimental Analysis of Discrete Algorithms), 1996.
17. M. R. Henzinger, V. King, and T. Warnow, *Constructing a tree from homeomorphic subtrees, with applications to computational evolutionary biology*, Algorithmica **24** (1999), 1–13.
18. Sampath Kannan, Tandy Warnow, and Shibu Yooseph, *Computing the local consensus of trees*, Symposium on Discrete Algorithms, 1995, pp. 68–77.

19. P. Kearney, M. Li, J. Tsang, and T. Jiang, *Recovering branches on the tree of life: An approximation algorithm*, SODA, 1999, pp. 537–5465.
20. F. G. R. Liu, M. M. Miyamoto, N. P. Freire, P. Q. Ong, M. R. Tennant, T. S. Young, and K. F. Gugel, *Molecular and morphological supertrees for eutherian (placental) mammals*, Science **291** (2001), 1786–1789.
21. A. Natanzon, R. Shamir, and R. Sharan, *Complexity classification of some edge modification problems*, Discrete Applied Mathematics **113** (2001), no. 1, 109–128.
22. A. Ortolani, *Spots, stripes, tail tips and dark eyes: predicting the function of carnivore colour patterns using the comparative method*, Biol. J. Linn. Soc. **67** (1999), 433–476.
23. I Pe'er, R. Shamir, and R. Sharan, *Incomplete directed perfect phylogeny.*, Proc. CPM 2000, 2000, pp. 143–153.
24. A. Purvis, *A modification to Baum and Ragan's method for combining phylogenetic trees*, Systematic Biology **44** (1995), 251–255.
25. M. A. Ragan, *Phylogenetic inference based on matrix representation of trees*, Molecular Phylogenetics and Evolution **1** (1992), 53–58.
26. M. J. Sanderson, A. Purvis, and C. Henze, *Phylogenetic supertrees: assembling the trees of life*, Trends Ecol. Evol. **13** (1998), 105–109.
27. C. Semple and M. Steel, *A supertree method for rooted trees*, Discrete Applied Mathematics **105** (2000), 147–158.
28. M. A. Steel, *The complexity of reconstructing trees from qualitative characters and subtrees*, Journal of Classification **9** (1992), 91–116.
29. M. F. Wojciechowski, M. J. Sanderson, K. P. Steele, and A. Liston, *Molecular phylogeny of the "temperate herbaceous tribes" of papilionoid legumes: a supertree approach, Adv. Legume Syst.*, in press.
30. M. Yannakakis, *Computing the minimum fill-in is NP-complete*, SIAM Journal on Algebraic and Discrete Methods **2** (1981), no. 1, 77–79.

A Space and Time Efficient Algorithm for Constructing Compressed Suffix Arrays

Tak-Wah Lam[1], Kunihiko Sadakane[2], Wing-Kin Sung[3,*], and Siu-Ming Yiu[1]

[1] Department of Computer Science, University of Hong Kong
Hong Kong
{twlam,smyiu}@csis.hku.hk

[2] Department of System Information Sciences
Graduate School of Information Sciences, Tohoku University
Sendai, Japan
sada@dais.is.tohoku.ac.jp

[3] Department of Computer Science, National University of Singapore
Singapore
ksung@comp.nus.edu.sg

Abstract. With the first Human DNA being decoded into a sequence of about 2.8 billion base pairs, many biological research has been centered on analyzing this sequence. Theoretically speaking, it is now feasible to accommodate an index for human DNA in main memory so that any pattern can be located efficiently. This is due to the recent breakthrough on compressed suffix arrays, which reduces the space requirement from $O(n \log n)$ bits to $O(n)$ bits. However, constructing compressed suffix arrays is still not an easy task because we still have to compute suffix arrays first and need a working memory of $O(n \log n)$ bits (i.e., more than 13 Gigabytes for human DNA). This paper initiates the study of constructing compressed suffix arrays directly from text. The main contribution is a new construction algorithm that uses only $O(n)$ bits of working memory, and more importantly, the time complexity remains the same as before, i.e., $O(n \log n)$.

1 Introduction

DNA sequences, which hold the code of life for living organisms, can be represented by strings over four characters A, C, G, and T. With the advance in bio-technology, the complete DNA sequences for a number of living organisms have been known. Even for human DNA, a draft which comprises about 2.8 billion characters, has been finished recently. This is, however, not the end, the next step is to analyse these sequences.

This paper is concerned with data structures for indexing a sequence over a fixed alphabet so that searching for an arbitrary pattern can be performed efficiently. Such tools are very useful in many areas. In particular, they find

* This research was supported in part by NUS Academic Research Grant R-252-000-119-112

O.H. Ibarra and L. Zhang (Eds.): COCOON 2002, LNCS 2387, pp. 401–410, 2002.

applications to many biological research activities on DNA, such as gene hunting, promoter consensus identification, and motif finding. Unlike English text, DNA sequences do not have word boundaries; suffix trees [9] and suffix arrays [8] are the most appropriate solutions in the literature for indexing DNA. For a DNA sequence with n characters, building a suffix tree takes $O(n)$ time, then a pattern P can be located in $O(|P| + occ)$ time, where occ is the number of occurrences. For suffix arrays, construction and searching takes $O(n \log n)$ time and $O(|P| \log n + occ)$ time, respectively. Both data structures requires $O(n \log n)$ bits; suffix arrays is associated with a smaller constant, though. For human DNA, the best known implementation of suffix tree and suffix array requires 40 Gigabytes and 13 Gigabytes, respectively [7]. Such memory requirement far exceeds the capacity of ordinary computers. Existing approaches for indexing human DNA include (1) using supercomputers with large main memory; and (2) storing the indexing data structure in the secondary storage [1, 6]. The first approach is expensive and inflexible, while the second one is slow. As more and more DNA are decoded, it is vital that individual biologists can eventually analyze different DNA sequences efficiently with their ordinary PCs.

Recent breakthrough results in compressed suffix arrays shed light on this direction [4, 5, 10]. It is now feasible to store a compressed suffix array in the main memory, which occupies only $O(n)$ bits, yet still supporting search efficiently. Precisely, the searching time increases only by a factor of at most $\log n$. For human DNA, such a compressed suffix array occupies about 2 Gigabytes. Nowadays a PC can have up to 4 Gigabytes of main memory and can easily accommodate such a data structure.

Theoretically speaking, a compressed suffix array can be constructed in $O(n \log n)$ time; however, the construction process requires much more than $O(n)$ bits of working memory. Among others, the original suffix array has to be built first, taking up at least $n \log n$ bits. In the context of human DNA, the working memory for constructing a compressed suffix array is at least 25 Gigabytes [11], far exceeding the capacity of ordinary PCs. This motivates us to investigate whether we can construct a certain kind of compressed suffix array in $O(n \log n)$ time while using $O(n)$ bits of memory. The latter requirement means construction directly from DNA sequences. This paper provides the first algorithm of such a kind, showing that the compressed suffix array proposed in [5, 10] can be built in a space and time efficient manner.

Experiments show that for human DNA, our space-efficient algorithm can run on a PC with 3 Gigabytes of memory and takes about 21 hours, i.e., about three times slower than the original algorithm implemented on a supercomputer with 64 Gigabytes of main memory [11] to accommodate the suffix array.

Technically speaking, our algorithm does not require much space other than for storing the compressed suffix array. This is based on an observation that the compressed suffix arrays of two consecutive suffixes are very similar. Thus, we can build the entire compressed suffix array directly from the DNA sequence in an incremental manner. The efficiency of our algorithm benefits from an elegant solution to an interesting data structure problem, which extends a balanced search tree to support selective increment over an arbitrary interval of up to n values in $O(\log n)$ time.

The rest of this paper is organized as follows. Section 2 gives a review of suffix arrays and compressed suffix arrays. Section 3 gives the technical background and a simple framework for computing the compressed suffix array incrementally. Sections 4 and 5 respectively detail the new data structure to support the incremental construction and the constructing algorithm.

2 Compressed Suffix Arrays in a Nutshell

Let Σ be an alphabet, and let \$ be a special character not in Σ. We assume that \$ is lexicographically smaller than any character in Σ. Intuitively, \$ is used to mark the end of a text. Consider a length-n text T, which is represented by an array $T[1..n+1] = T[1]T[2]\ldots T[n+1]$, where $T[n+1] = \$$. For $i = 1, 2, \ldots, n+1$, $T_i = T[i..n+1] = T[i]T[i+1]\ldots T[n+1]$ denotes a suffix of T starting from the i-th position.

i	$T[i]$	T_i
1	a	$acaaccg\$$
2	g	$caaccg\$$
3	a	$aaccg\$$
4	a	$accg\$$
5	g	$ccg\$$
6	g	$cg\$$
7	c	$g\$$
8	\$	\$

i	$SA[i]$	$T_{SA[i]}$
0	8	\$
1	3	$aaccg\$$
2	1	$acaaccg\$$
3	4	$accg\$$
4	2	$caaccg\$$
5	5	$ccg\$$
6	6	$cg\$$
7	7	$g\$$

i	$\Psi[i]$	$T[SA[i]]$
1	3	a
2	4	a
3	5	a
4	1	c
5	6	c
6	7	c
7	0	g

Fig. 1. Array, suffix array, and compressed suffix array of *acaaccg\$*

Suffix Arrays [8]: A suffix array is a sorted sequence of $n+1$ suffixes of T, denoted by $SA[0..n]$. Formally, $SA[0..n]$ is a permutation of the set of integers $\{1, 2, \ldots, n+1\}$ such that, according to the lexicographic order, $T_{SA[0]} < T_{SA[1]} < \ldots < T_{SA[n]}$. See Figure 1 for an example. Note that $SA[0] = n+1$. As each integer takes $\log n$ bits, the suffix array can be stored using $(n+1)\log n$ bits[1]. Given a text T together with the suffix array $SA[1..n]$, the occurrences of any pattern P in T can be found without scanning T again. Precisely, it takes $O(|P|\log n + occ)$ time, where occ is the number of occurrences.

Compressed Suffix Arrays [5, 10]: For every $i = 0, 1, 2, \ldots, n+1$, define $SA^{-1}[i]$ to be the integer j such that $SA[j] = i$. The major component of a compressed suffix array for a text T is an array $\Psi[1..n]$ where $\Psi[i] = SA^{-1}[SA[i]+1]$ for $i = 1, 2, \ldots, n$. See Figure 1 for an example.

Note that $\Psi[1..n]$ contains n integers. The trivial way to store it requires $n \log n$ bits. Nevertheless, $\Psi[1..n]$ can always be partitioned into $|\Sigma|$ strictly

[1] Throughout this paper, we assume that logarithm has a base of 2.

increasing sequences. This special property allows us to store it succinctly. This property is illustrated in the rightmost table in Figure 1 and its correctness is proved formally based on the following two lemmas.

Lemma 1. *For every $i < j$, if $T[SA[i]] = T[SA[j]]$, then $\Psi[i] < \Psi[j]$.*

Proof. Note that $i < j$ if $T_{SA[i]} < T_{SA[j]}$. If $i < j$ and $T[SA[i]] = T[SA[j]]$, then $T_{SA[i]+1} < T_{SA[j]+1}$. Let $p = \Psi[i]$ and $q = \Psi[j]$. Then $T_{SA[p]} = T_{SA[i]+1} < T_{SA[j]+1} = T_{SA[q]}$. Therefore, we have $p < q$. The lemma follows. □

Lemma 2. *$\Psi[1..n]$ can be partitioned into at most $|\Sigma|$ strictly increasing sequences S_c for all $c \in \Sigma$. For every integer $\Psi[i] \in S_c$, $T[SA[i]] = c$.*

Proof. Note that $[1..n]$ can be partitioned into at most $|\Sigma|$ intervals $[1 = i_1..j_1]$, $[i_2..j_2]$, ..., $[i_{|\Sigma|}..j_{|\Sigma|} = n]$ so that for every i, j in a particular interval, $T[SA[i]] = T[SA[j]]$.

Based on Lemma 1, for every $i < j$ within a particular interval $[i_k..j_k]$, $\Psi[i] < \Psi[j]$. Thus, $\Psi[i_k..j_k]$ is a strictly increasing sequence. The lemma follows. □

For DNA sequences, Σ has only four characters. Based on Lemma 2, Ψ is partitioned into four strictly increasing sequences. To represent an increasing sequence, we can store the difference between consecutive values instead of storing the actual values. Precisely, a sequence $(v_1, \ldots, v_l)$ of increasing values can be represented as $v_1, v_2 - v_1, v_3 - v_2, \ldots, v_l - v_{l-1}$. Furthermore, each individual difference is encoded using δ-coding [3], which is for representing integers with variable length. In summary, it is shown that Ψ can be represented using $O(n)$ bits. More importantly, with Ψ and some other auxiliary data structures, the occurrences of any pattern P in the text T can be found in $O(|P| \log n)$ time. See [10] for the algorithmic details.

3 Basic Properties and Incremental Construction

The existing approach to constructing the array Ψ is as follows: First, the suffix array SA is built. Next, the values of the array $\Psi[1..n]$ are computed one by one. Finally, based on δ-coding, a succinct representation of Ψ is generated. Since suffix array requires $n \log n$ bits to store, the working space for constructing a suffix array is at least $n \log n$ bits.

To reduce the working space required, we try to avoid creating the suffix array. Our construction algorithm works directly with the DNA sequence and is based on an incremental approach. Initially, we construct the compressed suffix array for only the last character of the text. Then, we repeatedly expand the text character by character and build a bigger and bigger compressed suffix array.

Let us give the mathematical background behind such an incremental approach. Suppose we are given the compressed suffix array for the text $T[1..n+1]$ where $T[n+1] = \$$. Let $SA_T[0..n]$ be the suffix array of T. Note that $SA_T[0] =$

i	$\Psi_T[i]$	$SA_T[i]$	$T_{SA_T[i]}$
0		8	$
1	3	3	*aaccg*$
2	4	1	*acaaccg*$
3	5	4	*accg*$
4	1	2	*caaccg*$
5	6	5	*ccg*$
6	7	6	*cg*$
7	0	7	*g*$

i	$\Psi_{T'}[i]$	$SA_{T'}[i]$	$T'_{SA_{T'}[i]}$
0		$8 \Rightarrow 9$	$
1	3	$3 \Rightarrow 4$	*aaccg*$
2	4	$1 \Rightarrow 2$	*acaaccg*$
3	$5 \Rightarrow 6$	$4 \Rightarrow 5$	*accg*$
4	1	$2 \Rightarrow 3$	*caaccg*$
5	$SA_T^{-1}[1] = 2$	1	*cacaaccg*$
5	$6 \Rightarrow 7$	$5 \Rightarrow 6$	*ccg*$
6	$7 \Rightarrow 8$	$6 \Rightarrow 7$	*cg*$
7	0	$7 \Rightarrow 8$	*g*$

Fig. 2. Suffix array and compressed suffix array of $T = acaaccg\$$ and $T' = cT$. Note that $x = 4$, that is, $T_{SA_T[x]} < T' < T_{SA_T[x+1]}$.

$n+1$. Let $\Psi_T[1..n]$ be the compressed suffix array where $\Psi_T[i] = SA_T^{-1}[SA_T[i]+1]$ for $i = 1, 2, \ldots, n$.

Suppose a character c is added to the beginning of $T[1..n+1]$, giving us a longer text $T'[1..n+2]$, which is equal to $cT[1..n+1]$. Let $SA_{T'}[0..n+1]$ be the suffix array of T'. Let $\Psi_{T'}[1..n+1]$ be the compressed suffix array where $\Psi_{T'}[i] = SA_{T'}^{-1}[SA_{T'}[i]+1]$ for $i = 1, 2, \ldots, n+1$.

Let us have a close look of the relationship between $T[1..n+1]$ and $T'[1..n+2]$. Suppose we search T' against the suffixes of T according to the order specified by SA_T. Let $x \in \{0, \cdots, n\}$ be the integer such that $T[SA_T[x]..n+1] < T' < T[SA_T[x+1]..n+1]$. The following lemma states the relationship between $SA_T[0..n]$ and $SA_{T'}[0..n+1]$. Figure 2 shows an example to illustrate the relationship between SA_T and $SA_{T'}$.

Lemma 3.

$$SA_{T'}[i] = \begin{cases} SA_T[i]+1 & \text{if } 0 \le i \le x \\ 1 & \text{if } i = x+1 \\ SA_T[i-1]+1 & \text{if } x+2 \le i \le n+1 \end{cases}$$

The above lemma states that $SA_{T'}[x+1] = 1$. Thus, $x = SA_{T'}^{-1}[1] - 1$. The next lemma states the relationship between $\Psi_T[1..n]$ and $\Psi_{T'}[1..n+1]$. See Figure 2 for an example.

Lemma 4. *For* $i \le x$,

$$\Psi_{T'}[i] = \begin{cases} \Psi_T[i] & \text{if } \Psi_T[i] \le x \\ \Psi_T[i]+1 & \text{if } \Psi_T[i] > x \end{cases}$$

For $i = x+1$,

$$\Psi_{T'}[i] = \begin{cases} SA_T^{-1}[1] & \text{if } SA_T^{-1}[1] \le x \\ SA_T^{-1}[1]+1 & \text{if } SA_T^{-1}[1] > x \end{cases}$$

For $i \ge x+2$,

$$\Psi_{T'}[i] = \begin{cases} \Psi_T[i-1] & \text{if } \Psi_T[i-1] \le x \\ \Psi_T[i-1]+1 & \text{if } \Psi_T[i-1] > x \end{cases}$$

Lemma 4 suggests the following algorithm to compute $\Psi_{T'}$ from Ψ_T.

Incremental Construction Algorithm:

Input: $\Psi_T[1..n]$ for $T[1..n+1]$, $SA_T^{-1}[1]$, and $T' = cT$ for some $c \in \Sigma$.
Output: $\Psi_{T'}$ for T' and $SA_{T'}^{-1}[1]$.

1. Locate the index x such that T' should be inserted between the suffices represented by $SA_T[x]$ and $SA_T[x+1]$.
2. Construct $\Psi_{T'}$ from Ψ_T by inserting the new suffix cT into Ψ and incremeting those values that are greater x as follows.
   - set $\Psi_{T'}[j] = \begin{cases} \Psi_T[j] & \text{if } j \leq x \\ SA_T^{-1}[1] & \text{if } j = x+1 \\ \Psi_T[j-1] & \text{if } j > x+1 \end{cases}$
   - for each $\Psi_{T'}[j]$, if $\Psi_{T'}[j] > x$, increment $\Psi_{T'}[j]$ by 1.
3. Set $SA_{T'}^{-1}[1] = x+1$.

The correctness of the above algorithm follows directly from Lemma 4. By executing the incremental construction algorithm for n times, the compressed suffix array for a length-n text can be built. Observe that each incremental construction step can be implemented in $O(n)$ time. Therefore, based on the compressed suffix array data structure in Section 2, we can construct the compressed suffix array Ψ_T for $T[1..n+1]$ using $O(n^2)$ time in $O(n)$ bits working space. This time complexity seems to be tight as each step may be required to increment $\Psi_{T'}$ for $\Omega(n)$ entries. Yet we show in the following section that the time complexity can be improved to $O(n \log n)$ without increasing the space complexity.

4 New Data Structure for Compressed Suffix Arrays

This section proposes a new data structure $\mathcal{D}$ for representing the compressed suffix array, which allows us to construct it in $O(n)$ bits working space while the time required remains $O(n \log n)$. The basic idea is to store the compressed suffix array in a balanced search tree like red-black tree. Furthermore, to save space, every node in the tree represents more than one value (standard trick). To support the selective increment operation efficiently as required in the previous section (see Lemma 6), each node stores relative search key instead of abslute search key. The details will be discussed in the rest of this section.

Recall from Lemma 2, $\Psi[1..n]$ can be partitioned into $|\Sigma|$ strictly increasing sequences $\{S_c | c \in \Sigma\}$. Note that $T[SA[i]] = c$ for all $\Psi[i] \in S_c$. Let $|S_c|$ be the number of integers in sequence S_c.

For a particular value $x \in S_c$, denote $rank(x)$ be its rank when the number in S_c is ordered in increasing order. The next lemma shows the relationship between the rank of S_c and the index of the Ψ function.

Lemma 5. *For $x \in S_c$, $\Psi^{-1}[x] = \sum_{d \in \Sigma, d<c} |S_d| + rank(x)$.*

Proof. Let $S' = \{y \in S_c \mid rank(y) < rank(x)\}$. We have $\Psi^{-1}[y] < \Psi^{-1}[x]$ if and only if $y \in S' \cup \bigcup_{d \in \Sigma, d<c} S_d$. Therefore, $\Psi^{-1}[x] = 1 + |S'| + |\bigcup_{d \in \Sigma, d<c} S_d|$. The lemma follows. □

Now, we describe the modified data structure $\mathcal{D}$ for the compressed suffix array. For each $c \in \Sigma$, the increasing sequence S_c is partitioned into segments of length $5 \log n$ up to $10 \log n$. Let the smallest value in each segment be the representative of the segment. To save space, for each segment, instead of storing the value of every integer, only their differences will be stored. More precisely, consider a segment with integers $v_1, \ldots, v_\ell$ where $v_1 < v_2 < \ldots < v_\ell$. v_1 is denoted as the representative of the segment and the data structure $\mathcal{D}$ stores only the differences $v_2 - v_1, v_3 - v_2, \ldots,$ and $v_\ell - v_{\ell-1}$.

For the space required, all segments can be stored using $5n$-bits. For the representatives, we store them in a modified red-black tree R_c. For every node w in R_c, $v(w)$ is denoted as the value of the representative represent by this node. $lp(w)$ is the first parent on the left when we go up the tree R_c starting from w; if such a parent does not exist, let it be the node u such that $v(u)$ is minimized.

Every node w in the modified red-black tree R_c has the following 6 fields:

- a color bit (red or black)
- a pointer to the left subtree.
- a pointer to the right subtree.
- a pointer to the segment represented by w.
- *size*(w), that is the total number of values represented by the subtree rooted at w.
- $d_w = v(w) - v(lp(w))$.

Although we do not store the value $v(w)$ for every node $w \in R_c$, its value can be recovered when we traverse down the tree R_c starting from the root. More precisely, when we traverse down the tree R_c, for every node w we met on the path, we can compute two information: $lp(w)$ and $v(w)$ in constant time.

Consider any node w of R_c. Suppose that for every node w' on the path from the root to w, we have already figured out the values of $lp(w')$ and $v(w')$. Let w_l and w_r be the left and right children of w, respectively. Note that the values of $lp(w_l), lp(w_r), v(w_l), v(w_r)$ can be computed as follows: When we go down to the node w_l, $lp(w_l) = lp(w)$ and $v(w_l) = d_{w_l} + v(lp(w_l))$. When we go down to the node w_r, $lp(w_r) = w$ and $v(w_r) = d_{w_r} + v(lp(w_r))$.

Observe that R_c looks very similar to a red-black tree. Thus, it has all the advantages of balanced binary search tree. In addition, it has the following property.

Lemma 6. *For any value x, let X be $\{w \in R_c \mid v(w) > x\}$. We can increase the value $v(w)$ for all $w \in X$ by some constant k using $O(\log n)$ time.*

Proof. Note that the absolute value of $v(w)$ is not stored for every node w in R_c. Instead, we store $d_w = v(w) - v(lp(w))$. This means that if the value of $lp(w)$ is increased by k, the value of w increases by k automatically.

Consider any node $w \in X$. There are two cases.

- Case 1. $v(lp(w)) > x$: In this case, since $lp(w) \in X$ and its value will be increased by k, the value of w will increase by k automatically.
- Case 2. $v(lp(w)) \leq x$: In this case, the value of $lp(w)$ will not be changed. The algorithm should increase d_w by k.

Let Y be $\{w \in X \mid v(lp(w)) \leq x\}$. Based on the above case analysis, we know that once d_w is increased for every $w \in Y$, the values of w for all $w \in X$ will increase by k automatically.

Let u be the node in R_c whose value is just larger than x. It can easily check that Y is in fact equal to $\{w \mid w$ is on the path of R_c from the root to u and $v(w) > x\}$. Since any path in R_c is of length $O(\log n)$, we can increase d_w by k for all $w \in Y$ in $O(\log n)$ time. □

For space complexity, note that the color (red or black) can be stored in 1 bit, the pointers to the left subtree or the right subtree can be stored in $\log(\frac{n}{5\log n})$-bits. Thus, each node can be stored using $5\log(\frac{n}{5\log n}+1)$-bits, which is smaller than $5\log n$. Since there are at most $\frac{n}{5\log n}$ representatives in all S_c for $c \in \Sigma$, the total space required by all modified red-black trees R_c is n.

In summary, the total space required by $\mathcal{D}$ to store the segments and the red-black trees is $5n + n = 6n$.

Lemma 7. *The total space required by the data structure $\mathcal{D}$ is $6n$ bits.*

5 An Efficient Incremental Construction

This section shows that, if the compressed suffix array is represented by data structure $\mathcal{D}$, the time required by the incremental construction algorithm (described in Section 3) is $O(\log n)$. Recall that the incremental construction step contains three sub-steps. This section discusses them one by one.

Step 1: This step tries to locate the index x such that T' should be inserted between the suffices represented by $SA_T[x]$ and $SA_T[x+1]$. Based on data structure $\mathcal{D}$, this step can be completed as follows.

1. We traverse the modified red-black tree R_c for S_c, return (a) a node p with value just smaller than $SA_T^{-1}[1]$; (b) *rank*(p). (Note the special case that $SA_T^{-1}[1]$ is smaller than the smallest value in red-black tree).
2. Sequential search the segment s corresponding to p and locate the value just greater than $SA_T^{-1}[1]$. Let q and r be the location and the rank of this value in the segment.
3. $x = \sum_{b<c, b\in\Sigma} |S_b| + rank(p) + r - 1$. (Note: $SA_{T'}^{-1}[1] = x+1$).

Note that step 1.1 just performs a search in the modified red-black tree R_c. Since R_c is a balanced tree, it requires $O(\log n)$ time. Note that during the traverse, *rank*(p) can be computed by means of *size*(). For step 1.2, since a segment contains at most $10\log n$ integers, a sequential search in this segment requires $O(\log n)$ time. Step 1.3 can be computed in $O(1)$ time. In total, step 1 requires $O(\log n)$ time.

Step 2: This step can be completed as follows.

1. Insert value $SA_T^{-1}[1]$ at location q of segment s.
2. Increment the total number of entries represented by the subtree at node p.
3. If the segment contains $\geq 10 \log n$ values, split the segment into two segments.
4. For every S_b where $b \in \Sigma$,
   (a) Increase all representatives in red-black tree whose values are greater than x by one. Finally, return the node p which is just smaller than or equal to x.
   (b) Sequential search the segment corresponding to p to find the value just greater than x. Increment this value (if exists) by one.

For Step 2.1, we need to insert the value $SA_T^{-1}[1]$ at location q of segment s. Assume that the segment s contains ℓ values $v_1, \ldots, v_\ell$. Also, recall that we only store the differences, that is, we store $v_2 - v_1, v_3 - v_2, \ldots, v_\ell - v_{\ell-1}$. To insert $SA_T^{-1}[1]$ into location q, Step 2.2 can be computed in $O(1)$ time. In Step 2.3, we need to split the segment, the algorithm is required to allocate a new segment and move $5 \log n$ values to the new segment. Such operation can be done in $O(\log n)$ as we need to move $5 \log n$ entries. Step 2.4(a) can be solved in $O(\log n)$ time based on Lemma 6 and the modified red-black tree data structure. Step 2.4(b) again can be solved in $O(\log n)$ time since a segment has less than $10 \log n$ values. In total, Step 2 requires $O(\log n)$ time.

Step 3: This step is simple. It takes $O(1)$ time.

In conclusion, the time required for each incremental construction is $O(\log n)$.

References

1. D. R. Clark and J. I. Munro. Efficient suffix trees on secondary storage. In *Proceedings of the Seventh Annual ACM-SIAM Symposium on Discrete Algorithms (SODA)*, pages 383–391. 1996.
2. Altschul S. F., Gish W., Miller W., Myers E. W., and Lipman D. J. Basic locol alignment search tool. *Journal of Molecular Biology*, pages 403–410, 1990.
3. P. Elias. Universal codeword sets and representation of the integers. *IEEE Transactions on Information Theory*, 21(2):194–203, 1975.
4. P. Ferragine and G. Manzini. Opportunistic data structures with applications. In *Proceedings of the 41st Annual Symposium on Foundations of Computer Science (FOCS)*, pages 390–398. 2000.
5. R. Grossi and J.S. Vitter. Compressed suffix arrays and suffix trees with applications to text indexing and string matching. In *Proceedings of the 32nd ACM Symposium on Theory of Computing*, pages 397-406, 2000.
6. E. Hunt, M. P. Atkinson, and R. W. Irving. A database index to large biological sequences. In *Proceedings of the 27th VLDB Conference*, pages 410–421. 2000.
7. S. Kurtz. Reducing the space requirement of suffix trees. *Software Practice and Experiences*, 29:1149–1171, 1999.
8. U. Manber and G. Myers. Suffix arrays: a new method for on-line string searches. *SIAM Journal on Computing*, 22(5):935–948, 1993.

9. E. M. MCreight. A space-economical suffix tree construction algorithm. *Journal of the ACM*, 23(2):262–272, 1976.
10. K. Sadakane. Compressed text databases with efficient query algorithms based on compressed suffix array. In *Proceedings of the 11th International Conference on Algorithms and Computation (ISAAC)*, pages 410–421. 2000.
11. K. Sadakane and T. Shibyya. Indexing huge genome sequences for solving various porblems. In *Genome Informatics*, pages 175–183. 2001.

Sharpening Occam's Razor
(Extended Abstract)

Ming Li[1,*], John Tromp[2,**], and Paul Vitányi[2,***]

[1] Department of Computer Science, Univ. California Santa Barbara, CA 93106, USA
mli@cs.ucsb.edu
[2] CWI, Kruislaan 413, 1098 SJ, Amsterdam, Netherlands
{tromp,paulv}@cwi.nl

Abstract. We provide a new representation-independent formulation of Occam's razor theorem, based on Kolmogorov complexity. This new formulation allows us to:

- Obtain better sample complexity than both length-based [4] and VC-based [3] versions of Occam's razor theorem, in many applications.
- Achieve a sharper reverse of Occam's razor theorem than that of [5]. Specifically, we weaken the assumptions made in [5] and extend the reverse to superpolynomial running times.

1 Introduction

Occam's razor theorem as formulated by [3,4] is arguably the substance of efficient pac learning. Roughly speaking, it says that in order to learn, it suffices to compress. A partial reverse, showing the necessity of compression, has been proved by Board and Pitt [5]. Since the theorem is about the relation between effective compression and pac learning, it is natural to assume that a sharper version ensues by couching it in terms of the *ultimate* limit to effective compression which is the Kolmogorov complexity. We present results in that direction.

Despite abundant research generated by its importance, several aspects of Occam's razor theorem remain unclear. There are basically two versions. The VC dimension-based version (Theorem 3.1.1 of [3]) gives the following upper bound on sample complexity: For a hypothesis space H with $VCdim(H) = d$, $1 \leq d < \infty$,

$$m(H, \delta, \epsilon) \leq \frac{4}{\epsilon}\left(d \log \frac{12}{\epsilon} + \log \frac{2}{\delta}\right). \quad (1)$$

* Supported in part by the NSERC Operating Grant OGP0046506, ITRC, and NSF-ITR Grant 0085801 at UCSB.
** Partially supported by an NSERC International Fellowship and ITRC.
*** Partially supported by the European Community through NeuroCOLT ESPRIT Working Group Nr. 8556. Affiliated with CWI and the University of Amsterdam.

O.H. Ibarra and L. Zhang (Eds.): COCOON 2002, LNCS 2387, pp. 411–419, 2002.

The following lower bound was proved by Ehrenfeucht *et al* [6].

$$m(H, \delta, \epsilon) > \max\left(\frac{d-1}{32\epsilon}, \frac{1}{\epsilon}\ln\frac{1}{\delta}\right). \tag{2}$$

The upper bound in Equation 1 and the lower bound in Equation 2 differ by a factor $\Theta(\log \frac{1}{\epsilon})$. It was show in [8] that this factor is, in a sense, unavoidable.

When H is finite, one can directly obtain the following bound on sample complexity for any consistent algorithm:

$$m(H, \delta, \epsilon) \leq \frac{1}{\epsilon}\ln\frac{|H|}{\delta}. \tag{3}$$

For a graded boolean space H_n, we have the following relationship between the VC dimension d of H_n and the cardinality of H_n,

$$d \leq \log|H_n| \leq nd. \tag{4}$$

When $\log|H_n| = O(d)$ holds, then the sample complexity upper bound given by Equation 3 can be seen to match the lower bound of Equation 2 up to a constant factor, and thus any consistent algorithm achieves optimal sample complexity for such hypothesis spaces.

The length based version of Occam's razor gives the sample complexity, for given ϵ and δ:

$$m = \max\left(\frac{2}{\epsilon}\ln\frac{1}{\delta}, \left(\frac{(2\ln 2)s^\beta}{\epsilon}\right)^{1/(1-\alpha)}\right), \tag{5}$$

when the *deterministic* occam algorithm returns a consistent hypothesis of length at most $m^\alpha s^\beta$ with $\alpha < 1$ and s is the length of the target concept.

In summary, the VC dimension based occam's razor theorem may be hard to use and it sometimes does not give the best sample complexity. The length-based Occam's razor is more convenient to use and often gives better sample complexity in the discrete case.

However, as we will demonstrate in this paper, the fact that the length-based Occam's razor theorem sometimes gives inferior sample complexity, can be due to the redundant representation format of the concept.

We believe Occam's razor theorem should be "representation-independent". That is, it should not be dependent on accidents of "representation format". (See [13] for other representation-independence issues.) In fact, the sample complexities given in Equations 1 and 3 are indeed representation-independent. However they are not easy to use and do not give optimal sample complexity.

In this paper, we give a Kolmogorov complexity-based Occam's razor theorem. We will demonstrate that our KC-based Occam's razor theorem is convenient to use (as convenient as the length based version), gives a better sample complexity than the length based version, and is representation-independent. In fact, the length based version can be considered as a specific computable approximation to the KC-based Occam's razor.

As one of the examples, we will demonstrate that the standard trivial learning algorithm for monomials actually often has a *better sample complexity* than the more sophisticated Haussler's greedy algorithm [7], using our KC-based Occam's razor theorem. This is contrary to the common belief that Haussler's algorithm is better.

Another issue related to Occam's razor theorem is the status of the reverse assertion. Although a partial reverse of Occam's razor theorem has been proved by [5], it applied only to the case of polynomial running time and sample complexity. They also required a property of closure under exception list. This latter requirement, although quite general, excludes some reasonable concept classes. Our new formulation of Occam's razor theorem allows us to prove a more general reverse of Occam's razor theorem, allowing the arbitrary running time and weakening the requirement of exception list of [5].

2 Occam's Razor

Let us assume the usual definitions, say Anthony and Biggs [1]. Also assume the notation of Board and Pitt [5]. For Kolmogorov complexity we assume the basics of [11].

In the following Σ is a finite alphabet. i.e. we consider only discrete learning problems in this paper.

First we define a pac-algorithm and a generalized notion of Occam-algorithm.

Definition 1. *A* pac-algorithm *for a class of representations* $\mathbf{R} = (R, \Gamma, c, \Sigma)$ *is a randomized algorithm* L *such that, for any* $s, n \geq 1, 0 < \epsilon, \delta < 1, r \in R^{\leq s}$, *and any probability distribution* D *on* $\Sigma^{\leq n}$, *if* L *is given* s, n, ϵ, δ *as input and has access to an oracle providing examples of* c(r) *according to* D, *then* L, *with probability at least* $1 - \delta$, *outputs a representation* r' *such that* $D(r' \oplus r) \leq \epsilon$. *The running time and sample complexity of the pac-algorithm are expressed as functions* $t(n, s, \epsilon, \delta)$ *and* $m(n, s, \epsilon, \delta)$.

Definition 2. *An* Occam-algorithm *for a class of representations* $\mathbf{R} = (R, \Gamma, c, \Sigma)$ *is a randomized algorithm which on input of a sample of length* m *of* $r \in R$, *and any* $\gamma > 0$, *with probability at least* $1-\gamma$ *outputs a representation* r' *consistent with the sample, such that* $K(r'|r, n, s) = m/f(m, n, s, \gamma)$, *with* $f(m, n, s, \gamma)$, *the compression achieved, being an increasing function of* m. *The running time of the Occam-algorithm is expressed as a function* $t(m, n, s, \gamma)$, *where* n *is the maximum length of the input examples.*

Our first theorem is a Kolmogorov complexity based Occam's Razor. We denote the minimum m such that $f(m, n, s, \gamma) \geq x$ by $f^{-1}(x, n, s, \gamma)$.

Theorem 1. *Suppose we have an Occam-algorithm for* $\mathbf{R} = (R, \Gamma, c, \Sigma)$ *with compression* $f(m, n, s, \gamma)$. *Write* f *as* $f(m, \gamma)$ *with the other parameters implicit. Then there is a pac-learning algorithm for* $\mathbf{R}$ *with sample complexity*

$$m(n,s,\epsilon,\delta) = \max\left(\frac{2}{\epsilon}\ln\frac{2}{\delta}, f^{-1}\left(\frac{2\ln 2}{\epsilon}, \delta/2\right)\right),$$

and running time $t(n,s,\epsilon,\delta) = t(m(n,s,\epsilon,\delta), n, s, \delta/2)$.

Proof. On input of ϵ, δ, s, n, the learning algorithm will take a sample of length $m = m(n,s,\epsilon,\delta)$ from the oracle, then use the Occam algorithm with $\gamma = \delta/2$ to find a hypothesis (with probability at least $1-\delta/2$) consistent with the sample and with low Kolmogorov complexity. In the proof we further abbreviate f to $f(m)$ with the other parameters implicit. Learnability follows in the standard manner from bounding (by the remaining $\delta/2$) the probability that all m examples of the target concept fall within the, probability ϵ or greater, symmetric difference with a bad hypothesis. Let $m \geq m(n,s,\epsilon,\delta)$. Then $m \geq f^{-1}(\frac{2\ln 2}{\epsilon}, \frac{\delta}{2})$ gives

$$\epsilon - \frac{\ln 2}{f(m)} \geq \frac{\epsilon}{2}.$$

Bounding $(1-\epsilon)^m$ by $e^{-\epsilon m}$ and taking negative logarithms,

$$2^{m/f(m)}(1-\epsilon)^m \leq \delta/2 \Leftrightarrow$$

$$m\left(\epsilon - \frac{\ln 2}{f(m)}\right) \geq \ln\frac{2}{\delta},$$

which follows from the above and the first lower bound on m. □

Corollary 1. *When the compression is of the form*

$$f(m,n,s,\gamma) = \frac{m^{1-\alpha}}{p(n,s,\gamma)},$$

one can achieve a sample complexity of

$$\max\left(\frac{2}{\epsilon}\ln\frac{2}{\delta}, \left(\frac{(2\ln 2)p(n,s,\delta/2)}{\epsilon}\right)^{1/(1-\alpha)}\right).$$

In the special case of total compression, where $\alpha = 0$, *this further reduces to*

$$\frac{2}{\epsilon}\left(\max\left(\ln\frac{2}{\delta}, (\ln 2)p(n,s,\delta/2)\right)\right). \tag{6}$$

For deterministic Occam-algorithms, we can furthermore replace $2/\delta$ *and* $\delta/2$ *in Theorem 1 by* $1/\delta$ *and* δ *respectively.*

Remark. Essentially, our new Kolmogorov complexity condition is a computationally universal generalization of the length condition in the original Occam's razor theorem of [4]. Here, in Theorem 1, we consider the shortest description length over all effective representations. This is representation-independent in the very strong sense of being an absolute and objective notion, which is recursively invariant by Church's thesis and the ability of universal machines to simulate each another.

Definition 3. *An* exception handler *for a class of representations* $\mathbf{R} = (R, \Gamma, c, \Sigma)$ *is an algorithm which on input of a representation* $r \in R$ *of length s, and an* $x \in \Sigma^*$ *of length n, outputs a representation* r' *of the concept* $c(r) \oplus \{x\}$*, of length at most* $e(s, n)$*, where e is the exception expansion function. The running time of the exception-handler is expressed as a function* $t(n, s)$ *of the representation and exception lengths. If* $t(n, s)$ *is polynomial in* n, s *and* $e(s, n)$ *is of the form* $s + p(n)$ *for some polynomial* $p()$ *then we say* $\mathbf{R}$ *is* polynomially closed under exceptions.

Theorem 2. *Let L be a pac-algorithm and E be an exception handler for* $\mathbf{R} = (R, \Gamma, c, \Sigma)$*. Then there is an Occam algorithm for* $\mathbf{R}$ *with compression* $\frac{1}{2\epsilon n}$*, where* ϵ*, depending on* m, n, s, γ*, is such that* $m(n, s, \epsilon, \gamma) = \epsilon m$ *holds.*

Proof. The proof is obtained in a fashion similar to Board and Pitt. Suppose we are given a sample of length m and confidence parameter γ. Assume without loss of generality that the sample contains m different examples. Define a uniform distribution on these examples with $\mu(x) = 1/m$ for each x in the sample. Let ϵ be as described. E.g. when $m(n, s, \epsilon, \gamma) = (\frac{1}{\epsilon})^b$ for some constant b, then $\epsilon = m^{-1/(b+1)}$. Apply L with $\delta = \gamma$ and above ϵ. It produces a concept which is correct with error ϵ, giving up to ϵm exceptions. We can just add these one by one using the exception handler. This will expand the concept size, but not the Kolmogorov complexity. The resulting representation can be described by the examples used plus the exceptions found, each taking n bits. This gives the claimed compression. □

Definition 4. *A* majority-of-3 algorithm *for a class of representations* $\mathbf{R} = (R, \Gamma, c, \Sigma)$ *is an algorithm which on input of 3 representation* $r_1, r_2, r_3 \in R^{\leq s}$*, outputs a representation* r' *of the concept* $MAJ(r_1, r_2, r_3)$ *of length at most* $e(s)$*, where e is the majority expansion function. The running time of the algorithm is expressed as a function* $t(s)$ *of the maximum representation length. If* $t(s)$ *and* $e(s)$ *are polynomial in s then we say* $\mathbf{R}$ *is* polynomially closed under majority-of-3.

Theorem 3. *Let L be a pac-algorithm with sample complexity* $m(n, s, \epsilon, \delta)$ *subquadratic in* $\frac{1}{\epsilon}$*, and let M be a majority-of-3 algorithm for* $\mathbf{R} = (R, \Gamma, c, \Sigma)$*. Then there is an Occam algorithm for* $\mathbf{R}$ *with compression* $m/3nm(n, s, \frac{1}{2\sqrt{m}}, \gamma/3)$*.*

Proof. Let us be given a sample of length m. Take $\delta = \gamma/3$ and $\epsilon = 1/\sqrt{m}$.
Stage 1: Define a uniform distribution on the m examples with $\mu(x) = 1/m$ for each x in the sample. Apply the learning algorithm. It produces (with probability at least $1 - \gamma/3$) a hypothesis r_1 which has error less than ϵ, giving up to $\epsilon m = \sqrt{m}$ exceptions. Denote this set of exceptions by E_1.

Stage 2: Define a new distribution on the m examples with $\mu(x) = \epsilon_2 = 1/(2\sqrt{m})$ for each x in E_1, and $\mu(x) = (1 - |E_1|/2\sqrt{m})/(m - |E_1|)$ for each x not in E_1. Apply the learning algorithm with error bound ϵ_2. It produces (with probability at least $1 - \gamma/3$) a hypothesis r_2 which is correct on all of E_1 and with error less than ϵ_2 on the remaining examples. This gives up to $\epsilon_2(m - |E_1|)/(1 -$

$|E_1|/2\sqrt{m}) < \sqrt{m}$ exceptions. Denote this set E_2. We have that E_2 is disjoint from E_1.

Stage 3: Define a new distribution on the m examples with $\mu(x) = 1/|E_1 \cup E_2|$ for each x in $E_1 \cup E_2$, and $\mu(x) = 0$ elsewhere. Apply the learning algorithm with error bound $\epsilon_3 = 1/2\sqrt{m}$. Note that $|E_1| \leq \sqrt{m}$ and $E_2 < \sqrt{m}$ gives that for x in $E_1 \cup E_2$, $\mu(x) > \epsilon_3$. Thus the algorithm produces (with probability at least $1 - \gamma/3$) a hypothesis r_3 which is correct on all of E_1 and E_2 and which might be totally wrong elsewhere (we don't care).

In total the number of examples consumed by the pac-algorithm is at most $3m(n, s, \frac{1}{2\sqrt{m}}, \gamma/3)$ each requiring n bits to describe. The three representations are combined into one representing the majority of the 3 concepts. This is necessarily correct on all of the m examples, since the 3 exception-sets are all disjoint. Furthermore, it can be described in terms of the examples fed to the pac-algorithm and thus achieves compression $f(m, n, s, \gamma) = m/3nm(n, s, \frac{1}{2\sqrt{m}}, \gamma/3)$. This is seen to be an increasing function of m given the assumed subquadratic sample complexity. □

The following corollaries use the fact that if a class is learnable, it must have finite VC-dimension and hence, according to Equation 1, they are learnable with sample complexity subquadratic in $\frac{1}{\epsilon}$.

Corollary 2. *Let a class* $\mathbf{R} = (R, \Gamma, c, \Sigma)$ *be closed under either exceptions or majority-of-3. Then* $\mathbf{R}$ *is pac-learnable iff there is an Occam algorithm for* $\mathbf{R}$.

Corollary 3. *Let a class* $\mathbf{R} = (R, \Gamma, c, \Sigma)$ *be polynomially closed under either exceptions or majority-of-3. Then* $\mathbf{R}$ *is polynomially pac-learnable iff there is a polynomial time Occam algorithm for* $\mathbf{R}$.

Example. Consider threshold circuits, acyclic circuits whose nodes compute threshold functions of the form $a_1x_1 + a_2x_2 + \cdots + a_nx_n \geq \delta$, $x_i \in \{0, 1\}, a_i, \delta \in N$ (note that no expressive power is gained by allowing rational weights and threshold). A simple way of representing circuits over the binary alphabet is to number each node and use *prefix-free encodings* of these numbers. For instance, encode i as $1^{|\text{bin}(i)|}0\text{bin}(i)$, the binary representation of i preceded by its length in unary. A complete node encoding then consists of the encoded index, encoded weights, threshold, encoded degree, and encoded indices of the nodes corresponding to its inputs. A complete circuit can be encoded with a node-count followed by a sequence of node-encodings. For this representation, a majority-of-3 algorithm is easily constructed that renumbers two of its three input representations, and combines the three by adding a 3-input node computing the majority function $x_1 + x_2 + x_3 \geq 2$. It is clear that under this representation, the class of threshold circuits are polynomially closed under majority-of-3. On the other hand they are not closed under exceptions, or under the exception lists of Board and Pitt [5].

Example. Let h_1, h_2, h_3 be 3 k-DNF formulas. Then $\text{MAJ}\{h_1, h_2, h_3\} = (h_1 \wedge h_2) \vee (h_2 \wedge h_3) \vee (h_3 \wedge h_1)$ which can be expanded into a $2k$-DNF formula. This is not good enough for Theorem 3, but it allows us to conclude that pac-learnability of k-DNF implies compression of k-DNF into $2k$-DNF.

3 Applications

We demonstrate how our KC-based Occam's razor theorem might be *conveniently* used, providing better sample complexity than the length-based version. In addition to giving better sample complexity, our new KC-based Occam's razor theorem, Theorem 1, is easy to use, as easy as the length based version, as demonstrated by the following two examples.

While it is easy to construct an artificial concept class with extremely bad representations such that our Theorem 1 gives *arbitrarily* better sample complexity than the length-based sample complexity given in Equation 5, we prefer to give real examples.

Application 1: Learning a String.
The DNA sequencing process can be modeled as the problem of learning a super-long string in the pac model [9,10]. We are interested in learning a target string t of length say 3×10^9 (length of a human DNA sequence). At each step, we can obtain as an example a substring of this sequence of length n, from a random location of t (Sanger's Procedure). In current practice, $n \approx 500$, and sampling is very expensive. Formally, the concepts we are learning are sets of possible length n substrings of a superstring, and these are naturally represented by the superstrings. We assume a minimal target representation (which may not hold in practice). Suppose we obtain a sample of m substrings (all positive examples). In biological labs, a Greedy algorithm which repeatedly merges a pair of substrings with maximum overlap is routinely used. It is conjectured that Greedy produces a common superstring t' of length at most $2s$, where s is the optimal length (NP-hard to find). In [2], we have shown that $s \leq |t'| \leq 4s$. Assume that $|t'| \approx 2s$.[1] Using the length-based Occam's razor theorem, this length of $2s$ would determine the sample complexity, as in Equation 6, with $p(n, s, \delta/2) = 2{\cdot}2s$ (the extra factor 2 is the 2-logarithm of the size of the alphabet $\{A, C, G, T\}$). Is this the best we can do? It is well-known that the sampling process in DNA sequencing is a very costly and slow process.

Let's now improve the sample complexity using our KC-based Occam's razor theorem.

Lemma 1. *Let t be the target string of length s and t' be the superstring returned by Greedy of length at most $2s$. Then*

$$K(t'|t, s, n) \leq 2s(2\log s + \log n)/n.$$

Proof. We will try to give t' a short description using some information from t. Let $S = \{s_1, \ldots, s_m\}$ be the set of m examples (substrings of t of length n). Align these substrings with the common superstring t', from left to right. Divide them into groups such that each group's leftmost string overlaps with every string in the group but does not overlap with the leftmost string of the previous group. Thus there are at most $2s/n$ such groups.

[1] Although only the $4s$ upper bound was proved in [2], it is widely believed that $2s$ holds.

To specify t', we only need to specify these $2s/n$ groups. After we obtain the superstring for each group, we re-construct t' by optimally merge the superstrings of the neighboring groups. To specify each group, we only need to specify the first and the last string of the group and how they are merged. This is because every other string in the group is a substring of the string obtained by properly merging the first and last strings. Specifying the first and the last strings requires $2 \log s$ bits of information to indicate their locations in t and we need another $\log n$ bits to indicate how they are merged. Thus $K(t'|t, n) \leq 2s(2 \log s + \log n)/n$. □

This lemma shows that Equation 6 can also be applied with $p(n, s, \delta/2) = 2 \cdot 2s(2 \log s + \log n)/n$, giving a factor $n/(2 \log s + \log n)$ improvement in sample-complexity. Note that in DNA practice, we have $n = 500$ and $s = 3 \times 10^9$. The sample complexity is reduced over "length based" Occam's razor by a multiplicative factor of $n/(2 \log s + \log n) \approx \frac{500}{2 \times 31 + 9} \approx 7$.

Application 2: Learning a Monomial.
Consider boolean space of $\{0, 1\}^n$. There are two well-known algorithms for learning monomials. One is the standard algorithm.

Standard Algorithm.

1. Initial Concept: $m = x_1 \overline{x_1} \ldots x_n \overline{x_n}$.
2. For each positive example, delete from m the variables that would make the example false.
3. Return the resulting monomial m.

Haussler [7] proposed a more sophisticated algorithm based on set-cover approximation as follows.

Haussler's Algorithm.

1. Use only negative examples. For each literal x, define S_x to be the set of negative examples such that x falsifies these negative examples. The sets associated with the literals in the target monomial form a minimum set cover of negative examples.
2. Run the approximation algorithm of set cover, this will use at most $k \log m$ sets or, equivalently, literals in our approximating monomial. Here k is the number of variables in the target monomial.

It is commonly believed that Haussler's algorithm has better sample complexity than the standard algorithm. We demonstrate that the opposite is sometimes true (in fact for most cases), using our KC-based Occam's razor theorem, Theorem 1. Let's assume that our target monomial M is of length $n - \sqrt{n}$. Then the length-based Occam's razor theorem gives sample complexity n/ϵ for both algorithms, by Formula 6. However, $K(M'|M) \leq \log 3\sqrt{n} + O(1)$, where M' is the monomial returned by the standard algorithm. This is true since the standard algorithm always produces a monomial M' that contains *all* literals of the target monomial M. Also, we only need $\log 3\sqrt{n} + O(1)$ bits to specify whether other literals are in or not in M'. Thus our Equation 6 gives the sample complexity of $O(\sqrt{n}/\epsilon)$. In fact, as long as $|M| > n/\log n$ (which is most likely to be the case if every monomial has equal probability), it makes sense to use the standard algorithm.

4 Conclusions

Several new problems are suggested by this research. If we have an algorithm that, given a length-m sample of a concept in Euclidean space, produces a consistent hypothesis that can be described with only $m^\alpha, \alpha < 1$ symbols (including a symbol for every real number; we're using uncountable representation alphabet), then it seems intuitively appealing that this implies some form of learning. However, as Board and Pitt noted in their paper [5], the standard proof of Occam's Razor does not apply, since we cannot enumerate these representations. The main open question is under what conditions (specifically on the real number computation model) such an implication would nevertheless hold.

Can we replace the exception element or majority of 3 requirement by some weaker requirement? Or can we even eliminate such closure requirement and obtain a complete reverse of Occam's razor theorem? Our current requirements do not even include things like k-DNF and some other reasonable classes.

Acknowledgements

We wish to thank Tao Jiang for many stimulating discussions.

References

1. M. Anthony and N. Biggs, *Computational Learning Theory*, Cambridge University Press, 1992.
2. A. Blum, T. Jiang, M. Li, J. Tromp, M. Yannakakis, Linear approximation of shortest common superstrings. *Journal ACM*, 41:4 (1994), 630-647.
3. A. Blumer and A. Ehrenfeucht and D. Haussler and M. Warmuth, Learnability and the Vapnik-Chervonenkis Dimension. *J. Assoc. Comput. Mach.*, 35(1989), 929-965.
4. A. Blumer and A. Ehrenfeucht and D. Haussler and M. Warmuth, Occam's Razor. *Inform. Process. Lett.*, 24(1987), 377-380.
5. R. Board and L. Pitt, On the necessity of Occam Algorithms. 1990 *STOC*, pp. 54-63.
6. A. Ehrenfeucht, D. Haussler, M. Kearns, L. Valiant. A general lower bound on the number of examples needed for learning. *Inform. Computation*, 82(1989), 247-261.
7. D. Haussler. Quantifying inductive bias: AI learning algorithms and Valiant's learning framework. *Artificial Intelligence*, 36:2(1988), 177-222.
8. D. Haussler, N. Littlestone, and, M. Warmuth. Predicting $\{0, 1\}$-functions on randomly drawn points. *Information and Computation*, 115:2(1994), 248–292.
9. T. Jiang and M. Li, DNA sequencing and string learning, *Math. Syst. Theory*, 29(1996), 387-405.
10. M. Li. Towards a DNA sequencing theory. *31st IEEE Symp. on Foundations of Comp. Sci.*, 125-134, 1990.
11. M. Li and P. Vitányi. *An Introduction to Kolmogorov Complexity and Its Applications.* 2nd Edition, Springer-Verlag, 1997.
12. L. G. Valiant. A Theory of the Learnable. *Comm. ACM*, 27(11), 1134-1142, 1984.
13. M.K. Warmuth. Towards representation independence in PAC-learning. In *AII-89*, pp. 78-103, 1989.

Approximating 3D Points with Cylindrical Segments*

Binhai Zhu

Department of Computer Science, Montana State University,
Bozeman, MT 59717-3880, USA
bhz@cs.montana.edu

Abstract. In this paper, we study a 3D geometric problem originated from computing neural maps in the computational biology community: Given a set S of n points in 3D, compute k cylindrical segments (with different radii, orientations, lengths and no segment penetrates another) enclosing S such that the sum of their radii is minimized. There is no known result in this direction except when $k = 1$. The general problem is strongly NP-hard and we obtain a polynomial time approximation scheme (PTAS) for any fixed $k > 1$ in $O(n^{3k-2}/\delta^{4k-3})$ time by returning k cylindrical segments with sum of radii at most $(1 + \delta)$ of the corresponding optimal value. Our PTAS is built upon a simple (though slower) approximation algorithm for the case when $k = 1$.

1 Introduction

Computing and simulating the behavior of a neuron is of particular interest to many researchers in computational biology and medical sciences. To do that, one needs to first study the functional characteristics of neural maps which is again based on the modeling of a neuron. In practice, what researchers in those areas do is to use electronic devices to obtain a dense sample of 3D points of a neuron and then try to approximate them using segments of cylinders (with different radii, orientations and lengths). Different segments with varying radii have different functional characteristics [JT96,PDJ99,JT00].

However, the problem seems to be difficult. In reality, researchers in those areas usually use available commercial software to reconstruct an approximate 3D polyhedron from the sample points. Then, from the polyhedron, people can manually compute the segments of cylinders which 'seems' to fit the polyhedron the best (Figure 1). In practice, this is a time-consuming process. In this paper, we try to study the problem from a computer scientist's point of view. We are especially interested in designing efficient approximation algorithms for solving the problem.

Theoretically, the problem is to compute k cylindrical segments which enclose a given set of 3D points P such that the sum of radii of the segments is minimized.

* This research is partially supported by Hong Kong RGC CERG grant CityU1103/99E, NSF CARGO grant DMS-0138065 and a MONTS grant.

O.H. Ibarra and L. Zhang (Eds.): COCOON 2002, LNCS 2387, pp. 420–429, 2002.

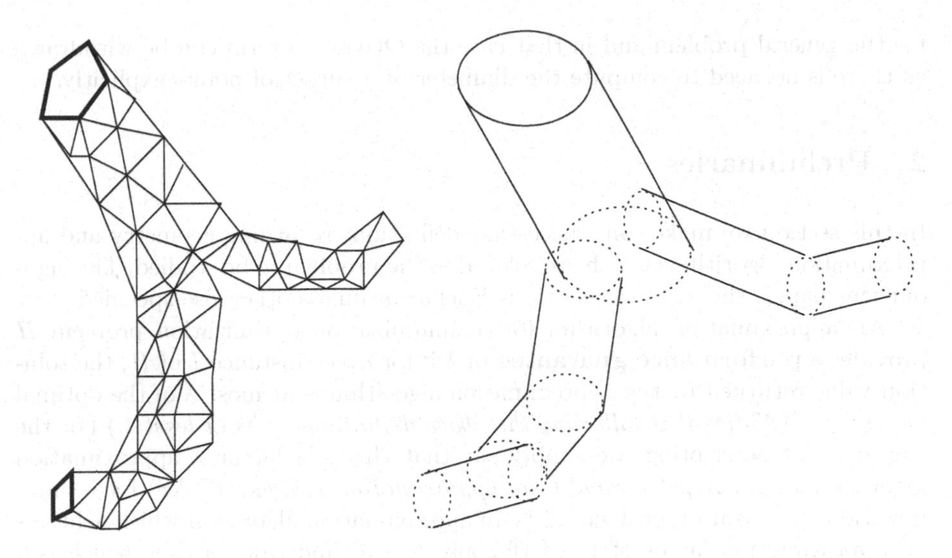

Fig. 1. Approximating a reconstructed polyhedron with cylindrical segments.

To the best of our knowledge, the only known related work uses m-flats (strips in 2D, cylinders in 3D) to cover a set of points in d-dimensions [AP00,HV02]; moreover, in our problem a cylindrical segment cannot penetrate another one by some small amount (to be defined later). Therefore, when $k > 1$, given same input the algorithms of [AP00,HV02] might generate completely different output compared with ours. (Of coures, when $k = 1$, using cylindrical segment or cylinder does not make much difference.) For $k = 1$, Schömer et al. presented an $O(n^4 \text{polylog}\, n)$ time algorithm [SSTY00] and Agarwal et al. obtained an $O(n^{3+\epsilon})$ time algorithm [AAS97]. When n is sufficiently large, both of the algorithms are not practical. Agarwal et al. also obtained a $(1+\delta)-$approximation which runs in $O(n/\delta^2)$ time [AAS97]. However, their algorithm used several subroutines (like computing the transversal of a set of 3D cubes and computing the smallest enclosing disks of a set of 2D points) which also make it hardly practical in terms of implementations. Recently, Chan [Ch00] obtained a $(1 + \delta)-$approximation which runs in $O(n/\delta)$ time [Ch00] using fixed dimensional convex programming.

In this paper, we show that it is possible to modify the proof in [MT82] to prove that the general problem when k is not part of the input is strongly NP-hard. We present a polynomial time approximation scheme (PTAS) for the problem for arbitrary fixed k. First, for $k = 1$ we present a factor-2 approximation which runs in $O(n \log n)$ time. Based on it, we then obtain a simple $(1 + \delta)-$approximation which runs in $O(n \log n + n/\delta^4)$ time. Although the running time is higher than those in [AAS97,Ch00], this algorithm is much simpler, it does not use any of those subroutines used in [AAS97] and it does use convex programming as in [Ch00], is hence practical. (The only complex subroutine is computing the diameter of a set of 3D points, which it calls only once.) Moreover, this algorithm can be used as the subroutine in the innermost loop of our PTAS for the general problem and in that case the $O(n \log n)$ term can be withdrawn as there is no need to compute the diameter of a subset of points explicitly.

2 Preliminaries

In this section we make some necessary definitions regarding geometry and approximation algorithms which are related to the problem to be studied. Throughout this paper, the distance metric is Euclidean unless otherwise specified.

An approximation algorithm for a minimization optimization problem Π provides a **performance guarantee** of λ if for every instance I of Π, the solution value returned by the approximation algorithm is at most λ of the optimal value for I. (*Notice that following the above definitions, λ is at least 1.*) For the simplicity of description, we simply say that this is a factor λ approximation algorithm for Π. A *polynomial time approximation scheme* (PTAS) for a minimization optimization problem Π is an approximation algorithm which achieves an approximation factor of $1+\delta$ (for any $\delta > 0$) and runs in time which is a polynomial of the input size n and $1/\delta$.

A cylinder C is an infinite set of points which have at most a distance R to a given line l in 3D. The line l is called the *center* of C. The sectional area of C which is vertical to l corresponds to a disk with radius R.

Given a line segment s_1s_2 in 3D, let the distance from a point q to the line through s_1s_2 be $d(q,r)$. The distance from q to s_1s_2 is $d(q,r)$ if r is on the line segment s_1s_2, otherwise the distance from q to s_1s_2 is infinite. A cylindrical segment S is an infinite set of points which have at most distance R to a given line segment s_1s_2 in 3D. Similarly, the line segment s_1s_2 is called the *center* of S. The two sectional areas through s_1, s_2 are called the *bases* of S. The length of s_1s_2, $d(s_1,s_2)$, is called the *length* of S and $2R$ is called the *width* of S. We denote them as *length*(S) and *width*(S) respectively and we assume that *length*$(S) \geq$ *width*(S). The ratio *length*(S)/*width*(S) is called the *aspect ratio* of S, denoted by $\alpha(S)$. It is easy to see that $\alpha(S) \geq 1$.

Two cylindrical segments S_1, S_2 intersects each other if $S_1 \cap S_2 \neq \emptyset$. Let the width of S_1, S_2 be w_1, w_2 ($w_1 \geq w_2$) and let the center of S_1, S_2 be l_1, l_2 respectively. S_1 *penetrates* S_2 if either the distance between l_1, l_2 is less than $(w_1 - w_2)/2$ or both of $S_1 - S_1 \cap S_2$ and $S_2 - S_1 \cap S_2$ are disconnected. This penetration constraint comes from the corresponding biological application as a branch of a neuron cannot penetrate another one. In practice, we simulate a neuron with approximate cylindrical segments so we allow one segment to penetrate another one by a small amount. (This makes sense as in practice even the sample points obtained contain small errors.) S_1 penetrates S_2 by an amount of δ if either the distance between l_1, l_2 is less than $|(w_1 - w_2)/2 - \delta|$ or both of $S_1 - S_1 \cap S_2$ and $S_2 - S_1 \cap S_2$ are disconnected. Clearly, it is easy to check whether one cylindrical segment penetrates another one by an amount of δ.

Given a set P of n points in 3D, the *diameter* of P is the maximum distance $d(p_1,p_2)$, $p_1, p_2 \in P$, over all points in P. We denote it as $D(P)$. The CSC (Cylinder Segments Cover) problem is defined as follows:

Instance: A set P of points $p_1, p_2, ..., p_n$ in the 3D space, integer k and real numbers K, σ.

Question: Does there exist a set of k cylindrical segments with radii $r_1, ..., r_k$ respectively, such that $r_1 + r_2 + ... + r_k \leq K$, no segment penetrates another by an amount of σ and each point in P is contained in at least one of the cylinder segments?

For the optimization version of the problem, we would like to minimize k and $r_1 + r_2 + ... + r_{k-1} + r_k$. Notice that the lengths of the segments do not quite matter as long as no one penetrates another one by an amount of σ — this latter condition naturally limits the lengths of some segments.

3 The Case when $k = 1$

In this section we present an efficient approximation algorithm for the case when $k = 1$. Our algorithm is elementary in the sense that it does not use any complex subroutines. The algorithm can also approximate the smallest enclosing cylinder of P by returning its length as infinity.

Given a set P of n points, let C^* be the smallest enclosing cylindrical segment of P (i.e., P is completely contained in C^*). We have the following simple approximation algorithm:

Algorithm 1.

Step 1 Compute the diameter $D(P)$ of P. Let $D(P)$ be $d(p_1, p_2)$.

Step 2 Use p_1p_2 as the center of the approximating cylindrical segment A. The maximum distance between point $q \in P$ and p_1p_2 is the radius of A. Also, return the length of A, $d(p_1, p_2)$.

Lemma 1. *Algorithm 1 presents a factor 2 approximation for the smallest enclosing cylindrical segment problem in $O(n \log n)$ time.*

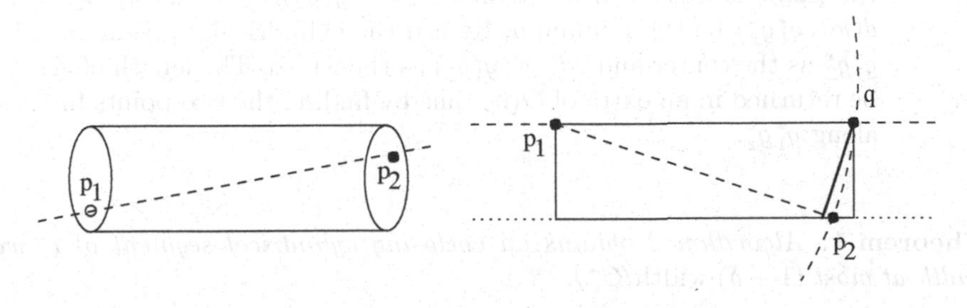

Fig. 2. Approximating a cylindrical segment along the diameter.

Proof: We refer to Figure 2. Let q be the furthest point from p_1p_2. The maximum value of $d(q; p_1p_2)$ is achieved when either p_1p_2 is on the surface of C^*, q is on the other side of C^* and the plane through $\triangle p_1p_2q$ contains the center of C^*, or, when p_1p_2 has the maximum angle to the center of C^*, the plane through $\triangle p_1p_2q$ contains the center of C^* and the height on p_1p_2 in $\triangle p_1p_2q$ is maximized. In both cases, this maximum distance is *width*(C^*) (Figure 2 shows one of the worst-case situation in 2D, exchanging p_2 with q shows the other worst case). The width of the approximating cylindrical segment is hence $2d(q; p_1p_2)$, which is bounded by $2\cdot$*width*(C^*). □

In practice, especially in biology related applications, a factor-2 approximation is hardly useful. So we must have better approximations if we do not want to use the exact solution (which is too expensive). In [AAS97,Ch00], two $(1+\delta)$-approximation algorithms for the smallest enclosing cylinder problem, were proposed. The running time are $O(n/\delta^2)$ and $O(n/\delta)$ respectively. We present below an approximation algorithm which achieves the same approximation factor but much easier. We do not need any complex subroutines used in [AAS97,Ch00].

Lemma 2. *Let u^*v^* be the center of the smallest enclosing cylindrical segment C^* of P. Then the smaller of the angles between p_1p_2 and u^*v^*, θ, is at most $\pi/4$.*

Proof: We have *width*(C^*)$\leq$*length*(C^*)$\cdot \sin\theta$. If $\theta > \pi/4$ then C^* would have a width greater than *width*(C^*). A contradiction to the optimality of C^*. □

The above two lemmas make the following simple factor-$(1+\delta)$ approximation possible.

Algorithm 2.

Step 1 Compute the approximate cylindrical segment A of P using Algorithm 1.

Step 2 Increase the radius of the two (circular) bases of A by a factor of $\sqrt{2}$, generate a grid of $2/\delta \times 2/\delta$ points on the enlarged bases of A. Let the sets of these grid points be G_1 and G_2 respectively.

Step 3 For each line g_1g_2 determined by two grid points $g_1 \in G_1, g_2 \in G_2$ find the point $w \in P$ which maximizes $d(w; g_1g_2)$. Over all g_1, g_2, w, let $d(w^*; g_1^*g_2^*)$ be the minimum. Return the cylindrical segment A^* with $g_1^*g_2^*$ as the center and $2d(w^*; g_1^*g_2^*)$ as the width. The length of A^* can be returned in an extra of $O(n)$ time by finding the two points furthest along $g_1^*g_2^*$.

Theorem 1. *Algorithm 2 obtains an enclosing cylindrical segment of P with width at most* $(1+\delta)\cdot$width*(C^*).*

Proof: Let the bases of A which contains p_1, p_2 be B_1, B_2 respectively. By Lemma 1, we know that *width*(C^*) $\leq$ *width*(A) $\leq 2\times$*width*(C^*). By Lemma 2, we know that u^*v^* and p_1p_2 have an angle of at most $\pi/4$. Therefore, the intersection of line u^*v^* and B_1, B_2 must be at most $\frac{1}{\cos\theta}$(*width*(C^*)$/2$)$\leq \sqrt{2}$(*width*(C^*)$/2$)

distance away from p_1, p_2, which is bounded by $\sqrt{2}(width(A)/2)$. So if we increase the radius of A by a factor of $\sqrt{2}$ and let the resulting bases be B_1', B_2' then the intersection of line u^*v^* and B_1, B_2 must be on or inside B_1', B_2'. As the distance between two adjacent grid points is $\frac{\sqrt{2}}{2}\delta \cdot width(A) \leq \sqrt{2}\delta \cdot width(C^*)$, clearly, these intersection points must be at most $\frac{1}{2}\delta \cdot width(A) \leq \delta \cdot width(C^*)$ distance away from the closest grid points. Following Algorithm 2, the returned width of A^*, $2d(w^*; g_1^* g_2^*)$, is at most $(1+\delta) \cdot width(C^*)$. □

The running time of the above algorithm is $O(n \log n + n/\delta^4)$, which is slower than those algorithms in [AAS97,Ch00]. However, our algorithm does not use any complex subroutines like computing the transversal of a set of cubes in 3D and computing the smallest enclosing disk of a set of points in 2D[1] as in [AAS97] or convex programming as in [Ch00]; therefore, algorithm is completely practical. The $O(n \log n)$ component is due to the computation of diameter of a set of 3D points [Ra00], which is not an easy task in terms of implementations. Fortunately, we can use any $O(n)$-time factor-c approximation for the diameter of P which only increases the running time of the corresponding algorithm by a constant factor (related to c). (Similar ideas are used in [AP00,Ch00] in which they use a factor-2 approximation for the diameter of P.) For the CSC problem, in each segment of a neuron (which can be approximated with a cylindrical segment) the number of points sampled is not too huge (usually a couple of thousand)[2]; therefore, we can even use the brute-force $O(n^2)$ algorithm in practice.

In the next section, we show how to use this algorithm as a subroutine in the innermost loop of our PTAS for any fixed $k > 1$.

4 The Case for Any Fixed $k > 1$

In this section, we shall proceed with the problem of covering P with k cylindrical segments such that no one penetrates another by an amount of δ and the sum of their radii is to be minimized. We present a PTAS for the problem, using the results in the previous section as an efficient subroutine.

Let $A_i^*, i = 1, ..., k$, be the k optimal approximating cylindrical segments of P. Let $S_i^*, i = 1, ..., k$ be the points of P which is contained in A_i^*. We have the following lemma.

Lemma 3. *For at least one i, the diameter of S_i^* contains at least one extreme point of P.*

Note that when $k = 2$ we can have a stronger version of the above lemma. In that case, the diameter of at least one of S_1^*, S_2^* is an antipodal pair of P. We present the following PTAS for the general problem and the objective is to return k approximating cylindrical segments $A_1, ..., A_k$.

[1] Computing the smallest enclosing disk might not be considered as complex now due to the package of Gärtner.

[2] The number of points sampled over a whole neuron is much larger, as it is composed of hundreds of segments.

Algorithm k**CSC.**

Step 1 Compute the convex hull of P, $CH(P)$.
Step 2 $i \leftarrow 1$.
Step 3 For each point $p \in CH(P)$, do the following.
3.1 For each point $q \in P, q \neq p$, let the points of P contained in the intersection of two balls centered at p, q and with radius $d(p,q)$, $B(p; d(p,q)) \cap B(q; d(p,q))$, be S_i'.
3.2 Use a slightly modified Algorithm 2 to compute an approximate cylindrical segment A_i as follows.
3.2.1 Compute the approximate cylindrical segment A_i' of S_i' using Algorithm 1.
3.2.2 Increase the radius of the two bases of A_i' by a factor of $\sqrt{2}$, generate a grid of $2/\delta \times 2/\delta$ points on the enlarged bases of A_i'. Let the sets of these grid points be G_1' and G_2' respectively.
3.2.3 For each line g_1g_2 determined by two grid points $g_1 \in G_1', g_2 \in G_2'$ and each point $r \in S_i'$ compute an approximate cylindrical segment A_i with radius $d(r; g_1g_2)$ which encloses a subset of points in S_i'. The length of A_i can be returned in an extra $O(n)$ time by finding the two points enclosed in A_i' which are the furthest along g_1g_2. Let the points of S_i' contained in A_i be S_i.
3.2.4 Update $P \leftarrow P - S_i$, $i \leftarrow i+1$. Repeat Step (3) until $i = k+1$.
Step 4 Identify all valid solutions by checking in each set of solution whether there exists a cylindrical segment which penetrates another one by δ. Among all valid solutions, compute the sum of the radii of the k cylindrical segments. Return the set of cylindrical segments with the minimum sum of radii.

From Lemma 3, for at least one i, the diameter of S_i^* contains at least one extreme point of P. The algorithm is basically to try all possible extreme point p and all possible point q such that pq is the diameter of S_i^*. By appling a slightly modified Algorithm 2 (i.e., to try all possible approximate cylindrical segments with radii $d(r; g_1g_2)$, over all $r \in S_i'$ and all grid points g_1, g_2 on different bases of A_i'), we can obtain an $(1+\delta)$ approximation A_i for A_i^* in $O(\frac{n^3}{\delta^4})$ time. Lemma 3 still holds after updating $P \leftarrow P - S_i$. Therefore, this recursive algorithm certainly can find a set of approximate cylindrical segments A_i for $A_i^*, i = 1, ..., k$ such that the sum of their radii is at most $(1+\delta)$ that of $A_i^*, i = 1, ..., k$.

Let $T(n,k)$ be the running time from step (2) to step (4) in the above algorithm, we have

$$T(n,k) = O(\frac{n^3}{\delta^4})T(n, k-1), k > 1,$$

$$T(n,1) = O(\frac{n}{\delta^4}),$$

which is $O(\frac{n^{3k-2}}{\delta^{4k}})$. Therefore the running time of the whole algorithm is $O(n \log n + \frac{n^{3k-2}}{\delta^{4k}})$. The $O(n \log n)$ term is the cost of step (1) for constructing 3D convex hull [PS85]. When $k > 1$, the running time is simply $O(\frac{n^{3k-2}}{\delta^{4k}})$. Note that at Step 3.2.4 if we stop at $i = k$ and simply use the algorithm in [Ch00] to compute the last approximate cylindrical segment containing all the remaining points then we can improve the running time of the above PTAS to $O(\frac{n^{3k-2}}{\delta^{4k-3}})$. (However, we cannot use Chan's algorithm at other steps in our algorithm.) We thus have the following theorem for CSC when $k > 1$ is fixed.

Theorem 2. *There is a polynomial time approximation scheme for the CSC problem when $k > 1$ is fixed: for any small $\delta > 0$, it runs in $O(\frac{n^{3k-2}}{\delta^{4k-3}})$ time and returns k minimum cylindrical segments whose sum of width are at most $(1+\delta)$ of the corresponding optimum.*

We remark that with simple modification, our PTAS for any fixed k would work for any function f over the k radii $r_1, ..., r_k$ as long as the function is monotone and $f(0, ..., 0) = 0$ (in our current setting $f(r_1, ..., r_k) = r_1 + ... + r_k$). For instance, our PTAS also works when the objective is to minimize the sum of volume of the cylindrical segments.

5 Hardness Result

In this section, we sketch how to modify the proof of Megiddo and Tamir [MT82] to show that the general problem of covering a set of 3D points with the minimum number of line segments (cylindrical segments with radii zero) in 3D is strongly NP-hard [GJ79]. Like in [MT82], we reduce the 3SAT problem to our problem. To simplify the presentation, we first sketch the proof of [MT82] and we then show how to modify the proof for our problem.

Given an instance of 3SAT $E_1 \wedge E_2 \wedge \cdots \wedge E_m$, where $E_j = a_j \vee b_j \vee c_j, \{a_j, b_j, c_j\} \subset \{v_1, \bar{v}_1, ..., v_n, \bar{v}_n\}$, $j = 1, ..., m$, Megiddo and Tamir construct m points in 2D such that each E_j is represented by a point p_j. For each pair of variable $v_i, \bar{v}_i$, they construct m^2 points. So in total $m + n \times m^2$ points (for m clauses and n variable pairs) are constructed. The reduction of [MT82] takes a pseudo-polynomial time to construct the $n \times m^2$ points which correspond to n variables. Each group of m^2 points (which corresponds to a variable) needs at least m lines to cover and there are only two sets of m lines to cover these m^2 points (which correspond to the true and false assignment of that variable). If a variable $v_i(\bar{v}_i)$ appears in E_j, then one line in the m lines corresponding to the true (false) assignment of v_i covers p_j. Therefore, $E_1 \wedge E_2 \wedge \cdots \wedge E_m$ is satisfiable if and only if the whole $m + n \times m^2$ points can be covered by mn lines.

In our problem, we need to use cylindrical segments in 3D to cover the points such that no segment penetrates another one by some amount σ. From now on we set $\sigma = 0$; moreover, each cylindrical segment is a line segment (i.e., with radius zero) and two segments can only intersect at a common endpoint. In the previous 2D construction, any covering line penetrates another one. To make the proof work for our problem, we first fix $p_1, ...p_m$ on a line L and then rotate each of the n groups of points around L such that the following property still holds. (Each group, which contains m^2 points, corresponds to a variable and

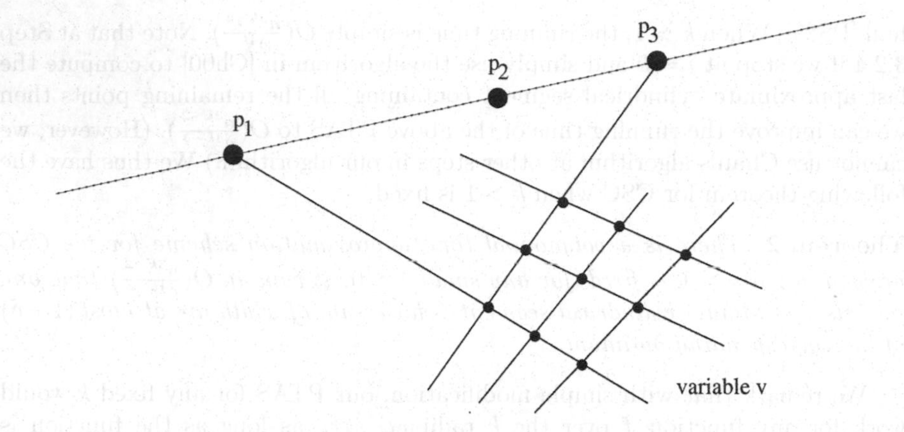

Fig. 3. v appears in E_1 and $\bar{v}$ appears in E_3.

each group of points always lie on the same plane after rotation.) Each group of m^2 points needs at least m 3D line segments to cover and there are only two sets of m 3D line segments to cover these m^2 points (which correspond to the true and false assignment of that variable). Also, none of the 3D line segments covers a point which belongs to different variable. If a variable $v_i(\bar{v}_i)$ appears in E_j, then one segment of the m segments which are corresponding to the true (false) assignment of v_i touches p_j. We show a brief example of the construction in Figure 3.

Therefore, $E_1 \wedge E_2 \wedge \cdots \wedge E_m$ is satisfiable if and only if the whole $m + n \times m^2$ 3D points can be covered by mn 3D line segments (cylindrical segments whose sum of radii is zero). The reduction, as in [MT82], takes a pseudo-polynomial time on determining the coordinates of the $n \times m^2$ points corresponding to the variables. Therefore, we have the following result.

Corollary 1. *It is strongly NP-hard to decide whether a set of 3D points can be covered by the minimum number of 3D cylindrical segments with the minimum sum of radii such that no segment penetrates another one.*

6 Concluding Remarks

In this paper, we investigate the problem of approximating a set P of n 3D points with k minimum cylindrical segments. We present a very simple algorithm for the case when $k = 1$. For the case when $k > 1$, we obtain a PTAS which runs in $O(\frac{n^{3k-2}}{\delta^{4k-3}})$ time. Here, we mention briefly our early empirical testing of the algorithms presented in this paper. Unfortunately, when $k \geq 2$ the algorithm is already impractical for $n = 10,000$ points. This probably explains in practice why practitioners first use commercial software to reconstruct a polyhedron from the input points — this will provide useful topological information for computing the approximating cylindrical segments. This implies, from another angle, that this PTAS is completely theoretical.

A lot of questions remain to be answered regarding this problem, both theoretically and practically. An immediate theoretical question is whether we can obtain a PTAS when k is not part of the input. Another problem is whether the objective of minimizing the sum of radii of a set of cylindrical segments is truely what the biologists want (they just want that the radius of each approximating segment is minimized). This question might not be able to be answered in satisfactory without practical simulations. Our feeling is that the true objective function should be minimizing the volume of the union of the cylindrical segments, which is not very easy to compute as the union has at least a quadratic combinatorial complexity.

References

AAS97. P. Agarwal, B. Aronov and M. Sharir. Line transversals of balls and smallest enclosing cylinders in three dimensions. In *Proc. 8th ACM-SIAM Symp on Discrete Algorithms (SODA'97)*, New Orleans, LA, pages 483-492, Jan, 1997.

AP00. P. Agarwal and C. Procopiuc. Approximation algorithms for projective clustering. In *Proc. 11th ACM-SIAM Symp on Discrete Algorithms (SODA'00)*, pages 538-547, 2000.

Ch00. T. Chan. Approximating the diameter, width, smallest enclosing cylinder, and minimum-width annulus. In *Proc. 16th ACM Symp on Computational Geometry (SCG'00)*, Hong Kong, pages 300-309, June, 2000.

GJ79. M. Garey and D. Johnson. *Computers and Intractability: A Guide to the Theory of NP-completeness.* Freeman, San Francisco, CA, 1979.

HV02. S. Har-Peled and K. Varadarajan. Projective clustering in high dimensions using core-sets. In *Proc. 18th ACM Symp on Computational Geometry (SCG'02)*, to appear, 2002.

JT96. G. Jacobs and F. Theunissen. Functional organization of a neural map in the cricket cercal sensory system. *J. of Neuroscience*, 16(2):769-784, 1996.

JT00. G. Jacobs and F. Theunissen. Extraction of sensory parameters from a neural map by primary sensory interneurons. *J. of Neuroscience*, 20(8):2934-2943, 2000.

MSW92. J. Matoušek, M. Sharir and E. Welzl. A subexponential bound for linear programming. *Algorithmica*, 16:498-516, 1992.

MT82. N. Megiddo and A. Tamir. On the complexity of locating linear facilities in the plane. *Operation Research Letters*, 1(5):194-197, 1982.

PDJ99. S. Paydar, C. Doan and G. Jacobs. Neural mapping of direction and frequency in the cricket cercal sensory system. *J. of Neuroscience*, 19(5):1771-1781, 1999.

PS85. F.P. Preparata and M.I. Shamos. *Computational Geometry: An Introduction.* Springer-Verlag, 1985.

Ra00. E. Ramos. Deterministic algorithms for 3-D diameter and some 2-D lower envelopes. In *Proc. 16th ACM Symp on Computational Geometry (SoCG'00)*, pages 290-299, June, 2000.

SSTY00. E. Schömer, J. Sellen, M. Teichmann and C.K. Yap. Smallest enclosing cylinders. *Algorithmica*, 27:170-186, 2000.

We91. E. Welzl. Smallest enclosing disks (balls and ellipsoids). In *New results and new trends in computer science*, LNCS 555, pages 359-370, 1991.

Algorithms for the Multicolorings of Partial k-Trees

Takehiro Ito, Takao Nishizeki, and Xiao Zhou

Graduate School of Information Sciences, Tohoku University
Aoba-yama 05, Sendai, 980-8579, Japan
take@nishizeki.ecei.tohoku.ac.jp, {nishi,zhou}@ecei.tohoku.ac.jp

Abstract. Let each vertex v of a graph G have a positive integer weight $\omega(v)$. Then a multicoloring of G is to assign each vertex v a set of $\omega(v)$ colors so that any pair of adjacent vertices receive disjoint sets of colors. A partial k-tree is a graph with tree-width bounded by a fixed constant k. This paper presents an algorithm which finds a multicoloring of any given partial k-tree G with the minimum number of colors. The computation time of the algorithm is bounded by a polynomial in the number of vertices and the maximum weight of vertices in G.

1 Introduction

Let $G = (V, E)$ be a graph with vertex set V and edge set E. A *vertex-coloring of a graph* G is to color all vertices so that any pair of adjacent vertices are colored with different colors. Let each vertex v of G have a positive integer weight $\omega(v)$. Let C be a set of colors, and let 2^C be the power set of C. Then a *multicoloring* Γ *of* G is a mapping from V to 2^C which assigns each vertex $u \in V$ a set $\Gamma(u)$ of $\omega(u)$ colors in C in a way that $\Gamma(v) \cap \Gamma(w) = \emptyset$ for any pair of adjacent vertices $v, w \in V$. Thus the ordinary vertex-coloring is merely a multicoloring for the special case where $\omega(v) = 1$ for every vertex v. The *multichromatic number* $\chi_\omega(G)$ of G is the minimum number of colors required for a multicoloring of G, that is,

$$\chi_\omega(G) = \min\{|C| : G \text{ has a multicoloring } \Gamma : V \to 2^C\}.$$

The *multicoloring problem* is to compute the multichromatic number $\chi_\omega(G)$ of a given graph G. Consider a graph G in Fig. 1(a) where $\omega(v)$ is attached to each vertex v. Since v_1, v_2 and v_5 are adjacent with each other, $\chi_\omega(G) \ge \omega(v_1) + \omega(v_2) + \omega(v_5) = 5$. Since G has a multicoloring with five colors $c_1, c_2, \cdots, c_5$ as illustrated in Fig. 1(b), $\chi_\omega(G) \le 5$. Thus $\chi_\omega(G) = 5$.

The multicoloring problem has a natural application in scheduling theory [9]. Consider a set V of jobs such that each job $v \in V$ needs a total of $\omega(v)$ units of time to be finished and there are several pairs of jobs which cannot be executed simultaneously. We wish to find a schedule of the minimum completion time. This problem can be modeled by a graph G in which a vertex corresponds to a job and an edge corresponds to a pair of jobs which cannot be executed simultaneously. A multicoloring of G with α colors corresponds to a preemptive schedule of

O.H. Ibarra and L. Zhang (Eds.): COCOON 2002, LNCS 2387, pp. 430–439, 2002.

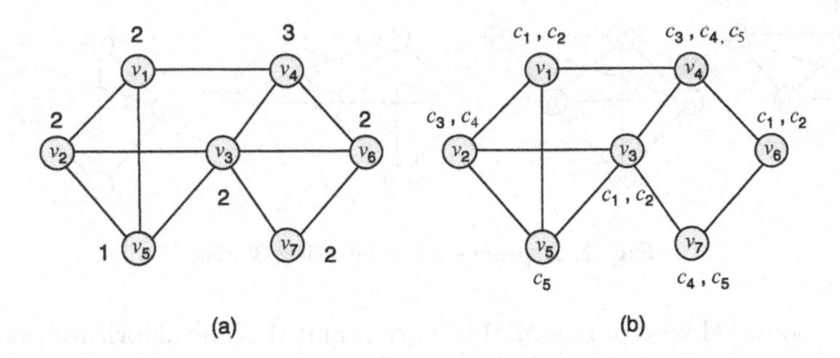

Fig. 1. (a) A partial 3-tree G and (b) a multicoloring Γ of G with five colors.

completion time α; if vertex v receives $\omega(v)$ colors, say $c_{i_1}, c_{i_2}, \cdots, c_{i_{\omega(v)}}$, then job v is executed in the i_1th, i_2th, $\cdots$, and $i_{\omega(v)}$th time slots, total in $\omega(v)$ time slots. The goal is to find a preemptive schedule of the minimum completion time of all jobs. Clearly, the minimum completion time is equal to the multichromatic number $\chi_\omega(G)$ of G.

Since the vertex-coloring problem is NP-hard, the multicoloring problem is of course NP-hard and hence it is very unlikely that the multicoloring problem can be efficiently solved for general graphs. However, there may exist an efficient algorithm to solve the multicoloring problem for a restricted class of graphs. Indeed, the problem can be solved for trees in time $O(n)$ [2,8], for triangulated graphs in time $O(n^2)$ [2,8], for perfect graphs in time $O(mn)$ [8], and for series-parallel graphs in time $O(nW)$ [12], where m is the number of edges, n is the number of vertices of a given graph G, and W is the *maximum vertex weight*, that is, $W = \max_{v \in V} \omega(v)$.

In this paper we consider another class of graphs, called "partial k-trees." A partial k-tree is a graph of "tree-width" bounded by a fixed constant k. The class of partial k-trees is fairly large; for example, trees, forests, outerplanar graphs and series-parallel graphs are all partial k-trees. A formal definition of a partial k-tree will be given in Section 2. It is thus desired to obtain an efficient algorithm to solve the multicoloring problem for partial k-trees. There are general methods to design algorithms for solving many problems on partial k-trees, including the vertex-coloring problem, the Hamiltonian cycle problem, the maximum independent vertex set problem, *etc.* [1,5,6,10]. However, a straightforward algorithm directly derived from the general methods takes time $O(n(2^{\chi_\omega(G)})^{2(k+1)}) = O(n2^{2(k+1)^2 W})$ to solve the multicoloring problem for partial k-trees G. Note that $\chi_\omega(G) \le (k+1)W$ since any partial k-tree G has a vertex-coloring with at most $k+1$ colors and hence G has a multicoloring with at most $(k+1)W$ colors. The computation time of the algorithm is not bounded by a polynomial in n and W. Thus it is desired to obtain a "parametric algorithm" of computation time bounded by a polynomial in both n and W [7].

In this paper we give an algorithm to solve the multicoloring problem for partial k-trees in time $O(nW^{2^{2k+3}} \log_2 W)$. Thus the algorithm is much faster than the straightforward algorithm above, and is the first one whose time com-

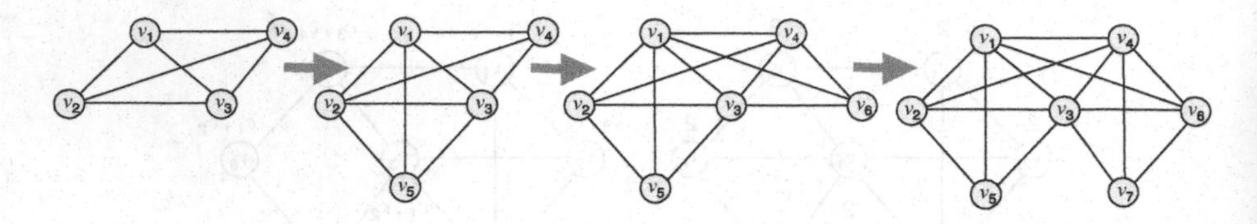

Fig. 2. A process of generating 3-trees.

plexity is bounded by a polynomial in both n and W. The algorithm takes linear time if W is bounded. It should be noted that an ordinary representation of a multicoloring of G requires space of size at least $\sum_{v\in V}\omega(v) = O(nW)$. Our algorithm is based on a clever and detailed formation of dynamic programming.

The rest of the paper is organized as follows. Section 2 includes basic definitions and notations. Section 3 gives an algorithm to solve the multicoloring problem for partial k-trees. Finally, Section 4 is a conclusion.

2 Terminology and Definitions

In this section we give some definitions. We deal with a *simple undirected* graph without multiple edges or self-loops. An edge joining vertices u and v is denoted by (u, v). We denote by n the number of vertices in G, and assume that k is a bounded positive integer.

A k-tree is defined recursively as follows [4]:

(1) A complete graphs with $k+1$ vertices is a k-tree.
(2) If G is a k-tree and k vertices induce a complete subgraph of G, then a graph obtained from G by adding a new vertex and joining it with each of the k vertices is a k-tree,

Any subgraph of a k-tree is called a *partial k-tree.* Thus a partial k-tree $G = (V, E)$ is a simple graph, and $|E| < kn$. Figure 2 illustrates a process of generating 3-trees. The graph in Fig. 1 is a partial 3-tree since it is a subgraph of the last 3-tree in Fig. 2.

A binary tree $T = (V_T, E_T)$ is called a *tree-decomposition of a partial k-tree* $G = (V, E)$ if T satisfies the following conditions (a) – (f):

(a) every node $X \in V_T$ of T is a subset of V, and $|X| \leq k+1$;
(b) $\bigcup_{X\in V_T} X = V$;
(c) for each edge $e = (u, v)$ of G, T has a leaf $X \in V_T$ such that $u, v \in X$;
(d) if node X_p lies on the path in T from node X_q to node X_r, then $X_q \cap X_r \subseteq X_p$;
(e) the number of nodes in T is $O(n)$; and
(f) each internal node X_i of T has exactly two children, say X_l and X_r, and either $X_i = X_l$ or $X_i = X_r$.

We will use notions as: leaf, node, child, and root in their usual meaning. Figure 3 illustrates a tree-decomposition T of the partial 3-tree in Fig. 1. We denote by X_0 the root of a tree-decomposition.

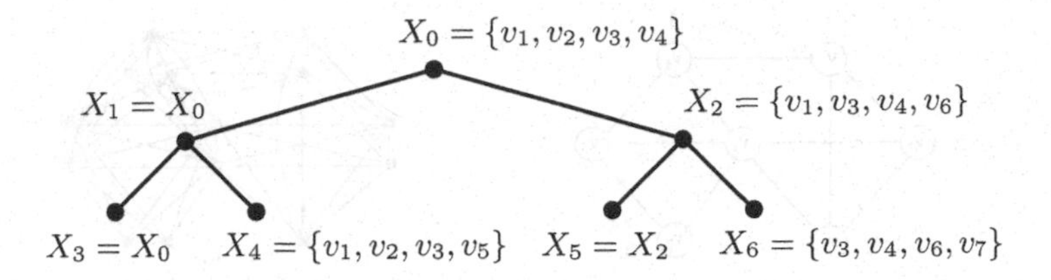

Fig. 3. Tree-decomposition of the partial 3-tree in Fig. 1.

Since a tree-decomposition of a partial k-tree can be found in linear time [3], we may assume that a partial k-tree G and its tree-decomposition T are given.

A multicoloring of a graph $G = (V, E)$ is an ordinary vertex-coloring of a new graph G_ω, constructed from G as follows: replace each vertex $u \in V$ with a complete graph $K_{\omega(u)}$ of $\omega(u)$ vertices, and join all vertices of $K_{\omega(v)}$ to all vertices in $K_{\omega(w)}$ for each edge $(v, w) \in E$, as illustrated in Fig. 4; the resulting graph is G_ω. Then a multicoloring of G induces an ordinary vertex-coloring of G_ω, and *vice versa.* Thus the multicoloring problem for G can be reduced to the vertex-coloring problem for G_ω. Since G_ω has $n' = \sum_{v \in V} \omega(v)$ ($\leq nW$) vertices and $O((n+m)W^2)$ edges, the reduction takes time $O((n+m)W^2)$, where $m = |E|$. Unfortunately, G_ω is not always a partial k-tree even if G is a partial k-tree. One cannot thus solve the multicoloring problem for a partial k-tree G by applying to G_ω the linear-time algorithm for finding a vertex-coloring of a partial k-tree in [1,5]. However, if G is a partial k-tree, then G_ω is a partial k'-tree for an integer $k' = (k+1)W$. The vertex-coloring problem can be solved for a partial k'-tree G_ω in time $O(n'(k'+1)^{2(k'+1)})$ [1,5]. Thus the multicoloring problem for a partial k-tree G can be solved in time $O(nW((k+1)W+1)^{2((k+1)W+1)})$. The computation time is not bounded by a polynomial in n and W. On the other hand, the computation time $O(nW^{2^{2k+3}} \log_2 W)$ of our algorithm is bounded by a polynomial in both n and W.

3 Algorithm

The main result of the paper is the following theorem.

Theorem 1. *Let k be a bounded positive integer, let G be any partial k-tree, and let α be any positive integer. Then one can know in time $O(n(\alpha+1)^{2^{2k+3}})$ whether G has a multicoloring with α colors.*

Using a binary search technique, one can compute the multichromatic number $\chi_\omega(G)$ of G by applying Theorem 1 for at most $\log_2((k+1)W)$ values of α, $1 \leq \alpha \leq (k+1)W$, since $\chi_\omega(G) \leq (k+1)W$. We thus have the following corollary.

Corollary 1. *The multichromatic number $\chi_\omega(G)$ of a partial k-tree G can be computed in time $O(nW^{2^{2k+3}} \log_2 W)$.*

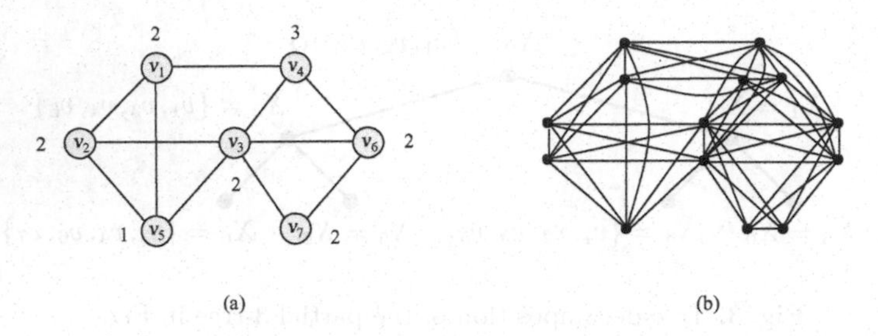

Fig. 4. Transformation of G to G_ω.

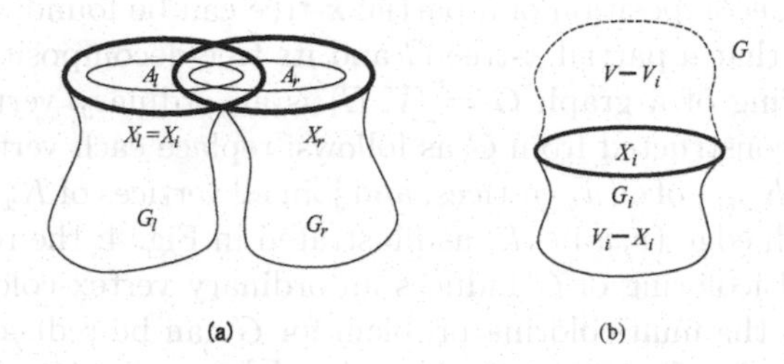

Fig. 5. (a) Graph $G_i = G_l \cup G_r$ (b) Graphs G and G_i.

In the remainder of this section we give a proof of Theorem 1. From now on we call a multicoloring simply a *coloring*. Let C be a set of α colors. Although we give an algorithm to examine whether a partial k-tree $G = (V, E)$ has a coloring $\Gamma : V \to 2^C$, it can be easily modified so that it actually finds a coloring Γ of G with α colors in C. Our idea is to extend a technique developed for the edge-coloring problem [4,11] to the multicoloring problem and to reduce the size of a Dynamic Programming (DP) table to $O((\alpha+1)^{2^{2k+3}})$ by considering "counts" and "pair-counts."

Let $G = (V, E)$ be a partial k-tree, and let $T = (V_T, E_T)$ be a tree-decomposition of G. Each node X_i of T corresponds to a subgraph $G_i = (V_i, E_i)$ of G. The vertex set V_i and edge set E_i of G_i are recursively defined as follows: if X_i is a leaf of T, then $V_i = X_i$ and $E_i = \{(u, v) \in E : u, v \in X_i\}$; if X_i is an internal node of T, the left child X_l of X_i corresponds to a subgraph $G_l = (V_l, E_l)$ of G, and the right child X_r corresponds to $G_r = (V_r, E_r)$, then $V_i = V_l \cup V_r$ and $E_i = E_l \cup E_r$, and hence G_i is a union of two graphs G_l and G_r as illustrated in Fig. 5 (a), where $X_i = X_l$ and X_r are indicated by ovals drawn by thick lines. Clearly $G = G_0$ for the root X_0 of T. One can easily observe that, for each node X_i of T, $X_i \subseteq V_i$ and any edge of G with both ends in X_i is contained in E_i. The condition (d) of a tree-decomposition implies that $V_l \cap V_r = X_l \cap X_r \subseteq X_i$ and $E_l \cap E_r = \{(u, v) \in E_i : u, v \in X_l \cap X_r\}$ as illustrated in Fig. 5 (a), and that no edge of G joins a vertex in $V_i - X_i$ and a vertex in $V - V_i$ for each node X_i of T. (See Fig. 5 (b).) Figure 6(a) depicts the partial 3-tree $G = G_0$ in Fig. 1(a); G_0 corresponds to the root $X_0 = \{v_1, v_2, v_3, v_4\}$ of a tree-decomposition T in Fig. 3. Figure 6(b) depicts G_1 corresponding to

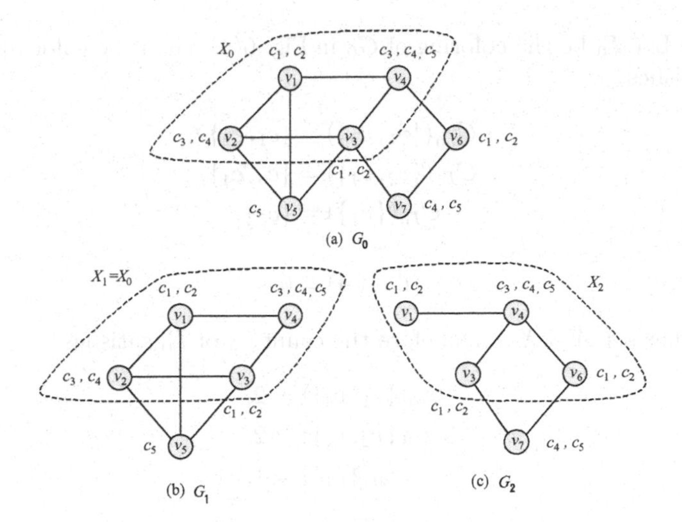

Fig. 6. (a) Colorings Γ_0 of $G = G_0$, (b) Γ_1 of G_1, and (c) Γ_2 of G_2.

the left child $X_1 = \{v_1, v_2, v_3, v_4\}$ of the root X_0, while Fig. 6(c) depicts G_2 corresponding to the right child $X_2 = \{v_1, v_3, v_4, v_6\}$. G_0 is the union of G_1 and G_2.

For a node $X_i \in V_T$ of T, a coloring Γ of G_i, and a color $c \in C$, we define a set $Y(X_i; \Gamma, c) \subseteq X_i$ as follows:

$$Y(X_i; \Gamma, c) = \{v \in X_i : c \in \Gamma(v)\}.$$

That is, $Y(X_i; \Gamma, c)$ consists of all vertices v in X_i that are assigned a color c by Γ. On the other hand, for a coloring Γ of G_i and a set $A \subseteq X_i$, we define a set $C_\Gamma(A) \subseteq C$ as follows:

$$C_\Gamma(A) = \{c \in C : A = Y(X_i; \Gamma, c)\}.$$

That is, $C_\Gamma(A)$ consists of all colors c in C such that $\{v \in X_i : c \in \Gamma(v)\} = A$. Probably $C_\Gamma(A) = \emptyset$ for many sets $A \subseteq X_i$. We call the mapping $C_\Gamma : 2^{X_i} \to 2^C$ the *color function of* Γ *on* X_i. Clearly, $\mathcal{F}_\Gamma = \{C_\Gamma(A) : A \in 2^{X_i}\}$ is a partition of set C.

We say that a coloring of G_i is *extensible* if it can be extended to a coloring of the whole graph $G = G_0$ without changing the coloring of G_i. Both the coloring Γ_1 of G_1 in Fig. 6(b) and the coloring Γ_2 of G_2 in Fig. 6(c) are extensible because both can be extended to the coloring Γ_0 of G_0 in Fig. 6(a).

Any mapping $\gamma : 2^{X_i} \to \{0, 1, 2, \cdots, \alpha\}$ is called a *count on a node* X_i. A count γ on X_i is defined to be *active* if G_i has a coloring Γ such that $\gamma(A) = |C_\Gamma(A)|$ for each set $A \in 2^{X_i}$. Such a count γ is called the *count of the coloring* Γ. Since $|C| = \alpha$ and $\mathcal{F}_\Gamma = \{C_\Gamma(A) : A \in 2^{X_i}\}$ is a partition of set C, any active count γ satisfies $\sum_{A \in 2^{X_i}} \gamma(A) = \alpha$.

Example Let Γ_0 be the coloring of G_0 in Fig. 6(a), then the color function C_{Γ_0} on X_0 satisfies

$$C_{\Gamma_0}(\{v_1, v_3\}) = \{c_1, c_2\},$$
$$C_{\Gamma_0}(\{v_2, v_4\}) = \{c_3, c_4\},$$
$$C_{\Gamma_0}(\{v_4\}) = \{c_5\},$$

and

$$C_{\Gamma_0}(A) = \emptyset$$

for any other set $A \subseteq X_0$. Therefore the count γ_0 of Γ_0 satisfies

$$\gamma_0(\{v_1, v_3\}) = 2,$$
$$\gamma_0(\{v_2, v_4\}) = 2,$$
$$\gamma_0(\{v_4\}) = 1,$$

and

$$\gamma_0(A) = 0$$

for any other set $A \subseteq X_0$. Similarly, the coloring Γ_1 of G_1 for $X_1 = \{v_1, v_2, v_3, v_4\}$ in Fig. 6(b) has a count γ_1 such that

$$\gamma_1(\{v_1, v_3\}) = 2,$$
$$\gamma_1(\{v_2, v_4\}) = 2,$$
$$\gamma_1(\{v_4\}) = 1,$$

and

$$\gamma_1(A) = 0$$

for any other set $A \subseteq X_1$. On the other hand, the coloring Γ_2 of G_2 for $X_2 = \{v_1, v_3, v_4, v_6\}$ in Fig. 6(c) has a count γ_2 such that

$$\gamma_2(\{v_1, v_3, v_6\}) = 2,$$
$$\gamma_2(\{v_4\}) = 3,$$

and

$$\gamma_2(A) = 0$$

for any other set $A \subseteq X_2$.

One can easily observe that the following lemma holds.

Lemma 1. *Let Γ and Γ' be colorings of G_i for a node X_i of T, and assume that Γ and Γ' have the same count. Then Γ is extensible if and only if Γ' is extensible.*

Define an equivalence relation $\cong$ on the set of all colorings of G_i, as follows: $\Gamma \cong \Gamma'$ if the colorings Γ and Γ' of G_i have the same (active) count. Then each active count on X_i characterizes an equivalence class of colorings of G_i. Lemma 1 implies that either all the colorings in an equivalence class are extensible or none of them is extensible. Since $|X_i| \leq k+1$, there are $(\alpha+1)^{2^{|X_i|}}$ $(\leq (\alpha+1)^{2^{k+1}})$

distinct counts $\gamma : 2^{X_i} \to \{0, 1, 2, \cdots, \alpha\}$ on X_i. The main step of our algorithm is to compute a table of all active counts on each node of T from leaves to the root X_0 of T by means of dynamic programming. From the table on the root X_0 one can easily know whether G has a coloring with α colors, as follows.

Lemma 2. *A partial k-tree G has a coloring with α colors if and only if the table on root X_0 has at least one active count.*

We first compute the table of all active counts on each leaf X_i of T as follows:

(1) enumerate all counts γ of X_i; and
(2) find all active ones from them.

There are at most $(\alpha + 1)^{2^{k+1}}$ distinct counts γ on X_i. Since $G_i = (V_i, E_i)$ and $V_i = X_i$, one can easily know that a count γ is active if and only if γ satisfies the following conditions (a), (b) and (c):

(a) $\sum\{\gamma(A) : A \in 2^{X_i}\} = \alpha$;
(b) $\sum\{\gamma(A) : A \in 2^{X_i}, v \in A\} = \omega(v)$ for each vertex $v \in X_i$; and
(c) $\gamma(A) = 0$ if $A \in 2^{X_i}$ and $u, v \in A$ for some edge (u, v) of G_i.

Since $|X_i| \le k + 1$, one can know in time $O(2^{k+1}) = O(1)$ whether a count γ satisfies Condition (a) above. Similarly, one can know in time $O(k2^{k+1}) = O(1)$ whether γ satisfies Condition (b). Since $|E_i| \le k(k + 1)/2 = O(k^2)$, one can know in time $O(k^2 2^{k+1}) = O(1)$ whether γ satisfies Condition (c). Thus, for each count γ, one can know in time $O(1)$ whether γ is active or not. Hence steps (1) and (2) above can be done for a leaf in time $O((\alpha + 1)^{2^{k+1}})$. Since T has $O(n)$ leaves, the tables on all leaves can be computed in time $O(n(\alpha + 1)^{2^{k+1}})$.

We next compute all active counts on each internal node X_i of T from all active counts on its children X_l and X_r. Either $X_i = X_l$ or $X_i = X_r$ by the condition (f) of a tree-decomposition. Therefore, one may assume without loss of generality that $X_i = X_l$. Any mapping $\rho : 2^{X_l} \times 2^{X_r} \to \{0, 1, 2, \cdots, \alpha\}$ is called a *pair-count on* X_i. Then there are $(\alpha + 1)^{2^{|X_l| + |X_r|}} (\le (\alpha + 1)^{2^{2(k+1)}})$ distinct pair-counts. For any coloring Γ of G_i, we denote by $\Gamma_l = \Gamma | G_l$ the *restriction* of Γ to G_l: $\Gamma_l(v) = \Gamma(v)$ for each vertex v of G_l. Similarly, we denote by $\Gamma_r = \Gamma | G_r$ the restriction of Γ to G_r. We denote by C_{Γ_l} the color function of Γ_l on X_l, and by C_{Γ_r} the color function of Γ_r on X_r. Then we define a pair-count ρ to be *active* if G_i has a coloring Γ such that, for each pair (A_l, A_r) of sets $A_l \subseteq X_l$ and $A_r \subseteq X_r$ (see Fig. 5),

$$\rho(A_l, A_r) = |C_{\Gamma_l}(A_l) \cap C_{\Gamma_r}(A_r)|.$$

Such a pair-count ρ is called the *pair-count of the coloring Γ of G_i*. We now have the following lemma, whose proof is omitted in this extended abstract due to the page limitation.

Lemma 3. *Let X_i be any internal node X_i of T, let X_l and X_r be the children of X_i, and let ρ be any pair-count on X_i. Then ρ is active if and only if ρ satisfies the following Conditions* (a) *and* (b):

(a) *if* $\rho(A_l, A_r) \geq 1$, *then* $A_l \cap X_r = A_r \cap X_l$; *and*
(b) *there is an active count* γ_l *on* X_l *such that for each set* $A_l \subseteq X_l$

$$\gamma_l(A_l) = \sum\{\rho(A_l, A) : A \subseteq X_r\}, \tag{1}$$

and there is an active count γ_r *on* X_r *such that for each set* $A_r \subseteq X_r$

$$\gamma_r(A_r) = \sum\{\rho(A, A_r) : A \subseteq X_l\}. \tag{2}$$

Using Lemma 3, we compute all active pair-counts ρ on X_i from all pairs of active counts γ_l on X_l and γ_r on X_r, as follows. There are at most $(\alpha+1)^{2^{2(k+1)}}$ pair-counts ρ on X_i. For each pair-count ρ of them, we examine whether ρ satisfies Conditions (a) and (b) in Lemma 3. For each pair-count ρ, one can know in $O(1)$ time whether ρ satisfies Condition (a), because there are at most $2^{2(k+1)} = O(1)$ distinct pairs (A_l, A_r). On the other hand, for each pair-count ρ, one can know in time $O\left((\alpha+1)^{2^{k+2}}\right)$ whether ρ satisfies Condition (b), because there are at most $((\alpha+1)^{2^{k+1}})^2 = (\alpha+1)^{2^{k+2}}$ pairs of active counts γ_l and γ_r, and one can know in time $O(1)$ for each of them whether it satisfies Eqs. (1) and (2). Thus all active pair-counts ρ on X_i can be found in time $O((\alpha+1)^{2^{2k+3}})$, since there are at most $(\alpha+1)^{2^{2(k+1)}}$ pair-counts ρ on X_i and $(\alpha+1)^{2^{2(k+1)}}(\alpha+1)^{2^{k+2}} \leq (\alpha+1)^{2^{2k+3}}$.

We then compute all active counts on an internal node X_i from all active pair-counts on X_i, as in the following Lemma 4, the proof of which is omitted due to the page limitation.

Lemma 4. *Let* X_i *be an internal node of* T, *let* X_l *and* X_r *be the two children of* X_i, *and let* $X_i = X_l$. *Then a count* γ *on* X_i *is active if and only if there exists an active pair-count* ρ *on* X_i *such that, for each set* $A \subseteq X_i$,

$$\gamma(A) = \sum\{\rho(A, A') : A' \subseteq X_r\}. \tag{3}$$

Using Lemma 4, we compute all active counts γ on X_i from all active pair-counts ρ on X_i. There are at most $(\alpha+1)^{2^{2(k+1)}}$ distinct active pair-counts ρ. From each ρ of them we compute an active count γ by Eq. (3). This can be done in $O(1)$ time since $|A|, |A'|, |X_i|, |X_l|, |X_r| \leq k+1 = O(1)$. We have thus shown that all active counts γ on X_i can be computed in time $O((\alpha+1)^{2^{2(k+1)}})$ from all active pair-counts ρ on X_i.

One can thus compute the DP table for an internal node X_i from the tables for the children X_l and X_r in time

$$O\left((\alpha+1)^{2^{2k+3}} + (\alpha+1)^{2^{2(k+1)}}\right) = O\left((\alpha+1)^{2^{2k+3}}\right).$$

Since T has $O(n)$ internal nodes, one can compute the DP tables for all internal nodes in time $O(n(\alpha+1)^{2^{2k+3}})$.

From the DP table for root X_0 one can know in time $O(1)$ by Lemma 2 whether G has a coloring with α colors.

This completes a proof of Theorem 1.

4 Conclusion

In this paper we obtained an algorithm to compute the multichromatic number $\chi_\omega(G)$ of a given partial k-tree G in time $O(nW^{2^{2k+3}} \log_2 W)$, where n is the number of vertices in G and W is the maximum weight of vertices in G. This is the first algorithm whose computation time is bounded by a polynomial in both n and W. If $W = O(1)$, the algorithm takes time $O(n)$. It is easy to modify the algorithm so that it actually finds a multicoloring of G with $\chi_\omega(G)$ colors. It is remaining as a future problem to obtain an algorithm which takes time polynomial in both n and $\log W$, and to obtain an algorithm for other classes of graphs.

References

1. S. Arnborg and J. Lagergren. Easy problems for tree-decomposable graphs. *Journal of Algorithms*, 12(2):308–340, 1991.
2. E. Balas and J. Xue. Minimum weighted colouring of triangulated graphs, with application to maximum weight vertex packing and clique finding in arbitrary graphs. *SIAM. J. Comput.*, 20:209–221, 1991.
3. H. L. Bodlaender. A linear time algorithm for finding tree-decompositions of small treewidth. *SIAM J. Comput.*, 25:1305–1317, 1996.
4. H. L. Bodlaender. Polynomial algorithms for graph isomorphism and chromatic index on partial k-trees. *Journal of Algorithms*, 11(4):631–643, 1990.
5. R. B. Borie, R. G. Parker and C. A. Tovey. Automatic generation of linear-time algorithms from predicate calculus descriptions of problems on recursively constructed graph families. *Algorithmica*, 7:555–581, 1992.
6. B. Courcelle and M. Mosbath. Monadic second-order evaluations on tree-decomposable graphs. *Theoretical Computer Science*, 109:49–82, 1993.
7. R. G. Downey and M. R. Fellows. Parameterized Complexity. *Springer-Verlag, New York*, 1998.
8. C. T. Hoáng. Efficient algorithms for minimum weighted colouring of some classes of perfect graphs. *Discrete Applied Mathematics*, 55:133–143, 1994.
9. D. Karger, C. Stein and J. Wein. Scheduling algorithms, in "Algorithms and Theory of Computation, Handbook" (Editor M. J. Atallah), CRC Press, 1998.
10. K. Takamizawa, T. Nishizeki and N. Saito. Linear-time computability of combinatorial problems on series-parallel graphs. *Journal of ACM*, 29(3):623–641, 1982.
11. X. Zhou, S. Nakano and T. Nishizeki. Edge-coloring partial k-trees. *Journal of Algorithms*, 21:598–617, 1996.
12. X. Zhou and T. Nishizeki. Efficient algorithms for weighted colorings of series-parallel graphs, In *Proc. of the 12th International Symposium on Algorithms and Computation*, Lect. Notes in Computer Science, Springer, 2223, 514-524, 2001.

A Fault-Tolerant Merge Sorting Algorithm

B. Ravikumar

Department of Computer Science
Sonoma State University
Rohnert Park, CA 94128
ravi@cs.sonoma.edu

Abstract. Sorting based on pairwise key comparisons is one of the most widely studied problems. We consider the problem of comparison based sorting in which some of the outcomes of comparisons can be faulty. We show how to modify merge-sorting to (nearly optimally) sort in the presence of faults. More specifically, we show that there is a variation of merge-sort that can sort n records with $O(n \log n)$ comparisons when upto $e = \Theta(\frac{\log n}{\log \log n})$ comparisons are faulty.

1 Introduction

The version of sorting problem studied here is the following: n keys stored in an array are to be sorted using pairwise comparisons. For convenience, we assume that all the keys are distinct. Among the responses to the comparison queries, some of them can be wrong. If the number of erroneous responses is a constant e, independent of n, then the standard query repetition strategy (repeat until one of yes/no answer is favored $e+1$ times). The number of key comparisons performed by this algorithm is upper-bounded by $(2e+1)T_{sort}(n)$ where $T_{sort}(n)$ is the number of comparisons performed by an error-free sorting algorithm. Thus $O(e\ n \log n)$ key comparisons are sufficient to sort n keys by query repetition as long as e is a constant. How large can e grow as a function of n so that it is still possible to sort n keys in $O(n \log n)$ time while there are e errors in comparisons? The main result of this paper is a modification of Merge-Sorting that sorts an array of n keys in $O(n \log n)$ time so long as e is $\Theta(\frac{\log n}{\log \log n})$.

This problem and its variations have received a lot of attention in the past ten years, see for example [1], [2], [4], [7], [6], etc. It was shown in [6] that $O(n \log n)$ comparisons are sufficient to sort so long as $e = \Theta(\log n)$. This result is tight within a constant factor in the sense that $\Omega(n \log n + e\ n)$ comparisons are necessary for any comparison based sorting algorithm to sort n keys if upto e of the pairwise comparisons can be erroneous. The algorithm presented in [6] is based on binary insertion sort and hence would require $\Omega(n^2)$ data movements to implement. Two improvements were proposed by [7] and [1], but both are modifications of the insertion sort presented in [6] and hence they do not improve the total number of operations significantly. It would be interesting to know if algorithms like heap sort or merge sort (which perform a total of $O(n \log n)$ operations including data movements) can be modified to work in the presence

O.H. Ibarra and L. Zhang (Eds.): COCOON 2002, LNCS 2387, pp. 440–447, 2002.

of errors. In this paper, we show how to modify Merge-Sorting in the presence of faults. Martin Farach-Colten in a recent talk (Alenex 2002) made a reference to the problem of sorting in the presence of faults as a real-world whose solution is required by the Google search engine and stated that (error-free version of) Merge-Sorting is not a good algorithm to use. As can be seen from this paper, a modified version of Merge Sorting solves the sorting problem in the presence of faults in a way that is significantly faster than query repetition.

It should be noted that for some comparison-based problems, when there are faults in comparisons, the best strategy is query repetition. Selection is one such problem. It is easy to see that merging of two equal length lists also falls into this category. This fact may explain why we could not get an adaptation of Merge-sorting with an upper-bound matching the number of comparisons performed by the asymptotic optimal algorithm of [6]. Our algorithm is off by a factor of $\Theta(log\ log\ n)$.

There are several reasons for interest in designing fault-tolerant Merge-Sorting or other sorting algorithms. The obvious one (e.g. mentioned in the above referenced talk) is that the hardware that implements the algorithm may generate erroneous outcomes occasionally. But there are more subtle applications as well. For example, given the outcomes of matches between various teams, we may want to rank the teams. Such ranking problems can be modeled as sorting with faulty comparisons. Another problem that can be modeled as fault-tolerant sorting is the problem of sorting (slowly) time-varying data. Here, the outcomes of some older comparisons correspond to errors. The problem of sorting in the presence of faults is also related to measures of pre-sortedness, an extensively studied topic. These connections and applications will be discussed in the final version of the paper. There, we will also describe a fault-tolerant adaptation of Heap-sort with a comparable upper-bound.

Although the complexity bounds in this paper can be presented in terms of n the number of keys and e the number of errors, we express our results in terms of the largest asymptotic growth on e that allows sorting to be done in $O(n\ log\ n)$ time.

2 A Simple Fault-Tolerant Merge Sorting

In this section, we present a simpler verion of fault-tolarant merge-sorting algorithm that can tolarate upto $O(\sqrt{log\ n})$ errors. We start with a fault-free version of merge-sort algorithm that takes as input a list of lists (each of which is known to be sorted) and merges them to form a sorted sequence provided there are no errors in comparisons. The algorithm is simply the "bottom-up" (or the non-resursive) version of merge-sort that merges lists 1 and 2, 3 and 4 etc., reducing the number of lists to half its original number, and repeats the process until a single list is created. We start with the array of n keys to be sorted, and make a list of n lists, each containing one item. We run the error-free merge-sort algorithm. Let L be the output produced. Now we check if every adjacent pair of keys in L is in the correct order by query repetition. Whenever a pair is found to be out of order, L is split. Thus after the checking phase, L gets split into potentially many lists. We repreat the cycle of sorting and checking again, until during the checking phase, the output passes as a sorted list. At each stage, we also maintain $rest$, the maximum number of errors allowable in future comparisons. This can save comparisons during the checking stage: it is enough to repeat queries sufficiently many times so that $rest$ flavorable votes are received for an yes or no outcome. We show that this algorithm has a total complexity of $\Theta(n \log n)$ so long as e is $O(\sqrt{\log n})$.

Consider the "bottom-up" (iterative) version of merge sorting which is the standard way to implement Merge-Sorting nonrecursively. The input is stored in an array A of size n. The algorithm proceeds in phases. During the first phase, sorted lists of length 2 are created by comparing $A[1]$ with $A[2]$, $A[3]$ with $A[4]$ etc. In the second phase, sorted lists of length 4 are created by merging $A[1], A[2]$ with $A[3], A[4]$ and so on until the whole list is sorted. In each of the merging steps, sequential merging algorithm (which performs $p + q - 1$ comparisons to merge two sorted lists of length p and q) [3] is used. We adapt this algorithm in the case of up to e errors as follows: The above algorithm will be run for a certain number of phases, without concren for possible errors that may have been introduced. Now the array is examined to determine if the sublists which would have been sorted if there were no errors are indeed sorted. This checking phase results in (correctly) partitioning some of these lists into shorter lists so that each of these shorter lists is sorted. We will now repeat the process by doing some more merging, followed by checking etc.

For a sequence $Z = < z_1, z_2, ..., z_n >$ define $runs(Z)$ as the minimum number of sorted segments into which Z can be divided.

Observation 1. Let X and Y be two sequences such that $runs(X) \leq p$ and $runs(Y) \leq q$. Suppose the standard sequential merging is used to merge the two lists and that during merging up to t errors can occur. If Z is the output, then $runs(Z) \leq p + q + t - 1$.

Proof. The proof is by induction on t. Let $X = < X_1, X_2, ..., X_p >$ which denotes the fact that the list X is the concatenation of p lists each of which is sorted. (When the list has a single item x, we may denote it by x, instead of $< x >$. Similarly, let $Y = < y_1, y_2, ..., y_q >$. Of course, when we run the standard merging algorithm on the list X, there will be no special markers that separate the X_i's. It will be a single list of length $\Sigma_{i=1}^{p} |X_i|$, and the same is true of the list Y.

Induction base $t = 0$. This case is shown again by induction on p. The base case of $p = 1$ is easy to show and is left as exercise. For the induction step, we can assume that both p and q are greater than 1. Suppose that when the pointer on X list moves for the first time past X_1 (to the first element of list X_2), the pointer on the Y-list is on some element of list Y_i for some i. The output list created thus far is due to the merging of X_1 and $< Y_1, ..., Y_{i-1}, P_i >$ where P_i is a prefix of Y_i which includes the keys to the left of the Y-pointer. (Let S_i denote the rest of Y_i.) By induction hypothesis, the runs on the output list produced thus far is at most $1 + i - 1 = i$. The remainder of the output list consists of merging the lists $< X_2, ..., X_p >$ with $< S_i, Y_{i+1}, ..., Y_q >$ and by induction hypothesis, the number of runs of this list is at most $(p-1)+(q-i+1)-1 = p+q-i-1$. Thus the total runs of the output list is at most $i+(p+q-i-1) = p+q-1$ and this completes the proof for $t=0$.

The induction on t follows a similar line and we omit the details. •

Observation 2. If the standard Merging algorithm is used to merge two sorted lists of combined length n, and if at most e errors occurred in comparisons, then $runs(Output)$ is at most $e+1$.

Proof. This result is a direct consequence of from Observation 1 by taking $p = q = 1$. •

Observation 3. If the standard merge-sorting algorithm produces a list Out on an input of length n and if at most e errors occurred in pairwise comparisons, then $runs(Out) \leq e+1$.

Proof. We can view the standard merge-sorting as the recursive (top-down) version. Suppose e_1 and e_2 are the errors occurring in the recursive calls, and e_3 is the number of errors occurring in the merging step. Suppose O_1 and O_2 are the outputs of the recursive calls. Then, Out is the result of merging O_1 and O_2. By induction hypothesis, $runs(O_1) \leq e_1+1$,$runs(O_2) \leq e_2+1$ and by Observation 1, $runs(Out) \leq e_1+1+e_2+1+e_3-1 = e+1$, and this completes the proof. •

Observation 4. Let $X = < X_1, X_2, ..., X_k >$ be such that each X_i is a sorted list and let $\Sigma_i|X_i| = n$. Suppose X_i's are merged by applying standard merging to merge X_1 with X_2, X_3 with X_4 etc. and repeating the process recursively until a sequence of m sorted lists are produced. Suppose there are no errors in comparisons. The total number of comparisons performed is at most $n\lceil log_2(k/m)\rceil$.

The proof of the above claim is obvious. We will now show the main result of this section.

Let $STANDARD\_MERGE(X)$ be a procedure which takes as input a list X of lists $X_1, X_2, ..., X_k$ and merges them pairwise and recursively repeats the process until a single sorted list is generated. This procedure proceeds as if there are no errors in comparisons. CHECK(L, *rest*) is a procedure which takes as input a list L and an integer *rest*. The lists L is not necessarily sorted; the goal of CHECK is to split L into a collection of (provably) sorted lists. CHECK returns two entities. The first one is a collection of lists $Y_1, Y_2, ..., Y_m$ where each Y_i is sorted. The other is an integer *current* which is the number of errors encountered during the comparisons performed by CHECK. CHECK will work under the assumption that there can be up to *rest* errors in comparisons when it starts. CHECK will repeat the comparison between two adjacent elements of the same list by repeating the query "Is $x < y$?" until one of the outcomes YES or NO is answered $rest+1$ times. At this point, the outcome becomes certain. If two adjacent elements are found to be out of place, the list is split at this point; otherwise the algorithm proceeds to the next comparison. In both cases, the value of *current* is updated by adding the number of erroneous responses. The process is repeated for all pairs of adjacent elements in each list Y_i and the split sequence of lists is output, along with *rest*. We now formally present the algorithm:

Input: A set of n keys $x_1, ..., x_n$ and e the error bound.
Step 1: $rest \leftarrow e$;
Step 2: $X = < x_1, x_2, ..., x_n >$
Step 3: $L = STANDARD\_MERGE(X)$;
Step 4: $< Y, current > = CHECK(L, rest)$;
while (Y is not a single list) do
begin
$rest = rest - current$;
$L = STANDARD\_MERGE(Y)$;
$< Y, current > = CHECK(L, rest)$;
end while;

The correctness of this algorithm is obvious. The next theroem presents the time complexity of this algorithm.

Theorem 1. *The algorithm presented above has a total time complexity of* $\Theta(n \log n)$ *if there are up to* $e = O(\sqrt{\log n})$ *errors in comparisons.*

Proof. Suppose $STANDARD\_MERGE$ is called by the algorithm t times. Then, clearly $t \leq e$ since the at least one error occurs during $STANDARD\_MERGE$ for the algorithm to proceed to the next iteration. Suppose the number of erroneous comparisons occurring during various calls to $STANDARD\_MERGE$ are e_2, ..., e_t. Then clearly, $e_1 + e_2 + ... + e_t \leq e$. The number of key comparisons performed by the t calls to $STANDARD\_MERGE$ is $n \log n + n \log(e_1+1) + n \log(e_2+1) + ... + n \log(e_t+1)$. The number of key comparisons performed by CHECK is
$en + (e - e_1)n + (e - e_1 - e_2)n + ... + (e - e_1 - ... - e_t)n$. From the arithmetic mean - geometric mean inequality, and from the inequality $e_1 + e_2 + ... + e_t \leq e$ it follows that $\{(e_1+1)(e_2+1)...(e_t+1)\}^{\frac{1}{t}} \leq (1 + \frac{e}{t})$. Thus the total number of comparisons performed by the algorithm is upper-bounded by $n \log n + (2t-1)ne + n\, t \log(1 + \frac{e}{t})$. If $e \leq \sqrt{\log n}$, this expression is upper-bounded by $O(n \log n)$.

3 A More Efficient Fault-Tolerant Merge Sorting

In this section, we present a more complex variation of standard merge-sort that can tolerate upto $O(\frac{\log n}{\log \log n})$ errors. This algorithm is similar to the simpler algorithm in the sense that it also repeats a standard merging followed by check and split, until the array is provably sorted. But instead of merging all the way to a single list, this algorithm stops the standard merging after a partial merging. The time to stop is chosen carefully so that an improved bound can be obtained. The algorithm and its analysis are presented below.

Theorem 2. *We can adapt Merge sorting to sort* n *keys in time* $\Theta(n \log n)$ *in the worst-case so long as* e *is* $O(\frac{\log n}{\log \log n})$.

Proof. We will first informally describe the algorithm. Let $STANDARD\_MERGE(X, t)$ be a procedure which takes as input a list X of lists $X_1, X_2, ..., X_k$

and merges them pairwise and them recursively repeats the process until it is reduced to a list of t lists. Thus the final list X' is $X_1', X_2', ..., X_t'$ where X_1' is the merge of $X_1, X_2, ..., X_k$, X_2' is the merge of $X_{k+1}, ..., X_{2k}$ etc. where $k = 2^{log\ (t/m)}$. CHECK($< X_1, ..., X_k >$, *rest*) is a procedure which takes as input a list X of lists $X_1, X_2, ..., X_k$ and an integer *rest*. The lists $X_1, X_2, \ldots, X_k$ are not necessarily sorted and the goal of CHECK is to split X_i's if necessary so that if the final output of CHECK is $X_1', X_2', ..., X_r'$ then each of the lists is sorted. CHECK will work under the assumption that there may be up to *rest* errors in comparisons when it starts. CHECK will repeat the comparison between two adjacent elements of the same list by repeating the query "Is $x < y$?" until one of the outcomes YES or NO is returned $rest + 1$ times. At this point, the outcome becomes certain. If two adjacent elements are found to be out of place, the list is split; otherwise the algorithm proceeds to the next comparison. In both cases, the value of *rest* is updated by subtracting the number of faulty outcomes. The process is repeated for all pairs of adjacent elements in all the lists and the split sequence of lists is output. Finally, let $SIMPLE\_MERGE(A, B, e)$ denote the standard merging algorithm to merge lists A and B, but it handles upto e errors in comparisons by query repetition. It performs each comparison required by the error-free simple merging algorithm (no fancy algorithm like Hwang and Lin [5]) but repeats each comparison sufficiently many times so that the outcome is determined with certainty. Thus the number of comparisons it performs is at most $O((|A| + |B|)e)$. The main algorithm, which is presented below uses the three algorithms described above as follows. It starts with the original sequence of n keys and merges them using $STANDARD\_MERGE$ until $\sqrt{n}$ lists are created. Then it uses CHECK. It repeats the process each time stopping with $\sqrt{t}$ lists if it started with t lists and then performs a CHECK. This phase concludes when CHECK returns a list with at most e lists. Now it switches to $SIMPLE\_MERGE$ to repeatedly merge adjacent pairs until a single list is created.

Input: A set of n keys $x_1, ..., x_n$ and e the error bound.
Step 1: for $i \leftarrow 1$ to n do $X_i = x_i$.
Step 2: $m \leftarrow n$; $rest \leftarrow e$;
Step 3: $X = < X_1, X_2, ..., X_m >$
Step 4: while $(m \geq e)$ do
  begin $t \leftarrow \lceil \sqrt{m} \rceil$;
  $STANDARD\_MERGE(X, t)$;
  $(< Y_1, Y_2, ..., Y_m >, current) \leftarrow CHECK(X, rest)$;
  $rest = rest - current$;
  $X = < Y_1, Y_2, ..., Y_m >$;
  end while;
Step 5: Merge the lists $Y_1, Y_2, ..., Y_m$ into a single list using $SIMPLE\_MERGE$.

Correctness of the algorithm: First observe that the following invariant "each X_i is sorted" holds for the output $< Y_1, Y_2, ..., Y_m >$ of CHECK. Partial correctness of the algorithm readily follows from this since when the while loop of Step 4 terminates, each Y_i is sorted. In Step 5, we use SIMPLE-MERGE to sort a set of sorted lists using an algorithm whose correctness is obvious. It is also easy to show that the while loop of Step 4 terminates. (In fact, there are at most e iterations of the while loop during which *runs* of X may increase. But in any case, the length can never increase beyond n. During the rest of the iterations, value of m decreases. From this it follows that the while loop terminates.

Analysis: From the observations presented in Section 2, it follows that there are at most $n^{\frac{1}{2^{i-1}}} + e$ sorted lists just prior to the i-th iteration of the while loop in Step 4. First we will show that the number of iterations of Step 4 performed by the algorithm is at most $O(lg\ lg\ n)$. The reason is as follows: If the number of lists at the beginning of an iteration is k, it is at most $\sqrt{k} + e$ at the end. Thus the number of iterations $i(n)$ is given by the recurrence inequality $i(n) \leq 1 + i(\sqrt{n} + e)$. Since e is $O(\frac{logn}{log\ log\ n})$, $i(n)$ is upper-bounded by $j(n)$ given by the recurrence inequality $j(n) \leq 1 + j(2\sqrt{n})$. The latter function $j(n)$ is easily seen to be upper-bounded by $O(log\ log\ n)$ and so is $i(n)$.

The number of comparisons performed by the call to *STANDARD_MERGE*(X, t) during the i-th iteration is given by:

$$nlog\left(\frac{n^{\frac{1}{2^{i-1}}} + e}{n^{\frac{1}{2^i}}}\right) \leq \frac{n\ log\ n}{2^i} + 1$$

Thus the total number of comparisons performed by **all** the calls to STANDARD-MERGE in Step 4 is upper-bounded by $n\ log\ n(1 + 1/2 + 1/2^2 + ...) + i$ (where i is the number of iterations) $= O(n\ log\ n)$.

The total number of comparisons performed by CHECK is easily seen as $(e + 1)\ n\ log\ log\ n) = O(n\ log\ n)$ since e is $O(\frac{logn}{log\ log\ n})$.

Finally, consider Step 5. Suppose we use SIMPLE-MERGE to merge k sorted lists with a total of n keys into a single sorted list during which e errors can occur, the total number of comparisons performed by the algorithm is $O(enlog\ k)$. In our case, $k = e = O(\frac{logn}{log\ log\ n})$, and so $O(enlog\ k)$ is $O(n\ log\ n)$. Thus the total number of key comparisons performed by the algorithm is $O(n\ log\ n)$.

4 Conclusions and Open Problems

The main problem that remains open is whether there is a natural adaptation of Merge-Sorting that can sort in $O(n\ log\ n)$ in the presence of $O(log\ n)$ faults. It would also be of interest to implement the two algorithms and compare them experimentally. Designing fault-tolerant versions of other comparison-based algorithms may also be a general area of interest. Finally, we would like to find applications for fault-tolerant algorithms for sorting besides the ones listed in the introduction.

References

1. A. Bagchi. On sorting in the presence of erroneous information. Information Processing Letters, 43(4):213-215, 28 September 1992.
2. R.S.Borgstrom and S.R. Kosaraju, Comparison-based search in the presence of errors, *Proc. of ACM Symposium on Theory of Computing* (1993), pp. 130-136.
3. T. Cormen, C. Leiserson, R. Rivest and C. Stein, *Introduction to Algorithms*, 2nd Edition, MIT Press 2001.
4. U.Feige, P.Raghavan, D.Peleg and E.Upfal, Computing with noisy information, *Proc. IEEE Symposium on Theoretical Computer Science* (1994), pp. 1001-1008.
5. D. Knuth, Art of Computer Programming, Volume 3. Sorting and Searching, 2nd Edition, Addison-Wesley, 1998.
6. K.B.Lakshmanan, B. Ravikumar and K. Ganesan, Coping with errors while sorting. *IEEE Transactions on Computers* 40 (9), (1991), pp. 1081-1084.
7. P.M.Long, Sorting and searching with a faulty comparison oracle. UCSC-CRL-92-15 (available electronically as ucsc-crl-92-15.ps.Z)

2-Compromise Usability in 1-Dimensional Statistical Databases

Ljiljana Branković[1] and Jozef Širáň[2]

[1] School of Electrical Engineering and Computer Science
The University of Newcastle
NSW 2308, Australia
lbrankov@cs.newcastle.edu.au

[2] Department of Mathematics, SvF
Slovak University of Technology
Radlinského 11
813 68 Bratislava, Slovakia
siran@lux.svf.stuba.sk

Abstract. Many believe that data mining poses the biggest privacy challenge in the next decade. In this paper we concentrate on threats to privacy arising from the possibility of combining aggregate data to deduce the confidential individual values. We determine the maximum number of sum totals that can be disclosed without leading to a 2-compromise in a 1-dimensional database for range queries.

Keywords: privacy in data mining, statistical database security, combinatorics, discrete mathematics.

1 Introduction

Ontario Privacy Commissioner Ann Cavoukian said in her report "Data mining: Staking a claim on your privacy" that data mining may be the most fundamental privacy challenge in the next decade.

Vast amounts of personal information are collected in the process of bank transactions, credit card payments in supermarkets, making phone calls, using reward cards, visiting doctors, renting videos and cars, to mention just a few. All this data is typically used for data mining and statistical analysis and is very often sold to other companies and organisations.

A breach of privacy occurs when individuals are not aware that the data has been collected in the first place, has been passed onto the other companies or has been used for the purposes other that the one for which it was originally collected.

Even when individuals approve of using their personal records for data mining and statistical analysis, for example for medical research on their diseases, it is still assumed that only aggregate values will be made available to researchers and that no individual values will be disclosed. To ensure this, it is not enough to disable access to individual records, as it is very often possible to compute

O.H. Ibarra and L. Zhang (Eds.): COCOON 2002, LNCS 2387, pp. 448–455, 2002.

individual values from a suitable combination of aggregate values. Possible solutions to this problem include adding noise to original data, so that disclosing a perturbed individual value does not imply a compromise. Another solution is to reject some statistical queries and disclose only those that, even when combined, do not lead to a compromise. By a compromise we mean a disclosure of a confidential individual value; more generally, by a k-compromise we mean a disclosure of an aggregate value based on k or less confidential individual values.

We present a new result regarding 2-compromise in 1-dimensional databases. We consider sum range queries, as they are the most common queries. Range queries are based on records with attribute values in a specified range. An example is a sum of salaries of all employees between 25 and 35 years of age. A 1-dimensional database of order n may be represented by a set $\{1,\ldots,n\}$ and range queries by subsets of the form $\{b, b+1,\ldots,e\}$, $1 \leq b \leq e \leq n$. In what follows we shall refer to range queries as intervals.

In this paper we answer the following question: To ensure that a 1-dimensional database is 2-compromise free, what percentage of range sum queries can we safely answer? Similar questions has been answered for k-compromise, where k is odd in [2]; for general queries in [4,3]; and for 1-compromise and multidimensional databases in [5,1].

2 Intervals and Threads

For any positive integer n let $[n]$ denote the set $\{1,2,\ldots,n\}$. Any subset of $[n]$ of the form $\{b, b+1,\ldots,e\}$, $b \leq e$, will be called an *interval* and denoted by $[b,e]$. The integers b and e are the *beginning* and the *ending* of the interval and $e-b+1$ is its *length*.

To each interval $I = [b,e] \subset [n]$ we now associate the n-dimensional vector $\omega_I = (\omega_1, \omega_2, \ldots, \omega_n)$ such that $\omega_i = 1$ for $b \leq i \leq e$ and $\omega_i = 0$ otherwise. A family $\mathcal{I}$ of intervals in $[n]$ is said to be *k-compromise free* if each linear combination of the vectors ω_I, $I \in \mathcal{I}$, is either a zero vector or contains more than k nonzero coordinates. In other words, the family $\mathcal{I}$ is k-compromise free if the only linear combination of the vectors ω_I $(I \in \mathcal{I})$ containing no more than k non-zero entries is the zero vector.

If $I_1 = [b_1, e_1]$ and $I_2 = [b_2, e_2]$ are intervals such that $b_2 = e_1 + 1$ then we say that I_1 and I_2 are *consecutive*. In such a case $I = I_1 \cup I_2$ is the *sum* of I_1 and I_2, denoted by $I = I_1 + I_2$. (Note that a union of two intervals is an interval also in the case when the two intervals overlap, but we will refer to a sum only in the sense defined above.) The definition of consecutiveness and sum can be extended in a natural way to more than two intervals. If $I = I_1 + I_2$ then we also write $I_2 = I - I_1$ and $I_1 = I - I_2$; these are the only instances when difference symbol for intervals will be used.

A family $\mathcal{I}$ of intervals in $[n]$ is *closed* if for any two intervals $I, J \in \mathcal{I}$ for which the sum (or the difference) is defined, we also have $I+J \in \mathcal{I}$ (or, $I-J \in \mathcal{I}$). For any such closed family $\mathcal{I}$ there exists a unique subfamily $\mathcal{I}_o$ of $\mathcal{I}$ with the following two properties:

(1) each interval in $\mathcal{I} \setminus \mathcal{I}_o$ can be obtained as a sum of intervals in $\mathcal{I}_o$; and
(2) no interval in $\mathcal{I}_o$ is a sum of two or more members of $\mathcal{I}_o$.

Indeed, for each positive integer $b \leq n$ that is a beginning of some interval in $\mathcal{I}$ let $I(b)$ be such an interval of the smallest length. Then $\mathcal{I}_o$ must contain all the intervals $I(b)$. On the other hand, it is easy to see that the set of such intervals satisfies (1) and is thus the entire $\mathcal{I}_o$. We will call $\mathcal{I}_o$ the *generating set* of $\mathcal{I}$.

Let $\mathcal{I}$ be a closed family of intervals in $[n]$. We will say that a subset $\mathcal{T}$ of $\mathcal{I}_o$ forms a *thread* if $\mathcal{T}$ consists of consecutive intervals (equivalently, if the disjoint union of all intervals in $\mathcal{T}$ is an interval again). We say that $\mathcal{S} = \{\mathcal{T}_1, \mathcal{T}_2, \ldots, \mathcal{T}_r\}$ is a *thread decomposition* of $\mathcal{I}_o$ if the set $\mathcal{I}_o$ is a disjoint union of the threads $\mathcal{T}_i \in \mathcal{S}$ and if no thread is a union of two or more threads in $\mathcal{S}$. (We note that a thread decomposition is uniquely determined up to the numbering of threads.)

From now on let $\mathcal{I}$ be a closed, 2-compromise free family of intervals in $[n]$. Let us fix a thread decomposition $\mathcal{S} = \{\mathcal{T}_1, \mathcal{T}_2, \ldots, \mathcal{T}_r\}$ of the generating set $\mathcal{I}_o$. Note that the assumption that $\mathcal{I}$ is 2-compromise free implies that the length of each interval in $\mathcal{I}_o$ is at least three and thus no interval in $\mathcal{I}_o$ can end with the element 1 or 2. For each thread $\mathcal{T}_i$, $1 \leq i \leq r$, let $B_i = \{b_1^i, b_2^i, \ldots, b_{t_i}^i\}$ denote the set of beginnings of intervals in $\mathcal{T}_i$, and let $E_i = \{e_1^i, e_2^i, \ldots, e_{t_i}^i\}$ denote the set of endings of intervals in $\mathcal{T}_i$, where t_i denotes the number of intervals in $\mathcal{T}_i$; we will call t_i the *length* of the thread $\mathcal{T}_i$. Let $E_i^* = \{b_1^i - 1\} \cup E_i$ and let $E^* = \cup_{i=1}^r E_i^*$. Similarly, let $B_i^* = B_i \cup \{e_{t_i}^i + 1\}$.

For each thread $\mathcal{T}_i$, $1 \leq i \leq r$, we define a set C_i as follows. For each j, $1 \leq j \leq t_i$, C_i contains e_j^i if every integer greater than e_j^i and less than or equal to n is in E^*; otherwise C_i contains $e_j^i + 1$. Additionally, if every integer greater than $b_1^i - 1$ and less than or equal to n is in E^*, then $b_1^i - 1$ is in C_i, otherwise b_1^i is in C_i. The reader is invited to check that for each i, $1 \leq i \leq r$, $C_i \subset [n]$ and $|C_i| = t_i + 1$.

Lemma 1. *For each $i, j \leq n$, $i \neq j$, the sets C_i and C_j are disjoint.*

Proof. As the intervals in $\mathcal{T}_i$ and $\mathcal{T}_j$ are all members of the generating set $\mathcal{I}_o$, no two of them can begin or end with the same element. Additionally, the first interval in $\mathcal{T}_i$ cannot start with $e_{t_j}^j + 1$ because then $\mathcal{T}_i \cup \mathcal{T}_j$ would be a thread itself, contrary to the definition of a thread decomposition. Similarly, the last interval in $\mathcal{T}_i$ cannot end with $b_1^i - 1$. Thus, if the sets C_i and C_j intersect in an element c then, without loss of generality, $c \in B_i^*$, that is, c is either $e_{t_i}^i + 1$ or a beginning of an interval in the thread $\mathcal{T}_i$ and, at the same time, $c \in E_j^*$, that is, c is either $b_1^j - 1$ or an ending of an interval in the thread $\mathcal{T}_j$. This implies that each integer greater than c (and less or equal to n) is an ending of some interval in $\mathcal{I}_o$. But then, by the definition of C_i, c is not in C_i, which contradicts our initial assumption that if the sets C_i and C_j intersect in an element c. □

Let s be the smallest integer with the property that every integer greater then or equal to s and less then or equal to n is in E^*. If such integer does not exist, let $s = n+1$. Note that every element $u < s$ in C_i, $1 \leq i \leq r$ is a beginning of an interval in $\mathcal{I}_o$, or $u = e^j_{t_j} + 1$ for some j, $1 \leq j \leq r$. Similarly, every element v, $v \geq s$ in C_i, $1 \leq i \leq r$ is an ending of some interval in $\mathcal{I}_o$, or $v = b^j_1 - 1$ for some j, $1 \leq j \leq r$. For $1 \leq i \leq r$ let $B^+_i = \{b+1;\ b \in B_i\}$ and let $E^-_i = \{e-1; e \in E_I\}$. Since the length of each interval in $\mathcal{I}_o$ is at least three, the sets B_i, B^+_i and E_i are mutually disjoint, and so are the sets B_i, E^-_i and E_i. For any j such that $1 \leq j \leq r$ let $F_j = B_j \cup (B^+_j \cap \{1, \ldots, s-1\}) \cup (E^-_j \cap \{s, \ldots, n\}) \cup E_j \cup (\{b^j_1 - 1, e^j_{t_j} + 1\} \cap [n])$.

Lemma 2. *For each $i \leq r$, $i \neq j$, we have $|C_i \cap F_j| \leq 1$.*

Proof. Let $b < b'$ be two elements in $C_i \cap F_j$.

Assume first that b, $b' < s$, that is, both b and b' are beginnings of intervals in $\mathcal{T}_i$, or $b \in B_i$ and $b' = e^i_{t_i} + 1$. As $\mathcal{T}_i$ is a thread, it follows in both cases that the interval $I = [b, b'-1]$ is a sum of intervals in $\mathcal{T}_i$. But at the same time we have $b, b' \in F_j$, which means that there exists an interval J that is a sum of intervals in $\mathcal{T}_j$ and has the form $J = [c, d]$ where $c \in \{b-1, b, b+1\}$ and $d \in \{b'-2, b'-1, b'\}$. It follows that the symmetric difference of the intervals I and J contains at most two elements, which contradicts the assumption that $\mathcal{I}$ is a 2-compromise free family (as the vector $\omega_I - \omega_J$ would have at most two non-zero coordinates).

We now assume that b, $b' \geq s$, that is, both b and b' are in E_i, or $b' \in E_i$ and $b = b^i_1 - 1$. It follows that there are intervals I and J in $\mathcal{I}$, where $I = [b+1, b']$ and $J = [c, d]$, $c \in \{b, b+1, b+2\}$ and $d \in \{b'-1, b', b+1'\}$, and we again arrive at the contradiction.

It remains to consider the case when $b < s$ and $b' \geq s$. We arrive at the same contradiction with considering the intervals $I = [b, b']$ and $J = [c, d]$, where $c \in \{b-1, b, b+1\}$ and $d \in \{b'-1, b', b'+1\}$. □

In our thread decomposition $\mathcal{S} = \{\mathcal{T}_1, \mathcal{T}_2, \ldots, \mathcal{T}_r\}$ of the generating set $\mathcal{I}_o \subset \mathcal{I}$ of intervals in $[n]$ we may assume without loss of generality that $\mathcal{T}_1$ contains an interval beginning with 1. Moreover, if n does not appear in an interval in $\mathcal{T}_1$ we may assume that the numbering is chosen in such a way that n appears as end of an interval in the thread $\mathcal{T}_2$. For the ease of formulation of the next result let $\varepsilon_1 = 2$ and $\varepsilon_2 = 0$ if the element n appears in the thread $\mathcal{T}_1$, and $\varepsilon_1 = \varepsilon_2 = 1$ if n appears in $\mathcal{T}_2$; let $\varepsilon_i = 0$ for $3 \leq i \leq r$.

Lemma 3. *Let the generating set $\mathcal{I}_o$ have a thread decomposition into $r \geq 2$ threads with the properties listed above. Then, for $1 \leq j \leq r$, the thread lengths t_i satisfy the following system of linear inequalities:*

$$3t_j + \sum_{i \neq j} t_i \leq n - 2 + \varepsilon_j \ .$$

Proof. Consider first the sets F_j introduced before Lemma 2 and observe that $|F_j| = 3t_j + 2 - \varepsilon_j$ for each j, $1 \leq j \leq r$. Recalling Lemma 1, we know that for $1 \leq i \leq r$, $i \neq j$, the sets C_i are pairwise disjoint. Moreover, by Lemma 2 we have $|F_j \cap C_i| \leq 1$ for any $i \neq j$. Using a routine set-theoretic estimate we obtain:

$$\begin{aligned} n \geq |F_j \cup (\cup_{i\neq j} C_i)| \geq |F_j| + \textstyle\sum_{i\neq j} |C_i| - \sum_{i\neq j} |F_j \cap C_i| \\ \geq |F_j| + \textstyle\sum_{i\neq j}(t_i+1) - \sum_{i\neq j} 1 \\ = 3t_j + 2 - \varepsilon_j + \textstyle\sum_{i\neq j} t_i \end{aligned}$$

which is equivalent with the inequality in the statement of the result. □

3 Main Result

Our goal is to find the largest cardinality $\mu_2(n)$ of a closed 2-compromise free family $\mathcal{I}$ of intervals in $[n]$. If $\mathcal{S} = \{\mathcal{T}_1, \mathcal{T}_2, \ldots, \mathcal{T}_r\}$ is a thread decomposition of the generating set $\mathcal{I}_o$ with properties as above, then clearly $|\mathcal{I}| = \sum_{i=1}^r \binom{t_i+1}{2}$ where t_i is the length of the thread $\mathcal{T}_i$. Let $\tau_2(n) = \max \sum_{i=1}^r \binom{t_i+1}{2}$ where the maximum is taken over all r and over all r-tuples $(t_1, t_2, \ldots, t_r)$ of non-negative integers satisfying the inequalities in Lemma 3 either for $\varepsilon_1 = 2$ or for $\varepsilon_1 = 1$. (In fact, the numbers t_i appearing in Lemma 3 are all positive, but allowing them to be equal to zero - which is equivalent to allowing "empty threads" - will not affect our computations.) Clearly, $\mu_2(n) \leq \tau_2(n)$; in what follows we determine $\tau_2(n)$ and show that $\tau_2(n) = \mu_2(n)$ for each odd $n \geq 1$ and each even $n \geq 52$.

Before we do so let us reformulate the problem of finding $\tau_2(n)$ in a more uniform way. Letting $s_i = t_i - \varepsilon_i/2$ for $1 \leq i \leq r$, the inequalities of Lemma 3 transform to

$$3s_j + \sum_{i\neq j} s_i \leq n - 3 \; . \tag{1}$$

Let $\varepsilon = \varepsilon_1 \in \{1, 2\}$; then $\varepsilon_2 = 2 - \varepsilon$. Further, let

$$f_\varepsilon = f_\varepsilon(s_1, s_2, \ldots, s_r) = \sum_{i\leq r} s_i(s_i+1) + \varepsilon s_1 + (2-\varepsilon)s_2 + \varepsilon/2 + 1 \; . \tag{2}$$

An easy computation shows that $2\tau_2(n)$ is determined by

$$2\tau_2(n) = \max_r \max_\varepsilon \max f_\varepsilon \tag{3}$$

where the third maximum is taken for $\varepsilon = 2$ over all r-tuples $s_i \geq 0$ which are integral solutions of (1), and for $\varepsilon = 1$ over all solutions of (1) such that $2s_1$ and $2s_2$ are odd integers and the remaining s_i are integers. We emphasize that the value of f_ε is an *even integer* for all such *admissible* r-tuples $(s_1, s_2, \ldots, s_r)$.

Theorem 1. *We have* $\tau_2(n) = \lfloor (n+1)^2/16 \rfloor$ *for each odd* $n \geq 1$ *and* $\tau_2(n) = \lceil n^2/16 \rceil$ *for each even* $n \geq 52$.

Proof. To outline our strategy, we first solve the following rational relaxation problem: Determine the maximum $2\theta_2(n)$ of the function f_2 from (2) over all r and over a region R consisting of all r-tuples $(s_1, s_2, \ldots, s_r)$ of *rational numbers* $s_i \geq 0$ that satisfy the inequalities (1). The function f_2 is convex, and as we know from the general theory, for any fixed r the maximum of f_2 taken over our (convex) region R occurs in one of the corners of R. Once we know $\theta_2(n)$, it is easy to see that we have an upper bound of the form $\tau_2(n) \leq \lfloor \theta_2(n) \rfloor$. We then show that equality holds there for all odd n, and finally take care about the more complicated case when n is even.

The form of the function f_2 suggests that for the purpose of determining $\theta_2(n)$ by means of computing corner coordinates and the corresponding values of f, we may without loss of generality assume that $s_1 \geq s_2 \geq \ldots \geq s_r$. Corners of the region R correspond to solutions of linear systems obtained from the $2r$ defining inequalities of R by choosing r of them and turning them into equations. Thus, if a corner $(s_1, s_2, \ldots, s_r)$ of R has exactly q nonzero coordinates $(0 \leq k \leq r)$ then $s_{q+1} = \ldots s_r = 0$ and $3s_j + \sum_{i \neq j} s_i = n - 3$ where $1 \leq j \leq q$. It follows that $s_1 = \ldots = s_q = (n-3)/(q+2)$, and then

$$f_2(s_1, s_2, \ldots, s_r) = (n-3)^2 q/(q+2)^2 + n - 1 \, . \tag{4}$$

The function $g(n, q) = (n-3)^2 q/(q+2)^2 + n - 1$ is maximised (for any fixed n) at $q = 2$, and $g(n, 2) = (n-3)^2/8 + n - 1 = (n+1)^2/8$. We therefore have $2\theta_2(n) = (n+1)^2/8$, and so

$$\tau_2(n) \leq \lfloor \theta_2(n) \rfloor = \lfloor (n+1)^2/16 \rfloor \, . \tag{5}$$

To show that, in fact, $\tau_2(n) = \lfloor (n+1)^2/16 \rfloor$ for all odd n we first consider the case when $n = 4m+3$, $m \geq 0$. Then for $r = 2$ the pair $s_1 = s_2 = m$ is an integer solution of the inequalities (1), and we have $f_2(m, m)/2 = (m+1)^2 = (n+1)^2/16$. If $n = 4m+1$ then, for $r = 2$ again, $s_1 = s_2 = m - 1/2$ is an admissible solution of (1) but this time for $\varepsilon = 1$, with $f_1(s_1, s_2) = m(m+1) = \lfloor (n+1)^2/16 \rfloor$.

The case when n is even requires a further analysis. It can be verified that both values $g(n, 1)/2$ and $g(n, 3)/2$ are strictly smaller than $\lceil n^2/16 \rceil + 1$ for all $n = 4m$, where $m \geq 13$ and all $n = 4m+2$ where $m \geq 10$. As $g(n, q)$ is for each fixed n a decreasing function of q when $q > 2$, to show that $\tau_2(n) = \lceil n^2/16 \rceil$ for even $n \geq 52$ it is sufficient (by (4) and the preceding fact) to restrict to $r = 2$ and to admissible pairs (s_1, s_2) with $s_1 \geq s_2 > 0$. We will now have a more close look at the set of admissible pairs for the maximisation problem (3) when $r = 2$.

If $n = 4m$, the admissible pairs (s_1, s_2) are determined by inequalities $3s_1 + s_2 \leq 4m - 3$ and $s_1 + 3s_2 \leq 4m - 3$, from which we have $s_1 + s_2 \leq 2m - 3/2$. But note that for an admissible pair the sum $s_1 + s_2$ is always an integer, and therefore $s_1 + s_2 \leq 2m - 2$. Assuming $s_1 \geq s_2$ we may replace the two original inequalities with a new set, namely, $3s_1 + s_2 \leq 4m - 3$ and $s_1 + s_2 \leq 2m - 2$ where $s_1 \geq s_2 \geq 0$. The admissible set for maximisation problem (3) for $r = 2$ now consists of all pairs (s_1, s_2) defined by this new set of inequalities (all other conditions are as described before the formulation of Theorem 1). Considering the corresponding rational relaxation problem for f_2 again, we may restrict to corners whose both coordinates are positive. With $s_1 \geq s_2$ there is just one such corner, namely, $(m - 1/2, m - 3/2)$, and it gives the value of $f_2 = 2m^2 + 3/2$. (Note that for the problem (3) this corner is admissible for f_1 but *not* for f_2; however, now we are dealing with the relaxation problem for f_2 where there is no guarantee that the values of f_2 are even integers.) Recalling the analysis in the previous paragraph, for $n = 4m \geq 52$ it now follows that $\tau_2(4m) \leq \lfloor f_2/2 \rfloor = m^2 = \lceil n^2/16 \rceil$. In fact, we have $\tau_2(n) = \lceil n^2/16 \rceil$ for $n = 4m \geq 52$, as is shown by evaluating f_1 at the above corner: $f_1(m - 1/2, m - 3/2) = 2m^2 = \lceil n^2/16 \rceil$.

Similarly, if $n = 4m + 2$ and $r = 2$ then the set of admissible pairs (s_1, s_2) for the problem (3) is determined by $3s_1 + s_2 \leq 4m - 1$ and $s_1 + s_2 \leq 2m - 1$ where $s_1 \geq s_2 \geq 0$. The rational relaxation problem for maximising f_2 over that region has solution at the corner $(m, m-1)$ with $f_2 = 2m^2 + 2m + 2$. Again, taking into account the analysis done before, for $m \geq 10$ we obtain $\tau_2(4m+2) \leq f_2/2 = \lceil n^2/16 \rceil$. The pair $(m, m-1)$ shows that, in fact, $\tau_2(n) = \lceil n^2/16 \rceil$ for $n = 4m + 2 \geq 42$. □

Recall that our goal is to find the largest cardinality $\mu_2(n)$ of a closed 2-compromise free family $\mathcal{I}$ of intervals in $[n]$. We saw earlier that $\mu_2(n) \leq \tau_2(n)$ for all n; now we prove that we have equality here in most cases.

Theorem 2. *We have* $\mu_2(n) = \lfloor (n+1)^2/16 \rfloor$ *for each odd* $n \geq 1$ *and* $\mu_2(n) = \lceil n^2/16 \rceil$ *for each even* $n \geq 52$.

Proof. By Theorem 1 it is sufficient to find instances of closed 2-compromise free families of cardinality $\tau_2(n)$ for the corresponding values of n. We will do so by describing thread decompositions of the generating sets $\mathcal{I}_o$; in all cases there will be just two threads $\mathcal{T}_1$ and $\mathcal{T}_2$.

If $n = 4m + 1$ then we set $\mathcal{T}_1 = \{[1,4], [5,8], \ldots, [4m-3, 4m]\}$ and $\mathcal{T}_2 = \{[3,6], \ldots, [4m-5, 4m-2], [4m-1, 4m+1]\}$. For $n = 4m + 3$ we define $\mathcal{T}_1 = \{[1,4], [5,8], \ldots, [4m-3, 4m], [4m+1, 4m+3]\}$ and $\mathcal{T}_2$ will be the same as before. If $n = 4m + 2$ then we let $\mathcal{T}_1 = \{[1,4], \ldots, [4m-7, 4m-4], [4m-3, 4m-1], [4m, 4m+2]\}$ and let $\mathcal{T}_2$ be as above. Finally, for $n = 4m$ we will have $\mathcal{T}_1 = \{[1,4], \ldots, [4m-7, 4m-4], [4m-3, 4m-1]\}$ and $\mathcal{T}_2 = \{[3,6], \ldots, [4m-9, 4m-6], [4m-5, 4m]\}$.

The reader is invited to check that the cardinalities of the (closed) families of intervals generated by the threads described above are equal to $\tau_2(n)$ and, more importantly, that all these families are 2-compromise free. □

We note that the above extremal families are by no means unique.

We define a usability for a range database as a ratio of the maximum number of range queries that can be answered without a compromise and a total number of range queries. Then, the usability for 2-compromise is $U(n) = \frac{2\lfloor (n+1)^2/16 \rfloor}{n(n+1)}$ for each odd $n \geq 1$ and $U(n) = \frac{2\lceil n^2/16 \rceil}{n(n+1)}$ for each even $n \geq 52$.

4 Conclusion

In this paper we determined the usability for 2-compromise in 1-dimensional range databases of order n for each odd n and each even $n \geq 52$. We note that the result is asymptotically equal to the one for 3-compromise. The question remains whether the usability for each $k = 2l$, $l \geq 1$ is asymptotically equal to the usability for $k = 2l + 1$.

Acknowledgements

The authors would like to thank an anonymous referee for pointing out a few inaccuracies in an earlier version of the paper.

References

1. L. Branković, P. Horak and M. Miller. An optimization problem in statistical database security. *SIAM Journal on Discrete Mathematics*, Volume 13, Number 3, pages 346–353, 2000.
2. L. Branković, M. Miller and J. Širáň. Range query usability of statistical databases. *To appear in International Journal of Computer Mathematics.*
3. J. R. Griggs. Database security and the distribution of subset sums in R^m. *Graph Theory and Combinatorial Biology, (Balatonlelle, 1996).*
4. J. R. Griggs. Concentrating subset sums at k points. *Bulletin of the ICA*, Volume 20, pages 65–74, 1997.
5. P. Horak, L. Branković and M. Miller. A combinatorial problem in database security. *Discrete Applied Mathematics*, Volume 91, Number 1-3, pages 119–126, 1999.

An Experimental Study and Comparison of Topological Peeling and Topological Walk

Danny Z. Chen[1,*], Shuang Luan[1,*], and Jinhui Xu[2,**]

[1] Department of Computer Science and Engineering, University of Notre Dame, Notre Dame, IN 46556, USA
{chen,sluan}@cse.nd.edu

[2] Department of Computer Science and Engineering, State University of New York at Buffalo, 201 Bell Hall Box 602000, Buffalo, NY, 14260, USA
jinhui@cse.buffalo.edu

Abstract. In this paper, we present an experimental study comparing two algorithms, *topological peeling* and *topological walk*, for traversing arrangements of planar lines. Given a set H of n lines and a convex region R on a plane, both topological peeling and topological walk sweep the portion A_R of the arrangement of H inside R in $O(K + n\log(n+r))$ time and $O(n + r)$ space, where K is the number of cells of A_R and r is the number of boundary vertices of R. In our study, we robustly implemented these two algorithms using the LEDA library. Based on the implementation, we carried out experiments to conduct several comparisons, such as the arrangement traversal fashions, memory consumption, and execution time. In general, topological peeling exhibits a better control on the propagation of its sweeping curve (called the *wavefront*). For memory consumption, two types of measures, logical and physical memory, were examined. Our experiments showed that although both algorithms use nearly the same amount of logical memory, topological peeling could use twice as much physical memory as topological walk. For execution time, experiments revealed an interesting phenomenon that topological peeling has a 10% to 25% faster execution time than topological walk in most cases. Our analysis of this phenomenon indicates that the execution times of topological peeling and topological walk are both sensitive to the ratio of the lower input lines to all input lines. When the ratio of the lower lines to all input lines is around 85%, the two algorithms have roughly the same amount of execution time. Under this ratio, topological peeling considerably outperforms topological walk; above this ratio, topological walk slightly outperforms topological peeling.

1 Introduction

The focus of this paper is on traversing arrangements of lines on the plane. Computing or sweeping the arrangement of a set of n planar lines is a fundamental

* The research was supported in part by the National Science Foundation under Grant CCR-9988468.

** The research was supported in part by a faculty start-up fund from the CSE dept., SUNY at Buffalo, and an IBM faculty partnership award.

O.H. Ibarra and L. Zhang (Eds.): COCOON 2002, LNCS 2387, pp. 456–466, 2002.

problem in computational geometry [11], and finds applications in solving many problems [3,4,9,10,11,12,21,23]. Extensive research has been done on sweeping the arrangement of planar lines, and quite a few interesting algorithms and techniques have been proposed [3,4,6,9,11,12,13,15,23]. A common feature of these techniques is to use a geometric curve (e.g., a polygonal line) to guide the sweep of the arrangement, and enumerate the faces of the arrangement (i.e., vertices, edges, or cells) based on the order of their intersections with the sweeping curve.

Edelsbrunner and Guibas [12] used a y-monotone curve to sweep the arrangement, resulting in a powerful *topological sweep* algorithm for computing the arrangement on the whole plane; their algorithm takes $O(n^2)$ time and $O(n)$ space for sweeping an arrangement A of n planar lines, which is optimal in both the time and space complexity. To efficiently traverse a portion of an arrangement inside a convex region on the plane, Asano, Guibas, and Tokuyama [3] presented an interesting output-sensitive *topological walk* algorithm, which computes the portion A_R of a planar arrangement A inside a convex region R in $O(K + n \log(n + r))$ time and $O(n + r)$ space, where K is the number of cells of A_R and r is the number of boundary vertices of R.

One unfavorable feature of topological walk [3] is that its traversal can visit the exterior of the region R; hence during the execution of the algorithm, if one removes the elements of the traversal associated with the outside of R, the traversed portion inside R can be disconnected (e.g., the swept cells may form multiple connected components). This feature can cause undesirable difficulty for solving certain optimization problems, such as the shortest path problem, on planar arrangements [8,9]. Another unfavorable feature is that topological walk reports only the lower chain of each cell (or the set of bounding lines below the cell) to identify the cell. In solving some CAD problems [22] and geometric optimization problems [2,7,20], the entire boundary of each cell in an arrangement is needed (e.g., used as the constraints for certain special non-linear optimization problem instance). Thus, such a "weak" representation of cells may not provide sufficient information for some applications.

Chen, Luan, and Xu recently presented another algorithm [9], called *topological peeling*, for sweeping the portion of an arrangement inside a convex region. Their algorithm achieves the same optimal time and space bounds as topological walk, and avoids the two drawbacks of topological walk mentioned above. More interestingly, topological peeling sweeps the arrangement in a wave-propagation fashion, i.e., the swept cells always form a single connected component, with a double-wriggle curve as the wavefront. This nice property enables them to define a convex region (called anchor region) in the traversed portion for each swept vertex, which allows the shortest path problem on the arrangement of n lines to be solved in $O(n^2)$ time and $O(n)$ space [8,9].

Since both topological peeling and topological walk can sweep a portion of an arrangement optimally, it will be interesting to determine which algorithm actually performs better in different scenarios. It is likely that such an understanding can be used to guide the applications of the two algorithms. In this paper, we report an experimental study on these two algorithms and compare their performance. We implemented the two algorithms by using the LEDA 4.1 Library, and conducted the comparison based on a large set of randomly generated arrangements (with up to 5000 lines for an arrangement). Our experiments mainly compare four aspects of the two algorithms: various coding issues, arrangement traversal fashions, execution time, and memory consumption. We also analyze the main reasons causing the difference in their performance.

Our experiments show that the executable files of both algorithms have roughly the same size, and their implementations need comparable efforts. A major difficulty encountered in our implementation is to make the programs numerically stable. After applying some interesting numerical methods to implicitly represent the arrangement vertices (i.e., intersection points of the input lines), we are able to make both programs numerically robust.

Based on our experiments, the execution time and memory consumption of both algorithms very much observe the bounds predicted by the theoretical analysis. That is, the execution time is roughly a linear function of the number of cells inside the convex region R and the memory consumption is linearly proportional to the number of lines and the size $|R|$ of R.

We measure two types of memory usage, namely the logical and physical memory used, in our experiments. By logical memory, we mean the sum of the basic space units (such as the number of tree nodes, pointers, the size of stacks, etc.) used by the algorithms. Experiments show that the maximum consumptions of the logical memory by the two algorithms during their executions are nearly the same as predicted by the theoretical analysis, while the physical memory used by topological peeling could be twice as much as that in topological walk due to the use of more pointers in the implementation.

For the execution time comparison, an interesting phenomenon observed in our experimentation is that although the topological peeling algorithm appears to be more sophisticated than the topological walk algorithm, the execution time of topological peeling is actually 10% to 25% faster than that of topological walk in most cases. We attribute the better execution time of topological peeling over topological walk to the ability of topological peeling to focus its search and exploration on relatively local structures of an arrangement. In topological peeling, two types of input lines, called upper lines and lower lines (see Section 2.1 for their definitions), are distinguished, and the induced arrangement of the upper (resp., lower) lines is swept by using a lower (resp., upper) horizon tree. Topological peeling repeatedly chops or "peels" a lower cell (i.e., a cell in the arrangement of the lower lines), called the *current cell*, and builds a local lower horizon tree for those upper lines intersecting the current cell. The sweep of the arrangement portion inside the current cell is thus restricted to searching a much smaller local tree, rather than a more complicated global tree as in topological walk. This difference, along with several other major differences (to be shown later), makes the execution time of topological peeling sensitive to the ratio of the number of lower lines over the number of all input lines in the arrangement. Our experimentation shows that when this ratio is around 85% (i.e., the number of lower lines is 85% of the total input lines), topological peeling has roughly the same execution time as topological walk. When the ratio is $< 85\%$, topological peeling in general considerably outperforms topological walk. When this ratio is $> 85\%$, topological walk often has a slightly better performance.

2 Overview of Topological Walk and Topological Peeling

In this section, we first give some preliminaries, and then outline the main steps of topological walk [3] and topological peeling [9]. For detailed discussions of these two algorithms, we recommend readers their original papers.

2.1 Preliminaries

Let H be a set of n planar lines, and $A(H)$ be the arrangement of H. Let R be a polygonal convex region, and A_R denote the portion of $A(H)$ inside R. The boundary $B(R)$ of R has r vertices. A line $l \in H$ is called an *upper* (resp., *lower*) line if the left intersection of l with $B(R)$ is on the upper (resp., lower) chain of $B(R)$. Let H_U (resp., H_L) denote the set of upper (resp., lower) lines in H. The arrangement of H_U (resp., H_L) inside R is called the upper (resp., lower) arrangement, denoted by A_R^U (resp., A_R^L).

A *cut* C of A_R induced by H [3,9], roughly speaking, is a list of edges on $B(R)$ or H such that (1) each line $l \in H$ contributes exactly one edge to C, (2) any two consecutive edges in C are incident to a cell of $A(H)$ with one edge above the other, and (3) the removal of C disconnects A_R. For any arrangement A_R, there always exists an initial cut C_0 such that each edge e of C contributed by a line $l \in H$ contains the left intersection of l with $B(R)$.

For a given cut C, one can construct a *horizon tree* by extending each edge in C to the right along its supporting line or the boundary $B(R)$. When two extending edges meet, one of the two edges stops, and this process results in a tree. The tree obtained by always stopping the extending edge with a smaller (resp., larger) slope is called an *upper* (resp., *lower*) horizon tree of C denoted by $T_U(C)$ (resp., $T_L(C)$). The horizon tree corresponding to the initial cut C_0 partitions the convex region R into a set of connected subregions, called *gulfs*. For a given A_R containing both the upper and lower lines, the upper horizon tree $T_U(C_0)$ contains some *dummy edges*, each of which is an edge outside and above R and connected to $T_U(C_0)$ at the left intersection of an upper line with $B(R)$. A twig in a horizon tree $T(C)$ is either the root of $T(C)$ or a branch node in $T(C)$ at which two cut edges of C meet.

In a horizon tree $T(C)$, given a twig w formed by two cut edges e_i and e_{i+1} of C with e_i stopping e_{i+1}, an *elementary step* on w is to replace e_i and e_{i+1} by two other edges e'_{i+1} and e'_i, where e'_{i+1} and e'_i are the right neighboring edges of e_i and e_{i+1} on their corresponding supporting lines. Also, the stopped edge e_{i+1} is extended to the right of w along its supporting line until it hits $T(C)$.

2.2 Main Steps of Topological Walk and Topological Peeling

To sweep A_R, topological walk [3] performs the following main steps.

1. Compute the initial cut C.
2. Construct the upper horizon tree $T(C)$.
3. Starting at the root of $T(C)$, use a dynamic depth-first search on $T(C)$ to find the leftmost twig w.
4. Perform an elementary step on w, and update C and $T(C)$.
5. Repeat Steps 3 and 4 until the cut C becomes empty.

Topological peeling [9] uses topological walk as a subroutine and performs the following main steps.

1. Partition H into two subsets: H_U of upper lines and H_L of lower lines.
2. Construct a global lower horizon tree T_L for the upper lines in H_U and a global upper horizon tree T_U for the lower lines in H_L.
3. Maintain a curve $Lfront$ to bound the swept cells of A_R^L.
4. Perform a topological walk on T_U, and for each (current) cell G in A_R^L thus visited for the first time, do the following:
   (a) extract the boundary $B(G)$ of G using $Lfront$ and T_U;
   (b) use $B(G)$ and T_L to build a local lower horizon tree $T_L(G)$ for those upper lines intersecting G;
   (c) simulate a topological walk on $T_L(G)$ to sweep $A_R^U \cap G$, and generate the intersections between upper and lower lines in an efficient way;

3 Implementation and Comparison Results

To further explore the behavior of the topological peeling and topological walk algorithms, we implemented these two algorithms and conducted a comparative study on them. The comparison results are presented in this section.

3.1 Experimental Environment

Our implementation is based on the C++ library LEDA 4.1. The source code of LEDA was obtained from *Algorithmic Solutions Software GmbH.* After the implementation, the two algorithms were tested and compared on a Sun Blade 1000 workstation with a 600 MHz UltraSPARC III Processor and 512 MB memory. The operating system is SunOS 5.8. Both programs were compiled by *g++* (GNU C++ version 2.7.2.1) with the same optimization options. The execution times were collected using UNIX user application *Top* (version 3.5 beta12), and the memory usage information was obtained by using special memory measuring procedures. To further reduce the machine variants, all data shown in this paper are based on the average of multiple runs, and all execution times were collected when memory measuring procedures were fully disabled.

3.2 Input Data Generation

The two algorithms have two kinds of input data: the straight lines for the arrangement and the convex region specifying the target portion of the arrangement. Every straight line is generated by randomly choosing two points inside a square. The convex regions are generated by computing the convex hull of m randomly chosen points. To reduce unnecessary overhead, all input data are generated off-line and fed to the algorithms during their executions.

3.3 Handling Numerical Errors

Numerical errors abound in geometric computing, and our implementation of the topological peeling and topological walk algorithms is no exception. A key here is how to handle properly the intersection points of the arrangements. Our solution for this numerical problem is to avoid using explicit coordinates of the intersection points since they can be easily corrupted by numerical errors. Instead, in our implementations, for each intersection point p, we use the two corresponding straight lines that give rise to that intersection at p to represent p. It appears that the programs based on this implicit representation of intersection points effectively prevented the propagation of numerical errors. Based on this implicit representation scheme, the mixed search procedure has accurate information to locate the starting intersection points, and the comparisons between intersection points are carried out by comparisons on their respective generating lines. Our experiments showed that this approach is very effective, and eliminate most of the abnormality caused by numerical errors.

3.4 Coding Issues

Overall, the efforts we made in implementing these two algorithms are comparable. The size of the final executable for topological peeling is 34.32MB, while that for topological walk is 33.52MB. This is because topological peeling and topological walk share many data structures and procedures (e.g. horizon tree and mixed search). Some differences come from the lockstep search for topological peeling and the handling of dummy edges for topological walk. The most difficult part of the implementation arose from the handling of a large number of pointers. These pointers are put into the programs to achieve fast execution time. For future implementations and applications, we would caution programmers to pay special attention to the maintenance and manipulation of these pointers.

3.5 Arrangement Traversal Fashions

A key difference between topological peeling and topological walk is the way they explore an arrangement. Figure 1 shows the traversal fashions of the two algorithms on the same arrangement within a convex polygon. The labels in Figures 1(b) and 1(c) denote the order in which each cell is traversed by each of the two algorithms. The cells to the above and left of the heavy polygonal lines (wavefront) form the regions that have been explored by the two algorithms after cell No. 7 is traversed. As seen from the figures, the region explored by topological walk can consist of different connected components, while the region traversed by topological peeling is always a single connected component bounded

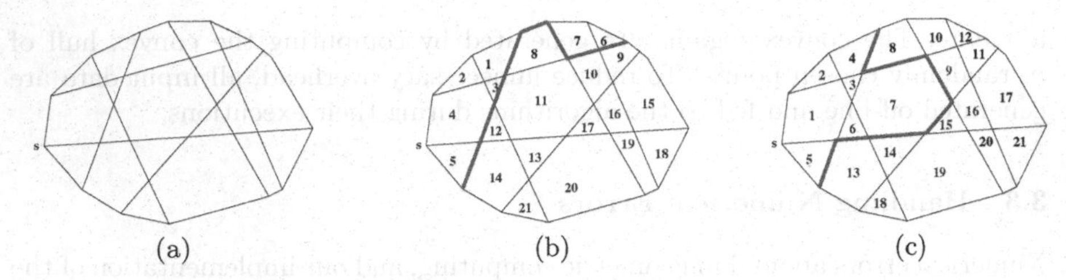

Fig. 1. (a) Arrangement in a convex region. (b) Topological walk traversal fashion. (c) Topological peeling traversal fashion.

by a wavefront whose shape is a *double wriggle curve* [9]. This is why topological peeling can be used to solve several problems (e.g., computing shortest paths) on planar arrangements, while topological walk is not directly applicable [9].

3.6 Execution Time Comparison

For the execution time comparison, we experimented with the programs for the two algorithms on randomly generated lines and convex polygons, and collected data on the execution times, the number of intersection points, and the number of cells reported. The results shown in Figure 2(a) and (b) are based on the data collected on arrangements of 2000 lines each, within a series of 10 convex polygons. The convex polygons are generated by computing the convex hull of 20 randomly generated points. Figure 2(a) shows the relation between the execution time and the total number of intersection points inside the convex polygon, and Figure 2(b) shows the relation between the execution time and the number of cells reported. Our experimental results demonstrate that the execution times of these two algorithms very much observe the time bounds predicted by the theoretical analysis, and the execution time increases almost linearly as the total number of intersection points or the total number of cells increases. More interestingly, these experimental results suggest that on average, topological peeling runs 10% to 25% faster than topological walk.

3.7 Memory Consumption Comparison

For the memory usage comparison, we collected two types of data: the actual physical size of the additional memory used, and the *logical units* of the additional memory used. The first type of memory usage is clear from its name; a discussion of the second type will be given later. The data on memory usage is obtained by putting a special memory measuring procedure in our codes. Each reported value represents the maximum amount of additional memory ever allocated during any time of a program execution. Now, we explain what we mean by *logical units* of memory. Precisely, it means the following. For topological peeling, it is the sum of: (1) the total number of nodes in the upper and lower horizon trees, (2) the total number of items in the stacks for traversing the upper

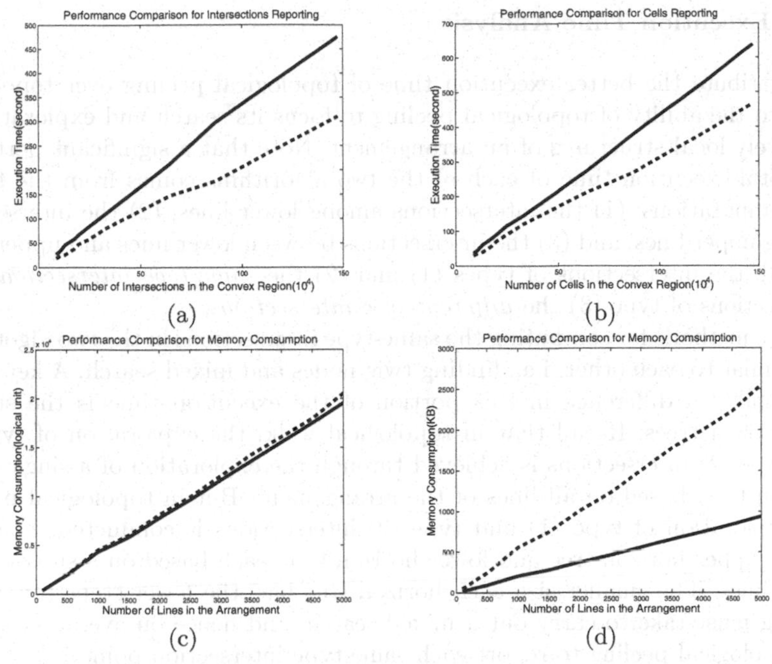

Fig. 2. Comparisons between topological peeling and topological walk. The solid line stands for topological walk and the dashed line stands for topological peeling.

and lower horizon trees, (3) the total number of nodes in the linked list representing the current lower cell, and (4) the total number of nodes in the linked list representing the wavefront. For topological walk, it is the sum of: (1) the total number of nodes in the upper horizon tree, and (2) the total number of items in the stack for traversing the upper horizon tree. We consider data on the logical memory usage because its value is closely related to the theoretical space bound and depends less on the choices made by specific programmers.

The experimental results shown in Figure 2(c) and (d) are based on randomly generated arrangements of up to 5000 lines each. Precisely, we considered arrangements of $500 * i$ lines, for every $i = 1, 2, \ldots, 10$ (i.e., from 500 lines to 5000 lines, with an interval size of 500). For every group size of lines, we ran 10 randomly generated examples on randomly generated convex polygons, and took the average. Figure 2(c) shows the relation between the physical size of additional memory used and the total number of lines in the arrangement, and Figure 2(d) shows the relation between the logical units of additional memory used and the total number of lines in the arrangement. Our experimental results suggest that both algorithms observe the space bound of the theoretical analysis, and the memory usage increases linearly with the number of lines in the arrangements. Although the two algorithms use nearly the same amount of logical memory, the topological peeling program uses over twice as much additional physical memory as topological walk. This is because more pointers are used in the topological peeling program, as suggested by the theoretical algorithm.

3.8 Execution Time Analysis

We attribute the better execution time of topological peeling over topological walk to the ability of topological peeling to focus its search and exploration on relatively local structures of an arrangement. Note that a significant portion of the total execution time of each of the two algorithms comes from the following computations: (1) the intersections among lower lines, (2) the intersections among upper lines, and (3) the intersections between lower lines and upper lines. We call the intersections of types (1) and (2) the *same-type intersections*, and intersections of type (3) the *different-type intersections*.

The methods for computing the same-type intersections in the two algorithms are similar to each other, i.e., finding twig nodes and mixed search. A key factor that makes a difference in this portion of the execution time is the sizes of the horizon trees. Recall that in topological walk, the exploration of type (1) and type (2) intersections is achieved through the exploration of a single upper horizon tree, based on all lines of the arrangement. But in topological peeling, the exploration of type (1) and type (2) intersections is conducted separately on the upper horizon tree and lower horizon tree, each based on a subset of the input lines. The smaller size each horizon tree has, the fewer steps topological peeling must take to carry out a mixed search, and hence on average less time for topological peeling to report each same-type intersection point.

Another difference-making factor lies in the way in which the two algorithms compute the different-type intersections. In topological walk, the methods for computing the different-type intersections are the same as for the same-type intersections. However, in topological peeling, things are somewhat different. Note that in topological peeling, arrangements A_R^L and A_R^U are swept by using their respective horizon trees T_U and T_L, in an interweaving manner. More precisely, topological peeling first performs topological walk on T_U to obtain a current lower cell G, then builds a partial lower horizon tree $T_L(G)$, and finally sweeps the portion of A_R^U inside the boundary $B(G)$ of G. Thus, topological peeling focuses its search on a local structure, i.e., the boundary $B(G)$ of the current upper cell G; furthermore, this local structure is very stable in the sense that it does not incur dynamic changes as occurred in topological walk. A key advantage of this stable local structure is to allow topological peeling to use an array (instead of a linked list) as data structure, which speeds up the search on this structure considerably. Consequently, instead of having to do mixed search on linked lists as in topological walk, topological peeling is able to do exponential search on arrays, yielding a faster execution time.

In a sense, we can view topological peeling as a "divide and conquer" version of topological walk. The "divide" stage is for separating the computation of type (1) and type (2) intersections onto upper horizon trees and lower horizon trees respectively; the "conquer" stage is for computing type (3) intersections in the partial lower horizon trees. To support our argument, we conducted the following experiments. We modified the arrangement generation procedure, so that we can control the ratio of the lower lines over all input lines in arrangements. The results in Figure 3 are based on the executions of both algorithms on randomly

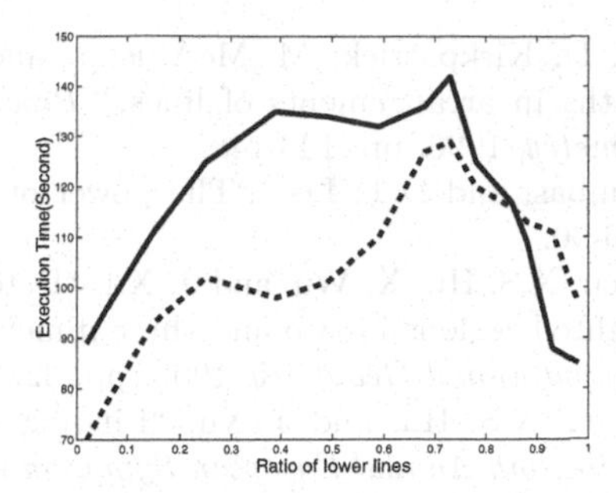

Fig. 3. The relation between the execution time and ratio of the lower lines over all input lines. The solid line stands for topological walk and the dashed line stands for topological peeling.

generated arrangements of 2000 lines each, within the same convex polygons. As shown in Figure 3, when the arrangements consist of mostly lower lines (i.e., the ratio of the lower lines over all input lines is close to 100%), topological walk outperforms topological peeling in execution time. This corresponds to the theoretical analysis, for in this special case, topological peeling is simply reduced to topological walk, except that topological peeling still needs to handle lockstep search. Hence it is natural that for this special case, topological peeling is slower. When the arrangements consist of mostly upper lines (i.e., the ratio of the lower lines over all input lines is close to 0%), topological peeling is also reduced to topological walk. However, unlike the previous case, in this special case topological peeling outperforms topological walk in execution time, because topological peeling need not handle dummy edges. The biggest advantage of topological peeling occurs in the more balanced situations in which the numbers of upper lines and lower lines are somewhat comparable. This is the case in which the "divide and conquer" version shows its power in execution time. As shown in Figure 3, for this more general case, topological peeling outperforms topological walk in execution time, by as much as 35%! The balance point at which both algorithms consume roughly the same amount of execution time appears when the ratio of the lower lines over all input lines is around 85%.

References

1. E.G. Anagnostou, L.J. Guibas, and V.G. Polimenis, "Topological sweeping in three dimensions," *Lecture Notes in Computer Science*, Vol. 450, *Proc. Int'l Symp. on Algorithms*, Springer-Verlag, 1990, pp. 310-317.
2. E.M. Arkin, Y.-J. Chiang, M. Held, J.S.B. Mitchell, V. Sacristan, S.S. Skiena, and T.-C. Yang, "On minimum-area hulls," *Algorithmica*, Vol. 21, 1998, pp. 119-136.
3. T. Asano, L.J. Guibas, and T. Tokuyama, "Walking in an arrangement topologically," *International Journal of Computational Geometry and Applications*, Vol. 4, No. 2, 1994, pp. 123-151.
4. T. Asano and T. Tokuyama, "Topological walk revisited," *Proc. 6th Canadian Conf. on Computational Geometry*, 1994, pp. 1-6.

5. P. Bose, W. Evans, D. Kirkpatrick, M. McAllister, and J. Snoeyink, "Approximating shortest paths in arrangements of lines," *Proc. 8th Canadian Conf. on Computational Geometry*, 1996, pp. 143-148.
6. B. Chazelle, L.J. Guibas, and D.T. Lee, "The power of geometric duality," *BIT*, Vol. 25, 1985, pp. 76-90.
7. D.Z. Chen, O. Daescu, X.S. Hu, X. Wu, and J. Xu, "Determining an optimal penetration among weighted regions in two and three dimensions," *Proc. 15th Annual ACM Symp. on Computational Geometry*, 1999, pp. 322-331.
8. D.Z. Chen, O. Daescu, X.S. Hu, and J. Xu, "Finding an optimal path without growing the tree," *Proc. 6th Annual European Symp. on Algorithms*, 1998, pp. 356-367.
9. D.Z. Chen, S. Luan, and J. Xu, "Topological peeling and implementation," *12th Annual Int. Symp. on Algorithms and Computation*, 2001, pp. 454-466.
10. D. Dobkin and A. Tal, "Efficient and Small Representation of Line Arrangements with Applications," *Proc. 17th Annual ACM symposium on Computational Geometry*, 2001, pp. 293-301.
11. H. Edelsbrunner, *Algorithms in Combinatorial Geometry*, Springer-Verlag, New York, 1987.
12. H. Edelsbrunner and L.J. Guibas, "Topologically sweeping an arrangement," *Journal of Computer and System Sciences*, Vol. 38, 1989, pp. 165-194.
13. H. Edelsbrunner, J. O'Rourke, and R. Seidel, "Constructing arrangements of lines and hyperplanes with applications," *SIAM J. Computing*, Vol. 15, 1986, pp. 341-363.
14. H. Edelsbrunner and D. Souvaine, "Computing median-of-squares regression lines and guided topological sweep," *Journal of the American Statistical Association*, Vol. 85, 1990, pp. 115-119.
15. H. Edelsbrunner and E. Welzl, "Constructing belts in two-dimensional arrangements with applications," *SIAM J. Computing*, Vol. 15, 1986, pp. 271-284.
16. D. Eppstein and D. Hart, "An efficient algorithm for shortest paths in vertical and horizontal segments," *Proc. 5th Int. Workshop on Algorithms and Data Structures*, 1997, pp. 234-247.
17. D. Eppstein and D. Hart, "Shortest paths in an arrangement with k line orientations," *Proc. 10th ACM-SIAM Symp. on Discrete Algorithms*, 1999, pp. 310-316.
18. K. Hoffman, K. Mehlhorn, R. Rosenstiehl, and R. Tarjan, "Sorting Jordan sequences in linear time using level-linked search trees," *Information and Control*, Vol. 68, 1986, pp. 170-184.
19. P.N. Klein, S. Rao, M.H. Rauch, and S. Subramanian, "Faster shortest-path algorithms for planar graphs," *Proc. 26th Annual ACM Symp. Theory of Computing*, 1994, pp. 27-37.
20. J. Majhi, R. Janardan, M. Smid, and P. Gupta, "On some geometric optimization problems in layered manufacturing," *Proc. 5th Int. Workshop on Algorithms and Data Structures*, 1997, pp. 136-149.
21. K. Miller, S. Ramaswami, P. Rousseeuw, T. Sellares, D. Souvaine, I. Streinu, and A. Struyf, "Fast implementation of depth contours using topological sweep," *Proc. 12th ACM-SIAM Symp. on Discrete Algorithms*, 2001, pp. 690-699.
22. J. Nievergelt and F.P. Preparata, "Plane-sweep algorithms for intersecting geometric figures," *Comm. of the ACM*, Vol. 25, No. 10, 1982, pp. 739-747.
23. F.P. Preparata and M.I. Shamos, *Computational Geometry: An Introduction*, Springer-Verlag, New York, 1985.

On-Line Maximizing the Number of Items Packed in Variable-Sized Bins

Leah Epstein[1,*] and Lene M. Favrholdt[2,**]

[1] School of Computer Science, The Interdisciplinary Center, Herzliya, Israel
lea@idc.ac.il

[2] Department of Mathematics and Computer Science, University of Southern Denmark
lenem@imada.sdu.dk

Abstract. We study an on-line bin packing problem. A fixed number n of bins, possibly of different sizes, are given. The items arrive on-line, and the goal is to pack as many items as possible. It is known that there exists a legal packing of the whole sequence in the n bins. We consider fair algorithms that reject an item, only if it does not fit in the empty space of any bin. We show that the competitive ratio of any fair, deterministic algorithm lies between $\frac{1}{2}$ and $\frac{2}{3}$, and that a class of algorithms including Best-Fit has a competitive ratio of exactly $\frac{n}{2n-1}$.

1 Introduction

The Problem. We consider the following bin packing problem. The input consists of n bins, possibly of different sizes, and a sequence of positively sized items. The bins as well as the sizes of the bins are denoted by $B_1, B_2, \ldots, B_n$. The items arrive on-line, i.e., each item must be packed before the next item is seen, and packed items cannot be moved between bins. The goal is to pack as many items as possible into the n bins. A bin is legally packed if the total size of the items assigned to it is at most the size of the bin. This problem of maximizing the number of items packed in a fixed number of bins is sometimes called *dual bin packing*, to distinguish it from the classical bin packing problem which is to pack *all* items in as *few* bins as possible. In [8] the problem is reported to have been named dual bin packing in [18]. Note that this name is also sometimes used for bin covering [2, 12, 13]. For a survey on classical bin packing in identical bins, see [14,11].

Throughout the paper, we restrict the input sequences to be *accommodating* [6, 7], i.e., sequences that an optimal off-line algorithm, which knows all items in advance, can pack completely. The reason for this restriction is that, for general sequences, no on-line algorithm can pack a constant fraction of the number of items that can be packed by an optimal off-line algorithm.

* Research supported in part by the Israel Science Foundation, (grant No. 250/01-1)

** Supported in part by the Danish Natural Science Research Council (SNF) and in part by the Future and Emerging Technologies program of the EU under contract number IST-1999-14186 (ALCOM-FT).

O.H. Ibarra and L. Zhang (Eds.): COCOON 2002, LNCS 2387, pp. 467–475, 2002.

The problem can also be seen as a scheduling problem with n uniformly related machines. In the basic scheduling problem, each job is to be assigned to one of the machines so as to minimize the makespan. This problem was first studied for the case of identical machines by Graham [15], and for uniformly related machines by [1, 10, 4]. For a survey on on-line scheduling problems, see [20]. Consider a scheduling problem with a deadline and assume that the aim is to schedule as many jobs as possible before this deadline. If an optimal off-line algorithm can schedule all jobs of any input sequence before the deadline, this problem is equivalent to our problem. Our problem can also be seen as a special case of the multiple knapsack problem (see [19, 9]), where all items have unit profit. (This problem was mainly studied in the off-line environment.)

The Algorithms. In this paper we study fair algorithms [3]. A *fair* algorithm rejects an item, only if the item does not fit in the empty space of any bin.

Some of the algorithms that are classical for the classical bin packing problem (where the whole sequence of items is to be packed in as few bins as possible) can be adapted to our problem. Such an adaptation for identical bins was already done in [7]: the n bins are all considered open from the beginning, and no new bin can be opened. We also use this adaptation. Since there is no unique way to define First-Fit for variable sized bins, we discuss this in Section 3.

The Quality Measure. The competitive ratio of an on-line algorithm $\mathbb{A}$ for the dual bin packing problem is the worst case ratio, over all possible input sequences, of the number of items packed by $\mathbb{A}$ to the number of items packed by an optimal off-line algorithm. Often an additive constant is allowed, yielding the following definition of the competitive ratio.

Definition 1. *For any algorithm $\mathbb{A}$ and any sequence I of items, let $\mathbb{A}(I)$ be the number of items packed by $\mathbb{A}$ and let $OPT(I)$ be the number of items packed by an optimal off-line algorithm. Furthermore, let $0 \le c \le 1$. An on-line algorithm $\mathbb{A}$ is* c-competitive *if there exists a constant b such that*

$$\mathbb{A}(I) \ge c \cdot OPT(I) - b, \text{ for any sequence } I \text{ of items.}$$

The competitive ratio *of $\mathbb{A}$ is*

$$C_{\mathbb{A}} = \sup\{c \mid \mathbb{A} \text{ is } c\text{-competitive}\}.$$

Note that since dual bin packing is a maximization problem, the competitive ratio lies between 0 and 1.

If the additive constant b is zero or negative, the algorithm is called *strictly c-competitive*. The bounds given in this paper are valid for the strict competitive ratio as well as for the competitive ratio in general.

For randomized algorithms, the competitive ratio is defined similarly, but $\mathbb{A}(I)$ is replaced by the expected value of $\mathbb{A}(I)$, $E(\mathbb{A}(I))$.

The Results. We show the following results for fair algorithms on accommodating sequences.

- Any fair algorithm has a competitive ratio of at least $\frac{1}{2}$, and the competitive ratio of Worst-Fit is exactly $\frac{1}{2}$.

– A class of algorithms that give preference to smaller bins has a competitive ratio of exactly $\frac{n}{2n-1}$. This class contains Best-Fit as well as the variant of First-Fit that sorts the bins in order of non-decreasing sizes.
– Any fair, deterministic algorithm has a competitive ratio of at most $\frac{2}{3}$, and any fair, randomized algorithm has a competitive ratio of at most $\frac{4}{5}$.

Previous Work. Dual bin packing in identical bins has been studied both in the off-line version [17, 16] and in the on-line version for accommodating sequences [6, 7, 3]. Even for identical bins, a restriction on the input sequences is needed in order to be able to achieve a constant competitive ratio [7]. In [7], fair algorithms are considered and it is shown that First-Fit has a competitive ratio of at least $\frac{5}{8}$ on accommodating sequences. An upper bound of $\frac{6}{7}$ for any fair or unfair randomized algorithm is also given. In [3], a $(\frac{2}{3} - \frac{2}{4n+1})$-competitive unfair algorithm is given, the negative result for fair deterministic algorithms is improved to 0.809, and the bound of $\frac{5}{8}$ for First-Fit is shown to be asymptotically tight (the upper bound approaches $\frac{5}{8}$ as n approaches infinity).

2 General Results on Fair Algorithms

In this section we show that, on accommodating sequences, the competitive ratio of any fair, deterministic algorithm lies between $\frac{1}{2}$ and $\frac{2}{3}$, and the competitive ratio of any randomized algorithm is at most $\frac{4}{5}$.

2.1 Positive Results

The main result of this section is that any fair algorithm is $\frac{1}{2}$-competitive on accommodating sequences. We need the following lemma which is adapted from a similar lemma for identical bins in [7].

Lemma 1. *For any fair algorithm, the number of rejected items is no larger than the number of accepted items, if the input sequence is accommodating.*

Proof. Given an instance of the dual bin packing problem with an accommodating sequence I, we define a sequence I' as follows. Each accepted item of size x is replaced by $\lfloor \frac{x}{s} \rfloor$ items of size s, where s is the minimum size of any rejected item. Each rejected item is decreased to have size s. Clearly, a packing of all items of I defines a legal packing of all items of I', hence I' is also an accommodating sequence.

Let P be the on-line packing of I and let P' be the packing of I' induced by P. Note that all items of I' have the same size. Thus, to calculate an upper bound on the number of items rejected we just need to find an upper bound on the number of items of size s that fit in the bins after doing the packing P'.

For each bin B_i, let k_i denote the number of items in bin B_i in the packing P. The empty space in B_i in the packing P' consists of the empty space in B_i in the packing P and the space freed by the rounding down of the items packed in B_i. The empty space in B_i in P is less than s, since the algorithm is fair, and the total size of each original item was decreased by less than s. Thus, the empty space in B_i in P' is strictly less than $s(k_i+1)$. We conclude that the number of rejected items is at most $\sum_{i=1}^{n} k_i$ which is the number of accepted items. □

Corollary 1. *Any fair algorithm has a competitive ratio on accommodating sequences of at least* $\frac{1}{2}$.

We close this section with an easy lemma that will be needed in Section 2.2 and Section 3. Let C be the set of non-empty bins in the optimal off-line packing. Let $N = |C|$.

Lemma 2. *Given an accommodating input sequence, any fair algorithm rejects at most* $N-1$ *items.*

Proof. If the on-line algorithm does not reject any items, its packing is optimal. Assume now, that at least one item is rejected. Let s be the minimum size of any rejected item. Since the algorithm is fair, the empty space in each bin is less than s. Another trivial upper bound on the empty space in any bin B is the size B of the bin. Thus, the total empty space in the on-line packing is strictly less than $Ns + \sum_{B \notin C} B$. The total empty space of OPT is at least $\sum_{B \notin C} B$. Hence, since OPT accepts all items, the total size of all rejected items is strictly less than Ns. Since all rejected items are of size at least s, there are at most $N-1$ rejected items. □

2.2 Negative Results

In this section we show an upper bound of $\frac{2}{3}$ for deterministic, fair algorithms and an upper bound of $\frac{4}{5}$ for randomized, fair algorithms.

We first prove the upper bound of $\frac{2}{3}$ for the strict competitive ratio. This is relatively easy for any $n \geq 2$. Consider for example the following instance with $n-2$ bins of size ε, $0 < \varepsilon < 1$, one bin of size 2, and one bin of size 3. The input sequence consists of two or three items that are all too large for the bins of size ε. The first item has size 1. If this first item is assigned to the bin of size 3, an item of size 3 arrives next. Otherwise, two items of size 2 will arrive. In the first case, only the first item is packed, since the second does not fit, and in the second case only two items are accepted, the third does not fit. It is easy to see that both sequences are accommodating. This gives an upper bound of $\frac{2}{3}$ on the strict competitive ratio, for $n \geq 2$. Applying Yao's inequality [21] as described in [5] on these two sequences gives an upper bound of $\frac{4}{5}$ on the strict competitive ratio for randomized algorithms. This can be seen in the following way. Consider the sequence where the first item of size 1 is followed by one item of size 3 with probability $p_1 = \frac{2}{5}$ and by two items of size 2 with probability $p_2 = \frac{3}{5}$. An algorithm that packs the first item in the bin of size 3 will have an expected performance ratio of at most $p_1 \cdot \frac{1}{2} + p_2 \cdot 1 = \frac{4}{5}$. Similarly, an algorithm that packs the first item in the bin of size 2 will have an expected performance ratio of at most $p_1 \cdot 1 + p_2 \cdot \frac{2}{3} = \frac{4}{5}$. Thus, no deterministic algorithm can have an expected performance ratio larger than $\frac{4}{5}$ on this sequence.

However, we are interested in negative results that hold for the competitive ratio in general, and not only for the strict competitive ratio. By Lemma 2, the number of rejected items is at most $n-1$. As long as there is only a constant number of bins, we can view the number of rejected items as just an additive constant, and hence any fair algorithm has competitive ratio 1. Thus, to prove the following theorem, we need to find arbitrarily long accommodating sequences with the property that only $\frac{2}{3}$ of the items are accepted.

Theorem 1. *Any fair, deterministic on-line algorithm for the dual bin packing problem has a competitive ratio of at most $\frac{2}{3}$ on accommodating sequences.*

Proof. For $\ell = 1, \ldots, \lfloor \frac{n}{2} \rfloor$, we give the pair of bins

$$B_{2\ell-1} = 2\ell + 4^{\ell}\varepsilon \text{ and } B_{2\ell} = 2\ell + 2 \cdot 4^{\ell}\varepsilon,$$

where $\varepsilon < \frac{1}{4^n}$ is a positive constant. Thus, $4^{\ell}\varepsilon < 1$, $1 \leq \ell \leq \lfloor \frac{n}{2} \rfloor$. If n is odd, the last bin is of size $\frac{\varepsilon}{2}$ (so that no items are packed in that bin for the sequence we define). The sequence contains $3 \cdot \lfloor \frac{n}{2} \rfloor$ items and is constructed so that exactly $2 \cdot \lfloor \frac{n}{2} \rfloor$ of them are accepted.

The sequence is defined inductively in steps $\lfloor \frac{n}{2} \rfloor, \lfloor \frac{n}{2} \rfloor - 1, \ldots, 1$. In step k, two large items are given and one small item is defined. The small items are given after all large items and are defined such that they will be rejected by the on-line algorithm. The sizes of the two large items are defined such that

- the on-line algorithm will pack them in B_{2k} and B_{2k-1}, one in each bin, and
- after packing the two items, the empty space in the two bins have the same size denoted E_k.

For convenience we define $E_{\lfloor \frac{n}{2} \rfloor+1} = 0$. As will be seen later, $E_{k+1} < E_k$, $1 \leq k \leq \lfloor \frac{n}{2} \rfloor$. Furthermore, we will prove that $E_1 < 1$.

The first large item given in step k has size $2k - E_{k+1}$. Thus, the very first item has size $2 \cdot \lfloor \frac{n}{2} \rfloor$, and the size of the first large item of each of the later steps depends on the empty space created in the previous step. Since $2k - E_{k+1} > 2k-1$ and all previous bins $B_n, \ldots, B_{2k+1}$ have less than one unit of empty space, this item fits only in B_{2k} and B_{2k-1}. What happens next depends on which of these two bins the algorithm chooses.

Case 1: The first large item is packed in B_{2k-1}. In this case, the next large item has size $2k - E_{k+1} + 4^k\varepsilon$. This item will be packed in B_{2k}. Now, the empty space in each of the bins B_{2k} and B_{2k-1} is $E_k = E_{k+1} + 4^k\varepsilon$. The small item defined in this step has size $S_k = E_k + 4^k\varepsilon$. Note that this item does not fit in B_{2k} or B_{2k-1}, but the off-line algorithm can pack the first large item in B_{2k} together with the small item and put the second large item in B_{2k-1}.

Case 2: The first large item is packed in B_{2k}. In this case, the next large item has size $2k - E_{k+1} - 4^k\varepsilon$. For $k \geq 2$, this item does not fit in B_{2k-2}, since $2k - E_{k+1} - 4^k\varepsilon > 2k - 1 - 4^k\varepsilon \geq 2k - 2 + 3 \cdot 4^k\varepsilon$, for $n \geq 2$, and $B_{2k-2} =$

$2k-2+2\cdot 4^{k-2}\epsilon$. Hence, this item must be packed in B_{2k-1}. Now, the empty space in each of the bins B_{2k} and B_{2k-1} is $E_k = E_{k+1}+2\cdot 4^k\varepsilon$. The small item defined in this step has size $S_k = E_k + 4^k\varepsilon$. This item does not fit in B_{2k} or B_{2k-1}, but the off-line algorithm can pack the first large item in B_{2k-1} and put the second large item in B_{2k-1} together with the small item.

Note that $E_{k+1}+4^k\varepsilon \leq E_k \leq E_{k+1}+2\cdot 4^k\varepsilon$, $1 \leq k \leq \lfloor \frac{n}{2} \rfloor$. The first inequality tells us that, to prove that none of the small items will be accepted, it suffices to prove that $S_k > E_1$, $2 \leq k \leq \lfloor \frac{n}{2} \rfloor$. This is easily done using the second inequality. For $2 \leq k \leq \lfloor \frac{n}{2} \rfloor$,

$$E_1 \leq E_k + 2\cdot \sum_{i=1}^{k-1} 4^i\varepsilon < E_k + 4^k\varepsilon = S_k.$$

Finally,

$$E_1 \leq E_{\lfloor \frac{n}{2} \rfloor+1} + 2\cdot \sum_{i=1}^{\lfloor \frac{n}{2} \rfloor} 4^i\varepsilon < 4^{\lfloor \frac{n}{2} \rfloor+1}\varepsilon < 4^{\lfloor \frac{n}{2} \rfloor+1-n} \leq 1.$$

□

We move on to randomized algorithms. Since the previous sequence was built step by step, we need to give a simpler sequence in order to prove the following theorem.

Theorem 2. *Any fair randomized algorithm has a competitive ratio on accommodating sequences of at most $\frac{4}{5}$.*

Proof. We use $\lfloor \frac{n}{2} \rfloor$ bins of size $1+\varepsilon$ and $\lfloor \frac{n}{2} \rfloor$ bins of size $2-\varepsilon$, where $0 < \varepsilon < \frac{1}{2}$. If n is odd, the last bin is of size ε. The sequence starts with $\lfloor \frac{n}{2} \rfloor$ items of size 1. We describe a proof for deterministic algorithms first. Since the algorithm is fair, all $\lfloor \frac{n}{2} \rfloor$ items are accepted. Let x be the number of bins of size $1+\varepsilon$ that received an item (no bin can receive more than one item). Then, exactly x bins of size $2-\varepsilon$ are empty. What happens next depends on the size of x.

Case $x \leq \frac{3}{5}\cdot\lfloor \frac{n}{2} \rfloor$. In this case, the sequence continues with $\lfloor \frac{n}{2} \rfloor$ items of size $2-\varepsilon$, and the on-line algorithm accepts $\lfloor \frac{n}{2} \rfloor + x$ items in total out of the $2\lfloor \frac{n}{2} \rfloor$. This gives a fraction of $\frac{\lfloor \frac{n}{2} \rfloor + x}{2\lfloor \frac{n}{2} \rfloor} \leq \frac{1+\frac{3}{5}}{2} = \frac{4}{5}$.

Case $x > \frac{3}{5}\cdot\lfloor \frac{n}{2} \rfloor$. In this case, the sequence continues with $\lfloor \frac{n}{2} \rfloor$ items of size $1+\varepsilon$ followed by $\lfloor \frac{n}{2} \rfloor$ items of size $1-\varepsilon$. After the arrival of items of size 1, there are $\lfloor \frac{n}{2} \rfloor$ empty bins. Thus, all items of size $1+\varepsilon$ are accepted and now each bin has exactly one item. Items of size $1-\varepsilon$ can only be assigned to bins of size $2-\varepsilon$ that contain an item of size 1, hence $\lfloor \frac{n}{2} \rfloor - x$ of them are accepted. Thus, the fraction $\frac{3\lfloor \frac{n}{2} \rfloor - x}{3\lfloor \frac{n}{2} \rfloor} < \frac{3-\frac{3}{5}}{3} = \frac{4}{5}$ of the items is accepted.

To get a randomized result, let x be the expectation of the number of bins of size $1+\varepsilon$ that got an item. The bound follows by linearity of expectation. □

3 Results on Specific Fair Algorithms

We now analyze specific algorithms. Some natural fair algorithms are First-Fit, Best-Fit, and Worst-Fit. The algorithm First-Fit is not a single algorithm, but a class of algorithms that give an order to the bins, and use the algorithm according to this order, i.e., assign an item to the first bin (in the ordered set of bins) that the item fits in. Among the various versions of First-Fit, two are most natural; *Smallest-Fit* assigns an item to the smallest bin it fits into, and *Largest-Fit* assigns an item to the largest bin it fits into. The other algorithms are used in their classical version, i.e., Best-Fit packs each item in a bin where it will leave the smallest possible empty space, and Worst-Fit packs it in the bin where it leaves the largest empty space. We refer to these four algorithms as SF, LF, BF, and WF.

We start the analysis by showing that $\frac{1}{2}$ is indeed the exact competitive ratio of WF and LF.

Theorem 3. *The competitive ratio of Worst-Fit and Largest-Fit on accommodating sequences is* $\frac{1}{2}$.

Proof. Let $\varepsilon > 0$ be a constant such that $\varepsilon \leq \frac{1}{n}$. Consider the following set of bins. One large bin of size n and $n-1$ small bins of size 1. The sequence consists of $n-1$ items of size 1 followed by $n-1$ items of size $1+\varepsilon$. Both algorithms LF and WF assign all items of size 1 to the large bin. As a result, all bins have a free space of size 1, hence none of the items of size $1+\varepsilon$ can be accepted. The optimal algorithm assigns each small item to a small bin, and all other items to the large bin; they all fit since

$$(1+\varepsilon)(n-1) \leq \frac{(n+1)(n-1)}{n} < n .$$

This example in combination with Corollary 1 proves the theorem. □

We further analyze a class of fair algorithms called *Smallest-Bins-First* to which SF and BF belong. This is the class of fair algorithms that whenever an item is assigned to an empty bin, this is the smallest bin in which the item fits. There are no additional rules, and the algorithm may use an empty bin even if the item fits in a non-empty bin, as long as it uses the smallest empty bin for that. SF belongs to this class according to its definition. BF belongs to this class since, among the empty bins that an item fits into, it fits better into the smaller bins than the larger bins. We give a tight analysis of this class as a function of n. Specifically we prove the following.

Theorem 4. *The competitive ratio of any Smallest-Bins-First algorithm on accommodating sequences is* $\frac{n}{2n-1}$.

Proof. If, after running the algorithm, all bins of the on-line algorithm are non-empty, then there are at least n accepted items and at most $n-1$ rejected items (by Lemma 2). Thus, in this case, the competitive ratio is at least $\frac{n}{2n-1}$.

Otherwise, consider the largest (last) bin b that remained empty after running the on-line algorithm. We consider items of size smaller than or equal to b, and items larger than b separately. Since a bin of size b is empty and no bin larger than b is empty, according to the definition of the class of algorithms, each bin of size more than b contains at least one item larger than b, namely the first item packed in the bin. Moreover, all items of size at most b are accepted. Let x_s be the number of items in bins of size at most b and let n_ℓ be the number of bins larger than b. Let N_s be the number of non-empty bins of OPT of size at most b and N_ℓ its number of non-empty bins larger than b. Clearly, $x_s \geq N_s$ (all those bins are of size at most b and contain at least one item). We get that the number of accepted items is at least $x_s + n_\ell \geq N_s + N_\ell = N$. Thus, by Lemma 2, the competitive ratio is at least $\frac{N}{2N-1} \geq \frac{n}{2n-1}$.

To show that the result is tight for this class of algorithms, let $\varepsilon < \frac{1}{n}$ be a positive constant. Consider the set of bins $B_i = 1+\varepsilon i, i = 1, \ldots, n$. The sequence consists of n items, one of size $1+\varepsilon(i-1)$ for each $i = 1, \ldots, n$, followed by $n-1$ items of size $\frac{n\varepsilon}{n-1}$. All algorithms in the class assign the item of size $1+\varepsilon(i-1)$ to B_i. All other items are rejected. The optimal off-line algorithm assigns each large item except the first one to a bin of its size. The first item and the $n-1$ small items are assigned to B_n. □

Note that when $n = 2$, the lower bound of $\frac{n}{2n-1}$ matches the general upper bound of $\frac{2}{3}$.

4 Conclusion

We have proven an upper bound of $\frac{2}{3}$ for all fair algorithms. We have also shown that any fair algorithm accepts at least half of the items, and that some algorithms do significantly better for very small n. It is left as an open problem to design a fair algorithm with a competitive ratio significantly larger than $\frac{1}{2}$ for any n, or prove that this is not possible. It is also unknown how much unfair algorithms can be better; the best negative result for those is $\frac{6}{7}$, which holds even for identical bins [7].

Acknowledgments

We would like to thank Joan Boyar for reading and commenting on the paper. We also thank Gerhard Woeginger for suggesting the title.

References

1. J. Aspnes, Y. Azar, A. Fiat, S. Plotkin, and O. Waarts. On-Line Routing of Virtual Circuits with Applications to Load Balancing and Machine Scheduling. *Journal of the ACM*, 44(3):486–504, 1997. Also in *Proc. 25th ACM STOC*, 1993, pp. 623-631.

2. S. F. Assmann, D. S. Johnson, D. J. Kleitman, and J. Y. Leung. On a Dual Version of the One-Dimensional Bin Packing Problem. *Journal of Algorithms*, 5:502–525, 1984.
3. Y. Azar, J. Boyar, L. Epstein, L. M. Favrholdt, K. S. Larsen, and M. N. Nielsen. Fair versus Unrestricted Bin Packing. *Algorithmica (to appear)*. Preliminary version at SWAT 2000, volume 1851 of *LNCS*: 200-213, Springer-Verlag, 2000.
4. P. Berman, M. Charikar, and M. Karpinski. On-Line Load Balancing for Related Machines. *Journal of Algorithms*, 35:108–121, 2000.
5. A. Borodin and R. El-Yaniv. *Online Computation and Competitive Analysis*. Cambridge University Press, 1998.
6. J. Boyar and K. S. Larsen. The Seat Reservation Problem. *Algorithmica*, 25:403–417, 1999.
7. J. Boyar, K. S. Larsen, and M. N. Nielsen. The Accommodating Function: A Generalization of the Competitive Ratio. *SIAM Journal on Computing*, 31(1):233–258, 2001.
8. J. L. Bruno and P. J. Downey. Probabilistic Bounds for Dual Bin-Packing. *Acta Informatica*, 22:333–345, 1985.
9. C. Chekuri and S. Khanna. A PTAS for the Multiple Knapsack Problem. In *Proc. 11th Annual ACM-SIAM Symposium on Discrete Algorithms*, pages 213–222, 2000.
10. Y. Cho and S. Sahni. Bounds for List Schedules on Uniform Processors. *SIAM Journal on Computing*, 9:91–103, 1988.
11. E. G. Coffman, Jr., M. R. Garey, and D. S. Johnson. Approximation Algorithms for Bin Packing: A Survey. In Dorit S. Hochbaum, editor, *Approximation Algorithms for NP-Hard Problems*, chapter 2, pages 46–93. PWS Publishing Company, 1997.
12. J. Csirik and J. B. G. Frenk. A Dual Version of Bin Packing. *Algorithms Review*, 1:87–95, 1990.
13. J. Csirik and V. Totik. On-Line Algorithms for a Dual Version of Bin Packing. *Discr. Appl. Math.*, 21:163–167, 1988.
14. J. Csirik and G. Woeginger. On-Line Packing and Covering Problems. In Amos Fiat and Gerhard J. Woeginger, editors, *Online Algorithms*, volume 1442 of *LNCS*, chapter 7, pages 147–177. Springer-Verlag, 1998.
15. R. L. Graham. Bounds for Certain Multiprocessing Anomalies. *Bell Systems Technical Journal*, 45:1563–1581, 1966.
16. E. G. Coffman Jr. and J. Y. Leung. Combinatorial Analysis of an Efficient Algorithm for Processor and Storage Allocation. *SIAM Journal on Computing*, 8(2):202–217, 1979.
17. E. G. Coffman Jr. J. Y. Leung, and D. W. Ting. Bin Packing: Maximizing the Number of Pieces Packed. *Acta Informatica*, 9:263–271, 1978.
18. J. Y. Leung. *Fast Algorithms for Packing Problems.* PhD thesis, Pennsylvania State University, 1977.
19. S. Martello and P. Toth. *Knapsack Problems.* John Wiley and Sons, Chichester, 1990.
20. J. Sgall. On-Line Scheduling. In *A. Fiat and G. J. Woeginger, editors,* Online Algorithms: The State of the Art, volume 1442 of *LNCS*, pages 196–231. Springer-Verlag, 1998.
21. A. C. Yao. Towards a Unified Measure of Complexity. *Proc. 12th ACM Symposium on Theory of Computing*, pages 222–227, 1980.

On-Line Grid-Packing with a Single Active Grid

Satoshi Fujita

Department of Information Engineering
Graduate School of Engineering, Hiroshima University
Higashi-Hiroshima, 739-8527, Japan

Abstract. In this paper, we study the problem of packing rectangular items into a minimum number of square grids in an on-line manner with a single active grid, where the size of each grid is $m \times m$ for some positive integer m, and the height and the width of each item are positive integers smaller than or equal to m, respectively. We first prove that the asymptotic competitive ratio of an optimal on-line algorithm is at least 23/11. We then propose an on-line algorithm that achieves a competitive ratio $O((\log\log m)^2)$.

1 Introduction

In this paper, we study the problem of packing rectangular items into a minimum number of square grids in an on-line manner. The size of each grid is $m \times m$ for some positive integer m, and the height and the width of each item are positive integers smaller than or equal to m, respectively. The term "on-line" implies that items are consecutively input and the packing of an item must be determined before the arrival of the next item. In the following, we refer to the above on-line packing problem as the grid-packing problem (GPP, for short). It is worth noting that GPP is a *discrete* version of the normal two-dimensional bin packing problem; i.e., in GPP, the size of each item can take discrete values instead of reals as in the normal bin-packing problem.

So far, the two-dimensional bin packing problem has been investigated extensively by many researchers, and several interesting results have been obtained during the past two decades [3,5,6,8,10]; e.g., it is shown that the asymptotic competitive ratio[1] of an optimal on-line algorithm is at least 1.907 [2] and at most 2.661 [9], and there is an off-line algorithm with approximation ratio 1.25 [1]. Although those bounds are based on several important and interesting techniques such as HARMONIC by Lee and Lee [7], at least in practice, it is not very realistic in the sense that it assumes each item can have *any* size in $(0,1] \times (0,1]$ and the difference to a particular value (e.g., $(1/2, 1/2)$) can be arbitrary small. In many real-world applications, however, the size of each item can take merely *discrete* values drawn from a given (finite) set, as in the packing of furniture into a room, packing of building blocks into a given area of VLSI chip, and so on. In addition, it is very common that an on-line packing problem should be solved

[1] A formal definition will be given in Section 2.

O.H. Ibarra and L. Zhang (Eds.): COCOON 2002, LNCS 2387, pp. 476–483, 2002.

by using a small working space, and in many cases, the opening of a new bin implies the closing of the current bin.

In this paper, we consider GPP with only one active grid for the packing. We measure the goodness of on-line algorithms in terms of the competitive ratio, that is typically represented as a function of the size of each grid. In the following, we first point out that the rotation of items is necessary for achieving a good competitive ratio by on-line algorithms (more precisely, we show that the ratio could not be smaller than $m/2$ if rotation of items is not allowed). We then prove that even when each item is rotatable, the competitive ratio of an optimal algorithm cannot be smaller than $23/11$. Finally, we propose an on-line algorithm with competitive ratio $O((\log \log m)^2)$, that is the main result of this paper.

The remainder of this paper is organized as follows. Section 2 introduces some basic definitions. Section 3 describes several elementary bounds and Section 4 describes our proposed algorithm with competitive ratio $O((\log \log m)^2)$. Section 5 concludes the paper with some future directions of research.

2 Preliminaries

Let $L = (a_1, a_2, \ldots, a_m)$ denote a sequence of input items. An item with width x and height y is denoted as (x, y) or $x \times y$. The competitive ratio of an on-line grid-packing algorithm A is defined as follows: Let OPT denote an optimal off-line algorithm and $A(L)$ the number of grids used by algorithm A for input sequence L. Then, the *asymptotic worst case ratio* (or *competitive ratio*) of algorithm A, denoted as R_A^∞, is defined as follows:

$$R_A^\infty \stackrel{\mathrm{def}}{=} \limsup_{n \to \infty} R_A^n$$
$$\text{where } R_A^n \stackrel{\mathrm{def}}{=} \max \left\{ \frac{A(L)}{OPT(L)} \mid OPT(L) = n \right\}.$$

In this paper, we will focus our attention to the following restricted class of on-line grid-packing algorithms; i.e., *an unused grid becomes active when it receives the first item, and once an active grid is declared to be closed, it never becomes active again.* Note that the above restriction plays an important role when we want to measure the space complexity of the algorithm in addition to the competitive ratio; i.e., the space complexity can be measured in terms of the number of active grids used by the algorithm. In the following, we say that an on-line grid-packing algorithm A is k-**space bounded** if the number of active grids used by the algorithm is bounded by k.

Note that the competitive ratio of k-space bounded on-line algorithms is defined as the ratio to an optimal off-line algorithm that can use an unbounded number of active bins, in this paper. It is a natural extension of the one-dimensional case, in which k-space bounded algorithms are compared with an off-line algorithm that can use any number of active bins (note that if $k = 1$, there is a unique way to use a single active bin efficiently in the one-dimensional bin-packing problem).

3 Bounds for 1-Space Bounded Algorithms

In this section, we derive several elementary bounds for 1-space bounded grid-packing algorithms. The following proposition claims that the competitive ratio of *any* 1-space bounded algorithm is provably bad, if the rotation of items is not allowed; i.e., the rotation of items is an essential factor to achieve a good competitive ratio by 1-space bounded algorithms.

Proposition 1. *Let GPP' denote a restricted version of GPP in which no rotation of items is allowed. Under GPP', the competitive ratio of any 1-space bounded algorithm is at least $m/2$, and there is a 1-space bounded algorithm with a competitive ratio $2m$.*

Proof. Let L_1 be a sequence of $2m^2$ items in which items $A = (m, 1)$ and $B = (1, m-1)$ are alternate m^2 times; i.e.,

$$L_1 = \underbrace{(A, B, \ldots, A, B)}_{2m^2}.$$

Given list L_1, any 1-space bounded on-line algorithm packs the $(2i-1)$st and the $(2i)$th items into the same grid, for $1 \leq i \leq m^2$; i.e., it consumes m^2 grids for L_1. On the other hand, an optimal off-line algorithm can pack L_1 into $2m-1$ grids as follows: 1) each of the first m grids accommodates m copies of B and one copy of A, and 2) each of the next $m-1$ grids accommodates m copies of A. A similar claim holds for any list L^* that is obtained by concatenating x L_1's for $x \geq 1$. Hence the competitive ratio of any 1-space bounded algorithm is at least

$$\frac{m^2}{2m-1} = \left(\frac{m}{2}\right) \times \left(\frac{1}{1-1/2m}\right) > \frac{m}{2},$$

which completes the former half of the proof.

The latter half can be proved by focusing on the occupation of the first row under a simple "bottom-left" algorithm, that proceeds as follows: When an input item is given, it is placed at the bottom of the current bin and is slided to the left-most position as much as possible. If there is no such space at the bottom of the bin, then it opens a new bin, and places the item at its bottom-left position. Under this "bottom-left" algorithm, the packing into the first row could be regarded as a greedy one-dimensional bin packing algorithm, that is known to have a competitive ratio 2. Q.E.D.

In the following, we consider cases in which the rotation of items is allowed. As the following proposition claims, the competitive ratio of 1-space bounded algorithms cannot be smaller than some constant value greater than two, even when we allow rotation of items.

Proposition 2. *The competitive ratio of any 1-space bounded algorithm is at least* $23/11$ (> 2.0909).

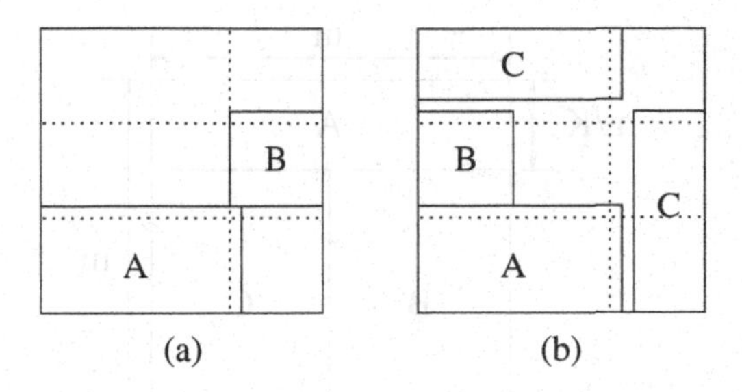

Fig. 1. Behavior of an on-line algorithm for L_2.

Proof. For simplicity, assume m is a multiple of 3 greater than or equal to 9. Let A, B, C, D be items defined as follows: $A = (2m/3+1, m/3+1)$, $B = (m/3, m/3)$, $C = (2m/3+1, m/3-2)$, and $D = (m/3-1, m/3-1)$. Let L_2 be a sequence of 63 items defined as follows:

$$L_2 = (\underbrace{A, B, \ldots, A, B}_{36}, \underbrace{C, \ldots, C}_{18}, \underbrace{D, \ldots, D}_{9}).$$

Given list L_2, any 1-space bounded algorithm packs items in L_2 as follows:

- the $(2i-1)$st and the $(2i)$th items are packed into the same grid, for $1 \leq i \leq 18$, i.e., it spends at least 18 grids for the first 36 items (Figure 1 (a)),
- the first two copies of C are packed into the 18th grid (recall that we allow rotation of items; see Figure 1 (b)),
- the remaining 16 copies of C are packed into 4 new grids, in such a way that each grid accommodates 4 copies, i.e., it spends 4 more grids, where the last grid can accommodate at most one copy of D, and
- 8 copies of D are packed into the 23rd grid.

As a result, any 1-space bounded on-line algorithm uses 23 grids for input L_2. On the other hand, an optimal off-line algorithm can pack L_2 into 11 grids as follows:

- each of the first 9 grids accommodates two copies of A, two copies of C, and one copy of D, and
- each of the next 2 grids accommodates 9 copies of B.

Since the 23rd grid cannot accommodate a new copy of A, by considering a sequence obtained by concatenating L_2 several times, we can conclude that the competitive ratio of any scheme is at least 23/11. Q.E.D.

It is worth noting that the above result is in contrast to the one-dimensional case, in which the competitive ratio is trivially at most two.

4 Proposed 1-Space Bounded Algorithm

In this section, we propose a 1-space bounded algorithm with competitive ratio $O((\log \log m)^2)$, for sufficiently large m's that is a power of two greater than or

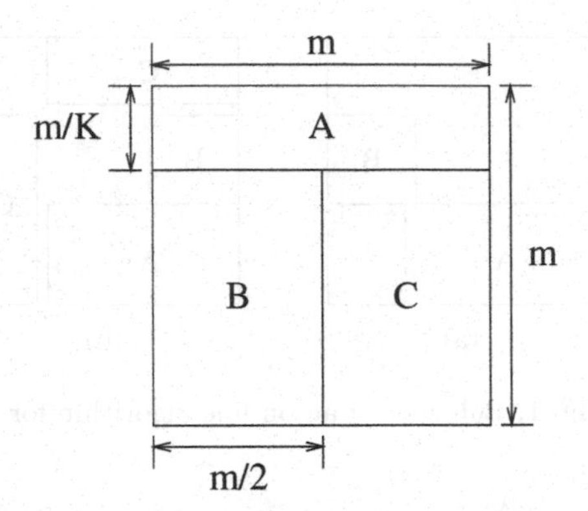

Fig. 2. Partition of a grid used in the proposed algorithm.

equal to 256. In the following, we assume that each item (x, y) satisfies $x \geq y$, without loss of generality. The algorithm consists of three procedures that are called separately depending on the width of the input item, i.e., it is packed by **procedure** A if $m/2 < x \leq m$; packed by **procedure** B if $m/\log_2 m < x \leq m/2$; and packed by **procedure** C if $0 < x \leq m/\log_2 m$. Note that the difficulty in designing on-line algorithms with a small competitive ratio resides in the treatment of medium sized items, that should be packed into the (current) active bin with several larger and/or smaller items, in a mixed manner. In the proposed algorithm, we overcome the difficulty by separating each bin into three parts, each of which is dedicated to large, medium, and small items, respectively.

Before describing each procedure in detail, we should define three subgrids used by those procedures. Let $K \stackrel{\text{def}}{=} \lceil \log_2 \log_2 m \rceil$. At first, each grid is divided into two subgrids of heights $\lceil m/K \rceil$ and $m - \lceil m/K \rceil$, respectively, and the second one is further divided into two subgrids of width $m/2$ each. In the proposed algorithm, the first subgrid (of height $\lceil m/K \rceil$ and width m) is dedicated to procedure A, and the last two subgrids are dedicated to procedures B and C, respectively. See Figure 2 for illustration.

4.1 Procedure A

Procedure A is called when the input item (x, y) satisfies $m/2 < x \leq m$. Recall that each grid is divided into three subgrids, and a subgrid of height $\lceil m/K \rceil$ and width m is dedicated for the packing by this procedure. More concretely, the packing of item (x, y) proceeds as follows:

Case 1: If $y > \lceil m/K \rceil$, then after closing the current grid G_1, it packs the input item into a new grid G_2. It then closes G_2, and opens a new grid G_3 as an active grid.

Case 2: If $y \leq \lceil m/K \rceil$, then the item is packed in a similar manner to the greedy one-dimensional bin packing algorithm; i.e., items are packed into the current subgrid in a bottom-left manner if the total height y' of the

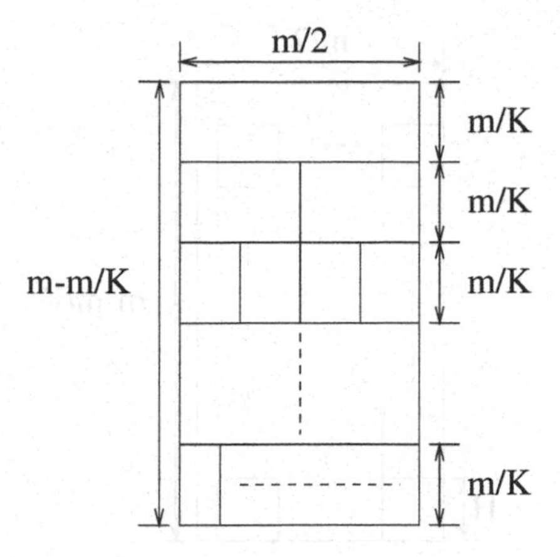

Fig. 3. Partition of the subgrid used by procedure B.

items in the current subgrid satisfies $y' + y \leq \lceil m/K \rceil$, but if not, it is packed into a new grid after closing the current grid.

Note that in both cases, two consecutive closed grids are filled with items with a total size at least $m^2/2K$.

4.2 Procedure B

Procedure B is called when the input item (x, y) satisfies $m/\log_2 m < x \leq m/2$. In procedure B, the dedicated subgrid of height $m - \lceil m/K \rceil$ and width $m/2$ is divided into $K-1$ strips of an (almost) equal height, and for each $1 \leq i \leq K-1$, the ith strip is further divided into 2^{i-1} substrips of width $m/2^i$ each; e.g., the second strip is split into two substrips of width $m/4$ each, the third strip is split into four substrips of width $m/8$ each, and so on (see Figure 3 for illustration). Note that the height of each strip is at least $\lfloor m/K \rfloor$ and the smallest width of the resultant substrips is at most $2m/\log_2 m$, since

$$m/2^{K-1} \leq m/2^{\log_2 \log_2 m - 1} = 2m/\log_2 m.$$

The packing of item (x, y) proceeds as follows:

Case 1: If $y > \lfloor m/K \rfloor$, then the item is packed into a new grid as in Case 1 of procedure A.

Case 2: If $y \leq \lfloor m/K \rfloor$ and $m/2^{i+1} < x \leq m/2^i$, then the item is packed into a substrip of the ith row
in a similar way to Case 2 of procedure A.

Note that in both cases, two consecutive closed grids are filled with items with a total size at least $\min\{m^2/K^2, m^2/4K\}$.

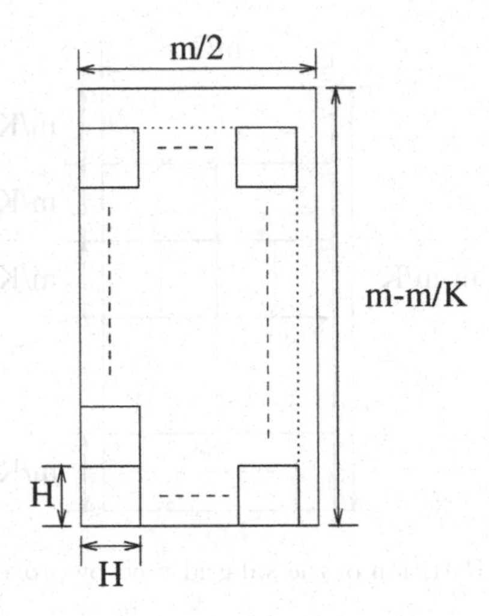

Fig. 4. Partition of the subgrid used by procedure C.

4.3 Procedure C

Procedure C is called when the input item (x, y) satisfies $0 < x \leq m/\log_2 m$. Let $H \stackrel{\text{def}}{=} \lfloor m/\log_2 m \rfloor$. In procedure C, the dedicated subgrid of height $m - \lceil m/K \rceil$ and width $m/2$ is divided into smaller subgrids of size $H \times H$ each. Note that when $m \geq 16$, we have $m - \lceil m/K \rceil \geq m/2$ since $\log_2 \log_2 m \geq 2$. Hence we can embed at least $(\log_2 m)^2/4$ such subgrids (of size $H \times H$) into the dedicated space, that is at least $2 \log_2 m$ for $m \geq 256$ (see Figure 4 for illustration).

The basic idea of the procedure is to use the following $\lceil \log_2 H \rceil$-space bounded algorithm for the packing into (sub)grids of size $H \times H$, in a repetitive manner; i.e., the current grid is closed and a new grid is opened when a half of $(\log_2^2 m)/4$ subgrids are (imaginary) closed by the algorithm.

The $\lceil \log_2 H \rceil$-space bounded algorithm uses $\lceil \log_2 H \rceil$ active (sub)grids numbered from 1 to $\lceil \log_2 H \rceil$. Let $b(i)$ denote the active (sub)grid with number i. (Sub)grid $b(1)$ is dedicated to the packing of items of width satisfying $2^{\lfloor \log_2 H \rfloor} < x \leq H$, and the other (sub)grids are used for the packing of items of width satisfying $0 < x \leq 2^{\lfloor \log_2 H \rfloor}$. In addition, (sub)grid $b(i)$ for $2 \leq i \leq \lceil \log_2 H \rceil$ is divided into a square grid of size $2^{\lfloor \log_2 H \rfloor} \times 2^{\lfloor \log_2 H \rfloor}$ and the remaining part, and the algorithm merely uses the first part for the packing (note that the size of the first part is at least one fourth of the given (sub)grid of size $H \times H$). Let $H' \stackrel{\text{def}}{=} 2^{\lfloor \log_2 H \rfloor}$ for brevity. During the execution of the algorithm, an active grid with number i ($2 \leq i \leq \lceil \log_2 H \rceil$) is partitioned into 2^{i-2} vertical strips of width $H'/2^{i-2}$ and height H', and they are dedicated for the packing of items with height x satisfying $H'/2^{i-1} < x \leq H'/2^{i-2}$. Note that the width of strips generated from the ($\lceil \log_2 H \rceil$)th (sub)grid is $H'/2^{\lceil \log_2 H \rceil - 2}$, that is either two or four.

For input item (x, y), the procedure proceeds as follows:

Case 1: If $2^{\lfloor \log_2 H \rfloor} < x \le H$, then it packs the item into the current $b(1)$ and it opens a new $b(1)$ after (imaginary) closing the current $b(1)$.
Case 2: If $H'/2^{i-1} < x \le H'/2^{i-2}$ for some $2 \le i < \lceil \log_2 H \rceil$, then it packs the item into a strip of $b(i)$ in a similar way to Case 2 of procedure A.
Case 3: If $1 \le x \le H'/2^{\lceil \log_2 H \rceil - 2}$, then it packs the item into a strip of $b(\lceil \log_2 H \rceil)$ in a similar way to Case 2 of procedure A.

By using a similar argument to the previous procedures, we can show that when a new grid is opened by procedure C after closing the current grid, the closed grid is filled with items with a total size m^2/c for some constant c, that is approximately eight for sufficiently large m.

As a result, we have the following proposition on the competitive ratio of the proposed algorithm.

Proposition 3. *The competitive ratio of the proposed 1-space bounded algorithm is* $O((\log \log m)^2)$.

5 Concluding Remarks

In this paper, we studied GPP with a single active grid, and proposed an on-line algorithm that achieves a competitive ratio $O((\log \log m)^2)$. An open problem is to find better upper and lower bounds for GPP. In particular, we are interested in finding an upper bound that is $O(\log \log m)$, or a constant. It is also an important open problem to find a good k-space bounded on-line algorithms for general $2 \le k < \log_2 m$.

References

1. B. S. Baker, D. J. Brown, and H. P. Katseff. A 5/4 Algorithm for Two-Dimensional Packing. *J. of Algorithms*, 2:348–368, 1981.
2. D. Blitz, A. van Vliet, and G. J. Woeginger. Lower bounds on the asymptotic worst-case ratio of on-line bin packing algorithms. unpublished manuscript, 1996.
3. D. Coppersmith and P. Raghavan. Multidimensional on-line bin packing: Algorithms and worst case analysis. *Oper. Res. Lett.*, 8:17–20, 1989.
4. J. Csirik and A. van Vliet. An on-line algorithm for multidimensional bin packing. *Oper. Res. Lett.*, 13:149–158, 1993.
5. S. Fujita and T. Hada. Two-Dimensional On-Line Bin Packing Problem with Rotatable Items. Proc. of COCOON, LNCS 1858, 210–220, 2000.
6. M. R. Garey and D. S. Johnson. *Computers and Intractability: A Guide for the Theory of NP-Completeness.* Freeman, San Francisco, CA, 1979.
7. C. C. Lee and D. T. Lee. A simple on-line bin packing algorithm. *J. Assoc. Comput. Mach.*, 32:562–572, 1985.
8. M. B. Richey. Improved bounds for harmonic-based bin packing algorithms. *Discrete Appl. Math.*, 34:203–227, 1991.
9. S. Seiden and R. van Stee. New bounds for multi-dimensional packing. Proc. of SODA, pp.486–495, 2002.
10. A. van Vliet. An improved lower bound for on-line bin packing algorithms. *Information Processing Letters*, 43:277–284, 1992.

Bend Minimization in Orthogonal Drawings Using Integer Programming

Petra Mutzel and René Weiskircher

Vienna University of Technology
Favoritenstraße 9-11 E186, A-1040 Vienna, Austria
{mutzel,weiskircher}@ads.tuwien.ac.at

Abstract. We consider the problem of minimizing the number of bends in an orthogonal planar graph drawing. While the problem can be solved via network flow for a given planar embedding of a graph G, it is NP-hard if we consider the set of all planar embeddings of G. Our approach combines an integer linear programming (ILP) formulation for the set of all embeddings of a planar graph with the network flow formulation for fixed embeddings. We report on computational experiments on a benchmark set containing hard problem instances that was already used for testing the performance of a previously published branch & bound algorithm for solving the same problem. Our new algorithm is about twice as fast as the branch & bound approach for the graphs of the benchmark set.

1 Introduction

Drawing graphs is important in many scientific and economic areas. Applications include the drawing of UML diagrams in software engineering and business process modeling as well as in the visualization of databases. A popular way of drawing graphs is representing the vertices as boxes and the edges as sequences of horizontal and vertical line segments connecting the boxes. This drawing style is called *orthogonal* drawing. A point where two segments of an edge meet is called a *bend*.

A well known approach for drawing general graphs is the topology-shape-metrics method. In the first step, the topology of the drawing is computed. The objective in this phase is to minimize the number of edge crossings. In the second step, the shape of the drawing is calculated. In the case of orthogonal drawings, the angles and the bends of the edges are computed. The objective is to minimize the number of bends for the given topology. Finally, the metrics of the drawing is computed while trying to achieve short edge lengths and small area for the given shape. In this paper, we focus on the bend minimization step (the seconds step). Given a planar graph, the task is to compute an orthogonal representation with the minimum number of bends.

The infinite set of different planar drawings of a graph can be partitioned into a finite set of equivalence classes called *embeddings* of a graph. An embedding defines the topology of a planar drawing without assigning lengths or shapes to the

O.H. Ibarra and L. Zhang (Eds.): COCOON 2002, LNCS 2387, pp. 484–493, 2002.

edges or fixing the shapes and positions of vertices. A *combinatorial embedding* fixes the sequence of incident edges around each vertex in clockwise order. This also fixes the list of *faces* of a drawing. The faces are the connected regions of the plane defined by a planar drawing. A *planar embedding* additionally defines the outer (unbounded) face of a planar drawing. *Orthogonal representations* are equivalence classes of orthogonal drawings, that fix the planar embedding and the bends and angles in an orthogonal drawing.

There are some results in the literature on the topic of optimizing certain functions over the set of all embeddings of a graph. Bienstock and Monma have studied the complexity of covering vertices by faces [BM88] and minimizing certain distance measures on the faces of a graph with respect to the outer face [BM89,BM90]. Garg and Tamassia have shown that optimizing the number of bends in an orthogonal drawing over the set of all embeddings of a planar graph is NP-hard [GT94].

Bertolazzi et al. [BBD00] have devised a branch & bound algorithm for solving the bend minimization problem over the set of all embeddings of a planar graph using SPQR-trees. In this paper, we attack the same problem using integer linear programming. To do this, we combine our integer linear program describing the set of all combinatorial embeddings of a planar biconnected graph [MW99,MW00] with a linear program that describes the set of all orthogonal representations of a planar graph with a fixed embedding. The result is a mixed integer linear program that represents the set of all orthogonal representations for a planar biconnected graph over the set of all embeddings. We use this new mixed integer linear program to optimize the number of bends in an orthogonal drawing over the set of all embeddings of a planar graph. Solving this program using a commercial solver (CPLEX) is significantly faster for large and difficult graphs than the branch & bound approach of Bertolazzi et al . as our computational results show.

Section 2 introduces SPQR-trees and summarizes the recursive construction of the integer linear program that describes the combinatorial embeddings of a graph. The linear program describing the orthogonal representations of a graph for a fixed embedding is the topic of Section 3. This is basically the formulation as a linear program of a minimum cost flow problem in a special network constructed from the graph and the embedding. In Section 4, we present the new mixed integer linear program that is the result of merging the integer linear program describing the embeddings of a graph with the linear program that describes the orthogonal representations for a graph where the embedding is fixed. The topic of Section 5, is the algorithm that we use to compute an orthogonal representation of a graph with the minimum number of bends over the set of all embeddings. The computational results we obtained by applying the algorithm to a set of hard benchmark graphs are given in Section 6. We compare the algorithm with a well known heuristic and with the branch & bound algorithm of Bertolazzi et al. The conclusion (Section 7) summarizes the main results and contains possible starting points for future work.

2 The ILP-Formulation Describing the Set of All Embeddings

The integer linear program (ILP) suggested in [MW99] describing the set of all combinatorial embeddings of a planar graph is constructed recursively using the SPQR-tree data structure. Because SPQR-trees are only defined for biconnected graphs, the same is true for the ILP. A graph is biconnected, if the number of its connected components can not be increased by deleting a vertex.

SPQR-trees have been defined by Di Battista and Tamassia [BT96]. They represent a decomposition of a biconnected graph into its triconnected components. A connected graph is triconnected, if there is no pair of vertices in the graph whose removal splits the graph into two or more components. An SPQR-tree has four types of nodes (Q-nodes, S-nodes, R-nodes and P-nodes) and with each node we associate a biconnected graph which is called the *skeleton* of that node. This graph can be seen as a simplified version of the original graph and its vertices are vertices of the original graph. The edges in a skeleton represent subgraphs of the original graph.

All leaves of the SPQR-tree are Q-nodes and all inner nodes S-, P or R-nodes. When we see the SPQR-tree as an unrooted tree, then it is unique for every biconnected planar graph. Another important property of these trees is that their size (including the skeletons) is linear in the size of the original graph and that they can be constructed in linear time [HT73,GM01]. As described in [BT96], SPQR-trees can be used to represent the set of all combinatorial embeddings of a biconnected planar graph. Every combinatorial embedding of the original graph defines a unique combinatorial embedding for each skeleton of a node in the SPQR-tree. Conversely, when we define an embedding for each skeleton of a node in the SPQR-tree, we define a unique embedding for the original graph.

The variables of the ILP correspond to directed cycles of the graph. Our recursive construction of the ILP guarantees that we only compute variables for cycles that form the boundary of a face in at least one embedding of the graph. So we generate the minimum set of variables needed to describe all embeddings. While the number of directed cycles in a graph grows exponentially with the size of the graph, our computational experiments in Section 6 show that the number of variables in our ILP grows only linearly.

We construct the program by splitting the SPQR-tree into smaller SPQR-trees, recursively constructing ILPs for these smaller trees, and then merging them into an ILP for the original graph. The basis of the recursive construction are SPQR-trees that have only one inner node. These graphs have a very simple structure and ILPs that describe their combinatorial embeddings are easy to construct. One type of constraints, similar to the subtour elimination constraints used in ILPs for the asymmetric travelings salesman problem (ATSP), are not explicitly added to the ILP because the number of these constraints is exponential. Instead we separate them in the optimization procedure using the same methods used for solving ATSP-problems with integer programming. We construct the ILPs of more complex graphs by merging the ILPs of the graphs generated by the splitting procedure and adding additional glue constraints. Using structural induction, we can show that the resulting ILP is correct and that the variables correspond exactly to the set of cycles that are face cycles in at least one embedding of the graph.

3 The Linear Program Describing Orthogonal Representations for a Fixed Embedding

Orthogonal representations not only fix the embedding of a graph but also the number, type and sequence of the bends on each edge in an orthogonal drawing. They do not fix the lengths of the edge segments in the drawing. The first efficient algorithm for computing an orthogonal representation of a graph with the minimum number of bends for a fixed planar embedding was presented by Tamassia [Tam87]. This algorithm constructs a flow network using the planar embedding and then computes a minimum cost flow in this network. This flow can be translated into an orthogonal representation of the graph with the minimum number of bends for the fixed embedding.

The drawback of the original method of Tamassia is that it can not deal with vertices of degree greater than four. Some modifications of the algorithm have been published that get over this constraint. The approach that we use implements the *podevsnef* drawing convention (planar orthogonal drawings with equal vertex size and non-empty faces) first mentioned in [FK96]. According to this convention, the vertices are drawn as boxes of the same size and the edges are positioned on a finer grid then the vertices. Because of this modification, more than one edge can be incident to each of the four sides of a vertex (see Fig. 1 for an example).

Bertolazzi et al. describe a minimum cost flow network N that can be used to compute an orthogonal representation in a simplified podevsnef model with the minimum number of bends for a fixed embedding [BBD00]. The network for G contains one node for every vertex of G (called *v-nodes*) and one vertex for every face cycle of the given embedding (called *c-nodes*).

Let f be the bijection that maps the vertices of G to the v-nodes of N and the face cycles of the planar embedding to the c-nodes. Then there is an arc between the v-node v_1 and the c-node v_2 if the vertex $f^{-1}(v_1)$ is on the cycle $f^{-1}(v_2)$. This arc is directed towards v_2 if the degree of $f^{-1}(v_1)$ is at most four and towards v_1 otherwise. There is an arc from the c-node v_3 to the c-node v_4 if the two cycles $f^{-1}(v_3)$ and $f^{-1}(v_4)$ share an edge.

The flow on arcs connecting v-nodes with c-nodes determines the angles between edges incident to the same vertex while the flow on arcs connecting two c-nodes determines the bends. Flow on an arc from a c-node to a v-node implies a zero-degree angle at the corresponding vertex between two incident edges and causes a bend on one of the edges. The amount of flow that each vertex in N produces or consumes together with the capacities for the edges guarantee that every feasible flow corresponds to an orthogonal representation. The cost per unit of flow on the arcs of the network are defined in such a way, that the cost of

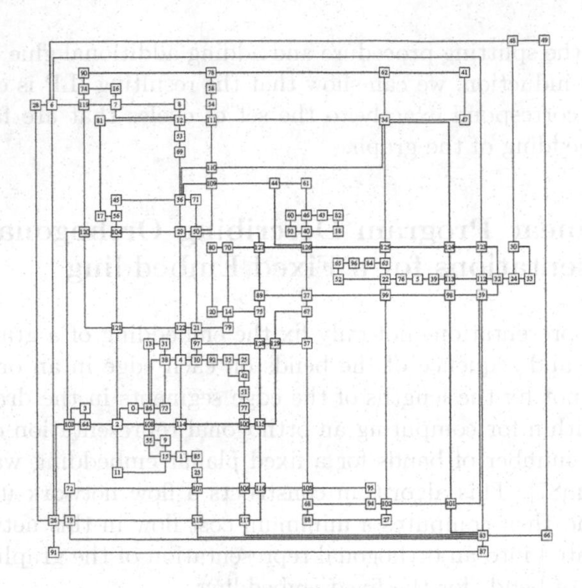

Fig. 1. A podevsnef drawing of a graph.

each feasible flow is equal to the number of bends in the represented orthogonal representation.

We used this network and transformed it into a linear program. There is one variable for each arc in the network that represents the amount of flow routed via this arc. One constraint for each vertex in the network makes sure that the number of incoming amount of flow minus the number of outgoing amount is equal to the demand of the node (some nodes have negative demand). We have one constraint for each arc that sets upper and lower bounds for the flow on the arc. The objective function minimizes the sum of the amount of flow over each arc multiplied by the cost of the arc. An optimal solution represents a minimum cost flow in N and thus an orthogonal representation with the minimum number of bends. Because of space constraints, we do not present the LP here, but the constraints are all contained in the mixed integer linear program of Section 4, that is used to compute an orthogonal representation with the minimum number of bends over all embeddings.

4 The Mixed Integer Linear Program Describing the Set of All Orthogonal Representations of a Graph

The flow network N of the last section describing the set of orthogonal representations of a graph with a fixed embedding contains one c-node for every face of the embedding. When we want to optimize over the set of all embeddings of a graph, we do not know which cycles will be face cycles in an optimal solution.

Therefore, we construct a new network N', where we have one c-node for every cycle in the graph, that is a face cycle in at least one embedding. The set of these cycles corresponds to the set of variables in our ILP from Section 2 that describes the set of all embeddings of a graph.

In a solution of the embedding ILP, the variables of the cycles that are face cycles in the represented embedding have value one while all other variables have value zero. Let A be the set of edges incident to the c-node for cycle c in N' and the variable for c in the embedding ILP be zero. Then all arcs in A must have flow zero. Therefore, the flow on the arcs of the network N' incident to c-nodes corresponding to cycles in G whose variable in the ILP is zero must also be zero.

To achieve this, we take the variables of the ILP into account when we compute the capacities of the edges and the amount of flow that each c-node consumes or produces. We first compute the capacities of the arcs and the demand of each c-node analogously to the corresponding values in the network N. Then we multiply the amount of flow produced or consumed by a c-node with the value of the corresponding variable in the ILP. This ensures that vertices in N' that correspond to cycles in G that are not face cycles do not produce or consume flow, because the corresponding variable in the ILP is zero.

Any arc that starts or ends at a c-node has capacity zero if the c-node corresponds to a cycle whose ILP-value is zero. If the capacity of the edge is limited even if the corresponding cycle is a face cycle, we can just multiply this limit with the ILP-value of the cycle. The arcs in the network N that connect two c-nodes have unlimited capacity. But we can easily compute an upper bound f_{max} for the flow produced in N' (we get a trivial upper bound by adding the supply of all nodes). This value can be used as the upper bound for the flow on any arc. For each arc a in N' connecting two c-nodes v_1 and v_2, we set the capacity to the minimum of the two products $f_{max}x_i$ where x_i is the the binary variable in the embedding ILP for the cycle corresponding to node v_i. This guarantees that the flow on a is zero if at least one of the cycles represented by the nodes v_i is not a face cycle in the chosen embedding.

The result is the network N', where the capacities of the edges and the amount of flow produced and consumed by the vertices depend on the values of the cycle variables in the ILP. We transform this network into a linear program and merge it with the ILP that represents the embeddings of the graph. The result is a mixed integer linear program (MILP), where an optimal solution corresponds to an orthogonal representation with the minimum number of bends over the set of *all* embeddings of the input graph.

MILP 1 is the resulting mixed integer linear program. We omitted the constraints that define the embedding because they are defined recursively and are not the main topic of this paper. The set C is the set of cycles in G that are face cycles in at least one embedding. The variable x_c is one if cycle c is a face cycle and variable o_c is one if it is the outer face cycle. The set E_{cc} is the set of arcs that connect two c-nodes. Arcs in E_{vc} start in a v-node and end in a c-node while the arcs in E_{cv} have the opposite direction. The expression $len(c)$ denotes the number of edges in cycle c.

MILP 1

$$\min \sum_{e \in E_N} cost(e) \cdot f_e$$

subject to

$$
\begin{array}{llll}
 & \sum_{c \in C} o_c = & 1 & \\
 & x_c - o_c \geq & 0 & \forall c \in C \\
\sum_{e=(v,w) \in E_N} f_e - & \sum_{e=(w,v) \in E_N} f_e = & 4 - deg(v) & \forall v \in V \\
\sum_{e=(c,w) \in E_N} f_e - & \sum_{e=(w,c) \in E_N} f_e = x_c(4 - len(c)) - 8 o_c & & \forall c \in C \\
 & f_e \leq & x_c(4 - deg(v)) & \forall e = (v,c) \in E_{vc} \\
 & f_e \leq & x_c & \forall e = (c,v) \in E_{cv} \\
 & f_e \leq & x_{c_1} f_{max} & \forall e = (c_1, c_2) \in E_{cc} \\
 & f_e \leq & x_{c_2} f_{max} & \forall e = (c_1, c_2) \in E_{cc} \\
 & f_e \geq & 0 & \forall e \in E_N \\
 & x_c, o_c \in & \{0,1\} & \forall c \in C
\end{array}
$$

5 The Algorithm for Minimizing the Number of Bends

The algorithm first computes the recursive ILP describing the set of all combinatorial embeddings of the graph. This also gives us the set of cycles of the graph that are face cycles in at least one embedding. This information is then used for computing the network N' and the corresponding MILP. We use CPLEX (version 6.5) to compute a solution and then separate subtour elimination constraints from the embedding ILP and re-optimize if necessary. When we have found a feasible solution, we transform it into an orthogonal representation of the graph.

To improve the performance of the algorithm, we modified the MILP slightly. For example, we only need outer face variables for half of the cycles. The orthogonal representations we exclude in this way are mirror images of other orthogonal representations that can still be represented. We also hard-coded a complete description of the set of embeddings for P-node skeletons with less than five vertices into our program to reduce the need for separating constraints.

6 Computational Results

Since we wanted to compare the performance of our approach with the branch & bound method for bend minimization by Bertolazzi [BBD00], we used the same set of graphs that they used for testing the performance of their algorithm.

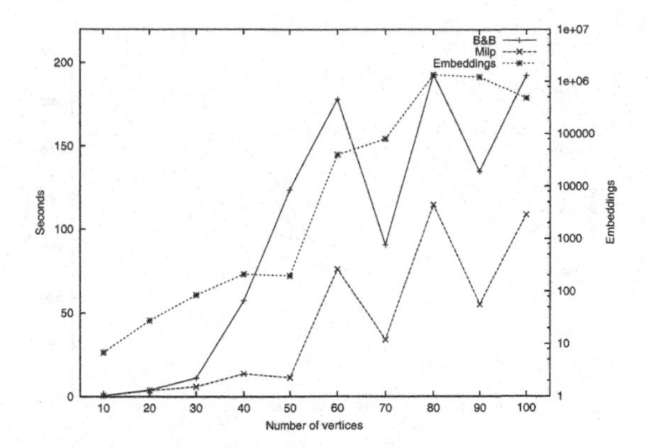

Fig. 2. Run time comparison with the branch & bound algorithm (linear scale) together with the average number of embeddings (logscale)

This set consists of 500 randomly generated graphs, 50 different graphs for each number of vertices from 10 to 100 in steps of 10.

Our algorithm and the branch & bound algorithm have the same limitations: They can only be applied to planar biconnected graphs, because they both use SPQR-trees. All the graphs in the benchmark set have these properties.

First, we compared the optimal results produced by our algorithm with the results computed by a popular heuristic. This heuristic chooses an arbitrary embedding for the graph and then computes a minimum cost flow in the network of section 3.

Let h be the number of bends in the orthogonal representation computed by the heuristic and o the number of bends in an orthogonal representation with the minimum number of bends. For each graph in the benchmark set, we computed the following value: $\frac{h-o}{h}100\%$. This is the percentage of the improvement we get using an optimal algorithm. Almost half of all graphs (246 out of 500) show a significant improvement (greater than 10%). The greatest absolute difference in the number of bends that we observed from the heuristic solution to the optimal solution was 12 bends. The average number of saved bends per graph using the optimal algorithm was 2.26. The average improvement over all graphs was 17.43%.

We compared the average running time for graphs with the same number of vertices of our new algorithm (MILP) and the branch & bound algorithm (B&B) from [BBD00]. Both algorithms were tested on a Sun Enterprise 450 Model 4400 with 4GB main memory. The running time of MILP includes the time needed for the recursive construction of the ILP that describes the embeddings of a graph.

Figure 2 shows the average running time of both algorithms. The x-axis shows the number of vertices in the graphs and the y-axis on the left the average running time in seconds for all graphs in the benchmark set with that number of vertices. The plot shows that our new algorithm needs on average only half the time needed by the branch & bound algorithm to compute the drawing with

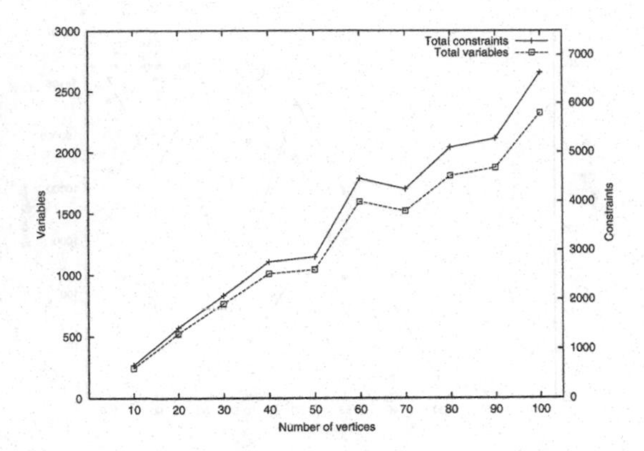

Fig. 3. The average number of constraints and variables grows only linearly with the size of the graphs

the minimum number of bends. The same plot contains the curve showing the average number of embeddings for graphs with the same number of vertices. As expected, the average number of embeddings grows exponentially with the size of the graphs (note that the y-axis on the right is logarithmic). However, the average number of constraints and variables in our mixed integer linear program grows only linearly with the size of the graphs (see Figure 3).

We also applied the branch & bound algorithm and our new algorithm to a set of 11529 graphs derived from graphs used in industrial applications. We created these graphs by planarizing the graphs in the benchmark set used in [BGL+97] and then adding edges to make them planar and biconnected. Because of space considerations, we can only mention a few statistics. The branch & bound algorithm failed to provide an optimal solution in one hour of computation time for 197 of the graphs, while our new algorithm exceeded this time limit for only 25 graphs. While the branch & bound algorithm is slightly faster on average for the graphs in the set with less than about 120 vertices, our algorithm has a significant speed advantage for the graphs with more than 150 vertices.

7 Conclusion

Using methods of integer linear programming to minimize the number of bends in an orthogonal drawing seems to be a promising approach. The main drawback is that at the moment, the algorithm only works for biconnected graphs. The reason is that SPQR-trees are only defined for biconnected graphs. A possible approach to get rid of this limitation is to work with the block tree of biconnected components of the graph. If it can be used to describe the set of all embeddings of a connected graph as an ILP, our approach can be easily extended to deal with any planar graph.

Acknowledgment

We thank Walter Didimo for providing the code of the branch & bound algorithm and the benchmark graphs.

References

BBD00. P. Bertolazzi, G. Di Battista, and W. Didimo. Computing orthogonal drawings with the minimum number of bends. *IEEE Transactions on Computers*, 49(8):826–840, 2000.

BGL+97. G. Di Battista, A. Garg, G. Liotta, R. Tamassia, E. Tassinari, and F. Vargiu. An experimental comparison of four graph drawing algorithms. *Comput. Geom. Theory Appl.*, 7:303–326, 1997.

BM88. D. Bienstock and C. L. Monma. On the complexity of covering vertices by faces in a planar graph. *SIAM Journal on Computing*, 17(1):53–76, 1988.

BM89. D. Bienstock and C. L. Monma. Optimal enclosing regions in planar graphs. *Networks*, 19(1):79–94, 1989.

BM90. D. Bienstock and C. L. Monma. On the complexity of embedding planar graphs to minimize certain distance measures. *Algorithmica*, 5(1):93–109, 1990.

BT96. G. Di Battista and R. Tamassia. On-line planarity testing. *SIAM Journal on Computing*, 25(5):956–997, 1996.

FK96. U. Fößmeier and M. Kaufmann. Drawing high degree graphs with low bend numbers. In F. J. Brandenburg, editor, *Graph Drawing (Proc. GD '95)*, volume 1027 of *LNCS*, pages 254–266. Springer-Verlag, 1996.

GM01. C. Gutwenger and P. Mutzel. A linear time implementation of SPQR-trees. In J. Marks, editor, *Graph Drawing (Proc. 2000)*, volume 1984 of *LNCS*, pages 77–90. Springer-Verlag, 2001.

GT94. A. Garg and R. Tamassia. On the computational complexity of upward and rectilinear planarity testing. In R. Tamassia and I. G. Tollis, editors, *Proceedings Graph Drawing '94*, volume 894 of *LNCS*, pages 286–297. Springer-Verlag, 1994.

HT73. J. E. Hopcroft and R. E. Tarjan. Dividing a graph into triconnected components. *SIAM Journal on Computing*, 2(3):135–158, 1973.

MW99. P. Mutzel and R. Weiskircher. Optimizing over all combinatorial embeddings of a planar graph. In G. Cornuéjols, R. Burkard, and G. Wöginger, editors, *Proceedings IPCO '99*, volume 1610 of *LNCS*, pages 361–376. Springer Verlag, 1999.

MW00. P. Mutzel and R. Weiskircher. Computing optimal embeddings for planar graphs. In *Proceedings COCOON '00*, volume 1858 of *LNCS*, pages 95–104. Springer Verlag, 2000.

Tam87. R. Tamassia. On embedding a graph in the grid with the minimum number of bends. *SIAM Journal on Computing*, 16(3):421–444, 1987.

The Conditional Location of a Median Path

Biing-Feng Wang[1], Shan-Chyun Ku[2], and Yong-Hsian Hsieh[1]

[1] National Tsing Hua University, Taiwan, ROC
bfwang@cs.nthu.edu.tw, eric@venus.cs.nthu.edu.tw
[2] Faraday Technology Corporation, Taiwan, ROC
scku@faraday.com.tw

Abstract. In this paper, we study the problem of locating a median path of limited length on a tree under the condition that some existing facilities are already located. The existing facilities may be located at any subset of vertices. Upper and lower bounds are proposed for both the discrete and continuous models. In the discrete model, a median path is not allowed to contain partial edges. In the continuous model, a median path may contain partial edges. The proposed upper bounds for these two models are $O(n\log n)$ and $O(n\log n\alpha(n))$, respectively. They improve the previous ones from $O(n\log^2 n)$ and $O(n^2)$, respectively. The proposed lower bounds are both $\Omega(n\log n)$. The lower bounds show that our upper bound for the discrete model is optimal and the margin for possible improvement on our upper bound for the continuous model is slim.

1 Introduction

Network location theory has been traditionally concerned with the optimal location of points. Let G=(V, E) be a graph. For any pair v, $u \in V$, let $d(v, u)$ be their distance. For each node $v \in V$, let $w(v)$ be its vertex weight. The well-known p-median problem is to find a subset H of p vertices minimizing $\Sigma_{v\in V}\{w(v)\times\min_{u\in H} d(v,u)\}$, which is called the *distance-sum* of H. The well-known p-center problem is to find a subset H of p vertices minimizing $\max_{v\in V}\{w(v)\times\min_{u\in H} d(v,u)\}$, which is called the *eccentricity* of H. Slater [8] firstly extended the network location theory to include a facility that is not merely a single-point but a path. He defined a *core* of a tree as a path of any length having the minimum distance-sum. Minieka [6] studied the problem of finding a median path of a specified length in a tree. In the problem, the length of a median path should be exactly equal to a given number l. Later, Hakimi *et al.* [3] studied the problem of locating a median path of limited length in a tree. In the problem, a median path should have length not larger than a given number l. Two models were discussed. If a median path is not allowed to contain partial edges, it is referred as the *discrete model*. Otherwise, it is referred as the *continuous model*. Recently, Alstrup *et al.* [2] gave an $O(n\log n)$ time algorithm for the discrete model and an $O(n\log n\alpha(n))$ time algorithm for the

O.H. Ibarra and L. Zhang (Eds.): COCOON 2002, LNCS 2387, pp. 494–503, 2002.

continuous model, where α is the inverse of Ackermann's function. With a little modification, their algorithm for the continuous model can be used to find a median path of a specified length as well. In [6], Minieka extended the location theory to include tree-shaped facilities. Results on locating tree-shaped facilities can be found in [6,9,10,12]. Path-shaped and tree-shaped facilities are also called *extensive facilities.* Parallel algorithms for locating different kinds of extensive facilities have also been examined in the literature [11,12].

Consider the situation that we are going to locate new facilities on a network in which some existing facilities are already located so that customers can be served by the closest facility whether exising or new. For example, consider a graph $G=(V, E)$ in which there are existing facilities located at the vertices of a subset $S \subseteq V$. *The conditional p-median problem* is to find a subset H of p vertices minimizing the distance-sum of $H \cup S$ and *the conditional p-center problem* is to find a subset H of p vertices minimizing the eccentricity of $H \cup S$. Minieka [5] coined the term *conditional location problem.* Since then many papers dealing with conditional location problems have appeared. Very recently, Tamir *et al.* [10] launched the study on finding the conditional location of an extensive facility of limited length on trees. They discussed both path-shaped and tree-shaped facilities, both eccentricity and distance-sum criteria, and both continuous and discrete models. Therefore, in total eight problems were examined. For most of the problems, sub-quadratic algorithms were presented.

In this paper, we study the problem of finding the conditional location a median path of limited length on a tree. Upper and lower bounds are proposed for both the discrete and continuous models. For the discrete model, the proposed upper bound is $O(n\log n)$. Tamir *et al.* [10] had proposed an $O(n\log^2 n)$ time algorithm for this model and conjectured that the time complexity of their algorithm can be improved to $O(n\log n)$ by using a more efficient implementation. Our upper bound is obtained by such an implementation. For the continuous model, the proposed upper bound is $O(n\log n\alpha(n))$, improving the previous one in [10] from $O(n^2)$. For the discrete model, the proposed lower bound is $\Omega(n\log n)$. It also holds for the unconditional case. Thus, for the discrete model, the upper bound proposed in this paper and the upper bound proposed in [2] for the unconditional case are both optimal. For the continuous model, the proposed lower bound is also $\Omega(n\log n)$. It shows that the margin for possible improvement on our upper bound for the continuous model is slim.

The remainder of this paper is organized as follows. In the next section, notation and preliminaries are presented. In Section 3, Alstrup *et al.*'s algorithms for the unconditional case are described. Then, in Sections 4 and 5, our algorithms for the conditional case are presented. Finally, in Section 6, lower bounds are proposed.

2 Notation and Preliminaries

Let $T = (V, E)$ be free tree. Let $n = |V|$. Each vertex $v \in V$ has a nonnegative weight $w(v)$ and each edge $e \in E$ has a nonnegative length $d(e)$. In this paper, we assume that T is drawn in the Euclidean plane so that each $e \in E$ is a line segment of length $d(e)$ and T is regarded as a closed and connected subset of points in the Euclidean plane. For any two points a, b of T, let $P(a, b)$ be the unique path from a to b, which is a collection of edges and at most two partial edges, and let $d(a, b)$ be the length of $P(a, b)$. A path $P(a, b)$ is *discrete* if both $a, b \in V$, and is *almost discrete* if at least one of a and b is a vertex. If two vertices $i, j \in V$ are neighbors, then by removing the edge (i, j) two subtrees are induced. We denote by T_j^i the subtree containing j. For convenience, for any $i \in V$, we call each T_j^i, $(i, j) \in E$, a *subtree* of i. For any subtree X of T, the vertex set and edge set of X are denoted, respectively, by $V(X)$ and $E(X)$. For easy description of algorithms, sometimes we will orient T into a rooted tree. In such a case, we denote by $p(i)$ the parent of a vertex $i \in V$.

For any vertex $i \in V$ and any subset Y of points of T, the *distance* from i to Y is the shortest distance from i to any point of Y. If $Y = \phi$, $d(i, Y) = \infty$. In this paper, we assume that there is a subset $S \subseteq V$ representing some existing facilities. For any subtree X and any subset Y of points of T, the *conditional weighted distance-sum* from X to Y is $D(X, Y) = \Sigma_{i \in V(X)} w(i) \times \min\{d(i, Y), d(i, S)\}$. If $X = T$, we simply write $D(Y)$ in place of $D(T, Y)$. Let $\lambda_i = D(i)$ for each $i \in V$ and let $\lambda_e = D(e)$ for each $e \in E$. For each $(i, j) \in E$, define a function $\delta_{(i,j)}$ as $\delta_{(i,j)}(x) = D(P(i, q))$, where $0 \leq x \leq d(i, j)$ and q is the point on (i, j) with $d(i, q) = x$. Note that δ is defined on ordered pairs of neighboring vertices and thus $\delta_{(i,j)} \neq \delta_{(j,i)}$. By definition, $\delta_{(i,j)}(0) = \lambda_i$ and $\delta_{(i,j)}(d(i, j)) = \lambda_{(i,j)}$.

The *median path problem* is to determine a path H of length $\leq l$ in T minimizing $D(H)$, where $l \geq 0$ is a real number. The determined path is called a *median path*. The problem is *unconditional* if $S = \phi$, and is *conditional* otherwise. When a median path is allowed to contain partial edges, we refer to the model as the *continuous model*, and otherwise we refer to it as the *discrete model*.

For any piece-wise linear function f, let $||f||$ be the number of linear segments of f. For a set F of functions, the *lower envelope* of F is the function Φ defined as $\Phi(x) = \min\{\infty, \min_{f \in F \text{ defined on } x}\{f(x)\}\}$ for all number $x \geq 0$. Hart and Sharir gave the following result.

Theorem 1 [4]: The lower envelope of n linear functions, each defining a line or a line segment, is a piece-wise linear function having $O(n\alpha(n))$ linear segments.

A *decomposition* of T by using a vertex $i \in V$ as a *decomposer* is a pair of two subtrees $T_1 = (V_1, E_1)$ and $T_2 = (V_2, E_2)$ such that $V_1 \cup V_2 = V$, $V_1 \cap V_2 = \{i\}$, and $E_1 \cup E_2 = E$. For any subtree X of T, we call a vertex in X having a neighbor in T outside X a *boundary vertex*. A *cluster* of T is a subtree having at most two boundary vertices. For any cluster C, $\beta(C)$ denotes the set of boundary vertices and $\pi(C)$ denotes the path between the boundary vertices. If C has only one boundary vertex, $\pi(C)$ denotes the vertex. Two clusters A and B can be *merged* if they intersect in a single vertex and $A \cup B$ is still a cluster. A *top tree* τ of T is a binary tree with the following properties [1,2].

1. Each node of τ represents a cluster of T.
2. The leaves of τ represent the edges of T.
3. Each internal node of τ represents the cluster merged from the two clusters represented by its children.
4. The root of τ represents T.
5. The height of τ is $O(\log n)$.

A top tree of T describes a way to recursively decompose T into subtrees having at most two boundary vertices, until each of the subtrees contains only a single edge [1,2].

3 Alstrup *et al.*'s Algorithms for the Unconditional Median Path Problem

3.1 The Continuous Model

For any subtree C of T, let $Cost(C)$ be the distance-sum of the best path of length$\leq l$ contained in C. That is, $Cost(C)=\min\{D(H)|\ H$ is a path of length$\leq l$ in $C\}$. Alstrup *et al.*'s algorithm firstly constructs a top tree τ of T in $O(n)$ time. Then, in a bottom-up fashion, it computes $Cost(C)$ for all clusters C represented by the nodes of τ. The computation for the root gets the distance-sum of a median path.

Besides $Cost(C)$, during the computation, for each cluster C, the value $D(\pi(C))$ and a function $\Phi_{b,C}$ for every $b\in\beta(C)$ are computed, where

$$\Phi_{b,C}(x)=\min\{D(P(b,\ q))|\ P(b,\ q) \text{ is contained in } C,\ d(b,\ q)\leq x\}$$

(i.e., $\Phi_{b,C}(x)$ is the distance-sum of the best path of length$\leq x$ that extends from b into C). Define for each $(i,\ j)\in E$ a function $f_{(i,j)}$ as

$$(1)\qquad f_{(i,j)}(x) = \begin{cases} \infty & \text{if} \quad x < 0, \\ \delta_{(i,j)}(x) & \text{if} \quad 0 \leq x < d(i,j), \text{ and} \\ \delta_{(i,j)}(d(i,j)) & \text{if} \quad x \geq d(i,j) \end{cases}$$

which is the distance-sum of the best path of length$\leq x$ that extends from i and is contained in the edge $(i,\ j)$. Then, we can express

$$\Phi_{b,C}(x) = min_{(i,j)\in E(C) \text{ and } i \text{ is on } P(b,j)}\{D(P(b,i)) + f_{(i,j)}(x - d(b,i)) - \lambda_i\}.$$

Clearly, since $S=\phi$, we have $\delta_{(i,j)}(x)=\lambda_i\text{-}x\times|V(T_j^i)|$ for $0\leq x\leq d(i,\ j)$. Thus, the function $\Phi_{b,C}$ is the lower envelope of $|E(C)|$ piece-wise linear functions. Since $\Sigma_{(i,j)\in E(C) \text{ and } i \text{ is on } P(b,j)}||f_{(i,j)}||=O(|E(C)|)$, we obtain the following.

Lemma 1 [2]: In the unconditional case, for any cluster C and $b\in\beta(C)$, $\Phi_{b,C}$ is a piece-wise linear function having $O(k\alpha(k))$ linear segments, where $k=|E(C)|$.

Alstrup *et al.*'s algorithm is as follows.

Algorithm 1. Median_Path(T, l)
Input: a tree T=(V, E) and a length l≥0
Output: the distance-sum of a median path having length≤l
begin
01 preprocess T to obtain λ_i, $\lambda_{(i,j)}$, and $f_{(i,j)}$ for all i∈V and (i, j)∈E
02 $\tau \leftarrow$ a top tree of T
03 **for** each leaf cluster e=(i, j)∈E **do**
04 $Cost(e) \leftarrow \min\{f_{(i,j)}(l), f_{(j,i)}(l)\}$
05 $D(\pi(e)) \leftarrow \begin{cases} \lambda_i & \text{if} \quad \beta(e) = \{i\} \\ \lambda_j & \text{if} \quad \beta(e) = \{j\} \\ \lambda_e & \text{if} \quad \beta(e) = \{i,j\} \end{cases}$
06 **for** each $b \in \beta(e)$ **do** $\Phi_{b,e} \leftarrow \begin{cases} f_{(i,j)} & \text{if} \quad b = i \\ f_{(j,i)} & \text{if} \quad b = j \end{cases}$
07 **for** each cluster C represented by an internal node of τ **do** (in a bottom-up fashion)
08 (A, B) ← the two clusters represented by the children of the internal node
09 c ← the intersection vertex of A and B
10 $H_c \leftarrow \min_{0 \le x \le l}\{\Phi_{c,A}(x)+\Phi_{c,B}(l\text{-}x)\text{-}\lambda_c\}$ // The distance-sum of the best path passing c in C //
11 $Cost(C) \leftarrow \min\{Cost(A), Cost(B), H_c\}$
12 $D(\pi(C)) \leftarrow \begin{cases} \lambda_b & \text{if } |\beta(C)| = 1 \text{ and } \beta(C) = \{b\} \\ D(\pi(A)) + D(\pi(B)) - \lambda_c & \text{otherwise} \end{cases}$
13 **for** each b∈$\beta(C)$ **do**
14 $\Phi_{b\,C}(\text{x}) \leftarrow$
$$\begin{cases} \min\{\Phi_{b,A}(x), \Phi_{b,B}(x)\} & \text{if} \quad b = c \\ \min\{\Phi_{b,A}(x), \Phi_{b,B}(x - d(b,c)) + D(\pi(A)) - \lambda_c\} & \text{if} \quad b \in \beta(A)\backslash\{c\} \\ \min\{\Phi_{b,A}(x - d(b,c)) + D(\pi(B)) - \lambda_c, \Phi_{b,B}(x)\} & \text{if} \quad b \in \beta(B)\backslash\{c\} \end{cases}$$
15 **return**($Cost(T)$)
end

Since S=ϕ, by using dynamic programming, it is easy to compute λ_i, $\lambda_{(i,j)}$, and $|V(T_j^i)|$ for all i∈V and (i, j)∈E in $O(n)$ time. Using the computed valuse, each $\delta_{(i,j)}$ and $f_{(i,j)}$ can be computed in $O(1)$ time. The computation for a leaf cluster takes $O(1)$ time. Since $\Phi_{c,A}$ and $\Phi_{c,B}$ are piece-wise linear, Step 10 determines $\min_{0 \le x \le l}\{\Phi_{c,A}(x)+\Phi_{c,B}(l\text{-}x)\text{-}\lambda_c\}$ in $O(||\Phi_{c,A}||+||\Phi_{c,B}||)$ time. In Step 14, we obtain $\Phi_{b,C}$ from $\Phi_{b,A}$ and $\Phi_{b,B}$ in $O(||\Phi_{b,A}||+||\Phi_{b,B}||)$ time. From Lemma 1, we conclude that the computation for a merged cluster C in Steps 8-14 requires $O(k\alpha(k))$ time, where k=$|E(C)|$. In each layer of τ, no two clusters C contain the same edge. Thus, the computation for a layer of τ takes $O(n\alpha(n))$ time. Since there are $O(\log n)$ layers, Algorithm 1 performs in $O(n\log n\alpha(n))$ time.

3.2 The Discrete Model

Algorithm 1 can be applied to the discrete model by using a different definition of $f_{(i,j)}$. Since only vertices can be the ends of a median path, we re-define $f_{(i,j)}$ as follows:

$$(2) \qquad f_{(i,j)}(x) = \begin{cases} \infty & \text{if } x < 0, \\ \lambda_i & \text{if } 0 \leq x < d(i,j), \text{and} \\ \lambda_{(i,j)} & \text{if } x \geq d(i,j) \end{cases}$$

With the new definition, each $f_{(i,j)}$ is a step function and thus $||\Phi_{b,C}|| = O(|E(C)|)$ for any cluster C and $b \in \beta(C)$. Therefore, the computation in Algorithm 1 for a cluster C takes $O(|E(C)|)$ time. We conclude that the problem of finding an unconditional discrete median path can be solved in $O(n\log n)$ time.

4 The Conditional Discrete Median Path Problem

It is easy to see that using the definition of $f_{(i,j)}$ in (2), Algorithm 1 solves the conditional discrete median path problem as well. By a careful check, it can be found that except Step 1, all steps in the algorithm are irrelevant to S. Thus, in the conditional case, those steps also takes $O(n\log n)$ time. In the remainder of this section, we show that Step 1 can be implemented in the same time.

Step 1 is to compute λ_i, $\lambda_{(i,j)}$, and $f_{(i,j)}$ for all $i \in V$ and $(i, j) \in E$. For each $(i, j) \in E$, let $m_{i,j} = D(T^i_j, i)$, which by definition is

$$\textstyle\sum_{v \in V(T^i_j)} w(v) \times \min\{d(v, S), d(v, i)\}.$$

Using $m_{i,j}$, we can compute $\lambda_i = \Sigma_{(i,j) \in E}\{m_{i,j}\}$ for all $i \in V$ and then compute $\lambda_{(i,j)} = \lambda_i + \lambda_j - m_{i,j} - m_{j,i}$ for all $(i, j) \in E$ in $O(n)$ time. Using λ_i and $\lambda_{(i,j)}$, the functions $f_{(i,j)}$ for all $(i, j) \in E$ can be computed in $O(n)$ time according to the definition in (2). Thus, our problem becomes the computation of all $m_{i,j}$, $(i, j) \in E$

We do the computaion of $m_{i,j}$ by using the divide-and-conquer strategy. For convenience, we also describe it on a top tree τ of T. For each cluster C represented by a node of τ, we will compute $m_{i,j}(C) = D(C^i_j, i)$ for each $(i, j) \in E(C)$, and we will compute two arrays $X_{b,C}$ and $Y_{b,C}$ for each $b \in \beta(C)$, where $X_{b,C}$ stores the ordering of the vertices $v \in V(C)$ by the distances $d(v, b)$ and $Y_{b,C}$ stores the ordering of the vertices $v \in V(C)$ by $d(v, S)$-$d(v, b)$. After the computation, we have $m_{i,j}(T) = m_{i,j}$ for every $(i, j) \in E$. The computation is as follows.

Procedure Computing_m(T)
begin
01 preprocess T to obtain $d(v, S)$ for all $v \in V$
02 $\tau \leftarrow$ a top tree of T
03 **for** each leaf cluster $e = (i, j) \in E$ **do**

04 $m_{i,j}(e) \leftarrow w(j)\times\min\{d(j, S), d(j, i)\}$
05 $m_{j,i}(e) \leftarrow w(i)\times\min\{d(i, S), d(i, j)\}$
06 **for** each $b\in\beta(e)$ **do**
07 $X_{b,e} \leftarrow \begin{cases}(i,j) & \text{if } b=i\\(j,i) & \text{if } b=j\end{cases}$
08 $Y_{b,e} \leftarrow \begin{cases}(i,j) & \text{if } d(i,S)-d(i,b)\le d(j,S)-d(j,b)\\(j,i) & \text{otherwise}\end{cases}$
09 **for** each cluster C represented by an internal node of τ **do** (in a bottom-up fashion)
10 $(A, B) \leftarrow$ the two clusters represented by the children of the internal node
11 $c \leftarrow$ the intersection vertex of A and B
12 **for** each $(i, j)\in E(C)$ **do** compute $m_{i,j}(C)$ and $m_{j,i}(C)$
13 **for** each $b\in\beta(C)$ **do**
14 compute $X_{b,C}$ and $Y_{b,C}$
end

The computation of $d(v, S)$ for all $v\in V$ in Step 1 takes $O(n\log n)$ time [10]. The computation for a leaf cluster takes $O(1)$ time. Thus, Steps 3-8 takes $O(n)$ time. Consider a fixed merged cluster C. Step 12 computes $m_{i,j}(C)$ and $m_{j,i}(C)$ for all $(i, j)\in E(C)$. Due to the symmetry between A and B, we only present the computation for $(i, j)\in E(A)$. For easy discussion, we assume that C is rooted at c. For each $(i, p(i))\in E(A)$, we have $C_i^{p(i)}=A_i^{p(i)}$ and $C_{p(i)}^i=A_{p(i)}^i\cup B$. Thus, for all $(i, j)=(i, p(i))\in E(A)$, we simply compute $m_{j,i}(C)=m_{j,i}(A)$ in $O(|V(C)|)$ time. Next, we describe the computation of $m_{i,j}(C)$ for all $(i, j)=(i, p(i))\in E(A)$. Assume that $X_{c,A}=(x_1, x_2,..., x_s)$ and $Y_{c,B}=(y_1, y_2,..., y_t\}$. For any vertex $x\in V(A)$, the ordering of the vertices $y\in V(B)$ by $d(y, S)$-$d(y, x)$ is the same with the ordering by $d(y, S)$-$d(y, c)$. Thus, for any vertex x_k in $X_{c,A}$, there exists an index z_k such that $d(y_q, S)$-$d(y_q, x_k)\le 0$ for $q<z_k$ and $d(y_q, S)$-$d(y_q, x_k)>0$ for $q\ge z_k$. By applying a process similar to merge, we compute all z_k in $O(|V(C)|)$ time from $X_{c,A}$ and $Y_{c,B}$. Then, since the indices z_k are nondecreasing, it is easy to compute $m_{i,j}(C)$ for all $(i, j)=(i, p(i))\in E(A)$ in $O(|V(C)|)$ time by using the following equation:

$$m_{x_k,p(x_k)}(C) = m_{x_k,p(x_k)}(A)+\sum\nolimits_{q<z_k} w(y_q)\times d(y_q, S)+\sum\nolimits_{q\ge z_k} w(y_q)\times d(y_q, x_k).$$

Therefore, Step 12 takes $O(|V(C)|)$ time. Now, consider the computation in Step 14. By symmetry, we may assume $b\in V(A)$. Clearly, we can obtain $X_{b,C}$ by merging $X_{b,A}$ and $X_{c,B}$ in linear time. Similarly, $Y_{b,C}$ can be obtained by merging $Y_{b,A}$ and $Y_{c,B}$ in linear time. Therefore, the computation in Steps 10-14 takes $O(|V(C)|)$ time in total, from which we conclude that the whole computation on τ takes $O(n\log n)$ time. We have the following theorem.

Theorem 2: The problem of finding a conditional discrete median path of length $\le l$ in a tree can be solved in $O(n\log n)$ time.

5 The Conditional Continuous Median Path Problem

By using the following important property, Tamir *et al.* presented an $O(n^2)$ time algorithm without applying the divide-and-conquer strategy.

Lemma 2 [10]: There is a conditional continuous median path that is almost discrete.

From now on, only almost discrete paths are considered. For easy description, in the remainder of this section, we assume that T is rooted at an arbitrary vertex $r \in V$. Let $P(v, q)$ be an almost discrete path, where v is a vertex and q is a point. The point q is possibly a vertex. For convenience, we say that $P(v, q)$ is of *type 1* if q is on the path from v to r, and is of *type 2* otherwise. For easy description, a discrete path between two vertices v_1, $v_2 \in V$ is regarded as two different almost discrete paths, one extending from v_1 to v_2 and the other extending from v_2 to v_1, such that any almost discrete path is either of type 1 or of type 2. Our strartegy for determining a median path is as follows. Firstly, we compute a best path among all paths of type 1. Then, we compute a best path among all paths of type 2. Finally, we determine a median path by simply comparing the two computed paths.

5.1 The Computation of a Best Path among All Paths of Type 1

Let $R=\{P(v, r)|\ v \in V,\ d(v, r) \leq l\}$ and $Q=\{P(v, q)|\ v \in V,\ d(v, r)>l,\ q$ is the point on $P(v, r)$ with $d(v, q)=l\}$. Our problem in this section is to compute the best path in $R \cup Q$. Since $|R \cup Q|=O(n)$, it is not difficult to derive an $O(n\log n)$ time algorithm for the computation. For easy analyzing, insteads of giving a new algorithm, we do the computation by performing an algorithm for the conditional discrete median path problem as follows. Firstly, we compute all the points q with $P(v, q) \in Q$, which requires $O(n\log n)$ time [9]. Then, repeatedly, for each computed point q, letting (i, j) be the edge containing q, we introduce a new vertex v_q at q. Finally, we perfom an algorithm for the conditional discrete median path problem. Note that the above computation may produce a path not in $R \cup Q$, which is of type 2.

5.2 The Computation of a Best Path among All Paths of Type 2

In Algorithm 1, the function $\delta_{(i,p(i))}$ is used to determine the distance-sum of a path having a partial edge extending form i to $p(i)$. Conversely, the function $\delta_{(p(i),i)}$ is used to determine the distance-sum of a path having a partial edge extending form $p(i)$ to i. Now, we are only interested in paths of type 2. Therefore, we obtain an algorithm by slightly modifying Algorithm 1 as follows. Firstly, for each $(i, p(i)) \in E$, we redefine $f_{(i,\ p(i))}(x) = \infty$ for all x, and

$$f_{(p(i),i)}(x) \quad = \quad \begin{cases} \infty & \text{if } \ x < 0, \\ \delta_{(p(i),i)}(x) & \text{if } \ 0 \leq x < d(p(i), i), \text{ and} \\ \lambda_{(p(i),i)} & \text{if } \ x \geq d(p(i), i) \end{cases}$$

Next, we modify Steps 10-11 of Algorithm 1 for the computation of $Cost(C)$. Since only almost discrete paths are considered, we compute

$$Cost(C) = \min\{Cost(A),\ Cost(B),\ H_{A,B},\ H_{B,A}\},$$

where

$$H_{A,B}=\min_{v\in V(A), d(c,v)\leq l}\ \{D(P(c,\ v))+\Phi_{c,B}(l\text{-}d(c,\ v))\text{-}\lambda_c\}$$

is the distance-sum of the best path of type 2 extending from a vertex in A to a point in B, and $H_{B,A}=\min_{v\in V(B), d(c,v)\leq l}\ \{D(P(c,\ v))+\Phi_{c,A}(l\text{-}d(c,\ v))\text{-}\lambda_c\}$ is the distance-sum of the best path of type 2 extending from a vertex in B to a point in A. It is not difficult to see that after the above modification, the computation for a cluster C can be implemented in $O(|V(C)|+k'\alpha(k'))$ time, where $k'=\Sigma_{(p(i),i)\in E(C)}||\delta_{(p(i),i)}||$.

Lemma 3: The total number of linear segments of all $\delta_{(p(i),i)}$, $(p(i),\ i)\in E$, is $O(n)$.

Lemma 4: All $\delta_{(p(i),\ i)}$, $(p(i),\ i)\in E$, can be computed in $O(n\log\ n)$ time.

Due to the page limitation, the proofs of Lemmas 3 and 4 are omitted. From the lemmas, it is not difficult to obtain the following theorem.

Theorem 3: The conditional continuous medina path problem can be solved in $O(n\log\ n\alpha(n))$ time.

6 Lower Bounds

Lemma 5 [7]: Solving the element uniqueness problem, which is to decide if any two of n given positive numbers are equal, requires $\Omega(n\log\ n)$ time in the comparison model.

Theorem 4: Finding an unconditional discrete median path of length$\leq l$ in a tree requires $\Omega(n\log\ n)$ time in the comparison model, even for a tree with all vertex weights being equal to 1.

Proof: Given an instance $A=(a_1,\ a_2,\ldots,\ a_n)$ of the element uniqueness problem, we construct in linear time a tree $T=(V,\ E)$ with $V=\{r,\ x_1,\ y_1,\ z_1,\ x_2,\ y_2,\ z_2,\ldots,\ x_n,\ y_n,\ z_n\}$ and $E=\{(r,\ x_1),\ (x_1,\ y_1),\ (x_1,\ z_1),\ (r,\ x_2),\ (x_2,\ y_2),\ (x_2,\ z_2),\ldots,\ (r,\ x_n),\ (x_n,\ y_n),\ (x_n,\ z_n)\}$. All vertex weights are 1. Let $t=3\times\max\{a_1,\ a_2,\ldots,\ a_n\}$. The length of $(r,\ x_i)$ is t. The lengths of $(x_i,\ y_i)$ and $(x_i,\ z_i)$ are a_i and $t\text{-}a_i$, respectively. Finally, let $l=3t$. With some efforts, we can show that there are two equal elements in A iff a discrete median path of T has distance-sum$\leq t(4n\text{-}7)$. Therefore the theorem holds. Q.E.D.

Theorem 5: Finding a conditional continuous median path of length$\leq l$ in a tree requires $\Omega(n\log\ n)$ time in the comparison model, even for a tree with all vertex weights being equal to 1.

Proof: Given an instance $A=(a_1, a_2, \ldots, a_n)$ of the element uniqueness problem, we construct in linear time a tree $T=(V, E)$ with $V=\{r, x_1, y'_1, y_1, z'_1, z_1, x_2, y'_2, y_2, z'_2, z_2, \ldots, x_n, y'_n, y_n, z'_n, z_n\}$ and $E=\{(r, x_1), (x_1, y'_1), (y'_1, y_1), (x_1, z'_1), (z'_1, z_1), (r, x_2), (x_2, y'_2), (y'_2, y_2), (x_2, z'_2), (z'_2, z_2), \ldots, (r, x_n), (x_n, y'_n), (y'_n, y_n), (x_n, z'_n), (z'_n, z_n)\}$. All vertex weights are 1. Let $t=3\times\max\{a_1, a_2, \ldots, a_n\}$. The length of (r, x_i) is t. The lengths of (x_i, y'_i), (y'_i, y_i), (x_i, z'_i), and (z'_i, z_i) are $a_i/2$, $a_i/2$, $(t\text{-}a_i)/2$, and $(t\text{-}a_i)/2$, respectively. The existing facility S includes all x_i, y'_i, and z'_i, $i=1, 2, \ldots, n$. Let $l=3t$. With some efforts, we can show that there are two equal elements in A iff a conditional continuous median path of T has distance-sum$\leq t(n\text{-}1)/2$. Q.E.D.

References

1. S. Alstrup, J. Holm, and M. Thorup, Maintaining median and center in dynamic trees, in *Proceedings of the SWAT 2000, Lecture Notes in Computer Science*, vol. 1851, Springer-Verlag, pp. 46-56, 2000.
2. S. Alstrup, P.W. Lauridsen, P. Sommerlund, and M. Throup, Finding cores of limited length, Technical Report, The IT University of Copenhagen, a preliminary version of this paper appeared in *Proceedings of the 5th International Workshop on Algorithms and Data Structures, Lecture Notes in Computer Science*, vol. 1272, Springer-Verlag, pp. 45-54, 1997.
3. S. L. Hakimi, E. F. Schmeichel and M. Labbé, On locating path- or tree- shaped facilities on networks, *Networks*, vol. 23, pp. 543-555, 1993.
4. S. Hart and M. Sharir, Nonlinearity of Davenport-Schinzel sequences and of general path compression schemes, *Combinatorica*, vol. 6, pp. 151-177, 1986.
5. E. Minieka, Conditional centers and medians on a graph, *Networks*, vol. 10, pp.265-272, 1980.
6. E. Minieka, The optimal location of a path or tree in a tree network, *Networks*, vol. 15, pp. 309-321, 1985.
7. F. P. Preparata and M. I. Shamos, *Computational Geometry: An Introduction*, Springer-Verlag, 1985.
8. P. J. Slater, Locating central paths in a network, *Transportation Science*, vol. 16, No. 1, pp. 1-18, 1982.
9. A. Tamir, Fully polynomial approximation schemes for locating a tree-shaped facility: a generalization of the knapsack problem, *Discrete Applied Mathematics*, vol. 87, pp. 229-243, 1998.
10. A. Tamir, J. Puerto, J.A. Mesa, and A.M. Rodriguez-Chia, Conditional location of path and tree shaped facilities on trees, manuscript, 2001.
11. B.-F. Wang, Finding a two-core of a tree in linear time, *SIAM Journal on Discrete Mathematics*, accepted.
12. B.-F. Wang, Efficient parallel algorithms for optimally locating a path and a tree of a specified length in a weighted tree network, *Journal of Algorithms*, vol. 34, pp. 90-108, 2000.

New Results on the k-Truck Problem*

Weimin Ma[1,2], Yinfeng Xu[1], Jane You[2], James Liu[2], and Kanliang Wang[1]

[1] School of Management, Xi'an Jiaotong University, Shaanxi 710049, PRC
[2] Dept. of Computing, Hong Kong Polytechnic Uinversity, Hung Hom, Kowloon, Hong Kong
cswmma@comp.polyu.edu.hk

Abstract. In this paper, some results concerning the k-truck problem are produced. First, the algorithms and their complexity concerning the off-line k-truck problem are discussed. Following that, a lower bound of competitive ratio for the on-line k-truck problem is given. Based on the *Position Maintaining Strategy* (PMS), we get some new results which are slightly better than those of [1] for general cases. We also use the *Partial-Greedy Algorithm* (PG) to solve this problem on a special line. Finally, we extend the concepts of the on-line k-truck problem to obtain a new variant: *Deeper On-line k-Truck Problem* (DTP).

1 Introduction

On-line problem and their *competitive analysis* have received considerable interest for about twenty years. S. Albers and S. Leonardi [2] coined out a comprehensive survey of this domain. On-line problems had been systematically investigated only when Sleator and Tarjian [3] suggested comparing an on-line algorithm to an optimal off-line algorithm and Karlin, Manasse, Rudolph and Sleator [4] coined the term *competitive analysis*. The task system, the k-server problem, and on-line/off-line games ([5], [6] and [7]) all attempt to model on-line problems and algorithms. In this paper, we first discussed the algorithms and its complexity concerning the off-line k-truck problem. Following that, a lower bound of competitive ratio for the on-line k-truck problem is given. Especially, based on the PMS, we get some new results for the general cases. In addition, we also use the PG to solve this problem on a special line and prove that PG is a $(1+(n-k)/\theta)$ -competitive algorithm for this case. Finally, we extend the concepts of the on-line k-truck problem to obtain a new variant: DTP.

2 Preliminaries

The k-truck problem can be stated as follows. We are given a metric space M, and k trucks which move among the points of M, each occupying one point of M.

* The authors would like to acknowledge the support of Central Research Grant GV-975 of the Hong Kong Polytechnic University and Research Grant from NSF of China. No.19731001

O.H. Ibarra and L. Zhang (Eds.): COCOON 2002, LNCS 2387, pp. 504–513, 2002.

Repeatedly, a request (a pair of points $x, y \in M$) appears. To serve a request, an empty truck must first move to x and then move to y with goods from x. How to minimize the total cost of all trucks? Obviously, the k-truck problem aims at minimizing the cost of all trucks. Because the cost of trucks with goods is different from that of trucks without goods on the same distance, the total distance cannot be considered as the objective to be optimized. For simplicity, we assume that the cost of a truck with goods is θ times that of one without goods on the same distance. We can then take $(1+\theta)$ times of the empty loaded distant as the objective of optimization.

The Model. Let $G = (V, E)$ denote an edge weighted graph with n vertices and the weights of edges satisfying the triangle inequality, where V is a metric space consisting of n vertices, and E is the set of all weighted edges. We assume that the weight of edge (x, y) is denoted by $d(x, y)$ and the weights are symmetric, i.e., for all $x, y, d(x, y) = d(y, x)$. We assume that k trucks occupy a k-vertexes which is a subset of V. A service request $r = (a, b), a, b \in V$ implies that there are some goods on vertex a that must be moved to vertex b (for simplicity, we assume that the weight of the goods is same all the time). A service request sequence R consists of some service request in turn, namely $R = (r_1, ..., r_m)$, where $r_i = (a_i, b_i), a_i, b_i \in V$. All discussion is based on the following essential assumptions: (1) Graph G is connected; (2) When a new service request occurs, k trucks are all free; (3) All trucks have the same load weight and the cost of a truck with goods is θ times that of one without goods on the same distance, and $\theta \geq 1$. For a known sequence $R = (r_1, ..., r_m)$, let $C_{\text{OPT}}(R)$ be the optimal total cost after finishing it. For a new service request r_i, if scheduling algorithm A can schedule without information regarding the sequence next to r_i, we call A an on-line algorithm. For on-line algorithm A, if there are constants α and β satisfying

$$C_{\text{A}}(R) \leq \alpha \cdot C_{\text{OPT}}(R) + \beta,$$

then for any possible R, A is called a competitive algorithm, where $C_{\text{A}}(R)$ is the total cost with algorithm A to satisfy sequence R.

If there is no limit for the R and θ, the on-line truck problem is called P. In problem P, if for any $r_i = (a_i, b_i), a_i, b_i$ and $\theta > 1$ holds, the problem is called $P1$. In problem P, if there is no limit for any $r_i = (a_i, b_i)$, but if $\theta = 1$, the problem is $P2$. In $P2$, if $d(a_i, b_i) > 0$, namely $a_i = b_i$, the problem is called $P3$. In problem P, if $d(a_i, b_i) = 0$, namely $a_i = b_i$, it is called $P4$.

Lemma 1. *[9] There exists an on-line algorithm for the k-server problem with the competitive ratio* $2k - 1$.

Lemma 2. *[1] Letting OPT be an optimal algorithm for an request sequence* $R = (r_1, ..., r_m)$, *then we have* $C_{OPT}(R) \geq C_{OPT}(\sigma) + \sum_{i=1}^{m} (\theta - 1) \cdot d(a_i, b_i)$, *where* $\sigma = ((a_1, a_1), ..., (a_m, a_m))$ *and* $r_i = (a_i, b_i)$.

Lemma 3. *[1] For any algorithm A for a request sequence* $R = (r_1, ..., r_m)$, *we have* $C_A(R) \geq \sum_{i=1}^{m} \theta \cdot d(a_i, b_i)$, *and* $C_{OPT}(R) \geq \sum_{i=1}^{m} \theta \cdot d(a_i, b_i)$.

Lemma 4. *[10] There exists an on-line algorithm for the k-server problem on a real line with the competitive ratio k.*

Position Maintaining Strategy (PMS) [8]

For the present request $r_i = (a_i, b_i)$, after a_i is reached, the truck reaching a_i must move from a_i to b_i with the goods to complete r_i. When the service for r_i is finished, the PMS moves the truck at b_i back to a_i (empty) before the next request arrives.

3 Off-Line Problem

In this section, two solutions for the off-line k-truck problem are discussed.

Definition (Configuration) *On the metric space M, a possible position of k trucks is called a configuration. That is, a configuration is a special k-multiset whose elements consist of at least one and at most k points of space M. Here, the special means that in the multiset the same node can be repeated from one to k times.*

3.1 Dynamic Programming (DP) Solution

In [6], a DP solution was given for the famous k-server problem. Similarly, we can develop a DP solution for the k-truck problem.

Lemma 5. *On a given graph G with n nodes, the number of possible configurations of all k trucks is $\binom{n+k-1}{n-1}$, where $k \leq n$.*

Proof. Assume that all k trucks and all n nodes line up along a line from left to right, thus there are $n + k$ locations on which there is either a truck or a node. Following that, we move all trucks between two nodes i and j (assuming that node i is right to node j and that there are not any other nodes between them) to node i. If there are not trucks between the two nodes, the meaning of this operation is that no truck is moved on to node i. In addition, in order to move all trucks on some nodes according to the above rules, we need to let the extreme right location be a node. The final task is to choose $n - 1$ locations, on which we will arrange the remaining $n - 1$ nodes, from the $n + k - 1$ locations. Obviously, we have $\binom{n+k-1}{n-1}$ choices. □

Let function $C_{\text{OPT}}(R, S)$ denote the cost of the minimum cost algorithm that handles request sequence R and ends up in configuration S. As in paper [6], we can compute this function recursively as follows, assuming that the trucks are initially in configuration S_0

$$C_{\text{OPT}}(\varepsilon, S) = \begin{cases} 0 & \text{if } S = S_0 \\ \text{undefine} & \text{otherwise} \end{cases}$$

$$C_{\text{OPT}}(Rr_i, S) = \begin{cases} \min_T (C_{\text{OPT}}(R, T) + d(T, \theta \cdot (a_i, b_i), S)) & \text{if } S = S_0 \\ \text{undefine} & \text{otherwise} \end{cases}$$

where $d(T, \theta \cdot (a_i, b_i), S)$ is the cost of transition from configuration T to configuration S and the last operation of transition is $a_i \to b_i$ (satisfying the request r_i at cost $\theta \cdot (a_i, b_i)$), T and S denote the configurations at time $i - 1$ and time i, respectively, and ε denotes the empty request sequence.

Theorem 1. *The above optimal off-line algorithm for the k-truck problem can give an optimal solution with time proportional to* $m \cdot \binom{n+k-1}{n-1}^2$, *where m is the length of the request sequence (the number of requests).*

Proof. Let $|R| = m$, we can develop a table-building method according to the above discussion. Build a table with $|R| + 1$ rows, each of which implies a subsequence of request sequence R, and $\binom{n+k-1}{n-1}$ columns each of which denote a possible configuration of trucks. Namely, the entry in row i and column j is $C_{\mathrm{OPT}}(R_i, S_j)$, where R_i is the subsequence of R of length i. Each row of the table can be built from the previous one within time $\binom{n+k-1}{n-1}^2$. Furthermore, only $|R| = m$ rows need these computations. The proof is completed. □

3.2 Minimum Cost Maximum Flow (MCMF) Solution

In [11], MCMF was used to resolve the off-line k-server problem. Our objective is to find an optimal strategy to serve a sequence of m requests with k trucks, if the request sequence is given in advance. Assume that the k-trucks initially occupy one point, the origin. And denote the i-th request by the binary-tuple (a_i, b_i). If there are m requests, the inputs to our problem are the superdiagonal entries of an $(m+1) \times (m+1)$ matrix, whose $(0, j)$ entry is the sum of cost from the original to the location of j-request start a_j (empty) and then to the request destination b_j (with the goods), $j = 1, 2, ..., m$, and whose (i, j) entry is the sum of cost from the location of i-request destination to the location of j-request start and then to the relevant destination with goods, $j, 1 \le i < j \le m$.

Theorem 2. *There is an* $O(km^2)$*-time off-line algorithm to find an optimal schedule for k trucks to serve a sequence of m requests (whether or not the triangle inequality holds).*

Proof. We can resolve the off-line the k-truck problem (with or without triangle inequality) by reducing it to the problem of finding a minimum cost flow of maximum value in an acyclic network. Suppose that there are k trucks $t_1, ..., t_k$ and m requests $r_1, ..., r_m$, where $r_i = (a_i, b_i)$, and $i = 1, ..., m$, we can build the following $(2 + k + 3m)$-node acyclic network: the vertex set is $V = \{s, s_1, ..., s_k, a_1, b_1, b'_1, ..., a_m, b_m, b'_m, t\}$. In that vertex set, nodes s and t are the source and sink, respectively. Each arc of our network has a capacity one. There is an arc of cost 0 from s to each s_i, an arc of cost 0 form each b'_i to t, as well as an arc to t from each s_i, of cost 0. ¿From each s_i, there is an arc to a_j of cost equal to the distance from the origin to the location of a_j. ¿From each a_j, there is only an arc to b_j of cost equal to $\theta \cdot d(a_i, b_i)$. For $i < j$, there is an arc from b'_i to a_j of cost equal to the distance between b_i to a_j. Moreover, form b_i to b'_i there is an arc of cost $-K$, where K is an extremely large positive real. The constructing of the network is completed.

It is easy to know that the value of the maximum flow in this network is k. Using minimum-cost augmentation [12], we can find an integral min-cost flow of value k in time $O(km^2)$, because all capacities are integral and the network is acyclic. An integral $s \to t$ flow of value k can be decomposed into k arc-disjoint $s \to t$ paths, the ith one passing through s_i. Obviously, this flow saturates all of the (b_i, b'_i) arcs, and hence corresponds to an optimal schedule for serving the requests, the ith server serving exactly those requests contained in the $s \to t$ path that passes through s_i, because $-K$ is so small. □

4 A Lower Bound

In this section we will give a lower bound of competitive ratio for the k-truck problem on a symmetric metric space. In other words, any general on-line algorithm for this problem, either a deterministic or a randomized algorithm, must have a competitive factor of at least $(\theta + 1) \cdot k/(\theta \cdot k + 2)$. In fact, we have actually proven a slightly more general lower bound on the competitive ratio. Suppose we wish to compare an on-line algorithm with k servers to an off-line one with $h \leq k$ servers. Naturally, the factor decreases when the on-line algorithm gets more servers than the off-line algorithm. We get the lower bound as $(\theta+1) \cdot k/((\theta+2) \cdot k - 2h + 2)$. A similar approach was taken in [6], where the lower bound and matching upper bound are given for the traditional k-server problem.

Theorem 3. *Let A be an on-line algorithm for the symmetric k-truck problem on a graph G with at least k nodes. Then, for any $1 \leq h \leq k$, there exist request sequences $R_1, R_2, \ldots$ such that: (1) For all i, R_i is an initial subsequence of R_{i+1}, and $C_A(R_i) < C_A(R_{i+1})$; (2) There exists an h-truck algorithm B (which may start with its trucks anywhere) such that for all i, $C_A(R_i) > (\theta+1) \cdot k \cdot C_B(R_i)/((\theta+2) \cdot k - 2h + 2)$.*

Proof. Without loss of generality, assume A is an on-line algorithm and that the k trucks start out at different nodes. Let H be a subgraph of G of size $k+2$, induced by the k initial positions of A's trucks and two other vertexes. Define R, A's nemesis sequence on H, such that $R(i)$ and $R(i-1)$ are the two unique vertexes in H not covered by A and a request $r_i = d(R(i), R(i-1))$ occurs at time i, for all $i \geq 1$. Then

$$C_A(R_t) = \sum_{i=1}^{t} (d(R(i+1), R(i)) + \theta \cdot d(R(i), R(i-1))) =$$

$$(1+\theta) \cdot \sum_{i=1}^{t-1} d(R(i+1), R(i)) + d(R(i+1), R(i)) + \theta \cdot d(R(1), R(0)),$$

because at each step R requests the node just vacated by A.

Let S be any h-element subset of H containing $R(1)$ but not $R(0)$. We can define an off-line h -truck algorithm $A(S)$ as follows: the trucks finally occupy the vertices in set S. To process a request $r_i = d(R(i), R(i-1))$, the following rule is applied: If Scontains $R(i)$, move the truck at node $R(i)$ to $R(i-1)$ with goods to satisfy the request, and update the S to reflect this change. Otherwise move the truck at node $R(i-2)$ to $R(i)$ without goods and then to $R(i-1)$ with goods, also to satisfy the request, and update S to reflect this change.

It is easy to see that for all $i > 1$, the set S contains $R(i-2)$ and does not contain $R(i-1)$ when step i begins. The following observation is the key to the rest of the proof: if we run the above algorithm starting with distinct equal-sized sets S and T, then S and T never become equal, for the reason described in the following paragraph.

Suppose that S and T differ before $R(i)$ is processed. We shall show that the versions of S and T created by processing $R(i)$, as described above, also differ. If both S and T contain $R(i)$, they both move the truck on node $R(i)$ to node $R(i-1)$, on which there is exactly not any truck. The other nodes have no changes, so S and T are still different and both S and T contain $R(i-1)$. If exactly one of S or T contains $R(i)$, then after the request exactly one of them contains $R(i-1)$, so they still differ. If neither of them contains $R(i)$, then both change by dropping $R(i-2)$ and adding $R(i-1)$, so the symmetric difference of S and T remains the same (non-empty).

Let us consider simultaneously running an ensemble of algorithms $A(S)$, starting from each h-element subset S of H containing $R(1)$ but not $R(0)$. There are $\binom{k}{h-1}$ such sets. Since no two sets ever become equal, the number of sets remains constant. After processing $R(i)$, the collection of subsets consists of all the h element subsets of H which contain $R(i-1)$.

By our choice of starting configuration, step 1 just costs $\theta \cdot d(R(1), R(0))$. At step i (for $i \geq 2$), each of these algorithms either moves the truck at node $R(i)$ to $R(i-1)$ (if S contains $R(i)$), at cost $\theta \cdot d(R(i), R(i-1))$, or moves the truck at node $R(i-2)$ to $R(i)$ and then to $R(i-1)$ (if S does not contain $R(i)$), at cost $d(R(i-2), R(i)) + \theta \cdot d(R(i), R(i-1))$. Of the $\binom{k}{h-1}$ algorithms being run, $\binom{k-1}{h-1}$ of them (the ones which contain $R(i-2)$ but not contain either $R(i)$) incur the cost of $d(R(i-2), R(i)) + \theta \cdot d(R(i), R(i-1))$. The remaining $\binom{k-1}{h-2}$ of algorithms incur the cost of $\theta \cdot d(R(i), R(i-1))$. Thus, for step i, the total cost incurred by all of the algorithms is

$$\binom{k}{h-1} \cdot \theta \cdot d(R(i), R(i-1)) + \binom{k-1}{h-1} \cdot d(R(i-2), R(i)).$$

The total cost of running all of these algorithms up to and including R(t) is

$$\sum_{i=1}^{t} \binom{k}{h-1} \cdot \theta \cdot d(R(i), R(i-1)) + \sum_{i=2}^{t} \binom{k-1}{h-1} \cdot d(R(i-2), R(i)).$$

Thus the expected cost of one of these algorithms chosen at random is

$$C_{\text{EXP}}(R_t) = \theta \cdot \sum_{i=1}^{t} d(R(i), R(i-1)) + \frac{\binom{k-1}{h-1}}{\binom{k}{h-1}} \cdot \sum_{i=2}^{t} d(R(i-2), R(i)) \leq \frac{(\theta+2)k-2h+2}{k} \cdot$$
$$\sum_{i=1}^{t-1} d(R(i), R(i+1)) + \frac{(\theta+1)k-h+1}{k} \cdot d(R(1), R(0)) - \frac{k-h+1}{k} \cdot d(R(t-1), R(t))$$

This inequality holds for the triangle inequality and expending of the binomial coefficients. Recall that the cost to A for the same steps was

$$C_{\text{A}}(R_t) = (1+\theta) \cdot \sum_{i=1}^{t-1} d(R(i+1), R(i)) + d(R(i+1), R(i)) + \theta \cdot d(R(1), R(0)),$$

Because the distances are symmetric, the two summations of the $C_{\text{EXP}}(R_t)$ and $C_{\text{A}}(R_t)$ are identical, except that both of the costs include some extra terms, which are bounded as a constant. Therefore, after some mathematical manipulation (e.g., let $t \to \infty$), we obtain

$$\frac{C_A(R_t)}{C_{EPT}(R_t)} \geq \frac{(\theta+1)\cdot k}{(\theta+2)\cdot k-2h+2}.$$

Finally, there must be some initial set whose performance is often no worse than the average of the costs. Let S be this set, and $A(S)$ be the algorithm starting from this set. Let R_i be an initial subsequence of R, for which $A(S)$ does no worse than average. □

Corollary 1. *For any symmetric k-truck problem, there is no c-competitive algorithm for* $c < (\theta+1)\cdot k/(\theta \cdot k+2)$.

Corollary 2. *For any symmetric k-taxi problem, there is no c-competitive algorithm for* $c < 2k/(k+2)$.

5 Competitve Ratios

5.1 Position Maintaining Strategy Solution

In [1], with the PMS, the case under which $\theta > (c+1)/(c-1)$ was studied, and a c-competitive algorithm was found to exist for the k-truck problem. In fact, we can get a somewhat better result for general cases.

Theorem 4. *For the on-line k-truck problem and a given graph G, if there is a c-competitive on-line algorithm for the k-server problem on G, then: (1) If* $\theta > (c+1)/(c-1)$, *then PMS is a c-competitivealgorithm; (2) If* $1 \leq \theta \leq (c+1)/(c-1)$, *then PMS is a* $(c/\theta + 1/\theta + 1)$*-competitive algorithm.*

Proof. For any $R = (r_1, ..., r_m)$, where $r_i = (a_i, b_i)$, considering the k-server problem's request sequence $\sigma = (a_1, ..., a_m)$, let A_σ be a c-competitive algorithm for the on-line k-server problem on graph G to satisfy the sequence. We design algorithm A as follows. For current service request $r_i = (a_i, b_i)$, first schedule a truck to a_i using algorithm A_σ, then complete the r_i with PMS. Thus total cost of A is

$$C_A(R) = \sum_{i=1}^{m} C_A(r_i) = \sum_{i=1}^{m} [C_A(a_i) + (\theta+1)\cdot d(a_i, b_i)] =$$

$$C_{A_\sigma}(\sigma) + (1 + 1/\theta)\cdot \sum_{i=1}^{m} \theta \cdot d(a_i, b_i)$$

where θ is defined above and $\theta \geq 1$. From lemma 2 and algorithm A_σ, we have

$$C_{A_\sigma}(\sigma) \leq c \cdot C_{OPT}(\sigma) + \beta \leq c \cdot [C_{OPT}(R) - \sum_{i=1}^{m} (\theta+1)\cdot d(a_i, b_i)] + \beta$$

Then we get

$$C_A(R) \leq c \cdot C_{OPT}(R) + [1 + 1/\theta - c\cdot(\theta-1)/\theta]\cdot \sum_{i=1}^{m} \theta \cdot d(a_i, b_i)] + \beta$$

If $\theta > (c+1)/(c-1)$, we get $C_A(R) \leq c \cdot C_{OPT}(R) + \beta$; if $1 \leq \theta \leq (c+1)/(c-1)$, and with lemma 3, $C_{OPT}(R) \geq \sum_{i=1}^{m} \theta \cdot d(a_i, b_i)$,we have $C_A(R) \leq (c/\theta + 1/\theta + 1)\cdot C_{OPT}(R) + \beta$, where c and β are some constants. □

Combining Theorem 4 and Lemma 1, the following corollary holds.

Corollary 3. *For the on-line k-truck problem on a given graph G, if $\theta > (c+1)/(c-1)$, holds, then there exists a $(2k-1)$-competitive algorithm; if $1 \leq \theta \leq (c+1)/(c-1)$, then there exists a $(2k/\theta+1)$-competitive algorithm.*

5.2 Comparison of Two Algorithms

In [1], an algorithm B, here we called it the PG, is given for the problem $P1$. The competitive ratio of algorithm B is $1+\lambda/\theta$, where $\lambda = d_{\max}/d_{\min}$, $d_{\max} = \max d(v_i, v_j)$, and $d_{\min} = \min d(v_i, v_j), i \neq j, v_i, v_j \in V$. We denote the PMS algorithm of subsection 5.1 by algorithm A. We may be confronted with the problem of choosing one algorithm from A' and B in different contexts. Respectively the competitive ratios of algorithms A and B are $c_A = \begin{cases} 2k-1 & \text{if } \theta > (c+1)/(c-1) \\ 2k/\theta+1 & \text{if } 1 \leq \theta \leq (c+1)/(c-1) \end{cases}$ and $c_B = 1+\lambda/\theta$. Letting $c_A = c_B$, we can get a k that makes the algorithm A and B equal as follows

$$k = \begin{cases} 1+\lambda/(2\theta) & \text{if } \theta > (c+1)/(c-1) \\ \lambda/2 & \text{if } 1 \leq \theta \leq (c+1)/(c-1) \end{cases}$$

Theorem 5. *For on-line k-truck problem P1, denoting the PMS and PG algorithms by A and B, respectively, at the aspect of the competitive ratio: if $\theta > (c+1)/(c-1)$ holds, if $k \leq 1+\lambda/(2\theta)$ then A is better than B, and contrarily if $k > 1+\lambda/(2\theta)$ then B is better than A; if $1 \leq \theta \leq (c+1)/(c-1)$ holds, if $k \leq \lambda/2$ then A is better than B, and contrarily if $k > \lambda/2$ then B is better than A.*

5.3 Partial-Greedy Algorithm on a Special Line

Let $G = (V, E)$ for the instance of an on-line k-truck problem consisting of a line of n vertices with $n-1$ edges whose lengths are equal to one. More formally, we have that $V = \{v_i | i = 1, ..., n\}$ and $E = \{v_i v_{i+1} | i = 1, ..., n-1\}$. All edge-weights are equal to one. It is natural to assume that no vertex has more than one truck (otherwise, we can get at this situation at most cost of $k \cdot (k+1)/2$). In addition, we assume that $n \geq k+2$ holds (otherwise the fourth case of the following algorithm does not exist).

Partial-Greedy Algorithm. *For the current request $r_i = (a_i, b_i)$ from the request sequence $R = (r_1, ..., r_m)$, schedule the k-truck problem P1 on the above special line with the following rules:*

(1) If there is a truck at a_i and also one at b_i, then PG moves the truck at a_i to b_i complete the request, and at the same time PG moves the truck at b_i to a_i with an empty load. The cost of PG for the r_i is $(1+\theta) \cdot d(a_i, b_i)$ and at present no vertex has more than one truck.

(2) If there is a truck at a_i and no truck at b_i, then PG moves the truck at a_i to b_i to complete the request. The cost of PG for the r_i is $\theta \cdot d(a_i, b_i)$, and at present no vertex has more than one truck.

(3) *If there is no truck at a_i and there is a truck at b_i, then PG moves the truck at b_i to a_i first without a load, and after that moves it from a_i to b_i to complete the request. The cost of PG for the r_i is $(1+\theta)\cdot d(a_i,b_i)$ and at present no vertex has more than one truck.*

(4) *If there is no truck at a_i and b_i, then PG moves the truck which is the closest to a_i (suppose that the truck is located at c_i) with an empty load and then moves to b_i to complete the request. The cost of PG for the r_i is $d(c_i,a_i)+\theta\cdot d(a_i,b_i)$, and again no vertex has more than one truck.*

Theorem 6. *PG is a $(1+(n-k)/\theta)$-competitive algorithm for the k-truck problem P1 on the above special line.*

Proof. For cases (1), (2) and (3), the cost of it PG is at most $(1+\theta)$ times the optimal cost for any request. For case (4), the extra cost is $d(c_i,a_i)$. Since c_i is the closest occupied vertex to a_i, we have $d(c_i,a_i)\le(n-k)\cdot d(a_i,b_i)$. Let $C_{\text{PG}}(R)$ denote the cost of algorithm *PG* for request sequence $R=(r_1,...,r_m)$, then we have

$$C_{\text{PG}}(R)=\sum_{i=1}^{m}\{max[d(b_i,a_i),d(c_i,a_i)]+\theta\cdot d(a_i,b_i)\}+\beta\le\sum_{i=1}^{m}\{(n-k)\cdot d(a_i,b_i)+\theta\cdot d(a_i,b_i)\}+\beta=(1+(n-k)/\theta)\cdot\sum_{i=1}^{m}\theta\cdot d(a_i,b_i)+\beta\le(1+(n-k)/\theta)\cdot C_{\text{OPT}}(R)$$

where β is the cost for preconditioning the truck such that each vertex has at most one truck and it is bounded by a constant related with G. The last inequality holds for the lemma 3. □

Similar to subsection 5.2, combining the lemma 4 and the above theorem 6, we have the following theorem.

Theorem 7. *For on-line k-truck problem P1 on the special line, denoting the PMS and PG algorithms by A and B, respectively, at the aspect of the competitive ratio: if $\theta>(c+1)/(c-1)$ holds, if $k\le(n+\theta)/(\theta+1)$ then A is better than B, and contrarily if $k>(n+\theta)/(\theta+1)$ then B is better than A; if $1\le\theta\le(c+1)/(c-1)$ holds, if $k\le(n-1)/2$ then A is better than B, and contrarily if $k>(n-1)/2$ then B is better than A.*

6 Deeper On-Line *k*-Truck Problem

We call the on-line *k*-truck problem studied in previous sections, the *Standard On-line k-truck problem* (STP). Here we will discuss another variant of it, the *Deeper On-line k-truck problem* (DTP). We formulate DTP as follows:
Given a metric space M, and k trucks which move among the points of M, each occupying one point of M, repeatedly, a request (a pair of points $x,y\in M$) appears. However, only the node x of request occurring is known when the information of the request is received, and the destination node y will not be known

until a truck has already been on the node of request occurring. To serve a request, an empty truck must first move to x and then move to y with goods from x. How to minimize the total cost of all trucks?

We easily know that the results of the competitive ratio of the PMS still hold for the DTP but these of the PG algorithm do not hold for the DTP.

Theorem 8. *For the DTP on a given graph G, if there is a c-competitive on-line algorithm for the k-server problem on G, then: (1) If $\theta > (c+1)/(c-1)$, then PMS is a c-competitive algorithm; (2) If $1 \leq \theta \leq (c+1)/(c-1)$, then PMS is a $(c/\theta + 1/\theta + 1)$-competitive algorithm.*

7 Concluding Remarks

Most of the results of this paper can be extended to the relevant cases of the k-taxi problem [8]. Although we get a lower bound of competitive ratio for the k-truck problem, the optimal lower bound of the competitive ratio for it is still open. Furthermore, whether there are some better on-line algorithms than PMS or PG needs further investigation.

References

1. W.M.Ma, Y.F.Xu, and K.L.Wang, On-line k-truck problem and its competitive algorithm. Journal of Global Optimization 21 (1): 15-25, September 2001.
2. S. Albers and S. Leonardi. Online algorithms. ACM Computing Surveys Vol.31. Issue 3 Sept. 1999.
3. D.D.Sleator, R.E.Tarjan, Amortized efficiency of list update and paging rules, Communication of the ACM, 28 (1985) 202-208.
4. A.Karlin, M.Manasse, L.Rudlph and D.D.Sleator. Competitive snoopy caching, Algorithmica, 3:79-119,1988.
5. M.S.Manasse, L.A.McGeoch, and D.D.Sleator, Competitive algorithms for on-line problems. In Proc. 20th Annual ACM Symp. on Theory of Computing, 322-33, 1988.
6. M.S.Manasse, L.A.McGeoch, and D.D.Sleator, Competitive algorithms for server problems, Journal of Algorithms,1990(11),208-230.
7. S.Ben-David, S.Borodin, R.M.Karp, G.Tardos, and A. Wigderson. On the power if randomization in on-line algorithms. In Proc. 22nd Annual ACM Symp. on Theory of Computing, 379-386, 1990.
8. Y.F.Xu, K.L.Wang, and B. Zhu, On the k-taxi problem, Information, Vol.2, No.4, 1999.
9. E.Koutsoupias, C.Papadimitriou, On the k-server conjecture, STOC.,507-511,1994.
10. M.Chrobak, L.Larmore, An optimal algorithm for the server problem on trees, SIAM Journal of Computing 20 (1991)144-148.
11. M.Chrobak, H.Karloff, T.Payne, S.Vishwanathan, New results on the server problem, SIAM Journal on Discrete Mathematics 4 (1991) 172-181.
12. R.Tarjan, Data Structures and Network Algorithms, SIAM, Philadelphia, 1983, 109-111.

Theory of Equal-Flows in Networks

K. Srinathan*, Pranava R. Goundan, M.V.N. Ashwin Kumar,
R. Nandakumar**, and C. Pandu Rangan

Department of Computer Science and Engineering, Indian Institute of Technology,
Madras, Chennai - 600036, India
{ksrinath,prg,mvnak}@cs.iitm.ernet.in,rangan@iitm.ernet.in

Abstract. The MAXIMUM-FLOW problem is a classical problem in combinatorial optimization and has many practical applications. We introduce a new variant of this well known Maximum-Flow problem, viz., the MAXIMUM-EQUAL-FLOW problem,wherein, for each vertex (other than the source) in the network, the actual flows along the arcs emanating from that vertex are constrained to be equal and integral. Surprisingly, unlike the Maximum-Flow problem that is known to admit a polynomial time solution, we prove that the Maximum-Equal-Flow problem is NP-Hard. Nevertheless, we provide an approximation algorithm for the Maximum-Equal-Flow problem. We develop a new (analogous) theory for Equal-Flows in networks and also illustrate the Maximum-Equal-Flow equivalents of the fundamental results in flow theory.

1 Introduction

Consider the problem of a company manufacturing a product P using certain resources $R_1, R_2, \ldots, R_n$. Each resource is in turn produced by other companies using other resources. Hence, for every product, it is possible to obtain a directed graph (network), wherein each node represents a product and a dircted arc (i, j) represents that the product i requires j as a resource in it's manufacture. Hence, in each such network, the *source* will be the final product P to be made and nature will be th e *sink*, providing all the required basic raw materials. But there may also be supply constraints on the availibility of resources for each of the products, which can be captured as capacity constraints on the flow in each of the corresponding arcs of the network. Such a graph is called a "Bill of Demand" in operations management terminology. The problem of finding the maximum number of units of the product P that a company can produce subject to these supply constraints, given the input requirements for each of the products, can be modeled mathematically as a problem similar to that of the MAXIMUM-FLOW in the "Bill of Demand" network, wherein each node distributes it's integral inflow among it's out-neighbours as per a fixed pre-specified ratio. Evidently, this problem cannot be solved using the algorithms for MAXIMUM-FLOW as they do not guarantee that the out-flows will be distributed as per the ratio. Interestingly,

* Financial support from Infosys Technologies Limited, India, is acknowledged.
** Currently at University of Chicago. E-mail: nanda@cs.uchicago.edu

O.H. Ibarra and L. Zhang (Eds.): COCOON 2002, LNCS 2387, pp. 514–524, 2002.

the above ratio-bound flow problem can be abstracted to the MAXI MUM-EQUAL-FLOW problem[1] in an analogous network. This could be done by applying the "conservation of units" principle to each node. Equivalently, we can use different measurement units to capture the required ratio for each of the out-arcs. Hence, the solution to the MAXIMUM-EQUAL-FLOW problem in the resulting network with capacity constraints in the modified units would give the optimal number of units of the final product P the company can make, subject to the supply constraints of other companies.

Informally, the EQUAL-FLOW problem in a network is a "uniform" view of the classical flows problem in a network, in the sense that in the former, each node distributes its in-flow equally and in integral quantities to all its neighbours unlike the latter, wherein each node is "random" and could distribute its in-flow in any manner that it wished. Hence, in situations like in the above problem of product maximization, where the maintenance of "equanimity" among the neighbours and the integrality constraint is of a higher priority than just a maximization of the overall flow, the "maximum-flow" sought for is in fact the maximum-equal-flow. In this work, we systematically study the problem of EQUAL-FLOWS in networks. We show that the MAXIMUM-EQUAL-FLOW problem is NP-hard, unlike the MAXIMUM-FLOW problem. We give the combinatorial and linear programming structure of this problem and design approximate algorithms for it. Furthermore, we illustrate the maximum-equal-flow equivalents of the fundamental results in flow theory viz., the *decomposition* theorem, the *augmenting path* theorem, and the *max-flow min-cut* theorem.

2 The MAXIMUM-EQUAL-FLOW Problem

We begin by introducing the well-known MAX-FLOW problem. Consider a network $N = (V, A, s, t, c)$ where (V, A) is a digraph with two specified nodes s (the *source*) and t (the *sink*). For each arc (i, j) the capacity $c(i, j)$, a positive integer, is given. A *flow* in N is an assignment of nonnegative integer value $f(i, j) \leq c(i, j)$ to each arc (i, j) such that for each node other than s and t, the sum of the flows of the incoming arcs is equal to the sum of flows along the outgoing arcs. The value of flow f is the sum of flows in the arcs leaving s. The problem of determining the maximum possible flow for s to t, subject to these constraints, is well known as the MAX-FLOW problem. A polynomial time algorithm can be obtained by building upon a fundamental fact about networks, known as the *max-flow min-cut theorem* [2,1]. Now consider the following variation of the MAX-FLOW problem. The flow function is integral and such that for all $i \in V$ and $(i, j_1), (i, j_2) \in A$, $j_1 \neq j_2$, we have $f(i, j_1) = f(i, j_2)$. In other words, the flow emanating out of a vertex is same in all the outgoing arcs (from that vertex, other than the source) and of an integral amount. Surprisingly, this problem is NP-hard (see Section 4) even though a polynomial time algorithm exists for the MAX-FLOW problem.

[1] By EQUAL-FLOW, we mean that each node distributes its in-flow equally to all its neighbours, without violating the integrality constraint of the actual flows.

Definition 1 (Maximum-Equal-Flow).
Instance: *A network* $N = (V, A, s, t, c)$*, positive integer* K*.*
Question: *Does there exist an equal-flow of value at least* K *in the network* N*?*

The algorithm for the MAX-FLOW problem would not work here. To see why the MAX-FLOW algorithm would not work in our case, let us have a brief look at the algorithm in [4]. If we want to find out if a given flow f is optimal, we have to check if a flow f' of value greater then f exists. If such a flow f' exists, then $\Delta f = f' - f$ is itself a positive flow. But Δf may have arcs with negative flows. This can be viewed as a positive flow in the reverse direction. This is equivalent to saying that Δf is a flow in a derived network $N(f) = (V, A', s, t, c')$, where $A' = \{A - \{(i,j) : f(i,j) = c(i,j)\}\} \cup \{(i,j) : (j,i) \in A, f(j,i) > 0\}$. So, finding if f is optimum is same as deciding if $N(f)$ has a positive flow, very similar to determining if a path exists between s and t. Hence, this can be done in polynomial time. Thus, starting with a network N of zero flow, we can repeatedly augment the flow. When no positive flow exists, we have arrived at the optimal. This algorithm seems to work even for the MAX-EQ-FLOW. But finding just a positive flow Δf is not enough; in fact we need a *positive equal path*. That is, the positive path should be such that *all* its "branches" have positive paths as well. Hence in the worst case, an analogous algorithm may end up potentially examining a super-polynomial number of paths. An even more grave problem is the fact that the non-existence of a positive flow path in the derived network implies that the current flow is the maximum (see Theorem 10). An analogous sufficiency condition may not always hold for positive equal flows (see Theorem 11). In the sequel (see Section 4), we show that the MAX-EQ-FLOW problem is in fact NP-hard. Nevertheless, in line with the efforts of [2], we develop a theory for equal-flows in networks that helps design approximation algorithms for the MAX-EQ-FLOW problem.

3 Equal Flow Theory

In this section, we state and prove the equal-flow equivalents of the classic flow decomposition, augmenting path and the max-flow min-cut theorems.

Observation 1 *For any (arbitrary) given network* $N = (V, A, s, t, c)$*, there exists a network* $N' = (V, A, s, t, c')$ *(i.e, a renaming of the capacities of the arcs in* N*) where* $c'(x) = c'(y)$ *for any two arcs* x *and* y *emanating from the same vertex (that is,* $\exists \alpha, \beta \in V$ *such that* $x = (v, \alpha), y = (v, \beta)$*), such that every maximum equal-flow through* N *is a maximum equal-flow through* N'*.*

Definition 2. *A network* $N = (V, A, s, t, c)$ *is said to be an* EQ-NETWORK *if the capacity function* $c : A \to Z_+$*, is such that for each vertex* $v \in V$*, all the arcs emanating from* v *have equal capacities.*

Observation 2 *Any* EQ-NETWORK $N = (V, A, s, t, c)$ *can be more succinctly represented as* $N = (V, A, s, t, c')$ *where* $c' : V \to Z_+$*, that is, each vertex can be* *assigned a capacity instead of all its out-going arcs assigned the* same *capacity. Hereafter (due to Observation 1), we will use* Networks *to mean* Eq-Networks.

3.1 Feasibility of Equal-Flows

This section involves the determination of answer to the following very basic question: *Does there exist any feasible (non-zero) equal-flow in the given network?* We initially assume the edge capacities to be infinity and answer the question and later study the effects of bounding the edge flows.

Observation 3 *Given a network N, there exists an ordering of the vertices such that $(v_i, v_j) \in A \Rightarrow i < j$ iff N contains no cycles.*

Observation 4 *For any network N without cycles, with infinite edge capacities, there exists a non-zero equal-flow iff there exists a non-zero (classical) flow.*

PROOF: Obviously if an equal-flow exists, then a classical flow exists. We provide an incremental construction of an equal-flow if a classical flow exists. From Observation 3, we know that the vertices can be ordered such that no back edges occur. Since classical flow exists, there exists a positive path from the source to the sink. Consider one such positive path, say $\mathcal{P}$. Now construct an equal-fractional-flow as follows. Assign a flow of 1 through the first edge of the path $\mathcal{P}$ and 0 to all the other edges from the source. We give below the transition from i^{th} vertex in the ordering to the $(i+1)^{th}$. Since there are no back edges from the $(i+1)^{th}$ vertex v_{i+1}, if v_{i+1} has a non-zero in-flow, then divide it equally among all its successors and proceed to the next vertex in the ordering. By this procedure, we have a valid equal-fractional-flow, which can be converted to an equal-flow by multiplying the individual flows by a suitable integer. □

Definition 3 (Splits). *We define a split $P \subseteq A$ of a network $N = (V, A, s, t, c)$ to be the set of arcs of the form $(V_1, V \backslash V_1)$ or $(V \backslash V_1, V_1)$ for some $V_1 \subseteq V$.*

Definition 4 (Forbidden Splits). *A split $P \subseteq A$ with arcs of the form $(X, V \backslash X)$ or $(V \backslash X, X)$, $X \subseteq V$, is said to be p-forbidden if the following properties hold, for any path p from s to t in N: (1) X or $V \backslash X$ contains both s and t. (2) All the arcs of P are similarly oriented from the set containing both s and t to its complement set of vertices. (3) For the path p from s to t, there exists a vertex v in the other set (complement to the one containing both s and t), such that there is a path q from s to v, with the first arc of p and q being the same.*

Theorem 5. *For an arbitrary network N with infinite edge capacities, there exists a non-zero equal-flow only if there exists a path p from s to t in N such that N contains no p-forbidden splits.*

PROOF: Let there exist a non-zero equal-flow in N. On the contrary, assume that every path p from s to t in N contains a p-forbidden split P such that every arc in P is of the form $(Y, V\backslash Y)$ or $(V\backslash Y, Y)$. Without loss of generality, we can assume that both s and t are in Y. Since P is a forbidden split, by definition there exists a path q from s to $v \in V\backslash Y$ such that first arc of p and q is the same. Let u be the last common vertex between p and q. Since there is a positive all-equal-flow through u, a non-zero flow enters $V\backslash Y$ but cannot reach the sink (condition 2 of Definition 4), contradicting the flow conservation property. □
The existence of an equal-flow through networks with bounded capacities requires the satisfaction of a few more conditions (in addition to the one mentioned in Theorem 5) that will be elaborated in the sequel.

Definition 5. *Given a network $N = (V, A, s, t, c)$, define the set $\mathcal{S} \subseteq A$ to be the set of all arcs that are emanating from the source, that is $\mathcal{S} = \{a \mid a \in A, a = (s, v)$ for some $v \in V\}$. We call the set $\mathcal{S}$ as the set of* SOURCE ARCS *in N.*

Definition 6. *The* CROSSING-INDEX *and the* THICKNESS-INDEX *of an arc $a \in A$ with respect to a source arc $a_s \in \mathcal{S}$ in a network $N = (V, A, s, t, c)$ is defined to be the values of the numerator and the denominator respectively, of the fractional flow (expressed as an irreducible fraction) through the arc a generated (see Subsection 3.2) by an unit flow across a_s, all the arc capacities being infinity. We denote the crossing-index (and the thickness-index) of an arc $a \in A$ with respect to a source arc $a_s \in \mathcal{S}$ by $\mathcal{C}(a, a_s)$ (and $\mathcal{T}(a, a_s)$ respectively).*

Observation 6 [FEASIBILITY CONDITION] *For an arbitrary network N with infinite edge capacities, there exists a non-zero all-equal-flow iff there exist a source arc a_s and a terminating arc*[2] *a_t such that $\mathcal{C}(a_t, a_s) > 0$.*

3.2 Feasible Values of Equal-Flows

In this section, we seek to answer to the following: (1)*Does there exist a feasible equal-flow in the given network of a stated magnitude (less than the value of the maximum-equal-flow)?* (2) *What are the characteristics of feasible equal-flows?*

Definition 7. *Basic-Equal-Fractional flow for a given network (if it exists) is the flow generated by assigning a flow of* 1 *to exactly one of the source arcs.*

Observation 7 *Any network contains at most $|\mathcal{S}|$ Basic-Equal-Fractional flows.*

We now give an algorithm to generate a Basic-Equal-Fractional flow w.r.t a source arc a_s, if it exists. Let $V = \{s, v_1, v_2, \ldots, v_n, t\}$ be the vertices of the network. Without loss of generality, we can assume that $a_s = (s, v_1)$. Assign variables f_i to denote the fractional flow in each of the outgoing arcs from the

[2] Terminating arcs are those that are in-coming arcs to the sink in the network.

vertex v_i, $1 \leq i \leq n$. Note that f_1 is a known linear function[3] of the variables $f_1, f_2, \ldots, f_n$ and hence by applying the flow conservation equation at the node v_1, we can eliminate the variable f_1. Similarly at node v_i we can eliminate f_i. Hence if a flow exists, all the variables will be eliminated at v_n, leading to the values of the fractional flows through each of the arcs. Note that the fractional flow (expressed as an irreducible fraction $\frac{n_a}{d_a}$) through an arc a is such that $n_a = \mathcal{C}(a, a_s)$ and $d_a = \mathcal{T}(a, a_s)$. We denote by $\mathcal{B}(a_s)$ the Basic-Equal-Fractional Flow generated by $a_s \in \mathcal{S}$.

Definition 8. *An Atomic-Equal-Flow is an integer valued equal-flow that cannot be expressed as a sum of smaller integer valued equal-flows.*

Lemma 1. *Any Atomic-Equal-Flow is the sum of an integral number of Basic-Equal-Fractional Flows.* □

Definition 9. *The* THICKNESS *of a source arc a_s, $\mathcal{J}(a_s)$, is the least integer such that the flow $\mathcal{J}(a_s) \cdot \mathcal{B}(a_s)$ is integral. Furthermore, the flow $\mathcal{J}(a_s) \cdot \mathcal{B}(a_s)$ is an Atomic-Equal-Flow. And, $\mathcal{J}(a_s) = LCM(\mathcal{T}(a, a_s)), \forall a \in A, \mathcal{T}(a, a_s) \neq 0$.*

3.3 Decomposition

Since any Equal-flow through N is (trivially) a flow through N, the classical decomposition theorem holds for Equal-flows too. In the sequel, we present a more tailor-made decomposition theorem for Equal-flows in networks.

Theorem 8 (Equal-Flow Decomposition). *Any integral equal-flow in a network $N = (V, A, s, t, c)$ can be expressed as a sum of atomic-equal-flows.* □

Theorem 9. *Given m irreducible fractions $\frac{n_1}{d_1}, \frac{n_2}{d_2}, \ldots, \frac{n_m}{d_m}$, and an expression $E = \sum_{i=1}^{m} \lambda_i \frac{n_i}{d_i}$, ($\lambda_i$'s are integers), E is an integer only if λ_i is an integral multiple of* $\text{LCM}\left[d_i, \left(\frac{\prod_{i=1}^{m} d_i}{d_i}\right)\right] \Big/ \left(\frac{\prod_{i=1}^{m} d_i}{d_i}\right)$.

PROOF: Note that $E = \sum_{i=1}^{m} \lambda_i \frac{n_i}{d_i} = \sum_{j=1}^{m} \left(n_j \lambda_j \frac{\prod_{i=1}^{m} d_i}{d_j}\right) / \prod_{i=1}^{m} d_i$. We require E to be an integer. We denote $\prod_{i=1}^{m} d_i$ by M. Let $\frac{M}{d_i} \cdot n_i \cdot \lambda_i \equiv b_i \ (mod\ M)$. Now, $n_i \cdot \lambda_i$ is an integer $\Longrightarrow \frac{M}{d_i} \Big| b_i$, $\left(b_i \text{ is an integral multiple of } \frac{M}{d_i}\right)$. E is an integer implies that the sum of b_i's ("remainders") is an integral multiple of M. Let us denote $r_i = \sum_{i=1}^{m} b_i - b_i$. $\Longrightarrow M|(r_i + b_i) \Rightarrow d_i|(r_i + b_i)$. It can be seen that $d_i|r_i$ (since each b_j, $j \neq i$ is a multiple of d_i) $\Longrightarrow d_i|b_i$. Since b_i is a multiple of d_i and $\frac{M}{d_i}$, it is also a multiple of $LCM\left(d_i, \frac{M}{d_i}\right)$. It follows that λ_i is a multiple of $LCM\left(d_i, \frac{M}{d_i}\right) \Big/ \frac{M}{d_i}$. Hence the theorem. □

[3] $\delta_{out}(v_i) \cdot f_i = \sum_{(j,i) \in A} f_j$ where δ_{out} denotes the outdegree.

Corollary 1. *Let $\mathcal{S} = \{a_{s_1}, a_{s_2}, \ldots, a_{s_t}\}$ be the set of source arcs in the network. The actual equal-flow through any source arc a_{s_i} is an integral multiple of*

$$\mathcal{G}(a_{s_i}) = \underset{\forall a \in A}{LCM} \left\{ LCM \left(\mathcal{T}(a, a_{s_j}), \frac{\prod_{j=1}^{t} \mathcal{T}(a, a_{s_j})}{\mathcal{T}(a, a_{s_j})} \right) \Big/ \frac{\prod_{j=1}^{t} \mathcal{T}(a, a_{s_j})}{\mathcal{T}(a, a_{s_j})} \right\} \square$$

3.4 Augmenting

Theorem 10 (Classical). *A flow is maximum iff there is no augmenting path.*

Since the non-existence of an augmenting path is both necessary and sufficient for the flow to be maximum, we have an algorithm to find the maximum flow through networks. Unfortunately, the corresponding theorem for equal-flows holds for maximal flows and not maximum flows (for which only the sufficiency holds). Hence an analogous algorithm would not always work.

Definition 10 (Augmenting Atomic-Equal-Flow). *An atomic-equal-flow $\mathcal{A}$ is said to be augmenting for a given equal-flow $\mathcal{F}$ if the equal-flow $\mathcal{A} + \mathcal{F}$ does not exceed the capacity in any of the arcs in the network.*

Observation 11 (Equal-Flow Augmenting) *Any equal-flow F is maximum only if there is no augmenting atomic-equal-flow.*

3.5 Max-Equal-Flow Min-Equal-Cut

In this section, we provide a theorem similar to the max-flow min-cut theorem for classical flows through networks. The results of this section, though not as elegant as their classical flow counterparts, help us design approximate algorithms for the maximum-equal-flow problem which is shown to be NP-hard in the sequel (Section 4). From Lemma 1 and Observation 8, we know that every arc $a \in A$ in N with the set of source arcs $\{a_{s_1}, \ldots, a_{s_t}\}$, is associated with a $t - tuple$, $\frac{n_1}{d_1}, \frac{n_2}{d_2}, \ldots, \frac{n_t}{d_t}$, of fractions[4] such that if $\lambda_1, \ldots, \lambda_t$ are the flows in each of the t source arcs respectively, then the flow through the arc a is $\sum_{i=1}^{t} \lambda_i \cdot \frac{n_i}{d_i}$. We use this fact to define the capacity of an equal-cut in the network.

Definition 11 (Equal-Cut). *An equal-cut is a split C where the arcs in C are of the form $V_1, V \backslash V_1$ or $V \backslash V_1, V_1$ and $s \in V_1$ and $t \in V \backslash V_1$.*

Definition 12 (Capability of an Equal-Cut). *The flow L_a through an arc $a \in A$ is a linear combination of the flows (λ_i's) through the source arcs. The flow is negative if the arc is into V_1, else it is positive. The flow through the equal-cut is the sum of the flows through all the arcs in the equal-cut, which is also a linear combination L_c of the λ_i's. The capability of an equal-cut is defined as the maximum value that L_c can take subject to each L_a being less than its capacity $c(a)$. Surprisingly, we will prove that any equal-cut has the same capability.*

[4] In fact, $n_i = \mathcal{C}(a, a_{s_i})$ and $d_i = \mathcal{T}(a, a_{s_i})$.

Theorem 12. *The Maximum-equal-flow through the network is equal to the capability of any equal-cut.*

PROOF: The max-equal-flow is less than or equal to any equal-cut. Since, otherwise the actual flow of at least one arc is greater than its capacity. Furthermore, the max-equal-flow is greater than or equal to the capability of any equal-cut, because the flow through any equal-cut defines a valid equal-flow in the sense that it does not violate the capacity constraints of any arc in the network. □

Corollary 2. *All equal-cuts have the same capability and this is equal to the maximum-equal-flow through the network.*

4 MAX-EQ-FLOW is NP-Hard

We give a polynomial reduction of the known NP-complete problem, viz. EXACT COVER BY 3-SETS [3] to that of the MAX-EQ-FLOW problem.

Definition 13. EXACT COVER BY 3-SETS (X3C)
Instance : *Set X with $|X| = 3q$ and a collection C of 3-element subsets of X.*
Question : *Does C contain an exact cover for X, i.e., a subcollection $C' \subseteq C$ such that every element of X occurs in exactly one member of C'?*

Theorem 13. MAX-EQ-FLOW *is NP-Hard.*

PROOF : We construct an instance of the MAX-EQ-FLOW problem from the given instance X (such that $|X| = 3q$) and C of the *Exact Cover by 3-Sets* problem as follows. Consider the network $N = (V, A, s, t, c)$, where $V = L \cup R \cup \{s, t\}$, where $|L| = |C|$, $|R| = |X|$ and there exist *bijective mappings* $h_l : L \to C$ and $h_r : R \to X$. The set of arcs is defined as $A = \{(\ell, r) \mid \ell \in L, r \in R, h_r(r) \in h_\ell(\ell)\} \cup \{(s, \ell) \mid \forall \ell \in L\} \cup \{(r, t) \mid \forall r \in R\}$. The capacity function $c : A \to Z$, where Z is the set of positive integers, is defined as $c(a) = \begin{cases} 3 \text{ if } a = (s, \ell) \in A, \text{ for some } \ell \in L. \\ 1 \text{ otherwise.} \end{cases}$.
Now the instance of the MAX-EQ-FLOW problem is: $N = (V, A, s, t, c)$, $|X|$, that is: *Is there an equal-flow of value $\geq |X|$ in N?*
<u>*Claim*</u>: If there exists an equal-flow of size $|X|$ in N, then there exists a $C' \subseteq C$ such that every element of X occurs in exactly one member of C'.
<u>*Proof*</u>: From the construction of the network N, since each vertex in R has an out-flow capacity of at most 1 (to the sink t), it is evident that exactly $3q$ vertices in R should be involved with non-zero actual equal-flows to produce an overall equal-flow of $3q$. Let $r_1, r_2, \ldots, r_{3q}$ be $3q$ vertices in R such that each of the arcs (r_i, t), $1 \leq i \leq 3q$ has a flow of 1 through it. Moreover, since the out-flow from each of the above $3q$ vertices is 1, it is essential that their in-flow was 1 too. Therefore, for each vertex r_i, $1 \leq i \leq 3q$, we have exactly one vertex $\ell_i \in L$ such that the actual flow in the arc (ℓ_i, r_i) is equal to 1. Also, there are exactly[5] q such vertices $\ell_1, \ell_2, \ldots, \ell_q$. We will now show that the sets $h_\ell(\ell_1), h_\ell(\ell_2), \ldots, h_\ell(\ell_q)$ are

[5] Due to the equal-flow constraint at ℓ_i $\forall i$

mutually disjoint sets in C and hence a solution to the given X3C problem. On the contrary, assume that there exists an element $r \in R$, such that $h_r(r) \in h_\ell(\ell_i)$ and $h_r(r) \in h_\ell(\ell_j)$, for some i, j, $1 \leq i, j \leq 3q$, $i \neq j$. This means that A contains the two arcs a_i and a_j, namely, $a_i = (\ell_i, r) \in A$ and $a_j = (\ell_j, r) \in A$. Since the actual flow is constrained to be an equal-flow of value $3q$, and it is already known that there exist arcs (ℓ_i, r_i) and (ℓ_j, r_j) with actual flows of 1 in them, it is necessary that the actual flows in the arcs a_i and a_j are respectively equal to 1 too (otherwise, it violates the equal-flow property at the nodes ℓ_i and ℓ_j respectively). But then the actual in-flow into the vertex r now stands at least 2 which contradicts the fact that the actual in-flow of any node $r \in R$ is ≤ 1.
Claim: If there exists no equal-flow of size $3q$ in N, then there exists no $C' \subseteq C$ such that every element of X occurs in exactly one member of C'.
Proof: On the contrary, let there exist q mutually disjoint sets, $c_1, c_2, \ldots, c_q$ in C. Then, in the network $N = (V, A, s, t, c)$ constructed as illustrated earlier, there exist q vertices $S = \{\ell_1, \ell_2, \ldots, \ell_q\}$, $S \subseteq L$ such that $h_\ell(\ell_i) = c_i$, $1 \leq i \leq q$. Let the actual flows $f : A \to Z$, where Z is the set of positive integers, to be:

$$f(a) = \begin{cases} 3 \text{ if } a = (s, \ell_i) \in A, \ell_i \in S, 1 \leq i \leq q. \\ 1 \text{ if } a = (\ell_i, r), \ell_i \in S \text{ for some } r \in R. \\ 1 \text{ if } a = (r, t), r \in R, \text{ such that } f((\ell, r)) = 1 \text{ for some } \ell \in S. \\ 0 \text{ otherwise.} \end{cases}$$

Since for any $r \in R$, there can exist at most one $\ell \in S$ such that $f((\ell, r)) = 1$ (otherwise $c_1, c_2, \ldots, c_q$ cannot be mutually disjoint sets), it is clear that the above assignment gives an actual equal-flow of value $3q$. □

5 Approximating the MAX-EQ-FLOW

We know that every flow can be expressed as the sum of atomic-equal-flows which themselves can be expressed as the sum of basic-equal-fractional-flows. In other words, any equal-flow through the network can be generated with the knowledge of the flows in the source arcs. We provide an elegant methodology to arrive at the flows in the source arcs that almost maximize the flow through the network, using certain combinatorial observations and linear programming.

5.1 Integer Linear Programming Formulation

From Theorem 1, it is clear that the flow through a source arc a_{s_i} can be expressed as a integral multiple of $\mathcal{G}(a_{s_i})$. So, let the flow through a_{s_i} be $x_i \cdot k_i$. Moreover, the flow through each arc $a_j \in A$ is expressible as a linear combination of the x_i's (see Subsection 3.5), say, $\sum_{a_{s_i} \in \mathcal{S}} k_{ji}.x_i$. Therefore, the maximum equal-flow problem is to: *Maximize* $\sum_{a_{s_i} \in \mathcal{S}} k_i x_i$ Subject to $\sum_{a_{s_i} \in \mathcal{S}} k_{ji}.x_i \leq c(a_j) \forall a_j \in A$ where the x_i's are integers and the k_i's and k_{ji}'s are known pre-computable constants.

5.2 Algorithm

Since in the worst case, solving the above ILP is computationally hard, we relax the condition of the x_i's being integers, and solve the resulting LP. We proceed to construct an all-integer based solution set by studying a sufficient condition on the actual flows through the source arcs in N to be integers.

Theorem 14. *It is sufficient to use*

$$k_i = \mathcal{K}(a_{s_i}) = \underset{\forall a \in A}{LCM} \left\{ \frac{LCM\left(\mathcal{T}(a, a_{s_j}), \frac{\prod_{j=1}^{t} \mathcal{T}(a,a_{s_j})}{\mathcal{T}(a,a_{s_j})}\right)}{\frac{\prod_{j=1}^{t} \mathcal{T}(a,a_{s_j})}{\mathcal{T}(a,a_{s_j})}} \times (1 + \Psi(a,i)) \right\}$$

in the integer linear programming formulation above, so that the solution to the ILP gives the maximum equal-flow in the corresponding network, where $\Psi(a,i)$ is as defined in the proof below.

PROOF: From Theorem 9 we know that given m irreducible fractions $\frac{n_1}{d_1}, \ldots, \frac{n_m}{d_m}$, and an expression $E = \sum_{i=1}^{m} \lambda_i \frac{n_i}{d_i}$, ($\lambda_i$'s are integers), E is an integer only if λ_i is an integral multiple of $g_i = LCM\left[d_i, \left(\frac{\prod_{i=1}^{m} d_i}{d_i}\right)\right] / \left(\frac{\prod_{i=1}^{m} d_i}{d_i}\right)$. But the above condition is not sufficient for E to be an integer. We give a procedure to find a sufficient condition for E to be an integer. First compute the g_i's for all $1 \le i \le m$. Get a modified set of fractions, viz. $\frac{n_1 g_1}{d_1}, \frac{n_2 g_2}{d_2}, \ldots, \frac{n_m g_m}{d_m}$ and express them as irreducible fractions[6], say, $\frac{n_1}{r_1}, \frac{n_2}{r_2}, \ldots, \frac{n_m}{r_m}$. Note that in general, not all the r_i's need to be 1 (in which case E is an integer). Let $I^{(i)} = \{i_1, i_2, \ldots, i_q\}$ be the set of all the indices such that $r_{i_1} = r_{i_2} = \cdots = r_{i_q} = r_i$. Hence, it is necessary that $\mathcal{N}_i = \sum_{j=1}^{q} n_{i_j}$ be a multiple of r_i. Let $\mathcal{N}_i = wr_i + s$ for some non-negative integers $w, s < r_i$. Let g be the least positive integer such that g can be expressed as a linear combination of the n_{i_j}'s with non-negative integer coefficients $\alpha_{i_1}, \ldots, \alpha_{i_q}$ such that $s + g$ is a multiple of r_i. Therefore, setting $\lambda_i = g_i(1 + \alpha_i)$ guarantees that E is an integer. In the case of the max-equal-flow in networks, define $\Psi(a,i) = \alpha_i$ when using for each arc in A, $n_i = \mathcal{C}(a, a_{s_i})$ and $d_i = \mathcal{T}(a, a_{s_i})$. The theorem follows. □

Note that the value of m^{lp} (the maximum equal-fractional-flow) is unaltered by the changing of the k_i's. Using Theorem 14 results in the solution set for the flow through each source arcs to be $x_i^{suf} = \left\lfloor \frac{x_i^{lp}}{\mathcal{K}(a_{s_i})} \right\rfloor \mathcal{K}(a_{s_i})$. The total flow in this case is $m^{suf} = \left(\sum_{i=1}^{t} \left\lfloor \frac{x_i^{lp}}{\mathcal{K}(a_{s_i})} \right\rfloor \mathcal{K}(a_{s_i})\right) \le m^{lp}$. From Theorem 14, it is clear that the feasible solution set of x_i^{suf}s (certainly) results in an integer equal-flow.

Theorem 15. *Given n positive integers $\lambda_1, \ldots, \lambda_n$ and n real numbers $r_1, \ldots, r_n$ such that* $(\min_{i=1}^{n} r_i) > 0$ *and* $(\max_{i=1}^{n} r_i) \ge 1$, $\frac{\sum_{i=1}^{n} \lambda_i \lfloor r_i \rfloor}{\sum_{i=1}^{n} \lambda_i r_i} > \frac{\lambda_{min}}{(n+1)\lambda_{max}}$, *where* $\lambda_{max} = \max_{i=1}^{n} \lambda_i$ *and* $\lambda_{min} = \min_{i=1}^{n} \lambda_i$.

[6] Note that the numerator remains the same.

PROOF: Left Hand Side =

$$\frac{\sum_{i=1}^{n}\lambda_i\lfloor r_i\rfloor}{\sum_{i=1}^{n}\lambda_i r_i} \geq \frac{\sum_{i=1}^{n}\lambda_{min}\lfloor r_i\rfloor}{\sum_{i=1}^{n}\lambda_{max} r_i} \geq \frac{\lambda_{min}}{\lambda_{max}}\frac{\sum_{i=1}^{n}\lfloor r_i\rfloor}{\sum_{i=1}^{n} r_i} \geq \frac{\lambda_{min}}{\lambda_{max}}\left(1 - \frac{\sum_{i=1}^{n} frac(r_i)}{\sum_{i=1}^{n} r_i}\right)$$

where $frac(r_i) = r_i - \lfloor r_i \rfloor$. Now, let $\sum_{i=1}^{n} frac(r_i) = q$. This, along with the fact that $(\max_{i=1}^{n} r_i) \geq 1$ would imply that $\sum_{i=1}^{n} r_i \geq q+1$. Therefore, $\frac{\lambda_{min}}{\lambda_{max}}\left(1 - \frac{\sum_{i=1}^{n} frac(r_i)}{\sum_{i=1}^{n} r_i}\right) \geq \frac{\lambda_{min}}{\lambda_{max}}\left(1 - \frac{q}{q+1}\right) \geq \frac{\lambda_{min}}{(q+1)\lambda_{max}}$. Since $q < n$, the theorem follows. ■

Corollary 3. *The above algorithm outputs a flow m^{suf} such that $\frac{m^*}{m^{suf}} < (t+1)\frac{\max_{i=1}^{t}\mathcal{K}(a_{s_i})}{\min_{i=1}^{t}\mathcal{K}(a_{s_i})}$, where m^* is the maximum-equal-flow, since $m^{lp} \geq m^*$.*

6 Conclusion

The problem of EQUAL-FLOW in a network is of immense theoretical and practical importance. In the absence of integrality constraints, this problem can be solved efficiently. However, there are practical problems (like the "Bill of Demand" problem discussed earlier) where the integrality constraint cannot be dispensed with. Surprisingly, integral MAXIMUM-EQUAL-FLOW, unlike integral MAXIMUM-FLOW turns out to be NP-Hard. Nevertheless, we provide an approximation algorithm, whose quality is proportional to the degree of the source node, to solve the problem.

References

1. P. Elias, A. Feinstein, and C. E. Shannon. Note on maximum flow through a network. *IRE Trans. Information Theory, IT-2*, 1956.
2. L. R. Ford and D. R. Fulkerson. *Flows in Networks.* Princeton University Press, Princeton, N. J., 1962.
3. M. R. Garey and D. S. Johnson. *Computers and Intractability: A Guide to the Theory of NP-Completeness.* W. H. Freeman and Company, 1979.
4. C. H. Papadimitriou. *Computational Complexity.* Addison-Wesley Publiction Company, 1994.

Minimum Back-Walk-Free Latency Problem
(Extended Abstract)

Yaw-Ling Lin*

Department of Comput. Sci. and Info Management, Providence University,
200 Chung Chi Road, Shalu, Taichung County, Taiwan 433, R.O.C.
yllin@pu.edu.tw

Abstract. Consider a graph with n nodes $V = \{1, 2, \ldots, n\}$ and let $d(i, j)$ denote the distance between nodes i, j. Given a permutation π on $\{1, 2, \ldots, n\}$ such that $\pi(1) = 1$, the back-walk-free latency from node 1 to node j is defined by $\ell_\pi(j) = \ell_\pi(j-1) + \min\{d(\pi(k), \pi(j)) \mid 1 \le k \le j-1\}$. Note that $\ell_\pi(1) = d(1,1) = 0$. Each vertex i is associated with a nonnegative weight $w(i)$. The (weighted) *minimum back-walk-free latency problem* (MBLP) is to find a permutation π such that the total back-walk-free latency $\sum_{i=1}^{n} w(i)\ell_\pi(i)$ is minimized.
In this paper, we show an $O(n \log n)$ time algorithm when the given graph is a tree. For a k-path trees, we derive an $O(n \log k)$ time algorithm; the algorithm is shown to be optimal in term of time complexity on any comparison based computational model. Further, we show that the optimal tour on weighted paths can be found in $O(n)$ time.
No previous hardness results were known for MBLP on general graphs. Here we settle the problem by showing that MBLP is NP-complete even when the given graph is a direct acyclic graph whose vertex weights are either 0 or 1.

1 Introduction

The *minimum latency problem* (MLP) [3], also known as the traveling repairman problem [2], or the deliveryman problem [10], is to find a minimum latency tour on a graph. Usually, the starting node of the tour is given and the goal is to minimize the sum of the arrival times at the other nodes. The arrival time is the distance traversed before reaching that node. Consider a distance matrix of a graph with n nodes, where $d(i, j)$ indicates the distance between nodes i, j. The minimum latency tour is a permutation π of vertices such that $\pi(1) = 1$, and the term

$$\sum_{i=2}^{n} \sum_{j=1}^{i-1} d(\pi(j), \pi(j+1))$$

is minimized. Note that $\ell_\pi(i) = \sum_{j=1}^{i-1} d(\pi(j), \pi(j+1)) = \ell_\pi(i-1) + d(\pi(i-1), \pi(i))$ is the traveling latency from node 1 to node i. In other words, the

* The work is supported in part by the National Science Council, Taiwan, R.O.C, grant NSC-89-2218-E-126-004.

O.H. Ibarra and L. Zhang (Eds.): COCOON 2002, LNCS 2387, pp. 525–534, 2002.

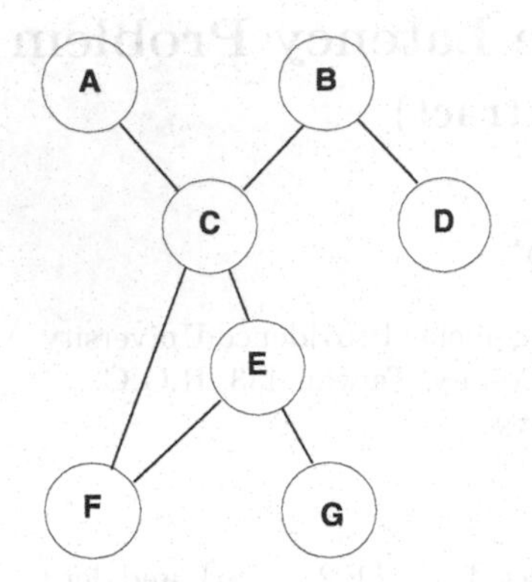

1. A is the starting vertex. Weights of vertices:

	A	B	C	D	E	F	G
$w(v)$	30	0	45	50	42	65	101

2. Distances of charted edges are 1's.

3. Distances of uncharted edges are 100's.

Optimal tour: <A,C,E,G,F,B,D> with latency:
0*30+1*45+2*42+3*101+4*65+5*0+6*50
= 992

Fig. 1. The minimum back-walk-free latency tour of a graph.

minimum latency tour is the permutation π that minimizes the total latency $\sum_{i=2}^{n} \ell_\pi(i)$. The problem arises in a number of applications in the operations research literatures, on-line search problems, and disk-head scheduling [2,10,12].

Despite its apparent similarity with the traveling salesperson problem (TSP), the MLP possesses a variety of aspects very different from the TSP. Small changes in the structure of a metric space can cause highly non-local changes in the structure of MLP. In fact, the problem is conjectured by D. West [8] to be NP-hard for edge-weighted trees, and even caterpillars, i.e., paths with edges sticking out. By contrast, the TSP can be optimally solved on a tree. However, both problems are known to be NP-hard when the points lie in a general metric space or even the (Euclidean) plane. Polynomial time algorithms for the optimal solutions are only known for the cases when the graph is a path [2,13], an edge-unweighted tree [10], a tree with diameter 3 [3], or a tree with a constant number of leaves [8].

The minimum latency problem can be generalized to the vertex-weighted case. The goal is to minimize the sum of latencies of each vertex multiplied by its weight. In other words, when each node i is given a real weight $w(i)$, the goal is to find a permutation π such that the term $\sum_{i=2}^{n} \ell_\pi(i) w(\pi(i))$ is minimized. To find an optimal tour that minimizes the vertex-weighted latency on a path, an $O(n^2)$ time algorithm can be easily deduced from the results in [2,3]. Wu [13] shows that finding the best starting vertex so as the minimum vertex-weighted latencies is minimized can also be done in $O(n^2)$ time.

In this paper, we discuss a variant of the vertex-weighted MLP problem that ignores the back-walking costs. Given a starting node 1 and a permutation π, the *back-walk-free travelling latency* from node 1 to node j is now defined by $\ell_\pi(j) = \ell_\pi(j-1) + \min\{d(\pi(k), \pi(j)) \mid 1 \le k \le j-1\}$. Given the distance matrix of a graph with n nodes, where $d(i,j)$ indicates the distance between nodes i, j, the (weighted) *minimum back-walk-free latency problem* (MBLP) is to find a permutation π that minimizes the total back-walk-free latency $\sum_{i=1}^{n} w(i)\ell_\pi(i)$. Figure 1 gives an instance of MBLP and its minimum latency tour.

Several variations of this problem focused on index and data allocation in a single broadcast channel using conventional techniques [6,7]. The issue of multiple broadcast channels using prune strategies and heuristic answers from single broadcast channel is addressed in [11,9].

When the given graph is a tree, Adolphson and Hu [1] gave an $O(n \log n)$ time algorithm; however, the technique used is difficult to be used for special cases of trees, where more efficient algorithms are possible. Thus, in this paper, we propose another $O(n \log n)$ time algorithm for the tree case. And, by exploiting the techniques we used on the algorithm, we derive an $O(n \log k)$ time algorithm for k-path trees. Further, we show that the optimal tour on weighted paths can be found in $O(n)$ time.

No previous hardness results were known for MBLP on general graphs. Here we settle the problem by showing that MBLP is NP-complete even when the given graph is a direct acyclic graph whose vertex weights are either 0 or 1, while each dag edge is associated with an unit distance, and each non-dag edge with a large distance.

2 Basic Notations and Properties

In this paper, we consider a rooted tree $T = (V, E)$ such that the vertex set $V = \{1, 2, \ldots, n\}$ and the root is node 1. Associate each vertex i with a nonnegative weight $w(i)$ and a nonnegative distance $d(i)$. Intuitively, $d(i, j) = d(j)$ if i is the parent node of j; otherwise, $d(i, j) = \infty$.

A permutation of nodes on subset of V is called a *subtour* of T. That is, a sequence $A = \langle a_1, a_2, \ldots, a_k \rangle$ is a subtour of T if $\{a_1, a_2, \ldots, a_k\} \subset V$ and $a_i \neq a_j$ whenever $i \neq j$. Denote $w(A) = \sum_{i=1}^{k} w(a_i)$; $d(A) = \sum_{i=1}^{k} d(a_i)$. The *cost ratio* of A is defined by $\rho(A) = w(A)/d(A)$.

Definition 1. *The* weighted latency *of a subtour* $A = \langle a_1, a_2, \ldots, a_k \rangle$ *is defined by* $L(A) = \sum_{i=1}^{k} (w(a_i) \cdot \sum_{j=1}^{i} d(a_j))$.

A subtour spanning every vertex of V, which defines a permutation on V, is called a *tour* on T. A tour π is *feasible* if the corresponding (weighted) latency $\ell(\pi) = \sum_{i=1}^{n} w(i)\ell_\pi(i) < \infty$. By assigning zero to $d(1)$, we have the following observation.

Proposition 1. *Let* $d(1) = 0$. *Given a feasible tour* A *with the corresponding permutation* π, *it follows that* $\ell(\pi) = \sum_{i=1}^{n} w(i)\ell_\pi(i) = L(A)$.

Let $A = \langle a_1, a_2, \ldots, a_i \rangle$ and $B = \langle b_1, b_2, \ldots, b_j \rangle$ be two disjoint subtour of T. The *concatenation* of A and B, denoted by $AB = \langle a_1, a_2, \ldots, a_i, b_1, b_2, \ldots, b_j \rangle$ is also a subtour of T.

Lemma 1 (Recursion formula of $L(\cdot)$). *Let* A, B *be two disjoint subtours of* T. *We have:* $L(AB) = L(A) + L(B) + d(A)w(B) = L(A) + L(B) + d(A)d(B)\rho(B)$.

Proof. The proof appears in the complete paper. □

We immediately have the following observation:

Proposition 2. *Let A, B be two disjoint subtours of T. It follows that $L(AB) \leq L(BA)$ if and only if $\rho(A) \geq \rho(B)$.*

Intuitively, it suggests us to place the heaviest vertex at the front of visited tour as early as possible.

Definition 2. *We say a subtour A of T is* atomic *if there exists a minimum (back-walk-free) latency tour P on T containing A contiguously. That is, $P = BAC$, such that B, C are two (possibly empty) subtours.*

Note that each vertex i defines an atomic subtour $\langle i \rangle$ with cost ratio $\rho(i) = w(i)/d(i)$.

3 Algorithm for General Trees

In this section we discuss properties of MBLP on general trees. Using these properties, an $O(n \log n)$ time algorithm for tree can be naturally deduced.

The idea here is to repeatedly identify the atomic parent-child pairs in the tree T, and *glue* them together until the final minimum latency tour is found. Initially, each node is, by definition, an atomic subtour of T; however, during the process, the original tree node is generalized to a super node, which contains atomic subtour of the T. Note that the topologically transformed tree structure reflects several interesting properties of T. For example, the subtour resides within each super node induces a (connected) subtree of T.

Consider two atomic subtours of T, namely A and B, and let T_A (T_B) denote the subtree of T induced by A (B). If there exists a node a of A and a node b of B such that a is a parent of b, then A is thus called a *parent* of B; conversely, B is a *child* of A.

First we show when to glue two atomic subtours into one:

Lemma 2. *Let A be an atomic subtour of T. The descendants of A consists of k mutually disjoint atomic subtours of T, namely, $A_1, A_2, \ldots, A_k$. Let $\rho(A_c) = \max\{\rho(A_j) \mid 1 \leq i \leq k\}$. If A_c is a child of A, and $\rho(A_c) \geq \rho(A)$, then AA_c is an atomic subtour of T.*

Proof. The proof appears in the complete paper. □

Corollary 1. *If A is an atomic subtour of T with largest ratio cost comparing to every other atomic subtours of T. Either it can be glued with its parent; otherwise, A can be appended to the output queue if A is a root.*

Theorem 1. *The minimum back-walk-free latency tour can be found in $O(n \log n)$ time for vertex-weighted trees.*

Proof. We propose an $O(n \log n)$ time algorithm, MBLP-TREE(T), shown in Figure 2. The algorithm essentially keeps clustering two atomic subtours into one. Each time the number of vertices is reduced by one. The correctness of the algorithm is easily verified by Lemma 2 and Corollary 1.

MBLP-TREE(T)
Input: A rooted tree T with weighted vertices $V = \{1, 2, \ldots, n\}$; node 1 is root.
Each vertex i has a nonnegative weight $w(i)$ and a nonnegative distance $d(i)$.
Output: A tour with minimum weighted back-walk-free latency.
1 **for** $i \leftarrow 2$ **to** n **do** ▷ Initializes the Fibonacci Queue Q
2 **new** v; $w[v] \leftarrow w(i); d[v] \leftarrow d(i); key[v] = w[v]/d[v]$; $list[v] \leftarrow \langle i \rangle$
3 ENQUEUE($Q, key[v]$) ; MAKE-SET(v)
4 **for** each $uv \in E(T)$ **do** $p[v] \leftarrow u$ ▷ u is a parent of v
5 **while** $Q \neq \emptyset$ **do**
6 $key[v] \leftarrow$ EXTRACT-MAX(Q); ▷ Pick the maximal cost ratio.
7 $u \leftarrow p[$FIND-SET$(v)]; w[u] \leftarrow w[u] + w[v]; d[u] \leftarrow d[u] + d[v]$
8 UNION$(u, v); p[v] \leftarrow p[u]$ ▷ Efficient clustering
9 $key[u] = w[u]/d[u]; list[u] = list[u] \circ list[v]$; INCREASE-KEY($Q, key[u]$)
10 **return** the output list of the root contains the minimum latency tour.

Fig. 2. Algorithm for finding minimum back-walk-free latency tour in tree.

Note that Step 1 to Step 4 of the algorithm takes $O(n)$ time to build a Fibonacci heap Q. Note that Q is a maximum-finding queue instead of the regular minimum-finding queue. Step 5 to Step 9 of MBLP-TREE(T) is to pick the maximum cost ratio node and glue it to its parent. The while loop executes exactly $n - 2$ times. The EXTRACT-MAX operation of Step 6 takes amortized $O(\log n)$ time. Thus totally $O(n \log n)$ time will be spent on Step 6. On the other hand, the INCREASE-KEY operations performed in Step 9 takes amortized $O(1)$ time; thus totally $O(n)$ time is spent on Step 9.

What is interesting in the algorithm is that we use the disjoint set data structure [4] in clustering the nodes contained within each subtour. A naive approach will need to locate all the pointers of the children of v, and let their parent pointer changed to the new parent u in Step 8 of the algorithm. Totally there are potentially $O(n^2)$ changes. Now with the help of the disjoint set data structure, Step 7 and Step 8 of the algorithm totally spends $n\alpha(n)$ time, where $\alpha(n)$ is the inverse of Ackermann's function. We thus conclude that the algorithm totally spends $O(n \log n)$ time. □

4 Algorithm for k-Path Trees

In this section, we discuss properties of MBLP on the k-path tree and propose an $O(n \log k)$ time algorithm. If k is a constant, the problem is linear time solvable.

A tree T is called a *k-path tree* if there is one vertex of T with degree k and all other vertices have degrees either 1 or 2. Note that a simple path is a 2-path. A *rooted k-path*, is the k-path rooted at the degree k vertex.

Let $P = \langle a_1, a_2 \ldots, a_m \rangle$ be a path of T, where vertices are listed from root to the leaf. It follows that any feasible subtour on P consists of consecutive vertices on P. Let $A_1, A_2, \ldots, A_k$ be disjoint subtours of P such that $\cup_{i=1}^{k} A_k = P$;

further, $P = A_1 A_2 \cdots A_k$. These A_i's is called a *path partition* of P; A_i is the ith path partition of P. Recall the definition in Section 2, the cost ratio of $A_i$is defined by $\rho(A_i) = w(A_i)/d(A_i)$.

Definition 3. *We call a subtour* $A = \langle a_1, a_2 \ldots, a_m \rangle$ right-skew *if and only if the cost ratio of any prefix* $\langle a_1, a_2 \ldots, a_j \rangle$ *is always smaller than or equal to the cost ratio of the remaining subpath* $\langle a_{j+1}, a_2 \ldots, a_m \rangle$. *A path partition* $P = A_1 A_2 \cdots A_k$ *is* decreasingly right-skew *if each* A_i *is right-skew and* $\rho(A_i) > \rho(A_j)$ *whenever* $i < j$.

Proposition 3. *Let* A, B *be two subtour of* T *with* $\rho(A) \geq \rho(B)$. *It follows that* $\rho(A) \geq \rho(AB) \geq \rho(B)$.

Lemma 3. *Let* A, B *be two right-skew subtour of* T *with* $\rho(A) \leq \rho(B)$. *It follows that* AB *is also right-skew.*

Proof. The proof appears in the complete paper. □

Here we first show that all single path P can be partition into decreasingly right-skew.

Lemma 4. *Any path can be uniquely partitioned into a decreasingly right-skew partition.*

Proof. Let $P = \langle v_1, v_2 \ldots v_n \rangle$. The argument obviously follows if $n = 1$. By induction, assume that a path $Q = A_1 A_2 \cdots A_k$, $|Q| = n$ is decreasingly right-skew partitioned. Now consider a path $P = \langle Q, a \rangle$.

The Lemma is proven if $\rho(a) < \rho(A_k)$. Otherwise, we find the largest i such that $\rho(A_i A_{i+1} \cdots A_k a) < \rho(A_{i-1})$; let $i = 1$ if such i can not be found. It suffices to show that $\langle A_i A_{i+1} \cdots A_k a \rangle$ is right-skew, and this can be done by observing that the single segment $\langle a \rangle$ is right-skew and by applying Lemma 3 repeatedly to the segments $A_i, A_{i+1}, \ldots, A_k, \langle a \rangle$ from right to left. That is, $\langle A_k a \rangle$ is right-skew because $\rho(A_k) \leq \rho(a)$, $\langle A_{k-1} A_k a \rangle$ is right-skew because $\rho(A_{k-1}) \leq \rho(A_k a)$, *etc.* Clearly, the partition is unique because other choices of i would not result in a decreasingly right-skew partition of A. □

Lemma 5. *Partitioning a path into a decreasingly right-skew partition can be done in* $O(n)$.

Proof. Consider the algorithm PATH-PARTI that produces a list of right-skew sequences in Figure 3. Note that i is the current working pointer scanning the vertices of P from right to left. Further, the pair $(i, p[i])$ represents a subsequence of P that is always right-skew throughout the entire algorithm. The observation is justified by Lemma 3, that two increasingly right-skew subtours can be grouped into one. Note that the condition checking and grouping is done by Step 3 to Step 6 of the algorithm.

The correctness of the algorithm follows from the fact that, after the execution of Step 1 to Step 6, each subsequence is right-skew, and the partition is increasing. Further, by Lemma 4, the partition must be unique.

```
PATH-PARTI(P)
Input: A path P = ⟨a_1, a_2, ..., a_n⟩
Output: A decreasingly right-skew partition of P.
1 for i ← n downto 1 do
2     p[i] ← i; w[i] ← w(a_i); d[i] ← d(a_i);       ▷ Each a_i alone is right-skew.
3     while (p[i] < n) and (w[i]/d[i] ≤ w[p[i] + 1]/d[p[i] + 1]) do
4         p[i] ← p[p[i] + 1]
5         w[i] ← w[i] + w[p[i] + 1]
6         d[i] ← d[i] + d[p[i] + 1]
7 i ← 1
8 while i ≤ n do        ▷ Reports (i, i) as a right-skew subtour ⟨a_i, ..., a_j⟩.
9     OUTPUT (i, p[i]); i ← p[i] + 1
```

Fig. 3. Partitioning a path into decreasingly right-skew in $O(n)$ time.

The $O(n)$ time complexity is justified by the amortized analysis. Note that total operations of the algorithm is clearly bounded by $O(n)$ except for the while loop body of Step 3 to Step 6. We define the potential of P to be k_i, the number of right-skew subsequences in P after the ith iteration of the for loop (Step 1 to Step 6.). Let us compute the amortized cost of operations done by Step 3 to Step 6. Let the actual cost of the operation is $c_i + 1$ in advancing c_i times of the $p[i]$ pointer. It follows that the potential difference is

$$\Phi(P_i) - \Phi(P_{i-1}) = (1 + k_{i-1} - c_i) - k_{i-1};$$

The amortized cost is therefore $\hat{c}_i = c_i + \Phi(P_i) - \Phi(P_{i-1}) = 1$ □

Let T be a rooted k-path. After deleting the root, T is decomposed into k paths: $P_1, \ldots, P_n$. For each P_i, we perform the PATH-PARTI(P_i) resulting $P_i = \langle A_1 \cdots A_j \rangle$ such that the partition $\langle A_1 \cdots A_j \rangle$ is decreasingly right-skew. It follows that each A_i is an atomic subtour in T.

Lemma 6. *A feasible right-skew subtour of a k-path, T, containing only vertices of one of its descendant path, is atomic in T.*

Proof. The proof appears in the complete paper. □

Theorem 2. *Finding the minimum back-walk-free latency tour in k-path can be done in $O(n \log k)$ time.*

Proof. We propose an $O(n \log k)$ time algorithm, MBLP-k-TREE(T), shown in Figure 4. The correctness of the algorithm follows from the fact that, The right-skew subsequence of each k-path is atomic in T as show by Lemma 6. Further, the algorithm reflects the spirit of MBLP-TREE(T) in extracting the maximal cost ratio and output.

The time complexity of the algorithm is easily shown to be $O(n \log k)$ by the fact that the size of the priority queue Q is at most k while there are totally $O(n)$ operations performed on Q. □

```
MBLP-k-PATH(T)
Input: A rooted k-path T.
Output: A tour with minimum weighted back-walk-free latency.
1 Decompose T into k paths: P_1, ..., P_k.
2 for i <- 1 to k do
3     list[i] <- PATH-PARTI(P_i)  ▷ Right-skew partition
4     ENQUEUE(Q, EXTRACT(list[i]))  ▷ Put the first element into priority queue Q.
5 while Q ≠ ∅ do
6      seq[i] <- EXTRACT-MAX(Q);  ▷ Pick the maximal cost ratio.
7      ENQUEUE(Q, EXTRACT(list[i]))
8      Append seq[i] to the output list.
9 return the output list.
```

Fig. 4. Algorithm for finding minimum back-walk-free latency tour in k-path.

Note that the output of the MBLP-TREE(T) can be viewed as a length $O(n)$ (multiple) permutation of the set $\{1, 2, \ldots, k\}$. Each output number represents the currently top vertex of one of the k paths being visited. Thus the number of all different possible output orderings can be as large as k^n. It follows that the solution space of the algorithm is therefore of the order $O(k^n)$. Since a comparison-based algorithm essentially searches the solution space in a binary (or ternary) manner, the underlying decision tree must have a depth of $\Omega(n \log k)$ by using simple counting arguments. It follows that

Proposition 4. *Any comparison-based algorithm for solving the MBLP on k-path spends $\Omega(n \log k)$ time in average.*

Corollary 2. *The algorithm* MBLP-k-PATH *spends $\Theta(n \log k)$ time, in average, in finding the minimum back-walk-free latency tour.*

Corollary 3. *Finding the minimum back-walk-free latency tour in a path (with any arbitrary starting vertex) can be done in $O(n)$ time.*

5 MBLP is NP-Complete

Despite its apparent simplicity comparing to the original MLP, here we show that MBLP is NP-complete even for a *directed acyclic graph* (usually briefed as dag) by reducing the 3-SAT problem to MBLP on a dag.

Theorem 3. *The decision version of MBLP for directed acyclic graphs is NP-complete.*

Proof. The decision version of MBLP is clearly in NP. Here we show that it is NP-hard by a reduction from 3-SAT [5] to MBLP.

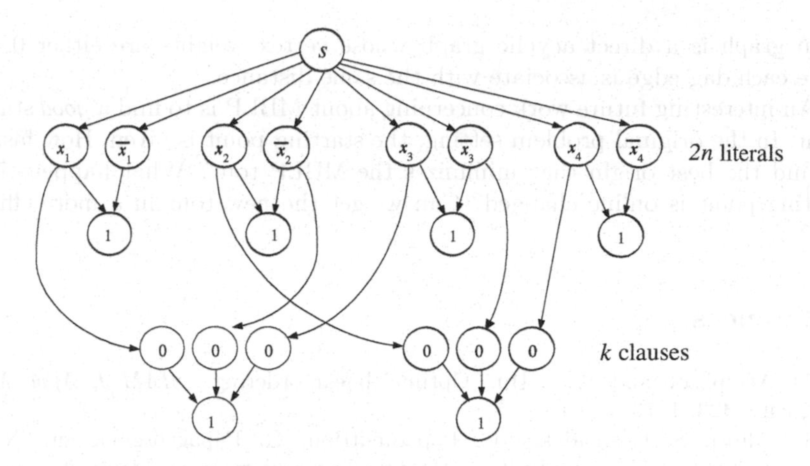

Fig. 5. Reducing the 3-SAT instance $(x_1 \vee \overline{x}_2 \vee x_3) \wedge (x_2 \vee \overline{x}_3 \vee x_4)$ into a dag.

Given an instance of 3-SAT, S, with literals $L = \{x_1, \overline{x}_1, \ldots, x_n, \overline{x}_n\}$ and clauses $C = \{c_1, \ldots, c_k\}$, we will construct a directed acyclic graph $G = (V, E)$ such that G has a back-walk-free tour with latency $(n+k)(n+k+1)$ if and only if S is satisfiable. The reduction is illustrated at Figure 5.

The constructed dag consists of one starting vertex s, and other three kinds of vertices: $2n$ *literal vertices* $V_L = \{x_1, \overline{x}_1, \ldots, x_n, \overline{x}_n\}$, $3k$ *clause vertices* $V_C = \cup_{i=1}^{k}\{c_{i1}, c_{i2}, c_{i3}\}$, and $n+k$ *sink vertices* $V_t = \{t_1, \ldots, t_n, t'_1, \ldots, t'_k\}$. The weights of vertices is defined by $w(v) = 1$ if $v \in V_t$; otherwise, $w(v) = 0$. The edge set of G consists of $4n$ *literal edges*, $E_L = \{(s, a) \mid a \in V_L\} \cup \{(x_i, t_i), (\overline{x}_i, t_i) \mid 1 \le i \le n\}$, k *clause edges*, $E_C = \cup_{i=1}^{k}\{(c_{i1}, t'_i), (c_{i2}, t'_i), (c_{i3}, t'_i)\}$, and finally the k *cross edges*, $E_X = \cup_{i=1}^{k}\{(a_{i1}, c_{i1}), (a_{i2}, c_{i2}), (a_{i3}, c_{i3})\}$ where $a_{i1}, a_{i2}, a_{i3} \in V_L$ are 3 vertices of the ith clause c_i. At last, we say that the distance of each (directed) edge is 1, and the distance of each unstated edge is $+\infty$.

The idea is that a low latency tour attempts to reach as many sink vertices as possible in the earliest manner, which constitutes the truth assignment of the given 3-SAT instance if it is satisfiable. The rest of the detailed arguments appears in the complete paper. □

6 Concluding Remarks

In this paper, a variation of the minimum latency problem, so called the minimum back-walk-free latency problem (MBLP) is considered. Several interesting properties of MBLP on trees are discussed, and we propose a simple and efficient $O(n \log n)$ time algorithm for general weighted tree, as well as an $O(n \log k)$ time algorithm for k-path trees. Further, we show that the optimal tour on weighted paths can be found in $O(n)$ time.

No previous hardness results were known for MBLP on general graphs. Here we settle the problem by showing that MBLP is NP-complete even when the given graph is a direct acyclic graph whose vertex weights are either 0 or 1, while each dag edge is associate with the same distance.

An interesting future work concerning about MBLP is to find a *good* starting point. In the original problem setting, the starting point is given. How fast can we find the best origin that minimizes the MBLP tour? What happens if the starting point is online changed? Can we get the new tour in a more efficient way?

References

1. D. Adolphson and T. C. Hu. Optimal linear ordering. *SIAM J. Appl. Math.*, 25:403–423, 1973.
2. F. Afrati, S. Cosmadakis, C. Papadimitriou, G. Papageorgiou, and N. Papakostantino. The complexity of the traveling repairman problem. *Informatique Theorique et Applications (Theoretical Informatics and Applications)*, 20(1):79–87, 1986.
3. A. Blum, P. Chalasani, D. Coppersmith, B. Pulleyblank, P. Raghavan, and M. Sudan. The minimum latency problem. In *Proceedings of the Twenty-Sixth Annual ACM Symposium on the Theory of Computing*, pages 163–171, Montréal, Québec, Canada, 23–25 May 1994.
4. T. Cormen, C. Leiserson, and R. Rivest. *Introduction to Algorithms.* MIT Press, 1990.
5. M.R. Garey and D.S. Johnson. *Computers and Intractability – A Guide to the Theory of NP-Completeness.* Freeman, New York, 1979.
6. T. Imielinski, S. Viswanathan, and B.R. Badrinath. Power efficient filtering of data on air. *4th International Conference on Extending Database Technoloy*, pages 245–258, March 1994.
7. T. Imielinski, S. Viswanathan, and B.R. Badrinath. Data on air: Organization and access. *IEEE Trans. on Knowledge and Data Engineering*, 9(3):353–372, May 1997.
8. E. Koutsoupias, C. Papadimitriou, and M. Yannakakis. Searching a fixed graph. In *Proc. 23rd Colloquium on Automata, Languages and Programming*, LNCS 1099, pages 280–289. Springer-Verlag, 1996.
9. Shou-Chih Lo and Arbee L.P. Chen. Index and data allocation in multiple broadcast channels. *IEEE Interantion Conference on Data Engineering 2000*, pages 293–302, 2000.
10. E. Minieka. The delivery man problem on a tree network. *Annals of Operations Research*, 18(1–4):261–266, February 1989.
11. N. Shivakumar and S. Venkatasubramanian. Energy-efficient indexing for information dissemination in wireless systems. *ACM, Journal of Wireless and Nomadic Application*, 1996.
12. Tsitsiklis. Special cases of traveling salesman and repairman problems with time windows. *NETWORKS: Networks: An International Journal*, 22, 1992.
13. Bang-Ye Wu. Polynomial time algorithms for some minimum latency problems. *Information Processing Letters*, 75(5):225–229, October 2000.

Counting Satisfying Assignments in 2-SAT and 3-SAT

Vilhelm Dahllöf*, Peter Jonsson**, and Magnus Wahlström***

Department of Computer and Information Science
Linköping University
SE-581 83 Linköping, Sweden
{vilda,petej,magwa}@ida.liu.se

Abstract. We present an $O(1.3247^n)$ algorithm for counting the number of satisfying assignments for instances of 2-SAT and an $O(1.6894^n)$ algorithm for instances of 3-SAT. This is an improvement compared to the previously best known algorithms running in $O(1.381^n)$ and $O(1.739^n)$ time, respectively.

1 Introduction

A number of recent papers, including [2,5,7,8,14], have studied various counting problems from different perspectives. This growing interest is probably due to the fact that decision problems – "is there a solution?" – have been thoroughly studied whereas our knowledge about counting problems – "how many solutions?" – is in comparison much more modest. Counting problems are not only mathematically interesting, they arise naturally when asking questions such as "what is the probability that a formula in the propositional calculus is true?" [11] or "what is the probability that a graph remains connected, given a probability of failure on each edge?" [12].

The subject of this paper is the #2-SAT and #3-SAT problems, i.e. counting the number of satisfying assignments to instances of 2-SAT and 3-SAT. These problems are known to be #P-complete [9,13]. Exact algorithms (with worst-case complexity analyses) have been studied earlier by Dubois [4], Zhang [14] and Littman *et al.* [10]. The algorithms by Dubois and Zhang run in $O(1.6180^n)$ time for #2-SAT and $O(1.8393^n)$ time for #3-SAT while the algorithms by Littman *et al.* run in $O(1.381^n)$ and $O(1.739^n)$ time, respectively.

We will here present an $O(1.3247^n)$ algorithm for the #2-SAT problem and an $O(1.6894^n)$ algorithm for the #3-SAT problem. Both algorithms run in polynomial space. These algorithms are (as the algorithms by Littman *et al.*) based

* The research is supported by CUGS – National Graduate School in Computer Science, Sweden.

** The research is partially supported by the *Swedish Research Council* (VR) under grant 221-2000-361.

*** The research is supported by CUGS – National Graduate School in Computer Science, Sweden.

O.H. Ibarra and L. Zhang (Eds.): COCOON 2002, LNCS 2387, pp. 535–543, 2002.

on the Davis-Putnam-Logemann-Loveland algorithm for satisfiability modified to count all satisfying assignments [2] and using a connected-components approach [1]. The basic idea is to choose a variable x and recursively count the number of satisfying assignments where x is true and where x is false. The main difference between our algorithms and those by Littman *et al.* lies in the choice of variable; our algorithms make a more elaborate analysis of the formula and are able to choose a "better" variable. The #2-SAT algorithm is also improved by using certain ideas from an algorithm for counting independent sets in graphs [3].

In the next section some definitions will be given and the notation used will be explained. Then the paper will use two sections to focus on the #2-SAT and #3-SAT algorithms, respectively.

2 Preliminaries

A *propositional variable* can have the values *true* and *false* and a *literal* is either a propositional variable x or its negation $\neg x$. A propositional formula on *conjunctive normal form* (CNF) is a conjunction of disjunctions of literals. A k-SAT formula ($k > 0$) is a propositional formula in CNF such that each clause contains at most k literals. A *satisfying assignment* is an assignment to every propositional variable of a formula making the entire formula true. #2-SAT is the problem of computing the number of satisfying assignments of 2-SAT formulae and #3-SAT is the corresponding problem for 3-SAT formulae. We note that an empty formula (i.e. a formula containing no clauses) has 1 model while a formula containing an empty clause (i.e. a clause containing no literals) is not satisfiable and has 0 models. We define the size of a formula $|F|$ as the number of variables F contains.

Given a formula F over the variables V, we define the *constraint graph* as an undirected graph (V, E) where $E = \{(x, y) \mid x, y \text{ appear in the same clause of } F\}$. We say that x and y are *constrained* iff $(x, y) \in E$ and the *constraints of* F are E. The *neighbourhood* of x ($N_F(x)$) is the set $\{y \mid (x, y) \in E\}$, the *degree* of x ($\delta_F(x)$) is $|N_F(x)|$ and the degree of F is $\delta(F) = \max\{\delta_F(x) \mid x \in Var(F)\}$. Note that if $\delta(F) = k$, then $|E| \leq k \cdot |F|/2$. We say that a formula is a *cycle* or a *path* whenever the constraint graph is a cycle or a path. A *connected component* of such a graph is a maximal subgraph such that for every pair of vertices x, y in the subgraph, there is a path from x to y. We say that the *connected components* of a formula are the subformulas corresponding to the connected components of the constraint graph.

Given a formula F and a literal l, we denote by $F|_l$ the result of making literal l true in F and simplifying the result by (1) removing any clause that contains l; (2) removing $\neg l$ whenever it appears; and (3) performing unit propagation as far as possible. By unit propagation, we mean to repeatedly apply rules of the following two types:

1. $(p \vee q \vee r) \wedge (p) \longrightarrow$ *true*; and
2. $(p \vee q \vee r) \wedge (\neg p) \longrightarrow (q \vee r)$.

We extend this notation to $F|_\Gamma$, where Γ is a set of literals.

Consider the following example: $F = (p \vee q \vee r)$ and note that #3-SAT$(F) = 7$. It seems reasonable that #3-SAT$(F|_q)$+#3-SAT$(F|_{\neg q})$ =#3-SAT(F) but this is not the case. We see that $(p \vee q \vee r)|_q$ is simplified to the empty formula (which has 1 model by definition) and #3-SAT$(F|_q)$+#3-SAT$(F|_{\neg q}) = 4$. The problem is that the variables q and r (which can be given arbitrary values) are "eliminated" in the simplification process. Thus, we introduce a function $\Psi(F, R, \Gamma)$, where F is the formula to be simplified, R the set of eliminated variables and Γ is the literals to be assigned *true*, it returns (F', R'), where F' is the simplified formula and R' is the new set of eliminated variables; R' is computed as $R \cup (Var(F) - (Var(F') \cup \Gamma))$, where $Var(F)$ denotes the set of variables of F.

The algorithms we present are based on the Davis-Putnam-Logemann-Loveland algorithm for satisfiability modified for counting satisfying assignments and they build on two straightforward principles: assume F is an arbitrary SAT formula, R a set of eliminated variables and let #SAT(F, R) denote the number of satisfying assignments of F modulo the eliminated variables in R.

Principle 1. Arbitrarily choose $x \in Var(F)$. Then,

$$\#\text{SAT}(F, R) = \#\text{SAT}(\Psi(F, R, \{x\})) + \#\text{SAT}(\Psi(F, R, \{\neg x\})).$$

Principle 2. Let F_1, F_2 be disjoint components of F. Then,

$$\#\text{SAT}(F, R) = 2^{|R|} \cdot \#\text{SAT}(F_1, \emptyset) \cdot \#\text{SAT}(F_2, \emptyset).$$

Note that these disjoint components can be found in polynomial time. The correctness proofs of the algorithms are omitted since they can trivially be inferred from these two principles.

3 Algorithm for #2-SAT

Before presenting the algorithm for #2-SAT, we need one more definition: given a formula F and a variable x, we define

$$L_F(x) = |\{u \mid (\sim u, \sim v) \in F \wedge u \in N_F(x) \wedge v \notin N_F(x) \cup \{x\}\}|$$

where $\sim u$ denotes either u or $\neg u$. Informally, one can say that $L_F(x)$, $x \in V$, tells how many of the neighbours of x that appear in a constraint together with some variables not in $N_F(x) \cup \{x\}$. If $L_F(x) = 1$, then $B(x)$ denotes the unique variable in $N_F(x)$ that is adjacent to one or more vertices outside $N_F(x) \cup \{x\}$. In Figure 1 there is an example where $L_F(x) = 1$.

Our algorithm for #2-SAT is presented in Figures 2 and 3. The function C_E computes #2-SAT by exhaustive search. It will be applied only to formulas of size ≤ 4 and can thus be safely assumed to run in $O(1)$ time. The correctness of the algorithm follows from the discussion in the previous section. When analysing the algorithm we will often encounter recurrences of the form

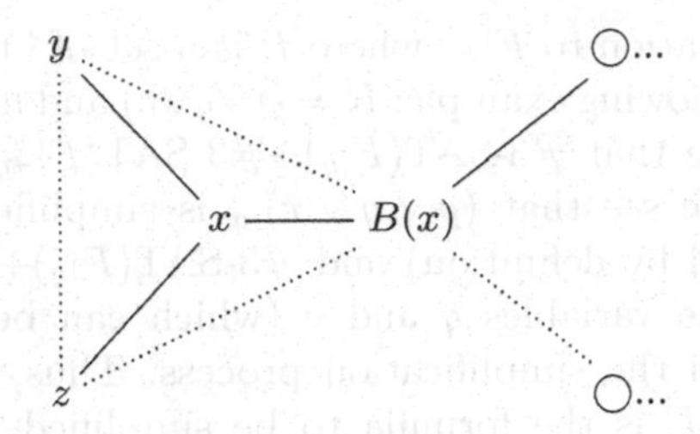

Fig. 1. A constraint graph where $L_F(x) = 1$; the filled lines indicate where there is a constraint by necessity, whereas a pointed line indicates a possible constraint

$$T(n) \leq \sum_{i=1}^{k} T(n - r_i) + \text{poly}(n).$$

They satisfy $T(n) \in O(\lambda(r_1, \ldots, r_k)^n)$ where $\lambda(r_1, \ldots, r_k)$ is the largest, real-valued root of the function $f(x) = 1 - \sum_{i=1}^{k} x^{-r_i}$. Note that we can ignore the polynomial factor so we will assume that all polynomial time computations take $O(1)$ time.

When inspecting C and C_3 one can see that every connected component of the input will be recursively solved by the algorithm. However, this will not increase the time required, since $\sum_{i=1}^{k} T(n_i) \leq T(n)$ when $n = \sum_{i=1}^{k} n_i$.

Given a 2-SAT formula F and a literal l, we define $Unit(F, l)$ to be the set of variables that appears in a clause of length 1 after having made literal l true in F and simplifying the result by (1) removing any clause that contains l; and (2) removing $\neg l$ whenever it appears (in other words, the usual simplification process employed by Ψ but without performing unit propagation). To illustrate $Unit$, consider the formula $F = (x \vee y) \wedge (x \vee z) \wedge (\neg x \vee w)$. It can easily be verified that $Unit(F, x) = \{w\}$ and $Unit(F, \neg x) = \{y, z\}$. In fact, if $\delta(x) = k$ in a 2-SAT formula F, then $|Unit(F, x)| + |Unit(F, \neg x)| \geq k$. This observation will be used several times in the complexity analysis.

Lemma 1. *Algorithm C_3 runs in $O(1.3045^n)$ time.*

Proof. In this lemma, we will measure the size of a formula G by the number m of constraints it contains. We will ultimately show that C_3 runs in $O(1.1939^m)$ time; by noting that a degree-3 formula with n vertices can contain at most $3n/2$ constraints, we get that the algorithm runs in $O(1.1939^{3n/2}) = O(1.3045^n)$ time.

We assume that the input formula F is connected and contains m constraints. Cases 0–2 are trivial and case 3 does not increase the running time.

Case 4: If F is a cycle, then F is immediately transformed into a path in both recursive calls. So, the interesting case is when the input is a path – then the time is bounded by the recurrence $T(m) \leq 4T(\lceil m/2 \rceil - 2)$ and straightforward calculations show that $T(m) \in O(4^{\log_2 m}) = O(m^2)$.

Case 5.1: $L_F(x) = 1$. We study the formula $F' = F|_{B(x)}$. First note that $\delta(B(x)) \geq 2$ since, otherwise, $|F| = 4$ and Case 2 would have applied. Now,

Algorithm $C_3(F, R)$
Case 0: F has an empty clause. Return 0.
Case 1: F is empty. Return $2^{|R|}$.
Case 2: $|F| \leq 4$. Return $C_E(F)$.
Case 3: F consists of two disjoint components F_1, F_2.
Return $2^{|R|} \cdot C_3(F_1, \emptyset) \cdot C_3(F_2, \emptyset)$.

Case 4: $\delta(F) \leq 2$. Pick a variable x in the following way:

1. If F is a path, choose x to be a variable that splits F into two paths of lengths $\lceil |F|/2 \rceil$ and $\lfloor |F|/2 \rfloor$.
2. If F is a cycle, choose x arbitrarily.

Return $C_3(\Psi(F, R, \{x\})) + C_3(\Psi(F, R, \{\neg x\}))$.

Case 5: $\delta(F) = 3$. Pick a variable x such that $\delta(x) = 3$ and do the following:

1. If $L_F(x) = 1$, then return $C_3(\Psi(F, R, \{B(x)\})) + C_3(\Psi(F, R, \{\neg B(x)\}))$.
2. If $L_F(x) > 1$, then return $C_3(\Psi(F, R, \{x\})) + C_3(\Psi(F, R, \{\neg x\}))$.

Fig. 2. Help function for computing #2-SAT

Algorithm $C(F, R)$
Case 0: F has an empty clause. Return 0.
Case 1: F is empty. Return $2^{|R|}$.
Case 2: F consists of two disjoint components F_1, F_2.
Return $2^{|R|} \cdot C(F_1, \emptyset) \cdot C(F_2, \emptyset)$.
Case 3: $\delta(F) \leq 3$. Return $C_3(F, R)$.
Case 4: Pick a variable x s.t. $\delta(x) \geq 4$ and return $C(\Psi(F, R, \{x\})) + C(\Psi(F, R, \{\neg x\}))$.

Fig. 3. Function for computing #2-SAT

$B(x)$ is given a fixed value and $\delta(B(x)) \geq 2$ implies that F' contains at least two constraints less than F. Furthermore, the neighbourhood of x forms a component F'' that is not connected to the rest of the formula. This component contains three variables and at least two constraints. Consequently, F'' will be taken care of in $O(1)$ time in the next recursive step of the algorithm (by component analysis and using C_E). The time needed in this case is thus bounded by $T(m) \leq 2T(m-4)$ since exactly the same situation arises when $B(x)$ is given the value false. It follows that $T(m) \in O(1.1892^m)$.

Case 5.2: $L_F(x) > 1$ – that is, of the variables $\{v, y, z\} = N_F(x)$, at least two are related to variables not in $\{v, x, y, z\}$. In this worst-case analysis, we assume that only v and y are related to variables not in $\{v, x, y, z\}$.

Let $S = Unit(F, x) \cap \{v, y\}$ and $S' = Unit(F, \neg x) \cap \{v, y\}$. The running time $T(m)$ of the algorithm satisfies the recursive relation

$$T(m) \leq T(m-3-|S|) + T(m-3-|S'|)$$

Here, three constraints are removed due to giving x a fixed value and at least one more constraint is removed for each variable v, y that appear in a unit clause. It is easy to see that $|S| + |S'| = 2$ and that the worst case occurs when $|S|$ or $|S'|$ equals 0. The time needed in this case is bounded by $T(m) \leq T(m-3)+T(m-5)$ and $T(m) \in O(1.1939^m)$.

Theorem 1. *Algorithm C runs in* $O(1.3247^n)$ *time.*

Proof. The result follows from Lemma 1 if $\delta(F) \leq 3$. Otherwise, let $S = Unit(F, x)$ and $S' = Unit(F, \neg x)$. The running time $T(n)$ of the algorithm satisfies the recursive relation

$$T(n) \leq T(n-1-|S|) + T(n-1-|S'|)$$

Since $\delta(x) \geq 4$, we know that $|S| + |S'| \geq 4$. Obviously, $\lambda(a,b) = \lambda(b,a)$ and $\lambda(a,b) \geq \lambda(a+1,b)$ so $\lambda(a,b) \leq \lambda(1,5) \approx 1.3247$ whenever $a, b \geq 5$. It can easily be checked that the small number of remaining cases also satisfy $\lambda(a,b) \leq \lambda(1,5)$ and $T(n) \in O(1.3247^n)$.

4 Algorithm for #3-SAT

In this section we present an algorithm D for #3-SAT and provide an upper bound on its running time.

To find this upper time bound, we will use a different method than the one described in Section 3. We will need to solve systems of recurrences like this:

$$f(n) = f(n-1) + g(n-1) \quad (1)$$
$$g(n) = f(n-1) + g(n-2) \quad (2)$$

To do this, we rewrite (2) so that the right hand side is only dependent upon $n-1$, by adding a new equation to the system:

$$f(n) = f(n-1) + g(n-1) \quad (3)$$
$$g(n) = f(n-1) + h(n-1) \quad (4)$$
$$h(n) = g(n-1) \quad (5)$$

We then write the equations into a matrix T, where each row represents a line in the system. An upper bound for the solution to the system is $f(n) \in O(\lambda^n)$, where λ is the largest eigenvalue of T. This can be seen by using generating functions for solving recurrence equations [6]. In this example:

$$T = \begin{pmatrix} 1 & 1 & 0 \\ 1 & 0 & 1 \\ 0 & 1 & 0 \end{pmatrix}$$

and $\lambda = 1.8019$.

Algorithm $D(F, R)$
Case 0: F has an empty clause. Return 0.
Case 1: F is empty. Return $2^{|R|}$.
Case 2: F consists of disjoint components F_1, F_2.
Return $2^{|R|} \cdot D(F_1, \emptyset) \cdot D(F_2, \emptyset)$.
Case 3: There exists some 1-clause $(x) \in F$. Return $D(\Psi(F, R, \{x\}))$.
Case 4: There exists some 2-clause in F. Pick a variable x in the following way:

1. If some variable v is a member of a 2-clause $(v \vee w)$ and $d(v) = 1$, pick w as x.
2. If some variable v is a member of a 2-clause $(v \vee w)$ and $d(v) = 2$, pick w as x.
3. Otherwise, pick as x a variable which is a member of as many 2-clauses as possible.

Return $D(\Psi(F, R, \{x\})) + D(\Psi(F, R, \{\neg x\}))$.

Case 5: Pick a variable x in the following way:

1. If $d(v) = 1$ for some variable v, pick a neighbour of maximum degree as x.
2. If $d(v) = 2$ for some variable v, pick a neighbour of maximum degree as x.
3. Otherwise, pick as x a variable of maximum degree.

Return $D(\Psi(F, R, \{x\})) + D(\Psi(F, R, \{\neg x\}))$.

Fig. 4. Function for computing #3-SAT

Some terminology is needed. A *k-clause* is a clause that contains k literals. The degree of x in F ($d_F(x)$), for a variable x, denotes the number of clauses in F that x appears in. Usually, we will use $d(x)$ when the formula F is clear from the context. Note that this is different from $\delta_F(x)$ as defined in Section 2. A formula where every variable has degree k is *k-regular*. $d_2(x)$ denotes the number of 2-clauses that x is a member of.

The algorithm is presented in Figure 4. The idea behind it is that if some variable v has a low degree ($d(v) = 1$ or 2), then we can pick its neighbours as branching variables to reach a situation where $d(v) = 0$ and eliminate v in this way. If no such variable exists, we simply pick a variable of high degree.

Lemma 2. *If F is a connected k-regular formula with only 3-clauses, then no k-regular formula $F' \neq F$ with only 3-clauses can be reached by assigning truth values to variables in F.*

Proof. Let F be a connected, k-regular formula which only contains 3-clauses. Assume that Γ is a set of assignments such that $F|_\Gamma$ is k-regular and only contains 3-clauses, and that $F|_\Gamma \neq F$. Note that since $F|_\Gamma$ only contains 3-clauses, $F|_\Gamma \subseteq F$. Let $F_1 = F|_\Gamma$ and $F_2 = F - F|_\Gamma$.

Since F is connected, there is some variable x which occurs in both F_1 and F_2. Since x occurs in F_1, it is not assigned a value by Γ. However, since $d_{F_1}(x) = d_F(x) - d_{F_2}(x) < k$, F_1 cannot be k-regular.

Theorem 2. *Algorithm D runs in $O(1.6894^n)$ time.*

Proof. We will analyze the time complexity of D case by case.

Cases 0, 1 and 3: These cases take $O(1)$ time.

Case 2: This case does not increase the time needed.

Case 4.1: When $w = true$, $d(v) = 0$ and v is removed from F; when $w = false$, $v = true$. $T(n) = 2T(n-2)$ so $T(n) \in O(1.4142^n)$.

Case 4.2: When $w = true$, 5.1 is met (or case 4.1, or some earlier case); when $w = false$, $v = true$. $T(n) = T(n-2)+T(n-3)+2T(n-4)$ so $T(n) \in O(1.5664^n)$.

Case 4.3: Even if the chosen variabel is only a member of one 2-clause, we have $T(n) = T(n-1) + T(n-2)$ with solution $T(n) \in O(1.6180^n)$. If it is a member of several 2-clauses, then the limit will be lower.

Case 5.1: This case takes $O(1.5214^n)$ time. Suppose $d(v) = 1$, $(u \vee v \vee w) \in F$ and that u is the chosen variable. When $u = true$, $d(v) = 0$ and v is removed from F; when $u = false$, case 4.1 is met. In total, the recurrence is $T(n) = T(n-2) + 2T(n-3)$ with solution $O(1.5214^n)$.

Case 5.2: This case takes $O(1.6560^n)$ time. Suppose $d(v) = 2$, $(u \vee v \vee w) \in F$ and that u is the chosen variable. When $u = true$, case 5.1 is met (or some earlier case); when $u = false$, case 4.2 is met (or some earlier case). The worst case is when no earlier case is met, and the recurrence for this case is $T(n) = 2T(n-3) + 3T(n-4) + 2T(n-5)$ with solution $O(1.6560^n)$.

Case 5.3: This case requires a more careful analysis. We shall formulate a set of recurrence equations that describes the behaviour of the algorithm in this case. Let x be the chosen variable. When we reach this case, F contains no 2-clauses. Setting $x = true$ creates k_1 2-clauses, and setting $x = false$ creates k_2 2-clauses, where $k_1 + k_2 = d(x)$. For the moment, assume that $d(x) = 4$.

Let $T(n, k)$ represent a situation where F contains n variables and k 2-clauses. The initial situation is $T(n, 0)$. Now, we enumerate a number of possible recurrences where x is the chosen variable, and the 2-clauses it appears in are $(x \vee v_i)$ or $(\neg x \vee v_i)$. Some cases that are clearly dominated by some listed case are ignored.

$$T(n,0) = T(n-1,k_1) + T(n-1,k_2) \quad (6)$$

$$T(n,k) = T(n-1,k-1) + T(n-2,k) \text{ if } d_2(x) = 1 \quad (7)$$

$$T(n,k) = T(n-2,k-1) + T(n-1,k) \text{ if } d_2(x) = 1 \quad (8)$$

$$T(n,k) = T(n-1,k-2) + T(n-3,k-d_2(v_1)-d_2(v_2)) \text{ if } d_2(x) = 2 \quad (9)$$

$$T(n,k) = T(n-2,k-1-d_2(v_1)) + T(n-2,k-1-d_2(v_2)) \text{ if } d_2(x) = 2 \quad (10)$$

The recurrences for $d_2(x) > 2$ are similar to (9)–(10).

The idea behind these recurrences is that when $x = true$ then every 2-clause $(x \vee v_i)$ is removed and for every 2-clause $(\neg x \vee v_i)$, $v_i = true$ and $d_2(v_i)$ 2-clauses are removed. In (7)–(8), we know that $d_2(v_1) = 1$; in (9)–(10) and similar recurrences, we know that $1 \leq d_2(v_i) \leq d_2(x)$ for all v_i involved.

In (7)–(8), $d(x) \geq 2$ is assumed, so that one new 2-clause will be created, either when $x = true$ or $x = false$. $d(x) = 1$ only occurs when $(x \vee v_1)$ is disjoint from the rest of F, so this assumption is allowed.

If we go through all possible combinations of rules like these, we find that the worst case is when only recurrences (6) and (7) are used, and $k_1 = 4, k_2 = 0$. The solution to this system is $T(n, 0) \in O(1.6894^n)$.

If $d(x) > 4$ then more 2-clauses are created and either more recurrences with low $d_2(x)$ or recurrences with a higher $d_2(x)$ are used. These cases are both easier than $d(x) = 4$.

The remaining case is when F is 3-regular. Then, as Lemma 2 shows, for the rest of the recursive calculations no 3-regular formula will appear, so there will always exist variables of degree 1 or 2. If $T(n)$ is the time required for a 3-regular formula and $T'(n)$ is the time required if variables of degree 1 or 2 are guaranteed to exist, we have $T(n) = 2T'(n-1)$ where $T'(n) \in O(1.6560^n)$, so $T(n) \in O(2 \cdot 1.6560^{n-1}) = O(1.6560^n)$.

In total, this shows that algorithm D runs in $O(1.6894^n)$ time.

References

1. R. J. Bayardo, Jr. and J. Pehoushek. Counting models using connected components. In *Proceedings of the Seventeenth National Conference on Artificial Intelligence (AAAI-2000)*, pages 157–162. AAAI Press, 2000.
2. E. Birnbaum and E. L. Lozinskii. The good old Davis-Putnam procedure helps counting models. *Journal of Artificial Intelligence Research*, 10:457–477, 1999.
3. V. Dahllöf and P. Jonsson. An algorithm for counting maximum weighted independent sets and its applications. In *Proceedings of the Thirteenth Annual ACM-SIAM Symposium on Discrete Algorithms (SODA-2002)*, pages 292–298, 2002.
4. O. Dubois. Counting the number of solutions for instances of satisfiability. *Theoretical Computer Science*, 81(1):49–64, 1991.
5. L. Goldberg and M. Jerrum. Counting unlabelled subtrees of a tree is #P-complete. *LMS Journal of Computation and Mathematics*, 3:117–124, 2000.
6. R. L. Graham, D. E. Knuth, and O. Patashnik. *Concrete Mathematics.* Addison-Wesley, Reading, MA, USA, second edition, 1994.
7. C. S. Greenhill. The complexity of counting colourings and independent sets in sparse graphs and hypergraphs. *Computational Complexity*, 9(1):52–72, 2000.
8. H. B. Hunt III, M. V. Marathe, V. Radhakrishnan, and R. E. Stearns. The complexity of planar counting problems. *SIAM Journal on Computing*, 27(4):1142–1167, 1998.
9. D. Kozen. *The design and analysis of algorithms.* Springer-Verlag, 1992.
10. M. L. Littman, T. Pitassi, and R. Impagliazzo. On the complexity of counting satisfying assignments. Unpublished manuscript.
11. D. Roth. On the hardness of approximate reasoning. *Artificial Intelligence*, 82(1–2):273–302, 1996.
12. S. P. Vadhan. The complexity of counting in sparse, regular, and planar graphs. *SIAM Journal on Computing*, 31(2):398–427, 2001.
13. L. Valiant. The complexity of enumeration and reliability problems. *SIAM Journal on Computing*, 8(3):410–421, 1979.
14. W. Zhang. Number of models and satisfiability of sets of clauses. *Theoretical Computer Science*, 155(1):277–288, 1996.

On the Maximum Number of Irreducible Coverings of an n-Vertex Graph by $n-3$ Cliques

Ioan Tomescu

Faculty of Mathematics, University of Bucharest, Bucharest, Romania
ioan@math.math.unibuc.ro

Abstract. In this paper the structure of the irreducible coverings by $n-3$ cliques of the vertices of a graph of order n is described. As a consequence, the number of such coverings for complete multipartite graphs is deduced. Also, it is proved that for sufficiently large n the maximum number of irreducible coverings by $n-3$ cliques of an n-vertex graph equals $3^{n-3} - 3 \cdot 2^{n-3} + 3$ and the extremal graph coincides (up to isomorphism) to $K_{3,n-3}$. This asymptotically solves a problem raised in a previous paper by the author (*J. Combinatorial Theory* B28, 2(1980), 127-141). The second extremal graph is shown to be isomorphic to $K_{3,n-3} - e$.

1 Definitions and Preliminary Results

For a graph G having vertex set $V(G)$ and edge set $E(G)$ a clique is a complete subgraph of G which is maximal relatively to set inclusion. The maximum number of vertices in a clique of G is the clique number of G and it is denoted $\omega(G)$ [1]. A k-clique is a clique containing k vertices. We say that a family of different cliques $c_1, c_2, \ldots, c_s$ of G is a covering of G by cliques if $\bigcup_{i=1}^{s} c_i = V(G)$. A covering C of G consisting of s cliques $c_1, \ldots, c_s$ of G will be called an irreducible covering of G if the union of any $s-1$ cliques from C is a proper subset of $V(G)$. This means that there exist s vertices $x_1, \ldots, x_s \in V(G)$ that are uniquely covered by cliques of C, i.e., $x_i \notin \bigcup_{\substack{k=1 \\ k \neq i}}^{s} c_k$ for every $1 \leq i \leq s$. The number of irreducible coverings by s cliques of G will be denoted $ic_s(G)$. $K_{p,q}$ will denote the complete bipartite graph whose parts contain p and q vertices, respectively and $K_{n_1,\ldots,n_r}$ the complete r-multipartite graph whose parts contain $n_1, \ldots, n_r$ vertices, respectively. A star is a graph $K_{1,k}$ where $k \geq 1$. $N(x)$ denotes the neighborhood of x, i.e., the set of vertices that are adjacent to x. The closed neighborhood of x, denoted $N^*(x)$ is $N(x) \cup \{x\}$. If graphs G and H are isomorphic we shall denote this by $G \cong H$.

The problem of determining the greatest number $\alpha(n)$, of cliques a graph with n vertices can have was solved by Miller and Muller [3] and independently by Moon and Moser [4] (see [2]): For every $n \geq 2$, $\alpha(n) = 3^{n/3}$ if $n \equiv 0 \pmod 3$, $\alpha(n) = 4 \cdot 3^{(n-4)/3}$ if $n \equiv 1 \pmod 3$, and $\alpha(n) = 2 \cdot 3^{(n-2)/3}$ if $n \equiv 2 \pmod 3$. For each $n \geq 2$ and $n \not\equiv 1 \pmod 3$ the extremal graph (which will be denoted $MMG(n)$) is unique (up to isomorphism) and it consists of n isolated vertices for

O.H. Ibarra and L. Zhang (Eds.): COCOON 2002, LNCS 2387, pp. 544–553, 2002.

$n = 2$ or 3 and for $n \geq 5$ it is a complete multipartite graph, whose parts contain 3 vertices ($n/3$ times) if $n \equiv 0 \pmod 3$ and 3 vertices ($(n-2)/3$ times) and 2 vertices if $n \equiv 2 \pmod 3$, respectively. If $n \equiv 1 \pmod 3$ there exist two extremal graphs: either it consists of one quadruple and $(n-4)/3$ triples of vertices or it consists of two pairs and $(n-4)/3$ triples. In this case we shall denote by $MMG(n)$ the multipartite complete graph whose parts are one quadruple and $(n-4)/3$ triples.

If $I(n, n-k)$ denotes the maximum number of irreducible coverings of the vertices of an n-vertex graph by $n-k$ cliques, in [8] it was shown that $\lim_{n\to\infty} I(n, n-k)^{1/n} = \alpha(k)$. Furthermore, $I(n, n-2) = 2^{n-2} - 2$ and $ic_{n-2}(G) = 2^{n-2} - 2$, where $|V(G)| = n \geq 4$ implies $G \cong K_{2,n-2}$.

There is a class of algorithms which yield all irreducible coverings for the set-covering problem, an example of an algorithm in this class being Petrick's algorithm [6]. Since every minimum covering is an irreducible one, this algorithm was intensively used for obtaining the minimal disjunctive forms of a Boolean function by using prime implicants of the function or for minimizing the number of states of an incompletely specified Mealy type automaton A by finding a minimum closed irreducible covering of the set of states of A by "maximal compatible sets of states", which are cliques in the graph of compatible states of A [5, 7]. Also, the chromatic number $\chi(G)$ of G equals the minimum number of cliques from an irreducible covering by cliques of the complementary graph $\overline{G}$. Hence an evaluation of the numbers like $I(n, n-k)$ may have implications on the worst-case complexity of these algorithms.

If $G = K_{p,q}$, every clique of G is an edge and an irreducible covering by edges of $K_{p,q}$ consists of a set of vertex-disjoint stars, centered in the part with p vertices or in the part with q vertices of $K_{p,q}$, which cover together all vertices of $K_{p,q}$. Some properties of the numbers $N(p,q)$ of all irreducible coverings by edges of $K_{p,q}$ were deduced in [8]; the exponential generating function of these numbers was obtained in [9]. The following lemma summarizes some properties observed in [8]:

Lemma 1. *A covering by edges C of $K_{p,q}$ is irreducible if and only if it does not contain three edges ab, bc and cd of a path P_4. If $1 \leq p \leq q$, the number of edges in an irreducible covering by edges of $K_{p,q}$ equals q if $p = 1$ and it lies between q and $p+q-2$ if $p \geq 2$.*

If $p \geq 2$ this number of edges is equal to q if all stars are centered in the part with p vertices and it is equal to $p+q-2$ if C consists of exactly two stars centered in different parts of $K_{p,q}$.

Lemma 2. *For every $n \geq 6$, $ic_{n-3}(K_{3,n-3}) = 3^{n-3} - 3 \cdot 2^{n-3} + 3$.*

Proof. If a star is an edge xy we can consider that its center is either x or y. It follows that all irreducible coverings by $n-3$ edges of $K_{3,n-3}$ consist of three stars centered in the part with three vertices of $K_{3,n-3}$, which implies that $ic_{n-3}(K_{3,n-3}) = s_{n-3,3} = 3!S(n-3,3) = 3^{n-3} - 3 \cdot 2^{n-3} + 3$, where $s_{n,k}$ denotes the number of all surjections from a set having n elements onto a set having k

elements and $S(n,k)$ is the Stirling's number of the second kind with parameters n and k, counting the number of all partitions of an n-element set into k classes. In [8, p. 138] it was conjectured that for $n \geq 6$, $I(n, n-3) = ic_{n-3}(K_{3,n-3})$ and the extremal graph is unique up to isomorphism, possibly for $n \geq 7$. In the last section it is shown that this property holds asymptotically, i.e., for every $n \geq n_0$ (n_0 fixed, $n_0 \geq 7$).

2 The Structure of the Irreducible Coverings by $n-3$ Cliques

From the definition of an irreducible covering we deduce that for any graph G of order n possessing an irreducible covering C by $n-3$ cliques there exists a set of $n-3$ vertices, $X = \{x_1, \ldots, x_{n-3}\}$, that are uniquely covered by cliques of C. Let $Y = V(G) \backslash X = \{y, z, t\}$ denote the remaining vertices of G. For every $A \subset Y$, $A \neq \emptyset$ let $N_A = \{x : x \in X \text{ and } N(x) \cap \{y,z,t\} = A\}$. For example, N_y is the set of vertices in X which are adjacent only to y, $N_{y,z,t}$ is the set of vertices in X which are adjacent to all y, z and t. It follows that all subsets N_A do not contain adjacent vertices, i.e., they induce independent subsets of vertices; also if $\emptyset \neq A \subset B \subset Y$ then $N_A \cup N_B$ induces an independent subset of vertices. To prove this property suppose that two distinct vertices x_i and x_j in the same N_A are adjacent. But x_i belongs to a clique c_i consisting of x_i and a subset $Z \subset A$. In this case $Z \cup \{x_i, x_j\}$ also is a clique, which contradicts the maximality of c_i relatively to set inclusion. The same argument applies whenever $x_i \in A$, $x_j \in B$, $\emptyset \neq A \subset B \subset Y$ and x_i and x_j are adjacent. If G does not contain triangles (3-complete subgraphs) then also $N_{y,z} \cup N_{y,t} \cup N_{z,t}$ induces an independent set of vertices. Indeed, if for example $ij \in E(G)$, where $i \in N_{y,z}$ and $j \in N_{y,t}$ then G would contain a triangle $\{y, i, j\}$, which contradicts the hypothesis.

We shall describe the structure of irreducible coverings by $n-3$ cliques of a graph of order $n \geq 4$ without isolated vertices by introducing five types $\alpha, \beta, \gamma, \delta, \varepsilon$ of such coverings:

α) Three vertex disjoint stars having centers denoted by u, v and w covering all vertices of G ($N^*(u) \cup N^*(v) \cup N^*(w) = V(G)$); all edges of these stars are 2-cliques.

β) r 3-cliques ($r \geq 1$) having an edge uv in common; p 2-cliques having an extremity in u; q 2-cliques having an extremity in v ($p, q \geq 0$) and a star whose edges are 2-cliques, centered in a vertex denoted w, vertex-disjoint from r triangles and $p+q$ edges, and covering together all vertices of G.

γ) r 3-cliques having a common edge uv; s 3-cliques having a common edge vw ($r, s \geq 1$); p 2-cliques having an extremity in u; q 2-cliques having an extremity in v and m 2-cliques having an extremity in w ($p, q, m \geq 0$), such that $N^*(u) \cup N^*(v) \cup N^*(w) = V(G)$.

δ) r 3-cliques, s 3-cliques and l 3-cliques having a common edge uv, vw and uw, respectively ($r, s, l \geq 1$); p 2-cliques, q 2-cliques and m 2-cliques with an extremity in u, v and w, respectively ($p, q, m \geq 0$) such that they cover together all vertices of G. It is important to note that the 3-clique $\{u, v, w\}$ does not belong to the covering (this would contradict the irreducibility and would increase the number of cliques to $n-2$).

ε) i 4-cliques having a triangle, denoted by $\{u, v, w\}$ in common ($i \geq 1$); r 3-cliques, s 3-cliques and l 3-cliques having in common the edge uv, vw and uw, respectively ($r, s, l \geq 0$); p 2-cliques, q 2-cliques and m 2-cliques with an extremity in u, v and w, respectively ($p, q, m \geq 0$) and covering together all vertices of G.

Lemma 3. *Every irreducible covering by $n-3$ cliques of a graph G of order $n \geq 4$ without isolated vertices is of one of the types $\alpha, \beta, \gamma, \delta, \varepsilon$.*

Proof. Let C be an irreducible covering by $n-3$ cliques of G. Since G has no isolated vertex it follows that

$$N_y \cup N_z \cup N_t \cup N_{y,z} \cup N_{y,t} \cup N_{z,t} \cup N_{y,z,t} = X,$$

the set of vertices that are uniquely covered by cliques of C. Every clique in C consists of a vertex of X and a nonempty subset of $Y = \{y, z, t\}$. If vertices y, z, t induce a subgraph isomorphic to $\overline{K}_3$ we obtain type α; if they induce a subgraph consisting of an edge and an isolated vertex we have type β; if they induce a subgraph isomorphic to $K_{1,2}$ we get type γ; if they induce a subgraph isomorphic to K_3 we obtain type δ or ε (depending upon the property that $N_{y,z,t}$ is empty or not).

If G of order n has an irreducible covering by $n-3$ cliques and possesses j isolated vertices $z_1, \ldots, z_j$, then by deleting these vertices we get another graph G_1 of order $n-j$ without isolated vertices, having an irreducible covering by $n-j-3$ cliques, since each of $z_1, \ldots, z_j$ is uniquely covered by a clique reduced to itself. It follows that any irreducible covering by $n-3$ cliques of G is of one of the types α, β, γ, δ, ε, where j isolated vertices were added in each case. From Lemma 3 it also follows that if a graph G of order n has an irreducible covering by $n-3$ cliques, then $\omega(G) \leq 4$. This characterization of the irreducible coverings by $n-3$ cliques of an n-vertex graph allows us to find the number of such coverings for all complete multipartite graphs.

Lemma 4. *Let $1 \leq a \leq b$ and $a+b=n \geq 2$. Then*

$$ic_{n-3}(K_{a,b}) = \begin{cases} 0 & \text{if } a = 1 \text{ or } 2\,; \\ 3^b - 3 \cdot 2^b + 3 & \text{if } a = 3\,; \\ a\binom{b}{2}(2^{a-1}-2) + b\binom{a}{2}(2^{b-1}-2) & \text{if } a \geq 4\,. \end{cases}$$

Proof. Since every clique of $K_{a,b}$ is an edge it follows that every irreducible covering by $n-3$ cliques has type α and $K_{a,b}$ cannot be covered by three vertex disjoint stars if $a = 1$ or 2. The case $a = 3$ was settled by Lemma 2. If $a \geq 4$ the covering condition implies that two centers belong to a part of $K_{a,b}$ and one center belongs to another part of $K_{a,b}$. If two centers u, v belong to the part with

a vertices and one center w belongs to another part (this can be done in $b\binom{a}{2}$ ways), then the stars centered in u and v can be chosen in $2S(b-1,2) = 2^{b-1}-2$ ways and the star centered in w can be uniquely chosen. A similar situation holds by permuting a and b and the formula in this case is justified.

The number of irreducible coverings by $n-3$ cliques of a complete tripartite graph $K_{a,b,c}$ $(a+b+c=n)$ is given by the following lemma.

Lemma 5. *Let $1 \le a \le b \le c$ and $a+b+c=n \ge 3$. Then*

$$ic_{n-3}(K_{a,b,c}) = \begin{cases} 0 & \text{if } a=b=1\ ; \\ 2^c-2 & \text{if } a=1 \text{ and } b=2\ ; \\ abc & \text{otherwise}\ . \end{cases}$$

Proof. $K_{a,b,c}$ has every clique isomorphic to K_3. Hence do not exist coverings of types α, β and ε. If $a=b=1$ also do not exist irreducible coverings of types γ and δ, thus implying $ic_{n-3}(K_{1,1,c})=0$. If there exists an irreducible covering of type γ then u and v being nonadjacent, they belong to a same class of $K_{a,b,c}$. Covering condition implies that the class of v consists only of v and the class of u and w is precisely $\{u,w\}$, thus implying $a=1$ and $b=2$. In this case $ic_{n-3}(K_{1,2,c}) = ic_{n-3}(K_{2,c}) = 2S(c,2) = 2^c-2$ since $c=n-3$ and by Lemma 1 every irreducible covering of $K_{2,n-3}$ has exactly $n-3$ edges. Otherwise ($a=1$ and $b \ge 3$ or $a \ge 2$) every irreducible covering by $n-3$ cliques of $K_{a,b,c}$ is of type δ. For every choice of u,v,w in different parts of $K_{a,b,c}$, there exists a unique irreducible covering C of type δ, where $\{u,v,w\} \notin C$. Since the number of choices of $\{u,v,w\}$ equals abc, it follows that in this case one has $ic_{n-3}(K_{a,b,c}) = abc$.

Lemma 6. *Let $1 \le a \le b \le c \le d$ and $a+b+c+d=n \ge 4$. Then*

$$ic_{n-3}(K_{a,b,c,d}) = \begin{cases} 1 & \text{if } a=b=c=1\ ; \\ 0 & \text{otherwise } (c \ge 2)\ . \end{cases}$$

Proof. Since $K_{a,b,c,d}$ has every clique isomorphic to K_4, every irreducible covering by $n-3$ cliques is of type ε, whenever $r=s=l=p=q=m=0$. If $a=b=c=1$ then we can choose uniquely $\{u,v,w\}$ to be the set of vertices in parts having a, b and c vertices, respectively and there exists a unique irreducible covering by $n-3$ cliques consisting of d 4-cliques having $\{u,v,w\}$ in common. If $c \ge 2$ then at least one vertex from $\{u,v,w\}$, let u, belongs to a part of $K_{a,b,c,d}$ together with another vertex x. In this case $\{x,u,v,w\}$ cannot be a 4-clique since ux is not an edge, hence $ic_{n-3}(K_{a,b,c,d})=0$.

Also note that $ic_{n-3}(G)=0$ for any complete t-multipartite graph G of order n for every $t \ge 5$, since otherwise $\omega(G) \le 4$, a contradiction. In the last section we also need the following result.

Lemma 7. *For every $n \ge 6$,*

$$ic_{n-3}(K_{3,n-3}-e) = 2 \cdot 3^{n-4} - 2^{n-2} + 2,$$

where $K_{3,n-3}-e$ is the graph obtained from $K_{3,n-3}$ by deleting any edge e.

Proof. Let u, v, w be the vertices in the part having three vertices of $K_{3,n-3}$ and the edge e which is deleted be xv. Every irreducible covering by $n-3$ edges of $K_{3,n-3} - e$ has type α. It is not posible that x be covered by a star consisting of edges xu and xw, since the remaining vertices, inducing a subgraph isomorphic to $K_{1,n-4}$ cannot be covered by two vertex disjoint stars. It follows that x is covered either by edge xu or by edge xw. In the first case, other two subcases hold: a) u is not covered by other edges of $K_{3,n-4}$. In this case the number of irreducible coverings by $n-3$ edges equals $ic_{n-4}(K_{2,n-4}) = 2^{n-4} - 2$;
b) u is also covered by other edges of $K_{3,n-4}$. Now the number of irreducible coverings by $n-3$ edges equals $ic_{n-4}(K_{3,n-4}) = 3^{n-4} - 3 \cdot 2^{n-4} + 3$. A similar situation occurs if x is covered by edge xw. Hence $ic_{n-3}(K_{3,n-3} - e) = 2(3^{n-4} - 2^{n-3} + 1)$.

Note that $ic_{n-3}(K_{3,n-3} - e) < ic_{n-3}(K_{3,n-3})$ for every $n \geq 6$.

3 Maximum Number of Irreducible Coverings by $n-3$ Cliques

Theorem 1. *There exists a fixed $n_0 (n_0 \geq 7)$ such that for every $n \geq n_0$ the following properties hold:*
i) $I(n, n-3) = 3^{n-3} - 3 \cdot 2^{n-3} + 3$;
ii) If a graph G of order n verifies $ic_{n-3}(G) = I(n, n-3)$ then $G \cong K_{3,n-3}$;
iii) If G reaches the maximum number of irreducible coverings by $n-3$ cliques in the class of graphs of order n that are not isomorphic to $K_{3,n-3}$, then $G \cong K_{3,n-3} - e$.

Proof. By Lemma 3 every irreducible covering by $n-3$ cliques of a graph G of order $n \geq 4$ is of one of types $\alpha, \dots, \varepsilon$, eventually with some isolated vertices added. The idea of proof is the following: we will prove that if $G \ncong K_{3,n-3}$ then either $G \cong K_{3,n-3} - e$ or $ic_{n-3}(G)$ is bounded above by $\alpha 3^{n-4} + P(n)2^n$, where $0 < \alpha < 2$ and $P(n)$ is a polynomial in n of a fixed degree, which is asymptotically smaller than $ic_{n-3}(K_{3,n-3} - e) = 2 \cdot 3^{n-4} - 2^{n-2} + 2$.

First the numbers of irreducible coverings C of types β, γ, δ and ε will be bounded above as follows: the number of irreducible coverings of types β or γ is less than or equal to $3\binom{n}{3}2^{n-3}$ and the number of irreducible coverings of types δ and ε is less than or equal to $\binom{n}{3}$. Indeed, suppose that C is of type β. Vertices u, v and w can be chosen in at most $3\binom{n}{3}$ ways. Covering condition implies that every vertex $x \notin \{u, v, w\}$ is adjacent to u, v or w. If x is adjacent to only one vertex of the set $\{u, v, w\}$, then x can be uniquely covered. The same situation holds if x is adjacent only to u and v; otherwise x can be covered in two ways by 2-cliques or 3-cliques. A similar situation holds for type γ: every $x \notin \{u, v, w\}$ can be covered uniquely or in two ways (whenever x is adjacent only to u and w or it is adjacent to all u, v and w). For types δ and ε vertices u, v and w induce K_3 and can be chosen in at most $\binom{n}{3}$ ways and any other vertex x can be uniquely covered since $\omega(G) \leq 4$. It follows that the number of irreducible coverings by $n-3$ cliques of G

of types β, γ, δ and ε is bounded above by $Q(n)2^n$, where $Q(n)$ is a polynomial in n of the third degree. Let $ic^{\alpha}_{n-3}(G)$ denote the number of irreducible coverings by $n-3$ cliques, of type α of G. It remains to show that if $G \not\cong K_{3,n-3}$ or $K_{3,n-3}-e$ then $ic^{\alpha}_{n-3}(G) \leq \alpha 3^{n-4} + R(n)2^n$, where $0 < \alpha < 2$ and $R(n)$ is a polynomial in n of a fixed degree. If G contains a triangle $\{a, b, c\}$ then by suppressing any edge of this subgraph the number of irreducible coverings by $n-3$ cliques, of type α is not decreasing. By repeating this procedure we finally get a triangle-free graph G' of order n such that $ic^{\alpha}_{n-3}(G') \geq ic^{\alpha}_{n-3}(G)$. So we shall suppose that $G \not\cong K_{3,n-3}$ or $K_{3,n-3} - e$ and G does not contain triangles, hence each clique is an edge. It is necessary to find a suitable upper bound of the number of irreducible coverings by $n-3$ edges of G of type α for all possible choices of the centers u, v, w of stars in the sets $N_y, N_z, N_t, N_{y,z}, N_{y,t}, N_{z,t}$ and $N_{y,z,t}$ such that the covering condition $N^*(u) \cup N^*(v) \cup N^*(w) = V(G)$ is satisfied. For example, if $u \in N_{y,z}, v \in N_{z,t}$ and $w \in N_{y,t}$ there is no vertex x adjacent to all u, v, w, hence the number of irreducible coverings by $n-3$ edges, of type α is bounded above in this case by $\binom{n}{3}2^{n-3}$; if $u \in N_{y,z}, v \in N_{z,t}, w \in N_z$, then z is the only vertex which is adjacent to all u, v, w and the upper bound is similar as above; if $u \in N_{y,z}, v \in N_y, w \in N_z$ then t cannot be covered, and so on.

Let $a = |N_y|, b = |N_z|, c = |N_t|, d = |N_{y,z}|, e = |N_{y,t}|, f = |N_{z,t}|, g = |N_{y,z,t}|$ and the functions: $\varphi(z,t) = 1$ if $zt \notin E(G)$ and $\varphi(z,t) = 2$ otherwise; $\psi(z,t) = 2$ if $zt \notin E(G)$ and $\psi(z,t) = 3$ if $zt \in E(G)$; $\eta(y,z,t) = 1$ if y, z, t induce $\overline{K}_3$ and $\eta(y,z,t) = 0$ otherwise; $\mu(y,z,t) = 1$ if y is not adjacent to z and t; $\mu(y,z,t) = 2$ if y is adjacent to exactly one vertex from the set $\{z,t\}$; $\mu(y,z,t) = 3$ if y is adjacent to both z and t. Note that $a+b+c+d+e+f+g \leq n-3$ and the inequality is strict when G contains isolated vertices.

By enumerating all possible cases (the verification of this procedure is left to the reader) one finds that only in the following five cases can exist more than one vertex which is adjacent to all centers u, v and w:

i) $u, v \in N_{y,z}$ and $w = t$, and other two cases deduced by circular permutations of y, z, t. Covering condition implies $N_{y,z} = \{u, v\}, N_y = N_z = \emptyset$ since u and v cannot be adjacent to any vertex in N_y or N_z, hence centers u, v, w can be chosen in a single way. Hence in this case $d = 2, a = b = 0$ and the upper bound has the form $\psi(y,t)\psi(z,t)3^c + \psi(t,z)\psi(y,z)3^b + \psi(t,y)\psi(z,y)3^a$ since for example for $N_{y,z} = \{u, v\}$ and $w = t$ only the c vertices of N_t can be covered in three ways. Other vertices can be uniquely covered, except y and z: if $yt \notin E(G)$ then y can be covered in two ways and if $yt \in E(G)$ we get three ways etc.

ii) $u \in N_y, v \in N_z, w = t$ and other two cases deduced by permutations. Covering condition implies $N_{y,z} = \emptyset$ (or $d = 0$) since u, v and w cannot be adjacent to any vertex in $N_{y,z}$. For any choice of $u \in N_y$ and $v \in N_z$ only vertices in N_t can be covered in at most three ways and vertices in $N_{y,t}$ and $N_{z,t}$ in at most two ways. Suppose that $N_y = \{u = u_0, u_1, \ldots, u_p\}$ and $N_z = \{v = v_0, v_1, \ldots, v_q\}$, where $p, q \geq 0$. Covering condition implies that $uv_1, \ldots, uv_q \in E(G)$ and $vu_1, \ldots, vu_p \in E(G)$. If a vertex $x \in N_t$ is adjacent to both u and v (and hence can be covered in three ways), the inexistence of triangles in G implies that $uv, xu_1, \ldots, xu_p, xv_1, \ldots, xv_q \notin E(G)$, hence x can be covered in a

unique way for all other choices of u and v. Hence by neglecting the terms of the form $S(n)2^n$, the upper bound in this case has the form $\varphi(y,t)\varphi(z,t)3^c2^{e+f} + \varphi(y,t)\varphi(y,z)3^a2^{e+d} + \varphi(y,z)\varphi(t,z)3^b2^{d+f}$.
iii) $u \in N_y, v = z, w = t$ and other two cases deduced by permutations. Covering condition implies $N_y = \{u\}$, hence $a = 1$ and centers u, v, w can be chosen in a single way. Only vertices of $N_{z,t}$ can be covered in three ways and vertices of N_t, N_z and $N_{y,z,t}$ in two ways. The upper bound is in this case $\mu(y,z,t)3^f2^{g+b+c} + \mu(z,y,t)3^e2^{g+a+c} + \mu(t,y,z)3^d2^{g+a+b}$.
iv) $u = y, v = z, w = t$. Since G does not contain triangles it follows that the upper bound is $\eta(y,z,t)3^g2^{d+e+f}$.
v) $u \in N_{y,z}, v \in N_y, w = t$ and other two cases deduced by permutations, or $u \in N_{y,z}, v \in N_z, w = t$ and other two cases deduced by permutations.
In the first case the covering condition implies $N_{y,z} = \{u\}$, $N_y = \{v\}$, hence $a = d = 1$ and centers u, v, w can be chosen in a single way. The vertices of N_t can be covered in three ways and the vertices of $N_{z,t}$ in two ways, hence the upper bound is

$$\varphi(z,t)\psi(y,t)3^c2^f + \varphi(z,y)\psi(t,y)3^a2^e + \varphi(y,z)\psi(t,z)3^b2^d$$
$$+\varphi(y,t)\psi(z,t)3^c2^e + \varphi(t,y)\psi(z,y)3^a2^d + \varphi(t,z)\psi(y,z)3^b2^f.$$

We will analyze two subcases: The centers are in the case i) (subcase I) or not (subcase II).

I. Note that if two cases occur at i) then the upper bound is not exponential since this implies $N_y = N_z = N_t = \emptyset$, hence $a = b = c = 0$. Suppose that $u, v \in N_{y,z}$ and $w = t$, which imply $|N_{y,z}| = 2$ and $N_y = N_z = \emptyset$. It follows that case ii) cannot hold since in this case at least two sets from N_y, N_z and N_t must be nonempty. Case iii) can hold only when $u \in N_t$. We have seen that only vertices of $N_{y,z}$ can be covered in three ways, but $|N_{y,z}| = 2$ and the upper bound produced in the case iii) is less than $P_1(n)2^n$ for large n. Case v) can also hold when $u \in N_{y,t}, v \in N_t, w = z$ or $u \in N_{z,t}, v \in N_t, w = y$. Since $N_y = N_z = \emptyset$ there is no vertex which can be covered in three ways and the upper bound corresponding to v) is less than $P_2(n)2^n$ for large n. It follows that by deleting terms less than $P_3(n)2^n$ the upper bound for $ic^{\alpha}_{n-3}(G)$ is equal to $S_1 = \psi(y,t)\psi(z,t)3^c + \eta(y,z,t)3^g2^{d+e+f}$. We have $d = |N_{y,z}| = 2, a = b = 0, c, e, f, g \geq 0$ and $c+e+f+g \leq n-5$. The global maximum of S_1 can be found by comparing local maxima of each term $T_1 = \psi(y,t)\psi(z,t)3^c$ and $T_2 = \eta(y,z,t)3^g2^{2+e+f}$. T_1 has a maximum for $c = n-5, d = 2, a = b = e = f = g = 0; \psi(y,t) = \psi(z,t) = 3$, hence $yt \in E(G), zt \in E(G)$ and all $n-5$ vertices of N_t are adjacent to u, v and t. In this case $G \cong K_{3,n-3}$ which contradicts the hypothesis. T_2 has a maximum for $\eta(y,z,t) = 1, d = 2, g = n-5, a = b = c = e = f = 0$ and in this case S_1 has the dominant term equal to $4 \cdot 3^{n-5} < 2 \cdot 3^{n-4}$. The second greatest value of T_1 occurs for $c = n-6, \psi(y,t) = \psi(z,t) = 3$ and in this case S_1 has the dominant term equal to $3^{n-4} < 2 \cdot 3^{n-4}$. The situation is similar if other two remaining cases occur at i).

II. If centers u, v and w are not in the case i), we have an upper bound deduced by adding the upper bounds obtained in the cases ii)–v): $S_2 = \varphi(y,t)\varphi(z,t)3^c2^{e+f} +$

$\varphi(y,t)\varphi(y,z)3^a2^{e+d} + \varphi(y,z)\varphi(t,z)3^b2^{d+f} + \mu(y,z,t)3^f2^{g+b+c} + \mu(z,y,t)3^e 2^{g+a+c} + \mu(t,y,z)3^d2^{g+a+b} + \eta(y,z,t)3^g2^{d+e+f} + (\varphi(z,t)\psi(y,t)2^f + \varphi(y,t)\psi(z,t)2^e)3^c + (\varphi(z,y)\psi(t,y)2^e + \varphi(t,y)\psi(z,y)2^d)3^a + (\varphi(y,z)\psi(t,z)2^d + \varphi(t,z)\psi(y,z)2^f)3^b$.

In fact there are some incompatibilities between some terms in S_2 and they cannot appear simultaneously. $T_3 = \varphi(y,t)\varphi(z,t)3^c2^{e+f}$ is incompatible with $T_4 = (\varphi(z,t)\psi(y,t)2^f + \varphi(y,t)\psi(z,t)2^e)3^c$ since the first appeared in the case ii) when $N_{y,z} = \emptyset$ and the second in the case v) when $|N_{y,z}| = 1$. A similar situation holds for terms containing factors 3^a and 3^b. By taking into account these incompatibilities and the conditions upon $a, b, \ldots, f \geq 0$ deduced from the properties of the sets $N_y, \ldots, N_{y,z,t}$ in the cases ii)-v), the local maxima of S_2 can occur for:

- $a = b = 1$, $c = n-5$, $d = e = f = g = 0$ and S_2 contains T_3, when the dominant term of S_2 is bounded above by $4 \cdot 3^{n-5} < 2 \cdot 3^{n-4}$, and other two similar cases are proved in the same way.
- $a = d = 1$, $c = n-5$, $b = e = f = g = 0$ and S_2 contains T_4. In this case G may have $ic^{\alpha}_{n-3}(G)$ maximum only if all $n-5$ vertices of N_t are adjacent to u, v and $w = t$, y is adjacent to u, v and w and z is adjacent only to u and w. It follows that $G \cong K_{3,n-3} - e$, a contradiction. The second greatest value of T_4 occurs e.g. for $a = d = f = 1$, $c = n-6$, $b = e = g = 0$. Since $f = |N_{z,t}| \geq 1$ and G does not contain triangles, it follows that $zt \notin E(G)$, which implies $\varphi(z,t) = 1, \psi(z,t) = 2$. We get $T_4 \leq (2 \cdot 2 + 2 \cdot 3)3^{n-6} = 10 \cdot 3^{n-6} < 2 \cdot 3^{n-4}$. Other two cases when a and b, respectively are maximized are proved similarly.
- $a = 1$, $f = n-4$, $b = c = d = e = g = 0$. In this case the dominant term of S_2 is equal to 3^{n-3} only if $\mu(y,z,t) = 3$, i.e., y is adjacent to t and z, and all $n-4$ vertices of $N_{z,t}$ are adjacent to $u, v = z$ and $w = t$. In this case $G \cong K_{3,n-3}$, which contradicts the hypothesis. The second greatest value of $\mu(y,z,t)3^f2^{g+b+c}$ may be reached for $a = b = 1$, $f = n-5$, $c = d = e = g = 0$ or $a = c = 1$, $f = n-5$, $b = d = e = g = 0$ (this case is similar to the previous one), or $a = g = 1$, $f = n-5$, $b = c = d = e = 0$. In the first case $\mu(y,z,t) = 3$ if $yz, yt \in E(G)$ and $n-5$ vertices are adjacent to $u, v = z$ and $w = t$, y is adjacent to u, v, w and the unique vertex from N_z is adjacent only to $z = v$ and u. Hence in this case $G \cong K_{3,n-3} - e$, a contradiction. In the third case $g = |N_{y,z,t}| = 1$, hence $yz, yt \notin E(G)$ since G does not contain triangles. This implies $\mu(y,z,t) = 1$ and $\mu(y,z,t)3^f2^{g+b+c} = 2 \cdot 3^{n-5}$.The third greatest value equals $4 \cdot 3^{n-5} < 2 \cdot 3^{n-4}$.
- $g = n-3$, $a = b = c = d = e = f = 0$ and $\eta(y,z,t) = 1$ (i.e., y, z, t induce $\overline{K}_3$) maximizes $\eta(y,z,t)3^g2^{d+e+f}$. But in this case $G \cong K_{3,n-3}$, which contradicts the hypothesis. The second greatest value of this term is reached for $\eta(y,z,t) = 1$ and $g = n-4$, d or e or f are equal to 1 and other variables are equal to 0. In this case $n-5$ vertices are adjacent to y, z and t, one vertex is adjacent to exactly two vertices from y, z and t and y, z, t induce $\overline{K}_3$. Hence $G \cong K_{3,n-3} - e$, a contradiction. The third greatest value equals $4 \cdot 3^{n-5} < 2 \cdot 3^{n-4}$.

Consequently, by neglecting terms of the form $S(n)2^n$, where $S(n)$ is a polynomial in n of a fixed degree, if G is not isomorphic to $K_{3,n-3}$ or $K_{3,n-3} - e$, the dominant term of the upper bound of $ic^{\alpha}_{n-3}(G)$ is at most $4 \cdot 3^{n-5}$ in both cases I and II. This completes the proof. Since $ic_3(K_{3,3}) = ic_3(K_{1,2,3}) = ic_3(K_{3,3}+e) = 6$ [8], where $K_{3,3}+e$ denotes $K_{3,3}$ plus one edge joining two nonadjacent vertices of $K_{3,3}$, it follows that $n_0 \geq 7$.

Let $G(k, n-k)$ be the graph of order n consisting of a set of $n-k$ pairwise nonadjacent vertices joined in all ways by edges to the vertices of $MMG(k)$, the extremal graph of order k having $\alpha(k)$ cliques. Since $\lim_{n\to\infty}(ic_{n-k}(G(k,n-k)))^{1/n} = \alpha(k)$ [8], the following conjecture seems to be plausible:

Conjecture 1. For every $k \geq 2$ there exists a fixed $n_0(k)$ such that for every $n \geq n_0(k)$ one has $I(n, n-k) = ic_{n-k}(G(k, n-k))$ and every graph G of order n such that $ic_{n-k}(G) = I(n, n-k)$ is isomorphic to $G(k, n-k)$.

This is true at least for $k = 2$ and 3.

References

1. Bollobás, B.: *Modern graph theory.* Springer-Verlag, New York (1998)
2. Even, S: *Algorithmic combinatorics.* Macmillan, New York (1973)
3. Miller, R. E., Muller, D. E.: A problem of maximum consistent subsets. *IBM Research Report RC*-240. J. T. Watson Research Center, Yorktown Heights, New York (1960)
4. Moon, J. W., Moser, L.: On cliques in graphs. *Israel J. Math.*, 3(1965) 23-28.
5. Paull, M. C., Unger, S. H.: Minimizing the number of states in incompletely specified sequential functions. *IRE Trans. Electronic Computers*, Vol. EC-8 (1959) 356-367
6. Petrick, S. R.: A direct determination of the irredundant forms of a Boolean function from the set of prime implicants. *AFCRC-TR*-56-110, Air Force Cambridge Research Center (1956)
7. Tomescu, I.: Combinatorial methods in the theory of finite automata (in French). In: *Logique, Automatique, Informatique*, Ed. Acad. R.S.R., Bucharest (1971) 269-423
8. Tomescu, I.: Some properties of irreducible coverings by cliques of complete multipartite graphs. *J. of Combinatorial Theory* B28, 2(1980) 127-141
9. Tomescu, I.: On the number of irreducible coverings by edges of complete bipartite graphs. *Discrete Mathematics* 150(1996) 453-456

On Reachability in Graphs with Bounded Independence Number

Arfst Nickelsen and Till Tantau*

Technische Universität Berlin
Fakultät für Elektrotechnik und Informatik
10623 Berlin, Germany
{nicke,tantau}@cs.tu-berlin.de

Abstract. We study the reachability problem for finite directed graphs whose independence number is bounded by some constant k. This problem is a generalisation of the reachability problem for tournaments. We show that the problem is first-order definable for all k. In contrast, the reachability problems for many other types of finite graphs, including dags and trees, are not first-order definable. Also in contrast, first-order definability does not carry over to the infinite version of the problem. We prove that the number of strongly connected components in a graph with bounded independence number can be computed using TC^0-circuits, but cannot be computed using AC^0-circuits. We also study the succinct version of the problem and show that it is Π_2^{P}-complete for all k.

1 Introduction

One of the most fundamental problems in graph theory is the reachability problem. For this problem we are asked to decide whether there exists a path from a given source vertex s to a given target vertex t in some graph G. For finite directed graphs this problem, which will be denoted REACH in the following, is well-known to be NL-complete [12,13]. It is thus easy from a computational point of view and efficient parallel algorithms are known for it. The complexity of the reachability problem drops if we restrict the type of graphs for which we try to solve it. The reachability problem $\mathrm{REACH_u}$ for finite undirected graphs is SL-complete [15] and thus presumably easier to solve. The even more restricted problem $\mathrm{REACH_{forest}}$ for undirected forests and the problem $\mathrm{REACH_{out\leq 1}}$ for directed graphs in which all vertices have out-degree at most 1 are L-complete [2].

In this paper we study the reachability problem for finite directed graphs whose independence number is bounded by some constant k. The independence number $\alpha(G)$ of a graph G is the maximum number of vertices that can be picked from G such that there is no edge between any two of these vertices. Thus we study the languages $\mathrm{REACH}_{\alpha\leq k} := \mathrm{REACH} \cap \{\langle G, s, t\rangle \mid \alpha(G) \leq k\}$ for constant k, where $\langle\rangle$ denotes a standard binary encoding. We show that, somewhat surprisingly, $\mathrm{REACH}_{\alpha\leq k}$ is first-order definable for all k.

* Work done in part while visiting the University of Rochester, New York. Supported by a TU Berlin Erwin-Stephan-Prize grant.

O.H. Ibarra and L. Zhang (Eds.): COCOON 2002, LNCS 2387, pp. 554–563, 2002.

First-order definability means the following. Let $\tau = (E^2, s, t)$ be the signature of directed graphs with two distinguished vertices. The binary relation symbol E represents an edge relation and the constant symbols s and t represent a source and a target vertex. We show that for each k there exists a first-order formula $\phi_{reach,\alpha \leq k}$ over the signature τ for which the following holds: for all *finite* directed graphs $G = (V, E)$ and all $s, t \in V$ the τ-structure (V, E, s, t) is a model of $\phi_{reach,\alpha \leq k}$ iff $\alpha(G) \leq k$ and there is path from s to t in G. The formulas will neither require an ordering on the universe nor the bit predicate [11].

The most prominent examples of graphs with bounded independence number are *tournaments* [18,20], which are directed graphs with exactly one edge between any two vertices. Their independence number is 1. Conditions for strong connectedness of tournaments (and thus, implicitly, for reachability) were proven in [9], but these conditions yield weaker bounds on the complexity of the reachability problem for tournaments than those shown in the present paper. A different example of graphs with bounded independence number, studied in [4], are directed graphs $G = (V, E)$ whose underlying undirected graph is claw-free, i. e., does not contain the $K_{1,m}$ for some constant m, and whose minimum degree is at least $|V|/3$. Their independence number is at most $3m - 3$.

Languages whose descriptive complexity is first-order are known to be very simple from a computational point of view. They can be decided by a family of AC^0-circuits (constant depth circuits) and also in constant parallel time on concurrent-read, concurrent-write parallel random access machines [16]. Since it is known that L-hard sets cannot be first-order definable [1,6], $\text{REACH}_{\alpha \leq k}$ is (unconditionally) easier to solve than REACH, REACH_{u}, and $\text{REACH}_{\text{forest}}$.

A problem closely related to the reachability problem is the problem of identifying the strongly connected components of a graph. We show that TC^0-circuits (constant depth circuits with threshold gates) can count the strongly connected components in graphs with bounded independence number, but AC^0-circuits cannot—not even in tournaments.

In hardware design one is often concerned with succinctly represented graphs, which are given implicitly via a program or a circuit that decides the edge relation of the graph. Papadimitriou, Yannakakis, and Wagner [19,23,24] have shown that the problems SUCCINCT-REACH, $\text{SUCCINCT-REACH}_{\text{u}}$, $\text{SUCCINCT-REACH}_{\text{forest}}$, and $\text{SUCCINCT-REACH}_{\text{out} \leq 1}$ are PSPACE-complete. Opposed to this, we show that $\text{SUCCINCT-REACH}_{\alpha \leq k}$ is Π_2^{P}-complete for all k.

Our results apply only to finite graphs. Let $\text{REACH}^{\infty}_{\alpha \leq k}$ be the class of all triples (G, s, t) such that G is a (possibly infinite) directed graph with $\alpha(G) \leq k$ in which there is a path from s to t. We show that there does not exist a set of first-order formulas (not even an uncountable one) whose class of models is exactly $\text{REACH}^{\infty}_{\alpha \leq k}$ for some k.

This paper is organised as follows. In Section 2 we study graph-theoretic definitions and results and prove a general theorem that shows how the independence number of a graph is connected to its different domination numbers. We believe this theorem to be of independent interest. In Section 3 we show that the problem $\text{REACH}_{\alpha \leq k}$ is first-order definable, by explicitly giving a defining formula. In Section 4 we study the circuit complexity of counting the number of strongly connected components in a graph. In Sections 5 we study the infinite version of our problem and in Section 6 the succinct version.

2 Graph-Theoretic Definitions and Results

In this section we first give definitions of basic graph-theoretic concepts. Then we prove a generalisation of the so-called lion king lemma, see Theorem 2.2. At the end of the section we prove Theorem 2.3, which will be the crucial building block of our first-order definition of $\text{REACH}_{\alpha \leq k}$.

A *graph* is a nonempty set V of vertices together with a set $E \subseteq V \times V$ of directed edges. Instead of $(x, y) \in E$ we will often write $x \to y$. The *out-degree* of a vertex u is the number of vertices v with $u \to v$. A *path of length* ℓ in a graph $G = (V, E)$ is a sequence $v_0, \dots, v_\ell$ of vertices with $v_0 \to v_1 \to \dots \to v_\ell$. A vertex t is *reachable* from a vertex s if there is a path from s to t. A *strongly connected component* is a maximal vertex set $U \subseteq V$ such that every vertex in U is reachable from every other vertex in U. A set $U \subseteq V$ is an *independent set* if there is no edge in E connecting vertices in U. The maximal size of independent sets in G is its *independence number* $\alpha(G)$. For $i \in \mathbb{N}$, a vertex $u \in V$ is said to *i-dominate* a vertex $v \in V$ if there is a directed path from u to v of length at most i. Let $\text{dom}_i(U)$ denote the set of vertices that are i-dominated by vertices in U. A set $U \subseteq V$ is an *i-dominating set* for G if $\text{dom}_i(U) = V$. The *i-domination number* $\beta_i(G)$ is the minimal size of an i-dominating set for G. A *tournament* is a graph with exactly one edge between any two different vertices and $(v, v) \notin E$ for all $v \in V$. Note that tournaments have independence number 1.

Lemma 2.1. *Let $G = (V, E)$ be a finite graph, $n := |V|$, $\alpha := \alpha(G)$. Then G has at least $\binom{n}{2} / \binom{\alpha+1}{2}$ edges and there exists a vertex with out-degree at least $(n-1) / 2\binom{\alpha+1}{2}$.*

Proof. The number of $(\alpha + 1)$-element subsets of V is $\binom{n}{\alpha+1}$. Every such set contains two vertices linked by an edge. Every such edge is in $\binom{n-2}{\alpha-1}$ different $(\alpha+1)$-element subsets of V. Therefore there are at least $\binom{n}{\alpha+1} / \binom{n-2}{\alpha-1} = \binom{n}{2} / \binom{\alpha+1}{2}$ edges in G. This also shows that the average out-degree in G is at least $\binom{n}{2} / n\binom{\alpha+1}{2} = (n-1) / 2\binom{\alpha+1}{2}$ and one vertex has at least this out-degree. □

Turán [21], referenced in [22], gives an exact formula for the minimal number of edges in a graph as a function of the graph's independence number. However, the simple bound from the above lemma will be more appropriate for our purposes.

Theorem 2.2. *Let $G = (V, E)$ be a finite graph, $n := |V|$, $\alpha := \alpha(G)$. Then $\beta_1(G) \leq \lceil \log_c n \rceil$ and $\beta_2(G) \leq \alpha$, where $c = (\alpha^2 + \alpha)/(\alpha^2 + \alpha - 1)$.*

Proof. We iteratively construct a 1-dominating set D_1 for G of size at most $\lceil \log_c n \rceil$. In each step we put a vertex v_i into D_1 that dominates as many vertices as possible of the subset $V_i \subseteq V$ not dominated so far. Formally, set $V_0 := V$

and for $i \geq 1$, as long as V_{i-1} is not empty, choose a vertex $v_i \in V_{i-1}$ such that $V_i := V_{i-1} \setminus \mathrm{dom}_1(\{v_i\})$ is as small as possible. Let $i_{\max}$ be the first i such that V_i is empty. By Lemma 2.1 the out-degree of v_i is at least $\bigl(|V_{i-1}|-1\bigr) \,/\, 2\binom{\alpha+1}{2}$ and thus

$$|V_i| \leq |V_{i-1}| - 1 - \frac{|V_{i-1}|-1}{2\binom{\alpha+1}{2}} < |V_{i-1}| - \frac{|V_{i-1}|}{2\binom{\alpha+1}{2}}$$
$$= |V_{i-1}|\biggl(1 - \frac{1}{2\binom{\alpha+1}{2}}\biggr) = |V_{i-1}|\biggl(\frac{\alpha^2+\alpha-1}{\alpha^2+\alpha}\biggr) = \frac{|V_{i-1}|}{c}.$$

This shows that the size of V_i decreases by at least the factor c in each step. Thus after at most $\lceil \log_c n \rceil$ iterations the set V_i is empty and $D_1 := \{v_1, \dots, v_{i_{\max}}\}$ is the desired 1-dominating set.

We next construct a 2-dominating set D_2 of size at most α by removing superfluous vertices from D_1. Formally, let $W_{i_{\max}} := \{v_{i_{\max}}\}$ and let $W_{i-1} := W_i$ if $v_i \in \mathrm{dom}_1(W_i)$, and $W_{i-1} := W_i \cup \{v_i\}$ otherwise. Clearly, $D_2 := W_1$ is a 2-dominating set. To prove $|D_2| \leq \alpha$, assume that D_2 contains at least $\alpha + 1$ vertices $v_{i_1}, \dots, v_{i_{\alpha+1}} \in D_1$. Since these vertices cannot be independent, there must exist indices i_r and i_s such that $(v_{i_r}, v_{i_s}) \in E$. By construction of the set D_1, this can only be the case if $i_s > i_r$. But then $v_{i_r} \notin D_2$ by construction of W_{i_r}, a contradiction. □

For tournaments G, Theorem 2.2 yields $\beta_1(G) \leq \log_2(n)$ and $\beta_2(G) = 1$. The first result was first proved by Megiddo and Vishkin in [17], where it was used to show that the dominating set problem for tournaments is not NP-complete, unless $\mathrm{NP} \subseteq \mathrm{DTIME}\bigl(n^{O(\log n)}\bigr)$. The second result is also known as the lion king lemma, which was first noticed by Landau in [14] in the study of animal societies, where the dominance relations on prides of lions form tournaments. It has applications in the study of P-selective sets [10] and many other fields.

Theorem 2.3. *Let $G = (V, E)$ be a finite graph, $n := |V|$, $\alpha := \alpha(G)$, $c := (\alpha^2+\alpha)/(\alpha^2+\alpha-1)$, and $s, t \in V$. Then the following statements are equivalent:*

1. *There is no path from s to t in G.*
2. *There is a subset $D_1 \subseteq V$ with $|D_1| \leq \lceil \log_c n \rceil$ such that* $\mathrm{dom}_1(D_1)$ *is closed under reachability, $s \in \mathrm{dom}_1(D_1)$ and $t \notin \mathrm{dom}_1(D_1)$.*
3. *There is a subset $D_2 \subseteq V$ with $|D_2| \leq \alpha$ such that* $\mathrm{dom}_2(D_2)$ *is closed under reachability, $s \in \mathrm{dom}_2(D_2)$ and $t \notin \mathrm{dom}_2(D_2)$.*

Proof. Both 2 and 3 imply 1, since no path starting at a vertex s inside a set that closed is under reachability can 'leave' this set to arrive at a vertex t outside this set. To show that 1 implies 2, consider the set S of vertices reachable from s in G. Then S is closed under reachability, $s \in S$ and $t \notin S$. The induced graph $G' := \bigl(S, E \cap (S \times S)\bigr)$ also has independence number at most α. Therefore, by Theorem 2.2, the graph G' has a 1-dominating set D_1 of size at most $\lceil \log_c n \rceil$. To show that 1 implies 3, consider the same graph G' once more. By Theorem 2.2 it also has a 2-dominating set D_2 of size at most α. □

3 First-Order Definability of the Problem

In this section we show that reachability in graphs with bounded independence number is first-order definable. We start with a review of some basic notions from descriptive complexity theory.

We use the *signature* or *vocabulary* $\tau = (E^2, s, t)$. It consists of a binary relation symbol E, representing an edge relation, and constant symbols s and t, representing a source and a target vertex. A *τ-structure* is a tuple (V, E, s, t) such that $E \subseteq V \times V$ and $s, t \in V$. We do not distinguish notationally between the symbols in the signature and their interpretation in a structure, because it is always clear from the context which of the two meanings is intended. The standardised binary code of a finite τ-structure (V, E, s, t) will be denoted $\langle V, E, s, t\rangle$. A set A of codes of finite τ-structures is *first-order definable* if there exists a first-order formula ϕ over the signature τ such that for all finite τ-structures (V, E, s, t) we have $(V, E, s, t) \models \phi$ iff $\langle V, E, s, t\rangle \in A$.

Theorem 3.1. *For each k, $\text{REACH}_{\alpha\leq k}$ is first-order definable.*

Proof. Let $k \geq 1$ be fixed. We give a stepwise construction of a formula $\phi_{reach,\alpha\leq k}$ such that $(V, E, s, t) \models \phi_{reach,\alpha\leq k}$ iff $\langle V, E, s, t\rangle \in \text{REACH}_{\alpha\leq k}$. Roughly spoken, the formula $\phi_{reach,\alpha\leq k}$ will say '$\alpha(G) \leq k$ and it is not the case that condition 3 of Theorem 2.3 holds for s and t'.

Let $\phi_{distinct}(v_1, \dots, v_k) \equiv \bigwedge_{i\neq j}[v_i \neq v_j]$. This formula expresses that vertices are distinct. The property '$\alpha(G) \leq k$' can be expressed as follows:

$$\phi_{\alpha\leq k} \equiv (\forall v_1, \dots, v_{k+1})\Big[\phi_{distinct}(v_1, \dots, v_{k+1}) \to \bigvee\nolimits_{i\neq j} E(v_i, v_j)\Big].$$

The next two formulas express that a vertex v, respectively a set $\{v_1, \dots, v_m\}$ of vertices, 2-dominates a vertex u:

$$\phi_{\text{2-}dom}(v, u) \equiv v = u \vee E(v, u) \vee (\exists z)\big[E(v, z) \wedge E(z, u)\big],$$

$$\phi_{\text{2-}dom}(v_1, \dots, v_m, u) \equiv \phi_{\text{2-}dom}(v_1, u) \vee \dots \vee \phi_{\text{2-}dom}(v_m, u).$$

Since $\beta_2(G) \leq \alpha(G) \leq k$, condition 3 of Theorem 2.3 can be expressed as follows:

$$\begin{aligned}\phi_{condition} \equiv{}& (\exists v_1, \dots, v_k)\\ &\Big[\phi_{\text{2-}dom}(v_1, \dots, v_k, s) \wedge \neg\phi_{\text{2-}dom}(v_1, \dots, v_k, t) \wedge\\ &\quad(\forall u, v)\big[(\phi_{\text{2-}dom}(v_1, \dots, v_k, u) \wedge \neg\phi_{\text{2-}dom}(v_1, \dots, v_k, v)) \to \neg E(u, v)\big]\Big].\end{aligned}$$

The desired formula $\phi_{reach,\alpha\leq k}$ is given by $\phi_{\alpha\leq k} \wedge \neg\phi_{condition}$. □

Note that the formula $\phi_{reach,\alpha\leq k}$ constructed in the proof has quantifier alternation depth three, beginning with a universal quantifier.

Theorem 3.1 be easily extended to the following larger class of graphs: define the *r-independence number* $\alpha_r(G)$ of a graph G as the maximal size of an r-independent set in G, which is a vertex subset such that there is no path of length at most r between any two different vertices in this subset. Then reachability in graphs with $\alpha_r(G) \leq k$ is first-order definable for all $k, r \in \mathbb{N}$.

4 Circuit Complexity of the Problem

In this section we study the circuit complexity of the problem $\text{REACH}_{\alpha\leq k}$, as well as the complexity of counting the number of strongly connected components in a graph with bounded independence number. We show that this number can be computed using TC^0-circuits, but cannot be computed using AC^0-circuits.

A family $\mathcal{C} = (C_n)_{n\in\mathbb{N}}$ of circuits is a *family of* AC^0*-circuits* if each C_n has n input gates, their size is bounded by a polynomial in n, their depth is bounded by a constant, and each C_n consist of unbounded fan-in/fan-out and-, or-, and not-gates. For TC^0-circuits we also allow threshold gates, whose output is 1 if the number of 1's at the input exceeds some threshold. For $x \in \{0,1\}^n$ we write $\mathcal{C}(x)$ for the output produced by C_n on input x. The output may be a bitstring since we allow multiple output gates. A circuit family $\mathcal{C}$ *decides* a set $A \subseteq \{0,1\}^*$, respectively *computes* a function $f\colon \{0,1\}^* \to \{0,1\}^*$, if for all $x \in \{0,1\}^*$ we have $x \in A$ iff $\mathcal{C}(x) = 1$, respectively $f(x) = \mathcal{C}(x)$.

As shown by Lindell [16], every first-order definable set can be decided by AC^0-circuits. In particular, by Theorem 3.1 there exists, for each k, an AC^0-circuit family $\mathcal{C}^k$ that decides $\text{REACH}_{\alpha\leq k}$. We now sketch how these families can be used to decrease the average case complexity of REACH, which is L-hard and thus does not have AC^0-circuits [1,6]. Suppose there exists a constant k for which we expect $\alpha(G) \leq k$ to hold with high probability for input graphs G. Then whenever $\alpha(G) \leq k$ holds, we can use $\mathcal{C}^k$ to decide in constant depth whether there is a path from s to t. For graphs with $\alpha(G) > k$ we use a slow standard reachability circuit to decide whether such a path exists. If the probability of $\alpha(G) \leq k$ is sufficiently large, the preprocessing will decrease the average time taken by the circuit to produce its output.

A problem closely related to the reachability problem is the problem of counting strongly connected components. The following theorem pinpoints the exact circuit complexity of this counting problem for graphs with bounded independence number. Let $\zeta_{\alpha\leq k}\colon \{0,1\}^* \to \{0,1\}^*$ be the function that maps the code $\langle G\rangle$ of a graph G to the binary representation of the number of strongly connected components in G if $\alpha(G) \leq k$, and that maps $\langle G\rangle$ to 0 if $\alpha(G) > k$.

Theorem 4.1. *For each k, $\zeta_{\alpha\leq k}$ can be computed by* TC^0*-circuits, but not by* AC^0*-circuits.*

Proof. Let k be fixed. Let $\phi_{reach,\alpha\leq k}(u,v)$ be the formula with two free variables expressing that v is reachable from u and that the underlying graph has independence number at most k. It is obtained from $\phi_{reach,\alpha\leq k}$ from the proof of Theorem 3.1 by replacing the constant symbols s and t by variables u and v. Consider the formula

$$\phi_{rep}(v) \equiv (\forall u)\big[u < v \to \big(\neg\phi_{reach,\alpha\leq k}(u,v) \vee \neg\phi_{reach,\alpha\leq k}(v,u)\big)\big],$$

where '<' is a relation that is interpreted as a total ordering of the set of vertices.

For a graph G with $\alpha(G) \leq k$, the formula $\phi_{rep}(v)$ will be true exactly for the smallest members (with respect to the ordering $<$) of each strongly connected component. Thus, the number of vertices v for which $\phi_{rep}(v)$ holds is exactly the number of strongly connected components in G. Since ϕ_{rep} is a first-order formula, there exists a family of AC^0-circuits that maps $\langle\{v_1,\dots,v_n\},E\rangle$ to a bitstring in which the i-th position is 1 iff $(\{v_1,\dots,v_n\},E) \models \phi_{rep}(v_i)$. Since the number of 1's in this bitstring can be computed in constant depth using threshold gates, $\zeta_{\alpha\le k}$ can be computed by TC^0-circuits.

Next, for the sake of contradiction, assume that there exists an AC^0-circuit family $\mathcal{C}$ that computes $\zeta_{\alpha\le k}$. We construct an AC^0-circuit for the parity function, contradicting the results of Ajtai et al. [1,6]. Let a bitstring $b = b_1 \dots b_n$ be given as input. Define a tournament $G = (\{1,\dots,n+1\},E)$ as follows: for $i+1<j$ there is an edge from j to i; for $i+1=j$ there is an edge from j to i if $b_i = 1$; otherwise there is an edge from i to j. If b contains no 1's, the tournament will form one big circle, thus having just one strongly connected component. Every additional 1 in b adds one strongly connected component. The parity of b is thus given by the toggled least-significant bit of $\mathcal{C}(\langle G\rangle)$. □

5 Infinite Version of the Problem

In this section we study the class $\text{REACH}^{\infty}_{\alpha\le k}$ and show that the results of Section 3 on the first-order definability of $\text{REACH}_{\alpha\le k}$ do not carry over to $\text{REACH}^{\infty}_{\alpha\le k}$. This class contains all triples (G,s,t) such that G is a (possibly infinite) graph with $\alpha(G)\le k$ in which there is a path from s to t. We start with a review of the relevant notions from model theory.

Let τ be a signature. A class K of τ-structures is called *elementary (over finite structures)* if there exists a first-order formula ϕ over τ such that for every (finite) τ-structure $\mathcal{A}$ we have $\mathcal{A}\models\phi$ iff $\mathcal{A}\in K$. (Some authors use 'finitely axiomatisable' instead of 'elementary'.) A class K of τ-structures is *Δ-elementary* if there exists a set Φ of first-order formulas over τ such that for every τ-structure $\mathcal{A}$ we have $\mathcal{A}\models\Phi$ iff $\mathcal{A}\in K$.

Fact 5.1 (Compactness Theorem). *Let Φ be a set of first-order formulas such that every finite $\Phi_0\subseteq\Phi$ has a model. Then Φ has a model.*

With these definitions, Theorem 3.1 simply states that $\text{REACH}^{\infty}_{\alpha\le k}$ is elementary over finite structures for all k. The below proof that $\text{REACH}^{\infty}_{\alpha\le k}$ is not even Δ-elementary follows the standard pattern of proofs applying the compactness theorem. The only essential part is the construction of appropriate model graphs for finite subsets of a hypothetical axiomatisation of $\text{REACH}^{\infty}_{\alpha\le k}$.

Theorem 5.2. *$\text{REACH}^{\infty}_{\alpha\le k}$ is not Δ-elementary for any k.*

Proof. Assume that there exists a set Φ of first-order formulas with $(V,E,s,t)\models\Phi$ iff $(V,E,s,t)\in\text{REACH}^{\infty}_{\alpha\le k}$. For each $n\in\mathbb{N}$ define the following formula ψ_n, which is fulfilled by a graph iff there is a path of length n from s to t.

$$\psi_n \equiv (\exists v_1,\dots,v_{n-1})\big[E(s,v_1)\wedge E(v_1,v_2)\wedge\cdots\wedge E(v_{n-2},v_{n-1})\wedge E(v_{n-1},t)\big].$$

Consider the set $\Psi := \Phi \cup \{\neg\psi_1, \neg\psi_2, \neg\psi_3, \dots\}$. We claim that every finite $\Psi_0 \subseteq \Psi$ has a model (V, E, s, t). To see this, let n be large enough such that for all $i \geq n$ we have $\neg\psi_i \notin \Psi_0$ and define a graph $G = (V, E)$ by $V := \{1, \dots, n+1\}$ and $(i, j) \in E$ iff $j \leq i + 1$. Then $\alpha(G) = 1 \leq k$ and the shortest path from $s := 1$ to $t := n + 1$ has length n. Thus (V, E, s, t) is a model of Ψ_0.

Since every finite subset of Ψ has a model, Ψ has a model (V, E, s, t) by the compactness theorem. Since this model fulfills $\neg\psi_n$ for all n, there can be no path of finite length from s to t in $G = (V, E)$. Thus Φ has a model that is not an element of $\text{REACH}^{\infty}_{\alpha \leq k}$. □

6 Succinct Version of the Problem

In this section we study succinctly represented graphs. Such graphs are given implicitly via a description in some description language. Since succinct representations allow one to encode large graphs into small codes, checking properties is (provably) harder for succinctly represented graphs than for graphs coded in the usual way. Papadimitriou et al. [19,24] have shown that most interesting problems for succinctly represented graphs are PSPACE-complete or even NEXP-complete. The following formalisation of succinct graph representations is due to Galperin and Wigderson [7], but others are also possible [24,8].

Definition 6.1. *A* succinct representation *of a graph $G = (\{0,1\}^n, E)$ is a $2n$-input circuit C such that for all $u, v \in \{0,1\}^n$ we have $(u, v) \in E$ iff $C(uv) = 1$.*

The circuit tells us for any two vertices of the graph whether there is a directed edge between them or not. Note that there is no need to bound the size of C.

Definition 6.2. *Let $A \subseteq \{\langle G, s, t\rangle \mid G = (V, E)$ is a finite graph, $s, t \in V\}$. Then SUCCINCT-A is the set of all codes $\langle C, s, t\rangle$ such that C is a succinct representation of a graph G with $\langle G, s, t\rangle \in A$.*

Theorem 6.3. *For each k, SUCCINCT-REACH$_{\alpha \leq k}$ is Π_2^{P}-complete.*

Proof. We first show SUCCINCT-REACH$_{\alpha \leq k} \in \Pi_2^{\mathrm{P}}$. Let $\langle C, s, t\rangle$ be an input and let C represent a graph $G = (V, E)$ with $V = \{0,1\}^n$. Note that $\log_2 |V| = n$. We first check whether $\alpha(G) \leq k$, which can easily be done using a coNP-machine. We then check whether there is path from s to t in G. By Theorem 2.3 this is case iff *for all* sets $D_1 \subseteq \{0,1\}^n$ of size at most $\beta_1(G)$ either $s \notin \text{dom}_1(D_1)$ or $t \in \text{dom}_1(D_1)$ or $\text{dom}_1(D_1)$ is not closed under reachability, i.e., *there exist* vertices $u \in \text{dom}_1(D_1)$ and $v \in \{0,1\}^n \setminus \text{dom}_1(D_1)$ such that $C(uv) = 1$. Since $\beta_1(G) \leq \lceil \log_c 2^n \rceil \leq \lceil n / \log_2 c \rceil$, the size of the D_1's that need to be checked is linear in n. Thus the 'for all ... exists ...' test is a Π_2^{P}-algorithm, since a membership test for the set $\text{dom}_1(D_1)$ can be performed in polynomial time.

We now prove that even the reachability problem SUCCINCT-REACH$_{\text{tourn}}$ for tournaments is Π_2^{P}-hard. Let $L \in \Pi_2^{\mathrm{P}}$ be any language. By the quantifier characterisation of the polynomial hierarchy [25] there exists a polynomial time decidable ternary relation R and a constant c such that

$$L = \{x \mid (\forall y, |y| = |x|^c)(\exists z, |z| = |x|^c)[R(x, y, z)]\}.$$

We construct a reduction from L to SUCCINCT-REACH$_{\text{tourn}}$. On input x we construct, in polynomial time, a circuit C and two bitstrings s, t such that $x \in L$ iff $\langle C, s, t\rangle \in$ SUCCINCT-REACH$_{\text{tourn}}$. Let n denote the length of x and let $\ell := n^c$.

The circuit C will represent a highly structured tournament G of exponential size. The vertex set of G is $V = \{0,1\}^{2\ell+1}$. Each vertex $v \in V$ can be split into a 'y-component' $y \in \{0,1\}^{\ell+1}$ and a 'z-component' $z \in \{0,1\}^{\ell}$ with $yz = v$. All vertices that have the same y-component form a *level.* All vertices on the same level are connected such that they form a strongly connected subtournament of G. We say a level is *above* another level if its y-component is lexicographically larger than the other level's y-component.

Edges between different levels generally point 'downwards', i. e., from higher levels to lower levels. The only exception are edges between a vertex with y-component $0\tilde{y}$ with $\tilde{y} \in \{0,1\}^{\ell}$ and the vertex with the same z-component on the level directly above. Such an edge points 'upwards' iff $R(x, \tilde{y}, z)$. The source is any vertex on the bottom level, the target is any vertex on level 10^{ℓ}.

The graph G is a tournament and the representing circuit C can be constructed in polynomial time. From each level y one can go (at best) only one level higher to the next level y', since all edges between non-neighbouring levels point downwards. Since all vertices on the same level are connected, if one can reach a vertex v on level $0\tilde{y}$, one can reach a vertex on the level directly above iff $R(x, y', z)$ holds for some $z \in \{0,1\}^{\ell}$. So in order to get from the source to the target, for all $\tilde{y} \in \{0,1\}^{\ell}$ there must exist a $z \in \{0,1\}^{\ell}$ such that $R(x, \tilde{y}, z)$. □

7 Conclusion and Open Problems

We showed that the complexity of the reachability problem for graphs with bounded independence number is lower than the complexity of the corresponding problem for, say, forests. However, we did not claim that is also easier to *actually find a path* in a tournament. While it is easily seen that there is a function in FL that maps every forest to a path from the first to the last vertex, provided such a path exists, we do not know whether such a function exists for tournaments. We recommend this problem for further research.

We do not know whether the three levels of quantifier alternation in the first-order formula for REACH$_{\alpha \le k}$ are necessary, but conjecture that this is the case. Since we do not refer to an ordering relation in our first-order formula, it seems promising to use an Ehrenfeucht-Fraïssé game [3,5] to prove this.

In the succinct setting, we proved that the problem SUCCINCT-REACH$_{\alpha \le k}$ is Π_2^{P}-complete for all k. Opposed to this, for $r > 1$ our arguments only show SUCCINCT-REACH$_{\alpha_r \le k} \in \Pi_3^{\mathrm{P}}$. In particular, we would like to know the exact complexity of SUCCINCT-REACH$_{\alpha_2 \le 1}$.

Acknowledgments

We would like to thank Mitsu Ogihara, Ken Regan, Alan Selman and Leen Torenvliet for helpful discussions.

References

1. M. Ajtai. Σ_1^1 formulae on finite structures. *Annals of Pure and Applied Logic*, 24:1–48, 1983.
2. S. Cook and P. McKenzie. Problems complete for deterministic logarithmic space. *J. Algorithms*, 8(3):385–394, 1987.
3. A. Ehrenfeucht. An application of games to the completeness problem for formalized theories. *Fundamenta Mathematicae*, 49:129–141, 1961.
4. R. Faudree, R. Gould, L. Lesniak, and T. Lindquester. Generalized degree conditions for graphs with bounded independence number. *J. Graph Theory*, 19(3):397–409, 1995.
5. R. Fraïssé. Sur quelques classifications des systèmes de relations. *Publ. Sci. Univ. Alger. Sér. A*, 1:35–182, 1954.
6. M. Furst, J. Saxe, and M. Sipser. Parity, circuits, and the polynomial-time hierarchy. *Math. Systems Theory*, 17(1):13–27, 1984.
7. H. Galperin and A. Wigderson. Succinct representations of graphs. *Inform. Control*, 56(3):183–198, 1983.
8. G. Gottlob, N. Leone, and H. Veith. Succinctness as a source of complexity in logical formalisms. *Annals of Pure and Applied Logic*, 97(1–3):231–260, 1999.
9. F. Harary and L. Moser. The theory of round robin tournaments. *Amer. Math. Monthly*, 73:231–246, 1966.
10. L. Hemaspaandra and L. Torenvliet. Optimal advice. *Theoretical Comput. Sci.*, 154(2):367–377, 1996.
11. N. Immerman. *Descriptive Complexity*. Springer-Verlag, 1998.
12. N. Jones. Space-bounded reducibility among combinatorial problems. *J. Comput. Syst. Sci.*, 11(1):68–85, 1975.
13. N. Jones, Y. Lien, and W. Laaser. New problems complete for nondeterministic log space. *Math. Systems Theory*, 10:1–17, 1976.
14. H. Landau. On dominance relations and the structure of animal societies, III: The condition for secure structure. *Bull. Mathematical Biophysics*, 15(2):143–148, 1953.
15. H. Lewis and C. Papadimitriou. Symmetric space-bounded computation. *Theoretical Comput. Sci.*, 19(2):161–187, 1982.
16. S. Lindell. A purely logical characterization of circuit uniformity. In *Proc. 7th Struc. in Complexity Theory Conf.*, pages 185–192. IEEE Computer Society, 1992.
17. N. Megiddo and U. Vishkin. On finding a minimum dominating set in a tournament. *Theoretical Comput. Sci.*, 61(2–3):307–316, 1988.
18. J. Moon. *Topics on Tournaments*. Holt, Rinehart, and Winston, 1968.
19. C. Papadimitriou and M. Yannakakis. A note on succinct representations of graphs. *Inform. Control*, 71(3):181–185, 1986.
20. K. Reid and L. Beineke. Tournaments. In *Selected Topics in Graph Theory*, pages 169–204. Academic Press, 1978.
21. P. Turán. Eine Extremalaufgabe aus der Graphentheorie (in Hungarian). *Matem. és Physikai Lapok*, 48:436–452, 1941.
22. P. Turán. On the theory of graphs. *Colloquium Math.*, 3:19–30, 1954.
23. K. Wagner. The complexity of problems concerning graphs with regularities. In *Proc. 7th Symposium on Math. Foundations of Comp. Sci.*, volume 176 of *Lecture Notes in Computer Science*, pages 544–552. Springer-Verlag, 1984.
24. K. Wagner. The complexity of combinatorial problems with succinct input representation. *Acta Informatica*, 23(3):325–356, 1986.
25. C. Wrathall. Complete sets and the polynomial-time hierarchy. *Theoretical Comput. Sci.*, 3(1):23–33, 1976.

On Parameterized Enumeration

Henning Fernau*

University of Newcastle, School of CS and EE
University Drive, NSW 2308 Callaghan, Australia
fernau@cs.newcastle.edu.au

Abstract. We study several versions of parameterized enumeration. The idea is always to have an algorithm which outputs all solutions (in a certain sense) to a given problem instance. Such an algorithm will be analysed from the viewpoint of parameterized complexity. We show how to apply enumeration techniques in a number of examples. In particular, we give a fixed parameter algorithm for the reconfiguration of faulty chips when providing so-called shared and linked spares.

1 Introduction

In classical complexity theory, there are three main ways to build complexity classes or classify computational problems, respectively: decision problems, functional problems, and counting problems.

As a running example, let us consider the *vertex cover problem* (VC) for undirected graphs. This yields the following decision problem (already formulated with a glimpse on parameterized complexity):

Instance: A graph $G=(V,E)$
Parameter: positive integer k
Question: Is there a vertex cover $C \subseteq V$ with $|C| \leq k$, i.e., each edge from E is incident to at least one vertex from C?

A vertex cover with k vertices will also be called k-vertex cover. Alternatively, one could ask for an algorithm that in fact yields a vertex cover $C \subseteq V$ with $|C| \leq k$ instead of merely stating its existence. This would be the functional version of the problem. One might also like to know how many different vertex covers $C \subseteq V$ with $|C| \leq k$ exist. This would be the counting version of VC.

Obviously, there is a fourth natural problem type, namely the functional version of the counting problem (which we will call *enumeration problem*): Output *all* k-vertex covers of a given graph. A variant would be to output *all minimum* k-vertex covers of a given graph, yielding the *optima enumeration problem.*

In classical complexity theory, it does not make much sense to ask for such an algorithm for vertex cover, since only the size of the graph is measured within complexity considerations. This means that, generally speaking, for NP-hard

---

* Most of the work was done while the author was with Wilhelm-Schickard-Institut für Informatik, Universität Tübingen, Sand 13, D-72076 Tübingen, Germany.

O.H. Ibarra and L. Zhang (Eds.): COCOON 2002, LNCS 2387, pp. 564–573, 2002.

problems, an exponential number of outputs is to be generated in the worst case. Even the counting problem is considerably hard. Notably, Goldberg, Spencer and Berque [9] published a low-exponential algorithm for counting vertex covers.

In contrast, the main idea of developing fixed parameter algorithms is to explicitly declare a part of the problem instance as a so-called parameter, expecting that this parameter tends to be small in practice, whereas the overall size of the instance might be huge. This means that one can afford (mildly) exponential behaviour of algorithms in terms of the parameter, as long as the overall running time is polynomial when considering the parameter as a fixed constant. More formally, a decision problem is called *fixed parameter tractable* if its running time is bounded by $f(k) \cdot n^{O(1)}$, where f is some arbitrary function, k is the parameter, and n is the size of the problem instance. In the case of VC, $O(c^k + kn)$ time algorithms have been developed, where $c < 1.3$, see [15].

We focus on the following forms of parameterized enumeration: generate *all* solutions, generate *all optimal* solutions, and generate *representative* solutions.

We will discuss all above-mentioned variants by means of examples in the following. This paper is intended to be a start-up of a theory of parameterized enumeration. The results obtained up to now are promising.

Why do we think that parameterized enumeration is important? There are a number of possible applications of such a theory, mainly dealing with the further processing of data. For example, Gramm and Niedermeier [10] developed a fixed parameter algorithm for the so-called minimum quartet inconsistency problem (MQI) which is important for constructing evolutionary trees in biology. An evolutionary tree is a rooted binary tree whose leaves are bijectively labelled by taxa from a set S. A quartet is an evolutionary tree with four leaves. A problem instance of MQI consists of an n-element set of taxa S and $\binom{n}{4}$ quartets such that, to each four-element subset S' of S, there is exactly one quartet whose leaves are labelled with taxa from S'. The aim is to construct an evolutionary tree T whose leaves are bijectively labelled by taxa from S such that the number of sub-trees of T with four leaves which are different from the input quartet with the same leaf labels is bounded by a given error bound, the parameter k of the problem. In this application, it is interesting for the human expert to see and check *all* reconstructed evolutionary trees (satisfying the given error bound) in order to choose the tree variants which appear to him to be the most reasonable choice, given his additional background knowledge on the subject. In fact, Gramm and Niedermeier already showed how to enumerate all such minimum solutions in time $O(4^k p(n))$.

The enumerated solutions could also be the basis of further computations, even as a kind of heuristic estimate. For example, some researchers interested in computing a k-dominating set of a graph heuristically assume that such a dominating set is included within a $2k$-vertex cover and use the known (comparatively fast) vertex cover algorithm (computing some $2k$-cover) in a preprocessing phase[1]. Similarly, one could start from all (minimum) vertex covers.

[1] U. Stege, personal communication about a Swedish bioinformatics group

Below, we will discuss an example from VLSI reconfiguration that shows the practical importance of knowing some representative of all kinds of uncomparable minimal solutions as a basis of further computations. Moreover, it is shown how these enumeration algorithms can be employed to solve practically relevant variants of decision problems in relation with VLSI reconfiguration.

More details are contained in the report version of this paper [6].

2 Generating All Solutions

Let $L \subseteq \Sigma^* \times \mathbb{N}$ be a parameterized language. In the vertex cover example, L would consist of pairs $(c(G), k)$, where G is some graph having a k-vertex cover and c is some natural coding function.

Let $L_f \subseteq \Sigma^* \times \Sigma^* \times \mathbb{N}$ be the "corresponding" functional language. A tuple (σ, x, k) is in L_f iff $(x, k) \in L$ and σ is a "solution witness" for (x, k). In the vertex cover example, L_f consists of triples $(c'(V'), c(G), k)$, where G is a graph having the k-vertex cover V', and c' and c are some coding functions.

Let us assume in the following that the parameterized problem we are considering is related to an optimization problem such that the parameter bounds the entity to be optimized. Then, we say that k is optimal for the given optimization problem instance x if the size of the optimal solution to x matches k.

L_f is *[optimally] fixed parameter enumerable* iff there is an algorithm which, given $(x, k) \in L$ [where k is optimal for x] generates all $\sigma \in \Sigma^*$ with $(\sigma, x, k) \in L_f$ in time $f(k) \cdot |x|^{O(1)}$.

L_f is of *[optimal] fixed parameter size* iff

$$|\{(\sigma, x, k) \mid \sigma \in \Sigma^*[, k \text{ optimal for } x]\}| \leq f(k) \cdot |x|^{O(1)}.$$

From the discussion in the introduction, we get a first example:

Example 1. MQI is minimally fixed parameter enumerable.

Lemma 1. *Let L_f be a functional parameterized language. If L_f is [optimally] fixed parameter enumerable, then L_f is of optimal fixed parameter size.*

Theorem 1. *VC is optimally fixed parameter enumerable in time $O(2^k k^2 + kn)$, where n is the number of vertices of the input graph and k is the parameter.*

Proof. (Sketch) This can be shown by using Buss' kernelization (see [5]) and a search-tree technique. More precisely, we use the following two kernelization rules as long as possible:

- If v is a vertex with no neighbours, v can be removed from the graph, since v will not be part of any minimum vertex cover.
- If v is a vertex of degree greater than k, v must be in any vertex cover, since otherwise all neighbours would be in the cover, which is not feasible, because we are looking for vertex covers with at most k vertices. Hence, we can remove v from the graph.

After having applied these kernelization rules exhaustively, we are left with a graph with at most k^2 vertices. Now, we can basically use the simple search-tree algorithm which already appeared in [13] before the advent of parameterized complexity to show the result. □

Remark 1. Essentially, there is no better minimum vertex cover enumeration algorithm than the one given in Theorem 1, since the graph

$$(\{1,\ldots,k\} \times \{1,2\}, \{\{(i,1),(i,2)\} \mid 1 \leq i \leq k\})$$

has 2^k many different minimum vertex covers.

Note that lower bounds are usually hard to obtain. This simple example is interesting, since it shows that there is no minimum vertex cover enumeration algorithm for planar vertex cover having running time of the form $c^{\sqrt{k}}n$, as it has been found for the decision problem [2].

Remark 2. On the contrary, vertex cover is *not* of fixed parameter size, since the n-vertex graph with no edges has $\binom{n}{k}$ many different k-vertex covers. Lemma 1 shows that VC is hence not fixed parameter enumerable.

The previous considerations show that the two notions of parameterized enumerability defined above are really different. On the other hand, we can prove the following general relationship between both notions:

Lemma 2. *If a minimization problem is fixed parameter enumerable, then it is optimally fixed parameter enumerable.*

If a maximization problem (where the size of the parameter is naturally bounded by a polynomial of the size of the problem instance) is fixed parameter enumerable, then it is optimally fixed parameter enumerable.

Proof. We consider the case of minimization problems. Maximization problems are treated similarly. One simply starts the enumeration algorithm with parameter 1, 2 through k and checks, for each output solution, whether it is minimal; the minimality is checked by going through all solutions generated by invocations of the enumeration algorithm with smaller parameter values. If the enumeration problem can be solved in time $f(k) \cdot |x|^{O(1)}$ on a problem instance (x, k), then the minima enumeration problem is solvable in time

$$f(k) \cdot |x|^{O(1)} \cdot \left(\sum_{j=0}^{k-1} f(j) \cdot |x|^{O(1)} \right) \leq k(f(k))^2 \cdot |x|^{O(1)}. \quad \square$$

The next remark shows that not all parameterized problems are optimally fixed parameter enumerable. Moreover, the given example proves again that the dominating set problem[2] appears to be harder than the vertex cover problem from a parameterized point of view, also see [5].

[2] A *dominating set* of a graph is a subset of vertices such that every vertex is either a member of the dominating set or a neighbour of a member of the dominating set.

Remark 3. Dominating set is even *not* of optimal fixed parameter size, as the k-fold disjoint graph union of K_n shows. By Lemma 1, this problem is not optimally fixed parameter enumerable.

Up to now, we only considered minimization problems. Let us briefly consider one maximization problem in the parameterized setting, namely, the problem of finding a maximum *independent set*, i.e., a set of vertices I of a given graph such that no vertex in I is neighbour of another vertex in I, of size (at least) k.

Remark 4. We first consider the independent set problem restricted to planar graphs. It is quite easy to see that this problem is optimally fixed parameter enumerable. Namely, construct a 4-colouring of the given planar graph G (which exists due to the famous four-colour theorem for planar graphs); each of the four such-obtained monochromatic vertex sets is independent and the largest one contains at least $n/4$ vertices. Hence, if $k < n/4$, we can always answer "no"; otherwise, we know that $n \leq 4k$ and, hence, there are at most $f(k) = \binom{4k}{k}$ many different independent sets of size k. Hence, enumerating all *maximum* independent sets would amount checking for at most $3kf(k)$ many vertex sets whether they are independent or not (the additional factor of $3k$ comes from the necessity of checking all possible extensions of a candidate set of size k).

As in Remark 2, one can see that planar independent set is *not* of fixed parameter size. As can be seen similarly to Remark 3, the independent set problem on general graphs is not of optimal fixed parameter size.

Finally, we observe that Theorem 1 can be used in order to show fixed parameter tractability of the decision problem mentioned in the introduction:

Remark 5. The following decision problem is fixed parameter tractable: Given a graph G and parameters k and ℓ, is there a k-dominating set included in some minimum $\ell \cdot k$-vertex cover of G? This can be seen by generating all minimum $\ell \cdot k$-vertex covers and then testing, for each k-element subset of such a cover, whether it forms a dominating set.

3 Generating All Representative Solutions

In the course of this section, we will mainly focus on parameterized minimization problems with two parameters, although the main ideas can be easily generalized to an arbitrary number of parameters. Similar notions can be coined for maximization problems, as well.

Let $L \subseteq \Sigma^* \times \mathbb{N}^2$ be a parameterized language with two parameters k_1, k_2 (stemming from a minimization problem). If $(\sigma, x, k_1, k_2) \in L_f$, where L_f is the functional problem corresponding to L as in the previous section, then (k_1, k_2) is called the *signature* of (σ, x) if $(\sigma, x, k_1', k_2') \in L_f$ and $(k_1', k_2') \leq (k_1, k_2)$ imply $(k_1', k_2') = (k_1, k_2)$, where we consider the partial order $(k_1', k_2') \leq (k_1, k_2)$ iff $k_1' \leq k_1$ and $k_2' \leq k_2$.

Table 1. Repairing chip arrays

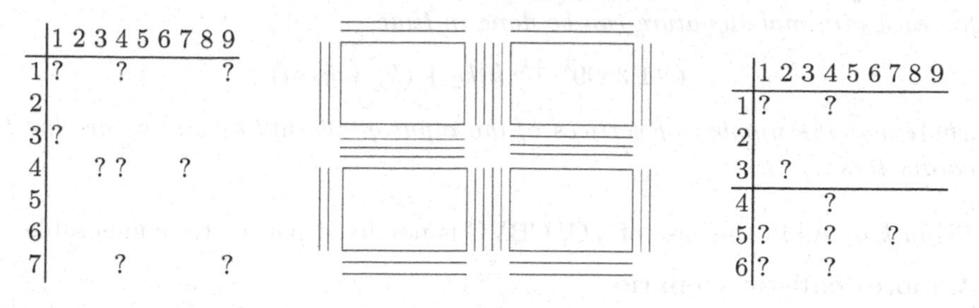

Lemma 3. *If we consider* k_1 *and* k_2 *as fixed, then there are at most* $\min\{k_1, k_2\}$ $+1$ *pairwise uncomparable (minimal) signatures.*

Hence, given a (codified) problem instance $x \in \Sigma^*$ and parameters k_1 and k_2, there are at most $\min\{k_1, k_2\} + 1$ elements in

$$\{(\sigma, x, k_1', k_2') \in L_f \mid (k_1', k_2') \leq (k_1, k_2) \wedge (k_1', k_2') \text{ is the signature of } (\sigma, x)\}$$

having different minimal signatures (k_1', k_2'). In some applications (as explained below), it is interesting to generate one *representative solution* for each minimal signature, given some problem instance. Due to the above lemma, there are at most $\min\{k_1, k_2\} + 1$ such representative solutions.

Detailed Example: Chip Reconfiguration

Kuo and Fuchs [12] provide a fundamental study of the *spare allocation problem*. Put concisely, this "most widely used approach to reconfigurable VLSI" uses spare rows and columns to tolerate failures in rectangular arrays of identical computational elements, which may be as simple as memory cells or as complex as processor units. If a faulty cell is detected, the corresponding entire row or column is replaced by a spare one.

The array on the left-hand side of Tab. 1 sketches a concrete small example of a 7×9 array, where faults are indicated by question marks. This array can be repaired, e.g., by using three spare rows (replacing rows number 1,4 and 7) and one spare column (replacing column number 1).

Equivalently, this reconfiguration problem can be formulated graph-theoretically as *Constraint Bipartite Vertex Cover (CBVC)* problem as follows: given a bipartite graph $G = (V_1, V_2, E)$ and *two* positive integers k_1 and k_2, are there two subsets $C_1 \subseteq V_1$ and $C_2 \subseteq V_2$ of sizes $|C_1| \leq k_1$ and $|C_2| \leq k_2$ such that each edge in E has at least one endpoint in $C_1 \cup C_2$?

In [8], a fixed parameter algorithm running in time less than $O(1.4^{k_1+k_2}n)$ was developed for this decision problem[3]. In fact, by analyzing the decision procedure developed in that paper one easily derives:

[3] A simpler algorithm for the CBVC problem with the additional restriction that only those bipartite covers are considered which also form a minimum vertex cover of the graph was established in [4].

Corollary 1. *For the CBVC problem, generating one representative solution for each minimal signature can be done in time*

$$O(1.3999^{k_1+k_2}k_1k_2 + (k_1 + k_2)n),$$

where n is the number of vertices of the input graph and k_1 and k_2 are the two parameters.

Remark 6. As in the case of VC, CBVC is not fixed parameter enumerable.

A more realistic scenario

As pointed out in, e.g., [11], there are several points due to which the problem formulated above is not a completely adequate model:

1. In the manufacturing process, the cost of repairing a chip by using vertical movements of the repair laser may be different from that of horizontal movements. This leads to a sort of weighted variant of CBVC.
2. As indicated in the middle figure of Tab. 1, a huge memory chip may be split into smaller blocks, each of them possibly having its own spare rows and columns. For reasons of economy, other designs are preferred in this case, e.g., each spare row depicted inbetween two memory blocks can be individually used to reconfigure either the block above or the block below it. In other words, in such complex designs, spares may be *shared*. Moreover, there may be spare rows or columns which are *linked*, which means that such a spare can only be used to reconfigure *one* certain row or column in several blocks. Obviously, the idea is here to reduce the costs of chip repair.

Combining Cor. 1 and Lemma 3, we conclude:

Theorem 2. *The weighted CBVC problem mentioned in point 1. above can be solved in time $O(1.3999^{k_1+k_2}k_1k_2+(k_1+k_2)n)$, where n is the number of vertices of the input graph and k_1 and k_2 are the two parameters.* □

Let us now consider the chip reconfiguration problem with memory blocks and shared spares.

Theorem 3. *Given a chip board with n elementary cells which is split into k_3 blocks each of which has at most k_1 neighbouring spare rows and k_2 neighbouring spare columns, then a reconfiguration strategy can be found in time*

$$O(k_3(1.3999^{k_1+k_2}k_1k_2 + (k_1 + k_2)n) + k_3(\min\{k_1, k_2\} + 1)^{\sqrt{k_3}+1})$$

if it exists.

Proof. (Sketch) At first, we run the representative enumeration procedure from Cor. 1 for each block. Then, all possible combinations of signatures for all blocks are examined to see whether the decision problem is solvable. This second step can be implemented more efficiently by using dynamic programming techniques in a sweep-line fashion. From a graph-theoretic point of view, we exploit the fact that a grid graph (representing the local dependencies between the blocks on the chip) with k vertices has treewidth of at most $\sqrt{k}+1$, see [3]. □

In other words, the parameterized enumeration of representative solutions can be used in order to show that another (related) decision problem is fixed parameter tractable, considering k_1, k_2 and k_3 as parameters of the problem.

The third mentioned variation which is also incorporating linked spares seems to be harder, since knowing only one representative solution per signature is of not much help here. Even worse, also the generation of all minimum solutions (which can be done as in the case of vertex cover elaborated above) would not help, since possibly non-optimal solutions (considered "locally" for each block) would be a better choice. For example, consider the chip depicted at the right-hand side of Tab. 1 with two blocks each containing three rows: For each of the two blocks, we have one spare row and, furthermore, there are two linked spare columns. If we use the linked spare columns in order to repair columns number 1 and 4, the array can be repaired by using the remaining two spare rows for row number 3 and row number 5. Only considering the first block, this solution is not minimal, since its signature $(1, 2)$ is outperformed by taking, e.g., a spare row for row number 1 and one of the two linked spare columns for column number 2. However, then the second block would be not repairable with the remaining spares (one spare row and one spare column).

Only at the expense of a considerable exponential blow-up, we can show the following fixed parameter tractability result:

Theorem 4. *Given a chip board with n elementary cells which is split into k_3 blocks each of which has at most k_1 neighbouring spare rows and k_2 neighbouring spare columns and assuming that there are, furthermore, at most k_4 linked spare rows and k_5 linked spare columns on the whole board, then a reconfiguration strategy can be found in time*

$$O\Bigg(k_3\Bigg((k_1+k_2+k_4+k_5)n+ \binom{k_3(k_1+k_4)}{k_4}\binom{k_3(k_2+k_5)}{k_5}\left[1.3999^{k_1+k_2}k_1k_2+(\min\{k_1,k_2\}+1)^{\sqrt{k_3}+1}\right]\Bigg)\Bigg)$$

if it exists.

Proof. Such a board can be reconfigured as follows: (1) Kernelize each block assuming that there are at most $k_1 + k_4$ spare rows and at most $k_2 + k_5$ spare columns per block. The size of the problem kernel such obtained is $k_3(k_1 + k_4)(k_2 + k_5)$. (2) Consider all possible assignments of the k_4 linked spare rows to one of the $k_3(k_1 + k_4)$ possibly faulty rows and all assignments of linked spare columns to possibly faulty columns and apply the algorithm sketched in the proof of the preceding theorem to each of the remaining "boards". □

Of course, the algorithm obtained in the previous theorem is only manageable for very small values of k_3, k_4 and k_5. A weighted variant of the last considered problem can be similarly tackled.

Remark 7. The example shown in this section proves that, from the point of view of applications, it might make perfect sense to consider problems with a certain number of parameters. The philosophy behind the development of fixed parameter algorithms is that the involved parameters should be small in practice, and this is exactly what we expect for all five parameters occurring in Theorem 4.

4 Conclusions

We considered the problem of enumerating all solutions of a given problem from the parameterized point of view. We coined different notions of parameterized enumeration and gave several examples, mainly from graph theory, with motivations from chip fabrication. We have shown that kernelizations as well as search trees (which are the most prominent ways to devise fixed parameter decision algorithms) are very useful techniques also for parameterized enumeration.

Remarkably, lower bounds and non-membership can be shown for several examples of enumeration problems and enumeration classes, whereas, in the classical area of decision problems, mostly relativized assertions of this kind are obtainable. We showed that answers to enumeration problems can be used in solutions of decision problems. In particular, we proved several more realistic scenarios of the chip reconfiguration problem [8] to be fixed parameter tractable.

Note that we deliberately focussed on considering the complexity of enumeration problems to include the time to output the solutions. Another variant where the size of the output solutions was considered as an extra sort of parameter (in the sense of providing output sensitive algorithms) was discussed by Grohe[4]. In this spirit, several papers on graph algorithms appeared, too, see, e.g., [14] and the references therein. When thinking about enumeration as some sort of preprocessing step for another algorithm which investigates all the obtained solutions, considering the size or number of output solutions as additional parameter does not seem to be reasonable.

It would be also interesting to consider the parameterized complexity of enumerating all (optimal) solutions *without repetitions*, as discussed in [14]. Of course, one could avoid repetitions by either examining all pairs of output solutions in a postprocessing phase (which would square the already exponential running time) or by additional bookkeeping (with tables of exponential size), but possibly better solutions can be found for concrete problems.

In [7], we showed that parameterized enumeration can be also used for proving parameterized tractability for maximization problems (in the sense elaborated in [7]), thus providing another sort of application of enumeration problems.

Finally, it would be interesting to see whether problems related to vertex cover are also fixed parameter enumerable. Here, the setting established in [16] might be helpful in order to prove enumerability results. More generally speaking, it would be interesting to develop enumeration techniques which are applicable not only to special situations. Such considerations might help answer the question whether or not the optima dominating set problem restricted to planar graphs is feasible or not, see [1] for the corresponding decision problem.

[4] in a talk on parameterized complexity and databases given at the Dagstuhl Workshop on Parameterized Complexity in August, 2001

Acknowledgments: We thank M. R. Fellows, R. Niedermeier, and U. Stege for some discussions. We are grateful to L. Brankovic for giving the conference talk.

References

1. J. Alber, H. L. Bodlaender, H. Fernau, and R. Niedermeier. Fixed parameter algorithms for planar dominating set and related problems. In M. M. Halldórsson, editor, *7th Scandinavian Workshop on Algorithm Theory SWAT 2000*, volume 1851 of *LNCS*, pages 97–110, 2000. Long version to appear in Algorithmica.
2. J. Alber, H. Fernau, and R. Niedermeier. Parameterized complexity: exponential speedup for planar graph problems. In F. Orejas, P. G. Spirakis, and J. v. Leeuwen, editors, *International Colloquium on Automata, Languages and Programming ICALP'01*, volume 2076 of *LNCS*, pages 261–272. Springer, 2001.
3. H. L. Bodlaender. A partial k-arboretum of graphs with bounded treewidth. *Theoretical Computer Science*, 209:1–45, 1998.
4. J. Chen and I. A. Kanj. On constrained minimum vertex covers of bipartite graphs: Improved algorithms. In A. Brandstädt and V. B. Le, editors, *Graph-Theoretic Concepts in Computer Science WG'01*, volume 2204 of *LNCS*, pages 55–65. Springer, 2001.
5. R. G. Downey and M. R. Fellows. *Parameterized Complexity*. Springer, 1999.
6. H. Fernau. On parameterized enumeration. Technical Report WSI–2001–21, Universität Tübingen (Germany), Wilhelm-Schickard-Institut für Informatik, 2001.
7. H. Fernau. Parameterized maximization. Technical Report WSI–2001–22, Universität Tübingen (Germany), Wilhelm-Schickard-Institut für Informatik, 2001.
8. H. Fernau and R. Niedermeier. An efficient exact algorithm for constraint bipartite vertex cover. *Journal of Algorithms*, 38(2):374–410, 2001.
9. M. K. Goldberg, T. H. Spencer, and D. A. Berque. A low-exponential algorithm for counting vertex covers. *Graph Theory, Combinatorics, Algorithms, and Applications*, 1:431–444, 1995.
10. J. Gramm and R. Niedermeier. Quartet inconsistency is fixed parameter tractable. In A. Amir and G. M. Landau, editors, *Proceedings of the 12th Annual Symposium on Combinatorial Pattern Matching (CPM 2001)*, volume 2089 of *LNCS*, pages 241–256. Springer, 2001.
11. R. W. Haddad, A. T. Dahbura, and A. B. Sharma. Increased throughput for the testing and repair of RAMs with redundancy. *IEEE Transactions on Computers*, 40(2):154–166, Feb. 1991.
12. S.-Y. Kuo and W. Fuchs. Efficient spare allocation for reconfigurable arrays. *IEEE Design and Test*, 4:24–31, Feb. 1987.
13. K. Mehlhorn. *Graph algorithms and NP-completeness*. Heidelberg: Springer, 1984.
14. S. Nakano. Efficient generation of triconnected plane triangulations. In J. Wang, editor, *Computing and Combinatorics, Proceedings COCOON 2001*, volume 2108 of *LNCS*, pages 131–141. Springer, 2001.
15. R. Niedermeier and P. Rossmanith. Upper bounds for vertex cover further improved. In C. Meinel and S. Tison, editors, *Proceedings of the 16th Symposium on Theoretical Aspects of Computer Science (STACS'99)*, volume 1563 of *LNCS*, pages 561–570. Springer, 1999.
16. N. Nishimura, P. Ragde, and D. M. Thilikos. Fast fixed-parameter tractable algorithms for nontrivial generalizations of vertex cover. In F. Dehne, J.-R. Sack, and R. Tamassia, editors, *Proc. 7th Workshop Algorithms and Data Structures WADS*, volume 2125 of *LNCS*, pages 75–86. Springer, 2001.

Probabilistic Reversible Automata and Quantum Automata

Marats Golovkins* and Maksim Kravtsev**

Institute of Mathematics and Computer Science, University of Latvia
Raiņa bulv. 29, Riga, Latvia
marats@latnet.lv, maksims@batsoft.lv

Abstract. To study relationship between quantum finite automata and probabilistic finite automata, we introduce a notion of probabilistic reversible automata (PRA, or doubly stochastic automata). We find that there is a strong relationship between different possible models of PRA and corresponding models of quantum finite automata. We also propose a classification of reversible finite 1-way automata.

1 Introduction

Here we introduce common notions used throughout the paper as well as summarize its contents.

We analyze two models of probabilistic reversible automata in this paper, namely, 1-way PRA and 1.5-way PRA.

If not specified otherwise, we denote by Σ an input alphabet of an automaton. Every input word is enclosed into *end-marker* symbols # and $. Therefore we introduce a *working alphabet* as $\Gamma = \Sigma \cup \{\#, \$\}$. By Q we normally understand the set of states of an automaton. By $\overline{L}$ we understand complement of a language L. Given an input word ω, by $|\omega|$ we understand the number of symbols in ω and with $[\omega]_i$ we denote i-th symbol of ω, counting from the beginning (excluding end-markers). By $q \xrightarrow{S} q'$, $S \subset \Sigma^*$, we denote that there is a positive probability to get to a state q' by reading some word $\xi \in S$, starting in q.

Let us consider A. Nayak's model of quantum automata with mixed states (QFA-N, [N 99]). (Evolution is characterized by a unitary matrix and subsequent measurements are performed after each step, POVM measurements not being allowed.) If a result of every measurement is a single configuration, not a superposition, and measurements are performed after each step, we actually get a probabilistic automaton. However, the following property applies to such probabilistic automata - their evolution matrices are *doubly* stochastic. This encourages us to give the following definition for probabilistic reversible automata:

* Research partially supported by the Latvian Council of Science, grant No. 01.0354 and grant for Ph.D. students; University of Latvia, K. Morbergs grant; European Commission, contract IST-1999-11234

** Research partially supported by the Latvian Council of Science, grant No. 01.0354 and European Commission, contract IST-1999-11234

O.H. Ibarra and L. Zhang (Eds.): COCOON 2002, LNCS 2387, pp. 574–583, 2002.

Definition 1.1. *A probabilistic automaton is called* reversible *if its linear operator can be described by a doubly stochastic matrix.*

At least two definitions exist, how to interpret word acceptance, and hence, language recognition, for reversible automata.

Definition 1.2. Classical acceptance. *(C-automata) We say that an automaton accepts (rejects) a word classically, if its set of states consists of two disjoint subsets: accepting states and rejecting states, and the following conditions hold:*
-the automaton accepts the word, if it is in accepting state after having read the last symbol of the word;
-the automaton rejects the word, if it is in rejecting state after having read the last symbol of the word.

Definition 1.3. "Decide and halt" acceptance. *(DH-automata) We say that an automaton accepts (rejects) a word in a decide-and-halt manner, if its set of states consists of three disjoint subsets: accepting states, rejecting states and non-halting states, and the following conditions hold:*
-the computation is continued only if the automaton enters a non-halting state;
-if the automaton enters an accepting state, the word is accepted;
-if the automaton enters a rejecting state, the word is rejected.

Having defined word acceptance, we define language recognition in an equivalent way as in [R 63]. We consider only bounded error language recognition in this paper. By $P_{x,A}$ we denote the probability that a word x is accepted by an automaton A.

Definition 1.4. *We say that a language L is recognized with bounded error by an automaton A with interval (p_1, p_2) if $p_1 < p_2$ and $p_1 = \sup\{P_{x,A} \mid x \notin L\}$, $p_2 = \inf\{P_{x,A} \mid x \in L\}$.*

We say that a language is recognized with a probability p if the language is recognized with interval $(1-p, p)$. We say that a language is recognized with probability $1-\varepsilon$, if for every $\varepsilon > 0$ there exists an automaton which recognizes the language with interval $(\varepsilon_1, 1-\varepsilon_2)$, where $\varepsilon_1, \varepsilon_2 \leq \varepsilon$.

In Section 2, we discuss properties of PRA C-automata (PRA-C). We prove that PRA-C recognize the class of languages $a_1^* a_2^* \dots a_n^*$ with probability $1-\varepsilon$. This class can be recognized by measure-many quantum finite automata [KW 97] (QFA-KW), with worse acceptance probabilities, however [ABFK 99]. This also implies that QFA-N recognize this class of languages with probability $1-\varepsilon$. Further, we show general class of regular languages, not recognizable by PRA-C. In particular, such languages as (a,b)*a and a(a,b)* are in this class. This class has strong similarities with the class of languages, not recognizable by QFA-KW [AKV 00]. We also show that the class of languages recognized by PRA-C is closed under boolean operations. In Section 3 we prove, that PRA DH-automata do not recognize the language (a,b)*a. In Section 4 we discuss some properties of 1.5-way PRA. We also present an alternative notion of probabilistic reversibility, not connected with quantum automata. In Section 5 we propose a classification of reversible automata (deterministic, probabilistic and quantum).

2 1-Way Probabilistic Reversible C-Automata

Definition 2.1. *1-way probabilistic reversible C-automaton (PRA-C) $A = (Q, \Sigma, q_0, Q_F, \delta)$ is specified by a finite set of states Q, a finite input alphabet Σ, an initial state $q_0 \in Q$, a set of accepting states $Q_F \subseteq Q$, and a transition function $\delta : Q \times \Gamma \times Q \longrightarrow \mathbb{R}_{[0,1]}$, where $\Gamma = \Sigma \cup \{\#, \$\}$ is the input tape alphabet of A and $\#$, $\$$ are end-markers not in Σ. Furthermore, transition function satisfies the following requirements:*

$$\forall(q_1, \sigma_1) \in Q \times \Gamma \sum_{q \in Q} \delta(q_1, \sigma_1, q) = 1 \tag{1}$$

$$\forall(q_1, \sigma_1) \in Q \times \Gamma \sum_{q \in Q} \delta(q, \sigma_1, q_1) = 1 \tag{2}$$

For every input symbol $\sigma \in \Gamma$, the transition function may be determined by a $|Q| \times |Q|$ matrix V_σ, where $(V_\sigma)_{i,j} = \delta(q_j, \sigma, q_i)$.

We define word acceptance as specified in Definition 1.2. The set of rejecting states is $Q \setminus Q_F$. We define language recognition as in Definition 1.4.

Now we present several results on the class of languages recognizable by PRA-C.

Lemma 2.2. *If a language is recognized by a PRA-C A with interval (p_1, p_2), exists a PRA-C which recognizes the language with probability p, where*

$$p = \begin{cases} \frac{p_2}{p_1+p_2}, & \text{if } p_1 + p_2 \geq 1 \\ \frac{1-p_1}{2-p_1-p_2}, & \text{if } p_1 + p_2 < 1. \end{cases}$$

Theorem 2.3. *If a language is recognized by a PRA-C, it is recognized by PRA-C with probability $1 - \varepsilon$.*

Proof. Idea of the proof. Assume that a language L is recognized by a PRA-C A. The language L is recognized with probability $1 - \varepsilon$, using a system of n identical copies of A. A system of n PRA-C automata may be simulated by a single PRA-C automaton. □

Lemma 2.4. *If a language L_1 is recognizable with probability greater than $\frac{2}{3}$ and a language L_2 is recognizable with probability greater than $\frac{2}{3}$ then languages $L_1 \cap L_2$ and $L_1 \cup L_2$ are recognizable with probability greater than $\frac{1}{2}$.*

Theorem 2.5. *The class of languages recognized by PRA-C is closed under intersection, union and complement.*

Proof. Let us consider languages L_1, L_2 recognized by some PRA-C automata. By Theorem 2.3, these languages is recognizable with probability $1 - \varepsilon$, and therefore by Lemmas 2.2 and 2.4, union and intersection of these languages are also recognizable. If a language L is recognizable by a PRA-C A, we can construct an automaton which recognizes a language $\overline{L}$ just by making accepting states of A to be rejecting, and vice versa. □

Lemma 2.6. *If A is a doubly stochastic matrix and X - a vector, then* $\max(X) \geq \max(AX)$ *and* $\min(X) \leq \min(AX)$.

Theorem 2.7. *For every natural positive n, a language $L_n = a_1^* a_2^* \ldots a_n^*$ is recognizable by some PRA-C with alphabet $\{a_1, a_2, \ldots, a_n\}$.*

Proof. We construct a PRA-C with $n+1$ states, q_0 being the initial state, corresponding to probability distribution vector $\begin{pmatrix} 1\, 0 \ldots 0 \end{pmatrix}^T$. The transition function is determined by $(n+1) \times (n+1)$ matrices

$$V_{a_1} = \begin{pmatrix} 1 & 0 & \ldots & 0 \\ 0 & \frac{1}{n} & \ldots & \frac{1}{n} \\ \vdots & \vdots & \ddots & \vdots \\ 0 & \frac{1}{n} & \ldots & \frac{1}{n} \end{pmatrix}, V_{a_2} = \begin{pmatrix} \frac{1}{2} & \frac{1}{2} & 0 & \ldots & 0 \\ \frac{1}{2} & \frac{1}{2} & 0 & \ldots & 0 \\ 0 & 0 & \frac{1}{n-1} & \ldots & \frac{1}{n-1} \\ \vdots & \vdots & \vdots & \ddots & \vdots \\ 0 & 0 & \frac{1}{n-1} & \ldots & \frac{1}{n-1} \end{pmatrix}, \ldots, V_{a_n} = \begin{pmatrix} \frac{1}{n} & \ldots & \frac{1}{n} & 0 \\ \vdots & \ddots & \vdots & \vdots \\ \frac{1}{n} & \ldots & \frac{1}{n} & 0 \\ 0 & \ldots & 0 & 1 \end{pmatrix}.$$

The accepting states are $q_0 \ldots q_{n-1}$, the only rejecting state is q_n. We prove, that the automaton recognizes the language L_n.

Case $\omega \in L_n$. All $\omega \in L_n$ are accepted with probability 1.

Case $\omega \notin L_n$. Consider k such that $\omega = \omega_1 \sigma \omega_2$, $|\omega_1| = k$, $\omega_1 \in L_n$ and $\omega_1 \sigma \notin L_n$. Since all one-letter words are in L_n, $k > 0$. Let $a_t = [\omega]_k$ and $a_s = \sigma$. So we have $s < t$, $1 \leq s \leq n-1$, $2 \leq t \leq n$. The word $\omega_1 a_s$ is accepted with probability $1 - \frac{t-s}{t(n-s+1)}$. By Lemma 2.6, since $\frac{t-s}{t(n-s+1)} < \frac{1}{t}$, reading the symbols succeeding $\omega_1 a_s$ will not increase accepting probability. Therefore, to find maximum accepting probability for words not in L_n, we have to maximize $1 - \frac{t-s}{t(n-s+1)}$, where $s < t$, $1 \leq s \leq n-1$, $2 \leq t \leq n$. We get that the automaton recognizes the language with interval $\left(1 - \frac{1}{\left[(\frac{n}{2})^2\right]+n+1},\ 1\right)$. (By Theorem 2.3, L_n can be recognized with probability $1 - \varepsilon$). □

Corollary 2.8. *Quantum finite automata with mixed states (model of Nayak, [N 99]) recognize $L_n = a_1^* a_2^* \ldots a_n^*$ with probability $1 - \varepsilon$.*

Proof. This comes from the fact, that matrices $V_{a_1}, V_{a_2}, \ldots, V_{a_n}$ from the proof of Theorem 2.7 (as well as tensor powers of those matrices) all have unitary prototypes (see Definition 5.1). □

Now we introduce a general class of regular languages not recognizable by PRA-C.

Definition 2.9. *We say that a regular language is of type $(*)$ if the following is true for the minimal deterministic automaton recognizing this language: Exist three states q, q_1, q_2, exist words x, y such that $q_1 \neq q_2$; $qx = q_1$, $qy = q_2$; $\forall t \in (x,y)^*\ \exists t_1 \in (x,y)^*\ q_1 t t_1 = q_1$; $\forall t \in (x,y)^*\ \exists t_2 \in (x,y)^*\ q_2 t t_2 = q_2$.*

We say that a regular language is of type $(*')$ if the following is true for the minimal deterministic automaton recognizing this language: Exist three states q,

q_1, q_2, exist words x, y such that $q_1 \neq q_2$; $qx = q_1$, $qy = q_2$; $q_1x = q_1$, $q_1y = q_1$; $q_2x = q_2$, $q_2y = q_2$.
We say that a regular language is of type $(*'')$ if the following is true for the minimal deterministic automaton recognizing this language: Exist two states q_1, q_2, exist words x, y such that $q_1 \neq q_2$; $q_1x = q_2$, $q_2x = q_2$; $q_2y = q_1$.

Type $(*'')$ languages are exactly those languages that violate the partial order condition of [BP 99].

Lemma 2.10. *If A is a deterministic finite automaton with a set of states Q and alphabet Σ, then $\forall q \in Q\ \forall x \in \Sigma^*\ \exists k > 0\ qx^k = qx^{2k}$.*

Lemma 2.11. *A regular language is of type $(*)$ iff it is of type $(*')$ or type $(*'')$.*

Proof. 1) If a language is of type $(*')$, it is of type $(*)$. Obvious.
2) If a language is of type $(*'')$, it is of type $(*)$. Consider a language of type $(*'')$ with states q_1'', q_2'' and words x'', y''. To build construction of type $(*)$, we take $q = q_1 = q_1''$, $q_2 = q_2''$, $x = x''y''$, $y = x''$. That forms transitions $qx = q_1$, $qy = q_2$, $q_1x = q_1$, $q_1y = q_2$, $q_2x = q_1$, $q_2y = q_2$. We have satisfied all the rules of $(*)$.
3) If a language is of type $(*)$, it is of type $(*')$ or $(*'')$. Consider a language whose minimal deterministic automaton has construction $(*)$. By Lemma 2.10, $\exists s \exists a\ q_1x^a = q_s$ and $q_sx^a = q_s$; $\exists t \exists b\ q_1y^b = q_t$ and $q_ty^b = q_t$; $\exists u \exists c\ q_2x^c = q_u$ and $q_ux^c = q_u$; $\exists v \exists d\ q_2y^d = q_v$ and $q_vy^d = q_v$. If $q_1 \neq q_s$, by the rules of $(*)$, $\exists z\ q_sz = q_1$. Therefore the language is of type $(*'')$. If $q_2 \neq q_u$, by the rules of $(*)$, $\exists z\ q_uz = q_2$, and the language is of type $(*'')$. Likewise, if $q_1 \neq q_t$ or $q_2 \neq q_v$, the language is of type $(*'')$. If $q_1 = q_s = q_t$ and $q_2 = q_u = q_v$, we have $qx^a = q_1$, $qy^d = q_2$, $q_1x^a = q_1y^b = q_1$, $q_2x^c = q_2y^d = q_2$. We get the construction $(*')$ if we take $x' = x^{ac}$, $y' = y^{bd}$. □

We are going to prove that every language of type $(*)$ is not recognizable by any PRA-C. For this purpose, we use several definitions from the theory of finite Markov chains ([KS 76]).

Definition 2.12. *A state q_j is accessible from q_i (denoted $q_i \to q_j$) if there is a positive probability to get from q_i to q_j (possibly in several steps).*

Definition 2.13. *States q_i and q_j communicate (denoted $q_i \leftrightarrow q_j$) if $q_i \to q_j$ and $q_j \to q_i$.*

Definition 2.14. *A Markov chain is called doubly stochastic, if its transition matrix is a doubly stochastic matrix.*

We recall the following theorem from the theory of finite Markov chains:

Theorem 2.15. *If a Markov chain with a matrix A is irreducible and aperiodic,*
a) it has a unique stationary distribution Z;
b) $\lim_{n\to\infty} A^n = (Z, \dots, Z)$;
c) $\forall X\ \lim_{n\to\infty} A^nX = Z$.

Several facts about doubly stochastic matrices follow from this theorem.

Corollary 2.16. *If a doubly stochastic Markov chain with an $m \times m$ matrix A is irreducible and aperiodic,*

$$a)\ \lim_{n\to\infty} A^n = \begin{pmatrix} \frac{1}{m} & \cdots & \frac{1}{m} \\ \cdots & \cdots & \cdots \\ \frac{1}{m} & \cdots & \frac{1}{m} \end{pmatrix};\ b)\ \forall X\ \lim_{n\to\infty} A^n X = \begin{pmatrix} \frac{1}{m} \\ \cdots \\ \frac{1}{m} \end{pmatrix}.$$

Lemma 2.17. *If M is a doubly stochastic Markov chain with a matrix A, then $\forall q\ q \to q$.*

Corollary 2.18. *Suppose A is a doubly stochastic matrix. Then exists $k > 0$, such that $\forall i\ (A^k)_{i,i} > 0$.*

Lemma 2.19. *If M is a doubly stochastic Markov chain and $q_a \to q_b$, then $q_a \leftrightarrow q_b$.*

Now, using the facts above, we can prove that any language of type $(*)$ is not recognizable by PRA-C.

Lemma 2.20. *If a regular language is of type $(*')$, it is not recognizable by any PRA-C.*

Proof. Assume from the contrary, that A is a PRA-C automaton which recognizes a language $L \subset \Sigma^*$ of type $(*')$.

Since L is of type $(*')$, it is recognized by a deterministic automaton D which has three states q, q_1, q_2 such that $q_1 \neq q_2$, $qx = q_1$, $qy = q_2$, $q_1x = q_1$, $q_1y = q_1$, $q_2x = q_2$, $q_2y = q_2$, where $x, y \in \Sigma^*$. Furthermore, exists $\omega \in \Sigma^*$ such that $q_0\omega = q$, where q_0 is an initial state of D, and exists a word $z \in \Sigma^*$, such that $q_1z = q_{acc}$ if and only if $q_2z = q_{rej}$, where q_{acc} is an accepting state and q_{rej} is a rejecting state of D. Without loss of generality we assume that $q_1z = q_{acc}$ and $q_2z = q_{rej}$.

The transition function of the automaton A is determined by doubly stochastic matrices $V_{\sigma_1}, \ldots, V_{\sigma_n}$. The words from the construction $(*')$ are $x = \sigma_{i_1} \ldots \sigma_{i_k}$ and $y = \sigma_{j_1} \ldots \sigma_{j_s}$. The transitions induced by words x and y are determined by doubly stochastic matrices $X = V_{\sigma_{i_k}} \ldots V_{\sigma_{i_1}}$ and $Y = V_{\sigma_{j_s}} \ldots V_{\sigma_{j_1}}$. Similarly, the transitions induced by words ω and z are determined by doubly stochastic matrices W and Z. By Corollary 2.18, exists $K > 0$, such that

$$\forall i\ (X^K)_{i,i} > 0 \text{ and } (Y^K)_{i,i} > 0. \tag{3}$$

Consider a relation between the states of the automaton defined as $R = \{(q_i, q_j) \mid q_i \xrightarrow{(x^K, y^K)^*} q_j\}$. By (3), this relation is reflexive. By Lemma 2.19, the relation R is symmetric.

Surely R is transitive. Therefore all states of A may be partitioned into equivalence classes $[q_0], [q_{i_1}], \ldots, [q_{i_n}]$. Let us renumber the states of A in such a way, that states from one equivalence class have consecutive numbers. First come the states in $[q_0]$, then in $[q_{i_1}]$, etc.

Consider the word $x^K y^K$. The transition induced by this word is determined by a doubly stochastic matrix $C = Y^K X^K$. We prove the following proposition. States q_a and q_b are in one equivalence class if and only if $q_a \to q_b$ with matrix C. Suppose $q_a \to q_b$. Then $(q_a, q_b) \in R$, and q_a, q_b are in one equivalence class. Suppose q_a, q_b are in one equivalence class. Then

$$q_a \xrightarrow{\xi_1} q_{i_1}, q_{i_1} \xrightarrow{\xi_2} q_{i_2}, \ldots, q_{i_{k-1}} \xrightarrow{\xi_k} q_b, \text{ where } \xi_s \in \{x^K, y^K\}. \quad (4)$$

By (3), $q_i \xrightarrow{x^K} q_i$ and $q_j \xrightarrow{y^K} q_j$. Therefore, if $q_i \xrightarrow{x^K} q_j$, then $q_i \xrightarrow{x^K y^K} q_j$, and again, if $q_i \xrightarrow{y^K} q_j$, then $q_i \xrightarrow{x^K y^K} q_j$. That transforms (4) to $q_a \xrightarrow{(x^K y^K)^t} q_b$, where $t > 0$. We have proved the proposition.

By the proved proposition, due to the renumbering of states, matrix C is a block diagonal matrix, where each block corresponds to an equivalence class of the relation R. Let us identify these blocks as $C_0, C_1, \ldots, C_n$. By (3), a Markov chain with matrix C is aperiodic. Therefore each block C_r corresponds to an aperiodic irreducible doubly stochastic Markov chain with states $[q_{i_r}]$. By Corollary 2.16, $\lim_{m \to \infty} C^m = J$, J is a block diagonal matrix, where for each $(p \times p)$ block C_r $(C_r)_{i,j} = \frac{1}{p}$. Relation $q_i \xrightarrow{(y^K)^*} q_j$ is a subrelation of R, therefore Y^K is a block diagonal matrix with the same block ordering and sizes as C and J. (This does not eliminate possibility that some block of Y^K is constituted of smaller blocks, however.) Therefore $JY^K = J$, and $\lim_{m \to \infty} Z(Y^K X^K)^m W = \lim_{m \to \infty} Z(Y^K X^K)^m Y^K W = ZJW$. So $\forall \varepsilon > 0\ \exists m\ \left\| \left(Z(Y^K X^K)^m W - Z(Y^K X^K)^m Y^K W\right) Q_0 \right\| < \varepsilon$. However, by construction $(*')$, $\forall k\ \forall m\ \omega(x^k y^k)^m z \in L$ and $\omega y^k (x^k y^k)^m z \notin L$. This requires existence of $\varepsilon > 0$, such that $\forall m\ \left\| \left(Z(Y^K X^K)^m W - Z(Y^K X^K)^m Y^K W\right) Q_0 \right\| > \varepsilon$. This is a contradiction. □

Lemma 2.21. *If a regular language is of type $(*'')$, it is not recognizable by any PRA-C.*

Proof. Proof is nearly identical to that of Lemma 2.20. □

Theorem 2.22. *If a regular language is of type $(*)$, it is not recognizable by any PRA-C.*

Proof. By Lemmas 2.11, 2.20, 2.21. □

We proved (Lemma 2.11) that the construction of type $(*)$ is a generalization the construction proposed by [BP 99]. Also it can be easily noticed, that the type $(*)$ construction is a generalization of construction proposed by [AKV 00]. (Constructions of [BP 99] and [AKV 00] characterize languages, not recognized by measure-many quantum finite automata of [KW 97].)

Corollary 2.23. *Languages (a,b)*a and a(a,b)* are not recognized by PRA-C.*

Proof. Both languages are of type $(*)$. □

3 1-Way Probabilistic Reversible DH-Automata

Definition 3.1. *The definition differs from one for PRA-C (Definition 2.1) by the following: languages are recognized according to Definition 1.3.*

It is easy to see that the class of languages recognized by PRA-C is a proper subclass of languages recognized by PRA-DH. For example, the language a(a,b)* is recognizable by PRA-DH. However, the following theorem holds:

Theorem 3.2. *Language (a,b)*a is not recognized by PRA-DH.*

Proof. Assume from the contrary that such automaton exists. While reading any sequence of a and b, this automaton can halt only with some probability p strictly less then 1, so accepting and rejecting probabilities may differ only by 1-p, because any word belonging to the language is not dependent on any prefix. Therefore for each $\varepsilon > 0$ we can find that after reading a prefix of certain length, the total probability to halt while continue reading the word is less then ε. In this case we can apply similar techniques as in the proof of Lemma 2.20. □

4 Alternative Approach to Finite Reversible Automata and 1.5-Way Probabilistic Reversible Automata

Let us consider an automaton $A' = (Q, \Sigma, q_0, Q_F, \delta')$ that can be obtained from a probabilistic automaton $A = (Q, \Sigma, q_0, Q_F, \delta)$ by specifying $\delta'(q, \sigma, q') = \delta(q', \sigma, q)$ for all q', σ and q. If A' is valid probabilistic automaton then we can call A and A' probabilistic reversible automata.

Definition 4.1. *An automaton of some type is called* ***weakly reversible*** *if the reverse of its transition function corresponds to the transition function of a valid automaton of the same type.*

Note: in case of deterministic automaton where $\delta : Q \times \Gamma \times Q \longrightarrow \{0, 1\}$ this property means that A' is still deterministic automaton, not nondeterministic. In case of one-way automata it is easy to check that this definition is equivalent to the one in Section 2. We give an example that illustrates that in case of 1.5-way automata these definitions are different.

Definition 4.2. *1.5-way probabilistic* ***weakly*** *reversible C-automaton $A = (Q, \Sigma, q_0, Q_F, \delta)$ is specified by Q, Σ, q_0, Q_F defined as in 1-way PRA-C Definition 2.1, and a transition function $\delta : Q \times \Gamma \times Q \times D \longrightarrow \mathbb{R}_{[0,1]}$, where Γ defined as in 1-way PRA-C definition and $D = \{0, 1\}$ denotes whether automaton stays on the same position or moves one letter ahead on the input tape. Furthermore, transition function satisfies the following requirements:*

$$\forall (q_1, \sigma_1) \in Q \times \Gamma \quad \sum_{q \in Q, d \in D} \delta(q_1, \sigma_1, q, d) = 1;$$

$$\forall (q_1, \sigma_1) \in Q \times \Gamma \quad \sum_{q \in Q, d \in D} \delta(q, \sigma_1, q_1, d) = 1$$

Definition 4.3. *1.5-way probabilistic reversible C-automaton* $A = (Q, \Sigma, q_0, Q_F, \delta)$ *is specified by* Q, Σ, q_0, Q_F *defined as in 1-way PRA-C Definition 2.1, and a transition function* $\delta : Q \times \Gamma \times Q \times D \longrightarrow \mathbb{R}_{[0,1]}$, *where* Γ *defined as in 1-way PRA-C definition and* $D = \{0, 1\}$ *denotes whether automaton stays on the same position or moves one letter ahead on the input tape. Furthermore, transition function satisfies the following requirements:*

$$\forall (q_1, \sigma_1) \in Q \times \Gamma \quad \sum_{q \in Q, d \in D} \delta(q_1, \sigma_1, q, d) = 1;$$

$$\forall (q_1, \sigma_1, \sigma_2) \in Q \times \Gamma^2 \sum_{q \in Q} \delta(q, \sigma_1, q_1, 0) + \sum_{q \in Q, \sigma \in \Gamma} \delta(q, \sigma_2, q_1, 1) = 1$$

Theorem 4.4. *Language (a,b)*a is recognizable by 1.5-way weakly reversible PRA-C.*

Proof. The $Q = \{q_0, q_1\}$, $Q_F = \{q_1\}$, δ is defined as follows; $\delta(q_0, a, q_0, 0) = \frac{1}{2}$, $\delta(q_0, a, q_1, 1) = \frac{1}{2}$, $\delta(q_1, a, q_0, 0) = \frac{1}{2}$, $\delta(q_1, a, q_1, 1) = \frac{1}{2}$, $\delta(q_0, b, q_0, 1) = \frac{1}{2}$, $\delta(q_0, b, q_1, 0) = \frac{1}{2}$, $\delta(q_1, b, q_0, 1) = \frac{1}{2}$, $\delta(q_1, b, q_1, 0) = \frac{1}{2}$, $\delta(q_0, \$, q_0, 1) = 1$, $\delta(q_1, \$, q_1, 1) = 1$. It easy to check that such automaton moves ahead according to the transition of the following deterministic automaton; $\delta(q_0, a, q_1, 1) = 1$, $\delta(q_1, a, q_1, 1) = 1$, $\delta(q_0, b, q_0, 1) = 1$, $\delta(q_1, b, q_0, 1) = 1$, $\delta(q_0, \$, q_0, 1) = 1$, $\delta(q_1, \$, q_1, 1) = 1$. So the probability of wrong answer is 0. The probability to be at the m-th position of the input tape after n steps of calculation for $m \leq n$ is C_n^m. Therefore it is necessary no more then $O(n * \log(p))$ steps to reach the end of the word of length n (and so obtain correct answer) with probability $1 - \frac{1}{p}$. □

5 A Classification of Reversible Automata

We propose the following classification for finite 1-way reversible automata:

	C-Automata	DH-Automata
Deterministic Automata	Permutation Automata [HS 66,T 68] (DRA-C)	Reversible Finite Automata [AF 98] (DRA-DH)
Quantum Automata with Pure States	Measure-Once Quantum Finite Automata [MC 97] (QRA-P-C)	Measure-Many Quantum Finite Automata [KW 97] (QRA-P-DH)
Probabilistic Automata	Probabilistic Reversible C-Automata (PRA-C)	Probabilistic Reversible DH-Automata (PRA-DH)
Quantum Finite Automata with Mixed States	not considered yet (QRA-M-C)	Enhanced Quantum Finite Automata [N 99] (QRA-M-DH)

Language class problems are solved for DRA-C, DRA-DH, QRA-P-C, for the rest types they are still open. Every type of DH-automata may simulate the corresponding type of C-automata.

In general, language classes recognized by C-automata are closed under boolean operations (though this is open for QRA-M-C), while DH-automata are not (though this is open for QRA-M-DH and possibly for PRA-DH).

Definition 5.1. *We say that a unitary matrix U is a* prototype *for a doubly stochastic matrix S, if $\forall i,j\ |U_{i,j}|^2 = S_{i,j}$.*

Not every doubly stochastic matrix has a unitary prototype. Such matrix is, for example, $\begin{pmatrix} \frac{1}{2} & \frac{1}{2} & 0 \\ \frac{1}{2} & 0 & \frac{1}{2} \\ 0 & \frac{1}{2} & \frac{1}{2} \end{pmatrix}$. In Introduction, we demonstrated some relation between PRA-C and QRA-M-DH (and hence, QRA-M-C). However, due to the example above, we do not know exactly, whether every PRA-C can be simulated by QRA-M-C, or whether every PRA-DH can be simulated by QRA-M-DH.

Theorem 5.2. *If all matrices of a PRA-C have unitary prototypes, then the PRA-C may be simulated by a QRA-M-C and by a QRA-M-DH.*
If all matrices of a PRA-DH have unitary prototypes, then the PRA-DH may be simulated by a QRA-M-DH.

References

[ABFK 99] A. Ambainis, R. Bonner, R. Freivalds, A. Ķikusts. Probabilities to Accept Languages by Quantum Finite Automata. *COCOON 1999, Lecture Notes in Computer Science*, 1999, Vol. 1627, pp. 174-183. http://arxiv.org/abs/quant-ph/9904066

[AF 98] A. Ambainis, R. Freivalds. 1-Way Quantum Finite Automata: Strengths, Weaknesses and Generalizations. *Proc. 39th FOCS*, 1998, pp. 332-341. http://arxiv.org/abs/quant-ph/9802062

[AKV 00] A. Ambainis, A. Ķikusts, M. Valdats. On the Class of Languages Recognizable by 1-Way Quantum Finite Automata. *STACS 2001, Lecture Notes in Computer Science*, 2001, Vol. 2010, pp. 75-86. http://arxiv.org/abs/quant-ph/0009004

[BP 99] A. Brodsky, N. Pippenger. Characterizations of 1-Way Quantum Finite Automata. http://arxiv.org/abs/quant-ph/9903014

[HS 66] J. Hartmanis, R. E. Stearns. Algebraic Structure Theory of Sequential Machines. *Prentice Hall*, 1966.

[KS 76] J. G. Kemeny and J. L. Snell. Finite Markov Chains. *Springer Verlag*, 1976.

[KW 97] A. Kondacs, J. Watrous. On The Power of Quantum Finite State Automata. *Proc. 38th FOCS*, 1997, pp. 66-75.

[MC 97] C. Moore, J. P. Crutchfield. Quantum Automata and Quantum Grammars. *Theoretical Computer Science*, 2000, Vol. 237(1-2), pp. 275-306. http://arxiv.org/abs/quant-ph/9707031

[N 99] A. Nayak. Optimal Lower Bounds for Quantum Automata and Random Access Codes. *Proc. 40th FOCS*, 1999, pp. 369-377. http://arxiv.org/abs/quant-ph/9904093

[R 63] M. O. Rabin. Probabilistic Automata. *Information and Control*, 1963, Vol. 6(3), pp. 230-245.

[T 68] G. Thierrin. Permutation Automata. *Mathematical Systems Theory*, Vol. 2(1), pp. 83-90

Quantum versus Deterministic Counter Automata

Tomohiro Yamasaki[1], Hirotada Kobayashi[1,2], and Hiroshi Imai[1,2]

[1] Department of Information Science, The University of Tokyo,
7-3-1 Hongo, Bunkyo-ku, Tokyo 113-0033, Japan
{yamasaki,hirotada,imai}@is.s.u-tokyo.ac.jp

[2] Quantum Computation and Information Project, ERATO, JST,
5-28-3 Hongo, Bunkyo-ku, Tokyo 113-0033, Japan

Abstract. This paper focuses on quantum analogues of various models of counter automata, and almost completely proves the relation between the classes of languages recognizable by bounded error quantum ones and classical deterministic ones in every model of counter automata. It is proved that (i) under some practically reasonable assumption, quantum ones are strictly stronger than deterministic ones in two-way one-counter automata, and (ii) for any fixed k, quantum ones and deterministic ones are incomparable in one-way k-counter automata.

1 Introduction

Quantum finite state automata were introduced by Moore and Crutchfield [8] and Kondacs and Watrous [4] independently. The latter showed that bounded error one-way quantum finite state automata (1QFAs) can recognize languages only in a proper subset of the class of regular languages, while the class of languages recognized by bounded error two-way quantum finite state automata (2QFAs) properly contains the class of regular languages.

One thing to be mentioned on 2QFAs is that, with respect to input length, they need logarithmically (not constantly) many qubits to be implemented. This is because the tape head on the input tape of a 2QFA is allowed in quantum superposition, and thus logarithmically many qubits are necessary to store the position of the tape head. In this context, quantum analogues of one-way counter automata and polynomial-time two-way counter automata need only logarithmically many qubits as well as their classical versions need logarithmically many bits (note that pushdown automata need polynomially many bits or qubits).

Another interesting property of counter automata was shown by Minsky [7] that two-way deterministic two-counter automata (2D2CAs) can simulate deterministic Turing machines, that is, 2D2CAs are universal. Morita [9] extended this to the universality of two-way reversible two-counter automata (2R2CAs). Hence, from the viewpoint of quantum computation, quantum analogues of two-way one-counter automata and one-way k-counter automata are of interest.

One-way quantum one-counter automata (1Q1CAs) were introduced by Kravtsev [5], and studied by Yamasaki, Kobayashi, Tokunaga, and Imai [11] and Bonner, Freivalds, and Kravtsev [1] in comparison with their various classical

O.H. Ibarra and L. Zhang (Eds.): COCOON 2002, LNCS 2387, pp. 584–594, 2002.

counterparts. In particular, it is known that 1Q1CAs can recognize several non-context-free languages [5,11] while there are regular languages that cannot be recognized by 1Q1CAs [11]. This implies the incomparability between 1Q1CAs and one-way deterministic one-counter automata (1D1CAs).

This paper gives the first formal treatments of two-way quantum one-counter automata (2Q1CAs) and one-way quantum k-counter automata (1QkCAs).

For two-way one-counter automata, it is proved that 2Q1CAs are at least as powerful as two-way deterministic one-counter automata (2D1CAs) in the following practical sense. That is, as far as we consider models with a counter tape of length bounded by some function with respect to input length (and this function is unknown to the finite control part of the automaton), we have a method of reversible simulation of 2D1CAs. Further it is proved that the non-context-free languages $L_{\mathrm{square}} = \{a^m b^{m^2} \mid m \geq 1\}$ and $L_{\mathrm{prod}} = \{a^{m_1} b^{m_2} c^{m_1 m_2} \mid m_1, m_2 \geq 1\}$, which cannot be recognized by 2D1CAs [3], can be recognized by polynomial-time 2Q1CAs with arbitrary small constant one-sided error. Other non-context-free languages such as $L_{\mathrm{power}} = \{a^m b^{2^m} \mid m \geq 1\}$ are also shown recognizable by one-sided error 2Q1CAs in polynomial time. These results of recognizability hold both in the usual model with a counter tape of unbounded length and in the restricted model with a counter tape of bounded length.

For one-way k-counter automata, this paper proves the existence of the family of languages $\{L_0^{k+1}\}$, where $L_0^{k+1} = \{a^{m_1} b a^{m_2} b \cdots a^{m_k} b a^{m_{k+1}} c a^{m_0} \mid m_i \geq 0, m_0 = m_j \text{ for some } 1 \leq j \leq k+1\}$ is known unrecognizable by 1DkCAs [2], can be recognized by 1QkCAs (actually by 1Q1CAs) with bounded error. It is also proved that, for any fixed integer k, the regular language $L_{\mathrm{last}} = \{\{a, b\}^* a\}$ cannot be recognized by bounded error 1QkCAs.

2 Definitions

Here we give formal definitions of two-way one-counter automata and one-way k-counter automata both in classical and quantum cases. It is assumed that every input x is written of the form ¢x\$ on the input tape, started by the left end-marker ¢ and terminated by the right end-marker \$. It is also assumed that each of ¢ and \$ does not appear in x. For convenience, let $\mathbb{Z}_{[a,b]}$ denote the set of integers in the interval of $[a, b]$.

2.1 Two-Way One-Counter Automata

In general, each two-way one-counter automaton is specified by $M = (Q, \Sigma, \delta, q_0, Q_{\mathrm{acc}}, Q_{\mathrm{rej}})$. Here Q is a finite set of states, Σ is the finite input alphabet, $q_0 \in Q$ is the initial state, $Q_{\mathrm{acc}} \subseteq Q$ is a set of accepting states, $Q_{\mathrm{rej}} \subseteq Q$ is a set of rejecting states, and δ is a transition function of the form

$$\delta \colon Q \times \Gamma \times S \times Q \times \{-1, 0, +1\} \times \{\leftarrow, \downarrow, \rightarrow\} \to \mathbb{C}, \tag{1}$$

where $\Gamma = \Sigma \cup \{¢, \$\}$ is the tape alphabet and $S = \{0, 1\}$. It is assumed that each two-way one-counter automaton has a tape served as a counter and the counter value is zero at the beginning of computation.

For the sake of reversible simulation discussed in Subsection 3.1, here we define two-way one-counter automata with a counter tape of bounded length. That is, for every input of length n, a counter tape is of length $2\xi(n)+1$ for some function $\xi \colon \mathbb{Z}^+ \to \mathbb{N}$ so that counter values are in the interval of $[-\xi(n), \xi(n)]$, and this function ξ is unknown to the finite control part of the automaton. Furthermore, it is assumed that the left-most and the right-most cells of the counter tape are indicated by symbols ¢ and \$ written on them, respectively, in order to prevent overflow and underflow of the counter. Thus the set S in (1) is redefined as $S = \{0, 1, ¢, \$\}$. Note that we have the usual model of two-way one-counter automata if we take ξ to be infinity independent of input length.

First we define two-way deterministic one-counter automata.

Definition 1. *A two-way deterministic one-counter automaton (2D1CA) $M = (Q, \Sigma, \delta, q_0, Q_{\mathrm{acc}}, Q_{\mathrm{rej}})$ is a two-way one-counter automaton whose transition function δ takes values in $\{0,1\}$ and satisfies that, for any $q \in Q$, $\sigma \in \Gamma$, and $s \in \{0, 1, ¢, \$\}$, there is a unique triplet of $q' \in Q$, $c \in \{-1, 0, +1\}$, and $d \in \{\leftarrow, \downarrow, \rightarrow\}$ such that $\delta(q, \sigma, s, q', c, d) = 1$.*

Assume that the input x is of length n and the counter tape is of length $2\xi(n) + 1$. For each counter value $z \in \mathbb{Z}_{[-\xi(n), \xi(n)]}$, let $s = \mathrm{sign}(z) \in \{0, 1, ¢, \$\}$, where $\mathrm{sign}(0) = 0$, $\mathrm{sign}(-\xi(n)) = ¢$, $\mathrm{sign}(\xi(n)) = \$$, and $\mathrm{sign}(z) = 1$ otherwise. At the beginning of computation, the automaton is in the initial state q_0 with its tape head scanning the left-most symbol ¢ of $w_x = ¢x\$$. At each step, it reads a symbol σ of w_x in a state q, checks $s = \mathrm{sign}(z)$ of a counter value z, and finds an appropriate transition $\delta(q, \sigma, s, q', c, d) = 1$ for some $q' \in Q$, $c \in \{-1, 0, +1\}$, and $d \in \{\leftarrow, \downarrow, \rightarrow\}$. Then it updates its state to q', changes the counter value z to $z + c$, and moves its tape head in direction d by a square (the tape head remains stationary if $d = \downarrow$). The automaton accepts x if it enters one of the final states in Q_{acc} and rejects x if it enters one of the final states in Q_{rej}.

Next we define two-way quantum one-counter automata (2Q1CAs). Assume that the counter tape is of length $2\xi(n) + 1$ for every input of length n. Then the number of configurations of a 2Q1CA M is precisely $(n + 2)(2\xi(n) + 1)|Q|$. For fixed M and ξ, let C_n denote this set of configurations. A computation on an input x of length n corresponds to a unitary evolution in the Hilbert space $\mathcal{H}_n = l_2(C_n)$. For each $(q, z, k) \in C_n$, where $q \in Q$, $z \in \mathbb{Z}_{[-\xi(n), \xi(n)]}$, and $k \in \mathbb{Z}_{[0, n+1]}$, let $|q, z, k\rangle$ denote the corresponding basis vector in $\mathcal{H}_n$. A transition operator U_x^δ for an input x on $\mathcal{H}_n$ is given by

$$U_x^\delta |q, z, k\rangle = \sum_{q', c, d} \delta(q, w_x(k), \mathrm{sign}(z), q', c, d) |q', z + c, k + \mu(d)\rangle,$$

where $w_x(k)$ denotes the kth symbol of $w_x = ¢x\$$ and $\mu(d) = -1(0)[+1]$ if $d = \leftarrow (\downarrow)[\rightarrow]$. It is assumed that U_x^δ is unitary, that is, $\left(U_x^\delta\right)^\dagger U_x^\delta = U_x^\delta \left(U_x^\delta\right)^\dagger = I$. After each transition, a state of a 2Q1CA is observed with the computational observable O, which is the orthogonal decomposition of $\mathcal{H}_n$ into $E_{\mathrm{acc}} \oplus E_{\mathrm{rej}} \oplus E_{\mathrm{non}}$, where $E_{\mathrm{acc}} = \mathrm{span}\{|q, z, k\rangle \mid q \in Q_{\mathrm{acc}}\}$, $E_{\mathrm{rej}} = \mathrm{span}\{|q, z, k\rangle \mid q \in Q_{\mathrm{rej}}\}$, and $E_{\mathrm{non}} = \mathrm{span}\{|q, z, k\rangle \mid q \in Q \setminus (Q_{\mathrm{acc}} \cup Q_{\mathrm{rej}})\}$. The outcome of any observation will be either "accepting" ($E_{\rm acc}$) or "rejecting" ($E_{\rm rej}$) or "non-halting" ($E_{\rm non}$). The probability of acceptance, rejection, and non-halting at each step is equal to the sum of the squared amplitude of each basis state in the new state for the corresponding subspace. In order to have U_x^δ be unitary, we have

$$\langle q_1, z_1, k_1 | q_2, z_2, k_2 \rangle = \sum_{\substack{q' \\ z_1+c_1=z_2+c_2 \\ k_1+\mu(d_1)=k_2+\mu(d_2)}} \delta^\dagger(q_1, w_x(k_1), \mathrm{sign}(z_1), q', c_1, d_1)\delta(q_2, w_x(k_2), \mathrm{sign}(z_2), q', c_2, d_2). \quad (2)$$

The *well-formedness conditions* of 2Q1CAs are ones derived from (2).

Definition 2. *A two-way quantum one-counter automaton (2Q1CA) $M = (Q, \Sigma, \delta, q_0, Q_{\rm acc}, Q_{\rm rej})$ is a two-way one-counter automaton whose transition function δ satisfies the well-formedness conditions.*

Finally, two-way reversible one-counter automata (2R1CAs) are simply defined as 2Q1CAs whose transition function δ only takes values in $\{0, 1\}$.

Now we give a definition of languages recognized by 2D1CAs and 2Q1CAs with a counter tape of bounded length.

Definition 3. *A language L is recognized by a 2D1CA M (resp. recognized by a 2Q1CA M with probability $p > 1/2$) if there exists a function $\xi_0 \colon \mathbb{Z}^+ \to \mathbb{N}$ such that, for any function $\xi \colon \mathbb{Z}^+ \to \mathbb{N}$ satisfying $\xi \geq \xi_0$ and any integer $n \geq 0$, M equipped with its counter tape of length $2\xi(n) + 1$ accepts any input $x \in L$ of length n (resp. accepts any input $x \in L$ of length n with probability at least p) and rejects any input $x \notin L$ of length n (resp. rejects any input $x \notin L$ of length n with probability at least p).*

To describe automata easily, we introduce the concept of *simple* 2Q1CAs. Consider the Hilbert space $l_2(Q)$ for the set Q of internal states of a 2Q1CA M.

Definition 4. *A 2Q1CA $M = (Q, \Sigma, \delta, q_0, Q_{\rm acc}, Q_{\rm rej})$ is simple, if there are unitary operators $V_{\sigma,s}$ on $l_2(Q)$ for each $\sigma \in \Gamma$ and $s \in \{0, 1, ¢, \$\}$, a counter function $C \colon Q \times \Gamma \to \{-1, 0, +1\}$, and a tape-head function $D \colon Q \to \{\leftarrow, \downarrow, \rightarrow\}$ such that, for any $q, q' \in Q$, $c \in \{-1, 0, +1\}$, and $d \in \{\leftarrow, \downarrow, \rightarrow\}$,*

$$\delta(q, \sigma, s, q', c, d) = \begin{cases} \langle q' | V_{\sigma,s} | q \rangle & \text{if } C(q', \sigma) = c \text{ and } D(q') = d, \\ 0 & \text{otherwise.} \end{cases}$$

If a 2Q1CA is simple, increase or decrease of a counter value is determined only by a symbol the automaton reads and a state it enters, while a move of the tape head is determined only by a state it enters. Thus it is easy to see that a simple 2Q1CA M satisfies the well-formedness conditions if there is a unitary operator $V_{\sigma,s}$ for each $\sigma \in \Gamma$ and $s \in \{0, 1, ¢, \$\}$ such that $\sum_{q' \in Q} (\langle q' | V_{\sigma,s} | q_1 \rangle)^\dagger \langle q' | V_{\sigma,s} | q_2 \rangle$ equals 1 if $q_1 = q_2$ and 0 otherwise.

For the case of 2D1CAs, simple 2D1CAs can be defined in a similar manner.

2.2 One-Way k-Counter Automata

Here we only deal with one-way two-counter automata. It is straightforward to extend our definitions to the k-counter cases.

In general, each one-way two-counter automaton is specified by $M = (Q, \Sigma, \delta, q_0, Q_{\mathrm{acc}}, Q_{\mathrm{rej}})$, whose transition function δ is of the form

$$\delta\colon Q \times \Gamma \times S \times S \times Q \times \{-1, 0, +1\} \times \{-1, 0, +1\} \to \mathbb{C},$$

where $\Gamma = \Sigma \cup \{¢, \$\}$ and $S = \{0, 1\}$. It is assumed that each one-way two-counter automaton has two counters, each of which initially contains zero.

First we define one-way deterministic two-counter automata.

Definition 5. *A one-way deterministic two-counter automaton (1D2CA) $M = (Q, \Sigma, \delta, q_0, Q_{\mathrm{acc}}, Q_{\mathrm{rej}})$ is a one-way two-counter automaton whose transition function δ takes values in $\{0, 1\}$ and satisfies that, for any $q \in Q$, $\sigma \in \Gamma$, and $s_1, s_2 \in \{0, 1\}$, there is a unique triplet of $q' \in Q$ and $c_1, c_2 \in \{-1, 0, +1\}$ such that $\delta(q, \sigma, s_1, s_2, q', c_1, c_2) = 1$.*

Transitions of a 1D2CA are almost same as those of a 2D1CA except that at every step the input tape head always moves right by a square and the automaton treats not only one but two counters.

Next we define one-way quantum two-counter automata. Given an input x of length n, the number of configurations of a 1Q2CA M is precisely $(2n + 5)^2 |Q|$. For fixed M, let C_n denote this set of configurations. A computation on x corresponds to a unitary evolution in the Hilbert space $\mathcal{H}_n = l_2(C_n)$. For each $(q, z_1, z_2) \in C_n$, where $q \in Q$ and $z_1, z_2 \in \mathbb{Z}_{[-n-2, n+2]}$, let $|q, z_1, z_2\rangle$ denote the corresponding basis vector in $\mathcal{H}_n$. For every $\sigma \in \Gamma$, a transition operator U_σ^δ on $\mathcal{H}_n$ is given by

$$U_\sigma^\delta |q, z_1, z_2\rangle = \sum_{q', c_1, c_2} \delta(q, \sigma, \mathrm{sign}(z_1), \mathrm{sign}(z_2), q', c_1, c_2) |q', z_1 + c_1, z_2 + c_2\rangle.$$

It is assumed that U_σ^δ is unitary. Similar to the case of a 2Q1CA, after each transition, a state of a 1Q2CA is observed with the computational observable O, which is the orthogonal decomposition of $\mathcal{H}_n$ into $E_{\mathrm{acc}} \oplus E_{\mathrm{rej}} \oplus E_{\mathrm{non}}$. In order to have the transition matrices U_σ^δ be unitary, we have

$$\langle q_1, z_{11}, z_{12} | q_2, z_{21}, z_{22}\rangle = \sum_{\substack{q' \\ z_{1j}+c_{1j} \\ = z_{2j}+c_{2j}}} \delta^\dagger(q_1, \sigma, \mathrm{sign}(z_{11}), \mathrm{sign}(z_{12}), q', c_{11}, c_{12}) \delta(q_2, \sigma, \mathrm{sign}(z_{21}), \mathrm{sign}(z_{22}), q', c_{21}, c_{22}). \quad (3)$$

The *well-formedness conditions* of 1Q2CAs are ones derived from (3).

Definition 6. *A one-way quantum two-counter automaton (1Q2CA) $M = (Q, \Sigma, \delta, q_0, Q_{\mathrm{acc}}, Q_{\mathrm{rej}})$ is a one-way two-counter automaton whose transition function δ satisfies the well-formedness conditions.*

3 2Q1CAs versus 2D1CAs

3.1 Reversible Simulation of 2D1CAs

First we show that an arbitrary 2D1CA is simulated by a 2R1CA as far as we consider the models with the counter tape of bounded length. Since 2R1CAs are special cases of 2Q1CAs, this implies that 2Q1CAs are practically at least as powerful as their classical deterministic counterpart.

We start with a useful property of 2D1CAs. The proof is easy and left to the reader.

Lemma 7. *Let L be a language that can be recognized by a 2D1CA. Then there exists a simple 2D1CA that recognizes L, whose counter function does not depend on a symbol it reads.*

Now we show the method of reversible simulation. Kondacs and Watrous [4] showed that any one-way deterministic finite state automaton can be simulated by a two-way reversible finite state automaton by using a technique due to Lange, McKenzie, and Tapp [6]. Our method is an extension of them.

Theorem 8. *Let L be a language that can be recognized by a 2D1CA. Then there exists a simple 2R1CA that recognizes L.*

Proof. Let $M = (Q, \Sigma, \delta, q_0, Q_{\mathrm{acc}}, Q_{\mathrm{rej}})$ be a 2D1CA for L that halts in at most $t(n)$ steps on a given input of length n. Then, for every function $\xi \geq t$ and every input x of length n, M equipped with its counter tape of length $2\xi(n)+1$ accepts $x \in L$ and rejects $x \notin L$. ¿From Lemma 7 we may assume that M is simple and its counter function does not depend on a symbol it reads. Without loss of generality, we assume that each Q_{acc} and Q_{rej} consists of only one state. We construct a simple 2R1CA $M' = (Q', \Sigma, \delta', q_0', Q_{\mathrm{acc}}', Q_{\mathrm{rej}}')$ for L such that for any function $\xi \geq t$ and every input x of length n, M' equipped with its counter tape of length $2\xi(n)+1$ accepts $x \in L$ and rejects $x \notin L$.

First, for each q, σ, s, and the transition matrix $V_{\sigma,s}$ of M, define $I_{q,\sigma,s} = \{q' \in Q \mid V_{\sigma,s}|q'\rangle = V_{\sigma,s}|q\rangle\}$ and $J_{q,\sigma,s} = \{q' \in Q \mid V_{\sigma,s}|q'\rangle = |q\rangle\}$, and fix an ordering of the set Q. Let $\max(\cdot)$ and $\min(\cdot)$ denote the maximum and minimum functions relative to this ordering. For any subset $R \subseteq Q'$, let $\mathrm{succ}(q, R)$ be the least element in R larger than q (assuming there is such an element).

Now we define M'. Let the state sets $Q' = Q \times \{-,+\}$, $Q_{\mathrm{acc}}' = Q_{\mathrm{acc}} \times \{-,+\}$, $Q_{\mathrm{rej}}' = Q_{\mathrm{rej}} \times \{-,+\}$, and let the initial state $q_0' = (q_0, +)$.

Define the transition matrices $V_{\sigma,s}'$ as, for each $(q,+), (q,-) \in Q', q \in Q$,

$$V_{\sigma,s}'|(q,+)\rangle = \begin{cases} |(\mathrm{succ}(q, I_{q,\sigma,s}), -)\rangle & \text{if } q \neq \max(I_{q,\sigma,s}), \\ |(r_{q,\sigma,s}, +)\rangle & \text{if } q = \max(I_{q,\sigma,s}), \end{cases}$$

$$V_{\sigma,s}'|(q,-)\rangle = \begin{cases} |(q,+)\rangle & \text{if } J_{q,\sigma,s} = \emptyset, \\ |(\min(J_{q,\sigma,s}), -)\rangle & \text{if } J_{q,\sigma,s} \neq \emptyset. \end{cases}$$

Here $r_{q,\sigma,s} \in Q$ is the state satisfying $V_{\sigma,s}|q\rangle = |r_{q,\sigma,s}\rangle$.

For $d = \leftarrow (\downarrow)[\rightarrow]$ let $-d = \rightarrow (\downarrow)[\leftarrow]$. For each $(q,+),(q,-) \in Q'$, $q \in Q$, and $\sigma \in \Gamma$, define the counter function C' and the tape-head function D' as $C'((q,\pm),\sigma) = \pm C(q,\sigma)$ and $D'((q,\pm)) = \pm D(q)$, where C and D are the counter function and tape-head function of M, respectively.

Let the transition function δ' of M' be defined from given $V'_{\sigma,s}$, C', and D'.

For a given M and every input x of length n, let G be an undirected graph with a set of vertices $Q \times \mathbb{Z}_{[-\xi(n),\xi(n)]} \times \mathbb{Z}_{[0,n+1]}$ and an edge between vertices (q_1, z, k) and $(q_2, z{+}C(q_2,\sigma), k{+}\mu(D(q_2)))$ if and only if $V_{w_x(k),\mathrm{sign}(z)}|q_1\rangle = |q_2\rangle$, where $w_x(k)$ denotes the kth symbol of $w_x = ¢x\$$, and $\mu(d) = -1(0)[+1]$ if $d = \leftarrow (\downarrow)[\rightarrow]$. Let G_0 be the connected component of G that contains the initial configuration $(q_0, 0, 0)$. Since M halts, there can be no cycles in G_0, and G_0 must contain exactly one vertex corresponding to a halting state $\in Q_{\mathrm{acc}} \cup Q_{\mathrm{rej}}$. Thus G_0 can be viewed as a tree with the single halting configuration vertex as the root. M' simulates M by traversing G_0 in a reversible manner.

Each configuration (q, z, k) of M corresponds to two configurations $((q,+), z, k)$ and $((q,-), z - C(q,\sigma), k - \mu(D(q)))$ of M', which are to be interpreted as follows (recall that C does not depend on σ). The configuration $((q,+), z, k)$ indicates that the subtree of G_0 rooted at the vertex (q, z, k) has just been traversed, and the configuration $((q,-), z - C(q,\sigma), k - \mu(D(q)))$ indicates that the subtree of G_0 rooted at (q, z, k) is now about to be traversed.

Consider the set $J_{q,\sigma,s} = I_{q'_i,\sigma,s} = \{q'_1, \ldots, q'_l\}$ for each $i = 1, \ldots, l$, where $\sigma = w_x(k)$ and $s = \mathrm{sign}(z)$. Assume that $q'_1 < q'_2 < \cdots < q'_l$ according to our ordering of Q. Suppose that M' is in a configuration $((q'_i,+), z, k)$ for $i < l$. Since $q'_i \neq \max(I_{q'_i,\sigma,s})$, the next configuration is $((q'_{i+1},-), z - C(q'_{i+1},\sigma), k - \mu(D(q'_{i+1})))$, and now the tree rooted at (q'_{i+1}, z, k) is about to be traversed. Now suppose that M' is in a configuration $((q'_l,+), z, k)$. Since $q'_l = \max(I_{q'_l,\sigma,s})$, the next configuration is $((q,+), z + C(q,\sigma), k + \mu(D(q)))$. Hence, for each $q \in Q$, $z \in \mathbb{Z}_{[-\xi(n),\xi(n)]}$, and $k \in \mathbb{Z}_{[0,n+1]}$, M' enters the configuration $((q,+), z + C(q,\sigma), k + \mu(D(q)))$ only after each of the subtrees rooted at its children has been traversed. Next, suppose that M' is in a configuration $((q,-), z, k)$. The next configuration is $((q'_1,-), z - C(q'_1,\sigma), k - \mu(D(q'_1)))$, and thus the subtree rooted at the vertex (q'_1, z, k) is now to be traversed. Finally, if $(q, z + C(q,\sigma), k + \mu(D(q)))$ has no predecessors, we have $J_{q,\sigma,s} = \emptyset$, and thus the configuration which immediately follows $((q,-), z, k)$ is $((q,+), z + C(q,\sigma), k + \mu(D(q)))$.

By traversing G_0 in this manner, M' eventually enters one of the configurations in $Q'_{\mathrm{acc}} \times \mathbb{Z}_{[-\xi(n),\xi(n)]} \times \mathbb{Z}_{[0,n+1]}$ or $Q'_{\mathrm{rej}} \times \mathbb{Z}_{[-\xi(n),\xi(n)]} \times \mathbb{Z}_{[0,n+1]}$, and clearly M' recognizes L. □

3.2 Recognizability by 2Q1CAs

Next we show that the non-context-free languages $L_{\mathrm{square}} = \{a^m b^{m^2} \mid m \geq 1\}$ and $L_{\mathrm{prod}} = \{a^{m_1} b^{m_2} c^{m_1 m_2} \mid m_1, m_2 \geq 1\}$, which cannot be recognized by 2D1CAs [3], can be recognized by polynomial-time 2Q1CAs with arbitrary small constant one-sided error. Other non-context-free languages $L_{m,m^2,\ldots,m^k} = \{a_1^m a_2^{m^2} \cdots a_k^{m^k} \mid m \geq 1\}$ and $L_{\mathrm{power}} = \{a^m b^{2^m} \mid m \geq 1\}$ are also shown recognizable by polynomial-time 2Q1CAs with arbitrary small constant one-sided error. The recognizability results shown in this subsection hold both in the usual model with a counter tape of unbounded length and in the restricted model with a counter tape of bounded length.

Proposition 9. *For a language $L \in \{L_{\mathrm{square}}, L_{\mathrm{prod}}, L_{m,m^2,\ldots,m^k}\}$ and for an arbitrary fixed integer $N \geq 2$, there exists a 2Q1CA M such that (i) for $x \in L$, M halts after $O(N|x|)$ steps and accepts x with certainty, and (ii) for $x \notin L$, M halts after at most $O(N|x|^2)$ steps and rejects x with probability at least $1 - 1/N$.*

Proof. We only show the case of L_{square}. The recognizability of L_{prod} can be shown in a similar manner, while that of $L_{m,m^2,\ldots,m^k}$ is shown by combining 2Q1CAs for L_{square} and L_{prod}.

We construct a 2Q1CA $M_{\mathrm{square}} = (Q, \Sigma, \delta, q_0, Q_{\mathrm{acc}}, Q_{\mathrm{rej}})$ such that for every input x of length n and any function ξ, $\xi(n) \geq n$, M_{square} equipped with its counter tape of length $2\xi(n) + 1$ recognizes L_{square}. For M_{square}, let the state set $Q = \{q_0, q_1, q_2, q_3, q_4, q^i_{5,j_1}, q^i_{6,j_2}, q^i_7 \mid 1 \leq i \leq N, 1 \leq j_1 \leq i, 1 \leq j_2 \leq N - i + 1\}$, the accepting state set $Q_{\mathrm{acc}} = \{q^N_7\}$, and the rejecting state set $Q_{\mathrm{rej}} = \{q^j_7 \mid 1 \leq j \leq N-1\}$. Define the transition matrices $V_{\sigma,s}$, the counter function C, and the tape-head function D as follows:

$$
\begin{aligned}
&V_{¢,0}|q_0\rangle = |q_0\rangle,\\
&V_{¢,0}|q^i_{5,0}\rangle = |q^i_{6,N-i+1}\rangle,\\
&V_{a,0}|q_0\rangle = |q_0\rangle,\\
&V_{a,0}|q_1\rangle = |q_2\rangle,\\
&V_{a,0}|q_3\rangle = |q_4\rangle,\\
&V_{a,s}|q^i_{5,j+1}\rangle = |q^i_{5,j}\rangle \quad \text{for } j \neq 0, i,\\
&V_{a,0}|q^i_{5,i+1}\rangle = |q^i_{5,i}\rangle,\\
&V_{a,1}|q^i_{5,i+1}\rangle = |q^i_{5,2i}\rangle,\\
&V_{a,1}|q^i_{5,1}\rangle = |q^i_{5,i}\rangle,\\
&V_{a,0}|q^i_{6,j+1}\rangle = |q^i_{6,j}\rangle \quad \text{for } 1 \leq j \leq N-i,\\
&V_{a,0}|q^i_{6,1}\rangle = |q^i_{6,N-i+1}\rangle,
\end{aligned}
$$

$$
\begin{aligned}
&V_{\$,0}|q_2\rangle = |q_3\rangle,\\
&V_{\$,0}|q^i_{6,N-i+1}\rangle = \frac{1}{\sqrt{N}} \sum_{k=1}^{N} e^{\frac{2\pi i k}{N}\sqrt{-1}} |q^k_7\rangle,\\
&V_{b,0}|q_0\rangle = |q_1\rangle,\\
&V_{b,0}|q_2\rangle = |q_2\rangle,\\
&V_{b,0}|q_3\rangle = |q_3\rangle,\\
&V_{b,0}|q_4\rangle = \frac{1}{\sqrt{N}} \sum_{i=1}^{N} |q^i_{5,0}\rangle,\\
&V_{b,1}|q^i_{5,i}\rangle = |q^i_{5,0}\rangle,\\
&V_{b,1}|q^i_{5,0}\rangle = |q^i_{5,2i}\rangle,\\
&V_{b,0}|q^i_{6,j+1}\rangle = |q^i_{6,j}\rangle,\\
&V_{b,0}|q^i_{6,1}\rangle = |q^i_{6,N-i+1}\rangle,
\end{aligned}
$$

$$
\begin{aligned}
&C(q^i_{5,2i}, a) = -1,\\
&C(q^i_{5,i}, a) = +1,\\
&C(q^i_{5,2i}, b) = -1,\\
&C(q^i_{5,0}, a) = +1,\\
&C(q, \sigma) = 0 \text{ otherwise},\\
&D(q_j) = \rightarrow \text{ for } j = 0, 2, 4,\\
&D(q_j) = \leftarrow \text{ for } j = 1, 3,\\
&D(q^i_{5,2i}) = \leftarrow,\\
&D(q^i_{5,i}) = \rightarrow,\\
&D(q^1_{6,N-i+1}) = \rightarrow,\\
&D(q) = \downarrow \text{ otherwise}.
\end{aligned}
$$

One can see that the computation of M_{square} consists of three phases. The first phase rejects any input not of the form a^+b^+. This phase is straightforward, similar to the case of a two-way reversible finite state automaton (without a counter) that recognizes every input of the form a^+b^+. For the input not of this form, the computation terminates with rejection. Otherwise, the second phase begins with the state q_4 with the tape head reading the left-most b.

At the start of the second phase, the computation branches into N paths, indicated by the states $q^1_{5,0}, \ldots, q^N_{5,0}$, each with amplitude $1/\sqrt{N}$. For each of these paths, M_{square} moves the tape head to left and right deterministically in the following way. Along the ith path, while the counter value is not zero, the automaton decreases the counter value by one and moves the tape head to left. In addition, every time the tape head reads the symbol a, it remains stationary for i steps. Upon the counter value being zero, it repeats the following until the tape head reads the left-most b: it increases the counter value by one and moves the tape head to right. If the tape head reads the symbol ¢, the computation enters the third phase with the state $q_{5,0}^{i}$. Thus, while M_{square} is scanning a's in the input during the second phase, the tape head requires precisely $(i+1)(\sum_{c=0}^{m_1-1}(2c+1)+m_1) = (i+1)(m_1^2+m_1)$ steps along the ith path, where m_1 is the number of a's.

Along the ith path on the third phase, every time the tape head reads the symbol a or b, it remains stationary for $N-i+1$ steps and then moves to right. Upon reading the symbol \$, each computation path again splits according to the quantum Fourier transformation, yielding the single accepting state q_7^N and the other rejecting states $q_7^1, \ldots, q_7^{N-1}$. Thus, while M_{square} is scanning a's and b's in the input during this phase, the tape head requires precisely $(N-i+1)(m_1+m_2)$ steps along the ith path, where m_2 is the number of b's. Therefore, it is easy to see that, under the assumption $i \neq i'$, $(i+1)(m_1^2+m_1)+(N-i+1)(m_1+m_2) = (i'+1)(m_1^2+m_1)+(N-i'+1)(m_1+m_2)$ if and only if $m_1^2 = m_2$.

First consider the case that $m_1^2 = m_2$. Since each of the N computation paths reaches the symbol \$ at the same time, the superposition immediately after performing the quantum Fourier transformation is $\frac{1}{N}\sum_{i=1}^{N}\sum_{k=1}^{N} e^{\frac{2\pi i k}{N}\sqrt{-1}}|q_7^k\rangle = |q_7^N\rangle$. Hence the accepting state q_7^N is entered with certainty.

Next suppose that $m_1^2 \neq m_2$. In this case, each of N paths reaches the symbol \$ at a different timing. Thus, there is no cancellation among the rejection states. For each path, the conditional probability that an observation results in q_7^N at the time is $1/N$. It follows that the total probability that an observation results in q_7^N is also $1/N$. Hence the input is rejected with probability $1-1/N$.

It is clear that each possible computation path has length $O(N|x|)$ for $x \in L_{\text{square}}$ and at most $O(N|x|^2)$ for $x \notin L_{\text{square}}$. □

Finally, we state the recognizability of $L_{\text{power}} = \{a^m b^{2^m} \mid m \geq 1\}$ without a proof. Interestingly, L_{power} can be recognized by 2Q1CAs in *linear-time*.

Proposition 10. *For the language L_{power} and for an arbitrary fixed integer $N \geq 2$, there exists a 2Q1CA M_{power} that accepts $x \in L_{\text{power}}$ with certainty and rejects $x \notin L_{\text{power}}$ with probability at least $1-1/N$. In either case, M_{power} halts after $O(N|x|)$ steps with certainty.*

4 1Q*k*CAs versus 1D*k*CAs

First it is shown the strength of 1QkCAs that there exists a language that can be recognized by 1QkCAs but not by 1DkCAs.

Fischer, Meyer, and Rosenberg [2] proved that the language $L_0^{k+1} = \{a^{m_1} ba^{m_2}b \cdots a^{m_k}ba^{m_{k+1}}ca^{m_0} \mid m_i \geq 0, m_0 = m_j \text{ for some } 1 \leq j \leq k+1\}$ cannot be recognized by 1DkCAs. In contrast to this, we prove that, for any fixed k, L_0^{k+1} is recognizable by bounded-error 1Q1CAs. Here one may consider a 1Q1CA as a 1Q2CA whose second counter never changes its value, or a 2Q1CA whose tape head moves to right by a square at every step. For the formal definition of 1Q1CAs, see [5,11,1].

Proposition 11. *There exists a 1Q1CA M_0^{k+1} that recognizes the language L_0^{k+1} with probability $1/2 + 1/(4k+2)$.*

Proof. We only show the recognizability of $L_0^3 = \{a^{m_1}ba^{m_2}ba^{m_3}ca^{m_0} \mid m_i \geq 0, m_0 = m_j \text{ for some } 1 \leq j \leq 3\}$ for the case of $k = 2$. It is straightforward to extend the proof to the case of general k.

Let $L_1^3 = \{a^m ba^*ba^*ca^m \mid m \geq 0\}, L_2^3 = \{a^*ba^m ba^*ca^m \mid m \geq 0\}$, and $L_3^3 = \{a^*ba^*ba^m ca^m \mid m \geq 0\}$. It is easy to prove that each of L_1^3, L_2^3, and L_3^3 is recognizable by a one-way reversible one-counter automaton (1R1CA). Let each M_1^3, M_2^3, and M_3^3 be the 1R1CA for L_1^3, L_2^3, and L_3^3, respectively. We construct a 1Q1CA M_0^3 by using M_1^3, M_2^3, and M_3^3. M_0^3 behaves as follows.

After reading the left end-marker ¢, the computation branches into four paths, path-1, path-2, path-3, and path-4, with amplitudes $\sqrt{1/5}$, $\sqrt{1/5}$, $\sqrt{1/5}$, and $\sqrt{2/5}$, respectively. In each path-i, $1 \leq i \leq 3$, M_0^3 behaves in the same manner as M_i^3 to check whether the input is in L_i^3. In the path-4, M_0^3 accepts any input.

Then, for every input $x \in L_0^3$, there is at least one path among the path-1, path-2, and path-3 in which x is accepted. Therefore, M_0^3 accepts $x \in L_0^3$ with probability $1/5 + 2/5 = 3/5$. On the other hand, every input $x \notin L_0^3$ is always rejected in all of the three paths, path-1, path-2, and path-3. Thus M_0^3 rejects $x \notin L_0^3$ with probability at least $1/5 + 1/5 + 1/5 = 3/5$.

Reversibility of this automaton is clear by its construction. □

Next, in contrast to Proposition 11, it is shown the weakness of 1QkCAs that there is a regular language that cannot be recognized by bounded-error 1QkCAs.

For the no-counter and one-counter cases, it is known that the language $L_{\text{last}} = \{\{a, b\}^*a\}$ cannot be recognized by 1QFAs and 1Q1CAs. Since L_{last} is regular, it is obviously recognizable by one-way deterministic finite state automata (and hence by 1DkCAs for any fixed k). These unrecognizability results for L_{last} are from the fact shown by Nayak [10], which states that, for each fixed $n \geq 0$, any general one-way quantum automaton that recognizes the language $\{wa \mid w \in \{a, b\}^*, |w| \leq n\}$ must have $2^{\Omega(n)}$ quantum basis states. It is easy to extend these to the k-counter case.

Proposition 12. *For any fixed $k \geq 0$, the language L_{last} cannot be recognized by 1QkCAs with bounded error.*

Proof. By the fact shown by Nayak, a 1QkCA for L_{last} must have at least $2^{\Omega(n)}$ quantum basis states for every input of length n. For every input of length n, however, the number of basis states a 1QkCA can have is at most $(2n+5)^k|Q|$, which is less than $2^{\Omega(n)}$ for sufficiently large n. This completes the proof. □

References

1. R. J. Bonner, R. Freivalds, and M. Kravtsev. Quantum versus probabilistic one-way finite automata with counter. In *Proceedings of the 28th Conference on Current Trends in Theory and Practice of Informatics (SOFSEM 2001)*, volume 2234 of *Lecture Notes in Computer Science*, pages 181–190, 2001.

2. P. C. Fischer, A. R. Meyer, and A. L. Rosenberg. Counter machines and counter languages. *Mathematical Systems Theory*, 2(3):265–283, 1968.
3. E. M. Gurari and O. H. Ibarra. Two-way counter machines and Diophantine equations. *Journal of the ACM*, 29(3):863–873, 1982.
4. A. Kondacs and J. Watrous. On the power of quantum finite state automata. In *Proceedings of the 38th Annual Symposium on Foundations of Computer Science*, pages 66–75, 1997.
5. M. Kravtsev. Quantum finite one-counter automata. In *Proceedings of the 26th Conference on Current Trends in Theory and Practice of Informatics (SOFSEM '99)*, volume 1725 of *Lecture Notes in Computer Science*, pages 431–440, 1999.
6. K.-J. Lange, P. McKenzie, and A. Tapp. Reversible space equals deterministic space. *Journal of Computer and System Sciences*, 60(2):354–367, 2000.
7. M. L. Minsky. Recursive unsolvability of Post's problem of 'tag' and other topics in the theory of Turing machines. *Annals of Mathematics*, 74(3):437–455, 1961.
8. C. Moore and J. P. Crutchfield. Quantum automata and quantum grammars. *Theoretical Computer Science*, 237(1–2):275–306, 2000.
9. K. Morita. Universality of a reversible two-counter machine. *Theoretical Computer Science*, 168(2):303–320, 1996.
10. A. Nayak. Optimal lower bounds for quantum automata and random access codes. In *Proceedings of the 40th Annual Symposium on Foundations of Computer Science*, pages 369–376, 1999.
11. T. Yamasaki, H. Kobayashi, Y. Tokunaga, and H. Imai. One-way probabilistic reversible and quantum one-counter automata. *Theoretical Computer Science*, in press. Preliminary version appeared in *Proceedings of the 6th Annual International Computing and Combinatorics Conference*, volume 1858 of *Lecture Notes in Computer Science*, pages 436–446, 2000.

Quantum DNF Learnability Revisited

Jeffrey C. Jackson[1,*], Christino Tamon[2,**], and Tomoyuki Yamakami[3]

[1] Duquesne University
Pittsburgh, PA 15282-1754, USA
jackson@mathcs.duq.edu
[2] Clarkson University
Potsdam, NY 13699-5815, USA
tino@clarkson.edu
[3] University of Ottawa
Ottawa, Canada
yamakami@site.uottawa.ca

Abstract. We describe a quantum PAC learning algorithm for DNF formulae under the uniform distribution with a query complexity of $\tilde{O}(s^3/\epsilon + s^2/\epsilon^2)$, where s is the size of DNF formula and ϵ is the PAC error accuracy[1]. If s and $1/\epsilon$ are comparable, this gives a modest improvement over a previously known classical query complexity of $\tilde{O}(ns^2/\epsilon^2)$. We also show a lower bound of $\Omega(s \log n/n)$ on the query complexity of any quantum PAC algorithm for learning a DNF of size s with n inputs under the uniform distribution.

1 Introduction

In this abstract we describe a quantum learning algorithm for DNF formulae under the uniform distribution using quantum membership queries. Although Bshouty and Jackson [2] have shown that it is possible to adapt Jackson's Harmonic Sieve algorithm [10] to the quantum setting, our goal is different. We will focus on reducing the number of quantum membership queries used by the DNF learning algorithm whereas their motivation was in showing that quantum examples are sufficient for learning DNF.

The Harmonic Sieve HS algorithm combines two crucial independent algorithms. The first algorithm is an inner algorithm for finding parity functions that weakly approximate the target DNF function. The second algorithm used in the Harmonic Sieve is an outer algorithm that is a boosting algorithm. A weak learning algorithm is an algorithm that produces hypotheses whose accuracy are slightly better than random guessing. Boosting is a method for improving the accuracy of hypotheses given by a weak learning algorithm.

For the inner algorithm, a Fourier-based algorithm given in [12] (called the KM algorithm) is used in HS for finding the weak parity approximators. The KM algorithm is based on a similar method given by Goldreich and Levin [6]

* This material is based upon work supported by the NSF Grant No. CCR-9877079.
** This work is supported by the NSF Grant No. DMR-0121146.
[1] The notation $\tilde{O}(f)$ hides extra logarithmic factors, i.e., $\tilde{O}(f) = O(f \log f)$.

O.H. Ibarra and L. Zhang (Eds.): COCOON 2002, LNCS 2387, pp. 595–604, 2002.

in their seminal work on hardcore bits in cryptography. Subsequently, Levin [13] and Goldreich [7], independently, gave highly improved methods for solving this so-called *Goldreich-Levin* problem. Their ideas were adapted by Bshouty et al. [3] to obtain a weak learning algorithm for DNF of n inputs with query and time complexity of $\tilde{O}(n/\gamma^2)$, where γ is the weak advantage of the parity approximator. By a result in [10], $\gamma = O(1/s)$ for DNF formula of size s.

For the outer algorithm, the original `HS` used a boosting method of Freund [5] called `F1`. Recently, it has been shown by Klivans and Servedio [11] that Freund's other boosting algorithm called $\mathtt{B}_{\mathtt{Comb}}$ gave superior running times. Let γ be the weak advantage of the weak learning algorithm and let ϵ be the target accuracy. Then Freund proved that `F1` is a $O(1/\epsilon^3)$-smooth $O(\gamma^{-2}\log(1/\epsilon))$-stage algorithm whereas $\mathtt{B}_{\mathtt{Comb}}$ is a $\tilde{O}(1/\epsilon)$-smooth $O(\gamma^{-2}\log(1/\epsilon))$-stage boosting algorithm. The fastest known algorithm for learning DNF is obtained by combining the two improved independent components that results in a total running time of $\tilde{O}(ns^4/\epsilon^2)$ and a query complexity of $\tilde{O}(ns^2/\epsilon^2)$ (see [3,11] and the references therein).

We describe an efficient *quantum* DNF learning algorithm by combining a quantum Goldreich-Levin algorithm `QGL` of Adcock and Cleve [1] with the superior boosting algorithm $\mathtt{B}_{\mathtt{Comb}}$. The quantum algorithm of Adcock and Cleve used only $O(1/\gamma)$ queries (beating a classical lower bound of $\Omega(n/\gamma^2)$ proved also in [1]). After adapting both algorithms for quantum PAC learning, we obtain a quantum Harmonic Sieve algorithm `QHS` with a sample complexity of $\tilde{O}(s^3/\epsilon + s^2/\epsilon^2)$. In contrast to the best known classical upper bound of $\tilde{O}(ns^2/\epsilon^2)$, this gives a modest improvement if s and $1/\epsilon$ are comparable.

As shown in [1], the quantum Goldreich-Levin algorithm has applications to quantum cryptography. In this work, we show one of its applications in computational learning theory.

For the sake of exposition, in this abstract we will describe our quantum DNF PAC learning algorithm using a conceptually simpler boosting algorithm `SmoothBoost` given by Servedio [14]. We describe a *boost-by-filtering* version of Servedio's `SmoothBoost` that is a $O(1/\epsilon)$-smooth $O(\gamma^{-2}\epsilon^{-1})$-stage boosting algorithm. So, we incur an extra $1/\epsilon$ factor in the sample complexity. We defer the details of using $\mathtt{B}_{\mathtt{Comb}}$ in `QHS` to the final version of this paper.

Finally, we prove a query lower bound of $\Omega(s \log n/n)$ on any quantum PAC learning algorithm for DNF under the uniform distribution with (quantum) membership queries.

2 Preliminaries

We are interested in algorithms for learning approximations to an unknown function that is a member of a particular class of functions. The specific function class of interest in this paper is that of DNF expressions, that is, Boolean functions that can be expressed as a disjunction of terms, where each term is a conjunction of Boolean variables (possibly negated). Given a *target* DNF expression $f : \{0,1\}^n \to \{-1,+1\}$ having s terms along with an *accuracy* parameter

Input: Parameters $0 < \epsilon < 1/2$, $0 \leq \gamma < 1/2$
Sample S of target f
Weak learning algorithm `WL`

Output: Hypothesis h

1. $U_S \equiv$ the uniform distribution over S
2. $M_1(x) \equiv 1, \forall x \in S$
3. $N_0(x) \equiv 0, \forall x \in S$
4. $\theta \leftarrow \gamma/(2+\gamma)$
5. $t \leftarrow 1$
6. **while** $\mathbf{E}_{x \sim U_S}[M_t(x)] > \epsilon$ **do**
7. $D_t(x) \equiv M_t(x)/(m\mathbf{E}_{x \sim U_S}[M_t(x)]), \forall x \in S$
8. $h_t \leftarrow \mathtt{WL}(S, D_t, \delta = \Omega(\epsilon^{-1}\gamma^{-2}))$
9. $N_t(x) \equiv N_{t-1}(x) + f(x)h_t(x) - \theta, \forall x \in S$
10. $M_{t+1}(x) \equiv [\![N_t(x) < 0]\!] + (1-\gamma)^{N_t(x)/2}[\![N_t(x) \geq 0]\!], \forall x \in S$
11. $t \leftarrow t+1$
12. **end while**
13. $T \leftarrow t-1$
14. $H \equiv \frac{1}{T}\sum_{i=1}^{T} h_i$
15. **return** $h \equiv \text{sign}(H)$.

Fig. 1. The `SmoothBoost` algorithm of Servedio [14].

$0 < \epsilon < 1/2$ and a *confidence* parameter $\delta > 0$, the goal is to with probability at least $1-\delta$ produce a *hypothesis* h such $\Pr_{x \sim U_n}[f(x) \neq h(x)] < \epsilon$, where U_n represents the uniform distribution over $\{0,1\}^n$. We will sometimes refer to such an h as an ϵ-approximator to f, or equivalently say that h has $\frac{1}{2} - \epsilon$ *advantage* (this represents the advantage over the agreement between f and a random function, which is 1/2). A learning algorithm that can guarantee only $\gamma > 0$ advantage in the hypothesis produced but can do so with arbitrarily small probability of failure δ is called a *weak learning algorithm*, and the hypothesis produced is a *weak approximator*.

The information our learning algorithm is given about the target function varies. One form is a *sample*, that is, a set S of input/output pairs for the function. We often use x to denote an input and $f(x)$ the associated output, and $x \in S$ to denote that x is one of the inputs of the pairs in S. Another type of information we sometimes use is a *membership oracle* for f, MEM_f. Such an oracle is given an input x and returns the function's output $f(x)$.

3 A Smoother Boost-by-Filtering Algorithm

A modification of Servedio's `SmoothBoost` boosting algorithm [14] is described in this section. A special case (discrete weak hypotheses and fixed margin) version of `SmoothBoost` sufficient for our purposes is shown in Figure 1. `SmoothBoost` is a boosting-by-sampling method that can be applied to a weak learning algorithm in order to produce a hypothesis that closely approximates the sample. Specifically, `SmoothBoost` receives as input a sample S of size m as well as accuracy parameter ϵ. It is also given a weak learning algorithm `WL`. The boosting algorithm defines a series of distributions D_t over S and successively calls the weak learning algorithm, providing it with the sample S and with one of the distributions D_t. In the end, the algorithm combines the weak hypotheses returned by the calls to the weak learner into a single hypothesis h.

Servedio proves three key properties of `SmoothBoost`:

Lemma 1 (Servedio). *Let f be a target function, and let S, ϵ, γ, h, and D_t be as defined in Figure 1. Then*

1. *If every weak hypothesis h_t returned by* `WL` *has advantage at least γ with respect to D_t, then* `SmoothBoost` *will terminate after $T = O(\epsilon^{-1}\gamma^{-2})$ stages.*
2. *If* `SmoothBoost` *terminates, then $\Pr_{x\sim U_S}[f(x) \neq h(x)] < \epsilon$, where U_S represents the uniform distribution over S (Servedio actually proves a stronger margin result that implies this).*
3. *$L_\infty(mD_t) \leq 1/\epsilon$ for all t, where $m = |S|$ (this is the* smoothness property *of* `SmoothBoost`*).*

Here we adapt this algorithm to obtain a boosting-by-filtering algorithm that will be used by the Harmonic Sieve. First, notice that Lemma 1 holds for the special case $S = \{0,1\}^n$. However, there are potential problems with running the SmoothBoost algorithm directly on such a large S. First, it is not computationally feasible to exactly compute $\mathbf{E}_{x\sim U_n}[M_t(x)]$, where U_n represents the uniform distribution over $\{0,1\}^n$. So instead we must estimate this quantity by sampling. This has a small impact on both the form of the loop condition for the algorithm (line 6), but also on the "distributions" D_t passed to the weak learner (line 7). In fact, the D_t that will be passed to the weak learner will generally not be a true distribution at all, but instead a constant multiplied by a distribution due to the constant error in our estimate of $\mathbf{E}_{x\sim U_n}[M_t(x)]$.

We will deal with the weak learner later, so for now let us assume that the weak learner produces the same hypothesis h_t given an approximation to D_t as it would given the actual distribution. Then notice that the computations for N_t and M_{t+1} are unchanged, so the only impact on the boosting algorithm has to do with the loop condition at line 6. This is easily addressed: let E_t represent an estimate of $\mathbf{E}_{x\sim U_n}[M_t(x)]$ to within additive error $\epsilon/3$ and change the loop condition to $E_t > 2\epsilon/3$. Then if the loop terminates it must be that $\mathbf{E}_{x\sim U_n}[M_t(x)] \leq \epsilon$, as before. It is easily verified that given this condition, Servedio's proof implies that h is an ϵ-approximator to f with respect to the uniform distribution. Furthermore, since $\mathbf{E}_{x\sim U_n}[M_t(x)] \geq \epsilon/3$ if the algorithm terminates, the other statements of Lemma 1 change only by constant factors. In particular, the smoothness condition of the lemma now becomes $L_\infty(2^n D_t) \leq 3/\epsilon$ for all t.

Finally, because $O \leq M_t(x) \leq 1$ for all t and x, the Hoeffding bound gives that taking the sample mean of $M_t(x)$ over a sample of size $\Omega(\epsilon^{-2})$ will, with

Input: Parameters $0 < \epsilon < 1/2$, $0 \leq \gamma < 1/2$
Membership oracle MEM_f
Weak learning algorithm WL

Output: Hypothesis h

1. Draw uniform sample R of $\Omega(\log(\epsilon^{-1}\gamma^{-1})/\epsilon^2)$ instances and label using MEM_f
2. $M_1(x) \equiv 1, \forall x \in \{0,1\}^n$, $N_0(x) \equiv 0, \forall x \in \{0,1\}^n$
3. $\theta \leftarrow \gamma/(2+\gamma)$
4. $t \leftarrow 1$
5. **while** $\mathbf{E}_{x \sim U_R}[M_t(x)] > 2\epsilon/3$ **do**
6. $\quad D_t(x) \equiv M_t(x)/(2^n \mathbf{E}_{x \sim U_R}[M_t(x)])$
7. $\quad h_t \leftarrow \mathtt{WL}(MEM_f, D_t, \delta = \Omega(\epsilon^{-1}\gamma^{-2}))$
8. $\quad N_t(x) \equiv N_{t-1}(x) + f(x)h_t(x) - \theta, \forall x \in \{0,1\}^n$
9. $\quad M_{t+1}(x) \equiv [\![N_t(x) < 0]\!] + (1-\gamma)^{N_t(x)/2}[\![N_t(x) \geq 0]\!], \forall x \in \{0,1\}^n$
10. $\quad t \leftarrow t+1$
11. **end while**
12. $T \leftarrow t-1$
13. $H \equiv \frac{1}{T}\sum_{i=1}^{T} h_i$
14. **return** $h \equiv \text{sign}(H)$.

Fig. 2. The SmoothBoost modified for boost-by-filtering.

constant probability, produce an estimate with additive error at most $\epsilon/3$. Furthermore, if the algorithm terminates in T steps, then a single uniform random sample R of size $\Omega(\log(T)/\epsilon^2)$ guarantees, with constant probability, that estimating the expected value of $M_t(x)$ by the sample mean over R at every step t will produce an $\epsilon/3$ accurate estimate at every step.

Figure 2 presents the modified SmoothBoost algorithm. Notice that in place of a sample S representing the target function f, we are assuming that we are given a membership oracle MEM_f. We will subsequently consider quantum versions of this algorithm and of the membership oracle. For this reason, we show the definitions of M and N as being over all of $\{0,1\}^n$, although for a classical algorithm the only values that would actually be used are those corresponding to $x \in R$.

While the SmoothBoost algorithm has been presented for illustration, Klivans and Servedio [11] have shown that one of Freund's boosting algorithms, which they call $\mathtt{B_{Comb}}$, is actually slightly superior to SmoothBoost for our purposes. Specifically, they note that $\mathtt{B_{Comb}}$ has properties similar to those of SmoothBoost given in Lemma 1, with the change that the number of stages T improves from $O(\epsilon^{-1}\gamma^{-2})$ to $O(\log(1/\epsilon)/\gamma^2)$ while the smoothness of each of the distributions D_t passed to the weak learner satisfies (when learning over all of $\{0,1\}^n$) $L_\infty(2^n D_t) = O(\log(1/\epsilon)/\epsilon)$. We will continue to use SmoothBoost in our analysis here, since $\mathtt{B_{Comb}}$ and its analysis are noticeably more complicated than SmoothBoost and its analysis. However, our final sample size bounds will be stated as if $\mathtt{B_{Comb}}$ is being used, and the final version of this paper will include details of the $\mathtt{B_{Comb}}$ analysis.

Input: Parameters n, $\gamma \in (0, 1/2)$, $\delta > 0$
Quantum membership oracle QMQ_f for Boolean function f
represented by a unitary tranformation U_{MQ}
Random uniform sample R of size $\tilde{\Omega}(\gamma^{-2}\log(1/\delta))$.
Output: A coefficient A with the property that $\Pr_D[f = \chi_A] \geq \frac{1}{2} + \gamma$
with probability at least $1 - \delta$.

1. Let C be defined as in Equation 1.
2. Label R using QMQ_f.
3. Define a sampling-based U_{EQ} as in Equation 2.
4. $|\varphi\rangle \leftarrow C|0_n\rangle_I|0_m\rangle_A|0\rangle_B$
5. **for** $k = 1, \ldots, O(1/\gamma)$ **do**
6. $\quad |\varphi\rangle \leftarrow -CU_0C^\dagger U_{EQ}|\varphi\rangle$
7. **end for**
8. Measure and return the contents of register I.

Fig. 3. The quantum weak learning algorithm QWDNF for uniform distribution.

4 A Query-Efficient Quantum WDNF Algorithm

In this section we describe a quantum weak learning algorithm WDNF for finding parity approximators of non-Boolean functions under smooth distributions. This algorithm is based on a quantum Goldreich-Levin algorithm given by Adcock and Cleve [1]. For completeness we describe the quantum Goldreich-Levin algorithm in the following. This algorithm uses the standard Pauli X (complement) and Z (controlled phase flip) gates as well as the Hadamard H gate (see [4]). Let $U_0(\sum_x \alpha_x|x\rangle) = \sum_{x\neq 0}\alpha_x|x\rangle - \alpha_0|0\rangle$ be the unitary transformation that flips the phase of the all-zero state.

The U_{MQ} transformation that represents a *noisy* membership oracle with respect to a parity function χ_A defined in [1] is given by

$$U_{MQ}|x\rangle|0_m\rangle = \alpha_x|x\rangle|u_x, A\cdot x\rangle + \beta_x|x\rangle|v_x, \overline{A\cdot x}\rangle,$$

where $\sum_x \alpha_x^2 \geq 1/2 + \gamma$ and $\sum_x \beta_x^2 \leq 1/2 - \gamma$. By a result of Jackson [10], for any DNF formula f with s terms, there is a parity function A such that $\Pr[f(x) = \chi_A(x)] \geq 1/2 + \gamma/2$, for $\gamma = 1/(2s+1)$. Thus, a noiseless DNF oracle QMQ_f is a noisy oracle U_{MQ} for some parity function χ_A. Thus, we may assume that U_{MQ} is a unitary transformation that represents a quantum membership oracle QMQ_f for a DNF formula f that maps $|x\rangle|0_m\rangle$ to $|x\rangle|u_x, f(x)\rangle$, for some string $u_x \in \{0,1\}^{m-1}$ that represents the work space of the oracle.

The quantum algorithm QGL of Adcock and Cleve is represented by the following unitary transformation

$$C = (H_n \otimes I_{m+1})(U_{MQ}^\dagger \otimes I_1)(I_{n+m-1} \otimes Z)(U_{MQ} \otimes I_1)(H_n \otimes I_m \otimes X) \quad (1)$$

applied to the initial superposition of $|0_n, 0_m, 0\rangle$. In [1] it was proved that the quantum algorithm QGL prepares a superposition of all n-bit strings such that the probability of observing the coefficient A is $4\gamma^2$. By repeating this for $O(1/\gamma^2)$ stages, we can recover A with constant probability.

The number of stages can be reduced to $O(1/\gamma)$ by using a technique called *amplitude amplification*. This amplification technique uses an iterate of the form $G = (-CU_0C^{\dagger}U_{EQ})^k C|0_n, 0_m, 0\rangle$, where k is approximately $O(1/\gamma)$, and U_{EQ} is a unitary transformation that represents a *quantum equivalence* oracle QEQ_f. The transformation U_{EQ} is defined as

$$U_{EQ}|a\rangle = \begin{cases} -|a\rangle & \text{if } |\mathbf{E}[f\chi_a]| \geq \theta \\ |a\rangle & \text{otherwise} \end{cases} \quad (2)$$

For the purpose of learning DNF, we need to simulate U_{EQ} using a sampling algorithm that has access to QMQ_f. A classical application of Hoeffding sampling requires $\Omega(1/\gamma^2)$ queries to QMQ_f [2]. To simulate U_{EQ}, we will simply use a sample R of size $\Omega(1/\gamma^2 \log(1/\delta))$ to obtain a good estimate with probability at least $1-\delta$.

Finally, recall that we will be applying boosting to this weak learning algorithm, which means that QWDNF will be called a number of times. However, it is not necessary to draw a new random sample R each time QWDNF is called, as the boosting algorithm merely wants a guarantee that the algorithm succeeds with high probability and does not require independence. The resulting quantum weak learning algorithm for DNF, which we denote QWDNF, is described in Figure 3.

4.1 Non-Boolean Functions over Smooth Distributions

Recall that in the Harmonic Sieve algorithm [10], we need to find weak Parity approximators for non-Boolean functions g that is based on the DNF formula f and the current boosting distribution D in SmoothBoost, i.e., we need to consider expressions of the form (we have dropped subscripts for convenience)

$$\mathbf{E}_D[f\chi_A] = \sum_x \frac{M(x)}{2^n \mathbf{E}[M(x)]} f(x)\chi_A(x) = \frac{\mathbf{E}[Mf\chi_A]}{\mathbf{E}[M]}.$$

This shows a reduction from finding a coefficient A such that $|\mathbf{E}_D[f\chi_A]|$ is large to finding a coefficient A so that $|\mathbf{E}_U[g\chi_A]|$, where $g(x) = M(x)f(x)$, is large. Assuming that $\mathbf{E}[M(x)] \geq \epsilon/3$, we will use the algorithm QWDNF to find a coefficient A such that for some constant c_2 $|\mathbf{E}[Mf\chi_A]| \geq \frac{c_2\epsilon}{3(2s+1)} = \Gamma$.

Note that $0 < M(x) \leq 1$, for all x. Thus we can use a technique of Bshouty and Jackson [2] that transforms the problem to the individual bits of $M(x)$. Let $d = \log(3/\Gamma)$, where Γ is as above. Let $\alpha(x) = \lfloor 2^d M(x) \rfloor / 2^d$, i.e., $M(x)$ truncated to include only d of its most significant bits. Assume that $\alpha = \sum_{j=1}^{d} \alpha_j 2^{-j} + k2^{-d}$, where $\alpha_j \in \{-1, 1\}$ and $k \in \{-1, 0, 1\}$. Thus

[2] Grover has proposed a quantum algorithm for estimating the mean that requires $O(\frac{1}{\gamma} \log\log \frac{1}{\gamma})$ queries. However, in our setting, we will use fewer queries if we estimate this value classically because we can use a single sample for all estimates, as discussed below.

$$|\mathbf{E}[Mf\chi_A]| - \frac{\Gamma}{3} \leq |\mathbf{E}[\alpha f\chi_A]| \leq \max_j |\mathbf{E}[\alpha_j f\chi_A]| + \frac{\Gamma}{3}$$

and therefore there exists j so that $|\mathbf{E}[\alpha_j f\chi_A]| \geq \Gamma/3$, assuming $|\mathbf{E}[Mf\chi_A]| \geq \Gamma$.

Note that to simulate U_{EQ} for verifying that the non-Boolean function $g(x) = M(x)f(x)$ has a Γ-heavy coefficient at A, i.e., $|\hat{g}(A)| \geq \Gamma$, we need a sample of size at least $1/\Gamma^2 \sim (s/\epsilon)^2$.

5 A Quantum Harmonic Sieve Algorithm

In this section, we describe a quantum version of the Harmonic Sieve algorithm obtained by combining the quantum Goldreich-Levin algorithm and the SmoothBoost boosting algorithm (see Figure 4).

The top level part of this algorithm involves $O(s^2/\epsilon)$ boosting rounds[3] and each round requires invoking the algorithm QWDNF that uses $\tilde{O}(s/\epsilon)$ queries. The "oracle" $QMQ_f \cdot D_t$ represents the procedure that will produce Boolean functions representing the bits of $M_t f$ and simulate quantum membership oracles to be passed to QWDNF. There is an additional cost of a random sample of size $\tilde{O}(s^2/\epsilon^2)$ for estimating the expression $\mathbf{E}[M_t]$ to within $O(\epsilon)$ and for simulating the equivalence oracle U_{EQ} used by QWDNF. The latter step requires estimating the expression $\mathbf{E}[M_t f\chi_A]$ to within $O(\epsilon/s)$ accuracy. This random sample is shared among all boosting stages and all calls to QWDNF. The key property exploited here is the oblivious nature of the sampling steps.

Thus the overall algorithm, if BComb is used as the boosting algorithm, requires $\tilde{O}(s^3/\epsilon + s^2/\epsilon^2)$ sample complexity. The best classical algorithm (also based on BComb) has complexity $\tilde{O}(ns^2/\epsilon^2)$. Thus, for $s = \Theta(1/\epsilon)$, the quantum algorithm is an improvement by a factor of n.

6 Lower Bounds

In this section, we prove a lower bound on the query complexity of any quantum PAC learning algorithm for DNF formulae.

Theorem 1. *Let $s \geq n/\log n$. Then any quantum PAC learning algorithm requires $\Omega(s\log n/n)$ queries to learn a DNF formula of size s over n variables under the uniform distribution, given $\epsilon < 1/4$ and any constant $\delta > 0$.*

Proof We use a construction given in Bshouty et al. [3]. Let $t = \log s$ and $u = n-t$. Consider the class of DNF formulae $C = \left\{ \bigvee_{a\in\{0,1\}^t} x^a y_a \ : \ \langle y_a\rangle_{a\in\{0,1\}^t} \right\}$, over the variable set of $V = \{x_1,\ldots,x_t\} \cup \{y_1,\ldots,y_u\}$, where $x^a = \bigwedge_{i=1}^t x_i^{a_i}$, with the convention $x_i^0 = x_i$ and $x_i^1 = \overline{x_i}$, and for each $a \in \{0,1\}^t$, y_a is a constant (0 or 1) or one of the variables y_i or its negation. Each $f \in C$ is specified uniquely by a word $y \in \Sigma^s$ over the alphabet $\Sigma = \{0,1,y_1,\overline{y_1},\ldots,y_u,\overline{y_u}\}$, i.e., we may denote f_y to be the DNF specified by the word $y \in \Sigma^s$. By the

[3] This could be improved to $O(s^2\log(1/\epsilon))$ rounds if Freund's BComb algorithm is used.

Input: Parameters $0 < \epsilon, \delta < 1$, n, a quantum membership oracle QMQ_f for a DNF formula f, s (the size of DNF f),
Output: h so that $\Pr[f \neq h] < \epsilon$.

1. Draw uniform sample R of $\Omega(s^2/\epsilon^2)$ instances and label using QMQ_f
2. $\gamma \leftarrow 1/(8s+4)$ (weak advantage)
3. $k \leftarrow c_1 \gamma^{-2} \epsilon^{-1}$ (number of boosting stages)
4. $M_1 \equiv 1$ (all-one function), $N_0 \equiv 0$ (all-zero function)
5. **for** $t = 1, \ldots, k$ **do**
6. $\quad E_t \leftarrow \mathbf{E}_{x \sim U_R}[M_t(x)]$
7. $\quad$ **if** $E_t \leq 2\epsilon/3$ **then break end if**
8. $\quad D_t \equiv M_t/(2^n E_t)$
9. $\quad h_t \leftarrow \mathtt{QWDNF}(n, \gamma\epsilon, \delta/2k, QMQ_f \cdot D_t, R)$ where $\Pr_{D_t}[h_t(x) \neq f(x)] \leq \frac{1}{2} - \gamma$.
10. $\quad N_t \equiv N_{t-1} + f h_t - \theta$
11. $\quad M_{t+1} \equiv [\![N_t < 0]\!] + (1-\gamma)^{N_t/2}[\![N_t \geq 0]\!]$
12. **end do**
13. $T = t - 1$
14. $H(x) \equiv \frac{1}{T}\sum_{i=1}^{T} h_i(x)$
15. **return** $h(x) = \mathrm{sign}(H(x))$

Fig. 4. The new **QHS** algorithm.

Gilbert-Varshamov bound, there is a code $L \subset \Sigma^s$ with minimum distance αs of size at least $\frac{|\Sigma|^s}{\sum_{k=0}^{\alpha s}\binom{s}{k}(|\Sigma|-1)^k} \geq \left(\frac{(2u+2)^{1-\alpha}}{2}\right)^s$ We focus on $C_L \subset C$ where the words y are taken from L. Note that for any distinct $y, z \in L$ we have $\Pr_U[f_y \neq f_z] = \mathbf{E}_U[f_{y \oplus z}] \geq \alpha s/2$, where the probability is taken over the uniform distribution on V. Letting $2\epsilon = \alpha s/2$, this implies that any two distinct DNF functions f_y, f_z, where $y, z \in L$, are (2ϵ)-separated. So any (ϵ, δ)-PAC algorithm for C_L must return exactly the unknown target function.

Now let A be any quantum (ϵ, δ)-PAC algorithm with access to a quantum membership oracle QMQ_f associated with a target DNF function f. Suppose that A makes T queries for any function $f \in C_L$. Following the notation in [8], let X^f be the truth table of the DNF function f, i.e., X^f is a binary vector of length $N = 2^n$. Let $P_h(X^f)$ be the probability function of A of returning as answer a DNF function h when the oracle is QMQ_f, for $h, f \in C_L$. By the PAC property of A, we have

- $P_f(X^f) \geq 1 - \delta$
- $\sum_{h: h \neq f} P_h(X^f) < \delta$

It is known that P_f is a multivariate polynomial of degree $2T$ over X^h, for any f, h. Let $N_0 = \sum_{t=0}^{2T}\binom{N}{t}$. For $X \in \{0,1\}^N$, let $\tilde{X} \in \{0,1\}^{N_0}$ be the vector obtained by taking all ℓ-subsets of $[N]$, $\ell \leq 2T$. The coefficients of P_h can be specified by a real vector $V_h \in \mathbb{R}^{N_0}$ and $P_h(X^f) = V_h^T X^f$. Let M be a matrix of size $|C_L| \times N_0$ whose rows are given by the vectors V_h^T for all $h \in C_L$. Let N be a matrix of size $|C_L| \times |C_L|$ whose columns are given by the vectors

MV_g for all $g \in C_L$. Observe that the (h, f) entry in the matrix N is given by $P_h(X^f)$. As in [8], we argue that since N is *diagonally dominant* (from the PAC conditions on δ above), it has full rank. Thus $N_0 \geq |C_L|$, which implies that $N^{2T} \geq |C_L| \geq \left(\frac{(2u+2)^{1-\alpha}}{2}\right)^s$. This implies that $4nT \geq s\log(n)(1-o(1))$ which gives $T \geq \Omega(s \log n/n)$. □

Acknowledgments

The second author thanks Richard Cleve for helpful discussions on the quantum Goldreich-Levin algorithm.

References

1. M. Adcock, R. Cleve. A Quantum Goldreich-Levin Theorem with Cryptographic Applications. *19th Int. Symp. Theor. Aspects of Comp. Sci.*, 2002.
2. N.H. Bshouty, J. Jackson. Learning DNF over the Uniform Distribution using a Quantum Example Oracle. *SIAM J. Computing*, 28(3):1136-1153, 1999.
3. N.H. Bshouty, J. Jackson, C. Tamon. More Efficient PAC-learning of DNF with Membership Queries under the Uniform Distribution. *12th Ann. ACM Conf. Comp. Learning Theory*, 286-295, 1999.
4. M. Nielsen, I. Chuang. *Quantum Computation and Quantum Information.* Cambridge University Press, 2000.
5. Y. Freund. Boosting a Weak Learning Algorithm by Majority. *Inf. Comp.*, **121**(2):256-285, 1995.
6. O. Goldreich, L. Levin. A Hardcore Predicate for all One-Way Functions. *21st Ann. ACM Symp. Theory of Computing*, 25-32, 1989.
7. O. Goldreich. *Modern Cryptography, Probabilistic Proofs and Pseudorandomness.* Springer-Verlag, 1999.
8. S. Gortler, R. Servedio. Quantum versus Classical Learnability. *16th Conf. on Comp. Complexity*, 473-489, 2001.
9. R. Impagliazzo. Hard-core distributions for somewhat hard problems. *36th Ann. Symp. Found. Comp. Sci.*, 538-545, 1998.
10. J. Jackson. An Efficient Membership-Query Algorithm for Learning DNF with Respect to the Uniform Distribution. *J. Comp. Syst. Sci.*, 55(3):414-440, 1997.
11. A. Klivans, R. Servedio. Boosting and Hardcore Sets. *40th Ann. Symp. Found. Comp. Sci.*, 624-633, 1999.
12. E. Kushilevitz, Y. Mansour. Learning Decision Trees using the Fourier Spectrum. *SIAM J. Computing*, 22(6): 1331-1348, 1993.
13. L. Levin. Randomness and Non-determinism. *J. Symb. Logic*, 58(3):1102-1103, 1993.
14. R. Servedio. Smooth Boosting and Linear Threshold Learning with Malicious Noise. *14th Ann. Conf. Comp. Learning Theory*, 473-489, 2001.

Author Index

Lecture Notes in Computer Science

For information about Vols. 1–2315
please contact your bookseller or Springer-Verlag

Vol. 2316: J. Domingo-Ferrer (Ed.), Inference Control in Statistical Databases. VIII, 231 pages. 2002.

Vol. 2317: M. Hegarty, B. Meyer, N. Hari Narayanan (Eds.), Diagrammatic Representation and Inference. Proceedings, 2002. XIV, 362 pages. 2002. (Subseries LNAI).

Vol. 2318: D. Bošnački, S. Leue (Eds.), Model Checking Software. Proceedings, 2002. X, 259 pages. 2002.

Vol. 2319: C. Gacek (Ed.), Software Reuse: Methods, Techniques, and Tools. Proceedings, 2002. XI, 353 pages. 2002.

Vol.2320: T. Sander (Ed.), Security and Privacy in Digital Rights Management. Proceedings, 2001. X, 245 pages. 2002.

Vol. 2321: P.L. Lanzi, W. Stolzmann, S.W. Wilson (Eds.), Advances in Learning Classifier Systems. Proceedings, 2002. VIII, 231 pages. 2002. (Subseries LNAI).

Vol. 2322: V. Mařík, O. Štěpánková, H. Krautwurmová, M. Luck (Eds.), Multi-Agent Systems and Applications II. Proceedings, 2001. XII, 377 pages. 2002. (Subseries LNAI).

Vol. 2323: À. Frohner (Ed.), Object-Oriented Technology. Proceedings, 2001. IX, 225 pages. 2002.

Vol. 2324: T. Field, P.G. Harrison, J. Bradley, U. Harder (Eds.), Computer Performance Evaluation. Proceedings, 2002. XI, 349 pages. 2002.

Vol 2326: D. Grigoras, A. Nicolau, B. Toursel, B. Folliot (Eds.), Advanced Environments, Tools, and Applications for Cluster Computing. Proceedings, 2001. XIII, 321 pages. 2002.

Vol. 2327: H.P. Zima, K. Joe, M. Sato, Y. Seo, M. Shimasaki (Eds.), High Performance Computing. Proceedings, 2002. XV, 564 pages. 2002.

Vol. 2328: R. Wyrzykowski, J. Dongarra, M. Paprzycki, J. Waśniewski (Eds.), Parallel Processing and Applied Mathematics. Proceedings, 2001. XIX, 915 pages. 2002.

Vol. 2329: P.M.A. Sloot, C.J.K. Tan, J.J. Dongarra, A.G. Hoekstra (Eds.), Computational Science – ICCS 2002. Proceedings, Part I. XLI, 1095 pages. 2002.

Vol. 2330: P.M.A. Sloot, C.J.K. Tan, J.J. Dongarra, A.G. Hoekstra (Eds.), Computational Science – ICCS 2002. Proceedings, Part II. XLI, 1115 pages. 2002.

Vol. 2331: P.M.A. Sloot, C.J.K. Tan, J.J. Dongarra, A.G. Hoekstra (Eds.), Computational Science – ICCS 2002. Proceedings, Part III. XLI, 1227 pages. 2002.

Vol. 2332: L. Knudsen (Ed.), Advances in Cryptology – EUROCRYPT 2002. Proceedings, 2002. XII, 547 pages. 2002.

Vol. 2333: J.-J.Ch. Meyer, M. Tambe (Eds.), Intelligent Agents VIII. Revised Papers, 2001. XI, 461 pages. 2001. (Subseries LNAI).

Vol. 2334: G. Carle, M. Zitterbart (Eds.), Protocols for High Speed Networks. Proceedings, 2002. X, 267 pages. 2002.

Vol. 2335: M. Butler, L. Petre, K. Sere (Eds.), Integrated Formal Methods. Proceedings, 2002. X, 401 pages. 2002.

Vol. 2336: M.-S. Chen, P.S. Yu, B. Liu (Eds.), Advances in Knowledge Discovery and Data Mining. Proceedings, 2002. XIII, 568 pages. 2002. (Subseries LNAI).

Vol. 2337: W.J. Cook, A.S. Schulz (Eds.), Integer Programming and Combinatorial Optimization. Proceedings, 2002. XI, 487 pages. 2002.

Vol. 2338: R. Cohen, B. Spencer (Eds.), Advances in Artificial Intelligence. Proceedings, 2002. XII, 373 pages. 2002. (Subseries LNAI).

Vol. 2340: N. Jonoska, N.C. Seeman (Eds.), DNA Computing. Proceedings, 2001. XI, 392 pages. 2002.

Vol. 2342: I. Horrocks, J. Hendler (Eds.), The Semantic Web – ISCW 2002. Proceedings, 2002. XVI, 476 pages. 2002.

Vol. 2345: E. Gregori, M. Conti, A.T. Campbell, G. Omidyar, M. Zukerman (Eds.), NETWORKING 2002. Proceedings, 2002. XXVI, 1256 pages. 2002.

Vol. 2346: H. Unger, T. Böhme, A. Mikler (Eds.), Innovative Internet Computing Systems. Proceedings, 2002. VIII, 251 pages. 2002.

Vol. 2347: P. De Bra, P. Brusilovsky, R. Conejo (Eds.), Adaptive Hypermedia and Adaptive Web-Based Systems. Proceedings, 2002. XV, 615 pages. 2002.

Vol. 2348: A. Banks Pidduck, J. Mylopoulos, C.C. Woo, M. Tamer Ozsu (Eds.), Advanced Information Systems Engineering. Proceedings, 2002. XIV, 799 pages. 2002.

Vol. 2349: J. Kontio, R. Conradi (Eds.), Software Quality – ECSQ 2002. Proceedings, 2002. XIV, 363 pages. 2002.

Vol. 2350: A. Heyden, G. Sparr, M. Nielsen, P. Johansen (Eds.), Computer Vision – ECCV 2002. Proceedings, Part I. XXVIII, 817 pages. 2002.

Vol. 2351: A. Heyden, G. Sparr, M. Nielsen, P. Johansen (Eds.), Computer Vision – ECCV 2002. Proceedings, Part II. XXVIII, 903 pages. 2002.

Vol. 2352: A. Heyden, G. Sparr, M. Nielsen, P. Johansen (Eds.), Computer Vision – ECCV 2002. Proceedings, Part III. XXVIII, 919 pages. 2002.

Vol. 2353: A. Heyden, G. Sparr, M. Nielsen, P. Johansen (Eds.), Computer Vision – ECCV 2002. Proceedings, Part IV. XXVIII, 841 pages. 2002.

Vol. 2355: M. Matsui (Ed.), Fast Software Encryption. Proceedings, 2001. VIII, 169 pages. 2001.

Vol. 2358: T. Hendtlass, M. Ali (Eds.), Developments in Applied Artificial Intelligence. Proceedings, 2002 XIII, 833 pages. 2002. (Subseries LNAI).

Vol. 2359: M. Tistarelli, J. Bigun, A.K. Jain (Eds.), Biometric Authentication. Proceedings, 2002. X, 197 pages. 2002.

Vol. 2360: J. Esparza, C. Lakos (Eds.), Application and Theory of Petri Nets 2002. Proceedings, 2002. X, 445 pages. 2002.

Vol. 2361: J. Blieberger, A. Strohmeier (Eds.), Reliable Software Technologies – Ada-Europe 2002. Proceedings, 2002 XIII, 367 pages. 2002.

Vol. 2362: M. Tanabe, P. van den Besselaar, T. Ishida (Eds.), Digital Cities II. Proceedings, 2001. XI, 399 pages. 2002.

Vol. 2363: S.A. Cerri, G. Gouardères, F. Paraguaçu (Eds.), Intelligent Tutoring Systems. Proceedings, 2002. XXVIII, 1016 pages. 2002.

Vol. 2364: F. Roli, J. Kittler (Eds.), Multiple Classifier Systems. Proceedings, 2002. XI, 337 pages. 2002.

Vol. 2366: M.-S. Hacid, Z.W. Raś, D.A. Zighed, Y. Kodratoff (Eds.), Foundations of Intelligent Systems. Proceedings, 2002. XII, 614 pages. 2002. (Subseries LNAI).

Vol. 2367: J. Fagerholm, J. Haataja, J. Järvinen, M. Lyly. P. Råback, V. Savolainen (Eds.), Applied Parallel Computing. Proceedings, 2002. XIV, 612 pages. 2002.

Vol. 2368: M. Penttonen, E. Meineche Schmidt (Eds.), Algorithm Theory – SWAT 2002. Proceedings, 2002. XIV, 450 pages. 2002.

Vol. 2369: C. Fieker, D.R. Kohel (Eds.), Algebraic Number Theory. Proceedings, 2002. IX, 517 pages. 2002.

Vol. 2370: J. Bishop (Ed.), Component Deployment. Proceedings, 2002. XII, 269 pages. 2002.

Vol. 2371: S. Koenig, R.C. Holte (Eds.), Abstraction, Reformulation, and Approximation. Proceedings, 2002. XI, 349 pages. 2002. (Subseries LNAI).

Vol. 2372: A. Pettorossi (Ed.), Logic Based Program Synthesis and Transformation. Proceedings, 2001. VIII, 267 pages. 2002.

Vol. 2373: A. Apostolico, M. Takeda (Eds.), Combinatorial Pattern Matching. Proceedings, 2002. VIII, 289 pages. 2002.

Vol. 2374: B. Magnusson (Ed.), ECOOP 2002 – Object-Oriented Programming. XI, 637 pages. 2002.

Vol. 2375: J. Kivinen, R.H. Sloan (Eds.), Computational Learning Theory. Proceedings, 2002. XI, 397 pages. 2002. (Subseries LNAI).

Vol. 2377: A. Birk, S. Coradeschi, T. Satoshi (Eds.), RoboCup 2001: Robot Soccer World Cup V. XIX, 763 pages. 2002. (Subseries LNAI).

Vol. 2378: S. Tison (Ed.), Rewriting Techniques and Applications. Proceedings, 2002. XI, 387 pages. 2002.

Vol. 2379: G.J. Chastek (Ed.), Software Product Lines. Proceedings, 2002. X, 399 pages. 2002.

Vol. 2380: P. Widmayer, F. Triguero, R. Morales, M. Hennessy, S. Eidenbenz, R. Conejo (Eds.), Automata, Languages and Programming. Proceedings, 2002. XXI, 1069 pages. 2002.

Vol. 2381: U. Egly, C.G. Fermüller (Eds.), Automated Reasoning with Analytic Tableaux and Related Methods. Proceedings, 2002. X, 341 pages. 2002 .(Subseries LNAI).

Vol. 2382: A. Halevy, A. Gal (Eds.), Next Generation Information Technologies and Systems. Proceedings, 2002. VIII, 169 pages. 2002.

Vol. 2383: M.S. Lew, N. Sebe, J.P. Eakins (Eds.), Image and Video Retrieval. Proceedings, 2002. XII, 388 pages. 2002.

Vol. 2384: L. Batten, J. Seberry (Eds.), Information Security and Privacy. Proceedings, 2002. XII, 514 pages. 2002.

Vol. 2385: J. Calmet, B. Benhamou, O. Caprotti, L. Henocque, V. Sorge (Eds.), Artificial Intelligence, Automated Reasoning, and Symbolic Computation. Proceedings, 2002. XI, 343 pages. 2002. (Subseries LNAI).

Vol. 2386: E.A. Boiten, B. Möller (Eds.), Mathematics of Program Construction. Proceedings, 2002. X, 263 pages. 2002.

Vol. 2387: O.H. Ibarra, L. Zhang (Eds.), Computing and Combinatorics. Proceedings, 2002. XIII, 606 pages. 2002.

Vol. 2389: E. Ranchhod, N.J. Mamede (Eds.), Advances in Natural Language Processing. Proceedings, 2002. XII, 275 pages. 2002. (Subseries LNAI).

Vol. 2391: L.-H. Eriksson, P.A. Lindsay *(Eds.), FME* 2002: Formal Methods – Getting IT Right. Proceedings, 2002. XI, 625 pages. 2002.

Vol. 2392: A. Voronkov (Ed.), Automated Deduction – CADE-18. Proceedings, 2002. XII, 534 pages. 2002. (Subseries LNAI).

Vol. 2393: U. Priss, D. Corbett, G. Angelova (Eds.), Conceptual Structures: Integration and Interfaces. Proceedings, 2002. XI, 397 pages. 2002. (Subseries LNAI).

Vol. 2398: K. Miesenberger, J. Klaus, W. Zagler (Eds.), Computers Helping People with Special Needs. Proceedings, 2002. XXII, 794 pages. 2002.

Vol. 2399: H. Hermanns, R. Segala (Eds.), Process Algebra and Probabilistic Methods. Proceedings, 2002. X, 215 pages. 2002.

Vol. 2401: P.J. Stuckey (Ed.), Logic Programming. Proceedings, 2002. XI, 486 pages. 2002.

Vol. 2402: W. Chang (Ed.), Advanced Internet Services and Applications. Proceedings, 2002. XI, 307 pages. 2002.

Vol. 2403: Mark d'Inverno, M. Luck, M. Fisher, C. Preist (Eds.), Foundations and Applications of Multi-Agent Systems. Proceedings, 1996-2000. X, 261 pages. 2002. (Subseries LNAI).

Vol. 2404: E. Brinksma, K.G. Larsen (Eds.), Computer Aided Verification. Proceedings, 2002. XIII, 626 pages. 2002.

Vol. 2405: B. Eaglestone, S. North, A. Poulovassilis (Eds.), Advances in Databases. Proceedings, 2002. XII, 199 pages. 2002.

Vol. 2407: A.C. Kakas, F. Sadri (Eds.), Computational Logic: Logic Programming and Beyond. Part I. XII, 678 pages. 2002. (Subseries LNAI).

Vol. 2408: A.C. Kakas, F. Sadri (Eds.), Computational Logic: Logic Programming and Beyond. Part II. XII, 628 pages. 2002. (Subseries LNAI).

Vol. 2409: D.M. Mount, C. Stein (Eds.), Algorithm Engineering and Experiments. Proceedings, 2002. VIII, 207 pages. 2002.